PHARMACOLOGICAL EFFECTS of ETHANOL on the NERVOUS SYSTEM

Pharmacology and Toxicology: Basic and Clinical Aspects

Mannfred A. Hollinger, Series Editor
University of California, Davis

Published Titles

Inflammatory Cells and Mediators in Bronchial Asthma, 1990, Devendra K. Agrawal and Robert G. Townley
Pharmacology of the Skin, 1991, Hasan Mukhtar
In Vitro *Methods of Toxicology*, 1992, Ronald R. Watson
Basis of Toxicity Testing, 1992, Donald J. Ecobichon
Human Drug Metabolism from Molecular Biology to Man, 1992, Elizabeth Jeffreys
Platelet Activating Factor Receptor: Signal Mechanisms and Molecular Biology, 1992, Shivendra D. Shukla
Biopharmaceutics of Ocular Drug Delivery, 1992, Peter Edman
Beneficial and Toxic Effects of Aspirin, 1993, Susan E. Feinman
Preclinical and Clinical Modulation of Anticancer Drugs, 1993, Kenneth D. Tew, Peter Houghton, and Janet Houghton
Peroxisome Proliferators: Unique Inducers of Drug-Metabolizing Enzymes, 1994, David E. Moody
Angiotensin II Receptors, Volume I: Molecular Biology, Biochemistry, Pharmacology, and Clinical Perspectives, 1994, Robert R. Ruffolo, Jr.
Angiotensin II Receptors, Volume II: Medicinal Chemistry, 1994, Robert R. Ruffolo, Jr.
Chemical and Structural Approaches to Rational Drug Design, 1994, David B. Weiner and William V. Williams
Biological Approaches to Rational Drug Design, 1994, David B. Weiner and William V. Williams
Direct Allosteric Control of Glutamate Receptors, 1994, M. Palfreyman, I. Reynolds, and P. Skolnick
Genomic and Non-Genomic Effects of Aldosterone, 1994, Martin Wehling
Human Growth Hormone Pharmacology: Basic and Clinical Aspects, 1995, Kathleen T. Shiverick and Arlan Rosenbloom
Placental Toxicology, 1995, B. V. Rama Sastry
Stealth Liposomes, 1995, Danilo Lasic and Frank Martin
TAXOL®: Science and Applications, 1995, Matthew Suffness
Endothelin Receptors: From the Gene to the Human, 1995, Robert R. Ruffolo, Jr.
Serotonin and Gastrointestinal Function, 1995, Timothy S. Gaginella and James J. Galligan
Drug Delivery Systems, 1995, Vasant V. Ranade and Mannfred A. Hollinger
Alternative Methodologies for the Safety Evaluation of Chemicals in the Cosmetic Industry, 1995, Nicola Loprieno
Phospholipase A_2 in Clinical Inflammation: Molecular Approaches to Pathophysiology, 1995, Keith B. Glaser and Peter Vadas
Experimental Models of Mucosal Inflammation, 1996, Timothy S. Gaginella
ImmunoPharmaceuticals, 1996, Edward S. Kimball

Pharmacology and Toxicology: Basic and Clinical Aspects

Mannfred A. Hollinger, Series Editor
University of California, Davis

Forthcoming Titles

Alcohol Consumption, Cancer and Birth Defects: Mechanisms Involved in Increased Risk Associated with Drinking, Anthony J. Garro
Antibody Therapeutics, William J. Harris and John R. Adair
Brain Mechanisms and Psychotropic Drugs, A. Baskys and G. Remington
Chemoattractant Ligands and Their Receptors, Richard Horuk
CNS Injuries: Cellular Responses and Pharmacological Strategies, Martin Berry and Ann Logan
Muscarinic Receptor Subtypes in Smooth Muscle, Richard M. Eglen
Neural Control of Airways, Peter J. Barnes
Pharmacological Regulation of Gene Expression in the CNS, Kalpana Merchant
Pharmacology in Exercise and Sports, Satu M. Somani
Pharmacology of Intestinal Secretion, Timothy S. Gaginella
Placental Pharmacology, B. V. Rama Sastry
Receptor Characterization and Regulation, Devendra K. Agrawal
Receptor Dynamics in Neural Development, Christopher A. Shaw
Ryanodine Receptors, Vincenzo Sorrentino
Therapeutic Modulation of Cytokines, M.W. Bodmer and Brian Henderson

PHARMACOLOGICAL EFFECTS of ETHANOL on the NERVOUS SYSTEM

Edited by

Richard A. Deitrich

V. Gene Erwin

CRC Press
Boca Raton New York London Tokyo

Library of Congress Cataloging-in-Publication Data

Pharmacological effects of ethanol on the nervous system / edited by
Richard A. Deitrich and V. Gene Erwin.
p. cm. -- (Pharmacology and toxicology)
Includes bibliographical references and index.
ISBN 0-8493-8389-7
1. Alcohol--Physiological effect. 2. Nervous system--Effect of
drugs on. I. Deitrich, Richard A. II. Erwin, V. Gene.
III. Series: Pharmacology & toxicology (Boca Raton, Fla.)
[DNLM: 1. Alcohol, Ethyl--pharmacology. 2. Central Nervous
System--drug effects. 3. Behavior--drug effects. QV 84 P5353
1995]
QP801.A3P46 1995
615′.7828--dc20
DNLM/DLC
for Library of Congress

95-9425
CIP

International Standard Book Number 0-8493-8389-7
Library of Congress Card Number 95-9425
Printed in the United States of America 1 2 3 4 5 6 7 8 9 0
Printed on acid-free paper

PREFACE

In selecting authors for this book we have concentrated heavily on those from the University of Colorado Alcohol Research Center. There are many other individuals who could have written equally high quality reviews, but these authors were chosen in an attempt to increase coordination between chapters. The book is designed to be useful for those entering the field, those already in the field but needing to expand their horizons, and those who would like a source of recent references in any given area.

We have tried to present the information in two ways. First are sections on the actions of ethanol on various receptor systems in the brain with references to the consequent behavioral effects. Next are behavioral effects with reference to the receptor systems that may mediate these actions. Finally there is a section on behavioral and structural consequences of ethanol damage in the central nervous system.

We would like to dedicate the book to two individuals who were important to the scientific careers of both the editors, the late Dr. Harold C. Heim, who was Professor and Dean of the School of Pharmacy of the University of Colorado and the late Dr. Leslie Hellerman, who was Professor of Physiological Chemistry at The Johns Hopkins School of Medicine.

INTRODUCTION

This book is intended to be an update on the recent developments in the actions of ethanol on the central nervous system. It is divided into three parts. First is a discussion of some of the cellular events that are perturbed by ethanol. Of necessity the authors of these chapters refer to behavioral effects of ethanol, but the focus is on the mechanisms by which ethanol alters cellular constituents presumably, to mediate behavioral events.

The second part is a discussion of the behavioral effects of ethanol, again referring back to the known cellular events that may underlie these behavioral events.

Finally there is a section of the possible mechanisms of structural damage to the central nervous system in fetal and adult brains.

The book makes no pretense to be all encompassing, and the discerning reader will notice gaps in the presentation; however, it concentrates on recent work and on areas where some reasonable correlations between cellular events and behavioral effects can be made. With this in mind we offer some criteria for making such correlations.

CRITERIA FOR INVOLVEMENT OF A NEUROTRANSMITTER-NEUROMODULATOR IN THE ACTIONS OF ETHANOL

1. Agonists and antagonists of the neurotransmitter system should interact with ethanol both *in vivo* and *in vitro* as would be predicted from the hypothesis (i.e., GABA agonist should be additive with ethanol, and GABA antagonists should block ethanol effects, whereas NMDA agonists should block ethanol effects, and NMDA antagonists should be additive). Effective ethanol concentrations should be in the range of pharmacologically relevant levels.
2. There should be a significant correlation between the interaction of agonists and antagonists with ethanol at the cellular level and their interactions with ethanol at the behavioral level.
3. Ethanol should reproduce or block many, but not necessarily all, of the effects of the neurotransmitter *in vivo* or *in vitro*. Likewise, neurotransmitter agonists or antagonists should reproduce some, but not necessarily all, of the effects of ethanol.
4. There should be cross tolerance and cross dependence between ethanol and agonists of the neurotransmitter system if the hypothesis involves chronic effects.
5. Ethanol and the agonists or antagonists should not interact metabolically, or if they do, this is controlled for by assay of ethanol levels and of the compound being studied.
6. Animals genetically selected for behavioral responses to ethanol should react as predicted to agonists and antagonists of the neurotransmitter system under study. This is a sufficient, but not necessary, condition since the base stock from which the animals were selected may have no polymorphism in the neurotransmitter system under study.

The reader can determine which, if any of the neurotransmitter system discussed in this book meet the criteria outlined above.

THE EDITORS

Richard A. Deitrich, Ph.D. is Professor of Pharmacology and the Principal Investigator of the Alcohol Research Center at the University of Colorado Health Sciences Center. He is also a faculty fellow of the Institute of Behavioral Genetics at the University of Colorado in Boulder, Colorado.

Dr. Deitrich graduated from the University of Colorado School of Pharmacy in 1953 with a B.S. degree. He then obtained his Master's degree at the same institution and his Ph.D. degree from the Department of Pharmacology at the University of Colorado Health Sciences Center in Denver. He spent two years as a postdoctoral fellow and later as Instructor at the Department of Biochemistry, The Johns Hopkins University School of Medicine. He returned to the University of Colorado Department of Pharmacology in 1963 where he has remained, except for a year in the laboratory of Dr. Jean-Pierre von Wartburg in Berne, Switzerland. He is a member of the American Society for Pharmacology and Experimental Therapeutics, American Society for Biochemistry and Molecular Biology, American Chemical Society, Research Society on Alcoholism, International Society for Biomedical Research on Alcoholism, the American Society for Neurochemistry, and Society for Behavioral Genetics.

He served as President of the Research Society on Alcoholism and in 1984 received their award for research excellence. He was Treasurer of the International Society for Biomedical Research on Alcoholism for several years.

Dr. Deitrich has presented numerous invited lectures at national and international meetings and at many universities. His research interests include the molecular mechanism of ethanol in the central nervous system and in the role of genetics in controlling behavioral reactions to ethanol in man and other animals.

Dr. V. Gene Erwin is Professor of Pharmaceutical Sciences at the University of Colorado Health Sciences Center and Scientific Director of the University of Colorado Alcohol Research Center. He obtained his B.S. and M.S. degrees from the University of Colorado School of Pharmacy and his Ph.D. degree from the Department of Pharmacology at the University of Colorado School of Medicine. He then spent several years as a postdoctoral fellow in the Department of Biochemistry at The Johns Hopkins School of Medicine. He returned to the School of Pharmacy at the University of Colorado in 1967 and eventually served as Dean of the School for 10 years from 1974 to 1984. He spent a sabbatical at the Salk Institute in San Diego. He is a faculty fellow at the Institute for Behavioral Genetics at the University of Colorado in Boulder, Colorado. He is a member of the American Society for Pharmacology and Experimental Therapeutics, Society for Neuroscience, and Society for Behavioral Genetics.

He has presented numerous seminars and talks at national and international meetings on his scientific, as well as administrative, interests. He serves on several editorial boards for scientific journals. His current interests are in the role of neurotensin in the actions of ethanol and in identifying specific genes that control the behavioral reactions to ethanol and other drugs such as cocaine.

CONTRIBUTORS

Ronald L. Alkana
University of Southern California
School of Pharmacy
Department of Molecular Pharmacology and Toxicology
Los Angeles, California

Nicolai A. Avdulov, Ph.D.
University of Minnesota School of Medicine
Department of Pharmacology
Minneapolis, Minnesota

Howard C. Becker, Ph.D.
Center for Drug and Alcohol Programs (CDAP)
Department of Psychiatry and Behavioral Sciences
Medical University of South Carolina and VA Medical Center
Charleston, South Carolina

George R. Breese
University of North Carolina at Chapel Hill
School of Medicine
Department of Psychiatry
Chapel Hill, North Carolina

Michael D. Browning
University of Colorado Health Science Center
Department of Pharmacology
Denver, Colorado

Michael E. Charness, M.D.
Harvard Medical School
Department of Neurology
Brigham and Women's Hospital
West Roxbury VA Medical Center
West Roxbury, Massachusetts

Wei-Jung A. Chen, Ph.D.
Texas A&M University Health Science Center
Department of Human Anatomy and Medical Neurobiology
College Station, Texas

Allan C. Collins
University of Colorado
Institute for Behavioral Genetics
Boulder, Colorado

John C. Crabbe, Ph.D.
Veterans Administration Medical Center and Oregon Health Sciences University
Portland, Oregon

Daryl L. Davies
University of Southern California
School of Pharmacy
Department of Molecular Pharmacology and Toxicology
Los Angeles, California

Richard A. Deitrich
University of Colorado Health Science Center
Alcohol Research Center and Department of Pharmacology
Denver, Colorado

Laura J. Draski
University of Colorado Health Science Center
Alcohol Research Center and Department of Pharmacology
Denver, Colorado

Thomas V. Dunwiddie
University of Colorado Health Science Center
Department of Pharmacology
and Veterans Administration Medical Center
Denver, Colorado

Kathryn T. Eggeman
University of Colorado Health Science Center
Department of Pharmacology
Denver, Colorado

V. Gene Erwin
University of Colorado Heath Center
Alcohol Research Center and School of Pharmacy
Denver, Colorado

R. Adron Harris, Ph.D.
University of Colorado Health Sciences Center
Department of Pharmacology and Alcohol Research Center
and Veterans Affairs Medical Center
Denver, Colorado

Steven J. Henriksen
The Scripps Research Institute
Department of Neuropharmacology and Alcohol Research Center
La Jolla, California

Matti E. Hillbom
University of Oulu
Department of Neurology
Oulu, Finland

Clyde W. Hodge, Ph.D.
Bowman Gray School of Medicine
Department of Physiology and Pharmacology
Winston-Salem, North Carolina

Paula L. Hoffman, Ph.D.
University of Colorado Heath Science Center
Department of Pharmacology
Denver, Colorado

Urule Igbavboa, D.Sc.
University of Minnesota School of Medicine
Department of Pharmacology
Minneapolis, Minnesota

George F. Koob, Ph.D.
The Scripps Research Institute
Department of Neuropharmacology
La Jolla, California

Anh Dzung Lê
Addiction Research Foundation
Biobehavioral Research Department
Toronto, Ontario, Canada

Francis Y.G. Lee, Sr., M.D.
The John Hopkins University School of Medicine
Department of Medicine
Baltimore, Maryland

Samuel Madamba
The Scripps Research Institute
Department of Neuropharmacology and Alcohol Research Center
La Jolla, California

Susan E. Maier, Ph.D.
Texas A&M University Health Science Center
Department of Human Anatomy and Medical Neurobiology
College Station, Texas

J.M. Mayer
Centre of Forensic Science
Toronto, Ontario, Canada

Thomas J. McCown
University of North Carolina
The Brain and Developmental Research Center
Chapel Hill, North Carolina

Pamela Metten, M.S.
Veterans Administration Medical Center and Oregon Health Sciences University
Portland, Oregon

S. John Mihic, Ph.D.
University of Colorado Health Sciences Center
Department of Pharmacology
Denver, Colorado

Debra Mullikin-Kilpatrick, Ph.D.
University of Massachusetts Medical Center
Department of Pharmacology
Worcester, Massachusetts

Zhiquo Nie
The Scripps Research Institute
Department of Neuropharmacology and
Alcohol Research Center
La Jolla, California

Roger Nordmann
Université René-Descartes
Biomédicale des Saints-Pères
Paris, France

A. M. Rao, Ph.D.
University of Kentucky Medical School
Department of Surgery
Lexington, Kentucky

Hélène Rouach
Université René-Descartes
Biomédicale des Saints-Pères
Paris, France

Herman H. Samson, Ph.D.
Bowman Gray School of Medicine
Department of Physiology and
Pharmacology
Winston-Salem, North Carolina

Friedhelm Schroeder, Ph.D.
Texas A&M University
Texas Veterinary Medical Center
Department of Physiology and
Pharmacology
College Station, Texas

Paul Schweitzer
The Scripps Research Institute
Department of Neuropharmacology and
Alcohol Research Center
La Jolla, California

George R. Siggins
The Scripps Research Institute
Department of Neuropharmacology and
Alcohol Research Center
La Jolla, California

Boris Tabakoff, Ph.D.
University of Colorado Health Science
Center
Department of Pharmacology
Denver, Colorado

Steven N. Treistman, Ph.D.
University of Massachusetts Medical
Center
Department of Pharmacology
Worcester, Massachusetts

Gary S. Wand, M.D.
The John Hopkins University School of
Medicine
Departments of Medicine and Psychiatry
Baltimore, Maryland

James R. West, Ph.D.
Texas A&M University Health Science
Center
Department of Human Anatomy and
Medical Neurobiology
College Station, Texas

W. Gibson Wood, Ph.D.
VA Medical Center
Geriatric Research, Education, and Clinical
Center
University of Minnesota School of
Medicine
Department of Pharmacology
Minneapolis, Minnesota

CONTENTS

Chapter 1

THE NEUROPHARMACOLOGY OF ETHANOL'S BEHAVIORAL ACTION: NEW DATA, NEW PARADIGMS, NEW HOPE

George F. Koob

TABLE OF CONTENTS

I. INTRODUCTION

The challenge addressed in this chapter is to attempt to bridge the gaps between brain mechanisms of ethanol action and its behavioral effects. The focus chosen has been on those behavioral effects that contribute to the reinforcement associated with various aspects of the ethanol dependence process. The field is moving at a rapid pace, and it is not at all clear at which level of analysis — molecular, cellular, or system — the breakthroughs will occur at any given time. What is clear, however, is that advances ultimately have to be made at all three levels to be able to generate a coherent picture of the neurobiological mechanisms for ethanol's behavioral effects. To address this issue, the following outline is proposed.

0-8493-8389-7/95/$0.00+$.50

First, the highlights of new advances in ethanol neuropharmacology relevant to the behavioral effects of ethanol will be reviewed. New paradigms important for exploring the interaction of neuropharmacology with behavior will then be discussed. Finally, an attempt will be made to integrate these observations into a coherent neurobiology of ethanol action based on common links both at the molecular and behavioral levels of analysis.

II. NEW ADVANCES IN THE STUDY OF THE NEUROPHARMACOLOGY OF ETHANOL ACTIONS

Several advances mark recent developments in the neuropharmacology of ethanol that have, or will have, significance for an understanding of the behavioral actions of ethanol. The discovery that ethanol, at meaningful concentrations (10–50 mM), can alter glutamate function, that low doses of ethanol may interact with the GABA receptor complex, that ethanol can interact with the neuromodulator adenosine, and that ethanol influences neuropeptide function are examples at the cellular level. At the biochemical level there is increasing evidence that ethanol may interact with calcium channels and G-proteins as well as the GABA receptor complex only in the presence of certain subunits at that receptor. These sites for the action of ethanol have been termed "ethanol receptive elements,"[1] and all of these neuropharmacological effects may contribute to the acute behavioral effects of ethanol including its "anxiolytic" or "tension reducing"-like effects.

A. THE RISE OF GLUTAMATE

As described thoroughly in Chapter 5, administered ethanol acutely inhibits NMDA postsynaptic function at a very low threshold concentration (5 mM). This effect has been seen both electrophysiologically and biochemically, as measured by inhibition of NMDA-induced ion currents and by inhibition of NMDA-stimulated calcium uptake, respectively. A possible mechanism for this effect is an inhibitory action of ethanol on the interaction of the co-agonist glycine with the NMDA receptor. Higher, but still relatively low, concentrations of ethanol (22 mM) also have been reported to inhibit glutamate actions at non-NMDA receptors. The hypothesized functional consequences of the glutamate receptor inhibition by ethanol are numerous and include ethanol-induced cognitive deficits, ethanol-induced anxiolytic-like effects, discriminative stimulus effects of ethanol, and mediation of some of the effects of chronic ethanol. There appears to be an up-regulation of NMDA receptors following chronic ethanol administration which could result in not only the withdrawal symptoms associated with cessation of chronic ethanol but also ethanol-induced excitotoxicity. Some evidence also exists which suggests that the NMDA receptor may play an important role in the development of tolerance to ethanol.

B. GABA REVISITED

GABA is the major inhibitory neurotransmitter in the vertebrate central nervous system, and it binds to a post-synaptic GABA-A receptor complex that is a member of the ligand-gated ion channel family of receptors. Based on pharmacological data, it has long been hypothesized that ethanol interacts with GABA to produce its behavioral effects. GABA mimetic agents have additive or synergistic effects with ethanol and GABA antagonists, as well as inverse agonists, and have been shown to antagonize the effects of GABA (see Chapters 4 and 14). However, the role of GABA in ethanol actions has undergone some reassessment in light of a number of observations. First, caution is urged in interpreting the role of the GABA complex in ethanol effects based on pharmacological interactions because co-application of GABAergic drugs with ethanol can alter the balance of excitatory and inhibitory neuronal influences independent of a direct action of ethanol on the GABA complex. Second, while ethanol can potentiate the effects of GABA agonists on chloride influx and can directly enhance chloride influx at high doses in some preparations, it is clear that the effectiveness of benzodiazepines and barbiturates is greater.

These concerns about the specificity and effectiveness of the GABA interaction in the actions of ethanol has led to some re-evaluations at the system, cellular, and molecular levels. First, it is clear that not all brain sites react in the same way to ethanol with regard to the GABA receptor interaction. Evidence suggests that brain sites that bind the benzodiazepine zolpidem are more likely to show GABA ethanol interactions. Second, in studies of ethanol's sedative effect involving selective breeding of long sleep versus short sleep mice, it is clear that ethanol potentiates GABA-mediated chloride uptake only in the long sleep mice. Finally, it appears that the requirement for the gamma 2 long subunit of the GABA receptor to observe ethanol potentiation of GABA-mediated conductance in oocytes is only apparent at low doses of ethanol and may involve the site for phosphorylation by protein kinase C contained in the gamma 2 long insert (see Chapter 4). In addition, it appears that the subunit specificity disappears with larger, more anesthetic-like doses of ethanol (>50 mM), suggesting a different mechanism of action for the moderate to high dose effects of ethanol. This suggests that the gamma subunit is necessary but not sufficient to promote ethanol sensitivity.

C. ADENOSINE AND A NEUROMODULATORY INTERACTION WITH ETHANOL

Adenosine has been hypothesized to have an important inhibitory neuromodulatory action in the central nervous system with which ethanol may interact. Adenosine can activate or inhibit adenylate cyclase, cause hyperpolarization of neurons, inhibit neurotransmitter release, and inhibit Ca^{2+} channels. There is no real evidence for specifically "purinergic" neurons; rather, there appears to be a generalized release of adenosine from brain tissue sufficient to activate extracellular adenosine receptors and probably produce a tonic inhibitory effect on neural activity (see Chapter 9). The activational effects of adenosine antagonists such as caffeine attest to this tonic inhibitory modulation. There appears to be three means by which ethanol can interact with this inhibitory neuromodulator. First, blood acetate concentrations can rise to the 1–2 mM range as a result of ethanol metabolism even with low levels of ethanol intake, providing a substrate for enhanced synthesis of adenosine. A second potential source of interaction is the potent ability of ethanol to inhibit adenosine transport back into neural tissue, leading to increased extracellular adenosine. The third potential source of adenosine/ethanol interaction is the G-protein coupled effector mechanisms through which the adenosine receptor functions. Ethanol enhances basal- and receptor-stimulated adenylate cyclase activity, possibly through an action on G-proteins. Adenosine has similar actions on G_S through the A2 receptor, and ethanol potentiates the increase in cyclic AMP that occurs with activation of A2 receptors in a cell culture system[2] (see Chapter 9). Some ethanol effects, but certainly not all, can be reversed by adenosine antagonists, and there is some evidence that the long sleep mice have an enhanced sensitivity to adenosine. Existing data suggest that the interaction of adenosine with ethanol appears to involve relatively low doses of ethanol, whereas high dose ethanol effects are likely mediated by other neuropharmacological actions.

D. ETHANOL AND NEUROPEPTIDES

Several neuropeptides, most notably neurotensin, opioid peptides, and corticotropin releasing factor, have been implicated in the actions of ethanol. Neurotensin has a number of pharmacological actions that resemble those of ethanol including hypothermia, analgesia, altered locomotor activity, and neuroendocrine effects (see Chapter 10). Neurotensin also potentiates the effects of ethanol, and ethanol in the intoxicating dose range can alter the levels of neurotensin in the brain. Chronic administration of ethanol is associated with a behavioral and neurochemical cross-tolerance to the effects of neurotensin. Finally, neurotensin appears to have differential effects in the long sleep and short sleep mouse lines, with the short sleep mice showing a greater sensitivity to neurotensin-induced potentiation of ethanol actions. Genetic analysis of these strain differences suggests significant correlations between neurotensin levels and receptor densities and ethanol sensitivity (see Chapter 10).

Pharmacological studies have long implicated opioid peptides in the reinforcing actions of ethanol. Naloxone and naltrexone decrease ethanol intake in non-dependent and dependent animals.[3-6] Recent evidence suggests that the delta receptor may be particularly important for this effect.[7] Opioid peptide antagonists also block fluid intake for sucrose and saccharin solutions, suggesting a more general role for opioid peptides in reinforcement processes. Nevertheless, these basic research findings have resulted in a number of significant clinical trials in humans showing that oral naltrexone can significantly decrease relapse in the detoxified alcoholic.[8,9]

Another neuropeptide recently implicated in the actions of ethanol is corticotropin-releasing factor (CRF). CRF is well established as the hypothalamic-releasing factor mediating the hypothalamic pituitary adrenal response to stress, but CRF is also localized independent of the neuroendocrine cascade where it is thought to mediate behavioral responses to stress (see section V.C.). CRF appears to be activated during ethanol withdrawal and may be involved in some of the behavioral symptoms associated with ethanol withdrawal (see section V.C.).

E. EFFECTS OF ETHANOL ON VOLTAGE-GATED ION CHANNELS — A SUBSTRATE FOR ETHANOL EFFECTS AT THE BIOCHEMICAL/MOLECULAR LEVEL

Ethanol has long been thought to act at the molecular level by disordering biological membranes, increasing their fluidity, and possibly perturbing the functions of proteins embedded in the membrane. However, the doses required to produce those effects have always been high (>50 mM) (see Chapter 2), and recent evidence has accumulated to suggest that lower doses of ethanol may act directly upon proteins and, more specifically, ligand-gated channels (GABA and glutamate). Other *in vitro* and *in vivo* work suggests that ethanol at concentrations at or below those considered intoxicating inhibits voltage-activated calcium channels (see Chapter 3) and may be a target for ethanol action in the central nervous system. Ethanol (<25 mM) inhibits calcium influx in brain synaptosome preparations, depending on the brain region being investigated. Ethanol also inhibits calcium-dependent neurotransmitter release, inhibits excitatory and inhibitory postsynaptic potentials in cultured spinal cord neurons that are attributable to increases in intracellular calcium, and inhibits potassium-stimulated calcium uptake in undifferentiated pheochromocytoma (PC12) cells. There are a number of different calcium channels that have been identified and defined by their voltage threshold of activation. Specific pharmacological agents that interact with the different calcium channels and new molecular biological approaches have been developed that will allow a determination of the specific subset or subunits of calcium channel that are sensitive to ethanol. For example, low doses of ethanol inhibit L-type and T-type calcium currents in whole-cell patch clamp studies in cultured cell lines. Also, chronic exposure to ethanol results in increases in calcium currents and an increase in the binding of the calcium antagonist dihydropyridine, suggesting the involvement of the L-type calcium channel in chronic ethanol effects. These changes may contribute to the adaptive responses of cells to ethanol, and this up-regulation of calcium channels may result in some of the symptoms of ethanol withdrawal. The question remains as to the significance of calcium channels for the behavioral effects of intoxicating doses of ethanol. As pointed out in Chapter 3, massive interference with all calcium channels would likely be fatal. It is therefore likely that ethanol at intoxicating doses only affects a small subset of calcium channels. Results reviewed above support this hypothesis, and the identification of specific pharmacological and molecular means of manipulating calcium channels should go far towards testing this hypothesis in future studies.

An additional highly sensitive effect of ethanol involves inhibition of another voltage-regulated channel, this one a K+ channel called the M channel (see Chapter 11). Ethanol at doses of 22–44 mM greatly reduced the M current amplitude; this effect was similar to that of muscarinic agonists, delta opioids, and somatostatin. These results are potentially very exciting because they lay a foundation for low-dose excitatory effects of ethanol at the cellular level and may provide the substrate for some of the selective delta opioid receptor interactions with ethanol at the behavioral level (see Section II.D.). Again, it is not yet clear exactly where ethanol acts to inhibit M channel function. One possibility is directly on the channel itself; alternatively, it may involve some G-protein interaction (see Chapter 11).

III. NEW PARADIGMS

The development of new paradigms for behavioral and molecular analyses has contributed significantly to the recent major advances in the understanding of ethanol action. At the behavioral level, a reliable validated animal model of ethanol reinforcement has been developed that leads to meaningful blood ethanol levels in the rat. In addition, advances have been made using drug discrimination and an operant conflict procedure. At the cellular level, *in vivo* microdialysis in awake moving animals allows, for the first time, a neurochemical window on events taking place during ethanol intoxication and dependence. Studies employing molecular biological techniques such as antisense, knockouts, and transgenic preparation to explore the neuropharmacology of ethanol intoxication are only just beginning.

A. OPERANT SELF-ADMINISTRATION OF ETHANOL — A REINFORCEMENT ANALYSIS

Early paradigms which assess the reinforcing effects of ethanol typically used an oral preference paradigm where animals were allowed to drink ethanol or water. This procedure suffered from numerous methodological problems, including failure to provide information about the frequency and magnitude of individual drinking episodes (typically 24-hour preference was recorded), failure to measure meaningful blood ethanol levels, and problems in determining if sufficient ethanol was ingested to produce pharmacological effects. More recently, Samson and colleagues have developed and validated an operant procedure for both limited access to ethanol and more prolonged access (see Chapter 13). A major breakthrough in this area was the development of a training procedure involving access to a sweetened solution and a subsequent fading-in of ethanol to avoid the aversiveness of the ethanol taste. This produced relatively rapid acquisition of responding for ethanol in non-deprived rats and allowed a choice for responding between ethanol and water. This ease of training combined with validation by measures of blood alcohol levels has established this procedure as a reliable means of measuring the reinforcing effects of ethanol and has been a boon to neuropharmacological analysis of ethanol reinforcement (see Chapter 13).

B. ANTI-CONFLICT EFFECTS OF ETHANOL

Ethanol has long been hypothesized to have anti-stress, tension-reducing, or anti-anxiety effects in humans. Early animal studies with ethanol established that ethanol was effective in reducing approach-avoidance behaviors in "conflict" situations,[10] and subsequently the tension-reducing properties of ethanol have been observed in a variety of behavioral situations.[11] However, these anti-conflict properties of ethanol have been difficult to establish in operant paradigms. A modification of the Geller-Seifter procedure[12] using incremental shock[13] and the lick suppression test[14] have proved particularly sensitive.[11] With the use of these procedures, ethanol reliably produces a dose-dependent release of punished responding as reflected as an increase in lever pressing during the conflict period, and these tests have proved very useful for exploring the neuropharmacological basis for the anti-conflict effects of ethanol.[11] Effective doses range between 0.5–1.0 g/kg intraperitoneally. These same doses of ethanol produce dose-dependent decreases in responding during the unpunished component, which probably reflects the acute motor-impairing effects of ethanol. Previous difficulties with the sensitivity of operant conflict tests may be related to this close correspondence between the anti-conflict effects of the drug and its motor-impairing abilities. Thus, a test that minimizes the response requirements in order to show an anti-conflict effect will be more likely to show sensitivity.

C. DRUG DISCRIMINATION

Two aspects of ethanol's discriminative properties have been successfully studied using drug discrimination procedures: the acute effects of ethanol and the stimulus effects associated with ethanol withdrawal. In studies of the acute effects of ethanol, animals are typically trained to discriminate ethanol from vehicle in a two-lever discrimination procedure.[15] Here, rats trained on a fixed ratio-20 schedule for food pellets are further trained to press for food on a water-appropriate lever after injection with water, or an ethanol-appropriate lever after injection with either 1.0, 1.5, or 2.0 g/kg of ethanol (separate groups). During testing, test sessions were imposed twice per week where 20 consecutive responses on either lever resulted in delivery of a food pellet and then termination of the session. The amount of responding on the drug-paired lever is an index of drug discrimination. These types of studies have produced substantial evidence that ethanol can serve as a mixed cue composed of several neuropharmacological components. These components can include GABA agonists, glutamate antagonists, neuroactive steroids, and subsets of the serotonin receptor agonists. While drug discrimination cues may not necessarily provide direct evidence for the reinforcing actions of ethanol, they provide a powerful means of assessing dose-dependent neuropharmacological interactions that can establish hypotheses for further testing in reinforcement models. Thus, the components of the mixed ethanol cue may change in prominence with changes in the training dose, and this may possibly reflect an important aspect of the neuropharmacology of ethanol reinforcement as well.[15]

Ethanol withdrawal can also be discriminated by rats.[16] When animals were made dependent by gavage or liquid diet, they selected a lever trained to pentylenetetrazol over saline during withdrawal before the onset of overt physical signs of withdrawal. In a subsequent study using an interesting two-drug discrimination procedure, rats were trained to discriminate between 20 mg/kg pentylenetetrazol and 3 mg/kg chlordiazepoxide and then were injected with a large dose of ethanol (3–4 g/kg). Pentylenetetrazol-appropriate responding occurred at 9–12 hours after the ethanol dosing.[17] These results suggest that there is a delayed rebound effect that may represent an acute withdrawal state. Study of such effects may provide clues as to the neuroadaptation that occurs during the development of ethanol dependence.

D. ETHANOL AND TEMPERATURE PHYSIOLOGY

A well known physiological response associated with intoxicating doses of ethanol is that ethanol produces a dose-dependent impairment of thermoregulation that at normal ambient temperatures results in a significant hypothermia (see Chapter 17). This hypothermia has important pharmacokinetic consequences in that ethanol metabolism is slowed by the temperature drop. However, new data suggests that hypothermia appears to have a role in mediating various aspects of dependence. Oral intake of ethanol may be limited by the aversive effects of ethanol hypothermia. Perhaps more importantly, hypothermia can serve as an internal cue for the intoxication associated with ethanol, and as such, may have a role in the conditioning of internal states important for relapse.[18]

E. *IN VIVO* MICRODIALYSIS AND *IN VIVO* ELECTROPHYSIOLOGY

Advances in technology in the past decade have made the dream of measuring functional changes in neurotransmitter levels during ongoing behavior a reality. The technique of *in vivo* microdialysis, where extracellular levels of neurotransmitters can be measured in awake, freely moving animals, has provided a powerful means of testing hypotheses regarding the neuropharmacological mechanism of action of ethanol, as well as generating hypotheses for testing. Systemically administered or self-administered ethanol in intoxicating doses produces an increase in extracellular dopamine in the nucleus accumbens.[19,20] More recent work has shown that extracellular levels of the neuropeptide CRF are increased during ethanol withdrawal.[21] The possibility of measuring monoamines as well as neuropeptides by *in vivo* microdialysis is a major new advance in the search for the neuropharmacological basis of the reinforcement associated with ethanol dependence.

While the techniques for *in vivo* extracellular recording of individual neurons in freely moving animals have been available for some time, only recently have they been applied to neuropharmacologically defined neuronal populations.[22] The use of techniques for the study of ethanol reinforcement has begun to be applied to date, but such studies have begun in earnest with other abused drugs.[23,24] The recording from neuropharmacologically defined neurons in freely behaving animals would provide a much needed bridge from the cellular to the behavioral level of analysis and would be an extremely valuable complement to the *in vivo* microdialysis approach that is well underway.

F. QUANTITATIVE GENETIC ANALYSIS

New advances in behavioral genetics have been utilizing a unique statistical-based analysis to map drug response to putative chromosomal locations.[25] In this procedure, recombinant inbred strains are tested for drug response phenotypes and these are correlated with known marker loci that have been previously mapped in each of the recombinant inbred strains. Significant correlations can identify specific chromosomal regions responsible for a given trait. Eventually, further analysis at the molecular level may provide identification of specific genes that lie within these regions. This technique has the power to identify multiple genes for a given drug-related trait.

G. MOLECULAR BIOLOGY MEETS BEHAVIOR

Advances in molecular biology have now reached the functional level in that means exist to actively interfere with synthesis of specific proteins resulting in changes in receptors or neurotransmitters either through developmental manipulations (transgenic overexpression or knockouts) or through interference with ongoing protein synthesis (knockdowns). Transgenic manipulations have resulted in overexpression of neurotransmitters such as CRF[26] or knockouts such as the dopamine D-1 knockouts.[27] Such manipulations can be used to test specific hypotheses regarding the necessary and critical role of a specific system but are limited by the fact the organism has the capability of adapting to this perturbation during the course of development. Nevertheless, one might still expect a neuropharmacological probe to expose some type of deficit. The knockdown procedure involving antisense oligonucleotides directed at the translation of the DNA message to RNA has the advantage of an acute perturbation but the disadvantage that large amounts of oligonucleotides must be directly injected into the brain. This technique does provide a powerful means of producing a highly selective knockdown of a specific receptor subtype implicated in a particular functional effect and has been successfully implemented in studies of NPY receptors[28] and dopamine receptors.[29] Another interesting *in vitro* approach that is beginning to be used to perform structure-function analysis of receptor subtypes is the domain swapping procedure where chimeras and point mutations are made to delineate regions of a protein responsible for a certain function (binding, signal transduction, targeting, enzyme activity, etc.).[30] Using this approach one can identify whether receptor subtypes share pockets of ligand binding activity or whether they are different. One can imagine such a technique being applied to different domains of some of the cloned receptors sensitive to ethanol such as the GABA complex and the glutamate complex.

IV. NEW HOPE

A. ETHANOL ACTIONS AT INOTROPIC RECEPTORS, VOLTAGE-REGULATED ION CHANNELS, AND SECOND MESSENGER SYSTEMS: NEW AND SENSITIVE SELECTIVITY OF NEUROPHARMACOLOGICAL ACTION

One of the striking observations that is evident from the chapters in this book is that ethanol has powerful and potent effects on inotropic receptors, calcium channels, and second messenger systems. This work generates the exciting hypothesis that ethanol at low doses has actions on brain neurotransmitter systems with a selectivity never dreamed possible and for the first time provides strong impetus for understanding the mechanism of intoxication from the bottom up approach. One obvious question is, are these actions in any way related to some of the earlier nonspecific membrane-perturbing hypotheses? The data suggest that this is not the case because even with more sophisticated measures of membrane function such as interdigitation, the doses required are significantly higher than those that produce acute intoxication or robust effects on neurotransmitter machinery (see Chapter 2). That ethanol can perturb subtle membrane function that may contribute to its anesthetic properties or even tolerance and dependence is not in question. The point is that much lower doses of ethanol produce robust effects on specific protein parts of neuropharmacological machinery or "receptive elements,"[1] and these observations have great promise for understanding the selectivity of its behavioral actions at intoxicating doses. However, as the dose of ethanol is increased, susceptible lipid domains in the membrane may begin to affect protein function that may ultimately contribute to the effects of higher doses of ethanol (see Chapter 2).

The action of low doses of ethanol on the function of the GABA complex in *in vitro* models appears to be related to phosphorylation by protein kinase C (see Chapter 4). Ethanol at low doses has been shown to directly inhibit inward Ca^{2+} current in nerve terminal preparations (see Chapter 3), and ethanol at low doses decreases Ca^{2+} flux in NMDA receptor complex, possibly through an interference with the interaction of the co-agonists glycine and glutamate (see Chapter 5). Also, acute ethanol can selectively enhance the activation of the G-protein G_S by agonists (see Chapter 8) and can potentiate the effects of adenosine to stimulate G_S. Putting all these observations together it seems clear that low doses of ethanol (<50 mM) can have powerful effects on the cellular machinery dedicated to producing excitation and inhibition. Rather than exerting a classical neuropharmacological action at any specific binding site, ethanol appears to act at sites related to opening ionophores for chloride ions resulting in enhanced inhibition or blocking ionophores or channels for Ca^{2+} ions resulting in a blockade of excitation. Furthermore, ethanol could potentiate certain adenosine-mediated effects such as stimulation of adenylate cyclase via enhancement of G_S activity. Ethanol also has selective effects on voltage-regulated ion channels, notably Ca^{2+} channels and the K^+ M channel. Thus, ethanol may alter the overall balance of excitatory and inhibitory neurotransmission via actions at the second messenger level, as well as at receptor ionophores that produce the classical dose-related disinhibition-sedation-hypnosis-anesthesia behavioral dose effect function. The existence of some common element(s) among all these actions that account for the low dose disinhibitory/reinforcing actions of ethanol seems unlikely. More likely is that ethanol has highly selective actions on "receptive elements" within the neuropharmacological machinery required for neurotransmission, and the challenge is to identify where and how these actions take place. Parallel and interactive advances at the molecular, cellular, and system level will be required.

B. ETHANOL REINFORCEMENT — FROM THE MESOLIMBIC (DA?) SYSTEM TO THE EXTENDED AMYGDALA

There is compelling evidence for an activation of the mesolimbic dopamine system with low doses of ethanol, and neuropharmacological evidence suggests that facilitation or inhibition of dopamine action in the mesolimbic dopamine system can influence ethanol reinforcement (see Chapter 13). However, it is clear that dopamine is not necessary for ethanol reinforcement in that several studies have shown that removal of the mesolimbic dopamine system does not alter ethanol self-administration.[31,32] Thus, other neurotransmitters have clearly been implicated in ethanol reinforcement including GABA, glutamate, serotonin, and opioid peptides. GABA antagonists, serotonin agonists, and opioid peptide antagonists are effective in decreasing ethanol self-administration in a variety of paradigms. Glutamate antagonists substitute for ethanol in drug discrimination studies, raising the possibility that glutamate may also be involved in this reinforcing effect. The role of these different neurotransmitter systems in ethanol reinforcement may vary. Discrete neurotransmitter systems may be differentially involved in various aspects of the reinforcement process (appetitive-consummatory) as elegantly outlined by Samson and Hodge (see Chapter 13).

However, the question remains as to what exactly constitutes the mesolimbic system and whether this system could be extended to include more classical neural substrates of emotional behavior. While the focus in Chapter 13 is on the mesolimbic dopamine system, this conceptualization results in some limitations in perspective. An alternative approach is to place the focus "upstream" at the region of the nucleus accumbens and its connections. Recent anatomical data suggest that a subregion of the nucleus accumbens, the shell of the nucleus accumbens, has cytoarchitectural similarities and intimate connections with certain other areas of the limbic forebrain, including the bed nucleus of the stria terminalis and the central nucleus of the amygdala.[33] This "extended amygdala" has important connections with the extrapyramidal motor system and the lateral hypothalamus and is rich in many of the neurotransmitters implicated in the reinforcing actions of ethanol such as serotonin, GABA, and opioid peptides.

Opioid peptide systems provide a case in point. Opiate antagonists have long been known to decrease ethanol drinking and operant responding for ethanol.[3] These opioid interactions with ethanol appear to involve delta receptors in that delta receptor antagonists are particularly potent in decreasing ethanol self-administration.[7] However, opioid antagonists also decrease consummatory behaviors for other reinforcers, including sweetened solutions, suggesting that these opioid antagonist effects may reflect a general action on reinforcement mechanisms with which ethanol interacts. However as noted above, recent data has shown that the opioid antagonist naltrexone is effective in decreasing relapse to alcoholism in detoxified alcoholics in several recent double-blind, placebo-controlled studies.[8,9] At this time the parts of the mesolimbic/extended amygdala systems which are involved in these opioid effects are unknown, the relationship to the control of other consummatory behavior such as food and water intake is unclear, and the interaction of opioid antagonists with the mesolimbic dopamine system remains to be elucidated. Nevertheless, the opioid system/ethanol interactions raise the possibility that multiple neurotransmitters are involved in the reinforcing actions of ethanol.

C. ETHANOL AND STRESS

Stress and ethanol have historically been closely linked, either through the tension reduction hypothesis or through the anxiety associated with chronic alcoholism or alcohol withdrawal and protracted abstinence. Both acute and chronic ethanol are powerful activators of the hypothalamic-pituitary adrenal axis, and this appears to be the result of activation of the neuropeptide CRF in the hypothalamus to, in turn, activate the classic neuroendocrine stress response.[34] However, recent evidence suggests that chronic ethanol may also interact with an extensive extrahypothalamic, extra-neuroendocrine CRF system implicated in behavioral responses to stress (for review, see Reference 35). CRF itself has "anxiogenic"-like effects, and CRF antagonists injected into the central nervous system can reverse many behavioral responses to stress. The anxiogenic-like effect of ethanol withdrawal can be reversed by intracerebroventricular administration of the CRF antagonist[36] and, in a more recent study, by intracerebral administration of the CRF antagonist into the central nucleus of the amygdala.[37]

Even more compelling is the observation of an increase in extracellular levels of CRF in the amygdala during ethanol withdrawal.[21]

D. MISSING LINKS

At the molecular level, research has begun to focus in on neuropharmacological changes that clearly are within the range of ethanol doses that produce intoxication. It is curious that two major systems, the GABA and glutamate receptor complexes, definitely affected by ethanol are inotropic but the net result of ethanol's action is a decrease in excitability. Also, ethanol's inhibition of Ca^{2+} function appears to be an important component of low-dose ethanol actions also resulting in decreases in neuronal excitability. Indeed, even where ethanol produces an activation of a second messenger system, such as the potentiation of ability of adenosine to activate G_S and subsequently adenylate cyclase, the ultimate functional effect appears to be one of a potentiation of inhibition. While at present it seems unlikely that there is some ethanol-protein interaction that is common to these effects, one wonders what future biochemical/molecular studies will reveal. The burgeoning field of functional molecular biology involving transgenic manipulations, antisense knockdowns, and knockouts should provide the means to address such questions by allowing the specific manipulation of cellular proteins involved in signal transduction.

What is also largely unknown is exactly which molecular neuropharmacological effect conveys the specificity associated with given neurotransmitter actions. For example, not all GABA receptors are affected by low doses of ethanol, and the mechanism by which ethanol selectively influences some, but not all, GABA receptor complexes is unknown. Furthermore, the precise localization of the ethanol-sensitive GABA receptor complexes which underlie ethanol's reinforcing effects is unknown. Similar questions can be posed regarding the opioid peptide and serotonin systems. Neuronal involvement of the mesolimbic dopamine system has been confirmed, but how ethanol specifically accesses this system is largely unknown. However, research will continue to address these questions, as data accumulates from both the "top down" and "bottom up" approaches. The brain sites important for the opioid antagonist actions will be identified, and the molecular mechanisms responsible for activation of the opioid peptide systems will be elucidated. The development of reliable measures of ethanol reinforcement, the cloning of opioid receptors, and the ability to measure extracellular levels of opioid peptides with *in vivo* microdialysis are only a small sample of the exciting new approaches to be applied to this field. A largely neglected but critical bridging area will be actual electrophysiological recording from neuropharmacologically identified cells during the various aspects of ethanol reinforcement.

Perhaps a more important area of research that has been neglected is how these reinforcement pathways are altered during the neuroadaptation that has long been hypothesized to be a part of ethanol tolerance and withdrawal. Neuroadaptation can take many forms and can occur at many levels, as is illustrated in a number of the chapters in this book. One approach to this issue has been to arbitrarily divide neuronal adaptation into two separate levels. A within-system adaptation was hypothesized to involve changes within the same neurotransmitter system involved in the acute reinforcing actions of the drug, and a between-system adaptation was hypothesized to involve a change between distinct neurotransmitter systems implicated in the acute reinforcing actions of ethanol.[38] An example of a within-systems adaptation could be the acute vs. chronic effects of ethanol on extracellular dopamine levels in the nucleus accumbens. Ethanol in intoxicating doses increased extracellular levels of dopamine in the nucleus accumbens during self-administration, and withdrawal from ethanol decreases dopamine levels in this brain region.[39] Alternatively, there could be a change at the second messenger level such that a G-protein stimulated by acute ethanol may no longer be influenced by chronic ethanol (see Chapter 8). For an example of a between-system adaptation, a neurotransmitter system not normally implicated in the acute reinforcing actions of ethanol may be recruited during the course of dependence to counteract the acute effects of ethanol. This activation may then be manifest during ethanol withdrawal. The response of central nervous system corticotropin releasing factor may be an example of a between-system adaptation (see Section IV.C.). However, regardless of these arbitrary distinctions, it is clear that adaptation can occur from the molecular-cellular to the system level of analysis, and the challenge for future research will be to determine how these coordinated changes mediate the motivation for ethanol dependence.

ACKNOWLEDGMENTS

This chapter was prepared with support from NIAAA Center Grant AA06420 (George F. Koob, Director) and Grant AA 08459 (GFK). I would like to thank Amanda Roberts and Brian Baldo for their helpful ideas and corrections on the manuscript and the Molecular and Experimental Medicine Word Processing Center for their help in manuscript preparation.

REFERENCES

1. Tabakoff, B. and Hoffman, P.L., Alcohol: Neurobiology, in *Substance Abuse. A Comprehensive Textbook*, Lowenstein, J.H., Ruiz, P., and Millman, R.B., Eds., Williams and Wilkins, Baltimore, 1992, 152–185.
2. Gordon, A.S., Collier, K., and Diamond, I., Ethanol regulation of adenosine receptor-stimulated cAMP levels in a clonal neural cell line: an *in vitro* model of cellular tolerance to ethanol, *Proc. Natl. Acad. Sci. U.S.A.*, 83:2105–2108, 1986.
3. Reid, L.D. and Hunter, G.A., Morphine and naloxone modulate intake of ethanol, *Alcohol*, 1:33–37, 1984.
4. Altshuler, H.L., Phillips, P.E., and Feinhandler, D.A., Alterations of ethanol self-administration by naltrexone, *Life Sci.*, 26:679–688, 1980.
5. Hyytia, P., Involvement of μ-opioid receptors in alcohol drinking by alcohol-preferring AA rats, *Alcoholism: Clin. Exp. Res.*, 17:697–701, 1993.
6. Weiss, F., Mitchiner, M., Bloom, F.E., and Koob, G.F., Free-choice responding for ethanol versus water in Alcohol-Preferring (P) and unselected Wistar rats is differentially altered by naloxone, bromocriptine and methysergide, *Psychopharmacology*, 101:178–186, 1990.
7. Froehlich, J.C., Zweifel, M., and Hart, J., Importance of delta opioid receptors in maintaining high alcohol drinking, *Psychopharmacology*, 103:467–472, 1991.
8. Volpicelli, J.R., Alterman, A.I., Hayashida, M., and O'Brien, C.P., Naltrexone in the treatment of alcohol dependence, *Arch. Gen. Psychiatry*, 49:876–880, 1992.
9. O'Malley, S.S., Jaffee, A.J., Chang, G., Schottenfeld, R.S., Meyer, R.E., and Rounsaville, B., Naltrexone and coping skills therapy for alcohol dependence, *Arch. Gen. Psychiatry*, 49:881–888, 1992.
10. Masserman, J.H. and Yum, K.S., An analysis of the influence of alcoholism experimental neuroses in cats, *Psychosom. Med.*, 8:36–52, 1946.
11. Koob, G.F. and Britton, K.T., Neurobiological substrates for the anti-anxiety effects of ethanol, in *Pharmacology and Neuropharmacology, Vol. 2: Alcohol and Alcoholisms*, Begleiter, H. and Kissin, B., Eds., Oxford University Press, 1995 (in press).
12. Geller, I. and Seifter, J., The effect of meprobamate, barbiturates, d-amphetamine and promazine on experimentally-induced conflict in the rat, *Psychopharmacologia*, 1:482–491, 1960.
13. Pollard, G.T. and Howard, J.V., The Geller-Seifter conflict paradigm with incremental shock, *Psychopharmacologia*, 62:117–171, 1979.
14. Vogel, J.R., Beer, B., and Clody, D.E., A simple and reliable conflict procedure for testing anti-anxiety agents, *Psychopharmacologia*, 21:1–7, 1971.
15. Grant, K.A. and Colombo, G., Discriminative stimulus effects of ethanol: effect of training dose on the substitution of N-methyl-d-aspartate antagonists, *J. Pharmacol. Exp. Ther.*, 264:1241–1247, 1993.
16. Lal, H., Harris, C.M., Benjamin, D., Springfield, A.C., Bladra, S., Emmett-Oglesby, M., Characterization of a pentylenetetrazol-like interoceptive stimulus produced by ethanol withdrawal, *J. Pharmacol. Exp. Ther.*, 247:508–518, 1988.
17. Gauvin, D.V., Harland, R.D., Criado, J.R., Michaelis, R.C., and Holloway, F.A., The discriminative stimulus properties of ethanol and acute ethanol withdrawal states in rats, *Drug Alcohol Depend.*, 24:103–113, 1989.

18. Cunningham, C.L., Niehus, J.S., and Bachtold, J.F., Ambient temperature effects on taste aversion conditioned by ethanol: contribution of ethanol-induced hypothermia, *Alcoholism: Clin. Exp. Res.*, 16:1117–1124, 1992.
19. Imperato, A. and DiChiara, G., Preferential stimulation of dopamine release in the nucleus accumbens of freely moving rats by ethanol, *J. Pharmacol. Exp. Ther.*, 239:219–239, 1986.
20. Weiss, F., Lorang, M.T., Bloom, F.E., and Koob, G.F., Ethanol self-administration stimulates dopamine release in the rat nucleus accumbens: genetic and motivational determinants, *J. Pharmacol. Exp. Ther.*, 267:250–258, 1993.
21. Pich, E.M., Lorang, MN., Yaganeh, M., DeFonseca, F.F., Koob, G.F., and Weiss, F., Increase of corticotropin-releasing factor release from the amygdala of awake rats during ethanol withdrawal as measured with microdialysis, *J. Neurosci.*, 1995 (in press).
22. Foote, S.L. and Aston-Jones, G.S., Pharmacology and physiology of central noradrenergic systems, in *Psychopharmacology: The Fourth Generation of Progress*, Bloom, F.E. and Kupfer, D.J., Eds., Raven Press, Ltd., New York, 1995, 335–345.
23. Chang, J.Y., Sawyer, S.F., Lee, R.-S., and Woodward, D.J., Electrophysiological and pharmacological evidence for the role of the nucleus accumbens in cocaine self-administration in freely moving rats, *J. Neurosci.*, 14:1224–1244, 1994.
24. Criado, J.R., Lee, R.-S., Berg, G., and Henriksen, S.J., Sensitivity of nucleus accumbens neurons *in vivo* to intoxicating doses of ethanol, *Alcoholism: Clin. Exp. Res.*, 19:164–169, 1995.
25. Crabbe, J.C., Belknap, J.K., and Buck, K.J., Genetic animal models of alcohol and drug abuse, *Science*, 264:1715–1723, 1994.
26. Stenzel-Poore, M.P., Heinrichs, S.C., Rivest, S., Koob, G.F., and Vale, W.W., Overproduction of corticotropin-releasing factor in transgenic mice: a genetic model of anxiogenic behavior, *J. Neurosci.*, 14:2579–2584, 1994.
27. Xu, M., Moratella, R., Gold, L.H., Hiroi, N., Koob, G.F., Graybiel, A.M., and Tonegawa, S., Dopamine D1 receptor mutant mice are deficient in striatal expression of dynorphin and in dopamine-mediated behavioral responses, *Cell*, 79:729–742, 1994.
28. Wahlestedt, C., Pich, E.M., Koob, G.F., Yee, F., and Heilig, M., Downregulation of brain neuropeptide Y (NPY) Y1 receptors by *in vivo* administration of an antisense oligodeoxy-nucleotide is associated with signs of anxiety in rats, *Science*, 259:528–530, 1993.
29. Zhou, L.W., Zhang, S.P., Qin, Z.H., and Weiss, B., *In vivo* administration of an oligodeoxynucleotide antisense to the D2 dopamine receptor messenger RNA inhibits D2 dopamine receptor-mediated behavior and the expression of D2 dopamine receptors in mouse striatum, *J. Pharmacol. Exp. Ther.*, 268:1015–1023, 1994.
30. Liggett, S.B., Freedman, N.J., Schwinn, D.A., and Lefkowitz, R.J., Structural basis for receptor subtype-specific regulation revealed by a chimeric beta 3/beta 2-adrenergic receptor, *Proc. Natl. Acad. Sci. U.S.A.*, 90:3665–3669, 1993.
31. Myers, R.D. and Quarfordt, S.D., Alcohol drinking attenuated by sertraline in rats with 6-OHDA or 5,7-DHT lesions of N. accumbens: a caloric response, *Pharmacol. Biochem. Behav.*, 40:923–928, 1991.
32. Rassnick, S., Stinus, L., and Koob, G.F., The effects of 6-hydroxydopamine lesions of the nucleus accumbens on oral self-administration of ethanol in the rat, *Brain Res.*, 623:16–24, 1993.
33. Heimer, L. and Alheid, G., Piecing together the puzzle of basal forebrain anatomy, in *The Basal Forebrain: Anatomy to Function*, Napier, T.C., Kalivas, P., and Hanin, I., Eds., Plenum Press, New York, 1991, 1–42.
34. Rivier, C., Bruhn, T., and Vale, W., Effect of ethanol on the hypothalamic-pituitary-adrenal axis in the rat: role of corticotropin-releasing factor (CRF), *J. Pharmacol. Exp. Ther.*, 229:127–131, 1984.
35. Koob, G.F., Heinrichs, S.C., Menzaghi, F., Pich, E.M., and Britton, K.T., Corticotrophin-releasing factor, stress and behavior, *Semin. Neurosci.*, 7:221–229, 1994.
36. Baldwin, H.A., Rassnick, S., Rivier, J., Koob, G.F., and Britton, K.T., CRF antagonist blocks alcohol withdrawal "anxiogenic" response, *Psychopharmacology*, 103:227–232, 1991.
37. Rassnick, S., Heinrichs, S.C., Britton, K.T., and Koob, G.F., Microinjection of a corticotropin-releasing factor antagonist into the central nucleus of the amygdala reverses anxiogenic-like effects of ethanol withdrawal, *Brain Res.*, 605:25–32, 1993.
38. Koob, G.F. and Bloom, F.E., Cellular and molecular mechanisms of drug dependence, *Science*, 242:715–723, 1988.
39. Smith, A., Parsons, L.H., Pich, E.M., Hyytia, P., Schulteis, G., Lorang, M.T., Yackey, M.F., Koob, G.F., and Weiss, F., Ethanol modifies extracellular levels of dopamine, serotonin, and corticotropin releasing factor in the limbic forebrain: studies in rats with different histories of ethanol exposure, *Neurosci. Abstr.*, 1994.

Chapter 2

MEMBRANES AND ETHANOL: LIPID DOMAINS AND LIPID-PROTEIN INTERACTIONS

W. Gibson Wood, Friedhelm Schroeder, A. M. Rao, Urule Igbavboa, and Nicolai A. Avdulov

TABLE OF CONTENTS

I. INTRODUCTION

The molecular mechanisms involved in producing ethanol intoxication and tolerance are not well-understood. An understanding of the behavior of ethanol at the molecular level is complicated by the fact that ethanol has a multi-faceted effect on cells in contrast to a localized site of action on a specific cellular target. The nonspecific effects of ethanol resulted in a large body of work that examined how ethanol affected bulk membrane lipid structure. One hypothesis was that ethanol fluidized the bulk membrane lipid and that the increase in fluidity was involved in intoxication (Chin and Goldstein, 1977a; Chin and Goldstein, 1977b; Chin and Goldstein, 1981; Goldstein et al., 1982). Moreover, ethanol tolerance was proposed to result from modification of bulk membrane lipid composition that reduced the fluidizing effects of ethanol (Chin and Goldstein, 1984; Daniels and Goldstein, 1982; Parsons et al., 1982; Chin et al., 1978; Taraschi et al., 1986; Rubin and Rottenberg, 1982; Waring et al., 1981). While there was substantial support for the bulk lipid hypothesis, there was also evidence that was not in agreement with the bulk lipid hypothesis (reviewed in Klemm, 1990; Deitrich et al., 1989; Goldstein, 1986; Sun and Sun, 1985; Wood and Schroeder, 1988; Wood et al., 1989a; Stubbs and Rubin, 1993). For example, a criticism of the bulk membrane lipid hypothesis was that a one or two degree change in temperature had a greater fluidizing effect on the bulk membrane than a concentration of ethanol that could not be tolerated by an individual *in vivo* (Hunt, 1985). It was reasoned that relatively small changes in body temperature did not produce intoxication. Therefore, ethanol-induced increases in bulk membrane fluidity could not account for membrane sensitivity to ethanol. Effects of temperature on bulk membrane fluidity was a common criticism of the bulk membrane lipid hypothesis and

0-8493-8389-7/95/$0.00+$.50

were based on the assumption that ethanol and temperature had equivalent effects on membranes. This assumption is not valid, and it is obvious that effects of ethanol and temperature on membranes are more complex than previously thought. Temperature and alcohols do not have equivalent effects on all aspects of membrane structure and function (Treistman and Wilson, 1987; Treistman et al., 1987; Sauerheber et al., 1982). For example, ethanol *in vitro* inhibits Na^+ + K^+-ATPase activity (Rangaraj and Kalant, 1982; Nhamburo et al., 1987; Levental and Tabakoff, 1980). Increasing temperature, on the other hand, increased activity of Na^+ + K^+-ATPase (Rangaraj and Kalant, 1982). Benzyl alcohol and ethanol inhibited glucose uptake in adipocytes (Sauerheber et al., 1982). Increasing temperature stimulated the uptake of glucose in adipocytes (Sauerheber et al., 1982).

Another argument against the bulk lipid hypothesis was effects of ethanol *in vitro* on membranes of aged animals (Strong and Wood, 1984; Armbrecht et al., 1983). Generally, aged animals respond differently to effects of ethanol *in vivo* when compared to younger animals (Wood and Armbrecht, 1982b). Aged animals sleep longer following an injection of ethanol and regain the righting reflex at a lower blood ethanol concentration when compared to younger animals (Wood and Armbrecht, 1982a; Ritzmann and Springer, 1980). However, fluidity of synaptic plasma membranes (SPM) and depolarized GABA release from synaptosomes of aged mice (22–28 months) were less perturbed by ethanol *in vitro* than mice 3–6 months of age (Strong and Wood, 1984; Armbrecht et al., 1983). The resistance of membranes of aged animals to ethanol perturbation was similar to resistance observed in membranes of young ethanol-tolerant animals. If an ethanol-induced increase in membrane bulk fluidity was an explanation for ethanol intoxication, then membranes of aged animals should have been more affected by ethanol *in vitro* in contrast to the observed findings of resistance to ethanol perturbation.

It has been well-established that plasma membranes of ethanol-tolerant animals and alcoholic patients were resistant to perturbation by ethanol *in vitro* (reviewed in Hunt, 1985; Wood et al., 1993b). Membrane tolerance to ethanol *in vitro* was attributed to changes in the bulk lipid composition of the membrane, e.g., total cholesterol content, specific phospholipids, and fatty acyl groups. Effects of chronic ethanol consumption on bulk membrane lipid composition have not been consistently observed (Hunt, 1985; Sun and Sun, 1985; Wood and Schroeder, 1988).

Ethanol-induced changes in membrane bulk lipid fluidity and lipid composition did not fully explain how ethanol produced intoxication or the development of tolerance. A simple conclusion is that membrane lipids are not involved in the actions of ethanol. However, membranes are complex structures whose environment consists of other properties in addition to bulk fluidity and lipid composition, e.g., dielectric constant, interdigitation, lipid domains, and lipid-protein interactions (Yeagle, 1992). Recent studies examining these different properties of membranes are revealing new and exciting data on effects of ethanol. This chapter will focus on the role of dielectric constant, interdigitation, lipid domains, and lipid-protein interactions as factors contributing to mechanisms of membrane sensitivity to ethanol and membrane tolerance to ethanol.

II. NEW APPROACHES TO UNDERSTANDING EFFECTS OF ETHANOL ON MEMBRANE PROPERTIES

A. MEMBRANE DIELECTRIC CONSTANT AND DEHYDRATION

The membrane interior is hydrophobic, and the surface is hydrophilic. The polarity of the membrane can be expressed by the dielectric constant. For example, the dielectric constant of water at 20°C is 80, and the dielectric constant of oleic acid at 20°C is 2.5. The inference is that the dielectric constant of the interior membrane core is much lower than the membrane surface. Orme, et al. (1988) studied the permeability of various ions and concluded that alcohols increased the dielectric constant and modification of the dielectric constant was the primary mechanism for effects of alcohols. Different alcohols were found to increase transmembrane flux of hydrophobic cations and anions. The hypothesis that alcohols increased the dielectric constant was based in part on the Born-Parsegian equation, which predicts that the permeability of small ions will increase over large ions as the dielectric constant is increased (Orme et al., 1988). In a more direct test of effects of alcohols on membrane dielectric constant, Rottenberg (1992) reported that alcohols, including ethanol, increased the membrane dielectric constant. These studies were conducted using model membranes and different biological membranes including SPM. Dielectric constant was measured using the fluorescent probe Prodan. Effects of ethanol on membrane dielectric constant were interpreted as a result of ethanol partitioning into the membrane surface.

Based on the findings of Orme et al. (1988) and Rottenberg (1992), we examined the role of membrane dielectric constant and ethanol in SPM of Long-Sleep (LS) and Short-Sleep (SS) lines of mice (Schroeder et al., 1994). LS mice sleep longer following an injection of ethanol as compared to SS mice (Tabakoff et al., 1980). Multifrequency phase and modulation fluorimetry of diphenylhexatriene (DPH) was used to resolve structural and dielectric differences in the membrane interior core. Fluorescence spectral peak ratios, fluorescence lifetime analysis, and initial rates of photoreaction of DPH in SPM provided sensitive measures of SPM interior core dielectric properties. Bulk membrane fluidity (i.e., limiting anisotropy and rotational relaxation time) did not significantly differ between the SPM of LS and SS mice. The membrane interior sensed by DPH was more polar in SPM of SS than LS mice. Ethanol *in vitro* (25–75 mM) increased the SPM interior core dielectric of LS, but not SS, mice. In addition, ethanol *in vitro* increased the fraction of the DPH short lifetime component in SPM of LS mice. The ethanol-induced change in lifetime reflects DPH photoreactivity, which increases with increasing dielectric. Ethanol increased the polar molecules (ethanol and/or water) in the SPM hydrophobic core sensed by DPH, and our findings were consistent with the earlier studies showing that alcohols increased dielectric constant (Orme et al., 1988; Rottenberg, 1992). Ethanol-induced effects on membrane dielectric constant may contribute to changes in neuronal function. Short-chain alcohols stimulate ion transport, and alcohols increase the membrane dielectric constant (Orme et al., 1988). Effects of chronic ethanol consumption on membrane dielectric constant have not been reported.

Ethanol increased the membrane dielectric constant in different biological membranes. This effect of ethanol on the membrane dielectric constant may have resulted from a weakening of hydrogen bonding allowing water and/or ethanol to hydrate the membrane hydrophobic core. However, there is evidence indicating that ethanol and other alcohols actually dehydrate the membrane (Chiou et al., 1990; Chiou et al., 1992; Klemm, 1990; Yurttas et al., 1992). Alcohols have been reported (Yurttas et al., 1992) to release water in reversed micelles determined by Fourier transform infrared spectroscopy (FTIR). The conclusion that alcohols dehydrate the membrane is not consistent with the findings showing that alcohols increased the membrane dielectric constant (Orme et al., 1988; Rottenberg, 1992; Schroeder et al., 1994). If alcohols remove water from the membrane, then the dielectric constant should be reduced even with ethanol present in the membrane. The dielectric constant of ethanol is 25, and the dielectric constant of water is 80 (Weast et al., 1987). A major difference between the membrane dielectric studies and the dehydration studies was the use of biological membranes and reverse micelles, respectively. The reason for using reverse micelles was that the detection of released water in a reversed micelle is more accurate than in a biological membrane where there is a large amount of water outside the membrane. A potential problem with the reverse micelle is that the FTIR method does not allow determination of where the water may go. Ethanol would appear to break hydrogen bonds, and it is possible that water at the membrane surface may move deeper into the membrane. Thus, free water could be increased, and the membrane dielectric would also increase as a result of ethanol being able to partition deeper into the membrane. It would seem that experiments need to be conducted using biological membranes or model unilamellar membranes to test whether ethanol removes water from membranes or redistributes water within the membrane.

B. MEMBRANE INTERDIGITATION

The traditional view of the membrane bilayer is that the exofacial and cytofacial leaflets meet in a well-defined midplane of the bilayer (Slater and Huang, 1992). However, it has been shown that the exofacial and cytofacial leaflets can interdigitate under certain conditions (Slater and Huang, 1992). Ethanol has been shown to induce interdigitation in model membranes. Rowe's group reported that ethanol could induce the interdigitated phase in phosphatidylcholines using fluorescence intensity of DPH (Nambi et al., 1988). An important aspect of that work was that the fluorescence method was calibrated with measurements made using X-ray diffraction (Nambi et al., 1988) on the same preparations. X-ray diffraction is a direct method to determine interdigitation. In a subsequent study, it was found that cholesterol blocked the effects of ethanol on inducing the interdigitated phase (Komatsu and Rowe, 1991). Ethanol-induced interdigitation was reported in large unilamellar vesicles of dipalmitoylphosphatidylcholine as measured by the ratio of excimer to monomer fluorescence intensity of pyrene-labeled phosphatidylcholine (Yamazaki et al., 1994). These results were confirmed by high-resolution electron cryomicroscopy.

Ethanol caused a shift from the noninterdigitated to the interdigitated phase in model membranes. The question was addressed as to whether ethanol would induce interdigitation in a biological membrane (Avdulov et al., 1994b). We examined effects of ethanol on SPM of high alcohol sensitivity (HAS) and low alcohol sensitivity (LAS) rats using energy transfer between the surface fluorescent probe 1-anilinonaphthalene-8-sulfonic acid (ANS) and the hydrophobic probe, benzo[*def*]phenanthrene (pyrene). ANS quenches the fluorescence of the pyrene monomer. The efficiency of the energy transfer was calculated by the following equation:

$$\left(I - I_{da}\right)/I_d \tag{1}$$

where I_d and I_{da} represent fluorescence intensity of donor (pyrene) in the absence and presence of acceptor (ANS), respectively. Interdigitation was considered to be proportional to the distance (D) between the donor and acceptor and was calculated using the following equation:

$$D = \left[\left(F'/F_{max}\right)\left(\langle E\rangle^{-1} - 1\right)\right]^{1/3} \tag{2}$$

where $\langle E\rangle$ was the average efficiency of energy transfer, F' was the fluorescence intensity of the concentration of acceptor utilized, and F_{max} was maximal fluorescence of acceptor in SPM. Both F' and F_{max} were obtained by ANS binding studies in the absence of pyrene. The ratio F':F_{max} provides a measure of binding of ANS to the SPM surface and allows correction for effects of ethanol on ANS binding.

The extent of interdigitation was similar in SPM of HAS and LAS rats in the absence of ethanol. SPM interdigitation was increased by 50 m*M* ethanol in HAS but not LAS rats. This study demonstrated that interdigitation could be determined in a biological membrane. In addition, effects of ethanol on SPM interdigitation were observed at a much lower ethanol concentration (50 m*M*) than effects seen in model membranes (>500 m*M*). The greater sensitivity of ethanol-induced interdigitation of SPM may have resulted from the different fluorescent techniques and/or inherent differences in model membranes and SPM. The model membrane studies used unilamellar vesicles that were homogenous in lipid composition. SPM are heterogenous in lipid composition, the lipids are asymmetrically distributed in the bilayer, and proteins are present. Such differences may contribute to greater effects of ethanol on interdigitation in a biological membrane as compared to a model membrane. Ethanol-induced partial interdigitation may occur as a result of lipid domains in a biological membrane.

There have been no reports on effects of chronic ethanol consumption on membrane interdigitation using fluorescence techniques or X-ray diffraction. However, data of a study on liver plasma membranes from chronic ethanol-treated rats were interpreted as resulting from ethanol-induced interdigitation (Lewis et al., 1989) Membrane order measured by vibrational infrared spectroscopy revealed that membranes of chronic ethanol-treated animals were more ordered and that ethanol *in vitro* also ordered the membrane. It was suggested that the presence of ethanol may have resulted in increased lipid packing associated with partial interdigitation (Lewis et al., 1989). Effects of ethanol *in vitro* were different on membrane order determined by vibrational infrared spectroscopy as compared to most studies which have reported that ethanol had a fluidizing effect on membranes. Vibrational infrared spectroscopy may report more on different properties of membranes than studies using probes such as DPH or spin-labeled probes.

Changes in membrane interdigitation have not only structural consequences but also can affect activity of proteins. Activity of Ca^{2+} + Mg^{2+}-ATPase inserted into phosphatidylcholines of different chain lengths was dependent on interdigitation (Johannsson et al., 1981). ATPase activity was increased as the chain length of phosphatidylcholine increased. Interdigitation is reduced as the chain length of the phospholipid is increased. In contrast, it was reported that Ca^{2+} + Mg^{2+}-ATPase activity required an optimal thickness or interdigitation and that phosphatidylcholines with longer or shorter chain lengths as compared to C18 reduced enzyme activity (Cornea and Thomas, 1994). An additional finding of that study was that mobility of Ca^{2+} + Mg^{2+}-ATPase was reduced, and it was suggested that changes in thickness affected protein aggregation. We have reported that ethanol *in vitro* induced protein clustering or aggregation in SPM of HAS rats (Avdulov et al., 1994b). Ethanol also induced interdigitation in SPM of HAS rats as described earlier in this section. Effects of ethanol on interdigitation have both structural and functional consequences and may help explain ethanol-induced changes in certain membrane proteins and receptors.

III. MEMBRANE LIPID DOMAINS

Lipids are not uniformly distributed in a membrane but occur in different domains or pools. Two large domains are vertical lipid domains and lateral lipid domains (Op den Kamp, 1979; Schroeder et al., 1991a; Devaux, 1991; Wood and Schroeder, 1992). Another type of domain that has been described is boundary or annular lipid domains associated with different membrane proteins (Lentz, 1988). Vertical domains refer to the two monolayers or leaflets of the membrane. Membrane leaflets differ in fluidity, lipid distribution, electrical charge, and regulatory activity of certain membrane proteins. The exofacial leaflets of different membranes (SPM, LM fibroblasts, and intestinal brush border membranes) were substantially more fluid than the cytofacial leaflet (Wood et al., 1989b; Schroeder et al., 1988; Sweet et al., 1987; Dudeja et al., 1991). Phosphatidylcholine is the predominant phospholipid in the exofacial leaflet whereas the cytofacial leaflet contains primarily phosphatidylserine, phosphatidylinositol, and phosphatidylethanolamine. Sphingomyelin is located exclusively in the exofacial leaflet of SPM, and the exofacial leaflet also contains approximately 20% of phosphatidylethanolamine. The exofacial leaflet is considered zwitterionic, and the cytofacial leaflet is negatively charged. Cholesterol, a major lipid of plasma membranes, is asymmetrically distributed with the greatest concentration of cholesterol in the cytofacial leaflet. Cholesterol asymmetry has been reported in SPM (Wood et al., 1990), LM membranes (Kier et al., 1986), and erythrocytes (Brasaemle et al., 1988; Schroeder et al., 1991b). Lateral lipid domains refer to the distribution of lipids across the horizontal plane of the membrane. Such domains have been identified for phosphatidylserine, phosphatidylethanolamine, and cholesterol (Treistman et al., 1987; Schroeder et al., 1991a). Lateral lipid domains may also occur within each membrane leaflet, but such data have not been reported. Annular lipid domains are lipids in close proximity to proteins and are involved in regulation of protein function (Lentz, 1988). The hypothesis that ethanol may act on membrane lipid domains had been proposed earlier by other investigators (Hunt, 1985; Harris, 1984; Levental and Tabakoff, 1980) but not directly examined until the past four or five years.

A. EXOFACIAL AND CYTOFACIAL LEAFLETS

Sheetz and Singer (1974) reported that drugs which differed in their charge properties had a differential effect on the shape of erythrocytes. Cationic drugs induced a cup-forming shape, and anionic drugs resulted in crenation. It was concluded in that study that the anionic drugs acted on the zwitterionic exofacial leaflet and the cationic drugs acted on the negatively charged cytofacial leaflet. Anionic drugs have been shown to have a greater effect on fluidity of the zwitterionic exofacial leaflet, and cationic drugs had a larger effect on the negatively charged cytofacial leaflet of LM plasma membranes (Sweet et al., 1987). Ethanol is not a highly charged drug, so it would not be expected that ethanol would preferentially act on one leaflet versus the other. However, it was known that there was a large difference in fluidity of the exofacial as compared to the cytofacial leaflet and that ethanol had a greater effect on fluid as compared to less fluid membranes. We made the simple prediction that ethanol would have a greater effect on the more fluid leaflet of the membrane (Wood and Schroeder, 1988; Schroeder et al., 1988). This hypothesis was tested in mouse SPM using trinitrobenzenesulfonic acid (TNBS) and the fluorescent probe DPH. The TNBS method had been used previously to determine fluidity of the individual membrane leaflets (Sweet et al., 1987; Sweet and Schroeder, 1988a; Kier et al., 1986; Schroeder, 1985). TNBS covalently binds to amino groups in the membrane. Trinitrophenyl groups quench fluorescence of DPH and other fluorescence probe molecules by energy transfer. Trinitrophenylation of intact synaptosomes at 4°C (nonpenetrating conditions) and subsequent isolation of SPM results in covalent attachment of trinitrophenyl quenching groups to the exofacial leaflet. An important requirement of the TNBS method is that TNBS does not penetrate into the cytofacial leaflet. We have shown that penetration of the cytofacial leaflet by TNBS at 37°C resulted in a reduction in fluorescence lifetime and a decrease in fluorescence intensity of DPH when compared to incubation at 4°C and SPM not incubated with TNBS.

The TNBS method is based on the assumption that the membrane is composed of fluorescing compartments of different accessibility to TNBS. When the fluorescence intensity, F, and anisotropy, r, are measured simultaneously, then

$$r = \Sigma F_i r_i \tag{3}$$

where F_i is the fraction of fluorescence intensity in compartment i. Therefore, values for the exofacial and cytofacial are solved using the following equation

$$r = F_c/F \;\; r_c + F - F_c/F \;\; r_c \tag{4}$$

where F and F_c are fluorescence of DPH in SPM isolated from synaptosomes that were incubated with buffer only or with TNBS at nonpenetrating conditions (4°C). Anisotropy, r, (both leaflets) and r_c (cytofacial leaflet anisotropy) were determined for DPH in SPM. The equation is then solved for r_e (exofacial leaflet).

The limiting anisotropy of DPH in SPM was markedly less in the exofacial leaflet than the cytofacial leaflet (Schroeder et al., 1988). The anisotropy value is proportional to fluidity with a lower value representing a more fluid membrane. Ethanol *in vitro* had an asymmetric effect on SPM. We found that ethanol had a greater effect on the more fluid exofacial leaflet than the less fluid cytofacial leaflet. Effects of ethanol on the SPM exofacial leaflet were observed beginning at a concentration of 25 m*M* ethanol. Earlier studies on effects of ethanol on bulk membrane fluidity had been criticized because of the use of high concentrations of ethanol. Effects of ethanol on the exofacial leaflet are well within the range that would be observed *in vivo*. It was subsequently shown by another group using TNBS quenching of DPH fluorescence that the exofacial leaflet of human intestinal brush border membranes was more fluid than the cytofacial leaflet and that benzyl alcohol had a much larger effect on the exofacial leaflet (Dudeja et al., 1991). These two studies demonstrated that alcohols had a specific effect on membrane structure and that this specificity was associated with the difference in fluidity of the exofacial and cytofacial leaflets.

In another approach used to examine effects of an alcohol on structure of the exofacial and cytofacial leaflets, the transbilayer movement or flipping of phospholipids from the exofacial to the cytofacial leaflet was measured using electron spin resonance and spin-labeled phosphatidylcholine, phosphatidylethanolamine, and phosphatidylserine (Basse et al., 1992). Benzyl alcohol *in vitro* increased the movement of phosphatidylethanolamine and phosphatidylserine from the exofacial to the cytofacial leaflet of erythrocytes when incubated at 4°C (Basse et al., 1992). Movement of phosphatidylcholine from the exofacial to the cytofacial leaflet in the presence of benzyl alcohol was observed at 37°C. It was suggested that the alcohol may have inhibited a protein (translocase/flippase) that had previously been proposed to be involved in regulation of the transbilayer distribution of phosphatidylethanolamine and phosphatidylserine.

Chronic ethanol consumption had the opposite effect on the transbilayer distribution of phosphatidylcholine (Wood et al., 1991). The transbilayer distribution of phosphatidylcholine was altered in erythrocytes of chronic ethanol-treated miniature swine. Phospholipid asymmetry was determined using phospholipase A_2 in which erythrocytes were incubated with the enzyme where only the exofacial leaflet was exposed to phospholipase A_2. This method for determining phospholipid asymmetry has been well-characterized in erythrocytes (Roelofsen, 1982). There was a significant increase in the amount of phosphatidylcholine in the erythrocyte exofacial leaflet of the chronic ethanol group as compared to the control group. The total amount of phosphatidylcholine in the erythrocyte membrane (exofacial + cytofacial leaflet) was not significantly different between the ethanol and control groups.

Alcohols *in vitro* have an asymmetric effect on fluidity of the two membrane leaflets. The exofacial leaflet was more affected by alcohols than the cytofacial leaflet. Chronic ethanol consumption also had an asymmetric effect on fluidity of the two membrane leaflets (Wood et al., 1989b). Fluidity of the SPM leaflets was determined using the TNBS method and DPH as described earlier in this chapter. Mice were administered ethanol or control liquid diets for 21 days, and SPM were prepared. Mice were not purposely withdrawn from ethanol prior to membrane preparation. The SPM exofacial leaflet of ethanol-treated mice was significantly less fluid as compared to the exofacial leaflet of pair-fed control mice. The cytofacial leaflet, on the other hand, was significantly more fluid in SPM of the ethanol-tolerant mice in contrast to the cytofacial leaflet of the control group. SPM bulk fluidity did not significantly differ between the two groups. Membrane tolerance to ethanol *in vitro* was observed in the SPM exofacial leaflet of the chronic ethanol group. The exofacial leaflet of the chronic ethanol group was resistant to effects of ethanol *in vitro* as compared to the exofacial leaflet of the control group. Significant fluidization of the exofacial leaflet in the chronic ethanol group was not observed until 200 m*M* ethanol. Ethanol *in vitro* did not have a significant effect on fluidity of the cytofacial leaflet of the chronic ethanol or control groups.

Chronic ethanol consumption had what appeared to be a paradoxical effect on the SPM leaflets. Whereas ethanol *in vitro* increased fluidity of the exofacial leaflet, ethanol *in vivo* reduced fluidity of the exofacial leaflet and increased fluidity of the cytofacial leaflet. Membrane fluidity is primarily a result of the lipid composition of the membrane. However, as discussed earlier in this chapter, chronic ethanol-induced changes in the total amount of membrane lipids have not been consistently observed. Membrane cholesterol has been one of the lipids most extensively studied as an explanation for ethanol-induced changes in membrane structure. Cholesterol above the phase transition temperature of membranes reduces membrane fluidity (Curtain et al., 1988), and it is one of the major lipids in plasma membranes, accounting for between 40 and 45 mol% of the total membrane lipid (Schroeder et al., 1991a; Wood et al., 1993b). Cholesterol is not evenly distributed in membranes but is located in different vertical and lateral pools or domains (Schroeder et al., 1991a). Cholesterol is asymmetrically distributed in membranes. Recent studies indicated that the cytofacial leaflet contained substantially more cholesterol than the exofacial leaflet in erythrocytes and LM fibroblast membranes (Brasaemle et al., 1988; Kier et al., 1986; Schroeder et al., 1991b). It is important to emphasize that the recent studies on cholesterol asymmetry of erythrocytes have used different methods to determine the transbilayer distribution of cholesterol (i.e., cholesterol oxidase, the fluorescent sterol dehydroergosterol, and fluorescent recovery after photobleaching), and data of each study showed that the exofacial contains substantially less cholesterol than the cytofacial leaflet.

We have used the fluorescent sterol dehydroergosterol and TNBS quenching to examine cholesterol distribution in SPM and effects of chronic ethanol consumption on the transbilayer distribution of cholesterol (Wood et al., 1990). Dehydroergosterol is a natural fluorescent sterol that occurs in different organisms (Schroeder et al., 1991a). Dehydroergosterol behaves similar to cholesterol with respect to uptake and exchange kinetics. The fluorescence of dehydroergosterol and its similarity to cholesterol make it an ideal probe for studying cholesterol domains. Membranes that have been treated with TNBS are incubated with dehydroergosterol contained in small unilamellar vesicles that allow exchange of cholesterol in the biological membrane with the fluorescent sterol. The fluorescence intensity of dehydroergosterol in the membrane is measured and the transbilayer distribution of cholesterol is determined based on the fractional fluorescence in TNBS treated versus untreated membranes. It is also necessary to quantify the total amount of dehydroergosterol that has been incorporated into the membrane so that the units of fluorescence intensity/amount of dehydroergosterol are equivalent for the TNBS labeled and control membranes. We have found that HPLC is a reliable and sensitive method for measuring small amounts of dehydroergosterol of membranes. Approximately 3–6 mol% dehydroergosterol in the membrane is sufficient for determination of transbilayer distribution. Higher amounts of dehydroergosterol can result in self-quenching of fluorescence of dehydroergosterol.

We showed that cholesterol is asymmetrically distributed in mouse SPM (Wood et al., 1990). The cytofacial leaflet had approximately 88% of the SPM cholesterol. The difference in the transbilayer distribution of cholesterol in SPM was consistent with the difference in fluidity of the two SPM leaflets. The exofacial leaflet was more fluid and contains substantially less cholesterol than the cytofacial leaflet. Chronic ethanol consumption altered SPM transbilayer cholesterol distribution. The SPM exofacial leaflet of the ethanol-tolerant group contained twice as much cholesterol when compared to the exofacial leaflet of the pair-fed control group. There was also a reduction of cholesterol in the cytofacial leaflet of the ethanol group. The total amount of SPM cholesterol was not affected by chronic ethanol consumption. Effects of chronic ethanol consumption on SPM transbilayer cholesterol distribution were in agreement with the effects of chronic ethanol consumption on SPM leaflet fluidity discussed earlier in this chapter. We concluded that the reduction in exofacial leaflet fluidity in the ethanol group resulted from the increase in cholesterol in the exofacial leaflet. Chronic ethanol consumption had a marked effect on the transbilayer distribution of cholesterol in SPM. Ethanol-induced changes in cholesterol distribution occurred in the absence of a change in the total amount of SPM cholesterol and underscores the conclusion that bulk membrane properties are not sensitive indicators of effects of ethanol (Wood and Schroeder, 1988).

B. LATERAL LIPID DOMAINS

Ethanol and other alcohols *in vitro* affected the lateral movement of specific phospholipids in *Aplysia* neuronal membranes (Treistman et al., 1987). The diffusion coefficient of fluorescent labeled phosphatidylethanolamine and phosphatidylcholine were differentially affected by ethanol and butanol as measured by fluorescence recovery after photobleaching. The diffusion coefficient was greater for phosphatidylethanolamine than for phosphatidylcholine. An additional important finding of that study was that temperature and alcohols had different effects on the diffusion rates of the phospholipids. As temperature increased, the diffusion coefficient of phosphatidylethanolamine was reduced and phosphatidylcholine was unaffected between 4°C and 25°C. Ethanol and butanol increased the diffusion coefficient of phosphatidylethanolamine. It was concluded that the membrane lipids were organized in domains and that alcohols had a selective effect on specific lipid domains (Treistman et al., 1987).

Lateral cholesterol domains have been identified in membranes (Schroeder et al., 1991a). Two domains are exchangeable and nonexchangeable pools of cholesterol. The kinetics and size of these cholesterol pools had been described in model membranes (McLean and Phillips, 1984; Thomas and Poznansky, 1988; Bar et al., 1987) and biological membranes (Gold and Phillips, 1990; Slotte and Bierman, 1988) and recently were reported in neuronal membranes (Rao et al., 1993). The synaptosomal membrane has approximately 50% of cholesterol that resides in an exchangeable pool (Rao et al., 1993).The exchangeable cholesterol pool has been attributed to cholesterol monomers and cholesterol-phospholipid complexes and the nonexchangeable pool to cholesterol aggregates (Nemecz et al., 1988). Cholesterol exchangeable and nonexchangeable pools have been shown to be affected by temperature, cholesterol/phospholipid ratio, and certain drugs (Schroeder et al., 1991a).

Effects of chronic ethanol consumption on cholesterol exchangeable and nonexchangeable pools have recently been reported (Wood et al., 1993a). Exchangeable and nonexchangeable cholesterol pools were determined using small unilamellar vesicles containing radiolabeled cholesterol that were incubated with synaptosomes of chronic ethanol-treated and pair-fed control mice. The rate of cholesterol exchange was significantly slower in synaptosomes of ethanol-tolerant mice as compared to synaptosomes of pair-fed control mice. The size of the cholesterol exchangeable pool was approximately 50% in both the ethanol and control groups and was similar to what had been reported previously in synaptosomal membranes (Rao et al., 1993). In the same study, it was also shown that the hydrolysis of sphingomyelin in the exofacial leaflet slowed cholesterol exchange and that this effect was greater in synaptosomes of the pair-fed control group than synaptosomes of the ethanol group. Sphingomyelin is exclusively an exofacial phospholipid and has been proposed to be involved in regulation of membrane cholesterol (Clejan and Bittman, 1984). The effects of chronic ethanol consumption on the rate of cholesterol exchange may be due to the differences in the transbilayer distribution of cholesterol. Increasing the cholesterol to phospholipid ratio of membranes has been shown to reduce the rate of cholesterol exchange (Phillips et al., 1987). Chronic ethanol consumption increased the amount of cholesterol in the exofacial leaflet, which will result in an increase in the cholesterol to phospholipid ratio of the exofacial leaflet of the ethanol group when compared to the exofacial leaflet of the control group.

Data on effects of ethanol *in vitro* and *in vivo* indicated that the exofacial leaflet is more affected by ethanol than the cytofacial leaflet. The transbilayer distribution of cholesterol would appear to be involved in the selective effects of ethanol on the exofacial leaflet. Chronic ethanol consumption altered the transbilayer distribution of cholesterol. Important issues are the understanding of how cholesterol is regulated in membranes and effects of ethanol on this regulation. These issues are discussed in the next section.

IV. MEMBRANE LIPID-PROTEIN INTERACTIONS

It has been well-established in the field of biochemistry that lipids are involved in regulating activity of certain proteins (Jones et al., 1988; Spector and Yorek, 1985; Sweet and Schroeder, 1988b; Lentz, 1988). For example, lipid-protein interactions have been described for ATPases (Yeagle, 1983; Missiaen et al., 1989), GABA receptor (Shouffani and Kanner, 1990), acetylcholine receptor (Berstein et al., 1989; Artigues et al., 1989; Leibel et al., 1987), as well as other proteins (Cserhati and Szogyi, 1992). However, the majority of studies on effects of ethanol on membranes and cell function have focused on either lipids or proteins, and little attention was directed toward examining effects of ethanol on lipid-protein interactions. An exception to this situation are the studies on the poly-phosphoinositol system, which have been recently reviewed in an excellent paper by Sun et al. (1993). This section will examine other recent studies on alcohols and lipid-protein interactions, as well as data on lipid-protein interactions that may have implications for understanding actions of ethanol on neuronal function. Emphasis will be on sterol carrier proteins and fatty acid binding proteins.

A. STEROL CARRIER PROTEINS

Chronic ethanol consumption altered the transbilayer distribution of cholesterol. The mechanism that regulates the transbilayer or asymmetric distribution of cholesterol in the exofacial and cytofacial membrane leaflets has not been elucidated. While there is evidence that a protein may mediate the transbilayer distribution of phosphatidylserine and phosphatidylethanolamine (Conner and Schroit, 1988; Zachowski and Gaudry-Talarmain, 1990; Zachowski, 1993), involvement of a protein in regulating cholesterol asymmetry has not been examined until recently. There have been several sterol carrier proteins identified, and studies on such proteins have been recently reviewed (Veerkamp et al., 1991). Sterol carrier protein-2 (SCP-2, also called nonspecific lipid transfer protein) and liver-fatty acid binding protein (L-FABP) bind cholesterol in a 1:1 molar ratio (Schroeder et al., 1991a). SCP-2 also binds phospholipids, and L-FABP binds fatty acids in addition to cholesterol. There is very little known about the occurrence of brain sterol carrier proteins or the effects of chronic ethanol consumption on sterol carrier proteins in general. A fatty acid binding protein (similar to L-FABP) from bovine brain was identified and found to bind fatty acids (Schoentgen et al., 1989). Binding of cholesterol was not examined in that study. Recently, we have identified two sterol carrier proteins in mouse brain homogenate: SCP-2 and a potentially new protein that cross-reacted with antibodies to heart-FABP (Myers-Payne et al., 1994a).

There have been three studies that examined effects of chronic ethanol consumption on sterol carrier proteins (Pignon et al., 1987; Myers-Payne et al., 1994a; Myers-Payne et al., 1994b). The amount of L-FABP was increased in the liver of chronic ethanol-treated rats (Pignon et al., 1987). The maximal binding capacity of palmitate was increased but the affinity for fatty acids was reported to be decreased in liver of the ethanol group. The reported decrease in binding affinity of fatty acids may have resulted from an increase in cholesterol binding to L-FABP in the chronic ethanol group. We have recently observed a significant increase in the amount of L-FABP in the liver of chronic ethanol-treated mice (Myers-Payne et al., 1994b). It was also found that the increase was due to membrane bound L-FABP in contrast to the cytosolic form of L-FABP. On the other hand, the amount of SCP-2 was significantly decreased in livers of the chronic ethanol-treated mice (Myers-Payne et al., 1994b). In a separate study, we found that the amount of SCP-2 was increased in brain of chronic ethanol-treated mice (Myers-Payne et al., 1994a). This increase in SCP-2 was attributed to an increase in the membrane-associated protein in contrast to the cytosolic form. It is tempting to speculate that the increase in SCP-2 in brain of chronic ethanol-treated mice contributed to modification of transbilayer cholesterol distribution of ethanol-treated mice discussed earlier in this chapter. An issue that needs to be resolved is an understanding of the differential effects of chronic ethanol consumption on brain SCP-2 and liver SCP-2.

B. PROTEIN FUNCTION AND LIPIDS

Some of the sterol carrier proteins bind fatty acids in addition to cholesterol. It has been shown that different fatty acids were involved in the regulation of certain ion channels (Ordway et al., 1991). Recently, it was reported that the NMDA receptor appeared to contain a fatty acid binding domain (Petrou et al., 1993). The amino acid sequence of several different fatty acid binding proteins were compared to the amino acid sequence of the NMDA receptor. There was significant similarity in a 131-residue domain of the NMDA receptor and FABPs. Regulation of proteins that have been shown to be affected by fatty acids may not be a function of the annular lipid environment but the result of direct binding of the fatty acid to the protein (Ordway et al., 1991). Another possibility is that cholesterol may directly bind to a protein such as the NMDA receptor. Evidence of potential binding sites for cholesterol on the acetylcholine receptor have been reported (Jones and McNamee, 1988; Fernandez-Ballester et al., 1994). There have been several studies on effects of ethanol on the NMDA receptor, but the mechanism of ethanol's action has not been determined (Michaelis et al., 1993). Fatty acids or cholesterol may be required for normal function of the NMDA receptor, and ethanol may interfere with the putative lipid binding sites, modification of the boundary lipids, or direct binding of ethanol to the receptor. A study showed that alcohols inhibited lipid-free protein kinase C (PKC) activity (Slater et al., 1993) and that may be an indication of binding of alcohols to a protein. It also was reported, however, that alcohol-induced inhibition of PKC was increased by the presence of lipids (Slater et al., 1993). Data from our laboratory using fluorescence and NMR techniques have shown that ethanol directly binds to lipid free bovine serum albumin beginning at a concentration of 50 m*M* ethanol (Avdulov et al., 1994a).

V. CONCLUSIONS

Almost 100 years ago Meyer and Overton independently reported that narcosis was associated with the partition coefficients of different compounds including alcohols (Meyer, 1899; Meyer, 1901; Overton, 1896). The pioneering work of Goldstein and her colleagues (reviewed in Goldstein, 1987), Rubin's group (reviewed in Rubin and Rottenberg, 1982), and many other investigators showed that ethanol altered the bulk lipid environment of the membrane. Based on recent evidence and the application of different techniques, it is now possible to expand upon that earlier body of work by looking at the specific effects of ethanol within the bilayer and actions of ethanol on lipid-protein interactions in membranes. The environment of a membrane is complex, with several different properties, and some of those properties may be involved in mechanisms of intoxication and tolerance. For example, the contribution of cholesterol asymmetry, membrane dielectric constant, and interdigitation is now only beginning to be understood regarding the actions of ethanol. Knowledge concerning lipid-protein interactions and effects of ethanol is at a most rudimentary stage in providing an explanation for intoxication and tolerance to ethanol.

It is reasonable to conclude that any explanation of the mechanisms responsible for ethanol intoxication and tolerance must include an interaction of lipids and proteins. However, it has been proposed that ethanol has a direct effect on proteins and that membrane lipids do not contribute to the actions of ethanol (Moss et al., 1991; Franks and Lieb, 1984). This idea is flawed for several different reasons. There is an extensive amount of data in the field of biochemistry which have shown that different lipids were involved in the regulation of various proteins. Studies reviewed in this chapter supported the involvement of membrane lipid domains in intoxication and tolerance to ethanol. Finally, it has never been reported that a protein per se could explain ethanol intoxication or tolerance. A parsimonious approach is that it is possible that acute effects of ethanol may result from a direct action of ethanol on certain membrane proteins which might involve direct binding to the hydrophobic portion of a protein, displacement of protein bound lipids, or modification of lipids in close proximity to the protein. As the concentration of ethanol is increased, effects on susceptible lipid domains begin to affect protein function. Ethanol-induced changes in annular lipids or interdigitation may modify the hydrophobic area of proteins. For example, ethanol-induced effects on interdigitation may alter exposure of the hydrophobic portion of a protein or produce protein clustering or aggregation. Repeated exposure to ethanol may trigger a homeostatic response which would reduce the amount of ethanol that partitioned into the membrane. Such a homeostatic mechanism may be increased synthesis and incorporation of a sterol carrier protein into the membrane that would translocate cholesterol from the cytofacial to the exofacial leaflet resulting in membrane tolerance.

ACKNOWLEDGMENTS

This work was supported in part by NIH Grants AA07292 and AG11056 (W.G.W.), the Medical Research Service, and the Geriatric Research, Education, and Clinical Center of the Department of Veterans Affairs.

REFERENCES

Armbrecht, H.J., Wood, W.G., Wise, R.W., Walsh, J.B., Thomas, B.N., and Strong, R. (1983). Ethanol-induced disordering of membranes from different age groups of C57BL/6NNIA mice. *J. Pharmacol. Exp. Ther.,* 226, 387–391.

Artigues, A., Villar, M.T., Fernandez, A.M., Ferragut, J.A., and Gonzalez-Ros, J.M. (1989). Cholesterol stabilizes the structure of the nicotinic acetylcholine receptor reconstituted in lipid vesicles. *Biochim. Biophys. Acta,* 985, 325–330.

Avdulov, N.A., Chochina, S.V., Schroeder, F., Daragan, V., Mayo, K., and Wood, W.G. (1994a). Direct binding of ethanol to bovine serum albumin using fluorescence probes and NMR. Unpublished manuscript.

Avdulov, N.A., Wood, W.G., and Harris, R.A. (1994b). Effects of ethanol on structural parameters of rat brain membranes: relationship to genetic differences in ethanol sensitivity. *Alcoholism: Clin. Exp. Res.,* 18, 53–59.

Bar, L.K., Barenholz, Y., and Thompson, T.E. (1987). Dependence on phospholipid composition of the fraction of cholesterol undergoing spontaneous exchange between small unilamellar vesicles. *Biochemistry,* 26, 5460–5465.

Basse, F., Sainte-Marie, J., Maurin, L., and Bienvenue, A. (1992). Effect of benzyl alcohol on phospholipid transverse mobility in human erythrocyte membrane. *Eur. J. Biochem.,* 205, 155–162.

Berstein, G., Haga, T., and Ichiyama, A. (1989). Effect of the lipid environment on the differential affinity of purified cerebral and atrial muscarinic acetylcholine receptors for pirenzepine. *Mol. Pharmacol.,* 36, 601–607.

Brasaemle, D.L., Robertson, A.D., and Attie, A.D. (1988). Transbilayer movement of cholesterol in the human erythrocyte membrane. *J. Lipid Res.,* 29, 481–489.

Chin, J.H. and Goldstein, D.B. (1977a). Effects of low concentrations of ethanol on the fluidity of spin-labeled erythrocyte and brain membranes. *Mol. Pharmacol.,* 13, 435–441.

Chin, J.H. and Goldstein, D.B. (1977b). Drug tolerance in biomembranes: a spin label study of the effects of ethanol. *Science,* 196, 684–685.

Chin, J.H. and Goldstein, D.B. (1981). Membrane-disordering action of ethanol. Variation with membrane cholesterol content and depth of the spin label probe. *Mol. Pharmacol.,* 19, 425–431.

Chin, J.H. and Goldstein, D.B. (1984). Cholesterol blocks the disordering effects of ethanol in biomembranes. *Lipids,* 19, 929

Chin, J.H., Parsons, L.M., and Goldstein, D.B. (1978). Increased cholesterol content of erythrocyte and brain membranes in ethanol-tolerant mice. *Biochim. Biophys. Acta,* 513, 358–363.

Chiou, J., Krishna, P.R., Kamaya, H., and Ueda, I. (1992). Alcohols dehydrate lipid membranes: an infrared study on hydrogen bonding. *Biochim. Biophys. Acta,* 1110, 225–233.

Chiou, J., Ma, S., Kamaya, H., and Ueda, I. (1990). Anesthesia cutoff phenomenon: interfacial hydrogen bonding. *Science,* 248, 583–585.

Clejan, S. and Bittman, R. (1984). Decreases in rates of lipid exchange between *Mycoplasma gallisepticum* cells and unilamellar vesicles by incorporation of sphingomyelin. *J. Biol. Chem.,* 259, 10823–10826.

Conner, J. and Schroit, A. (1988). Transbilayer movement of phosphatidylserine in erythrocytes: inhibition of transport and preferential labeling of a 31000–Dalton protein by sulfhydryl reactive reagents. *Biochemistry,* 27, 848–951.

Cornea, R.L. and Thomas, D.D. (1994). Effects of membrane thickness on the molecular dynamics and enzymatic activity of reconstituted Ca-ATPase. *Biochemistry,* 33, 2912–2920.

Cserhati, T. and Szogyi, M. (1992). Interaction of phospholipids with proteins and peptides. New Advances 1990. *Int. J. Biochem.,* 24, 525–537.

Curtain, C.C., Gordon, L.M., and Aloia, R.C. (1988). Lipid domains in biological membranes: conceptual Development and Significance, in *Lipid Domains and the Relationship to Membrane Function.* R.C. Aloia, C.C. Curtain, and L.M. Gordon, eds. (New York: Alan R. Liss), 1–15.

Daniels, C.K. and Goldstein, D.B. (1982). Movement of free cholesterol from lipoproteins or lipid vesicles into erythrocytes. *Mol. Pharmacol.,* 21, 694–700.

Deitrich, R.A., Dunwiddie, T.V., Harris, R.A., and Erwin, V.G. (1989). Mechanism of action of ethanol: initial central nervous system actions. *Pharmacol. Rev.,* 41, 489–537.

Devaux, P.F. (1991). Static and dynamic lipid asymmetry in cell membranes. *Biochemistry,* 30, 1163–1173.

Dudeja, P.K., Harig, J.M., Wali, R.K., Knaup, S.M., Ramaswamy, K., and Braistus, T.A. (1991). Differential modulation of human small intestinal brush-border membrane hemileaflet fluidity affects leucine aminopeptidase activity and transport of d-glucose and l-glutamate. *Arch. Biochem. Biophys.,* 284, 338–345.

Fernandez-Ballester, G., Castresana, J., Fernandez, A.M., Arrondo, J.R., Ferragut, J.A., and Gonzalez-Ros, J.M. (1994). A role for cholesterol as a structural effector of the nicotinic acetylcholine receptor. *Biochemistry,* 33, 4065–4071.

Franks, N.P. and Lieb, W.R. (1984). Do general anaesthetics act by competitive binding to specific receptors? *Nature,* 310, 599–601.

Gold, J.C. and Phillips, M.C. (1990). Effects of membrane lipid composition on the kinetics of cholesterol exchange between lipoproteins and different species of red blood cells. *Biochim. Biophys. Acta,* 1027, 85–92.

Goldstein, D.B., Chin, J.H., and Lyon, R.C. (1982). Ethanol disordering of spin-labeled mouse membranes; correlation with genetically determined ethanol sensitivity of mice. *Proc. Natl. Acad. Sci. U.S.A.,* 79, 4231–4233.

Goldstein, D.B. (1986). Effect of alcohol on cellular membranes. *Ann. Emergency Med.,* 15, 1013–1018.

Goldstein, D.B. (1987). Ethanol-induced adaptation in biological membranes. *Ann. N.Y. Acad. Sci.,* 492, 103–111.

Harris, R.A. (1984). Defining the membrane pathology produced by chronic alcohol consumption. *Lab. Invest.,* 50(2), 113–114.

Hunt, W.A. (1985). Alcohol and Biological Membranes (New York: The Guilford Press).

Johannsson, A., Keightley, C.A., Smith, G.A., Richards, C.D., Hesketh, T.R., and Metcalfe, J.C. (1981). The effect of bilayer thickness and *n*-alkanes on the activity of the (Ca^{2+} + Mg^{2+})-dependent ATPase of sarcoplasmic reticulum. *J. Biol. Chem.,* 256, 1643–1650.

Jones, O.T., Eubanks, J.H., Earnest, J.P., and McNamee, M.G. (1988). A minimum number of lipids are required to support the functional properties of the nicitinic acetylcholine receptor. *Biochemistry,* 27, 3733–3742.

Jones, O.T. and McNamee, M.G. (1988). Annular and nonannular binding sites for cholesterol associated with the nicotinic acetylcholine receptor. *Biochemistry,* 27, 2364–2374.

Kier, A.B., Sweet, W.D., Cowlen, M.S., and Schroeder, F. (1986). Regulation of transbilayer distribution of a fluorescent sterol in tumor cell plasma membranes. *Biochim. Biophys. Acta,* 861, 287–301.

Klemm, W.R. (1990). Dehydration: a new alcohol theory. *Alcohol,* 7, 49–50.

Komatsu, H. and Rowe, E.S. (1991). Effect of cholesterol on the ethanol-induced interdigitated gel phase in phosphatidylcholine: use of fluorophore pyrene-labeled phosphatidylcholine. *Biochemistry,* 30, 2463–2470.

Leibel, W.S., Firestone, L.L., Legler, D.C., Braswell, L.M., and Miller, K.W. (1987). Two pools of cholesterol in acetylcholine receptor-rich membranes from *Torpedo. Biochim. Biophys. Acta,* 897, 249–260.

Lentz, B.R. (1988). Organization of membrane lipids by intrinsic membrane proteins, in *Lipid Domains and the Relationship to Membrane Function.* R.C. Aloia, C.C. Curtain, and L.M. Gordon, eds. (New York: Alan R. Liss), 141–161.

Levental, M. and Tabakoff, B. (1980). Sodium-potassium activated adenosine triphosphatase activity as a measure of neuronal membrane characteristics in ethanol-tolerant mice. *J. Pharmacol. Exp. Ther.,* 212, 315–319.

Lewis, E.N., Levin, I.W., and Steer, C.J. (1989). Infrared spectroscopic study of ethanol-induced changes in rat liver plasma membrane. *Biochim. Biophys. Acta,* 986, 161–166.

McLean, L. and Phillips, M.C. (1984). Cholesterol transfer from small and large unilamellar vesicles. *Biochim. Biophys. Acta,* 776, 21–26.

Meyer, H. (1899). Welche Eigenschaft der Anasthetica bedingt ihre narkitische Wirkung? *Naunyn-Schmiedebergs Arch. Exp. Pathol. Pharmakol.,* 42, 109–118.

Meyer, H. (1901). Zur Theorie der Alkolnarkose: der Einfuss wechselnder Temperatur auf Wirkungsstarke und Theilungscoefficient der Narcotica. *Naunyn-Schmiedebergs Arch. Exp. Pathol. Pharmakol.,* 46, 338–346.

Michaelis, E.K., Michaelis, M.L., and Kumar, K.N. (1993). Molecular Properties of an NMDA Receptor Complex and Effects of Ethanol on this Complex, in *Alcohol, Cell Membranes, and Signal Transduction in Brain.* C. Alling, I. Diamond, S.W. Leslie, G.Y. Sun, and W.G. Wood, eds. (New York: Plenum Press), 139–149.

Missiaen, L., Raeymaekers, L., Wuytack, F., Vrolix, M., De Smedt, H., and Casteels, R. (1989). Phospholipid-protien interactions of the plasma membrane Ca2+-transporting ATPase. Evidence for a tissue-dependent functional difference. *Biochem. J.,* 263, 687–694.

Moss, G.W.J., Franks, N.P., and Lieb, W.R. (1991). Modulation of the general anesthetic sensitivity of a protein: a transition between two forms of firefly luciferase. *Proc. Natl. Acad. Sci. U.S.A.,* 88, 134–138.

Myers-Payne, S.C., Fontaine, R.N., Hubbell, T., Spener, F., Borchers, T., Wood, W.G., and Schroeder, F. (1994a). The effects of chronic ethanol consumption on lipid transfer proteins in mouse brain. unpublished manuscript.

Myers-Payne, S.C., Fontaine, R.N., Langland, G., Wood, W.G., and Schroeder, F. (1994b). Chronic ethanol induces translocation of liver fatty acid-binding protein to membranes. unpublished manuscript.

Nambi, P., Rowe, E.S., and McIntosh, T.J. (1988). Studies of the ethanol-induced interdigitated gel phase in phosphatidylcholines using the fluorophore 1,6–diphenyl-1,3,5–hexatriene. *Biochemistry,* 27, 9175–9182.

Nemecz, G., Fontaine, R.N., and Schroeder, F. (1988). A fluorescence and radiolabel study of sterol exchange between membranes. *Biochim. Biophys. Acta,* 943, 511–521.

Nhamburo, P.T., Salafsky, B.P., Tabakoff, B., and Hoffman, P.L. (1987). Effects of ethanol on ouabain inhibition of mouse brain (Na+,K+)ATPase activity. *Biochem. Pharmacol.,* 36, 2027–2033.

Op den Kamp, J.A.F. (1979). Lipid asymmetry in membranes. *Annu. Rev. Biochem.,* 48, 47–71.

Ordway, R.W., Singer, J.J., and Walsh, J.V., Jr. (1991). Direct regulation of ion channels by fatty acids. *TINS,* 14, 96–100.

Orme, F.W., Moronne, M.M., and Macey, R.I. (1988). Modification of erythrocyte membrane dielectric constant by alcohols. *J. Membrane Biol.,* 104, 57–68.

Overton, E. (1896). Über die osmotischen Eigenschaften der in ihrer Betdeutung für die Toxikologie und Pharmakologie. *Z. Phys. Chem.*, 22, 189–209.

Parsons, L.M., Gallaher, E.J., and Goldstein, D.B. (1982). Rapidly developing functional tolerance to ethanol is accompanied by increased erythrocyte cholesterol in mice. *J. Pharmacol. Exp. Ther.*, 223(2), 472–476.

Petrou, S., Ordway, R.W., Singer, J.J., and Walsh, J.V., Jr. (1993). A putative fatty acid-binding domain of the NMDA receptor. *Trends Biochem. Sci.*, 18, 41–42.

Phillips, M.C., Johnson, W.J., and Rothblat, G.H. (1987). Mechanisms and consequences of cellular cholesterol exchange and transfer. *Biochim. Biophys. Acta*, 906, 223–276.

Pignon, J., Bailey, N.C., Baraona, E., and Lieber, C.S. (1987). Fatty acid-binding protein: a major contributor to the ethanol-induced increase in liver cytosolic proteins in the rat. *Hepatology*, 7(5), 865–871.

Rangaraj, N. and Kalant, H. (1982). Effect of chronic ethanol treatment on temperature dependence and on norepinephrine sensitization in rat brain (Na^+ + K^+)-adenosine triphosphatase. *J. Pharmacol. Exp. Ther.*, 223, 536–539.

Rao, A.M., Igbavboa, U., Semotuk, M., Schroeder, F., and Wood, W.G. (1993). Kinetics and size of cholesterol lateral domains in synaptosomal membranes: modification by sphingomyelinase and effects on membrane enzyme activity. *Neurochem. Int.*, 23, 45–52.

Ritzmann, R.F. and Springer, A. (1980). Age differences in brain sensitivity and tolerance to ethanol in mice. *Age*, 3, 15–17.

Roelofsen, B. (1982). Phospholipases as tools to study the localization of phospholipids in biological membranes. A critical review. *J. Toxicol.*, 1(1), 87–197.

Rottenberg, H. (1992). Probing the interactions of alcohols with biological membranes with the fluorescent probe Prodan. *Biochemistry*, 31, 9473–9481.

Rubin, E. and Rottenberg, H. (1982). Ethanol-induced injury and adaptations in biological membranes. *Fed. Proc.*, 41, 2465–2471.

Sauerheber, R.D., Esgate, J.A., and Kuhn, C.E. (1982). Alcohols inhibit adipocyte basal and insulin-stimulated glucose uptake and increase the membrane lipid fluidity. *Biochim. Biophys. Acta*, 691, 115–124.

Schoentgen, F., Pignede, G., Bonanno, L.M., and Jolles, P. (1989). Fatty acid-binding protein from bovine brain. *Eur. J. Biochem.*, 185, 35–40.

Schroeder, F. (1985). Fluorescence probes unravel asymmetric structure of membranes. *Sub-Cell. Biochem.*, 11, 51–100.

Schroeder, F., Morrison, W.J., Gorka, C., and Wood, W.G. (1988). Transbilayer effects of ethanol on fluidity of brain membrane leaflets. *Biochim. Biophys. Acta*, 946, 85–94.

Schroeder, F., Jefferson, J.R., Kier, A.B., Knittel, J., Scallen, T.J., Wood, W.G., and Hapala, I. (1991a). Membrane cholesterol dynamics: cholesterol domains and kinetic pools. *Proc. Soc. Exp. Bio. Med.*, 196, 235–252.

Schroeder, F., Nemecz, G., Wood, W.G., Morrot, G., Ayraut-Jarrier, M., and Devaux, P.F. (1991b). Transmembrane distribution of sterol in the human erythrocyte. *Biochim. Biophys. Acta*, 1066, 183–192.

Schroeder, F., Colles, S.M., Kreishman, G.P., Heyliger, C.E., and Wood, W.G. (1994). Synaptic plasma membrane structure and polarity of long-sleep and short-sleep mice. *Arch. Biochem. Biophys.*, 309, 369–376.

Sheetz, M.P. and Singer, S.J. (1974). Biological membranes as bilayer couples. A molecular mechanism of drug-erythrocyte interactions. *Proc. Natl. Acad. Sci. U.S.A.*, 71, no.11., 4457–4461.

Shouffani, A. and Kanner, B.I. (1990). Cholesterol is required for the reconstitution of the sodium- and chloride-coupled, GABA-aminobutyric acid transporter from rat brain. *J. Biol. Chem.*, 265, 6002–6008.

Slater, J.L. and Huang, C.-H. (1992). Lipid bilayer interdigitation, in *The Structure of Biological Membranes*. P. Yeagle, ed. (Boca Raton, FL: CRC Press), 175–210.

Slater, S.J., Cox, K.J.A., Lombardi, J.V., Ho, C., Kelly, M.B., Rubin, E., and Stubbs, C.D. (1993). Inhibition of protein kinase C by alcohols and anaesthetics. *Nature*, 364, 82–84.

Slotte, J.P. and Bierman, E.L. (1988). Depletion of plasma-membrane sphingomyelin rapidly alters the distribution of cholesterol between plasma membranes and intracellular cholesterol pools in cultured fibroblasts. *Biochem. J.*, 250, 653–658.

Spector, A.A. and Yorek, M.A. (1985). Membrane lipid composition and cellular function. *J. Lipid Res.*, 26, 1015–1035.

Strong, R. and Wood, W.G. (1984). Membrane properties and aging: *in vivo* and *in vitro* effects of ethanol on synaptosomal GABA-aminobutyric acid release. *J. Pharmacol. Exp. Ther.*, 229, 726–730.

Stubbs, C.D. and Rubin, E. (1993). Molecular mechanisms of ethanol and anesthetic actions: lipid and protein-based theories, in *Alcohol, Cell Membranes, and Signal Transduction in Brain*. C. Alling, I. Diamond, S.W. Leslie, G.Y. Sun, and W.G. Wood, eds. (New York: Plenum Press), 1–11.

Sun, G.Y. and Sun, A.Y. (1985). Ethanol and membrane lipids. *Alcoholism: Clin. Exp. Res.*, 9, 164–180.

Sun, G.Y., Zhang, J.-P., and Lin, T.-A. (1993). Effects of acute and chronic ethanol administration on the polyphosphoinositide signaling activity in brain, in *Alcohol, Cell Membranes, and Signal Transduction in Brain*. C. Alling, I. Diamond, S.W. Leslie, G.Y. Sun, and W.G. Wood, eds. (New York: Plenum Press), 205–218.

Sweet, W.D., Wood, W.G., and Schroeder, F. (1987). Charged anesthetics selectively alter plasma membrane order. *Biochemistry*, 26, 2828–2835.

Sweet, W.D. and Schroeder, F. (1988a). Polyunsaturated fatty acids alter sterol transbilayer domains in LM fibroblast plasma membranes. *FEBS Lett.*, 229, 188–192.

Sweet, W.D. and Schroeder, F. (1988b). Lipid domains and enzyme activity, in *Lipid Domains and the Relationship to Membrane Function*. R.C. Aloia, C.C. Curtain, and L.M. Gordon, eds. (New York: Alan R. Liss), 17–42.

Tabakoff, B., Ritzmann, R., Raju, T.S., and Deitrich, R.A. (1980). Characterization of acute and chronic tolerance in mice selected for inherent differences in sensitivity to ethanol. *Alcoholism: Clin. Exp. Res.*, 4, 70–73.

Taraschi, T.F., Ellingson, J.S., Wu, A., Zimmerman, R., and Rubin, E. (1986). Phosphatidylinositol from ethanol-fed rats confers membrane tolerance to ethanol. *Proc. Natl. Acad. Sci. U.S.A.*, 83, 9398–9402.

Thomas, P.D. and Poznansky, M.J. (1988). Cholesterol transfer between lipid vesicles. Effect of phospholipids and gangliosides. *Biochem. J.*, 251, 55–61.

Treistman, S., Moynihan, M.M., and Wolf, D. (1987). Influence of alcohols, temperature, and region on the mobility of lipids in neuronal membrane. *Biochim. Biophys. Acta*, 898, 109–120.

Treistman, S. and Wilson, A. (1987). Alkanol effects on early potassium currents in aplysia neurons depend on chain length. *Proc. Natl. Acad. Sci. U.S.A.*, 84, 9299–9303.

Veerkamp, J.H., Peeters, R.A., and Maatman, R.G.H.J. (1991). Structural and functional features of different types of cytoplasmic fatty-acid-binding proteins. *Biochim. Biophys. Acta*, 1081, 1–24.

Waring, A.J., Rottenberg, H., Ohnishi, T., and Rubin, E. (1981). Membranes and phosphlipids of liver mitochondria from chronic alcoholic rats are resistant to membrane disordering by alcohol. *Proc. Natl. Acad. Sci. U.S.A.*, 78(4), 2582–2586.

Weast, R.C., Astle, M.J., and Beyer, W.H. (1987). CRC Handbook of Chemistry and Physics (Boca Raton, FL: CRC Press, Inc.).

Wood, W.G. and Armbrecht, H.J. (1982a). Age differences in ethanol-induced hypothermia and impairment in mice. *Neurobiol. Aging*, 3, 243–246.

Wood, W.G. and Armbrecht, H.J. (1982b). Behavioral effects of ethanol in animals: age differences and age changes. *Alcoholism: Clin. Exp. Res.*, 6, 3–12.

Wood, W.G., Gorka, C., Johnson, J.A., Sun, G.Y., Sun, A.Y., and Schroeder, F. (1991). Chronic ethanol consumption alters transbilayer distribution of phosphatidylcholine in erythrocytes of Sinclair (S-1) Miniature Swine. *Alcohol*, 8, 395–399.

Wood, W.G., Gorka, C., Rao, A.M., and Schroeder, F. (1989a). Specific action of ethanol on lateral and vertical membrane domains, in *Molecular Mechanisms of Alcohol*. G.Y. Sun, P.K. Rudeen, W.G. Wood, Y. Wei, and A.Y. Sun, eds. (Clifton, NJ: Humana Press), 3–13.

Wood, W.G., Gorka, C., and Schroeder, F. (1989b). Acute and chronic effects of ethanol on transbilayer membrane domains. *J. Neurochem.*, 52, 1925–1930.

Wood, W.G., Rao, A.M., Igbavboa, U., and Semotuk, M. (1993a). Cholesterol exchange and lateral cholesterol pools in synaptosomal membranes of pair-fed control and chronic ethanol-treated mice. *Alcoholism: Clin. Exp. Res.*, 17, 345–350.

Wood, W.G., Rao, A.M., Schroeder, F., and Igbavboa, U. (1993b). Membrane cholesterol and ethanol: domains, kinetics, and protein function, in *Alcohol, Cell Membranes, and Signal Transduction in Brain*. C. Alling, I. Diamond, S.W. Leslie, G.Y. Sun, and W.G. Wood, eds. (New York: Plenum Press), 13–32.

Wood, W.G., Schroeder, F., Hogy, L., Rao, A.M., and Nemecz, G. (1990). Asymmetric distribution of a fluorescent sterol in synaptic plasma membranes: effects of chronic ethanol consumption. *Biochim. Biophys. Acta*, 1025, 243–246.

Wood, W.G. and Schroeder, F. (1988). Membrane effects of ethanol: bulk lipid versus lipid domains. *Life Sci.*, 43, 467–475.

Wood, W.G. and Schroeder, F. (1992). Membrane exofacial and cytofacial leaflets: a new approach to understanding how ethanol alters brain membranes, in *Alcohol and Neurobiology: Receptors, Membranes, and Channels*. R.R. Watson, ed. (Boca Raton, FL: CRC Press), 161–184.

Yamazaki, M., Miyazu, M., Asano, T., Yuba, A., and Kume, N. (1994). Direct evidence of induction of interdigitated gel structure in large unilamellar vesicles of dipalmitoylphosphatidylcholine by ethanol: studies by excimer method and high-resolution electron cryomicroscopy. *Biophys. J.*, 66, 729–733.

Yeagle, P. (1992). *The Structure of Biological Membranes*. (Boca Raton, FL: CRC Press).

Yeagle, P.L. (1983). Cholesterol modulation of (Na^{+}+K^{+})-ATPase ATP hydrolyzing activity in the human erythrocyte. *Biochim. Biophys. Acta*, 727, 39–44.

Yurttas, L., Dale, B.E., and Klemm, W.R. (1992). FTIR evidence for alcohol binding and dehydration in phospholipid and ganglioside micelles. *Alcoholism: Clin. Exp. Res.*, 16, 863–869.

Zachowski, A. (1993). Phospholipids in animal eukaryotic membranes: transverse asymmetry and movement. *Biophys. J.*, 294, 1–14.

Zachowski, A. and Gaudry-Talarmain, Y.M. (1990). Phospholipid transverse diffusion in synaptosomes: evidence for the involvement of the aminophospholipid translocase. *J. Neurochem.*, 55(4), 1352–1356.

Chapter 3

VOLTAGE-GATED CALCIUM CHANNELS

Debra Mullikin-Kilpatrick and Steven N. Treistman

TABLE OF CONTENTS

I. INTRODUCTION

For many years, the molecular mechanism of action for alcohol was thought to reflect its ability to disorder biological membranes, making them more fluid, and possibly perturbing the functions of excitable membrane proteins embedded in the lipid matrix (Metcalfe et al., 1968; Chin and Goldstein, 1977a; Chin and Goldstein, 1977b; Johnson et al., 1979). Often, the concentrations of ethanol necessary to produce measurable changes in lipid properties were

0-8493-8389-7/95/$0.00+$.50

well above those which cause behavioral consequences. Recently, a body of evidence has been accumulating suggesting that ethanol and general anesthetics may exert their actions directly upon proteins. Currently, it is not possible to state unequivocally that either of these alternatives is the unique basis for ethanol action in the nervous system. However, since the actions of ethanol produce a wide array of behavioral effects ranging from the more mundane, such as the need for urination, to more complex, such as both sedation and stimulation, it is not unreasonable to expect that a wide range of excitable membrane proteins will be affected, either directly or indirectly, at concentrations of ethanol which are pharmacologically relevant. In this review, we focus on the actions of ethanol on voltage-gated calcium channels. Modulation of this ubiquitous class of proteins will likely affect every aspect of physiology and behavior.

Ethanol's effects on voltage-gated ion channels were first described for sodium channels in the squid giant axon, which were found to be relatively insensitive to the drug (Armstrong and Binstock, 1964; Moore et al., 1964). A series of experiments performed in our laboratory on voltage-gated potassium channels in identified neurons of *Aplysia* suggested that ethanol was not acting by perturbation of a bulk lipid environment and that the biophysical state of the channel might influence its response to ethanol exposure (Treistman and Wilson, 1987a,b). Once again, relatively high ethanol concentrations were necessary to see significant effects on ionic current parameters, most notably in this case, on the kinetics of current decay. A number of ligand-gated channels have more recently been shown to respond to ethanol at more relevant concentrations. Studies using hippocampal neurons to examine NMDA mediated currents (Lovinger et al., 1989) and cerebral cortical neurons to examine GABA mediated currents (Reynolds and Prasad, 1991) demonstrated that these channel proteins were inhibited and augmented, respectively, by ethanol. Results from our laboratory using both rat neurohypophysial terminals and rat pheochromocytoma (PC12) cells have shown that voltage-activated calcium channels are inhibited by acute exposure to ethanol using concentrations which are at or below those considered to be legally intoxicating in the USA (~0.08% or 17 m*M*) (Wang et al., 1991a; Grant et al., 1993; Mullikin-Kilpatrick and Treistman, 1994a,b,c) and within the range of sensitivity of the ligand-gated channels mentioned.

Voltage-activated calcium channels are highly selective for calcium. These proteins represent a major pathway for calcium entry into cells and, as such, regulate intracellular calcium homeostasis. Calcium channels provide for rapid changes in potential across the plasma membrane which are required for neuronal signalling. Changes in intracellular calcium concentrations in response to membrane depolarization or activation of neurotransmitter receptors can affect many calcium-dependent processes such as neurotransmitter release, excitation-contraction coupling, cellular proliferation, and neuronal survival. The effects of both acute (minutes) and chronic (days) exposure to ethanol on calcium channels has been examined using biochemical as well as electrophysiological techniques in several systems including nonneuronal and neuronal preparations. Knowledge of the molecular basis for alcohol's action in these different systems could provide insight into alcohol-related disorders such as fetal alcohol syndrome. The relationship between the effects of exposure to ethanol on calcium channels in the laboratory and the development of alcoholism in humans is unknown. However, the finding that a reduction in release of arginine vasopressin from rat neurohypophysial terminals is directly related to a reduction of calcium current in these terminals in the presence of low concentrations of ethanol (Wang et al., 1991a,b,c) suggests that calcium channels are targets of ethanol action *in vivo* and are intimately involved in the behavioral effects of ethanol.

A. APPROACHES

Several techniques have been used to study the mechanism of ethanol action on voltage-activated calcium channels. In this review, we will focus on electrophysiological, biochemical, and molecular biological approaches.

1. Electrophysiological

Since the mode of signalling within the nervous system is by electricity, it is clear that the most direct measure of ethanol's actions on the nervous system will involve monitoring the electrical activity of nerve cells. These studies can include an examination of multicelluar circuit properties or can focus on the electrophysiology of individual neurons using either extracellular or intracellular recording techniques. Voltage clamp recordings provide a great deal of power in understanding the actions of ethanol on particular channel populations, including voltage-gated channels, and single channel recording techniques offer the promise of highly sophisticated analyses of the interaction between ethanol and a putative target protein. These techniques can provide significant information in addition to that obtained from biochemical studies, such as characterization of ethanol action at the single cell level and identification of the channel type affected by ethanol.

In most voltage clamp studies, calcium channels are isolated from other voltage-activated channels by the use of certain ions in the bath (e.g., barium as charge carrier and tetraethylammonium chloride to block potassium channels) and in the electrode (e.g., cesium chloride to block potassium channels), as well as with particular voltage protocols (e.g., high-voltage activated calcium channels can be evoked by a step to 0 or +10 mV from a holding potential of –90 mV, while low-voltage activated calcium channels can be evoked by a step to –20 mV from a similar holding potential). Calcium channel types can be further distinguished using pharmacological agents and inorganic ions. One can then examine the effects of ethanol on various channel properties such as the voltage-dependency and kinetics of channel activation and inactivation. Some of the first studies examining the effects of acute exposure to ethanol on calcium channels in *Aplysia* (Camacho-Nasi and Treistman, 1986; Camacho-Nasi and Treistman, 1987) involved the use of the two electrode voltage clamp technique. Since 1981 and the introduction of the patch clamp technique by Neher, Sakmann, and colleagues (1981), conventional whole-cell (ruptured patch), perforated patch (using the pore forming reagents nystatin or amphotericin B), and cell-attached patch clamp (single-channel recording) techniques have been used to examine the effects of ethanol on calcium channels in a number of biological systems. Excised patch recording is rarely used to record calcium channel activity, since the channels run down within minutes after excision, presumably due to the loss of an essential cytoplasmic element. An alternative to excised patch recording is to examine calcium channels reconstituted into planar lipid bilayers. In the presence of both the activated α subunit of the stimulatory guanine nucleotide regulatory protein (G protein) and a dihydropyridine (DHP) agonist, the activity of calcium channels in bilayers can be maintained for several minutes (see Rosenberg et al., 1986). As yet, the effects of ethanol on calcium channels reconstituted into planar lipid bilayers have not been examined.

2. Biochemical

Classic biochemical techniques using radioactive calcium to measure depolarization- (by electrical stimulation or elevated potassium) induced calcium flux or calcium-sensitive dyes such as fura-2 to determine changes in intracellular calcium concentration have been used to study the effects of acute as well as chronic exposure to ethanol on calcium channels in many systems (Daniell and Leslie, 1986; Friedman et al., 1980; Harris and Hood, 1980; Messing et al., 1986; Davidson et al., 1988; Daniell et al., 1987). One shortcoming of these techniques is that identification of the type of calcium channel affected by ethanol is made more difficult by not being able to use channel kinetics and voltage dependencies as identifiers. Taking advantage of DHPs, which are highly specific ligands for high voltage-activated L-type calcium channels, binding studies using radiolabeled DHP antagonists have been utilized to study the effects of both acute and chronic exposure to ethanol in freshly dissociated and primary neuronal cultures as well as cultured cell lines (Greenberg and Cooper, 1984; Harper et al., 1989; Messing et al., 1986).

3. Molecular Biology

As yet, molecular biological techniques have not been exploited in the study of ethanol's action on calcium channels. This is probably due to the fact that preparations in which calcium channels have been shown to be sensitive to relevant concentrations of ethanol (5–50 m*M*) have only recently been identified (Wang et al., 1991a; Mullikin-Kilpatrick and Treistman, 1994a,b,c; Huang and McArdle, 1994). The isolation of messenger RNA (mRNA) from these sources, and its subsequent microinjection into model expression systems, could facilitate the identification of the channel subunit(s) or site where ethanol interacts with the calcium channel. Expression systems such as the *Xenopus laevis* oocyte (Biel et al., 1990; Mikami et al., 1989; Mori et al., 1991); the fiberblastoid L-cell, which does not express any calcium channels (Peres-Reyes et al., 1989; Varadi et al., 1991); the murine muscular dysgenesis myotube, which is defective in the expression of the α_1 subunit (the fundamental subunit containing the calcium ionophore and the DHP binding site) of the calcium channel (Tanabe et al., 1987); a human embryonic kidney cell line (Williams et al., 1992a); and the IMR 32 neuroblastoma cell line (Tarroni et al., 1994) have all been developed and provide a means for functional analysis. Complete complementary DNA (cDNA) clones of calcium channel subunits that direct the expression of functional calcium channels in *Xenopus* oocytes have very recently been isolated from cDNA libraries obtained from nonneuronal (Ellis et al., 1988; Yaney et al., 1992) and neuronal preparations (Starr et al., 1991; Pragnell et al., 1991; Soong et al., 1993). These cloned and expressed channels will provide the means to examine, in isolation, particular subunits and combinations thereof which may be affected by ethanol. The coexpression of antisense oligonucleotides directed against specific calcium channel subunits or injection of antibodies generated against specific regions of the channel subunits will also aid researchers in determining the site(s) of action for ethanol.

B. CALCIUM CHANNELS

To date, calcium channels have been divided into two distinct classes as defined by their voltage threshold of activation: low-voltage activated (LVA) and high-voltage activated (HVA) (see Table 1 for a full description of calcium channel nomenclature). The LVA channels are termed "T" (for transient), while HVA channels are further divided into three subclasses: "L" (for long-lasting), "N" (for neither L nor T), and "P" (for Purkinje cell) depending upon their response to particular pharmacological agents. Other types of calcium channels such as P-type (Llinás et al., 1989) and Q-type (Sather et al., 1993), which may be similar to the P-type, are functionally defined types and have been recently classified by pharmacological characterization in hippocampal neurons (Wheeler et al., 1994). In addition, R-type channels (for resistant) have also been reported (Ellinor et al., 1993), but this channel type is not well established.

The T-type calcium channels activate at relatively hyperpolarized step potentials from very hyperpolarized holding potentials (close to the resting membrane potential) and carry a transient current which inactivates quickly during long pulses. These channels, which are equally permeable to Ba^{2+} and Ca^{2+}, have a single-channel conductance of 8–10 picosiemens (pS). These channels are found in muscle and neuronal tissue and are responsible for spontaneous membrane potential fluctuations (for review, see Bertolino and Llinás, 1992). The L-type channels activate at depolarized step potentials from relatively depolarized holding potentials (more depolarized than –40 mV) and carry a long-lasting current, which shows little voltage-dependent inactivation during long pulses. These channels are more permeable to Ba^{2+} than to Ca^{2+} and exhibit a single-channel conductance of 25–28 pS. A particular characteristic of L-type channels is that they must be phosphorylated in order to open upon membrane depolarization (Armstrong and Eckert, 1987). L-type channels, which are found in both neuronal and non-neuronal tissue, seem to be associated with the propagation of action potentials, hormone release (Wang et al., 1993), and with signal transduction events which occur at the membrane (Kostyuk, 1989). The identification of N-type channels, which are also

TABLE 3.1
Calcium Channel α_1 Subunits

Gene Product			Functional Correlates	
Consensus Name(s)	**Original Name(s) if Different**	**Sites of Expression**	**Current**	**Drug Sesitivity of Native Currents**
α_{1S}	Skeletal muscle CaCh1 α_{1skm}	Skeletal muscle, BC3H1 cells	HVA L type	Sensitive to DHPs, diltiazem and verapamil Insensitive to sub-µM ω-CTx-GVIA and funnel web spider venoms (ω-Aga-IVA, FTX)
α_{1A}	BI	Brain, cerebellum,	HVA Q type?	w-CTx-MVIIC (>100 n*M*); ω-Aga-IVA (>10 n*M*)
	CaCh4	Purkinje and granule cells, kidney,		DHP insensitive
	rbA	PC12 cells, C cells	HVA P type?	Sensitive to ω-Aga-IVA (<10 n*M*) and low sFTX DHP insensitive
α_{1B}	BIII CaCh5 rbB	Brain, peripheral neurons, PC12 cells, C cells	HVA N type	Sensitive to ω-CTx-GVIA (100–500 n*M*) and ω-CTx-NVIIC (>100 n*M*) DHP insensitive
α_{1C}	Cardiac Smooth muscle/ lung CaCh2 rbC	Heart, HIT cells, GH3 cells, brain, aorta, lung kidney, fibroblasts, PC12 cells, C cells	HVA L type	DHP sensitive Insensitive to low concentrations of ω-CTx-GVIA, ω-Aga-IVA, or sFTX
$\alpha_{1C\text{-}a}$	CaCh2a CaCh2-I	Heart		
$\alpha_{1C\text{-}b}$	CaCh2b CaCh2-II	Smooth muscle, lung		
$\alpha_{1C\text{-}c}$	rbC CaCh2-III	Brain		
α_{1D}	CaCh3 Neuroendocrine rbD	Brain, pancreas, HIT cells, GH3 cells, PC12 cells, C cells	HVA L type	DHP sensitive Reversibly sensitive to ω-CTX-GVIA, ω-Aga-IVA, or FTX
α_{1E}	CaCh6 BII rbE	Brain, heart, C cells	HVA R type?	Sensitive to low Ni Insensitive to DHPs or ω-CTx-NVIIC, or to low concentrations of ω-CTx-GVIA, ω-Aga-IVA or sFTX

Note: This table is intended as a guide and refers only to mammalian calcium channels. Not all previously used names are listed. Vertebrate doe-1 and doe-4 α_1 subunits, cloned from the marine ray Discopyge ommata, are orthologs of mammalian α_{1E} and α_{1B}, respectively. HVA and LVA, high and low voltage activated; DHP, dihydropyridine; ω-CTx-G and ω-CTx-M, ω-conotoxins from marine snails Conus geographus and Conus magus, respectively; Aga, agatoxin (funnel web spider Agelenopsis aperta toxin); sFTX, syhthetic funnel web spider toxin. Q-type calcium channel: current in cerebellar granule cells sensitive to ω-CTx-MVIIC but insensitive to DHPs, low ω-CTx-GVIA, and low ω-Aga-IVA; R-type calcium channel: residual in cerebellar granule cells after blocking with DHP, ω-Aga-IVA, ω-CTx-GVIA, and ω-CTx-MVIIC.

From: Birnbaumer, L., Campbell, K.P., Catterall, W.A., Harpold, M.M., Hofmann, F., Horne, W.A., Mori, Y., Schwartz, A., Snutch, T.P., Tanabe, T., and Tsien, R.W. (1994) The naming of voltage-gated calcium channels. *Neuron,* 13:505–506. With permission.

found in neuronal and non-neuronal tissue, is more difficult, as currents tend to activate at a similar range of potentials as L-type channels. In addition, N-type channels have both a slow-inactivating and a long-lasting component of current (Plummer et al., 1989). These channels are more permeable to Ba^{2+} than to Ca^{2+} and exhibit a single-channel conductance of 11–15 pS. The function of N-type channels may include an involvement in the generation of action potentials as well as some forms of neurotransmitter release (Miller, 1987; Wang et al., 1993). The P-type channels, which are the most widely distributed calcium channel in the mammalian central nervous system (Hillman et al., 1991; Llinás et al., 1992), activate slowly during depolarizing step potentials from relatively depolarized holding potentials (more depolarized than –60 mV). These channels display three distinct levels of single-channel conductance, these being 9–10, 13–14, and 18–19 pS. Because of the similarity of the single-channel conductance of P-type channels to L- and N-type channels, one cannot use purely operational definitions to classify the different calcium channel types and must rely upon pharmacological tools as well. The P-type calcium channel appears to play a role in neurotransmitter release (Artalejo et al., 1994).

1. Pharmacology

The classification of calcium channels has been assisted by the discovery and use of specific pharmacological agents. Blockers of LVA T-type channels include amiloride (Tang et al., 1988), high molecular weight alcohols (Llinás and Yarom, 1986), and nickel (Byerly and Hagiwara, 1988). Dihydropyridines which exhibit agonistic (causing an increase in macroscopic current or longer single-channel openings) or antagonistic (causing a decrease in macroscopic current or shorter single-channel openings) actions are voltage-dependent and highly selective for HVA L-type channels (Catterall, 1988; Reuter, 1985). These compounds are thought to exert their effects by binding to a site on the extracellular surface between the third and fourth internal homologous repeats (Nakayama et al., 1991) as well as to a site on the intracellular portion (membrane spanning region S6) of the fourth repeat of the α_1 subunit of the L-type channel (Regulla et al., 1991). High voltage-activated N-type channels are inhibited by omega-conotoxin GVIA, a toxin isolated from the marine snail *Conus geographus* (Olivera et al., 1985). Low concentrations of the toxin are highly selective for N-type channels and have been used successfully to identify them in a heterogenous population of calcium channels in neurohypophysial terminals (Wang et al., 1992). A neurotoxin, which has been isolated from the venom of the funnel-web spider *Agelenopsis aperta*, referred to as funnel-web spider toxin (FTX), is highly selective for HVA P-type channels (Llinás et al., 1989) as is a peptide, omega-agatoxin-IVA, which was isolated from the same venom (Adams et al., 1990). The P-type channel is insensitive to DHP ligands and omega-conotoxin GVIA. All HVA calcium channel types can be blocked by cadmium.

2. Molecular Biology

Much of the knowledge about calcium channel structure and function and the primary sequence of the protein(s) (inferred from cloning the corresponding cDNAs) originated from studies of calcium channels in skeletal muscle. The L-type calcium channel, which has been purified from skeletal muscle, is a large polypeptide of 430 kDaltons, made up of four distinct subunits: α_1 subunit, the major subunit of calcium channels; β; γ, which is specific for calcium channels in skeletal muscle; and the highly conserved α_2/δ.

The α_1 subunit, which is thought to be comprised of four internally homologous segments each made up of six membrane spanning regions, contains the channel pore, DHP binding sites (Striessnig et al., 1991), and several sites which are phosphorylated *in vitro* by cAMP-dependent protein kinase (PKA). Full-length cDNA clones for the α_1 subunits have been isolated from skeletal muscle, cardiac tissue, smooth muscle, endocrine glands, and brain. These cDNAs direct the functional expression of calcium channels in tissue culture cell lines and *Xenopus laevis* oocytes. These fairly homologous (41–70%) cDNAs are encoded by six different genes (for review, see Hofmann et al., 1994). The product of the S gene has been cloned from rabbit skeletal muscle (Tanabe et al., 1987) and encodes the L-type α_{1S} subunit. Splice products of the C gene cloned from cardiac (Mikami et al., 1989), smooth muscle (Biel et al., 1990), and brain (Snutch et al., 1991) encode for L-type $\alpha_{1C\text{-}a}$, $\alpha_{1C\text{-}b}$, and $\alpha_{1C\text{-}c}$ subunits, respectively. The gene product of the D gene was isolated from neuronal and endocrine tissues and encodes the neuroendocrine-specific L-type α_{1D} subunit (Williams et al., 1992b; Seino et al., 1992). The gene product of the A gene was cloned from brain cDNA libraries and encodes

P- and/or Q-type α_{1A} subunits (Mori et al., 1991). The cDNA of the B gene was cloned exclusively from brain (Williams et al., 1992a), and the gene product encodes the N-type α_{1B} subunit. The gene product of the E gene from brain encodes the α_{1E} subunit (Niidome et al., 1992; Soong et al., 1993). It is uncertain whether this gene product is a novel channel type, R-type, or is analogous to the T-type channel. Two cDNAs (doe-1 and doe-4) have been cloned from the marine ray (Horne et al., 1993). The sequences of these gene products are similar to the sequences of other gene products encoding α subunits of possibly R-type or N-type channels, respectively.

There are four different genes with tissue-specific transcripts, β_{1-4}, which give rise to the intracellularly located β subunit (for review, see Hofmann et al., 1994). This subunit appears to be required for optimal expression and function of the calcium channel and also contains sites for phosphorylation by PKA. In addition, this subunit has been shown to increase both macroscopic current and the rate of activation of current, as well as shift the voltage dependency of steady-state inactivation of calcium current (Singer et al., 1991; Varadi et al., 1991; Lacerda et al., 1991).

The cDNA for the transmembrane γ subunit was cloned from, and is specific for, skeletal muscle (Bosse et al., 1990; Jay et al., 1990). This subunit appears to be required for the voltage dependency of the L-type channel in skeletal muscle and increases the speed of channel inactivation.

Transcripts of the highly conserved α_2/δ subunit (the extracellular α_2 subunit is disulfide bridged to the transmembrane δ subunit) also appear to be required for optimal expression and function of the channel (for review, see Hofmann et al., 1994). This subunit has been found to increase current density of expressed α_1 and β subunits with no effect on current kinetics (Williams et al., 1992b; Singer et al., 1991). This subunit is found in skeletal muscle, heart, brain, and smooth muscle.

Another subunit, a 95-kiloDalton glycoprotein, has been shown to be an integral part of omega-conotoxin-sensitive N-type channels in neurons (Witcher et al., 1993). It is not known at this time whether there are ancillary subunits, in addition to those mentioned above, which may be involved in calcium channel function.

3. Localization in Brain

Subtypes of the voltage-dependent calcium channel are heterogeneously distributed throughout the brain. Gene products of the A (P- and/or Q-type α_1 subunit), B (N-type α_1 subunit), C (L-type α_1 subunit isolated from cardiac, smooth muscle, or brain), and D (neuroendocrine L-type α_1 subunit) genes, respectively, have been identified by Northern blot analysis in rat cerebellum, hippocampus, olfactory bulb, and spinal cord (Starr et al., 1991). Using blot hybridization analysis, it has been shown that the α_1 subunit mRNA derived from the E gene is expressed in the neocortex, striatum, lateral septum, amygdala, hippocampus, and the dentate gyrus (Niidome et al., 1992; Soong et al., 1993). This gene product has also been reported to show high levels of expression in the cerebellar cortex, medial habenula, substantia nigra, pars compacta, caudal brain stem, and dorsal raphe, using *in situ* hybridization (Soong et al., 1993). Autoradiographic mapping using labeled DHP antagonists showed that L-type calcium channels were localized in various regions of the rat brain (Cortes et al., 1983). Dihydropyridine-sensitive gene products of the C gene as well as the related gene products of the A, B, and D genes have also been identified in rabbit cerebellum (Perez-Reyes et al., 1990). In particular, high levels of expression of α_{1C-a} transcripts have been identified, using PCR amplification, in the cerebellum, trigeminals, hypothalamus, and olfactory bulb, while transcripts of the alternatively spliced variant α_{1C-b} have been shown in cerebellum, trigeminals, striatum, hypothalamus, hippocampus, cortex, and olfactory bulb (Snutch et al., 1991). High levels of N-type α_1 subunit mRNA have been identified in rat cerebellum, thalamus, and hippocampus, while lower levels of expression were shown in the striatum, cortex, and olfactory bulb by Northern blot analysis (Dubel et al., 1992). Using both blot hybridization analysis (Mori et al., 1991) and immunohistochemical analysis (Cherksey et al., 1991), the gene product of the A gene, the P/Q-type α_1 subunit, has been localized in the central nervous system to the Purkinje and granule cells of the cerebellum. The gene product of the E gene has been shown to be abundant in the cerebral cortex, hippocampus, and corpus striatum (Niidome et al., 1992). Within the neuronal cell bodies and proximal dendrites of the rat cerebal cortex and hippocampus, immunoreactivity has been detected for the α_1 subunits of both L-type calcium channel subtypes C and D, suggesting that these channels are intimately involved in regulating calcium-dependent processes which are located in those compartments (Hell et al., 1993).

4. Functional Consequences of Modulation by Ethanol

It is probably safe to say that alteration of calcium flux through voltage-gated calcium channels would affect virtually all physiological processes, either directly or indirectly. The release of transmitters and neurohormones from presynaptic fibers is dependent upon calcium entry, and thus, synaptic transmission would be seriously compromised by perturbation of function of this subset of calcium channels. It is especially important to note that, since release of transmitter is an exponential function of the rise in internal calcium, even small changes in channel flux can lead to large changes in synaptic transmission. Calcium entry through voltage-gated channels contributes to the action potential in many neurons and also plays a role as current carrier in subthreshold currents which control firing patterns such as bursting pacemaker activity. Calcium entering through voltage-gated channels also activates other conductances, such as calcium-activated potassium channels and calcium-activated chloride channels, as well as calcium-activated second messenger cascades. Given the severe consequences of any interference with the operation of calcium channels, it is reasonable to assume that they are not globally affected by exposure to relevant concentrations of ethanol. However, as we will see, a reduction in calcium current and calcium-mediated release, as a consequence of acute exposure to ethanol, has been shown using biochemical as well as electrophysiological methodology in several neuronal- and endocrine-like systems (Rabe and Weight, 1988; Oakes and Pozos, 1982; Camacho-Nasi and Treistman, 1986; Camacho-Nasi and Treistman, 1987; Koski et al., 1991; Daniell and Leslie, 1986; Brennan et al., 1989a,b; Messing et al., 1986; Wang et al., 1991a,b,c; Twombly et al., 1990). As our techniques for the study of calcium channels become more sophisticated, especially with respect to advances in molecular biology of these channels, we will likely find that only a small subset of calcium channel subunit combinations, or only channels predisposed by residence in particular channel states, are sensitive to ethanol at relevant concentrations.

In contrast to the effects of acute exposure to ethanol, chronic exposure to ethanol has been reported by a number of laboratories to result in an increase in potassium-stimulated calcium uptake (Messing et al., 1986), calcium-dependent release (Harper et al., 1989; Harper and Littleton, 1991), and DHP antagonist binding sites (Skattebol and Rabin, 1987; Messing et al., 1986). An electrophysiological study from our laboratory has shown that there is a significant increase in voltage-activated calcium current after chronic exposure to ethanol (Grant et al., 1993). The increase in binding sites and the up-regulation of calcium currents have been shown to involve a genetic up-regulation of the calcium channel protein (Harper et al., 1989) and protein kinase C (Messing et al., 1990). The increase in calcium current observed after chronic exposure suggests that calcium channels may contribute to the adaptive response of cells to ethanol. The consequences of functional up-regulation of calcium channels would be expected to be as significant for nervous system function as the acute inhibition of these channels. Indeed, there is evidence, which will be discussed later in this review, that many of the symptoms of withdrawal from chronic ethanol may be attributable to the upregulation of voltage-gated calcium channels. Once again, it is likely that given the severe consequences of a global alteration in calcium channel function, adaptive changes in response to chronic ethanol exposure will be limited to a subset of calcium channels.

II. EFFECTS OF ACUTE EXPOSURE TO ETHANOL

In this section, we present evidence which indicates that both neuronal and nonneuronal calcium channels as well as related functions of these channels are targets for acute exposure to ethanol.

A. PRIMARY CULTURES OR ACUTELY DISSOCIATED PREPARATIONS

1. Calcium Flux Studies

Over the years, a number of studies have used mammalian brain synaptosomal preparations to examine the effects of acute exposure to ethanol on depolarization-dependent calcium uptake *in vitro*. Initially, several groups showed that concentrations as low as 80 m*M* ethanol significantly enhanced calcium accumulation into depolarized and nondepolarized whole brain synaptosomes isolated from naive mice (Blaustein and Ector, 1975; Friedman et al., 1980), while other groups reported that ethanol, at concentrations equal to or less than 100 mM, inhibited the depolarization-dependent influx of calcium into synaptosomal preparations (Harris and Hood, 1980; Stokes and Harris, 1982). All of these studies examined the effects of ethanol during the slow-phase (30 s to 16 min time periods) of calcium uptake. Subsequently, studies were done to examine the effect of similarly low concentrations of ethanol on the fast-phase of calcium uptake, with significant inhibitory effects reported at periods of 5 s or less (Leslie et al., 1983; Daniell and Leslie, 1986). Results from most of these studies showed that 25–200 m*M* ethanol inhibited uptake in a dose-dependent fashion (maximal inhibition of 28 and 33% at 150 and 200 m*M* ethanol, respectively) but that the time-dependent inhibition of calcium uptake depended upon the brain region being investigated. Inhibition of slow-phase, potassium-stimulated calcium uptake was greater with synaptosomes prepared from cerebellum and striatum than from cortex or brain stem, whereas inhibition of fast-phase uptake was greater with preparations from brainstem or midbrain. It was suggested that these differences in calcium uptake in brain regions could be the result of differences in the calcium channels expressed in the various brain regions. A study (Hawthorn et al., 1992) examining the effect of alcohols of varying chain length on calcium uptake into synaptosomes reported that uptake was inhibited by the alcohols, with the potency increasing with chain length and that the inhibition could be blocked by FTX, a toxin specific for P-type calcium channels. These results suggested that P-type channels were targets for alcohol action in mammalian synaptosomal preparations. In another study, high concentrations of ethanol (100–800 m*M*) inhibited the fast-phase of depolarization-dependent calcium uptake by 25–90% in cerebrocortical synaptosomes, while the slow-phase of uptake was blocked 40 and 60% by 400 and 800 m*M* ethanol, respectively (Skattebol and Rabin, 1987). Davidson and co-workers (1988) reported that an acute exposure to 100, 200, and 500 m*M* ethanol reduced uptake into depolarized rat forebrain synaptosomes by approximately 17, 25, and 50%, respectively, suggesting that ethanol decreases the rate of calcium entry.

Fluorescent calcium-sensitive dyes such as the calcium indicator fura-2 have been used to examine ethanol-induced changes in intracellular calcium levels in synaptosomal preparations (Davidson et al., 1988; Rezazadeh et al., 1989). Daniell and co-workers (1987) showed that very high concentrations of ethanol (350–700 m*M*) caused an initial rapid rise (<10 s) in resting intracellular calcium levels in synaptosomes isolated from mouse whole brain. The levels of intracellular calcium remained elevated over the next 14 min and decreased the effectiveness of depolarization to further increase intracellular calcium levels. Acute exposure to concentrations of ethanol 100 m*M* or higher has also been shown to induce changes in intracellular calcium levels by releasing calcium from intracellular stores in brain microsomes (Daniell and Harris, 1989). Another study using fura-2 demonstrated that lower concentrations of ethanol caused a significant increase in intracellular calcium levels in forebrain synaptosomes (Davidson et al., 1988). Ethanol (100 m*M*) also caused an increase in basal levels of intracellular calcium in rat pinealocytes loaded with fura-2, and reduced subsequent DHP agonist- and potassium-stimulated calcium uptake (Chik et al., 1992). In isolated rat liver hepatocytes, ethanol has been shown to cause an increase in the influx of extracellular calcium into quin 2-loaded hepatocytes over a period of 2 min (Rubin and Hoek, 1988), while no ethanol-induced increase in either resting intracellular calcium or vasopressin-mediated increases in intracellular calcium was noted in another study utilizing hepatocytes (Daniell et al., 1987).

2. Neurotransmitter Release Studies

One of the first reports that examined the inhibitory effects of acute exposure to ethanol on calcium-dependent neurotransmitter release involved the use of electrically stimulated rat brain cortical slices (Carmichael and Israel, 1975). Subsequent studies, using rat synaptosomes isolated from various regions of the brain, demonstrated that depolarization-induced release of endogenous norepinephrine was unaltered by a 20 min exposure to 25–200 m*M* ethanol (Daniell and Leslie, 1986). In contrast, basal and carbachol-induced secretion of catecholamines from primary cultures of bovine adrenal chromaffin cells was potently inhibited by an acute exposure to 50 m*M* ethanol (Brennan et al., 1989a,b; Brennan and Littleton, 1991). More recently, a preparation of isolated rat neurohypophysial nerve terminals has been used to show that either electrically or potassium-stimulated release of arginine-vasopressin from these isolated nerve endings is significantly reduced by 10 m*M* ethanol (Wang et al., 1991a,b,c). This decrease in neuropeptide release correlated well with ethanol-induced decreases in macroscopic calcium currents in these terminals.

3. Dihydropyridine Antagonist Binding Studies

High concentrations of ethanol (400 m*M*) were required to inhibit specific [^{3}H]nitrendipine binding, characterized by a decrease in binding affinity without alteration in the number of binding sites (B_{max}) in rat brain membranes (Greenberg and Cooper, 1984). Another study involving rat brain membranes demonstrated that an acute oral administration of ethanol also produced an increase of 40% in antagonist binding sites, suggesting that ethanol had an effect on L-type calcium channels (Bergamaschi et al., 1988).

4. Electrophysiological Studies

In CA1 cells of isolated rat hippocampus, acute exposure of the terminal region of the neurons to ethanol (5–20 m*M*) has been shown to increase both inhibitory and excitatory postsynaptic potentials; 20 m*M* ethanol significantly depressed calcium-dependent spikes in these neurons (Carlen et al., 1982). These results were attributed to an ethanol-induced increase in intracellular calcium. Excitatory and inhibitory postsynaptic potentials have also been shown to be reduced by 10–40 m*M* ethanol in cultured spinal cord neurons (Gruol, 1982). Acute exposure of freshly dissociated dorsal root ganglion neurons to ethanol generated a decrease in electrical excitability, specific membrane capacitance, and action potential overshoot, whereas specific membrane resistance was increased (Bunting and Scott, 1989). It was proposed that ethanol caused the changes in passive electrical properties of the membrane by penetrating the lipid bilayer and expanding the membrane. In another report using dissociated dorsal root ganglion neurons, recordings demonstrated that acute exposure to ethanol increased the spike threshold and decreased the duration of the evoked action potentials in a dose-dependent fashion but did not alter the amplitude, rise time, or resting potential of the neurons (Eskuri and Pozos, 1987). Interestingly, the potency and reversibility of ethanol's actions in this study depended upon the temperature of the bath solution. Dissociated dorsal root ganglion neurons from embryonic rats exposed to 106, 320 and 639 m*M* ethanol for 1 h displayed significant reversible alterations in the repolarization phase of the action potential (Oakes and Pozos, 1982). This alteration was attributed to a decrease in potassium conductance. Because of the similar effects generated by cobalt (a calcium channel blocker), these authors reasoned that the observed decrease in potassium conductance was due to a decrease in the calcium current which activated calcium-dependent potassium channels.

Two lines of mice known as long-sleep and short-sleep have been bred for differences in behavioral sensitivity to ethanol (Erwin et al., 1976). Examination of L-type calcium channels in isolated dorsal root ganglion neurons from these lines has shown no difference in their sensitivity to acute exposure to ethanol (for review, see McArdle et al., 1992). In both lines, acute exposure to 5 m*M* ethanol caused an increase in L-type calcium current, whereas 43 m*M* ethanol produced a significant decrease in current amplitude. Single-channel recordings revealed an increase in the probability of single-channel openings with 5 m*M* ethanol and a decrease in channel openings with 43 m*M* ethanol. In addition, the facilitating effect of ethanol was not due to a specific effect on L-type channels, as the frequency of opening of channels carrying outward current was also increased. A very recent study has revealed that the inhibitory effect of ethanol on calcium channels requires activation of a pertussis-sensitive guanine nucleotide regulatory protein and that this activation causes a reduction in single-channel open time (Huang and McArdle, 1994).

Acutely dissociated rat neurohypophysial terminals have been used to determine the ethanol sensitivity of high-voltage activated calcium channels (Wang et al., 1991a,b,c). Whole-cell patch clamp recordings showed that both L- and N-type current amplitudes were significantly reduced by an acute exposure to ethanol (10–100 m*M*) in a dose-dependent manner. This reduction was not due to a shift in the voltage-dependency of either activation or steady-state inactivation. In the same preparation, 100 m*M* ethanol had no effect on a fast, transient potassium current. More recently, terminal-attached single-channel recordings have shown that ethanol (25–100 m*M*) inhibits DHP-sensitive L-type calcium channels by reducing the open probability of the channel, primarily by decreasing the duration of openings, without altering channel conductance (Wang et al., 1994). The presence of a DHP agonist enhanced the inhibitory effect of ethanol on the channel open time, indicating the possibility that the ethanol sensitivity of the channel is dependent upon gating mode. The calculated half-maximal inhibition by ethanol was 9.1 m*M*. These effects on channel gating characteristics are consistent with ethanol interacting directly with the L-type channel.

Whole-cell patch clamp recordings of calcium currents in rat pinealocytes have shown that as little as 25 m*M* ethanol inhibited peak amplitude of L-type currents in a time-dependent fashion (Chik et al., 1992). The maximum inhibitory effect of 40% occurred with 100 m*M* ethanol and was not due to a shift in the activation current-voltage relationship.

B. CELL LINES

Rat pheochromocytoma (PC12) cells, a cell line which was derived from a rat adrenal tumor, are an established model for the study of neuronal differentiation and, as such, have been used extensively to examine the effects of acute exposure to ethanol on calcium channels. Initial studies demonstrated that 50 m*M* ethanol inhibited potassium-stimulated calcium uptake by 16% in undifferentiated cells, while half-maximal inhibition occurred at 211 m*M* (Messing et al., 1986). This reduction in calcium influx was not due to an increase in calcium efflux. Results from a study performed in our laboratory indicated that potassium-stimulated calcium uptake was reduced by 48% when undifferentiated PC12 cells were acutely exposed to 200 m*M* ethanol (Grant et al., 1993). In addition, patch clamp recordings indicated that 200 m*M* ethanol significantly inhibited the macroscopic calcium currents in these cells (Grant et al., 1993). Undifferentiated PC12 cells acutely exposed to 100 and 200 m*M* ethanol showed a decrease in potassium-stimulated dopamine release, but no change in bradykinin-stimulated release, suggesting that acute exposure to ethanol can differentially affect neurotransmitter release associated with different pathways in these cells (Oldham et al., 1989). Acute exposure to 25–100 m*M* ethanol also inhibited carbachol- and potassium-stimulated calcium uptake in these cells (Koski et al., 1991). Inhibition of the cholinergic mediated response was evident within 5 s, whereas inhibition of the potassium mediated response required more than 30 s. Several studies (Rabe and Weight, 1988; Rabe et al., 1987) have probed the effects of ethanol on neurotransmitter release and intracellular calcium levels in PC12 cells. Ethanol inhibited muscarine-stimulated release of norepinephrine from these cells, and this effect was independent of calcium influx. In addition, 400 m*M* ethanol caused an elevation in intracellular calcium.

Whole-cell patch clamp studies from our laboratory have shown that ethanol, at concentrations equal to or less than those used for calcium uptake or neurotransmitter release studies (5, 25, and 50 m*M*), inhibits L-type calcium currents in undifferentiated PC12 cells independent of the current carrying ion and without causing any consistent shift in the activation current-voltage relationship (Mullikin-Kilpatrick and Treistman, 1994a). More recently, our laboratory has shown that calcium channels in nerve growth factor (NGF)-treated PC12 cells (resulting in differentiated, sympathetic neuron-like cells) are significantly less sensitive to the effects of ethanol than are channels in undifferentiated cells (Mullikin-Kilpatrick and Treistman, 1995a). Ethanol produced a hyperpolarizing shift in steady-state inactivation in calcium channels of PC12 cells which accounted for a significant portion of inhibition in both cell types. This is in contrast to ethanol's effect on calcium channels in neurohypophysial terminals, in which ethanol did not cause a shift in steady-state inactivation. Recent results from our laboratory indicate that the difference in ethanol sensitivity of L-type calcium currents in undifferentiated cells and NGF-treated cells is due to the involvement of the pertussis toxin-sensitive G protein, G_{i2}, in undifferentiated cells and not in NGF-treated cells (Mullikin-Kilpatrick and Treistman, 1994c; 1995b). This G protein does not appear to be involved in the ethanol-induced shift in steady-state inactivation of the channels in undifferentiated cells. These results suggest that ethanol can inhibit L-type calcium channels by multiple mechanisms or pathways. The time course of the NGF-induced reduction in ethanol inhibition of calcium channels in PC12 cells occurred in two phases and was reversible by removal of NGF (Mullikin-Kilpatrick and Treistman, 1994b).

PC12 cell line is not the only cell line which has been used to examine the effects of ethanol. In a hybrid neural cell line, NCB-20, derived from mouse neuroblastoma and Chinese hamster brain cells, 400 m*M* ethanol elevated intracellular calcium levels approximately 2-fold as assessed by the fluorescent indicator fura-2 (Chan and Greenberg, 1991). Both L- and T-type calcium currents have been shown to be reversibly inhibited by concentrations of ethanol above 30 m*M* in differentiated N1E-115 neuroblastoma and in NG108-15 neuroblastoma X glioma hybrid cells (Twombly et al., 1990). The voltage dependence of inactivation was unaltered by ethanol, and there was no indication of use-dependent inhibition in these neuroblastoma cell lines.

C. INVERTEBRATE PREPARATIONS

There have been a number of studies which examined the effects of acute exposure to ethanol on calcium channels in isolated neurons of the marine mollusk *Aplysia californica*. One of the first studies, employing two electrode voltage clamp techniques, showed that ethanol at high concentrations (110–1100 m*M*) reduced the amplitude of the action potential in bursting pacemaker R15 neurons, in R2 cells, and in bursting pacemaker cells in the L2–L6 group (Bergmann et al., 1974). This study also demonstrated that acute exposure to ethanol, at these concentrations, also reduced calcium currents and sodium currents. In another study using bursting pacemaker neurons, Schwartz (1983) reported that 685 m*M* ethanol decreased slow inward calcium current but increased an outward potassium current. In an ensuing report, Schwartz (1985) found that the decrease in calcium current was not due to an increase in intracellular calcium concentration. Subsequently, a study from our laboratory reported that 50 and 500 m*M* ethanol reversibly inhibited peak current amplitude by 20 and 90%, respectively, in identified neurons of *Aplysia* (Camacho-Nasi and Treistman, 1986). It was concluded from a later study that ethanol inhibition of current did not occur as a result of ethanol occluding the pore but by increasing the inactivation rate of channel currents when calcium was used as charge carrier (Camacho-Nasi and Treistman, 1987). In addition, it was noted that a significant time lag was observed in the ethanol-mediated reduction of current. Additional studies from our laboratory have confirmed that exposure to 100 and 300 m*M* ethanol reduces calcium currents by 20 and 60% in identified neurons of *Aplysia* (Treistman and Wilson, 1991).

The nervous system of *Helix aspersa*, a garden snail, has also been used as a model system to examine the effects of acute exposure to ethanol on calcium channels (Oyama et al., 1986). Ethanol, at concentrations equal to or less than 200 m*M*, reduced current amplitude and altered the time constant of decay of the calcium current. Results similar to these were also found by Camacho-Nasi and Treistman (1987) in *Aplysia* neurons.

III. EFFECTS OF CHRONIC EXPOSURE TO ETHANOL

One component of tolerance is the capacity to show a decreased sensitivity to a substance upon repeated exposure to that substance. This section provides evidence which indicates that chronic exposure to ethanol affects calcium channels and their related functions in many systems and that these channels appear to be involved in the development of tolerance to ethanol.

A. PRIMARY CULTURES OR ACUTELY DISSOCIATED PREPARATIONS

1. Calcium Flux Studies

To the best of our knowledge, there are no reports of the effects of chronic exposure to ethanol on calcium uptake in primary cultures or acutely dissociated preparations.

2. Neurotransmitter Release Studies

Bovine adrenal chromaffin cells displayed an approximate 3-fold increase in potassium-stimulated release of catecholamines after being maintained in culture for 6 d in the presence of 200 m*M* ethanol (Harper et al., 1989). Acute ethanol inhibition of stimulated release was reduced after chronic exposure, suggesting the development of tolerance in these cells. In addition, the increase in catecholamine release after chronic exposure could be inhibited by a 10 min exposure to a DHP antagonist prior to eliciting release. These data suggested that DHP-sensitive channels may be involved in the development of tolerance.

When compared to control cells, carbachol- and veratrine-induced release of catecholamines from bovine adrenal chromaffin cells cultured in the presence of 200 m*M* ethanol for 6 d was resistant to the inhibitory effects of an acute exposure to ethanol (Harper and Littleton, 1990). The results showed that complete tolerance to ethanol had developed for the carbachol-induced release over the whole range of ethanol concentrations used (25–400 m*M*) during chronic exposure, further supporting the idea that this model system could be useful in examining the mechanism(s) by which tolerance develops.

3. Dihydropyridine Antagonist Binding Studies

The ability of channel populations to be up-regulated in response to chronic ethanol exposure is variable. Cultured bovine adrenal chromaffin cells maintained in the presence of 200 m*M* ethanol for 4 d showed an approximate 2-fold increase in [^{3}H] nitrendipine binding sites on membranes prepared from these cells (Brennan et al., 1989a,b). These results suggest that the increase in binding sites represents an adaptation of the cells to overcome ethanol's inhibitory effects. A subsequent report demonstrated that the ethanol-induced increase in DHP binding sites could be prevented by blocking mRNA synthesis, suggesting that *de novo* protein synthesis was involved in up-regulating the calcium channel proteins (Harper et al., 1989). A later article showed that an elevated concentration of external calcium during chronic exposure to ethanol reduced the increase in DHP-sensitive binding sites observed in chronic-treated cells. Similar treatment with elevated calcium had no effect on binding sites in control cells (Brennan and Littleton, 1990). An overnight treatment of cells with pertussis toxin (thought to prevent activation of inhibitory guanine nucleotide regulatory proteins) caused an increase in binding sites similar to chronic ethanol treatment, suggesting that inactivation of pertussis toxin-sensitive G proteins may be involved in the effects of long-term ethanol exposure.

B. CELL LINES

Potassium-stimulated calcium uptake into undifferentiated PC12 cells cultured in the presence of 150–200 m*M* ethanol for 4–6 d was increased 25–80% over uptake levels in control cells (Messing et al., 1986; Skattebol and Rabin, 1987; Grant et al., 1993). Sensitivity to an acute ethanol challenge was approximately the same in control and ethanol-treated cells; the half-maximal inhibitory concentration for ethanol inhibition of uptake was 194 m*M* (Messing et al., 1986), suggesting that tolerance to an acute ethanol challenge with ethanol had not occurred. In contrast, Skattebol and Rabin (1987) observed that chronic ethanol treatment of PC12 cells enhanced the inhibitory effect of an acute ethanol challenge on calcium uptake. This difference in reported tolerance development was attributed to different methodologies used as well as clonal variabilities.

To our knowledge, only one study has examined the effects of chronic exposure to ethanol on neurotransmitter release. In this study, undifferentiated PC12 cells chronically treated with 200 m*M* ethanol showed an enhancement of dopamine release in the presence of high potassium but no change in release in response to bradykinin (Oldham et al., 1989). These results indicated that chronic exposure to ethanol can selectively affect neurotransmitter release mechanisms within the same cell.

Ligand binding studies have demonstrated that DHP antagonist binding sites on undifferentiated PC12 cell membranes (Messing et al., 1986; Skattebol and Rabin, 1987) or intact PC12 cells (Marks et al., 1989) were increased 1.5–2-fold after chronic exposure to ethanol, without an appreciable change in binding affinity. Chronic exposure of the cells to calcium channel blockers such as nifedipine, verapamil, or diltiazem did not cause an increase in DHP-sensitive binding sites, even though these drugs acutely inhibited calcium uptake in PC12 cells (Marks et al., 1989). A subsequent study by Messing and colleagues (1990) using both inhibitors and activators of protein kinase C indicated that this calcium-phospholipid-dependent kinase participates in the up-regulation of DHP-sensitive binding sites in PC12 cells.

An electrophysiological study from our laboratory indicated that the increase in DHP-sensitive binding sites is manifested as a 1.5-fold increase in voltage-activated calcium currents in undifferentiated PC12 cells (Grant et al., 1993). In addition, calcium channel currents in undifferentiated PC12 cells cultured in the presence of 180 m*M* ethanol for 6 d were reduced only 15% by an acute exposure to 25 m*M* ethanol, compared to a reduction of 42% for cells cultured in the absence of ethanol (Mullikin-Kilpatrick and Treistman, 1993). Thus, both an increase in DHP-sensitive L-type calcium current and a change in acute ethanol sensitivity of these channels after chronic exposure suggests that these calcium channels may contribute to the adaptive response of PC12 cells to ethanol.

In an abstract published in 1991 (Bergamaschi et al.), it was reported that exposure of undifferentiated NG108-15 cells to 200 m*M* ethanol for 3 d resulted in an increase in DHP-sensitive binding sites, similar to that seen in PC12 cells. In contrast, omega-conotoxin GVIA binding sites were not altered after chronic treatment of undifferentiated NG108-15 cells, suggesting that these two channel types were differentially modified by chronic exposure to ethanol. When differentiated NG108-15 cells were exposed to chronic ethanol, a reduction in omega-conotoxin binding sites was observed. Thus, both calcium channel types may contribute to an adaptative response of neuroblastoma cells to ethanol.

C. INVERTEBRATE PREPARATIONS

Calcium currents in *Aplysia californica* neurons are unaltered as a function of chronic exposure to ethanol. Electrophysiological studies performed in our laboratory demonstrated that exposure to 200 m*M* ethanol for periods up to 3 weeks did not alter calcium channel characteristics of R15 neurons, nor did chronic exposure alter the inhibition of currents produced by an acute challenge with 100 or 300 m*M* ethanol (Treistman and Wilson, 1991). Thus, calcium channels in different systems appear to show a different propensity for adaptation to chronic ethanol.

D. WHOLE ANIMAL STUDIES

1. Calcium Flux and Neurotransmitter Release Studies

In depolarized whole brain synaptosomes from mice kept on a diet containing ethanol for 10 d, calcium uptake was decreased compared to synaptosomal preparations from naive animals (Friedman et al., 1980). Since an acute challenge with 80 m*M* ethanol increased calcium uptake into synaptosomes from naive animals but did not cause an increase in uptake in synaptosomes from chronic-treated animals, it was concluded that tolerance had developed. Another study which used synaptosomes isolated from cerebral cortex of rats showed that chronic ethanol treatment of the animals resulted in an adaptive response such that an acute challenge with ethanol no longer inhibited depolarization-dependent calcium uptake (Leslie et al., 1983). Another report showed that chronic ethanol treatment produced an adaptative response to the effect of an acute challenge with ethanol on calcium uptake into hypothalamic synaptosomes but did not produce this adaptive response in brainstem or cerebellar synaptosomes (Daniell and Leslie, 1986). In addition, voltage-dependent release of norepinephrine was reduced by an acute exposure to ethanol in hypothalamic and cerebellar synaptosomes but not in preparations from brainstem, suggesting that the effects of ethanol on uptake and release can differ depending upon the brain region. Another study involving mice chronically exposed to ethanol demonstrated that depolarized brain synaptosomes from these animals accumulated significantly less calcium than depolarized synaptosomes from naive animals and were less sensitive (developed tolerance to ethanol) to the inhibitory effects of acute exposure to ethanol on potassium-stimulated calcium uptake (Harris and Hood, 1980). Neither calcium influx into nor dopamine release from depolarized brain synaptosomes of rats maintained on an ethanol-containing diet for 8 weeks were inhibited by an acute challenge with 200 m*M* ethanol, in contrast to results obtained with synaptosomes from control animals (Woodward et al., 1990). Bay K 8644 increased uptake and release in control synaptosomes but had no effect in synaptosomes isolated from ethanol-treated animals. Omega-conotoxin had no effect on depolarization-dependent calcium uptake in either preparation but inhibited dopamine release from both preparations. These results suggest that neuronal DHP- and conotoxin-sensitive calcium channels can be affected differently by chronic exposure to ethanol.

Depolarization-induced catecholamine release is enhanced in brain slices from ethanol-dependent animals (Lynch and Littleton, 1983). A report by Harper and Littleton (1991), using bovine adrenal chromaffin cells, showed that this release was inhibitable by DHP antagonists, leading them to conclude that DHP-sensitive channels might be involved in the adaptative response to ethanol.

In freshly isolated Kupffer cells from rats fed a liquid ethanol-containing diet for 5–8 weeks, depolarization-induced calcium uptake was increased and the addition of nitrendipine completely prevented this increase, suggesting that L-type calcium channels could be involved in alcoholic liver injury (Goto et al., 1993).

2. Dihydropyridine Antagonist Binding Studies

Repeated intraperitoneal injections of ethanol given to rats for 10 d resulted in a significant increase in the number of [^{3}H] nimodipine binding sites in cerebrocortical membranes (Dolin et al., 1987; Dolin and Little, 1989). This increase in antagonist binding sites was prevented by treatment with nitrendipine concomitant with chronic ethanol, and this result was confirmed by other workers (Whittington et al., 1991) and used to support the hypothesis that an increase in DHP-sensitive calcium channels in neurons may be involved in the physical dependence upon ethanol. Another study involving rats which were fed an ethanol-containing diet for 8 weeks also showed a slight yet significant increase in [^{3}H] nitrendipine binding sites in striatal membranes compared to membranes prepared from control animals (Woodward et al., 1990).

In a study using mice selectively bred for greater or lesser propensity for ethanol-withdrawal seizures, which were treated for 3 d with ethanol, it was found that there was an approximate 109% increase in [^{3}H] nitrendipine binding sites in withdrawal seizure prone mice compared to only a 30% increase in withdrawal seizure resistant mice (Brennan et al., 1990). This study suggested a role for L-type calcium channels in alcohol withdrawal seizures.

3. Electrophysiological Studies

No change in acute ethanol sensitivity of calcium channels was observed in hippocampal neurons isolated from long-sleep and short-sleep mice which were maintained on an ethanol-containing diet (for review, see McArdle et al., 1992). There was, however, a significant increase in L-type current amplitude in the neurons of long-sleep mice but not in neurons from short-sleep mice. Isolated mouse hippocampal slices were also used to investigate withdrawal hyperexcitability of animals physically dependent on ethanol (Whittington and Little, 1993). Intracellular recordings showed that a DHP agonist increased and a DHP antagonist decreased withdrawal hyperexcitability, presumably reflecting the effects of activation and inhibition of DHP-sensitive channels on the development of hyperexcitability. This study also demonstrated that an increase in Ca^{2+} channel activity in CA3 and CA1 areas was crucial for production of epileptiform activity during ethanol withdrawal, suggesting a role for DHP-sensitive calcium channels in alcohol withdrawal.

IV. CONCLUSIONS

Intracellular calcium affects numerous physiological processes, and cells have developed an effective system of mechanisms to maintain desirable intracellular concentrations. The fact that both NMDA receptors and voltage-gated calcium channels, which contribute significantly to calcium entry into nerve cells, are affected by relevant concentrations of ethanol is thus of major interest. Since massive interference with the functioning of voltage-gated calcium channels would likely be fatal, it is probable that intoxicating concentrations of ethanol affect only a small subset of the calcium channels in the nervous system, defined by particular subunit combinations, by "state" of the channel, or by the selective presence of cellular mediators of ethanol's actions. Some actions of ethanol, such as the decreased plasma levels of vasopressin after alcohol ingestion, seem clearly attributable to the inhibitory actions of ethanol on L-type calcium channels in posterior pituitary nerve terminals. Emerging knowledge of the molecular biology of calcium channels should help in elucidating the fingerprint of sensitive channels or neurons. The consequence of massive interference with calcium channels may be most evident in the withdrawal syndrome following chronic ethanol exposure, which seems to be at least partially attributable to an up-regulation of calcium channels and can prove fatal. A fuller understanding of the nature of both the acute and chronic actions of ethanol on voltage-gated calcium channels is clearly necessary if we are to understand the basis for the short- and long-term consequences of alcohol use.

ACKNOWLEDGMENTS

This work was supported by NIH grants AA05542 and AA08003.

REFERENCES

Adams, M.E., V.P. Bindokas, L. Hasegawa, and V.J. Venema (1990) Omega-agatoxins: novel calcium channel antagonists of two subtypes from funnel web spider (Agelenopsis aperta) venom. *J. Biol. Chem.,* 265:861–867.

Armstrong, C.M. and L. Binstock (1964) The effects of several alcohols on the properties of the squid giant axon. *J. Gen. Physiol.,* 48:265–276.

Armstrong, D. and R. Eckert (1987) Voltage-activated calcium channels that must be phosphorylated to respond to membrane depolarization. *Proc. Natl. Acad. Sci. U.S.A.,* 84:2518–2522.

Artalejo, C.R., M.E. Adams, and A.P. Fox (1994) Three types of Ca^{2+} channel trigger secretion with different efficacies in chromaffin cells. *Nature,* 367:72–76.

Bergamaschi, S., S. Govoni, R.A. Rius, and M. Trabucchi (1988) Acute ethanol and acetaldehyde administration produce similar effects on L-type calcium channels in rat brain. *Alcohol,* 5(4):337–340.

Bergamaschi, S., C. Lopez, F. Battaini, M. Parenti, and M. Trabucchi (1991) Omega-conotoxin and dihydropyridine labelled calcium channels are differentially affected by ethanol in NG 108–15 cells. *Soc. Neurosci. Abs.,* 151.12.

Bergmann, M.C., M.R. Klee, and D.S. Faber (1974) Different sensitivities to ethanol of three early transient voltage clamp currents of Aplysia neurons. *Pflugers Arch.,* 548:139–153.

Bertolino, M. and R.R. Llinás (1992) The central role of voltage-activated and receptor-operated calcium channels in neuronal cells. *Annu. Rev. Pharmacol. Toxicol.,* 32:399–421.

Biel, M., P. Ruth, E. Bosse, R. Hullin, W. Stuhmer, V. Flockerzi, and F. Hofmann (1990) Primary structure and functional expression of a high voltage activated calcium channel from rabbit lung. *FEBS Lett.,* 269:409–412.

Blaustein, M.P. and A.C. Ector (1975) Barbiturate inhibition of calcium uptake by depolarized nerve terminals in-vitro. *Mol. Pharmacol.,* 11:369–378.

Bosse, E., S. Regulla, M. Biel, P. Ruth, H.E. Meyer, V. Flockerzi, and F. Hofmann (1990) The cDNA and deduced amino acid sequence of the gamma subunit of the L-type calcium channel from rabbit skeletal muscle. *FEBS Lett.,* 267:153–156.

Brennan, C.H., J. Crabbe, and J.M. Littleton (1990) Genetic regulation of dihydropyridine-sensitive calcium channels in brain may determine susceptibility to physical dependence on alcohol. *Neuropharmacology,* 29:429–432.

Brennan, C.H., L.J. Guppy, and J.M. Littleton (1989a) Chronic exposure to ethanol increases dihydropyridine-sensitive calcium channels in excitable cells. *Ann. NY Acad. Sci.,* 560:467–469.

Brennan, C.H., A. Lewis, and J.M. Littleton (1989b) Membrane receptors, involved in up-regulation of calcium channels in bovine adrenal chromaffin cells, chronically exposed to ethanol. *Neuropharmacology,* 28:1303–1307.

Brennan, C.H. and J.M. Littleton (1990) Second messengers involved in genetic regulation of the number of calcium channels in bovine adrenal chromaffin cells in culture. *Neuropharmacology,* 29:689–693.

Brennan, C.H. and J.M. Littleton (1991) Chronic exposure to anxiolytic drugs, working by different mechanisms causes up-regulation of dihydropyridine binding sites on cultured bovine adrenal chromaffin cells. *Neuropharmacology,* 30:199–205.

Bunting, T.A. and B.S. Scott (1989) Aging and ethanol alter neuronal electric membrane properties. *Brain Res.,* 501:105–115.

Byerly, L. and S. Hagiwara (1988) Calcium channel diversity, in *Calcium and Ion Channel Modulation,* A.D. Grinnell, D. Armstrong, and M.B. Jackson, eds., New York, Plenum Press, 3–18.

Camacho-Nasi, P. and S.N. Treistman (1986) Ethanol effects on voltage-dependent membrane conductances: comparative sensitivity of channel populations in Aplysia neurons. *Cell. Mol. Neurobiol.,* 6:263–279.

Camacho-Nasi, P. and S.N. Treistman (1987) Ethanol-induced reduction of neuronal calcium currents: an examination of possible mechanisms. *Cell. Mol. Neurobiol.,* 7:191–207.

Carlen, P.L., N. Gurevich, and D. Durand (1982) Ethanol in low doses augments calcium-mediated mechanisms measured intracellularly in hippocampal neurons. *Science,* 215:306–309.

Carmichael, F.J. and Y. Israel (1975) Effects of ethanol on neurotransmitter release by rat brain cortical slices. *J. Pharm. Exp. Ther.,* 193:824–834.

Catterall, W.A. (1988) Structure and function of voltage-sensitive ion channels. *Science,* 242:50–61.

Chan, J. and D. Greenberg (1991) Intracellular calcium in NCB-20 cells: elevation by depolarization and ethanol but not by glutamate. *Brain Res.,* 539:328–331.

Cherksey, B.D., M. Sugimori, and R.R. Llinás (1991) Properties of calcium channels isolated with spider toxin, FTX. *Ann. NY Acad. Sci.,* 635:80–89.

Chik, C.L., Q. Liu, M. Girard, E. Karpinski, and A.K. Ho (1992) Inhibitory action of ethanol on L-type Ca^{2+} channels and Ca^{2+}-dependent guanosine 3′,5′-monophosphate accumulation in rat pinealocytes. *Endocrinology,* 131:1895–1902.

Chin, J.H. and D.B. Goldstein (1977a) Effects of low concentrations of ethanol on the fluidity of spin-labeled erythrocyte and brain membranes. *Mol. Pharmacol.,* 13:435–441.

Chin, J.H. and D.B. Goldstein (1977b) Drug tolerance in biomembranes: a spin label study of the effects of ethanol. *Science,* 196:684–685.

Cortes, R., P. Supavilai, M. Karobath, and J.M. Palacios (1983) The effects of lesions in the rat hippocampus suggest the association of calcium channel blocker binding sites with specific neuronal population. *Neurosci. Lett.,* 42(3):249–254.

Daniell, L.C., E.P. Brass, and R.A. Harris (1987) Effect of ethanol on intracellular ionized calcium concentrations in synaptosomes. *Mol. Pharmacol.,* 32:831–837.

Daniell, L.C. and S.W. Leslie (1986) Inhibition of fast phase calcium uptake and endogenous norepinephrine release in rat brain region synaptosomes by ethanol. *Brain Res.,* 377:18–28.

Daniell, L.C. and R.A. Harris (1989) Ethanol and inositol 1,4,5–triphosphate release calcium from separate stores of brain microsomes. *J. Pharmacol. Exp. Ther.,* 250:875–881.

Davidson, M., P. Wilce, and B. Shanley (1988) Ethanol increases synaptosomal free calcium concentration. *Neurosci. Lett.,* 89:165–169.

Dolin, S., H. Little, M. Hudspeth, C. Pagonis, and J. Littleton (1987) Increased dihydropyridine-sensitive calcium channels in rat brain may underlie ethanol physical dependence. *Neuropharmacology,* 26:275–279.

Dolin, S.J. and H.J. Little (1989) Are changes in neuronal calcium channels involved in ethanol tolerance. *J. Pharmacol. Exp. Ther.,* 250:985–991.

Dubel, S.J., T.V.B. Starr, J. Hell, M.K. Ahlijanian, J.J. Enyeart, W.A. Catterall, and T.P. Snutch (1992) Molecular cloning of the α1 subunit of an omega-conotoxin-sensitive calcium channel. *Proc. Natl. Acad. Sci. U.S.A.,* 89:5058–5062.

Ellinor, P.T., J.-F. Zhang, A.D. Randall, M. Zhou, T.L. Schwarz, R.W. Tsien, and W.A. Horne (1993) Functional expression of a rapidly inactivating neuronal calcium channel. *Nature,* 363:455–458.

Ellis, S.B., M.E. Williams, N.R. Ways, R. Brenner, A.H. Sharp, A.T. Leung, K.P. Campbell, E. McKenna, W.J. Koch, A. Hui, A. Schwartz, and M.M. Harpold (1988) Sequence and expression of mRNAs encoding of α_1 and α_2 subunits of a DHP-sensitive calcium channel. *Science,* 241:1661–1664.

Erwin, V.G., W.D. Heston, and G.D. McClearn (1976) Effects of hypnotics in mice genetically selected for sensitivity to ethanol. *Pharmacol. Biochem. Behav.,* 4:679–681.

Eskuri, S.A. and R.S. Pozos (1987) The effect of ethanol and temperature on calcium-dependent sensory neuron action potentials. *Alc. Drug Res.,* 7:153–162.

Friedman, M.B., C.K. Erickson, and S.W. Leslie (1980) Effects of acute and chronic ethanol administration on whole mouse brain synaptosomal calcium influx. *Biochem. Pharmacol.,* 28:1903–1908.

Goto, M., J.J. Lemasters, and R.G. Thurman (1993) Activation of voltage-dependent calcium channels in Kupffer cells by chronic treatment with alcohol in the rat. *J. Pharm. Exp. Ther.,* 267:1264–1268.

Grant, A.J., G. Koski, and S.N. Treistman (1993) Effect of chronic ethanol on calcium currents and calcium uptake in undifferentiated PC12 cells. *Brain Res.,* 600:280–284.

Greenberg, D.A. and E.C. Cooper (1984) Effects of ethanol on [^{3}H]nitrendipine binding to calcium channels in brain membranes. *Alcohol,* 8:568–571.

Gruol, D.L. (1982) Ethanol alters synaptic activity in cultured spinal cord neurons. *Brain Res.,* 243:25–33.

Hamill, O.P., A. Marty, E. Neher, B. Sakmann, and F.J. Sigworth (1981) Improved patch-clamp techniques for high-resolution current recording from cells and cell-free membrane patches. *Pflugers Arch.,* 391:85–100.

Harper, J.C., C.H. Brennan, and J.M. Littleton (1989) Genetic up-regulation of calcium channels in a cellular model of ethanol dependence. *Neuropharmacology,* 28:1299–1303.

Harper, J.C. and J.M. Littleton (1990) Development of tolerance to ethanol in cultured adrenal chromaffin cells. *Alcoholism: Clin. Exp. Res.,* 14:508–512.

Harper, J.C. and J.M. Littleton (1991) Characteristics of catecholamine release from adrenal chromaffin cells cultured in medium containing ethanol. 1 Spontaneous and K^+ induced release. *Alc. Alcohol.,* 26:25–32.

Harris, R.A. and W.F. Hood (1980) Inhibition of synaptosomal calcium uptake by ethanol. *J. Pharm. Exp. Ther.,* 213:562–568.

Hawthorn, M.H., J.N. Ferrante, Y.W. Kwon, A. Rutledge, E. Luchowski, R. Bangalore, and D.J. Triggle (1992) Effect of an homologous series of aliphatic alcohols on neuronal and smooth muscle voltage-dependent calcium channels. *Eur. J. Pharmacol.,* 229:143–148.

Hell, J.W., R.E. Westenbroek, C. Warner, M.K. Ahlijanian, W. Prystay, M.M. Gilbert, T.P. Snutch, and W.A. Catterall (1993) Identification and differential subcellular localization of the neuronal class C and class D L-type calcium channel α1 subunits. *J. Cell Biol.,* 123:949–962.

Hillman, D., S. Chen, T.T. Aung, B. Cherksey, M. Sugimori, and R.R. Llinás (1991) Localization of P-type calcium channels in the central nervous system. *Proc. Natl. Acad. Sci. U.S.A.,* 88:7076–7080.

Hofmann, F., M. Biel, and V. Flockerzi (1994) Molecular basis for CA^{2+} channel diversity. *Annu. Rev. Neurosci.,* 17:399–418.

Horne, W.A., P.T. Ellinor, I. Inman, M. Zhou, R.W. Tsien, and T.L. Schwarz (1993) Molecular diversity of Ca^{2+} channel α_1 subunits from the marine ray Discopyge ommata. *Proc. Natl. Acad. Sci. U.S.A.,* 90:3787–3791.

Huang, G-J. and J.J. McArdle (1994) Role of the GTP-binding Protein G_o in the suppressant effect of ethanol on voltage-activated calcium channels of murine sensory neurons. *Alcoholism: Clin. Exp. Res.,* 18:608–615.

Jay, S.D., S.B. Ellis, A.F. McCue, M.E. Williams, T.S. Vedvick, M.M. Harpold, and K.P. Campbell (1990) Primary structure of the gamma subunit of the DHP-sensitive calcium channel from skeletal muscle. *Science,* 248:490–492.

Johnson, D.A., N.M. Lee, R. Cooke, and H.H. Loh (1979) Ethanol-induced fluidization of brain lipid bilayers: required presence of cholesterol in membranes for the expression of tolerance. *Mol. Pharmacol.,* 15:739–746.

Koski, G., K. Lawrence, and D. Righi (1991) Ethanol-induced inhibition of carbachol-stimulated uptake of calcium in PC12 pheochromocytoma cells. *Neuropharmacology,* 30:267–274.

Kostyuk, P.G. (1989) Diversity of calcium ion channels in cellular membranes. *Neuroscience,* 28:253–261.

Lacerda, A.E., H.S. Kim, P. Ruth, E. Perez-Reyes, V. Flockerzi, F. Hofmann, L. Birnbaumer, and A.M. Brown (1991) Normalization of current kinetics by interaction between the α_1 and β subunits of the skeletal muscle dihydropyridine-sensitive Ca_{2+} channel. *Nature,* 352:527–530.

Leslie, S.W., E. Barr, J. Chandler, and R.P. Farrar (1983) Inhibition of fast- and slow-phase depolarization-dependent synaptosomal calcium uptake by ethanol. *J. Pharmacol. Exp. Ther.,* 225:571–575.

Llinás, R., M. Sugimore, J.W. Lin, and B. Cherksey (1989) Blocking and isolation of a calcium channel from neurons in mammals and cephalopods utilizing a toxin fraction (FTX) from funnel-web spider poison. *Proc. Natl. Acad. Sci. U.S.A.,* 86:1689–1693.

Llinás, R., M. Sugimori, D.E. Hillman, and B. Cherksey (1992) Distribution and functional significance of the P-type,voltage-dependent Ca^{2+} channels in the mammalian central nervous system. *TINS,* 15:351–355.

Llinás, R. and Y. Yarom (1986) Specific blockage of the low-threshold calcium channel by high molecular weight alcohols. *Soc. Neurosci. Abstr.,* 12:174.

Lovinger, D.M., G. White, and F.F. Weight (1989) Ethanol inhibits NMDA-activated ion current in hippocampal neurons. *Science,* 243:1721–1724.

Lynch, M.A. and J.M. Littleton (1983) Possible association of alcohol tolerance with increased synaptic calcium sensitivity. *Nature,* 303:175–176.

Marks, S.S., D.L. Watson, C.L. Carpenter, R.O. Messing, and D.A. Greenberg (1989) Comparative effects of chronic exposure to ethanol and calcium channel antagonists on calcium channel antagonist receptors in cultured neural (PC12) cells. *J. Neurochem.,* 53:168–172.

McArdle, J.J., S.P. Aiken, J.J. Choi, G.-J. Huang, X. Ming, E. Shifik, L. Tsiokas, S. Vincent, S. Zhang, and M. Watson (1992) Insights into the molecular mechanism of action of ethanol on brain calcium channels and muscarinic receptors derived from long- and short-sleep mice, in *Alcohol and Neurobiology Receptors, Membranes and Channels,* R.R. Watson, ed., CRC Press, Boca Raton, FL, 127–139.

Messing, R.O., C.L. Carpenter, I. Diamond, and D.A. Greenberg (1986) Ethanol regulates calcium channels in clonal neural cells. *Proc. Natl. Acad. Sci. U.S.A.,* 83:6213–6215.

Messing, R.O., A.B. Sneade, and B. Savidge (1990) Protein kinase C participates in up-regulation of dihydropyridine-sensitive calcium channels by ethanol. *J. Neurochem.,* 55:1383–1389.

Metcalfe, J.C., P. Seeman, and A.S.V. Burgen (1968) The proton relaxation of benzyl alcohol in erythrocyte membranes. *Mol. Pharmacol.,* 4:87–95.

Mikami, A., K. Imoto, T. Tanabe, T. Niidome, Y. Mori, H. Takeshima, M. Narumiya, and S. Numa (1989) Primary structure and functional expression of the cardiac dihydropyridine sensitive calcium channel. *Nature,* 340:230–233.

Miller, R.J. (1987) Multiple calcium channels and neuronal function. *Science,* 235:46–52.

Moore, J.W., W. Ulbricht, and M. Takata (1964) Effect of ethanol on the sodium and potassium conductances of the squid axon membrane. *J. Gen. Physiol.,* 48:279–295.

Mori, Y., T. Friedrich, K. Mansuk, A. Mikami, J. Nakai, P. Ruth, E. Bosse, F. Hofmann, V. Flockerzi, T. Furuichi, K. Mikoshiba, K. Imoto, T. Tanabe, and S. Numa (1991) Primary structure and functional expression from complementary DNA of a brain calcium channel. *Nature,* 350:398–402.

Mullikin-Kilpatrick, D. and S.N. Treistman (1993) Electrophysiological studies on calcium channels in naive and ethanol-treated PC12 cells. *Alc. Alcohol., Suppl.,* 2:385–389.

Mullikin-Kilpatrick, D. and S.N. Treistman (1994a) Ethanol inhibition of L-type Ca channels in PC12 cells: role of permeant ions. *Eur. J. Pharmacol.,* 270:17–25.

Mullikin-Kilpatrick, D. and S.N. Treistman (1994b) Time course of nerve growth factor modulation of ethanol inhibition of Ca^{2+} currents in PC12 cells. *Neurosci. Lett.,* 176:101–104.

Mullikin-Kilpatrick, D. and S.N. Treistman (1994c) Inhibitory action of ethanol in dihydropyridine-sensitive L-type C^{2+} channels in undifferentiated (UND) PC12 cells is mediated by G_i. *Soc. Neurosci. Abs,* 375.12.

Mullikin-Kilpatrick, D. and S.N. Treistman (1995b) G_{i2} is involved in ethanol inhibition of L-type calcium channels in undifferentiated but not differentiated PC12 cells. *Mol. Pharmacol.,* 47:997–1005.

Mullikin-Kilpatrick, D. and S.N. Treistman (1995a) Inhibition of dihydropyridine-sensitive Ca^{2+} channels by ethanol in undifferentiated and nerve growth factor-treated PC12 cells[1]: interaction with the inactivated state. *J. Pharm. Exp. Ther.,* 295:705–711.

Nakayama, H., M. Taki, J. Striessnig, H. Glossmann, W.A. Catterall, and Y. Kanoaka (1991) Identification of 1,4-dihydropyridine binding regions within the α_1 subunit of skeletal muscle Ca^{2+} channels by photoaffinity labeling with diazepine. *Proc. Natl. Acad. Sci. U.S.A.,* 88:9203–9207.

Niidome, T., M.-S. Kim, T. Friedrich, and Y. Mori (1992) Molecular cloning and characterization of a novel calcium channel from rabbit brain. *FEBS Lett.,* 308:7–13.

Oakes, S.G. and R.S. Pozos (1982) Electrophysiologic effects of acute ethanol exposure. I. Alterations in the action potentials of dorsal root ganglia neurons in dissociated culture. *Dev. Brain Res.,* 5:243–249.

Oldham, S.B., T. Ritchie, and E.P. Noble (1989) Ethanol exposure alters K^+- but not bradykinin-induced dopamine release in PC12 cells. *Alc. Alcohol.,* 24:43–54.

Olivera, B.M., W.R. Gray, R. Zeikus, J.M. McIntosh, J. Varga, J. Rivier, V. de Santos, and L.J. Cruz (1985) Peptide neurotoxins from fish-hunting cone snails. *Science,* 230:1338–1343.

Oyama, Y., N. Akaike, and K. Nishi (1986) Effects of N-alkanols on the calcium current of intracellularly perfused neurons of Helix aspersa. *Brain Res.,* 376:280–284.

Peres-Reyes, E., H.S. Kim, A.E. Lacerda, W. Horne, W. Wei, D. Rampe, K.P. Campbell, A.M. Brown, and L. Birnbaumer (1989) Induction of calcium currents by the expression of the α1–subunit of dihydropyridine receptor from skeletal muscle. *Nature,* 340:233–236.

Perez-Reyes, E., X.Y. Wei, A. Castellano, and L. Birnbaumer (1990) Molecular diversity of L-type calcium channel. Evidence for alterative splicing of the transcripts of three non-allelic genes. *J. Biol. Chem.,* 265:20430–20436.

Plummer, M.R., D.E. Logothetis, and P. Hess (1989) Elementary properties and pharmacological sensitivities of calcium channels in mammalian peripheral neurons. *Neuron,* 2:1453–1463.

Pragnell, M., J. Sakamoto, S.D. Jay, and K.P. Campbell (1991) Cloning and tissue-specific expression of the brain calcium channel β-subunit. *FEBS Lett.,* 291:253–258.

Rabe, C.S., E. Delorme, and F.F. Weight (1987) Muscarine stimulated neurotransmitter release from PC12 cells. *J. Pharm. Exp. Ther.,* 243:534–541.

Rabe, C.S. and F.F. Weight (1988) Effects of ethanol on neurotransmitter release and intracellular free calcium in PC12 cells. *J. Pharm. Exp. Ther.,* 244:417–422.

Regulla, S., T. Schneider, W. Nastainczyk, H.E. Mayer, and F. Hofmann (1991) Identification of the site of interaction of the dihydropyridine channel blockers nitrendipine and azidopine with calcium channel α_1. *EMBO J.,* 10:45–49.

Reuter, H. (1985) A variety of calcium channels. *Nature,* 6027:316–391.

Reynolds, J.N. and A. Prasad (1991) Ethanol enhances $GABA_A$ receptor-activated chloride currents in chick cerebral cortical neurons. *Brain Res.,* 564:138–142.

Rezazadeh, S.M., J.J. Woodward, and S.W. Leslie (1989) Fura-2 measurement of cytosolic free calcium in rat brain cortical synaptosomes and the influence of ethanol. *Alcohol,* 6:341–345.

Rosenberg, R.L., P. Hess, J.P. Reeves, H. Smilowitz, and R.W. Tsien (1986) Calcium channels in planar lipid bilayers: insights into mechanisms of ion permeation and gating. *Science,* 231:1564–1566.

Rubin, R. and J.B. Hoek (1988) Alcohol-induced stimulation of phospholipase C in human platelets requires G-protein activation. *Biochem. J.,* 254:147–153.

Sather, W.A., T. Tanabe, J.-F. Zhang, Y. Mori, M.E. Adams, and R.W. Tsien (1993) Distinctive biophysical and pharmacological properties of class A (B1) calcium channel α_1 subunits. *Neuron,* 11:291–303.

Schwartz, M.H. (1983) Voltage clamp analysis of ethanol effects on pacemaker currents of Aplysia neurons. *Brain Res.,* 278:341–345.

Schwartz, M.H. (1985) Effect of ethanol on subthreshold currents of Aplysia pacemaker neurons. *Brain Res.,* 332:337–353.

Seino, S., L. Chen, M. Seino, O. Blondel, J. Takeda, J.H. Johnson, and G. Bell (1992) Cloning of the α1 subunit of a voltage-dependent calcium channel expressed in pancreatic β cells. *Proc. Natl. Acad. Sci. U.S.A.,* 89:584–588.

Singer, D., M. Biel, I. Lotan, V. Flockerzi, F. Hofmann, and N. Dascal (1991) The roles of the subunits in the function of the calcium channel. *Science,* 253:1553–1557.

Skattebol, A. and R.A. Rabin (1987) Effects of ethanol on Ca uptake in synaptosomes and PC12 cells. *Biochem. Pharmacol.,* 36:2227–2229.

Snutch, T.P., W.J. Tomlinson, J.P. Leonard, and M.M. Gilbert (1991) Distinct calcium channels are generated by alternative splicing and are differentially expressed in the mammalian CNS. *Neuron,* 7:45–57.

Soong, T.W., A. Stea, C.D. Hodson, S.J. Dubel, S.R. Vincent, and T.P. Snutch (1993) Structure and functional expression of a member of the low voltage-activated calcium channel family. *Science,* 260:1133–1136.

Starr, T.V.B., W. Prystay, and T.P. Snutch (1991) Primary structure of a calcium channel that is highly expressed in the rat cerebellum. *Proc. Natl. Acad. Sci. U.S.A.,* 88:5621–5625.

Stokes, J.A. and R.A. Harris (1982) Alcohols and synaptosomal calcium transport. *Mol. Pharmacol.,* 22:99–104.

Striessnig, J., B.J. Murphy, and W.A. Catterall (1991) Dihydropyridine receptor of L-type Ca^{2+} channels: identification of binding domains for [^{3}H](+)-PN200–110 and [^{3}H] azidopine within the α1 subunit. *Proc. Natl. Acad. Sci. U.S.A.,* 88:10769–10773.

Tanabe, T., H. Takeshima, A. Mikami, V. Flockerzi, H. Takahashi, K. Kangawa, M. Kojima, H. Matsuo, T. Hirose, and S. Numa (1987) Primary structure of the receptor for calcium channel blockers from skeletal muscle. *Nature,* 328:313–318.

Tang, C-M., F. Presser, and M. Morad (1988) Amiloride selectively blocks the low threshold (T) calcium channel. *Science,* 240:213–215.

Tarroni, P., M. Passafaro, A. Pollo, M. Popoli, F. Clementi, and E. Sher (1994) Anti-β2 subunit antisense oligonucleotides modulate the surface expression of the α1 subunit of the N-type omega-CTX sensitive Ca^{2+} channels in IMR 32 human neuroblastoma cells. *Biochem. Biophys. Res. Commun.,* 201:180–185.

Treistman, S.N. and A. Wilson (1987a) Effects of ethanol on early potassium currents in Aplysia: cell specificity and influence of channel state. *J. Neurosci.,* 7:3207–3214.

Treistman, S.N. and A. Wilson (1987b) Alkanol effects on early potassium currents in Aplysia neurons depend on chain length. *Proc. Natl. Acad. Sci. U.S.A.,* 84:9299–9303.

Treistman, S.N. and A. Wilson (1991) Effects of chronic ethanol on currents carried through calcium channels in Aplysia. *Alcoholism: Clin. Exp. Res.,* 15:489–493.

Twombly, D.A., M.D. Herman, C.H. Kye, and T. Narahashi (1990) Ethanol effects on two types of voltage-activated calcium channels. *J. Pharmacol. Exp. Ther.,* 254:1029–1037.

Varadi, G., P. Lory, D. Schultz, M. Varadi, and A. Schwartz (1991) Acceleration of activation and inactivation by the β subunit of the skeletal muscle calcium channel. *Nature,* 352:159–162.

Wang, X., G. Dayanithi, J.R. Lemos, J.J. Nordmann, and S.N. Treistman (1991a) Calcium currents and peptide release from neurohypophysial terminals are inhibited by ethanol. *J. Pharm. Exper. Ther.,* 259:705–711.

Wang, X., J.R. Lemos, G. Dayanithi, J.J. Nordmann, and S.N. Treistman (1991b) Ethanol reduces vasopressin release by inhibiting calcium currents in nerve terminals. *Brain Res.,* 551:338–341.

Wang, X., S.N. Treistman, and J.R. Lemos (1991c) Direct identification of individual vasopressin-containing nerve terminals of the rat neurohypophysis after 'whole-cell' patch-clamp recordings. *Neurosci. Lett.,* 124:125–128.

Wang, X., S.N. Treistman, and J.R. Lemos (1992) Two types of high threshold calcium currents inhibited by omega-conotoxin in nerve terminals of rat neurohypophysis. *J. Physiol.,* 445:181–199.

Wang, X., S.N. Treistman, A. Wilson, J.J. Nordmann, and J.R. Lemos (1993) Calcium channels and peptide release from neurosecretory terminals. *NIPS,* 8:64–68.

Wang, X., G. Wang, J.R. Lemos, and S.N. Treistman (1994) Ethanol directly modulates gating of a dihydropyridine-sensitive Ca^{2+} channel in neurohypophysial terminals. *J. Neurosci.,* 14:5453–5460.

Wheeler, D.B., A. Randall, and R.W. Tsien (1994) Roles of N-type and Q-type Ca^{2+} channels in supporting hippocampal synaptic transmission. *Science,* 264:107–111.

Whittington, M.A., S.J. Dolin, T.L. Patch, R.J. Siarey, A.R. Butterworth, and H.J. Little (1991) Chronic dihydropyridine treatment can reverse the behavioural consequences of and prevent adaptations to, chronic ethanol treatment. *Br. J. Pharmacol.,* 103:1669–1676.

Whittington, M.A. and H.J. Little (1993) Changes in voltage-operated calcium channels modify ethanol withdrawal hyperexcitability in mouse hippocampal slices. *Exper. Physiol.,* 78:347–370.

Williams, M.E., P.F. Brust, D.H. Feldman, S. Patthi, S. Simerson, A. Maroufi, A.F. McCue, G. Velicelebi, S.B. Ellis, and M.M. Harpold (1992a) Structure and functional expression of an omega-conotoxin-sensitive human N-type calcium channel. *Science,* 257:389–395.

Williams, M.E., D.H. Feldman, A.F. McCue, R. Brenner, G. Velicelebi, S.B. Ellis, and M.M. Harpold (1992b) Structure and functional expression of α_1, α_2, β subunits of a novel human neuronal calcium channel subtype. *Neuron,* 8:71–84.

Witcher, D.R., M. DeWaard, J. Sakamoto, C. Franzini-Armstrong, M. Pragnell, S.D. Kahl, and K.P. Campbell (1993) Subunit identification and reconstitution of the N-type Ca^{2+} channel complex purified from brain. *Science,* 261:486–489.

Woodward, J.J., T. Machu, and S.W. Leslie (1990) Chronic ethanol treatment alters omega-conotoxin and Bay K 8644 sensitive calcium channels in rat striatal synaptosomes. *Alcohol,* 7:279–284.

Yaney, G.C., M.B. Wheeler, X. Wei, E. Perez-Reyes, L. Birnbaumer, A.E. Boyd, and L.G. Moss (1992) Cloning of a novel alpha-1 subunit of the voltage-dependent calcium channel from the beta-cell. *Mol. Endocrin.,* 6:2143–2152.

Chapter 4

ALCOHOL ACTIONS AT THE $GABA_A$ RECEPTOR/CHLORIDE CHANNEL COMPLEX

S. John Mihic and R. Adron Harris

TABLE OF CONTENTS

I. INTRODUCTION

GABA is the major inhibitory neurotransmitter in the vertebrate central nervous system (Krnjević, 1974). It binds to post-synaptic $GABA_A$ receptors linked to integral chloride channels (Barker and Ransom, 1978), decreasing both cellular excitability and the likelihood of action potential generation. The $GABA_A$ receptor is a member of the ligand-gated ion channel family of receptors (Schofield et al., 1987), related to the nicotinic acetylcholine (Noda et al., 1983; Boulter et al., 1987) and glycine (Grenningloh et al., 1987) receptors. A number of subunits assemble to form a hetero-oligomeric complex, with the membrane-spanning portions of the subunits forming the walls of an ion channel (Olsen and Tobin, 1990). Use of molecular cloning techniques has provided the amino acid sequences of the α, β, γ, δ, and ρ subunits, most of which have a number of subtypes (for a recent review, see Dunn et al., 1994).

The function of $GABA_A$ receptors can be modulated by a number of classes of sedative, hypnotic, or anesthetic agents. Benzodiazepines, barbiturates, volatile and steroid anesthetics, divalent and trivalent metals, and alcohols have all been reported to affect the function of $GABA_A$ receptors (see Sieghart, 1992, for a review). Benzodiazepine receptor ligands can act as agonists or inverse agonists, respectively enhancing or inhibiting GABAergic currents, in a manner that can be blocked by benzodiazepine receptor antagonists. The subunit composition of $GABA_A$ receptors plays a great role in determining their pharmacological properties. For example, the γ subunit, in combination with α and β subunits, is required to observe functional effects of benzodiazepines (Pritchett et al., 1989), although barbiturates potentiate

0-8493-8389-7/95/$0.00+$.50

GABAergic currents in receptors composed solely of α and β subunits. Furthermore, zinc preferentially inhibits the effects of GABA in αβ, and not αβγ, receptors (Draguhn et al., 1990).

This chapter will review recent findings (since 1990) on the actions of ethanol at the $GABA_A$ receptor. Earlier work was reviewed by Deitrich et al. (1989) and Leidenheimer and Harris (1992). Acute and chronic effects of ethanol, assessed using behavioral, biochemical, electrophysiological, or molecular biology techniques, will be discussed, and a comparison will be made between the actions of alcohols and volatile anesthetics. Finally, we will attempt to integrate recent developments in the field and to highlight important advances.

II. BEHAVIORAL STUDIES

Numerous studies have attempted to determine which behavioral effects of ethanol are mediated by changes in $GABA_A$ receptor function, but this has proven difficult. Since the $GABA_A$ receptor is the primary mediator of inhibitory neurotransmission in the mammalian CNS, any potentiation or inhibition of this receptor system will alter the balance of neuronal excitatory and inhibitory influences and thus alter behavior. For example, the duration of ethanol-induced loss of righting reflex is increased by the $GABA_A$ agonist muscimol and decreased by the $GABA_A$ receptor antagonist bicuculline (Liljequist and Engel, 1982; Martz et al., 1983). However, this does not necessarily mean that ethanol produces its behavioral effects by an action at the $GABA_A$ receptor. It may be that ethanol acts at a biochemical site distinct from the $GABA_A$ receptor to promote inhibitory neurotransmission (or inhibit excitatory processes) and that the effect of bicuculline at the $GABA_A$ site merely restores some sort of balance. Thus, reports of ethanol having enhancing or antagonizing effects in combination with agents known to interact with the $GABA_A$ receptor provide only indirect evidence that ethanol acts there. The interaction of ethanol with GABAergic agents *in vivo* could be physiological (involving different biochemical systems) rather than pharmacological (at the same biochemical site).

Recent work showed that ethanol may interact with modulators of $GABA_A$ receptor function but only in certain brain regions. Muscimol microinjected into the medial septal area of rats enhanced, while bicuculline microinjection antagonized, impairment of the aerial righting reflex by ethanol (Givens and Breese, 1990). Since this was not seen in the lateral septum, the authors suggested that specific brain regions may play a critical modulatory role in the sedative actions of ethanol. These findings are in agreement with electrophysiological findings of ethanol enhancement of GABAergic function in the medial septal area but not in the lateral septum (Givens and Breese, 1990). However, Soldo et al. (1994), using brain slices, found ethanol effects in both brain regions.

Since benzodiazepine receptor agonists are anxiolytic, while inverse agonists are anxiogenic, one would expect that if ethanol enhances GABAergic function, it could also decrease anxiety. The indirect GABA agonist valproate and the chloride channel blocker picrotoxin were tested for their effects on the anxiolytic effects of ethanol in a behavioral conflict task (negative contrast) which quantified how animals respond to an abrupt, unexpected reduction in reward (Becker and Anton, 1990). Valproate, when administered together with a sub-effective dose of ethanol (0.5 g/kg), dose-dependently potentiated the anxiolytic action of ethanol. Picrotoxin antagonized the anxiolytic effects of a larger dose of ethanol (1.0 g/kg) and also impaired the ability of valproate to enhance the anxiolytic effects of the smaller dose of ethanol. In the elevated plus-maze test, 2 g/kg ethanol increased both the percentage of entries into and time spent in the open arms, indicating an anxiolytic action which was reversed by the benzodiazepine receptor inverse agonist Ro 15–4513 (Prunell et al., 1994). In contrast, Ro 15–4513 did not antagonize the anxiolytic effect of 2 g/kg ethanol in the acquisition of shuttlebox avoidance; however, animals treated with 4 g/kg of ethanol showed impaired shuttlebox avoidance acquisition, which was completely reversed by Ro 15–4513. Ro 15–4513 had no effect on its own in any of the parameters examined.

Earlier work showed that Ro 15–4513 antagonized both the low-dose anticonflict effects of ethanol as well as the intoxication produced by higher doses (Suzdak et al., 1986b). More recently, Ro 15–4513 was also shown to decrease alcohol consumption (June et al., 1991, 1992a; Rassnick et al., 1993) and ethanol-induced hypoactivity (Wood et al., 1989). It also altered ethanol discrimination but only for brief periods when injected after ethanol and at doses and postinjection times which also disrupted responding (Middaugh et al., 1991). This time and dose dependence may explain why Rees and Balster (1988) saw attenuation of ethanol discrimination by Ro 15–4513 in mice, while Hiltunen and Jarbe (1988) failed to do so in rats. Not all effects of ethanol are reversed by Ro 15–4513 administration. Ro 15–4513 failed to antagonize ethanol-induced taste aversion (Jeffreys et al., 1990; June et al., 1992b), hypothermia (Wood et al., 1989), or conditioned place preference (Risinger et al., 1992).

Behavioral differences between the alcohol-sensitive (ANT) and alcohol-resistant (AT) lines of rats have been correlated with a difference in a single amino acid of the $GABA_A$ receptor α_6 subunit. The ANT rat line displays a high acute sensitivity to the motor-impairing effects of ethanol as well as unusual cerebellar $GABA_A$ receptor pharmacology, compared to the relatively alcohol-insensitive AT rat line. The α_6 subunit, which is expressed exclusively in cerebellar granule cells, when combined with β_2 and γ_2 subunits normally yields receptors which bind the benzodiazepine inverse agonist [^{3}H]Ro15–4513 with high affinity but do not bind other benzodiazepines or β-carbolines (Lüddens et al., 1990). This α_6 subunit is part of a cerebellar receptor subtype, selective for Ro15–4513, which antagonizes a number of the behavioral effects of ethanol (Syapin et al., 1990). The increased benzodiazepine sensitivity of the ANT rats was shown to be due to a single amino acid difference in the α_6 sequences of ANT and AT rats (Korpi et al., 1993). The α_6 subunit of ANT rats was expressed normally but contained a point mutation generating an arginine-to-glutamine substitution at amino acid 100. This resulted in $\alpha_{6(Q100)}\beta_2\gamma_2$ receptors which showed diazepam enhancement of GABA-mediated currents as well as diazepam-sensitive binding of [^{3}H]Ro15–4513 in ANT rats (Korpi et al., 1993). However, the importance of this amino acid difference to ethanol sensitivity is likely minor since, in the second generation of AT/ANT cross-breeding, no clear correlation was found between cerebellar receptor binding and alcohol sensitivity (Korpi et al., 1992).

Withdrawal following chronic ethanol administration has been shown to lead to increased seizure susceptibility. This is believed to be due, at least in part, to decreased GABAergic function. McCown and Breese (1993) found that when rats experienced 6 or 10 withdrawals from a 5 d ethanol liquid diet, there was a significant decrease in the threshold frequency for seizure genesis, 6–7 d after ethanol withdrawal. In the same animals, there was also a withdrawal-related increase in the effectiveness of bicuculline to reduce the seizure threshold current within the inferior collicular cortex, suggesting a possible local decrease in $GABA_A$ receptor function. Increased seizure susceptibility following ethanol withdrawal was also seen by Kokka et al. (1993) who found that rats on a chronic intermittent ethanol dosing regimen had a decreased seizure threshold to the convulsant, pentylenetetrazol. This reduction in seizure threshold persisted for at least 40 d of alcohol abstinence following the 60th dose and resembled the "kindling" phenomenon produced by the chronic administration of pentylenetetrazol. The authors postulated that this "kindling" phenomenon contributes to the development of dependence on central nervous system depressants acting on the $GABA_A$ receptor (Kokka et al., 1993). Frye et al. (1991) showed that bilateral microinfusions of bicuculline methiodide and picrotoxinin, as well as kainic acid (an excitatory amino acid agonist used as a positive control) produced similar increases in the frequency of wild-running seizures. However, ethanol dependence did not significantly change susceptibility to seizure induction by any of the convulsants tested, although susceptibility to the more severe, clonic seizures was increased for each convulsant, suggesting that the inhibition of $GABA_A$-mediated induction of wild-running seizures was not selectively increased by ethanol dependence (Frye et al., 1991). The authors concluded that their results did not support the hypothesis of down-regulation of $GABA_A$ receptor-chloride channel complex activity during the development of physical dependence to ethanol. Furthermore, recent work using recombinant inbred (RI) strains of long-sleep (LS) and short-sleep (SS) mice suggests that seizure susceptibility to bicuculline, in 24 LS × SS RI strains, does not significantly correlate with sensitivity to ethanol as measured by the sleep-time response (Wehner et al., 1991).

Behavioral evidence linking the $GABA_A$ receptor to ethanol dependence was obtained by Buck et al. (1991b) who found that a single injection of the benzodiazepine receptor antagonist flumazenil (10 mg/kg, 14 h before withdrawal) given to mice receiving an ethanol-containing liquid diet for 10 d, reduced seizure severity during withdrawal from ethanol. Furthermore, tolerance and diazepam-induced cross-tolerance to the ataxic effects of ethanol were also reversed by a single injection of flumazenil given 2–26 h before tolerance testing, suggesting that even brief occupation of benzodiazepine receptors by an antagonist may reset cellular mechanisms responsible for the development of ethanol tolerance and dependence.

In conclusion, the results presented above suggest that the $GABA_A$ receptor/chloride channel complex may be an important site of action of ethanol. Evidence is most compelling for ataxia, antianxiety, reinforcement, and the withdrawal seizures produced by ethanol. However, as mentioned earlier, caution must be taken to prevent over-interpretation of these findings, since any coapplication of $GABA_A$ergic drugs with ethanol will alter the balance of excitatory and inhibitory neuronal influences. This could very well be responsible for the enhancement or antagonism of the effects of ethanol produced by $GABA_A$ receptor agonists and antagonists, respectively. Furthermore, these behavioral findings do not allow us to distinguish a direct effect of ethanol on the $GABA_A$ receptor from an effect on some other biochemical process that itself modulates $GABA_A$ receptor function. Differentiating direct from indirect effects of ethanol on the $GABA_A$ receptor requires the use of biochemical and electrophysiological techniques which will be discussed in the next section.

III. FUNCTIONAL STUDIES

A. ACUTE EFFECTS OF ETHANOL

1. Studies Using *In Vivo* Preparations

A number of recent studies have examined the effects of ethanol on GABAergic function *in vivo*. Givens and Breese (1990) iontophoresed GABA onto rhythmically bursting neurons of the medial septal area and assessed the inhibition produced before and after the administration of systemic ethanol. They found that ethanol markedly enhanced GABA-mediated neuronal firing rates. This enhancement of the effects of GABA reversed to control levels while the blood ethanol concentration remained elevated, suggesting that some form of acute tolerance might be occurring. Under the same experimental conditions, there was no potentiation by ethanol in the lateral septum, illustrating that ethanol may only have enhancing effects in specific brain regions (Givens and Breese, 1990). Systemic administration of ethanol (0.5 g/kg) also potentiated GABAergic responses in the inferior colliculus (Simson et al., 1991). In addition, local application of ethanol in the inferior colliculus potentiated the inhibitory effects of GABA on neuronal activity but did not enhance the inhibitory effects of glycine. In confirmation of earlier work, the local applications of ethanol failed to enhance GABAergic inhibition in the lateral septum (Simson et al., 1991).

Criswell et al. (1993) studied a series of brain regions *in vivo* in an attempt to better understand what factors are responsible for ethanol enhancement of GABAergic responses. They found that ethanol enhanced the effects of iontophoretically applied GABA in the medial septal area, inferior colliculus, substantia nigra reticulata ventral pallidum, and the diagonal band of Broca. Ethanol had no enhancing effect in the lateral septum, ventral tegmental area, and the CA1 region of the hippocampus. Interestingly, there was an excellent correlation between positive ethanol effects and high [^{3}H]zolpidem (a benzodiazepine receptor agonist) binding among these brain regions. The authors suggest that zolpidem binding may be predictive of ethanol enhancement and that this requires the colocalized presence of α_1, β_2, and γ_2 subunits of the $GABA_A$ receptor. Pritchett and Seeburg (1990) found that zolpidem bound preferably to $GABA_A$ receptors containing an α_1 (and not α_2) subunit, although Puia et al. (1991) found that zolpidem efficacy was only decreased when the α_5 subunit was expressed with β_1 and γ_2. It should be noted that not all investigators have seen differential effects of ethanol in the medial and lateral septum. A recent study by Soldo et al. (1994), using rat brain slices, found that 80 m*M* ethanol enhanced $GABA_A$-mediated inhibitory postsynaptic currents in both medial and lateral septal neurons.

At least one study has failed to confirm ethanol enhancement of the effects of GABA in the medial septum. Frye et al. (1994), testing acutely dissociated neurons, failed to see significant mean ethanol (3–300 m*M*) effects in Sprague Dawley or High Alcohol Sensitivity (HAS) rats. However $GABA_A$ currents were enhanced by 300 m*M* EtOH in Fischer 344, ACI, and Wistar Kyoto rats. Some neurons from rats of all five strains displayed ethanol potentiation, while others displayed inhibition, of GABA-mediated currents. There was no additional enhancing effect of preincubating cells with ethanol for 2–3 min. In contrast to ethanol, pentobarbital, midazolam, lanthanum, and zinc effects were observed in all cells.

There is some evidence suggesting that receptor systems other than just the $GABA_A$ receptor itself can play a role in the enhancement of GABAergic currents by ethanol. Lin et al. (1991) showed that systemically applied ethanol enhanced $GABA_A$-mediated inhibition of Sprague Dawley rat cerebellar Purkinje cell firing but only if norepinephrine or isoproterenol were concomitantly applied with GABA. This suggests that EtOH potentiation of GABAergic activity in the cerebellar cortex is enhanced by β-adrenergic receptor activation. Isoproterenol could also inhibit neuronal firing rates when applied alone. When β-adrenergic agonists were not exogenously applied, there was little or no effect of ethanol on GABAergic currents, leading the authors to suggest that ethanol does not potentiate GABA's effects by an interaction directly at the $GABA_A$ receptor (Palmer and Hoffer, 1990). Freund et al. (1993) found minor ethanol enhancement of GABA-mediated inhibition of cerebellar Purkinje cell firing rates, although locally pressure-ejected ethanol produced decreased firing rates on its own. This ethanol effect was antagonized by preapplication of bicuculline, suggesting that it is a $GABA_A$-mediated event.

Prolonged local application of ethanol also produced an enhancement of isoproterenol-modulated $GABA_A$ responses (Lin et al., 1993a); this enhancement was lessened by the β-adrenergic antagonist timolol. Furthermore, the α-adrenergic agonist phenylephrine did not produce ethanol potentiation of $GABA_A$ responses (Lin et al., 1993a). There are at least three plausible mechanisms for this β-adrenergic receptor/ethanol interaction:

1. isoproterenol could produce an increased sensitivity of the $GABA_A$ receptor to ethanol
2. ethanol may be interacting with the β-adrenergic receptor to increase its ability to modulate $GABA_A$ receptor function (Lin et al., 1993a)
3. ethanol may potentiate catecholaminergic enhancement of GABAergic function by inhibiting reuptake of norepinephrine from the extracellular space (Lin et al., 1993c)

As Fischer 344 rats age, they exhibit decreases in β-adrenergic function. Lin et al. (1993b) demonstrated significantly greater ethanol enhancement of isoproterenol potentiation of $GABA_A$ responses in young (4–6 months) than old (24–25 months) Fischer 344 rats. In contrast, in the older animals, they frequently found ethanol potentiation of isoproterenol *inhibition* of $GABA_A$ responses, while this was rarely seen in young rats. Both of these opposing effects of ethanol on the actions of isoproterenol were antagonized by timolol.

2. Chloride Influx Studies

Ethanol has been shown to stimulate GABA-mediated $^{36}Cl^-$ uptake into cerebral cortical microsacs (Allan and Harris, 1985, 1986), synaptoneurosomes (Suzdak et al., 1986a), and cultured spinal cord neurons (Ticku et al., 1986). Reviews of the earlier extensive literature on the effects of ethanol in potentiating GABA-mediated $^{36}Cl^-$ uptake into cultured cells and vesicular preparations, particularly in different lines of ethanol-sensitive and -resistant animals can be found in Deitrich et al. (1989), Harris (1990), Allan and Harris (1991), and Leidenheimer and Harris (1992). This chapter details more recent developments in the field (since 1990).

Recent work has shown that the mechanism of ethanol enhancement of $GABA_A$ responses may significantly differ from that of benzodiazepines and barbiturates. Allan et al. (1991) found that ethanol (10–30 m*M*) in the presence of the $GABA_B$ agonist baclofen could promote chloride influx into mouse cortical microsacs; however, neither compound applied alone had any effect. This influx could be prevented by coapplication of either the $GABA_A$ antagonist bicuculline or the $GABA_B$ receptor antagonists phaclofen and saclofen. These findings raise the possibility that, unlike benzodiazepines and barbiturates, ethanol enhancement of $GABA_A$ receptor function may require coactivation of $GABA_B$ receptors and may not necessarily occur through a direct action on the $GABA_A$ receptor itself. This could occur through $GABA_B$ receptor interactions with G proteins resulting in the activation of kinases which have been shown to modulate $GABA_A$ receptor function. However, Mehta and Ticku (1990) failed to observe any effects of either baclofen or phaclofen on GABA-stimulated $^{36}Cl^-$ influx into mouse spinal cord cultured neurons in the presence or absence of 20 m*M* ethanol. This difference in findings may be due to the duration of exposure to phaclofen; Mehta and Ticku (1990) exposed their cells to phaclofen for 7 s while Allan et al. (1991) preincubated their microsacs with phaclofen for 2 min. In addition, Mehta and Ticku (1990) used spinal cord neurons while Allan et al. (1991) used cerebral cortical microsacs. Ethanol enhancement does not appear to involve changes in intracellular calcium levels, since pretreatment of cultured mammalian cortical cells with the intracellular Ca^{2+}-depleting agent dantrolene or with the calcium ionophore A 23187 did not affect ethanol potentiation of GABA-mediated $^{36}Cl^-$ influx (Mehta and Ticku, 1994).

Arguing in favor of a direct action of ethanol on the $GABA_A$ receptor/chloride channel complex, Ueha and Kuriyama (1991) studied the effects of ethanol on purified $GABA_A$ receptors in a reconstituted vesicle preparation. This study reported that the EC_{50} of ethanol enhancement of GABA-stimulated $^{36}Cl^-$ influx was approximately 0.2 μ*M*. However, the effects of concentrations of EtOH in the nanomolar range are of dubious physiological relevance. Allan, also studying ethanol enhancement of GABAergic chloride influx into reconstituted vesicles, has only seen effects at millimolar concentrations of ethanol (Andrea Allan, personal communication).

There is still some disagreement in the literature about whether ethanol can directly stimulate $^{36}Cl^-$ uptake in the absence of exogenously added GABA. Harris and colleagues observed enhancement of basal flux only when the assay was carried out at 0°C (Allan and Harris, 1986, 1987; McQuilkin and Harris, 1990). However, Suzdak et al. (1986a), Mehta and Ticku (1990, 1994), and Engblom and Åkerman (1991) did observe ethanol effects in the absence of exogenously added GABA, although usually at concentrations higher than those required to potentiate GABA-mediated chloride uptake.

The potentiating effect of ethanol on Cl^- uptake has not been universally observed. Morelli et al. (1988) saw no enhancement of GABA-mediated $^{36}Cl^-$ uptake by ethanol (0.5–300 m*M*), although they did observe pentobarbital potentiation. Mihic et al. (1991, 1992a) also saw pentobarbital and diazepam enhancement but no effects of ethanol at pharmacologically relevant concentrations. The ethanol potentiation seen by Harris and colleagues was not as robust as that produced by benzodiazepines and barbiturates and was seen only under certain conditions. McQuilkin and Harris (1990), for example, reported that ethanol potentiation was seen in mouse microsac preparations only when the mice were sacrificed by decapitation and not by cervical dislocation. Furthermore, leaving the tissue on ice for 2 h completely abolished the ethanol potentiation. Neither of these experimental variables decreased the degree of potentiation seen with barbiturates and benzodiazepines. Assay temperature is likely not an important variable, unless it is taken to an extreme. McQuilkin and Harris (1990) found ethanol potentiation of GABA's effect at 34°C but not 0°C, while Suzdak et al. (1986a) saw their effect at 30°C. Mihic et al. (1992a) failed to see enhancement using temperatures of 32, 34, or 37°C; they also sacrificed animals by decapitation and used the tissue immediately after preparation.

Even among the studies showing ethanol effects on chloride influx, the results vary among laboratories. For example, Engblom and Åkerman (1991), using the chloride-sensitive dye 6-methoxy-*N*-(3-sulfopropyl)quinolinium (SPQ), showed that, separately, both ethanol and GABA increased Cl^- efflux from Sprague Dawley rat brain synaptoneurosomes. However, their effects were infra additive when both were added together. Furthermore, there was no apparent concentration-dependent saturation of this effect of ethanol, in contrast to the results of Suzdak et al. (1986a) and Allan and Harris (1986) who saw biphasic effects of ethanol. Proctor et al. (1992b) compared the effects of ethanol in cerebral cortical slices using electrophysiological techniques with the effects in microsacs using the chloride influx assay. They found that maximal enhancement of GABA-mediated chloride flux occurred with an ethanol concentration of 60 m*M*; since the effect was biphasic, 160 m*M* EtOH was without effect. In contrast, the electrophysiological study showed a robust potentiation by 160 m*M* EtOH and a weaker effect using 80 m*M* ethanol.

3. Electrophysiological Studies in Brain Slices and Isolated or Cultured Cells

Like the chloride influx assays described above, not all electrophysiological studies agree on whether ethanol enhances GABAergic currents. Some investigators (Takada et al., 1989; Aguayo, 1990; Nishio and Narahashi, 1990; Wafford et al., 1990, 1991) have observed ethanol potentiation of GABA-mediated neurotransmission, while others (Osmanovic and Shefner, 1990; White et al., 1990) have not. Furthermore, even in laboratories which have reported stimulation by ethanol, differences in the nature of the enhancement still remain. However, although not all the reasons behind these discrepancies are fully understood, progress is being made and the patterns of ethanol action are becoming clearer.

Nishio and Narahashi (1990) studied the effects of ethanol on the $GABA_A$-activated current in rat dorsal root ganglion neurons in primary culture. GABA produced an inward chloride current which was composed of an initial transient peak followed by a sustained phase. Ethanol (30–300 m*M*) enhanced the peak current produced by 100 μ*M* GABA (a near-maximal concentration) in a concentration-dependent manner without affecting the sustained current. Further work by this group showed that the potencies of alcohols in enhancing GABAergic currents relate to their carbon chain lengths and their membrane/buffer partition coefficients (Nakahiro et al., 1991). The alcohols tested were 30–300 m*M* ethanol, 1–30 m*M* n-butanol, 30–1000 m*M* n-hexanol, and 3–100 m*M* n-octanol. Ethanol (300 m*M*) and n-octanol (30 and 100 μ*M*) both shifted GABA concentration-response curves to the left and had no effects at the highest GABA concentration (300 μ*M*). An EtOH concentration of approximately 250 m*M* was required in order to show 25% potentiation, while for n-butanol the EC_{25} value was about 4.5 m*M* (Nakahiro et al., 1991). In addition, preapplication of ethanol for 1–3 min before applying GABA did not increase the magnitude of potentiation seen (Nakahiro et al., 1991). These concentrations of ethanol and n-octanol also significantly increased the desensitization rates of $GABA_A$ responses (Nakahiro et al., 1991), as anesthetics have been shown to do previously (Nakahiro et al., 1989). Arakawa et al. (1992) then showed that alcohols could induce chloride currents in cultured rat dorsal root ganglion neurons without exogenously applied GABA, but the concentrations required (e.g., 1 *M* for ethanol) were very high.

Ethanol enhancement of GABAergic currents has also been observed in primary cultures of 7–8 d old chick embryo cerebral cortical neurons (Reynolds and Prasad, 1991). The ethanol concentration-response curve appeared to be biphasic with 20 m*M* ethanol producing less enhancement than 5 m*M* EtOH; this potentiation was seen in approximately 60% of the cells tested. GABA was applied by very brief (20–100 ms) pressure pulses using a Picospritzer, while ethanol was bath-applied. Interestingly, ethanol enhancement depended on the particular culture used; in four cultures, all cells tested displayed ethanol enhancement of GABAergic currents, while no cells in three other cultures showed this effect. In a more recent study Reynolds et al. (1992) examined acutely dissociated neurons from rat neocortical slices and primary cultures of chick, mouse, and rat brain regions (hippocampus, cerebral cortex, cerebellum, and spinal cord). Ethanol (1–50 m*M*) potentiated $GABA_A$ responses in some cells (28–100%) from each brain region studied, with variability being seen among rat litters used. The mouse hippocampal neurons seemed particularly sensitive to EtOH (1–10 m*M*); in addition, this potentiation was seen only at lower GABA concentrations (Reynolds et al., 1992). Cells which did not respond to ethanol still did show significant responses to flurazepam, suggesting that more than just the presence of a γ_2 subunit is required to see ethanol enhancement. Aguayo (1990), studying $GABA_A$-activated chloride currents in cultured mouse hippocampal and cortical neurons found ethanol effects similar to those of Reynolds and Prasad (1991). Whole-cell patch-clamp recordings showed a biphasic effect of ethanol with a maximal effect at a concentration of 10 m*M* and less potentiation using 80 m*M* EtOH. This biphasic ethanol concentration-response curve may be the result of slow (>1 s) ethanol application, since it was not seen when GABA and ethanol were applied quickly (<100 ms) (Aguayo and Pancetti, 1994). In agreement with Reynolds et al. (1992), Aguayo (1990) still saw significant pentobarbital and flurazepam enhancement of GABAergic currents in cells not showing an ethanol effect (30% of cells).

In isolated rat superior cervical ganglion neurons, 20–100 m*M* EtOH produced noncompetitive inhibition of GABAergic currents, with a greater inhibitory effect being seen in neurons from adult animals compared to newborns (Aguayo and Alarcón, 1993). However, a high concentration of ethanol (850 m*M*) produced potentiation and decreased the GABA EC_{50} in both animal groups. Inhibition of GABAergic currents by ethanol has also been reported by Grant and Treistman (1991) who found inhibitory effects of ethanol (18.3% with 100 m*M*) in the abdominal ganglion of *Aplysia californica*. The significance of this finding is difficult to assess because *Aplysia* $GABA_A$ receptors are thought to be quite different from those found in vertebrates (Ikemoto et al., 1988).

Recently, Wafford et al. (1990) injected whole brain mRNA from LS and SS mice in *Xenopus* oocytes and demonstrated functional $GABA_A$ receptor expression, using a voltage-clamp technique. These receptors displayed typical $GABA_A$ receptor pharmacology: potentiation by barbiturates and benzodiazepines and inhibition by picrotoxinin. Wafford et al. (1990) also found that 20 m*M* ethanol could only potentiate the effects of GABA in oocytes expressing LS, but not SS, mouse brain mRNA. This is consistent with studies showing ethanol potentiation of GABA-mediated chloride uptake into brain microsacs prepared from LS mice, with little or no effect in SS mouse microsacs (Allan and Harris, 1986; Zahniser et al., 1992). Diazepam potentiation was not significantly different for receptors expressed in oocytes from the two lines, although $^{36}Cl^-$ uptake experiments in brain microsacs had shown a greater effect of flunitrazepam in preparations from LS than from SS mice (Harris and Allan, 1989). Wafford et al. (1990) also found, in oocytes injected with LS or SS mRNA, that ethanol inhibition of NMDA-activated cation currents did not differ between the mouse lines, suggesting that actions of ethanol at the NMDA receptor do not appear to be important in determining the differences between LS and SS mice in the duration of loss of righting reflex.

There appears to be no concordance in the literature regarding ethanol effects in the hippocampus. Allan and Harris (1986) first showed that ethanol stimulated GABA-mediated chloride influx into cerebral cortical and cerebellar, but not hippocampal, microsacs. Proctor et al. (1992a) tested ethanol in both hippocampal microsacs (chloride flux) and slices (electrophysiological recordings). There were no effects of ethanol in microsacs, on synaptically evoked inhibitory postsynaptic potentials (see also Soldo et al., 1994) or on responses evoked by GABA locally released from a micropipette. However, they found that 80 and 160 m*M* ethanol did enhance synaptically evoked GABAergic responses in cerebral cortical neurons in rat brain slices (Proctor et al., 1992b; Soldo et al., 1994) and that 80 m*M* ethanol also potentiated $GABA_A$-mediated inhibitory postsynaptic currents (IPSCs) in intermediate lateral and medial septal neurons (Soldo et al., 1994). Since Wafford et al. (1991) showed that ethanol enhancement of GABAergic currents in *Xenopus* oocytes depended on the presence of the γ_{2L} subunit of the $GABA_A$ receptor, the lack of ethanol effect in the hippocampus was initially thought to be due to a lack of expression of this subunit in this brain region (Whiting et al., 1990). However, more recent studies have shown expression of the γ_{2L} subunit in the hippocampus (Zahniser et al., 1992) and electrophysiological evidence of ethanol enhancement of hippocampal GABAergic currents has also been shown (Aguayo, 1990; Reynolds et al., 1992; Aguayo and Pancetti, 1994). A recent report by Weiner et al. (1994a) examined the mechanism responsible for ethanol enhancement of $GABA_A$ currents in hippocampal CA1 neurons. Ethanol (10–50 m*M*) was shown to potentiate $GABA_A$-mediated inhibitory postsynaptic currents with a requirement for the presence of intracellular ATP; in contrast, diazepam enhancement did not require ATP. Similar to the studies of Aguayo (1990) and Reynolds et al. (1992), ethanol potentiation was biphasic with peak potentiation being observed using a concentration of 20 m*M* EtOH. Furthermore, the ethanol enhancement of GABAergic currents was blocked by the protein kinase C (PKC) inhibitors Ro 31–8220 and PKC_{19-31}. Weiner et al. (1994a) suggested that ethanol enhancement may be due to an inhibition of PKC-mediated phosphorylation, thus decreasing the phosphorylation state of $GABA_A$ receptors. However, Machu et al. (1991) and Messing et al. (1991) failed to find any significant effects of ethanol on PKC function in cell-free preparations. Furthermore, Aguayo and Pancetti (1994) reported that the phorbol ester PMA, which activates PKC, antagonized ethanol enhancement of GABAergic responses. Since there are contradictory data on whether PKC activates or inhibits $GABA_A$ receptor function (see Chapter 7), it is not surprising that the nature of ethanol/PKC/$GABA_A$ receptor interactions is currently unclear.

Weiner et al. (1994b) further characterized their ethanol effects in the hippocampal slice by showing that intracellular dialysis of neurons with (GDP-βS), a guanine nucleotide phosphate which inhibits G-protein-mediated cellular events, decreased ethanol but not pentobarbital enhancement of IPSCs. Moreover, dialysis with GTP-γS, which irreversibly activates G-proteins, enhanced ethanol effects on $GABA_A$ receptors. These results are consistent with those of other studies demonstrating that the effects of ethanol on $GABA_A$ receptors could be modified by G-protein-linked receptors. For example, β-adrenergic (Lin et al., 1993a) and $GABA_B$ (Allan et al., 1991) receptors have both been reported to modulate the enhancement of $GABA_A$ receptor function by ethanol.

One must also consider the time required for $GABA_A$ receptors to become dephosphorylated by ethanol inhibition of tonic PKC activity and compare it to how quickly ethanol has been reported to enhance GABAergic currents. Reynolds et al. (1992) observed ethanol potentiation of GABAergic responses in cultured mouse hippocampal neurons within 50–100 ms of the application of GABA + EtOH.

It might be argued that the enhancing effects of ethanol in hippocampal cells seen by Aguayo (1990) and Reynolds et al. (1992) could be due to their use of cultured cells, unlike Proctor et al. (1992a) and Soldo et al. (1994) who used brain slices. Neurons in culture could conceivably develop altered $GABA_A$ pharmacology. However, the data of Weiner et al. (1994a) argue against this possibility since they saw their positive effects of ethanol in hippocampal slices.

Very recent work from our laboratory provides evidence that PKC modulates the effects of ethanol on the $GABA_A$ receptor. Null-mutant mice lacking the γ isoform of PKC (Abeliovich et al., 1993) display a reduced sensitivity to ethanol on several behavioral tests (Harris et al., 1995a). In microsacs prepared from these PKC "knock-out" mice, we found that ethanol no longer enhanced GABA-mediated chloride flux (Harris et al., 1995a), further strengthening the hypothesis that ethanol enhancement of GABAergic currents involves more than merely the presence of specific subunits of the $GABA_A$ receptor.

Several groups have not been able to show ethanol potentiation of GABAergic currents. Osmanovic and Shefner (1990) found that 40–60 m*M* ethanol did not potentiate GABA-activated currents in rat locus coeruleus brain slice preparations while pentobarbital had a clear effect. White et al. (1990) also found no enhancement by 10, 50, or 100 m*M* ethanol on the GABA-activated chloride current in isolated adult rat sensory neurons, despite seeing an inhibition of NMDA-induced current. Furthermore, the results of Morrisett and Swartzwelder (1993) indicate that ethanol does not enhance GABAergic transmission (but midazolam does) in dentate slices, although it does inhibit NMDA responses. These data suggest that some factors, other than merely the presence of γ_{2L} subunits, are required in order to see ethanol potentiation of the function of $GABA_A$ receptors. Furthermore, the work of Criswell et al. (1993) shows that mRNA for both the γ_{2L} and γ_{2S} subunits have been found in brain regions which they found to be either sensitive or insensitive to ethanol.

Electrophysiological measurements allow one to study GABAergic neurons from a wide variety of brain regions, *in vivo* or *in vitro*. However, often one cannot be sure of the exact concentrations of applied agonists and drugs at the neuron under study, since drugs are introduced to a bath perfusate or are injected from the tip of a micropipette in the vicinity of the target cell, from which they diffuse away at an unknown rate. Furthermore, since there is heterogeneity of $GABA_A$ receptor subunit expression among, and within, brain regions, one may find a particular effect in one neuron but not necessarily in another. For example, Study and Barker (1981) showed that $GABA_A$-activated conductance was potentiated by diazepam in 82% of cultured mouse spinal cord neurons while pentobarbital had an effect in 85% of the cells; however, only 53% of the neurons showed enhancement by both diazepam and pentobarbital.

4. Electrophysiological Studies Using Recombinant Receptors

Within the last three years a number of studies have been published on the effects of ethanol on defined subunits of $GABA_A$ receptors. Whiting et al. (1990) and Kofuji et al. (1991) recently discovered that the γ_2 subunit of the $GABA_A$ receptor has an alternatively spliced variant, containing an additional eight amino acids. The original γ_2 subunit is now referred to as γ_{2S} (for short) while the spliced variant is called γ_{2L} (for long). Wafford et al. (1991) reported that it is the eight amino acids in the γ_{2L} subunit that are required to show ethanol potentiation of GABA-mediated conductance in oocytes. Receptors containing the γ_{2S} subunit were insensitive to ethanol but retained their sensitivity to pentobarbital and diazepam. The ethanol enhancement of GABAergic currents was prevented by injection of γ_{2L}-specific antisense nucleotides along with brain mRNA into *Xenopus* oocytes. Furthermore, $GABA_A$ receptors formed after the injection of $\alpha_1\beta_1\gamma_{2S}$ or $\alpha_1\beta_1\gamma_{2L}$ cRNAs were sensitive to pentobarbital and diazepam, but only the $\alpha_1\beta_1\gamma_{2L}$ receptors showed sensitivity to 20 m*M* ethanol (Wafford et al., 1991). Wafford and Whiting (1992) next mutated the consensus site for phosphorylation by protein kinase C contained in the γ_{2L} insert. This resulted in the loss of ethanol modulation of GABAergic currents in oocytes but did not affect benzodiazepine pharmacology. Preincubation of oocytes with the nonspecific PKA/PKC inhibitor, H-7, also blocked ethanol enhancement. Wafford and Whiting (1992) concluded that phosphorylation events at this site on the $GABA_A$ receptor could act as a control mechanism for neuronal responses to alcohol exposure.

One recent study used mammalian cells stably transfected with $\alpha_1\beta_1$ or $\alpha_1\beta_1\gamma_{2L}$ $GABA_A$ receptor subunit DNAs to study effects of ethanol on the $GABA_A$ receptor (Harris et al., 1995b). Use of stably transfected cells allowed comparison of $^{36}Cl^-$ flux and whole-cell patch-clamp measurements of $GABA_A$ receptor function. Both techniques detected a similar modulation of the GABA receptor by ethanol, flunitrazepam, and pentobarbital. The potentiating action of ethanol required the γ subunit and was maximal at a concentration of 10 m*M*. Analysis of data from individual cells expressing $\alpha_1\beta_1\gamma_{2L}$ subunits showed considerable variability in sensitivity to ethanol, and the ethanol potentiation was also influenced by the type of matrix used for cell culture. These results suggest that a γ subunit is necessary but not sufficient to observe ethanol sensitivity in this cell system. Actions of ethanol observed in these stably transfected mammalian cells are generally consistent with those discussed above for neuronal cultures, brain membrane vesicles, and specific subunits expressed in *Xenopus* oocytes. In particular, this work confirms the requirement of a γ subunit for enhancement of GABA action by ethanol (Wafford et al., 1991), although the γ_{2S} splice variant was not used, so the necessity of the γ_{2L} splice variant could not be addressed.

Two laboratories have been unable to obtain the potentiating effects of ethanol after expressing specific subunits of GABA$_A$ receptors. Kleingoor et al. (1991) did not find ethanol potentiation of GABA-mediated conductance using rat $\alpha_6\beta_2\gamma_2$ or $\alpha_1\beta_2\gamma_2$ subunits expressed in human embryonic kidney (HEK) 293 cells. However, since they did not specify which γ_2 subunit they were expressing (γ_{2L}, γ_{2S}, or both), it is not possible to determine if this failure to see an effect of ethanol was due to the choice of γ subunit. It may, instead, have been because they chose to use a β_2 subunit, unlike Wafford et al. (1991) who used a β_1 subunit, or because they used a mammalian expression system instead of *Xenopus* oocytes. Kleingoor et al. (1991) did, however, see flunitrazepam enhancement and DMCM inhibition of the GABA-mediated current in cells transfected with the $\alpha_1\beta_2\gamma_2$ subunits. Sigel et al. (1993) studied the effects of 20–100 m*M* EtOH on rat $\alpha_1\beta_1\gamma_{2L}$, $\alpha_1\beta_1\gamma_{2S,}$ and $\alpha_1\beta_2\gamma_{2S}$ receptors expressed in *Xenopus* oocytes. Ethanol was either preapplied for 30 s or coapplied with GABA. They found minor, but significant, stimulation of GABAergic responses at the higher ethanol concentrations tested (>60 m*M*) and that this enhancement did not depend on the presence of the γ_{2L} subunit. The fact that Sigel et al. (1993) and Kleingoor et al. (1991) used rat clones, while Wafford et al. (1991) used bovine clones is not likely to be important; the rat and bovine γ_{2L} sequences differ only in three amino acids, all in the N-terminal domain.

We have also observed the small enhancement of GABAergic currents reported by Sigel et al. (1993) and have also found it not to depend on the presence of the γ_{2L} subunit (Mihic et al., 1994a). Anesthetic concentrations of ethanol (50–400 m*M*) and butanol (1–20 m*M*) produced significant potentiation of GABAergic currents in *Xenopus* oocytes expressing human GABA$_A$ receptor cDNAs. The threshold for ethanol enhancement was approximately 100 m*M*. These enhancing effects of alcohols were seen in receptors expressing just $\alpha\beta$ receptors and those expressing $\alpha\beta\gamma$ receptors. Substituting the α_5 subunit for the α_1, or the β_2 for the β_1, did not affect the degree of ethanol potentiation (Mihic et al., 1994a). We believe that other laboratories have also seen manifestations of the same phenomenon. GABA responses of GT1–7 cells expressing α and β, but not γ, subunits were potentiated by 172 and 344 m*M* EtOH, while 17.2 m*M* EtOH was without effect (Hales et al., 1992). Kurata et al. (1993) transfected HEK 293 cells with rat $\alpha_1\beta_2$ or $\alpha_1\beta_2\gamma_{2S}$ cDNAs and found that 100 μ*M* octanol significantly shifted to the left the GABA concentration-response curves of both receptor types. Further work by this same group found that 0–300 m*M* EtOH, in contrast to 100 μ*M* octanol, did not potentiate GABA-mediated current amplitudes in $\alpha_1\beta_2\gamma_{2S}$, $\alpha_6\beta_2\gamma_{2S,}$ or $\alpha_6\beta_2\gamma_{2L}$ receptors expressed in HEK cells (Marszalec et al., 1994). However, 100 m*M* ethanol did increase the rate of GABA current desensitization in $\alpha_6\beta_2\gamma_{2S}$, but not $\alpha_1\beta_2\gamma_{2S}$, receptors in a manner that was not affected by the broad-spectrum kinase inhibitor H-7, The authors concluded that expression of the γ_{2L} subunit, by itself, is not sufficient to produce potentiation of GABAergic currents (Marszalec et al., 1994).

The phenomenon of anesthetic concentrations of alcohols potentiating GABA$_A$ responses, with no apparent subunit specificity, appears to be distinct from the low (e.g., 5–50 m*M*) ethanol potentiation discussed above. Furthermore, this high-dose ethanol phenomenon has likely already been reported in the literature without being identified as being distinct from the low-dose effect. Some of the work cited earlier, particularly by Narahashi's group (Nishio and Narahashi, 1990; Nakahiro et al., 1991), should now be re-evaluated as a study of the high-dose effect of ethanol. For example, Nakahiro et al. (1991) found that an EtOH concentration of approximately 250 m*M* was required in order to show 25% potentiation of GABAergic currents in rat dorsal root ganglion neurons maintained in primary culture, with lower concentrations having minimal effects. This is very similar to the effects we found with ethanol using $\alpha_1\beta_1$ and $\alpha_1\beta_1\gamma_{2S}$ receptors expressed in *Xenopus* oocytes (Mihic et al., 1994a). Aguayo and Pancetti (1994) recently noted that cultured hippocampal neurons displayed two types of sensitivity to ethanol. Some cells showed clear enhancement using ethanol concentrations below 100 m*M*, while others required concentrations of up to 850 m*M*.

When studying homomeric ρ_1 receptors expressed in *Xenopus* oocytes, we were surprised to find significant *inhibition* of GABAergic currents by a number of alcohols (Mihic and Harris, submitted). This inhibition was seen using low (400 n*M* - EC_{10}) but not high (5 μ*M* - EC_{90}) concentrations of GABA. Using 400 n*M* GABA, ethanol had an IC_{50} of approximately 100 m*M*, with a concentration of 25 m*M* producing significant (~25%) inhibition. In comparison, 20 m*M* butanol inhibited these GABAergic currents by 70% (Mihic and Harris, submitted).

B. CHRONIC EFFECTS OF ETHANOL

Compared to the acute studies, far less work has been done on the effects of chronic ethanol exposure on $GABA_A$ receptor function. For this reason the electrophysiological and biochemical data will be reviewed together in this section. Effects of chronic ethanol on receptor subunit mRNA levels are reviewed later (in Section IV).

Early work by Allan and Harris (1987) showed that treatment of DBA/2J mice with an ethanol-containing liquid diet for 7 d did not affect muscimol-stimulated $^{36}Cl^-$ uptake into cerebellar microsacs. Other studies using mice (Buck and Harris, 1990b) or rats (Frye et al., 1991; Mihic et al., 1992b) also found that chronically treating animals with ethanol did not affect basal or muscimol-stimulated chloride flux. However, Morrow et al. (1988) found a 26% decrease in the maximal effect of muscimol after chronic ethanol treatment but only if animals attained blood alcohol levels of 150 mg/dL or more. These animals were sacrificed within 15 min of being removed from an ethanol inhalation chamber. Sanna et al. (1993) also noted that chronically treating rats with ethanol, followed by only 1–3 h of withdrawal, produced a significant inhibition of the ability of GABA to induce chloride influx into cortical microsacs. However, withdrawal for 9–24 h or 3–6 d resulted in no decrement in GABA's ability to stimulate $^{36}Cl^-$ uptake. Kuriyama and Ueha (1992) found that exposing male ddy mice to ethanol in a vapor chamber for 7 d also decreased GABA-mediated $^{36}Cl^-$ uptake in nonwithdrawn animals. However, when animals were withdrawn from ethanol for 8 h, GABA-mediated chloride influx returned to control levels. Time after ethanol withdrawal did not appear to play a role in the studies of Buck and Harris (1990b) and Mihic et al. (1992b), although Frye et al. (1991) did use withdrawn animals (for 3–5 h), which might explain their negative results. When *Xenopus* oocytes expressing mouse brain cortical mRNA were exposed to 50 m*M* EtOH for 2–4 d without a withdrawal period, there was a decrease in the amount of current produced by a maximally effective concentration of muscimol (Buck and Harris, 1991).

A direct stimulation of chloride uptake by pentobarbital has also been reported to be reduced after chronic ethanol treatment (Morrow et al., 1988). In addition, pentobarbital stimulation of GABA-mediated chloride influx was also reduced (Morrow et al., 1990). However, no changes in pentobarbital's direct or GABA-potentiating effects were seen by Buck and Harris (1990b) or Mihic et al. (1992b).

Allan and Harris (1987) and Morrow et al. (1988) found that ethanol failed to enhance muscimol-mediated chloride flux after chronic ethanol treatment, in contrast to the potentiation that was seen in ethanol-naive animals. However, the direct effect of ethanol on basal chloride influx remained unchanged after the chronic treatment (Morrow et al., 1988). Cross tolerance from ethanol to benzodiazepines on the biochemical level can be expressed as a decreased benzodiazepine enhancement of GABA-mediated chloride influx. Buck and Harris (1990b) found that flunitrazepam efficacy in enhancing muscimol-activated $^{36}Cl^-$ uptake was markedly reduced in mice fed an ethanol-containing (6% v/v) liquid diet for 10 d. However, the decreased flunitrazepam potentiation observed in animals only occurred if they were sacrificed without undergoing withdrawal; if the animals were withdrawn for 24 h before sacrifice, no differences from pair-fed controls were found. No changes in the binding parameters of [^{3}H]flunitrazepam, [^{3}H]Ro 15–4513, or [^{3}H]SR 95531 accompanied the functional changes observed. Mihic et al. (1992b) also found decreased diazepam enhancement of GABA-mediated chloride flux in rats undergoing chronic ethanol treatment, but only if the rats received an i.p. injection of ethanol before sacrifice; i.e., they were nonwithdrawn. Sanna et al. (1993) also saw decreased flunitrazepam enhancement in brain vesicles prepared from chronically ethanol-treated rats, but only if animals were withdrawn for 1–3 h and not 9–24 h or 3–6 d. Flunitrazepam potentiation and DMCM inhibition of GABA-mediated conductance was also decreased in *Xenopus* oocytes exposed to 50 m*M* EtOH for 2–4 d (Buck and Harris, 1991). It was also found that the concomitant presence of the protein synthesis inhibitor, cycloheximide, with chronic ethanol could prevent the changes in flunitrazepam and muscimol function following chronic ethanol treatment. This suggests that post-translational modifications of $GABA_A$ receptors, rather than changes in gene expression, are responsible for the decreased sensitivity to muscimol and flunitrazepam following chronic ethanol treatment (Buck and Harris, 1991).

Buck and Harris (1990b) also found that chronic ethanol treatment of mice enhanced DMCM and Ro 15–4513 inhibition of GABA-mediated $^{36}Cl^-$ uptake by brain microsacs, leading to the conclusion that the chronic ethanol treatment was decreasing the coupling of benzodiazepine receptor agonists while at the same time favoring the coupling of inverse agonists. This might contribute to seizure susceptibility during ethanol withdrawal. However, DMCM inhibition of GABAergic currents was decreased by chronic ethanol treatment in *Xenopus* oocytes (Buck and Harris, 1991).

Allan et al. (1992) showed that cross tolerance from benzodiazepines to ethanol can also be demonstrated biochemically. Microsacs prepared from lorazepam-tolerant mice showed cross-tolerance to both ethanol and phenobarbital through decreased stimulation of GABA-mediated chloride uptake. However, pentobarbital-stimulation of the effect of GABA was not changed by chronic lorazepam treatment.

C. COMPARISONS WITH ANESTHETICS

Many behavioral, biochemical, and electrophysiological studies suggest that potentiation of chloride conductance through the $GABA_A$ receptor/chloride channel complex is involved in producing anesthesia. The $GABA_A$ analog, THIP, produces sedation, analgesia, and loss of righting reflex (Cheng and Brunner, 1985). Muscimol and THIP (Cheng and Brunner, 1985) increase, while the $GABA_A$ antagonist bicuculline (Sivam et al., 1982) decreases, the duration of pentobarbital-induced loss of righting reflex. Barbiturates (Nicoll, 1975; Study and Barker, 1981), diethyl ether (Huidobro-Toro et al., 1987), steroid anesthetics (Harrison and Simmonds, 1984; Turner et al., 1989), and propofol (Peduto et al., 1991) all potentiate the effects of GABA in electrophysiological and chloride uptake assays. In addition, diethyl ether, halothane, enflurane, and methoxyflurane increase $^{36}Cl^-$ uptake into cerebral cortical brain vesicles in a concentration-dependent, picrotoxin-reversible manner (Moody et al., 1988). Halothane also potentiates GABA-mediated neuronal depression in electrophysiological studies (Gage and Robertson, 1985; Nakahiro et al. 1989). Lin et al. (1992, 1993d) demonstrated that the structurally diverse anesthetics enflurane, halothane, isoflurane, diethyl ether, ketamine, pentobarbital, phenobarbital, 3α-hydroxy-5α-dihydroprogesterone, and propofol could all enhance GABAergic currents in *Xenopus* oocytes.

Volatile anesthetics potentiate the effects of GABA in $\alpha_1\beta_1$ and $\alpha_1\beta_1\gamma_{2S}$ receptors (Lin et al., 1993d; Harrison et al., 1993; Mihic et al., 1994b). This lack of subunit specificity contrasts with the absolute necessity for the γ_{2L} subunit required for the 20 m*M* ethanol potentiation as seen by Wafford et al. (1991) and Wafford and Whiting (1992). High concentrations of alcohols (Mihic et al., 1994a) also share with volatile and steroid anesthetics and barbiturates the ability to potentiate currents in $GABA_A$ receptors lacking a γ_2 subunit (Shingai et al., 1991; Porter et al., 1992; Harrison et al., 1993) and may have a similar mechanism of action. This similarity leads us to believe that the high-dose ethanol effects reported by Mihic et al. (1994a) have a different mechanism of action than those reported by Wafford et al. (1991).

D. INTEGRATED MECHANISM AND OVERVIEW

To our knowledge, the potentiation of GABA's effects by benzodiazepines and barbiturates has been observed by all laboratories that have studied these compounds, although it is clear that not all neurons that respond to GABA show effects of benzodiazepines or pentobarbital (Study and Barker, 1981). However, as reviewed above, some laboratories have failed to observe consistent actions of ethanol on $GABA_A$ responses. Benzodiazepines and barbiturates are known to interact directly with defined binding sites on the $GABA_A$ receptor/chloride channel complex, which may explain the consistency of these results among laboratories. Ethanol may also directly affect the receptor itself, but in such a way that only under specific conditions (e.g., receptor phosphorylation status) does enhancement of GABAergic function occur (Wafford et al., 1991; Wafford and Whiting, 1992; Harris et al., 1995a). Alternatively, ethanol may instead act at another distinct site, with a change in the function of that site then affecting GABA-mediated currents (Allan et al., 1991; Lin et al., 1991, 1993a). Regardless of which hypothesis is correct, it is undeniable that enhancement of $GABA_A$-mediated events by ethanol is much more sensitive to external influences than that produced by benzodiazepines or barbiturates. The fact that ethanol potentiation does not correlate well with the presence or absence of the γ_{2L} subunit in specific brain regions and that some studies using this subunit have not found ethanol effects further suggest that the mere presence of particular $GABA_A$ receptor subunits is insufficient to guarantee positive effects of ethanol. In contrast, the expression of the γ subunit with α and β subunits, produces universally reproducible expression of benzodiazepine pharmacology (Pritchett et al., 1989).

We believe that the positive effects of ethanol on $GABA_A$ receptors should be divided into two groups: the low-dose effects that appear to require a γ subunit (Wafford et al., 1991; Harris et al., 1995b) and the high-dose effects that we recently found not to depend on any particular subunit class (Mihic et al., 1994a). Another way of differentiating between these two phenomena is to examine their ethanol concentration-response curves. The low-dose effect exhibits a biphasic concentration-response curve, seen in both biochemical (Allan and Harris, 1986; Suzdak et al., 1986a) and electrophysiological (Aguayo, 1990; Reynolds and Prasad, 1991; Weiner et al., 1994a) studies, with a maximal effect at approximately 10–30 m*M* ethanol. As the concentration is increased, the potentiating effect of ethanol decreases. In contrast, the high-dose effect begins to appear at about 100 m*M* ethanol and does not appear to be biphasic (Mihic et al., 1994a). Furthermore, the latter effect is likely to be similar in nature to the potentiation of GABAergic currents produced by anesthetics (Harrison et al., 1993; Mihic et al., 1994b).

IV. EXPRESSION OF $GABA_A$ RECEPTOR SUBUNITS

One mechanism for effecting ethanol tolerance on the cellular level is to up- or down-regulate the numbers of receptors that are being affected by ethanol. In this model, the cell attempts to compensate for the continued disruptive effects of ethanol and consequently to re-establish homeostasis. One, albeit indirect, method to monitor changes in $GABA_A$ receptor subunit levels is to determine mRNA levels after chronic ethanol consumption. A number of recent studies have investigated the effects of chronic ethanol administration on $GABA_A$

receptor subunit mRNA levels. Fourteen days of ethanol vapor administration led to substantial decreases in α_1 (Morrow et al., 1990; Montpied et al., 1991) and α_2 but not α_3 (Montpied et al., 1991) GABA subunit transcript concentrations in rat cerebral cortex. Decreases of 40–50% in α_1 mRNA and 29% in α_2 mRNA levels were reported. However, β-actin mRNA, total RNA, rRNA, and poly(A)+ RNA levels were not significantly altered, suggesting that chronic ethanol exposure specifically alters the amounts of mRNAs coding for the α subunits of the $GABA_A$ receptor (Morrow et al., 1990; Montpied et al., 1991) rather than having a nonspecific effect on transcription or mRNA turnover in general.

Mhatre and Ticku (1992) saw essentially the same pattern of decreases in α subunit mRNA levels as did Morrow et al. (1990). The levels of α_1, α_2, and α_5 transcripts were down 45–61% while α_3 levels remained unchanged. These decreases persisted for 24 h after withdrawal and returned to normal 12 h later. Interestingly, α_6 mRNA levels in the cerebellum increased 76% following chronic ethanol (Mhatre and Ticku, 1992). Coexpression of the α_6 subunit with β and γ subunits produces $GABA_A$ receptors which bind the inverse agonist Ro 15–4513 but not benzodiazepine receptor agonists. An earlier paper by the same group (Mhatre et al., 1988) reported that the density of [^{3}H]Ro 15–4513 binding sites was increased following chronic ethanol administration. Mhatre et al. (1993), using polyclonal antibodies raised against specific α subunits, also found that the decreases in α_1 and α_2 mRNA levels following chronic ethanol treatment were paralleled by 61% and 47% decreases in the α_1 and α_2 subunit protein levels, respectively. However, they saw a 30% decrease in α_3 protein levels despite reporting no decreases in α_3 mRNA levels earlier (Mhatre and Ticku, 1992). In general agreement with these findings on subunit protein levels, Freund and Ballinger (1991) noted that the numbers of benzodiazepine receptors in alcoholics were decreased by 30% in the hippocampus and by 25% in the frontal cortex but were not altered in the temporal cortex or putamen. These changes occurred in histologically normal brains in subjects not taking benzodiazepines.

Buck et al. (1991a) studied the effects of chronic ethanol administration on a number of different $GABA_A$ receptor subunits (α_1, α_3, α_6, γ_{2S}, γ_{2L}, and γ_3) in withdrawal seizure-prone (WSP) and seizure-resistant (WSR) mice. Decreases in α_1 levels were seen only in the WSP mice, and α_6 mRNA levels were decreased in WSR but not WSP mice. No changes were seen in γ_{2L} or γ_{2S} mRNA levels, but γ_3 transcript amounts were increased in both lines. Since ethanol potentiation of $GABA_A$-mediated chloride influx is no longer seen in mice which have undergone chronic treatment with ethanol (Allan and Harris, 1987), one might have expected to see some changes in γ_{2L} subunit expression, because the γ_{2L} subunit is reported to be required to see ethanol enhancement of GABAergic currents (Wafford et al., 1991).

In contrast to the other studies, Hirouchi et al. (1993) found that continuous ethanol inhalation by mice for more than 5 d significantly *increased* α_1 subunit mRNA amounts without significantly altering β-actin or total poly(A)+ RNA content. This increase in the expression of $GABA_A$ receptor α_1 subunit mRNA returned to normal 8 h after the ethanol inhalation was terminated. The fact that some of the studies described above used animals that had been withdrawn from ethanol, while others did not employ withdrawal, cannot fully explain the discrepancy in α_1 mRNA levels. Montpied et al. (1991) sacrificed their rats immediately after terminating the ethanol treatment while Mhatre and Ticku (1992) withdrew their animals for 24 h; despite this major experimental difference, the two groups obtained identical patterns of changes in α subunit mRNA transcription. It is unlikely that mice and rats respond differently to the chronic ethanol treatment since Buck et al. (1991a) saw decreases in α_1 mRNA levels in WSP mice.

The decreases in α_1 and α_2 transcript concentrations found by Morrow et al. (1990) appear to be specific to ethanol since chronic pentobarbital administration had no effect on the levels of α subunit mRNAs in rat cerebral cortex. Decreases in α subunit mRNA levels after chronic benzodiazepine treatment have also been reported (Heninger et al., 1990; Kang and Miller, 1991; Zhao et al., 1994), although the patterns of decreases among α subunits varied from study to study; this may depend on the benzodiazepine receptor agonists used and the chronic administration paradigms employed. It is also possible that the levels of $GABA_A$ receptor subunit mRNA transcripts do not accurately predict $GABA_A$ receptor protein composition and consequently also function. Williamson and Pritchett (1994), observing low levels of messenger RNA for α_2, α_3, and α_5 subunits to be inconsistent with the high levels of type II benzodiazepine binding (that mediated by non-α_1-containing receptors) in the rat cerebellum at birth, concluded that models attempting to explain differences in $GABA_A$ receptor pharmacology merely as a result of changes in α subunit gene expression were inadequate. In conclusion, the most consistent results obtained by several groups appears to be the decrease in α_1 subunit transcript levels following chronic ethanol treatment, although even this was not universally seen. For a detailed review, see Morrow (1995).

An important question is whether the changes in receptor subunit mRNA and protein levels produced by chronic alcohol treatment actually produce a "subunit substitution" (Morrow, 1995) that can account for the observed changes in $GABA_A$ receptor function. One approach to answering this question is to use cells stably transfected with defined receptor subunits (e.g., $\alpha_1\beta_1\gamma_{2L}$) where subunit substitution is not possible (Harris et al., 1995b). Preliminary studies indicate that chronic exposure of these cells to ethanol reduces the coupling between GABA and benzodiazepine receptors as measured by GABA enhancement of [^{3}H]flunitrazepam binding and flunitrazepam enhancement of GABA-activated chloride flux (Klein et al., 1994; Klein et al., unpublished results). Thus, at least some of the effects of chronic ethanol observed in animals and cultured neuronal cells can be produced in transfected cells where subunit substitution is not possible. It is likely that these changes are post-translational (e.g., receptor phosphorylation) and such cell systems should be useful for elucidating mechanisms of receptor adaptation that do not require changes in subunit expression.

V. OVERVIEW AND FUTURE DIRECTIONS

Currently the most critical research question is what factors determine whether a $GABA_A$ receptor will respond to ethanol. One of the first reports of ethanol enhancement of GABA-activated chloride flux noted marked genetic differences in this action; that is, ethanol enhancement was found using membranes from LS but not SS mice (Allan and Harris, 1986). Reports since then have made it abundantly clear that there are many instances in which ethanol does not enhance GABA action, but these have provided little information about the molecular determinants of ethanol action on this receptor. Almost five years ago, it was suggested that both the proper subunits and the proper phosphorylation state was required for ethanol sensitivity of the receptor (Wafford et al., 1991). This is consistent with a series of studies suggesting that ethanol enhancement of GABA action on cerebellar Purkinje cells requires activation of cAMP production and cAMP-dependent protein kinase activity (Lin et al., 1991, 1993a; Freund et al., 1993) as well as studies of mutant mice lacking a specific kinase (Harris et al., 1995b). However, there is no direct evidence that ethanol changes the phosphorylation state of $GABA_A$ receptors or that all ethanol-insensitive receptors can be converted to an ethanol-sensitive state by changes in phosphorylation. (Some of these issues are discussed in more detail in chapter 7.) Importance of the mechanism of ethanol action is not restricted to $GABA_A$ receptors but applies to all members of the ligand-gated superfamily of ion channels. All of these channels are affected by ethanol, but the mechanism is not known for any channel. Because of the homology among these channels, it is likely that if the mechanism of ethanol action is defined for one channel, it will apply to all channels. This currently represents a critical gap in our knowledge.

Effects of chronic ethanol exposure on $GABA_A$ receptors appear more consistent than those of acute exposure, but the mechanism is no clearer. Some effects occur within minutes (Allan and Harris, 1987; Buck and Harris, 1990a) and may be related to acute tolerance while others are observed after days of exposure and show genetic correlation with severity of withdrawal seizures (Buck et al. 1991a). Some of these changes must require post-translational modification of the receptor whereas others may be due to changes in expression of receptor subunits. Elucidation of these mechanisms will likely give considerable insight regarding normal regulation of $GABA_A$ receptor function as well as the processes of tolerance and dependence.

Most of our knowledge of mechanisms of receptor regulation (e.g., internalization, coupling with G-proteins, and phosphorylation) has come from studies of cultured cells, and it is likely that such approaches will be essential for understanding actions of ethanol on $GABA_A$ receptors. However, cultured cells are model systems, and the importance of changes observed in isolated cells must be ascertained in brain and extended to behavior. This is much more difficult, but recent advances in genetic manipulation *in vivo*, including transgenic and null mutant animals as well as *in vivo* administration of antisense oligonucleotides and ribozymes to selectively reduce gene expression, should prove quite useful. Many of these techniques are beginning to be applied by alcohol researchers, and we can hope that they will resolve some of the current controversies regarding ethanol action on $GABA_A$ (and related) receptors.

ACKNOWLEDGMENTS

The authors would like to thank Drs. Andrea Allan, Ron Klein, and Jeff Weiner for providing us with unpublished data. This work was supported by the Department of Veteran Affairs and an NIH/NIAAA grant (AA06399) to R.A.H. and by a Medical Research Council of Canada postdoctoral fellowship to S.J.M.

REFERENCES

Abeliovich, A., Chen, C., Goda, Y., Silva, A.J., Stevens, C.F., and Tonegawa, S., Modified hippocampal long-term potentiation in PKCγ-mutant mice. *Cell,* 75:1253–1262 (1993).

Aguayo, L.G., Ethanol potentiates the $GABA_A$-activated Cl^- current in mouse hippocampal and cortical neurons. *Eur. J. Pharmacol.,* 187:127–130 (1990).

Aguayo, L.G. and Alarcón, J.M., Modulation of the developing rat sympathetic $GABA_A$ receptor by Zn^{++},benzodiazepines, barbiturates and ethanol. *J. Pharmacol. Exp. Ther.,* 267:1414–1422 (1993).

Aguayo, L.G. and Pancetti, F.C., Ethanol modulation of the γ-aminobutyric $acid_A$- and glycine-activated Cl^- current in cultured mouse neurons. *J. Pharmacol. Exp. Ther.,* 270:61–69 (1994).

Allan, A.M. and Harris, R.A., Ethanol and barbiturates enhance GABA-stimulated influx of ^{36}Cl in isolated brain membranes. *Pharmacologist,* 27:125 (1985).

Allan, A.M. and Harris, R.A., Gamma-aminobutyric acid and alcohol actions: neurochemical studies of long sleep and short sleep mice. *Life Sci.,* 39:2005–2015 (1986).

Allan, A.M. and Harris, R.A., Acute and chronic ethanol treatments alter GABA receptor-operated chloride channels. *Pharmacol. Biochem. Behav.,* 27:665–670 (1987).

Allan, A.M. and Harris, R.A., Neurochemical studies of genetic differences in alcohol action. In: *The genetic basis of alcohol and drug action.* (J.C. Crabbe and R.A. Harris, eds.), Plenum Press, New York (1991).

Allan, A.M., Burnett, D., and Harris, R.A., Ethanol-induced changes in chloride flux are mediated by both GABA(A) and GABA(B) receptors. *Alcoholism: Clin. Exp. Res.,* 15:233–237 (1991).

Allan, A.M., Baier, L.D., and Zhang, X., Effects of lorazepam tolerance and withdrawal on $GABA_A$ receptor-operated chloride channels. *J. Pharmacol. Exp. Ther.,* 261:395–402 (1992).

Arakawa, O., Nakahiro, M., and Narahashi, T., Chloride current induced by alcohols in rat dorsal root ganglion neurons. *Brain Res.,* 578:275–281 (1992).

Barker, J.L. and Ransom, B.R., Amino acid pharmacology of mammalian central neurons grown in tissue culture. *J. Physiol.,* 280:331–354 (1978).

Becker, H.C. and Anton, R.F., Valproate potentiates and picrotoxin antagonizes the anxiolytic action of ethanol in a nonshock conflict task. *Neuropharmacology,* 29:837–843 (1990).

Boulter, J., Connolly, J., Deneris, E., Goldman, D., Heinemann, S., and Patrick, J., Functional expression of two neuronal nicotinic acetylcholine receptors from cDNA clones identifies a gene family. *Proc. Natl. Acad. Sci. U.S.A.,* 84:7763–7767 (1987).

Buck, K.J. and Harris, R.A., Benzodiazepine agonist and inverse agonist actions on $GABA_A$ receptor-operated chloride channels. I. Acute effects of ethanol. *J. Pharmacol. Exp. Ther.*, 253:706–712 (1990a).

Buck, K.J. and Harris, R.A., Benzodiazepine agonist and inverse agonist actions on $GABA_A$ receptor-operated chloride channels. II. Chronic effects of ethanol. *J. Pharmacol. Exp. Ther.*, 253:713–719 (1990b).

Buck, K.J. and Harris, R.A., Chronic ethanol exposure of *Xenopus* oocytes expressing mouse brain mRNA reduces $GABA_A$ receptor-activated current and benzodiazepine modulation. *Mol. Neuropharmacol.*, 1:59–64 (1991).

Buck, K.J., Hahner, L., Sikela, J., and Harris, R.A., Chronic ethanol treatment alters brain levels of γ-aminobutyric $acid_A$ receptor subunit mRNAs: relationship to genetic differences in ethanol withdrawal seizure severity. *J. Neurochem.*, 57:1452–1455 (1991a).

Buck, K.J., Heim, H., and Harris, R.A., Reversal of alcohol dependence and tolerance by a single administration of flumazenil. *J. Pharmacol. Exp. Ther.*, 257:984–989 (1991b).

Cheng, S.-C. and Brunner, E.A., Inducing anesthesia with a GABA analog, THIP. *Anesthesiology*, 63:147–151 (1985).

Criswell, H.E., Simson, P.E., Duncan, G.E., McCown, T.J., Herbert, J.S., Morrow, A.L., and Breese, G.R., Molecular basis for regionally specific action of ethanol on γ-aminobutyric $acid_A$ receptors: generalization to other ligand-gated ion channels. *J. Pharmacol. Exp. Ther.*, 267:522–537 (1993).

Deitrich, R.A., Dunwiddie, T.V., Harris, R.A., and Erwin, V.G., Mechanism of action of ethanol: initial central nervous system actions. *Pharmacol. Rev.*, 41:489–537 (1989).

Draguhn, A., Verdorn, T.A., Ewart, M., Seeburg, P.H., and Sakmann, B., Functional and molecular distinction between rat $GABA_A$ receptor subtypes by Zn^{2+}. *Neuron*, 5:781–788 (1990).

Dunn, S.M.J., Bateson, A.M., and Martin, I.L., Molecular neurobiology of the $GABA_A$ receptor. In: *International Review of Neurobiology*, Vol. 36. (R.J. Bradley and R.A. Harris, eds.) Academic Press, San Diego (1994).

Engblom, A.C. and Åkerman, K.E.O., Effect of ethanol on γ-aminobutyric acid and glycine receptor-coupled fluxes in rat brain synaptoneurosomes. *J. Neurochem.*, 57:384–390 (1991).

Freund, G. and Ballinger, W.E., Loss of synaptic receptors can precede morphologic changes induced by alcoholism. *Alcohol Alcohol. Suppl.*, 1:385–391 (1991).

Freund, R.K., van Horne, C.G., Harlan, T., and Palmer, M.R., Electrophysiological interactions of ethanol with GABAergic mechanisms in the rat cerebellum *in vivo*. *Alcoholism: Clin. Exp. Res.*, 17:321–328 (1993).

Frye, G.D., Mathew, J., and Trzeciakowski, J.P., Effect of ethanol dependence on $GABA_A$ antagonist-induced seizures and agonist-stimulated chloride uptake. *Alcohol*, 8:453–459 (1991).

Frye, G.D., Fincher, A.S., Grover, C.A., and Griffith, W.H., Interaction of ethanol and allosteric modulators with $GABA_A$-activated currents in adult medial septum/diagonal band neurons. *Brain Res.*, 635:283–292 (1994).

Gage, P.W. and Robertson, B., Prolongation of inhibitory postsynaptic currents by pentobarbitone, halothane and ketamine in CA1 pyramidal cells in rat hippocampus. *Br. J. Pharmacol.*, 85:675–681 (1985).

Givens, B.S. and Breese, G.R., Site-specific enhancement of gamma-aminobutyric acid-mediated inhibition of neural activity by ethanol in the rat medial septal area. *J. Pharmacol. Exp. Ther.*, 254:528–538 (1990).

Grant, A.J. and Treistman, S.N., Ethanol sensitivity of postsynaptic receptors in the abdominal ganglion of *Aplysia californica*. *Brain Res.*, 546:217–221 (1991).

Grenningloh, G., Rienitz, A., Schmitt, B., Methfessel, C., Zensen, M., Beyreuther, K., Gundelfinger, D.E., and Betz, H., The strychnine-binding subunit of the glycine receptor shows homology with nicotinic acetylcholine receptors. *Nature*, 328:215–220 (1987).

Hales, T.G., Kim, H., Longoni, B., Olsen, R.W., and Tobin, A.J., Immortalized hypothalamic GT1–7 neurons express functional γ-aminobutyric acid type A receptors. *Mol. Pharmacol.*, 42:197–202 (1992).

Harris, R.A., Distinct actions of alcohols, barbiturates and benzodiazepines on GABA-activated chloride channels. *Alcohol*, 7:273–275 (1990).

Harris, R.A. and Allan, A.M., Genetic differences in coupling of benzodiazepine receptors to chloride channels. *Brain Res.*, 490:26–32 (1989).

Harris, R.A., McQuilkin, S.J., Paylor, R., Abeliovich, A., Tonegawa, S., and Wehner, J.M., PKCγ null mutant mice show decreased behavioral actions of ethanol and altered $GABA_A$ receptor function. *Proc. Natl. Acad. Sci. U.S.A.*, 92:3633–3635 (1995a).

Harris, R.A., Proctor, W.R., McQuilkin, S.J., Klein, R.L., Mascia, M.P., Whatley, V.J., Whiting, P.J., and Dunwiddie, T.V., Ethanol increases $GABA_A$ responses in cells stably transfected with receptor subunits. *Alcoholism: Clin. Exp. Res.*, 19:226–232 (1995b).

Harrison, N.L. and Simmonds, M.A., Modulation of the GABA receptor complex by a steroid anesthetic. *Brain Res.*, 323:287–292 (1984).

Harrison, N.L., Kugler, J.L., Jones, M.V., Greenblatt, E.P., and Pritchett, D.B., Positive modulation of human gamma-aminobutyric acid type A and glycine receptors by the inhalation anesthetic isoflurane. *Mol. Pharmacol.*, 44:628–632 (1993).

Heninger, C., Saito, N., Tallman, J.F., Garrett, K.M., Vitek, M.P., Duman, R.S., and Gallager, D.W., Effects of continuous diazepam administration on $GABA_A$ subunit mRNA in rat brain. *J. Mol. Neurosci.*, 2:101–107 (1990).

Hiltunen, A.J. and Jarbe, T.U., Ro 15–4513 does not antagonize the discriminative stimulus- or rate-depressant effects of ethanol in rats. *Alcohol,* 5:203–207 (1988).

Hirouchi, M., Hashimoto, T., and Kuriyama, K., Alteration of $GABA_A$ receptor alpha 1-subunit mRNA in mouse brain following continuous ethanol inhalation. *Eur. J. Pharmacol.,* 247:127–130 (1993).

Huidobro-Toro, J.P., Bleck, V., Allan, A.M., and Harris, R.A., Neurochemical actions of anesthetic drugs on the γ-aminobutyric acid receptor-chloride channel complex. *J. Pharmacol. Exp. Ther.,* 242:963–969 (1987).

Ikemoto, Y., Akaike, N., and Kijima, H., Kinetic and pharmacological properties of the GABA-induced chloride current in Aplysia neurones: a 'concentration clamp' study. *Br. J. Pharmacol.,* 95:883–895 (1988).

Jeffreys, R.D., Pournaghash, S., Glowa, J.R., and Riley, A.L., The effects of Ro 15–4513 on ethanol-induced taste aversions. *Pharmacol. Biochem. Behav.,* 35:803–806 (1990).

June, H.L., Lummis, G.H., Colker, R.E., Moore, T.O., and Lewis, M.J., Ro 15–4513 attenuates the consumption of ethanol in deprived rats. *Alcoholism: Clin. Exp. Res.,* 15:406–411 (1991).

June, H.L., Colker, R.E., Domangue, K.R., Perry, L.E., Hicks, L.H., June, P.L., and Lewis, M.J., Ethanol self-administration in deprived rats: effects of Ro 15–4513 alone, and in combination with flumazenil (Ro 15–1788). *Alcoholism: Clin. Exp. Res.,* 16:11–16 (1992a).

June, H.L., June, P.L., Domangue, K.R., Hicks, L.H., Lummis, G.H., and Lewis, M.J., Failure of Ro 15–4513 to alter an ethanol-induced taste aversion. *Pharmacol. Biochem. Behav.,* 41:455–460 (1992b).

Kang, I. and Miller, L.G., Decreased $GABA_A$ receptor subunit mRNA concentrations following chronic lorazepam administration. *Br. J. Pharmacol.,* 103:1285–1287 (1991).

Klein, R.L., Whiting, P.J., and Harris, R.A., Benzodiazepine treatment causes uncoupling of recombinant $GABA_A$ receptors expressed in stably transfected cells. *J. Neurochem.,* 63:2349–2352 (1994).

Kleingoor, C., Ewert, M., von Blankenfeld, G., Seeburg, P.H., and Kettenmann, H., Inverse but not full benzodiazepine agonists modulate recombinant $\alpha_6\beta_2\gamma_2$ $GABA_A$ receptors in transfected human embryonic kidney cells. *Neurosci. Lett.,* 130:169–172 (1991).

Kofuji, P., Wang, J.B., Moss, S.J., Huganir, R.L., and Burt, D.R., Generation of two forms of the γ-aminobutyric acid receptorA γ2-subunit in mice by alternative splicing. *J. Neurochem.,* 56:713–715 (1991).

Kokka, N., Sapp, D.W., Taylor, A.M., and Olsen, R.W., The kindling model of alcohol dependence: similar persistent reduction in seizure threshold to pentylenetetrazol in animals receiving chronic ethanol or chronic pentylenetetrazol. *Alcoholism: Clin. Exp. Res.,* 17:525–531 (1993).

Korpi, E.R., Uusi-Oukari, M., Castren, E., Suzdak, P.D., Seppala, T., Sarviharju, M., and Tuominen, K., Cerebellar $GABA_A$ receptors in two rat lines selected for high and low sensitivity to moderate alcohol doses: pharmacological and genetic studies. *Alcohol,* 9:225–231 (1992).

Korpi, E.R., Kleingoor, C., Kettenmann, H., and Seeburg, P.H., Benzodiazepine-induced motor impairment linked to point mutation in cerebellar $GABA_A$ receptor. *Nature,* 361:356–359 (1993).

Krnjević, K., Chemical nature of synaptic transmission in vertebrates. *Physiol. Rev.,* 54:418–540 (1974).

Kurata, Y., Marszalec, W., Hamilton, B.J., Carter, D.B., and Narahashi, T., Alcohol modulation of cloned $GABA_A$ receptor-channel complex expressed in human kidney cell lines. *Brain Res.,* 631:143–146 (1993).

Kuriyama, K. and Ueha, T., Functional alterations in cerebral $GABA_A$ receptor complex associated with formation of alcohol dependence: analysis using GABA-dependent $^{36}Cl^-$ influx into neuronal membrane vesicles. *Alcohol Alcohol.,* 27:335–343 (1992).

Leidenheimer, N.J. and Harris, R.A., Acute effects of ethanol on $GABA_A$ receptor function: molecular and physiological determinants. In: *GABAergic Synaptic Transmission* (G. Biggio, A. Concas and E. Costa, eds.), Raven Press, New York (1992).

Liljequist, S. and Engel, J.A., Effects of GABAergic agonists and antagonists on various ethanol-induced behavioral changes. *Psychopharmacology,* 78:71–75 (1982).

Lin, A.M.-Y., Freund, R.K., and Palmer, M.R., Ethanol potentiation of GABA-induced electrophysiological responses in cerebellum: requirement for catecholamine modulation. *Neurosci. Lett.,* 122:154–158 (1991).

Lin, A.M.-Y., Freund, R.K., and Palmer, M.R., Sensitization of γ-aminobutyric acid-induced depressions of cerebellar Purkinje neurons to the potentiative effects of ethanol by *beta* adrenergic mechanisms in rat brain. *J. Pharmacol. Exp. Ther.,* 265:426–432 (1993a).

Lin, A.M.-Y, Bickford, P.C., and Palmer, M.R., The effects of ethanol on γ-aminobutyric acid-induced depressions of cerebellar Purkinje neurons: influence of *beta* adrenergic receptor action in young and aged Fischer 344 rats. *J. Pharmacol. Exp. Ther.,* 264:951–957 (1993b).

Lin, A.M.-Y., Bickford, P.C., Palmer, M.R., and Gerhardt, G.A., Ethanol inhibits the uptake of exogenous norepinephrine from the extracellular space of the rat cerebellum. *Neurosci. Lett.,* 164:71–75 (1993c).

Lin, L-H., Chen, L.L., Zirrolli, J.A., and Harris, R.A., General anesthetics potentiate GABA actions on $GABA_A$ receptors expressed by *Xenopus* oocytes: lack of involvement of intracellular calcium. *J. Pharmacol. Exp. Ther.,* 263:569–578 (1992).

Lin, L-H., Whiting, P., and Harris, R.A., Molecular determinants of general anesthetic action: role of $GABA_A$ receptor structure. *J. Neurochem.,* 60:1548–1553 (1993d).

Lüddens, H., Pritchett, D.B., Köhler, M., Killisch, I., Keinänen, K., Monyer, H., Sprengel, R., and Seeburg, P.H., Cerebellar $GABA_A$ receptor selective for a behavioural alcohol antagonist. *Nature,* 346:648–651 (1990).

Machu, T.K., Olsen, R.W., and Browning, M.D., Ethanol has no effect on cAMP-dependent protein kinase-, protein kinase C-, or Ca^{2+}-calmodulin-dependent protein kinase II-stimulated phosphorylation of highly purified substrates *in vitro. Alcoholism: Clin. Exp. Res.,* 15:1040–1044 (1991).

Marszalec, W., Kurata, Y., Hamilton, B.J., Carter, D.B., and Narahashi, T., Selective effects of alcohols on γ-aminobutyric acid A receptor subunits expressed in human embryonic kidney cells. *J. Pharmacol. Exp. Ther.,* 269:157–163 (1994).

Martz, A., Deitrich, R.A., and Harris, R.A., Behavioural evidence for the involvement of gamma-aminobutyric acid in the actions of ethanol. *Eur. J. Pharmacol.,* 89:53–62 (1983).

McCown, T.J. and Breese, G.R., A potential contribution to ethanol withdrawal kindling: reduced GABA function in the inferior collicular cortex. *Alcoholism: Clin. Exp. Res.,* 17:1290–1294 (1993).

McQuilkin, S.J. and Harris, R.A., Factors affecting actions of ethanol on GABA-activated chloride channels. *Life Sci.,* 46:527–541 (1990).

Mehta, A.K. and Ticku, M.K., Are $GABA_B$ receptors involved in the pharmacological effects of ethanol? *Eur. J. Pharmacol.,* 182:473–480 (1990).

Mehta, A.K. and Ticku, M.K., Ethanol enhancement of GABA-induced $^{36}Cl^-$ influx does not involve changes in Ca^{2+}. *Pharmacol. Biochem. Behav.,* 47:355–357 (1994).

Messing, R.O., Petersen, P.J., and Henrich, C.J., Chronic ethanol exposure increases levels of protein kinase C delta and epsilon and protein kinase C-mediated phosphorylation in cultured neural cells. *J. Biol. Chem.,* 266:23428–23432 (1991).

Mhatre, M.C., Mehta, A.K., and Ticku, M.K., Chronic ethanol administration increases the binding of the benzodiazepine inverse agonist and alcohol antagonist [^{3}H]Ro 15–4513 in rat brain. *Eur. J. Pharmacol.,* 153:141–145 (1988).

Mhatre, M.C. and Ticku, M.K., Chronic ethanol administration alters gamma-aminobutyric $acid_A$ receptor gene expression. *Mol. Pharmacol.,* 42:415–422 (1992).

Mhatre, M.C., Pena, G., Sieghart, W., and Ticku, M.K., Antibodies specific for $GABA_A$ receptor alpha subunits reveal that chronic alcohol treatment down-regulates alpha-subunit expression in rat brain regions. *J. Neurochem.,* 61:1620–1625 (1993).

Middaugh, L.D., Bao, K., Becker, H.C., and Daniel, S.S., Effects of Ro 15–4513 on ethanol discrimination in C57BL/6 mice. *Pharmacol. Biochem. Behav.,* 38:763–767 (1991).

Mihic, S.J., Wu, P.H., and Kalant, H., Differences among effects of sedative-hypnotic drugs on GABA-mediated chloride flux: quench flow studies. *Brain Res.,* 555:259–264 (1991).

Mihic, S.J., Wu, P.H., and Kalant, H., Potentiation of GABA-mediated chloride flux by pentobarbital and diazepam but not ethanol. *J. Neurochem.,* 58:745–751 (1992a).

Mihic, S.J., Kalant, H., Liu, J.-F., and Wu, P.H., Role of the GABA-receptor/chloride-channel complex in tolerance to ethanol and cross tolerance to diazepam and pentobarbital. *J. Pharmacol. Exp. Ther.,* 261:108–113 (1992b).

Mihic, S.J., Whiting, P.J., and Harris, R.A., Anaesthetic concentrations of alcohols potentiate $GABA_A$ receptor-mediated currents: lack of subunit specificity. *Eur. J. Pharmacol. — Mol. Pharmacol.,* 268:209–214 (1994a).

Mihic, S.J., McQuilkin, S.J., Eger, E.I., II, Ionescu, P., and Harris, R.A., Potentiation of $GABA_A$ receptor-mediated chloride currents by novel halogenated compounds correlates with their abilities to induce general anesthesia. *Mol. Pharmacol.,* 46:851–857 (1994b).

Montpied, P., Morrow, A.L., Karanian, J.W., Ginns, E.I., Martin, B.M., and Paul, S.M., Prolonged ethanol inhalation decreases gamma-aminobutyric $acid_A$ receptor alpha subunit mRNAs in the rat cerebral cortex. *Mol Pharmacol.,* 39:157–163 (1991).

Moody E.J., Suzdak, P.D., Paul, S.M., and Skolnick, P., Modulation of the benzodiazepine/gamma-aminobutyric acid receptor chloride channel complex by inhalation anesthetics. *J. Neurochem.,* 51:1386–1393 (1988).

Morelli, M., Deidda, S., Garau, L., Carboni, E., and Di Chiara, G., Ethanol: lack of stimulation of chloride influx in rat brain synaptoneurosomes. *Neurosci. Res. Commun.,* 2:77–84 (1988).

Morrisett, R.A. and Swartzwelder, H.S., Attenuation of hippocampal long-term potentiation by ethanol: a patch-clamp analysis of glutamatergic and GABAergic mechanisms. *J. Neurosci.,* 13:2264–2272 (1993).

Morrow, A.L., Montpied, P., Lingford-Hughes, A., and Paul, S.M., Chronic ethanol and pentobarbital administration in the rat: effects on $GABA_A$ receptor function and expression in brain. *Alcohol,* 7:237–244 (1990).

Morrow, A.L., Regulation of $GABA_A$ receptor function and gene expression in the central nervous system. In: *International Review of Neurobiology,* Vol. 38. (R.J. Bradley and R.A. Harris, eds.), Academic Press, San Diego (1995).

Morrow, A.L., Suzdak, P.D., Karanian, J.W., and Paul, S.M., Chronic ethanol administration alters gamma-aminobutyric acid, pentobarbital and ethanol-mediated $^{36}Cl^-$ uptake in cerebral cortical synaptoneurosomes. *J. Pharmacol. Exp. Ther.,* 246:158–164 (1988).

Nakahiro, M., Yeh, J.Z., Brunner, E., and Narahashi, T., General anesthetics modulate GABA receptor channel complex in rat dorsal root ganglion neurons. *FASEB J.,* 3:1850–1854 (1989).

Nakahiro, M., Arakawa, O., and Narahashi, T., Modulation of gamma-aminobutyric acid receptor-channel complex by alcohols. *J. Pharmacol. Exp. Ther.,* 259:235–240 (1991).

Nicoll, R.A., Presynaptic action of barbiturates in the frog spinal cord. *Proc. Natl. Acad. Sci. U.S.A.*, 72:1460–1463 (1975).

Nishio, M. and Narahashi, T., Ethanol enhancement of GABA-activated chloride current in rat dorsal root ganglion neurons. *Brain Res.*, 518:283–286 (1990).

Noda, M., Takahashi, H., Tanabe, T., Toyosato, M., Kikyotani, S., Furutani, Y., Hirose, T., Takashima, H., Iayama, S., Miyata, T., and Numa, S., Structural homology of *Torpedo californica* acetylcholine receptor subunits. *Nature*, 302:528–532 (1983).

Olsen, R.W. and Tobin, A.J., Molecular biology of $GABA_A$ receptors. *FASEB J.*, 4:1469–1480 (1990).

Osmanovic, S.S. and Shefner, S.A., Enhancement of current induced by superfusion of GABA in locus coeruleus neurons by pentobarbital, but not ethanol. *Brain Res.*, 517:324–329 (1990).

Palmer, M.R. and Hoffer, B.J., GABAergic mechanisms in the electrophysiological actions of ethanol on cerebellar neurons. *Neurochem. Res.*, 15:145–151 (1990).

Peduto, V.A., Concas, A., Santoro, G., Biggio, G., and Gessa, G.L., Biochemical and electrophysiologic evidence that propofol enhances GABAergic transmission in the rat brain. *Anesthesiology*, 75:1000–1009 (1991).

Porter, N.M., Angelotti, T.P., Twyman, R.E., and Macdonald, R.L., Kinetic properties of $\alpha_1\beta_1\gamma$-aminobutyric $acid_A$ receptor channels expressed in Chinese hamster ovary cells: regulation by pentobarbital and picrotoxin. *Mol. Pharmacol.*, 42:872–881 (1992).

Pritchett, D.B. and Seeburg, P.H., Gamma-aminobutyric $acid_A$ receptor alpha 5-subunit creates novel type II benzodiazepine receptor pharmacology. *J. Neurochem.*, 54:1802–1804 (1990).

Pritchett, D.B., Sontheimer, H., Shivers, B.D., Ymer, S., Kettenmann, H., Schofield, P.R., and Seeburg, P.H., Importance of a novel $GABA_A$ receptor subunit for benzodiazepine pharmacology. *Nature*, 338:582–585 (1989).

Proctor, W.R., Allan, A.M., and Dunwiddie, T.V., Brain region-dependent sensitivity of $GABA_A$ receptor-mediated responses to modulation by ethanol. *Alcoholism: Clin. Exp. Res.*, 16:480–489 (1992a).

Proctor, W.R., Soldo, B.L., Allan, A.M., and Dunwiddie, T.V., Ethanol enhances synaptically evoked $GABA_A$ receptor-mediated responses in cerebral cortical neurons in rat brain slices. *Brain Res.*, 595:220–227 (1992b).

Prunell, M., Escorihuela, R.M., Fernandez-Teruel, A., Nunez, J.F., and Tobena, A., Differential interactions between ethanol and Ro 15–4513 on two anxiety tests in rats. *Pharmacol. Biochem. Behav.*, 47:147–151 (1994).

Puia, G., Vicini, S., Seeburg, P.H., and Costa, E., Influence of recombinant γ-aminobutyric acid-A receptor subunit composition on the action of allosteric modulators of γ-aminobutyric acid-gated Cl^- currents. *Mol. Pharmacol.*, 39:691–696 (1991).

Rassnick, S., D'Amico, E., Riley, E., and Koob, G.F., GABA antagonist and benzodiazepine partial inverse agonist reduce motivated responding for ethanol. *Alcoholism: Clin. Exp. Res.*, 17:124–130 (1993).

Rees, D.C. and Balster, R.L., Attenuation of the discriminative stimulus properties of ethanol and oxazepam, but not of pentobarbital, by Ro 15–4513 in mice. *J. Pharmacol. Exp. Ther.*, 244:592–598 (1988).

Reynolds, J.N. and Prasad, A., Ethanol enhances $GABA_A$ receptor-activated chloride currents in chick cerebral cortical neurons. *Brain Res.*, 564:138–142 (1991).

Reynolds, J.N., Prasad, A., and MacDonald, J.F., Ethanol modulation of GABA receptor-activated Cl^- currents in neurons of the chick, rat and mouse central nervous system. *Eur. J. Pharmacol.*, 224:173–181 (1992).

Risinger, F.O., Malott, D.H., Riley, A.L., and Cunningham, C.L., Effect of Ro 15–4513 on ethanol-induced conditioned place preference. *Pharmacol. Biochem. Behav.*, 43:97–102 (1992).

Sanna, E., Serra, M., Cossu, A., Colombo, G., Follesa, P., Cuccheddu, T., Concas, A., and Biggio, G., Chronic ethanol intoxication induces differential effects on $GABA_A$ and NMDA receptor function in the rat brain. *Alcoholism: Clin. Exp. Res.*, 17:115–123 (1993).

Schofield, P.R., Darlison, M.G., Fujita, N., Burt, D.R., Stephenson, F.A., Rodriguez, H., Rhee, L.M., Ramachandran, J., Reale, V., Glencorse, T.A., Seeburg, P.H., and Barnard, E.A., Sequence and functional expression of the $GABA_A$ receptor shows a ligand-gated receptor super-family. *Nature*, 328:221–227 (1987).

Shingai, R., Sutherland, M.L., and Barnard, E.A., Effects of subunit types of the cloned $GABA_A$ receptor on the response to a neurosteroid. *Eur. J. Pharmacol., — Mol. Pharmacol.*, 206:77–80 (1991).

Sieghart, W., $GABA_A$ receptors: ligand-gated Cl^- ion channels modulated by multiple drug-binding sites. *Trends Pharmacol.*, 13:446–450 (1992).

Sigel, E., Baur, R., and Malherbe, P., Recombinant $GABA_A$ receptor function and ethanol. *FEBS Lett.*, 324:140–142 (1993).

Simson, P.E., Criswell, H.E., and Breese, G.R., Ethanol potentiates γ-aminobutyric acid-mediated inhibition in the inferior colliculus: evidence for local ethanol/γ-aminobutyric acid interactions. *J. Pharmacol. Exp. Ther.*, 259:1288–1293 (1991).

Sivam, S.P., Nabeshima, T., and Ho, I.K., Acute and chronic effects of pentobarbital in relation to postsynaptic GABA receptors: a study with muscimol. *J. Neurosci. Res.*, 7:37–47 (1982).

Soldo, B.L., Proctor, W.R., and Dunwiddie, T.V., Ethanol differentially modulates $GABA_A$ receptor-mediated chloride currents in hippocampal, cortical and septal neurons in rat brain slices. *Synapse*, 18:94–103 (1994).

Study, R.E. and Barker, J.L., Diazepam and (-) pentobarbital: fluctuation analysis reveals different mechanisms for potentiation of γ-aminobutyric acid responses in cultured central neurons. *Proc. Natl. Acad. Sci. U.S.A.*, 78:7180–7184 (1981).

Suzdak, P.D., Schwartz, R.D., Skolnick, P., and Paul, S.M., Ethanol stimulates gamma-aminobutyric acid receptor-mediated chloride transport in rat brain synaptoneurosomes. *Proc. Natl. Acad. Sci. U.S.A.,* 83:4071–4075 (1986a).

Suzdak, P.D., Glowa, J.R., Crawley, J.N., Schwartz, R.D., Skolnick, P., and Paul, S.M., A selective imidazobenzodiazepine antagonist of ethanol in the rat. *Science,* 234:1243–1247 (1986b).

Syapin, P.J., Jones, B.L., Kobayashi, L.S., Finn, D.A., and Alkana, R.L., Interactions between benzodiazepine antagonists, inverse agonists, and acute behavioral effects of ethanol in mice. *Brain. Res. Bull.,* 24:705–709 (1990).

Takada, R., Saito, K., Matsuura, H., and Inoki, R., Effect of ethanol on hippocampal GABA receptors in the rat brain. *Alcohol,* 6:115–119 (1989).

Ticku, M.K., Lowrimore, P., and Lehoullier, P., Ethanol enhances GABA-induced ^{36}Cl influx in primary spinal cord cultured neurons. *Brain. Res. Bull.,* 17:123–126 (1986).

Turner, D.M., Ransom, R.W., Yang, J.S., and Olsen, R.W., Steroid anesthetics and naturally occurring analogs modulate the gamma-aminobutyric acid receptor complex at a site distinct from the barbiturates. *J. Pharmacol. Exp. Ther.,* 248:960–966 (1989).

Ueha, T. and Kuriyama, K., Direct action of ethanol on cerebral $GABA_A$ receptor complex: analysis using purified and reconstituted $GABA_A$ receptor complex. *Neurochem. Intl.,* 19: 319–325 (1991).

Wafford, K.A., Burnett, D.M., Dunwiddie, T.V., and Harris, R.A., Genetic differences in the ethanol sensitivity of $GABA_A$ receptors expressed in *Xenopus* oocytes. *Science,* 249:291–293 (1990).

Wafford, K.A., Burnett, D.M., Leidenheimer, N.J., Burt, D.R., Wang, J.B., Kofuji, P., Dunwiddie, T.V., Harris, R.A., and Sikela, J.M., Ethanol sensitivity of the $GABA_A$ receptor requires 8 amino acids contained in the γ2L subunit. *Neuron,* 7:27–33 (1991).

Wafford, K.A. and Whiting, P.J., Ethanol potentiation of $GABA_A$ receptors requires phosphorylation of the alternatively spliced variant of the gamma 2 subunit. *FEBS Lett.,* 313:113–117 (1992).

Wehner, J.M., Pounder, J.I., and Bowers, B.J., The use of recombinant inbred strains to study mechanisms of drug action. *J. Addict. Dis.,* 10:89–107 (1991).

Weiner, J.L., Zhang, L., and Carlen, P.L., Potentiation of $GABA_A$-mediated synaptic current by ethanol in hippocampal CA1 neurons: possible role of protein kinase C. *J. Pharmacol. Exp. Ther.,* 268:1388–1395 (1994a).

Weiner, J.L., Zhang, L., and Carlen, P.L., Guanine nucleotide phosphate analogs modulate ethanol potentiation of $GABA_A$-mediated synaptic current in hippocampal CA1 neurons. *Brain Res.,* 665:307–310 (1994b).

White, G., Lovinger, D.M., and Weight, F.F., Ethanol inhibits NMDA-activated current but does not alter GABA-activated current in an isolated adult mammalian neuron. *Brain Res.,* 507:332–336 (1990).

Whiting, P., McKernan, R.M., and Iversen, L.L., Another mechanism for creating diversity in γ-aminobutyrate type A receptors: RNA splicing directs expression of two forms of γ2 subunit, one of which contains a protein kinase C phosphorylation site. *Proc. Natl. Acad. Sci. U.S.A.,* 87:9966–9970 (1990).

Williamson, R.E. and Pritchett, D.B., Levels of benzodiazepine receptor subtypes and $GABA_A$ receptor alpha-subunit mRNA do not correlate during development. *J. Neurochem.,* 63:413–418 (1994).

Wood, A.L., Healey, P.A., Menendez, J.A., Verne, S.L., and Atrens, D.M., The intrinsic and interactive effects of RO 15–4513 and ethanol on locomotor activity, body temperature, and blood glucose concentration. *Life Sci.,* 45:1467–1473 (1989).

Zahniser, N.R., Buck, K.J., Curella, P., McQuilkin, S.J., Wilson-Shaw, D., Miller, C.L., Klein, R.L., Heidenreich, K.A., Keir, W.J., Sikela, J.M., and Harris, R.A., $GABA_A$ receptor function and regional analysis of subunit mRNAs in long-sleep and short-sleep mouse brain. *Mol. Brain Res.,* 14:196–206 (1992).

Zhao, T.J., Chiu, T.H., and Rosenberg, H.C., Reduced expression of gamma-aminobutyric acid type A/benzodiazepine receptor gamma 2 and alpha 5 subunit mRNAs in brain regions of flurazepam-treated rats. *Mol. Pharmacol.,* 45:657–663 (1994).

Chapter 5

ETHANOL AND GLUTAMATE RECEPTORS

Boris Tabakoff and Paula L. Hoffman

TABLE OF CONTENTS

I. INTRODUCTION

The last decade has seen a revolution in the conceptualization of the mechanisms by which ethanol produces its neuropharmacologic actions. Although there continues to be no evidence of a proteinaceous receptor for ethanol, the selective effects of ethanol, at concentrations <100 mg% (approximately 22 m*M*), on ligand-gated ion channels can certainly qualify these proteins as "receptive elements" (Tabakoff and Hoffman, 1987) for the expression of ethanol's actions. This chapter will be devoted to an examination of the acute and chronic effects of ethanol on ionotropic glutamate receptors, but a demonstration of a chemical's (ethanol's) actions on a receptor system does not, in and of itself, demonstrate that ethanol produces its physiologic/pharmacologic effects via such actions.

The hypothesis of involvement of the glutamate receptor-activated ion channels in the actions of ethanol within the CNS arises from two lines of reasoning. The first relates to the fact that glutamatergic transmission is the major mechanism for generating excitatory signals within the brain, and these signals are critical in functions, such as the initial stages of memory (Collingridge 1985; Morris et al., 1986), and in CNS neuronal development in the early stages of life (McDonald and Johnston, 1990; Collingridge and Lester, 1989). Excessive signalling through glutamatergic mechanisms has been linked to pathologic events, such as seizures and

0-8493-8389-7/95/$0.00+$.50

neuronal damage, and death (Isokawa and Levesque, 1991; Choi, 1992). Ethanol interferes with establishment of memory (Walker and Hunter, 1978; Lister et al., 1991) and with proper neuronal development (Abel, 1980). After chronic ethanol ingestion, withdrawal seizures are evidenced (Tabakoff and Rothstein, 1983), and chronic ethanol ingestion and withdrawal produce neuronal damage (Walker et al., 1980; Charness, 1993). This spectrum of ethanol's actions well fits the events mediated by glutamatergic transmission, and one can speculate that ethanol is modulating glutamatergic transmission to produce effects on memory, neuronal development, CNS withdrawal hyperexcitability, and neuronal damage.

The second line of reasoning that implicates glutamate systems in the actions of ethanol derives from experiments in which the discriminative stimulus properties of certain glutamate receptor antagonists (e.g., the NMDA receptor antagonists, dizocilpine, ketamine, and phencyclidine) were investigated in animals trained to discriminate moderate doses of ethanol from vehicle (Grant et al., 1991; Schechter et al., 1993; Sanger, 1993). The NMDA receptor antagonists fully "substituted" for ethanol in experiments with pigeons, mice, and rats, and such results were interpreted to indicate that ethanol and the NMDA receptor antagonists may have a common mechanism of action. The data on discriminative stimulus properties of NMDA receptor antagonists and ethanol complement the data regarding similarities in physiologic and behavioral actions of the NMDA antagonists and ethanol. For instance, ketamine and phencyclidine at relatively low doses produce ataxia and have anxiolytic effects, and at higher doses these drugs have anticonvulsant properties (Balster, 1987). More recently the effects of ethanol and NMDA receptor antagonists were studied (Balster et al., 1993) in lines of mice selectively bred for sensitivity and resistance to ethanol withdrawal-induced convulsions (WSP, withdrawal seizure prone mice; WSR, withdrawal seizure resistant mice [Phillips et al., 1989]). WSP mice have been shown to have greater numbers of NMDA receptors in certain brain areas (e.g., hippocampus) than WSP mice (Valverius et al., 1990). Ethanol and competitive and noncompetitive NMDA receptor antagonists reduced responding on a fixed-ratio performance task in WSR mice at doses which produced no effect on responding in WSP mice (Balster et al., 1993). The authors explained their results by postulating similar inhibitory actions of ethanol and NMDA antagonists at the NMDA receptors and used the evidence for greater numbers of NMDA receptors in brains of WSP mice to explain the greater resistance of the WSP mice to the effects of ethanol and the NMDA antagonists. The similarities of the actions of ethanol and NMDA antagonists in discriminative stimulus experimental paradigms (Grant et al., 1991) and in animals selectively bred for differential responses to ethanol (Balster et al., 1993) provide further evidence for commonalities in mechanism of action of NMDA antagonists and ethanol and additionally link ethanol's effects to glutamatergic neurotransmission.

II. GLUTAMATE RECEPTORS

A. STRUCTURAL AND FUNCTIONAL FEATURES OF NON-NMDA RECEPTORS

At least four subtypes of glutamate receptor have been described based on relative affinities for various agonists. These include the NMDA, kainate, quisqualate/AMPA, and 4-aminophosphonobutyrate (APB) receptors (Collingridge and Lester, 1989). These glutamate receptors can be classified as ionotropic (NMDA, kainate, AMPA receptors: ligand-gated ion channels) or metabotropic (the APB receptor appears to be one form of metabotropic receptor; these are G protein-coupled receptors linked to phospholipase C and adenylyl cyclase) (see Hollmann and Heinemann, 1994). Most work with ethanol to date has focused on the ionotropic glutamate receptors. Of these receptors, the non-NMDA receptors mediate fast excitatory neurotransmission. These receptor subunits were the first to be cloned and have now been characterized and organized into three structural classes (Heinemann et al., 1991; Dingledine, 1991; Hollmann and Heinemann, 1994). The first class consists of the subunits called GluR1 through GluR4. The pharmacological characteristics of these subunits (i.e., responses of expressed subunits to kainate and alpha-amino-3-hydroxy-5-methyl-4-isoxazole propionic acid [AMPA], an agonist at the "quisqualate" subtype of receptor [Collingridge and Lester, 1989]), indicated that receptors formed from these subunits are "AMPA-preferring": AMPA is the most potent agonist, although kainate produces a greater maximal response. Another class of subunits consists of GluR5–7, which share 75–80% sequence homology among themselves and only about 40% with the other four GluR subunits (Hollmann and Heinemann, 1994). The ligand binding and electrophysiological properties of receptors formed from these expressed subunits suggest that they represent the "low-affinity kainate receptor," as characterized in brain, at which AMPA is not an agonist or only a weak agonist (see Hollmann and Heinemann, 1994). High-affinity kainate receptors, designated KA-1 and KA-2, have also been cloned and apparently represent a third subfamily of receptors. These proteins do not form functional homomeric channels when expressed, but membrane preparations from cells transfected with KA-1 or KA-2 display high-affinity kainate binding. The combined expression of KA-2 with GluR5 and/or GluR6 produces functional receptors; receptors composed of KA-2 and GluR6 respond to AMPA.

While these data might suggest that there are distinct kainate and AMPA receptors, as was originally proposed on the basis of pharmacological data, only homomeric receptors have been found to display a selective response to a single agonist (e.g., GluR6 homomers respond only to kainate, not AMPA). Hollmann and Heinemann (1994) suggest that, if heteromeric receptors are the rule, there may be no "pure kainate" or "pure AMPA" receptors *in vivo*. It is not known how the non-NMDA glutamate receptor subunits are assembled *in vivo*, although immunoprecipitation studies have demonstrated association of KA2, GluR6, and/or GluR7 (see Hollmann and Heinmann, 1994).

There is some evidence that the non-NMDA glutamate receptors can be modified by phosphorylation and that this modification may affect receptor function. For example, one splice variant of GluR1, expressed in human embryonic kidney (HEK) 293 cells, was found to be phosphorylated on serine residues under basal conditions (Moss et al., 1993). Although activators of protein kinase A (PKA) have been reported to alter glutamate receptor responses in brain (Greengard et al., 1991; Wang et al., 1991; Cerne et al., 1992) and in cells expressing GluR1 and GluR3 (Keller et al., 1992), there was no change in phosphorylation of GluR1 upon treatment of cells with activators of either protein kinase A or C. However, cotransfection of GluR1 with protein tyrosine kinase produced phosphorylation of tyrosine residues in GluR1 (Moss et al., 1993). In a study in which the GluR6 subunit was expressed in HEK 293 cells, it was found that PKA enhanced kainate responses and that elimination of two serine residues in GluR6 by site-directed mutagenesis could block the effect of PKA (Wang et al., 1993; Raymond et al., 1993). In cultured hippocampal neurons, GluR subunits 1–3 (AMPA-preferring receptors) were phosphorylated by Ca^{++}-calmodulin-dependent protein kinase II (CaM-kinase II), which in turn was activated by Ca^{++} entering through the NMDA receptor-coupled ion channel (see below) (McGlade-McCulloh et al., 1993; Tan et al., 1994). It has also been reported that phorbol ester treatment inhibited the response to kainate in *Xenopus* oocytes expressing GluR1 plus 3, GluR2 plus 3, or GluR6, with the magnitude of inhibition greatest in the latter case (Dildy-Mayfield and Harris, 1994a), although phorbol ester treatment did not affect kainate responses in cultured cerebellar granule cells (Snell et al., 1994b). Overall, these data suggest the possibility that the function of the non-NMDA glutamate receptors can be regulated by phosphorylation (see Chapter 7).

B. STRUCTURAL AND FUNCTIONAL FEATURES OF NMDA RECEPTORS

Recently, a number of subunits of the NMDA receptor have also been cloned. The first subunit identified in rat was NR1, and the homolog in mouse was called ζ (Nakanishi 1992; Moriyoshi et al., 1991; Yamazaki et al., 1992). These subunits are widely distributed in brain (Nakanishi 1992), and when mRNA encoding NR1 or ζ is expressed in *Xenopus* oocytes, a homomeric receptor is formed which displays many of the characteristics of the native NMDA receptor but which produces small amplitude currents in the presence of agonist (Nakanishi 1992; Moriyoshi et al., 1991; Yamazaki et al., 1992). A second family of NMDA receptor subunits, designated NR2 in the rat and ε in the mouse (Kutsuwada et al., 1992; Monyer et al., 1992) has also been characterized. There are four subtypes of this subunit (NR2 A-D; ε 1–4 [Nakanishi, 1992; Kutsuwada et al., 1992; Monyer et al., 1992; Ikeda et al., 1992; Ishii et al., 1993]), and the distribution of mRNA for these subunits is localized to particular brain areas (Nakanishi, 1992; Kutsuwada et al., 1992; Monyer et al., 1992). The expression in *Xenopus* oocytes of NR2 or ε subunit mRNA alone does not produce functional receptors; however, expression of NR1 with various members of the NR2 family results in receptors with characteristics that more nearly resemble those of native receptors (Kutsuwada et al., 1992; Monyer et al., 1992). The various heteromeric receptors have different properties (e.g., affinity for agonists or antagonists; sensitivity to glycine [see below]) which depend on the member of the NR2 or ε family which is expressed along with NR1 or ζ (Kutsuwada et al., 1992; Monyer et al., 1992). It has also been reported that eight splice variants of NR1 exist, adding to the complexity of receptor structure and function (Sugihara et al., 1992; Hollmann et al., 1993). In addition to these NMDA receptor subunits, a "glutamate binding protein" has been cloned and has been suggested to be one subunit of an NMDA receptor that consists of several proteins (Kumar et al., 1991). There is no homology between the glutamate binding protein and the other NMDA receptor subunits, and it has been postulated that the complex of which this protein is a part represents a distinct type of NMDA receptor (Kumar et al., 1991). A further complexity has recently been added to studies of NMDA receptors. Smirnova et al. (1993a) have identified and cloned a presynaptic glutamate (NMDA) receptor protein. This protein has been considered important for modulating release of neurotransmitters (Smirnova et al., 1993b) but has structural features (e.g., a single transmembrane region) and certain functional characteristics (e.g., incomplete Mg^{++} block) which significantly distinguish it from the earlier identified NMDA receptor subunits.

The pharmacological characteristics of the postsynaptic NMDA receptor have already been well described (Collingridge and Lester, 1989). The receptor is coupled to an ion channel which, when activated, is permeable to calcium as well as monovalent cations. The function of the NMDA receptor is voltage-dependent, meaning that the response to NMDA is increased as the cell is depolarized. Thus, for example, synaptically released glutamate interacting with ionotropic non-NMDA receptors will result in rapid depolarization of the postsynaptic cell, promoting activation of the NMDA receptor by glutamate. The voltage dependence is a result of the binding of Mg^{++} within the NMDA receptor-coupled channel. Mg^{++} blocks the channel but is released upon cellular depolarization. The NMDA receptor-channel complex also contains binding sites for a number of compounds that can "allosterically" affect receptor function. Glycine binds to a strychnine-insensitive site and enhances the action of NMDA; in fact, glycine is postulated to be *required* for NMDA action, and glycine and NMDA or glutamate may be thought of as coagonists at the receptor (Kleckner and Dingledine, 1988). Within the ion channel there is a binding site for the dissociative anesthetic phencyclidine (PCP), which blocks the channel, and the complex also contains binding sites for Zn^{++}, which inhibits NMDA action, and for polyamines, which enhance the response to NMDA (Collingridge and Lester, 1989).

Protein kinase activity also modulates NMDA receptor function. The NR1 and 2 subunits (and the corresponding subunits in the mouse) all contain a number of consensus sequences for phosphorylation by Ca^{++}-calmodulin-dependent protein kinase and PKC (Nakanishi 1992; Kutsuwada et al., 1992; Monyer et al., 1992; Ikeda et al., 1992). It has been reported that direct (phorbol ester) or indirect (through agonists that increase polyphosphoinositide metabolism and diacylglycerol production) stimulation of PKC can increase electrophysiological responses to NMDA in trigeminal neurons (Chen and Huang, 1991), in embryonic striatal neurons (Murphy et al., 1994), and in *Xenopus* oocytes expressing whole brain mRNA or mRNAs coding for particular NMDA receptor subunits (Kutsuwada et al., 1992; Urushihara et al., 1992). Both PKC and CaM-kinase II were found to enhance glutamate or NMDA stimulation of the binding of the NMDA receptor antagonist, dizocipline, which recognizes a site within the ion channel, in postsynaptic densities of rat brain (Kitamura et al., 1993). The effect of PKC stimulation can vary depending on alternative splicing and on the subunit composition of the receptor. For example, the greatest potentiation of NMDA responses by phorbol ester was observed with the splice variant of NR1 that contains an N-terminal insertion and the shortest COOH-terminal region (lacking C-terminal inserts) (Durand et al., 1993). The importance of receptor subunit composition was demonstrated by the findings that, while NMDA activation of expressed heteromeric ζ/ε1 or ζ/ε2 receptors was increased by phorbol ester treatment, activation of the ζ/ε3 or ζ/ε4 receptors was not (Kutsuwada et al., 1992; Mori et al., 1993). We and others have recently found that phorbol ester treatment of cerebellar granule cells in primary culture *decreases* the response to NMDA (Snell et al., 1994a; Courtney and Nicholls 1992). It has also been reported that phorbol esters decrease the electrophysiological response to NMDA in rat CA1 hippocampal neurons (Makram and Segal 1992).

Tingley et al. (1993) demonstrated that PKC phosphorylated NR1 and that most of the phosphorylation sites occurred on the C-terminal portion of the protein, which had been postulated to be localized extracellularly. This finding necessitated the generation of a new model of receptor topography, in which the C-terminal portion of the protein is intracellular (Tingley et al., 1993). However, although the NR1 protein can be phosphorylated, there is evidence that this phosphorylation does not account for the modulation of NMDA receptor function by PKC. When a mutated form of the ζ1 subunit, from which the phosphorylation sites had been eliminated, was expressed alone or in combination with the ε2 subunit in *Xenopus* oocytes, the response to NMDA was still potentiated by treatment with a phorbol ester activator of PKC (Yamakura et al., 1993). Similarly, when a mutant form of NR1 lacking four consensus sites for PKC phosphorylation in the C-terminal region was expressed in oocytes in combination with NR2A, receptor function was potentiated by phorbol ester to the same degree as the function of the expressed NR1/NR2A receptor that contained the consensus sites (Sigel et al., 1994). The elimination of consensus sites within the putative intracellular region between the third and fourth transmembrane segments of NR1 or NR2A also did not influence the response of the expressed receptor to phorbol ester treatment (Sigel et al., 1994). These experiments suggest that phosphorylation of the NMDA receptor subunits per se is not the basis for modulation of receptor activity by PKC. On the other hand, there is evidence from studies of receptor chimeras that the C-terminal region of the NR2 subunits *is* important for potentiation of NMDA responses by PKC (Mori et al., 1993), and the role of this protein region needs to be further investigated (see Chapter 7).

The activity of NMDA receptors can also be potentiated by reducing agents such as dithiothreitol (DTT) and inhibited by oxidizing agents such as dithionitrobenzoic acid (DTNB), consistent with the presence of a redox site on the receptor which may modulate NMDA receptor function in response to *in vivo* oxidizing or reducing agents (Aizenman et al., 1989; Reynolds et al., 1990). Heteromeric receptors (containing both NR1 and NR2 subunits) are more sensitive to DTT and DTNB than homomeric receptors, and differences in sensitivity among NR2 subunits may also exist (see McBain and Mayer, 1994).

III. ACUTE EFFECTS OF ETHANOL

A. POSTSYNAPTIC NMDA RECEPTORS

Biochemical and electrophysiological studies have demonstrated a very consistent effect of ethanol, acutely, to inhibit NMDA receptor function. Whole cell patch clamp studies in dissociated hippocampal neurons in culture showed that ethanol at low concentrations (threshold, 5 m*M*) inhibited NMDA-induced ion currents (Lovinger et al., 1989; Lima-Lindman and

Albuquerque, 1989). The electrophysiological studies have since been extended to neurons and brain slices from adult animals, with similar results (Lovinger et al., 1990). In addition, it has been demonstrated that ethanol can inhibit the activation of some medial septal neurons by iontophoretically applied NMDA *in vivo* (Simson et al., 1991).

Ethanol also inhibits biochemical responses to NMDA. In cerebellar granule cells in culture, ethanol inhibited NMDA-stimulated calcium uptake, measured both with fura-2 and with $^{45}Ca^{++}$, as well as cyclic GMP production (Hoffman et al., 1989a,b). The effects of ethanol occurred at very low concentrations, with a threshold of 10–25 m*M*. Similar results have been found in dissociated whole brain cells from neonatal rats (Dildy-Mayfield and Leslie, 1989). Thus, diverse studies from a number of laboratories have clearly shown that ethanol is a potent inhibitor of NMDA responses in the brain and in neuronal cells in culture. It has also been reported that ethanol inhibits the response to glutamate at a peripheral NMDA receptor in the guinea pig ileum myenteric plexus (Frye, 1991).

The sensitivity of the NMDA receptor to inhibition by ethanol has led to a search for the mechanism of action of ethanol at this receptor. Originally, in studies of NMDA-stimulated cyclic GMP production in primary cultures of cerebellar granule cells, it was reported that ethanol did not interact with the phencyclidine (PCP) binding site (Hoffman et al., 1989a). In a recent study using dissociated brain cells, however, the inhibition produced by ethanol and dizocilpine, which binds to the PCP site, was not completely additive, suggesting a possible interaction of ethanol with the PCP site (Dildy-Mayfield and Leslie, 1991). Studies of calcium uptake in cerebellar granule cells, as well as electrophysiological investigations in cultured hippocampal neurons, indicated a lack of interaction between the inhibition caused by ethanol and that caused by Mg^{++} (Rabe and Tabakoff, 1990; Lovinger et al., 1990). On the other hand, when electrophysiological measurements were made using slices of rat hippocampus, ethanol inhibition of NMDA responses was found to be enhanced in the presence of Mg^{++} (Martin et al., 1991). The data obtained when ethanol and Mg^{++} were covaried in these experiments, however, suggested different mechanisms of action for the two compounds (Morrisett et al., 1991). Enhancement of ethanol inhibition of NMDA responses (stimulation of nitric oxide synthase activity) by Mg^{++}, as well as by Zn^{++}, has also recently been reported in experiments with cultured rat cerebral cortical cells (Chandler et al., 1994). Ethanol does not appear to act directly at the NMDA binding site, since, in most instances, the inhibition does not follow a "competitive" pattern (Rabe and Tabakoff, 1990; Dildy-Mayfield and Leslie, 1991). Furthermore, analysis of ligand binding to various sites on the NMDA receptor complex in mouse hippocampal and cerebral cortical membranes (glutamate, dizocilpine, and glycine binding sites) showed no effect of 100 m*M* ethanol on the K_D or B_{max} for these ligands when equilibrium binding studies were carried out (Snell et al., 1993). There is one report that very low concentrations of ethanol (0.1–10 μ*M*) *increased* the affinity for dizocipline in rat cerebral cortical membranes in the absence of added glutamate or glycine; however, the physiological significance of the effect of these low ethanol concentrations is not clear (DeMontis et al., 1991). Polyamine enhancement of NMDA receptor responses, measured as an increase in NMDA-stimulated c-*fos* expression *in vivo,* was reduced by acute ethanol administration at a dose of 2 g/kg, suggesting a role for the polyamine site in receptor sensitivity to ethanol (Matsumoto et al., 1993). In a recent study, the oxidizing agent DTNB, which had little effect *per se* on NMDA-stimulated norepinephrine release from rat hippocampal slices, increased the inhibitory potency of ethanol, suggesting that the redox site on the receptor might play a role in the action of ethanol (Woodward, 1994a).

The most convincing data at present regarding a site of action of ethanol at the NMDA receptor, however, come from studies of the effect of glycine on ethanol-induced inhibition. In primary cultures of cerebellar granule cells, glycine was shown to reverse the ethanol-induced inhibition of NMDA-stimulated calcium uptake (Rabe and Tabakoff, 1990; Snell et al., 1994b). A structure-activity analysis confirmed that glycine interaction with the "coagonist" site was involved in the reversal of the ethanol effect (Rabe and Tabakoff, 1990).

Analysis of the glycine dose-response curve revealed that ethanol significantly decreased the potency of glycine to enhance the response to NMDA (Snell et al., 1994b). Thus, the inhibitory effect of ethanol on NMDA receptor responses at low glycine concentrations may be attributed to the decreased effect of glycine. Glycine has also been found to reverse ethanol inhibition in studies of NMDA-stimulated increases in intracellular calcium in dissociated neonatal brain cells (Dildy-Mayfield and Leslie, 1991). Furthermore, in mouse brain membranes, ethanol was shown to interfere with ligand binding to the NMDA receptor-coupled channel, when binding was analyzed under *nonequilibrium* conditions (Snell et al., 1993). In the absence of glycine, ethanol inhibited dizocipline binding to mouse cerebral cortical and hippocampal membrane preparations, and this inhibition could be overcome by increasing concentrations of glycine (Snell et al., 1993). These data suggest a specific site of action of ethanol at the NMDA receptor, which could involve interference with the action of the coagonists.

The mechanism of ethanol inhibition of NMDA receptor function and interaction with the glycine site was further examined using cerebellar granule cells. Although, as mentioned, ethanol reduced glycine potency in these cells (Snell et al., 1994b), ethanol did not directly affect *equilibrium* binding of ligands to the glycine site on the NMDA receptor in brain membranes (Snell et al., 1993). It was observed that treatment of cerebellar granule cells with a phorbol ester, which, as noted above, inhibited the response to NMDA in these cells, also reduced the potency of glycine to enhance NMDA-stimulated increases in intracellular Ca^{++} (Snell et al., 1994a). Both the inhibitory effect of ethanol and that of the phorbol ester could be reversed by PKC inhibitors, and the inhibitory effects of ethanol and phorbol ester were not additive (Snell et al., 1994a,b). These results suggested a similar mechanism of action for ethanol and phorbol esters in the inhibition of NMDA receptor responses in cerebellar granule cells and an involvement of PKC in the response to ethanol.

However, it appears that interference with the glycine-NMDA interaction is not the only mechanism by which ethanol can inhibit NMDA receptor function. Glycine did not reverse the effect of ethanol on NMDA-stimulated neurotransmitter release from cerebral cortical or hippocampal slices (Gonzales and Woodward, 1990; Woodward, 1994b) and did not alter ethanol inhibition of NMDA-activated current in cultured mouse hippocampal neurons (Weight et al., 1991) or ethanol inhibition of electrophysiological responses to NMDA in slices of adult rat hippocampus (Martin et al., 1991). In primary cultures of cerebral cortical cells, obtained from embryonic rats, glycine did not overcome the inhibitory effect of ethanol on NMDA-stimulated increases in intracellular Ca^{++} (Bhave et al., 1994). Furthermore, in these cells, phorbol ester exposure *increased* the response to NMDA at low glycine concentrations. The data suggest the possibility that factors such as the subunit composition of the NMDA receptor, and/or the characteristics of PKC present in different cell types, may influence the mechanism of ethanol inhibition of NMDA receptor function.

The suggestion that ethanol may inhibit NMDA receptor function via different mechanisms in different cell types is also supported by the finding that sensitivity to ethanol varies in brain regions. For example, ethanol was a much less potent inhibitor of NMDA receptor function in cerebral cortical cells in primary culture than in cerebellar granule cells (Bhave et al., 1994). Electrophysiological studies in rats have also demonstrated that systemic administration of ethanol produces variable amounts of inhibition of NMDA responses among cells in the medial septum and that locally applied ethanol inhibits NMDA-evoked responses in the inferior colliculus and hippocampus but not the lateral septum (Simson et al., 1991; 1993). There have been some investigations of the contribution of structural characteristics of the NMDA receptor to its sensitivity to ethanol. Four of the splice variants of NR1 were expressed in *Xenopus* oocytes, and inhibition of NMDA-activated current by ethanol was examined. It was reported that, of the homomeric receptors tested, the receptor formed by the splice variant "LL" (containing both N- and C-terminal insertions) was most sensitive to ethanol. This sensitivity was the result of ethanol inhibition of a "hump" in the current, which slowly developed after the initial response. This hump was eliminated when barium was substituted for Ca^{++} in the extracellular medium, and, under these conditions, there was little difference among the splice variants in sensitivity to ethanol. Thus it was suggested that a calcium-dependent reduction of current by ethanol is most apparent in receptors composed of the LL variant, while a calcium-independent inhibition was common to all variants (Koltchine et al., 1993).

The effect of heteromeric subunit composition on sensitivity to ethanol has also been examined. Koltchine et al. (1993) reported that cotransfection of the ε1 subunit with the NR1 subunit did not alter sensitivity to ethanol. Masood et al. (1994) found that receptors consisting of ζ1/ε1 (homologous to NR1/NR2A) or ζ1/ε2 (NR1/NR2B) were more sensitive to inhibition by ethanol than homomeric ζ1 receptors or heteromeric ζ1/ε3 (NR1/NR2C) receptors expressed in *Xenopus* oocytes. Kuner et al. (1993) reported that, in Ca^{++}-containing medium, NR1/NR2B receptors expressed in oocytes were most sensitive to ethanol inhibition and NR1/NR2A receptors were least sensitive, while, when Ca^{++} was removed from the medium, the NR1/NR2C receptor was least sensitive to ethanol. However, their interpretation of the results was that, although the receptor containing NR2C was slightly less sensitive to ethanol, the inhibition of function of all three recombinant receptors by ethanol was essentially the same. These authors also did not observe differences in sensitivity to ethanol when different splice variants of NR1 were expressed in combination with NR2A. It should be noted that the currents recorded by Kuner et al. (1993) were much greater than those recorded by Masood et al. (1994) or Koltchine et al. (1993) (for homomeric receptors only). It appears that, while the subunit composition of NMDA receptors, including splice variants, may have some influence on sensitivity to ethanol, the molecular structure of NMDA receptors is unlikely to be the only determinant of the response to ethanol.

B. PRESYNAPTIC NMDA RECEPTOR-MEDIATED NEUROTRANSMITTER RELEASE

One also needs to consider ethanol's actions on presynaptic receptor-gated Ca^{++} channels with characteristics of the NMDA subtype of glutamate receptor (Göthert and Fink, 1991; Smirnova et al., 1993a). Activation of these receptors stimulates the release of dopamine (DA) or norepinephrine (NE). NMDA receptors have also been shown to modulate the release and/or turnover of acetylcholine (Göthert and Fink, 1989) and serotonin (Löscher et al., 1991) in certain areas of brain, but the use of brain slices for the demonstration of certain of these effects (Göthert and Fink, 1989) precludes the specific assignment of these NMDA receptors to a presynaptic locale. Glutamatergic (NMDA) receptors have been shown to *promote* the release of acetylcholine in the striatum (Göthert and Fink, 1989), while the release of serotonin in the caudate nucleus (Becquet et al., 1990), in the cortex and in the nucleus accumbens (Löscher et al., 1991) is *inhibited* by the NMDA receptor-coupled systems.

The effects of ethanol on NMDA receptor modulated neurotransmitter release (Göthert and Fink, 1989) were initially described at the time that the first reports appeared on the inhibitory effects of ethanol on the function of the postsynaptic NMDA receptor-gated ion channels (Hoffman et al., 1989a; Lovinger et al., 1989; Lima-Landman and Albuquerque, 1989). The initial report (Göthert and Fink, 1989) and those that followed (Gonzales and Woodward, 1990; Woodward and Gonzales, 1990) indicated that low concentrations of ethanol (<50 m*M*) inhibited NMDA-stimulated dopamine, norepinephrine, and acetylcholine release from brain tissue slices. Catecholamine release was inhibited at concentrations of ethanol ranging from 10–60 m*M*, and the alcohol-induced inhibition was shown to be related to the lipophilic nature of various alcohols (Gonzales et al., 1991). Gonzales et al. (1991) demonstrated that alcohols of increasing chain length (e.g., propanol and butanol) were more potent than ethanol in their effects on NMDA-stimulated norepinephrine release. Studies on the mechanism of ethanol's actions on NMDA receptor-mediated transmitter release have focused additional attention on the ethanol-induced disruption of the coagonist (glycine) interactions with glutamate in the activation of the ion channel but indicate brain regional differences. For instance, Woodward and Gonzales (1990) have shown that increasing concentrations of glycine can reverse the ethanol-induced inhibition of dopamine release in striatal slices but not of norepinephrine release in cerebral cortical slices (Gonzales and Woodward, 1990).

The *in vitro* data on ethanol's actions on the presynaptic receptor-gated ion channels such as the NMDA receptor/channel complex suggest that catecholamine release would be *inhibited* or unaffected by ethanol. Measurements of catecholamine release *in vivo*, however, demonstrate the difficulties of extrapolating from the *in vitro* results and accentuate the importance of 1) the individual characteristics of a particular neuron, and 2) neuronal system *interactions* in altering transmitter release and metabolism in the alcohol-intoxicated animal. The *in vivo* studies of dopamine release and metabolism are instructive in this regard. Methodologic advances have allowed for *direct* measures of DA release in brain areas of freely moving rats. These techniques involve the use of dialysis tubing inserted into a particular brain area to collect released DA and its metabolites (Imperato and Di Chiara, 1986). When these techniques were used, low doses of ethanol (0.25–0.5 g/kg, i.p.) were shown to *stimulate* DA release in rat brain. The effect of these low doses was confined to the nucleus accumbens, and higher ethanol doses (>1 g/kg) were necessary to stimulate DA release in the caudate nucleus (Imperato and Di Chiara, 1986). Electrophysiologic measurements of the activity of DA neurons in the ventral tegmental area and the substantia nigra after systemic administration of ethanol in doses of 0.5–2.0 g/kg also demonstrated an increased firing rate of these neurons (Gessa et al., 1985). The disparate effects of ethanol noted on DA release from slices of striatum *in vitro* (Woodward and Gonzales, 1990) and the results seen in the studies where ethanol was administered systemically may be due to indirect effects of ethanol arising from *other* brain areas and neuronal systems that modulate the sensitivity of DA neurons to the direct effects of ethanol.

The proposal that changes in the activity of DA neurons occur *in vivo* via ethanol's actions on other neurotransmitter systems gains prominence because of the recent studies of the effects of 5-HT_3 receptor antagonists in altering DA neuron activity (Blandina et al., 1989), and the demonstration that NMDA receptor *antagonists* (i.e., dizocilpine administered *in vivo*) can produce an *increase* in dopamine release (Imperato et al., 1990) and metabolism (Löscher et al., 1991) resembling that seen with *in vivo* administration of ethanol. One can assume that both ethanol and dizocipline are reducing the activity of NMDA receptor-gated ion channels, albeit by somewhat different mechanisms. However, inhibition of a *presynaptic* NMDA receptor/channel on DA neurons would be difficult to reconcile with an *increase* in DA release and turnover, and a parsimonious explanation of the obtained results (Löscher et al., 1991; Imperato et al., 1990) would involve an NMDA receptor-containing interneuronal system which would modulate DA release. Serotonergic neurons are candidates for the role of modulators of DA neuron activity. Dizocilpine (Löscher et al., 1991) and ethanol (Tabakoff and Boggan, 1974) both *increase* serotonin turnover. If the serotonin (5-HT_3 receptor) *antagonist* ICS-205–930 was administered prior to the systemic administration of ethanol to rats, the stimulation of DA release by ethanol was greatly diminished (Wozniak et al., 1990). Interestingly, ICS-205–930 also reduced the effects of ethanol on DA release when ethanol was directly administered into the brain, but significantly higher doses of ICS-205–930 were needed to produce this reduction in the action of intracerebrally administered ethanol (Wozniak et al., 1991). The interactions between serotonergic and dopaminergic systems have recently been well detailed, particularly in the striatum (Blandina et al., 1989). Serotonin (5-HT_3) receptor *agonists* have been shown to increase DA release. One could postulate that some of the effects of ethanol in promoting the release of DA *in vivo* are mediated indirectly *via* ethanol's actions (through the NMDA receptor?) on serotonin neurons.

C. RECEPTOR RESPONSES TO KAINATE

Although in initial studies it was reported that the NMDA receptor, as compared to other glutamate receptors, was selectively sensitive to the action of ethanol (Lovinger et al., 1989;

Hoffman et al., 1989a), more recent studies have demonstrated significant inhibition of the response to kainate as well (Dildy-Mayfield and Harris, 1992a,b). When rat brain mRNA was expressed in *Xenopus* oocytes, ethanol was found to reduce the maximal response to kainate, and ethanol inhibition increased as the kainate concentration decreased. A similar result was found in primary cultures of cerebellar granule cells (Snell et al., 1994b). This kainate concentration-dependent phenomenon may underlie initial observations that NMDA receptor function was more sensitive to ethanol inhibition than non-NMDA receptor function (Lovinger et al., 1989; Hoffman et al., 1989a) since, in the earlier studies, only a single concentration of kainate (one that produced a maximal response) was tested. Relatively low concentrations of ethanol (22 m*M*) were also reported to reduce the amplitude of glutamate-induced EPSPs in a rat nucleus accumbens slice preparation, and this inhibition was attributed to ethanol's interference with glutamate actions at non-NMDA receptors (Nie et al., 1993). Although, in cerebellar granule cells, phorbol ester treatment did not alter the response to kainate (increase in intracellular Ca^{++} [Snell et al., 1994b]), Dildy-Mayfield and Harris (1994a,b) recently reported that treatment of oocytes expressing recombinant non-NMDA glutamate receptors with phorbol ester reduced kainate responses. These authors also found that ethanol inhibition of kainate responses was enhanced when oocytes were perfused with buffer containing high Ca^{++} concentrations and that inhibition of PKC by injection of an inhibitor peptide prevented the enhanced actions of ethanol. However, the inhibition of PKC did not alter the response to ethanol in normal, low-Ca^{++} buffer, leading to the conclusion that there may be two types of ethanol inhibition of kainate responses, PKC-dependent and PKC-independent.

D. POSITIVE AND NEGATIVE CONSEQUENCES OF THE ACUTE EFFECTS OF ETHANOL ON GLUTAMATE RECEPTORS

Acute ethanol inhibition of glutamate receptor function may have significant pharmacological consequences. Inhibition of responses at non-NMDA receptors would be expected to depress synaptic transmission. Inhibition of NMDA receptor responses could have more diverse effects. For example, the NMDA receptor is believed to play an important role in neurotoxicity, due to its ability to allow large amounts of calcium to enter the cell. Although ethanol itself is often thought of as a neurotoxin, it has been shown that ethanol can reduce the toxicity caused by NMDA in cultured rat brain cells (Lustig et al., 1992; Chandler et al., 1991; Takadera et al., 1990). It has been hypothesized that NMDA-induced neurotoxicity is a result of the activation of nitric oxide synthase and the consequent production of nitric oxide, a short-lived, highly reactive free radical, upon exposure of neurons to NMDA (Dawson et al., 1991). Thus, ethanol inhibition of the NMDA response, including inhibition of nitric oxide production, may have important clinical implications with respect to neurotoxicity caused by hypoxia and/or stroke.

The inhibition of NMDA receptor function by acute exposure to ethanol may also result in ethanol-induced cognitive deficits. This postulate is supported by studies showing that ethanol, at low concentrations, inhibits long-term potentiation (LTP) in the hippocampus (Blitzer et al., 1990). There is a great deal of evidence implicating NMDA receptor function in the induction of LTP, which is a model for the synaptic plasticity believed to be involved in learning and memory (Collingridge and Lester, 1989). In one study, the dose-dependence of the effect of ethanol on LTP was found to correlate well with the inhibition by ethanol of responses to NMDA (Blitzer et al., 1990).

IV. CHRONIC EFFECTS OF ETHANOL ON NMDA RECEPTORS

The NMDA receptor-coupled channel appears to undergo adaptive changes in animals or cells exposed chronically to ethanol. In mice that ingested ethanol in a liquid diet for seven days, dizocilpine binding was increased in hippocampus and several other brain areas, as determined by membrane binding and autoradiographic analyses (Grant et al., 1990; Gulya et al., 1991; Sanna et al., 1993). These data were interpreted to indicate an increase in the number of NMDA receptor complexes or an "up-regulation" of receptors in the brains of ethanol-fed animals. More recent membrane binding studies have demonstrated an increase in NMDA-sensitive glutamate binding but not in strychnine-insensitive glycine binding sites or in binding of a competitive NMDA receptor antagonist in hippocampal tissue of ethanol-fed mice (Snell et al., 1993). Similarly, an early investigation reported an increase in synaptosomal glutamate binding in rats after chronic ethanol ingestion (Michaelis et al., 1978), and glutamate binding has also been reported to be increased in hippocampal membranes from human alcoholics (Michaelis et al., 1990). The finding that there are changes in ligand binding at one of the coagonist recognition sites on the NMDA receptor and not at the other suggests the possibility that the various compounds may bind to sites on different NMDA receptor subunits and that there may be changes in subunit composition following chronic ethanol exposure. Early evidence indicating that various ligands bind to different sites on the NMDA receptor included the finding that NMDA-sensitive glutamate and glycine binding sites demonstrated differential rates of development (McDonald et al., 1990). More recently, ligand binding to recombinant NMDA receptors expressed in HEK 293 cells showed that homomeric receptors consisting of NR1a subunits (the splice variant of NR1 that does not contain the N-terminal insertion but does contain the C-terminal insertions) could bind a glycine site antagonist, while expression of both NR1a and NR2a was necessary to allow binding of glutamate antagonists and channel blocking compounds (Lynch et al., 1994). It has recently been reported that NR1 protein was increased in hippocampus of rats fed ethanol in a liquid diet for 12 weeks (Trevisan et al., 1994). Based on the recombinant receptor study (Lynch et al., 1994) such a change might not alone account for the observed increase in dizocilpine binding after chronic ethanol treatment, and measurement of the levels of other receptor subunits will be necessary to interpret the above described changes in ligand binding produced by chronic administration of ethanol.

It is of interest to note that both stress and hypoxia have been reported to alter NMDA receptor characteristics in brain. Immobilization stress increased dizocilpine and glutamate binding in certain brain areas of rats but did not alter binding of glycine site ligands, and treatment of rats with corticosterone also increased the binding of glutamate and a competitive antagonist (Yoneda et al., 1994). Chronic ethanol treatment has been shown to result in increased circulating levels of corticosteroids and a perturbation of the circadian rhythm of these hormones (Tabakoff et al., 1978). The changes in NMDA receptor ligand binding are similar (although not identical) for stress and ethanol treatments, and it is possible that hormonal alterations during ethanol ingestion and/or withdrawal may contribute to changes in NMDA receptor characteristics. Similarly, a brief *in vitro* hypoxic-hypoglycemic episode was reported to produce an increase in mRNA for NR2C in slice preparations of rat hippocampus and cortex (Perez-Velazquez and Zhang, 1994). Similar hypoxic-hypoglycemic episodes might occur during chronic ethanol treatment, leading to alterations in NMDA receptor subunit expression. Alternatively, ethanol itself or an adaptation to the initial inhibition of NMDA receptor function by ethanol could lead to altered characteristics of the NMDA receptor. Another possibility is that chronic ethanol ingestion could produce post-translational changes in the NMDA receptor that affect ligand binding. Such modifications have been shown to occur with other treatments. For example, exposure of neuronal membranes to DTT has been reported to increase dizocilpine binding (Reynolds et al., 1990).

It is important to know whether changes in ligand binding to the NMDA receptor or alterations in receptor subunit expression are accompanied by changes in receptor *function*. In rats that were exposed chronically to ethanol, there was no change in the ability of NMDA to stimulate neurotransmitter release in several brain areas (Brown, et al., 1991). However, the presence of physical dependence on ethanol was not demonstrated in these rats; furthermore, the NMDA receptors controlling neurotransmitter release in brain slices (presynaptic, see above) may respond differently from postsynaptic receptors to chronic ethanol ingestion. In contrast to the results on neurotransmitter release, when primary cultures of rat cerebellar granule cells were exposed chronically to ethanol (100 m*M* for 1–4 days; 20 m*M* for three or more days), there was an increased response to NMDA, measured as an increase in intracellular calcium using fura-2 (Iorio et al., 1992). A similar change was recently reported in primary cultures of rat cerebral cortical cells (Ahern et al., 1994). In the cerebellar granule cells, there was also an increased response to glycine, in the presence of constant levels of NMDA, at all concentrations of glycine tested. Other characteristics of the response to NMDA, including inhibition by dizocilpine and the competitive antagonist 5-aminophosphonovalerate (APV), as well as inhibition by ethanol added *in vitro*, did not change, suggesting that there was an increased number of NMDA receptors in the chronically ethanol-exposed cells, rather than a change in receptor properties. Ligand binding studies using intact cerebellar granule cells have confirmed that chronic ethanol exposure increases the number of NMDA receptors (measured with dizocilpine binding) (Hoffman et al., 1995). These data are consistent with the changes in ligand binding seen in brains of mice fed ethanol chronically and suggest an adaptive up-regulation of the NMDA receptor after chronic exposure to ethanol.

A. INVOLVEMENT OF NMDA RECEPTOR UP-REGULATION IN ETHANOL WITHDRAWAL SEIZURES

In the intact animal, this up-regulation may be related to certain symptoms of ethanol withdrawal, since administration of competitive and noncompetitive antagonists of NMDA receptors can reduce ethanol withdrawal seizures in mice and rats (Grant et al., 1990; Liljequist, 1991; Morrisett et al., 1990). Furthermore, the time course for the increase in hippocampal dizocilpine binding sites paralleled the time course for the appearance of ethanol withdrawal seizures (Gulya et al., 1991). In particular, the number of binding sites was increased at the time of withdrawal (after one week of ethanol feeding) and at eight hours after withdrawal, when withdrawal seizures peak, but had returned to control values by 24 hours after withdrawal, when withdrawal signs had dissipated. In a longer-term ethanol feeding study (28 weeks), in which dizocilpine binding was measured at 48 hours after ethanol withdrawal, no changes were observed (Tremwel et al., 1994), consistent with the hypothesis that NMDA receptor up-regulation after ethanol ingestion is a transient phenomenon that contributes to withdrawal signs. In addition, mice that have been selectively bred to be susceptible to ethanol withdrawal seizures (WSP mice) have a greater number of hippocampal dizocilpine binding sites than mice bred to be resistant to ethanol withdrawal seizures (WSR mice) (Valverius et al., 1990). This difference remained after chronic ethanol ingestion by the mice, although the number of dizocilpine binding sites was increased by the chronic ethanol exposure both in WSP and WSR mice (Valverius et al., 1990). The finding of a biochemical difference in two selected lines of mice, such as WSP and WSR, strongly supports a role for that biochemical system in mediating the selected behavioral trait. It has also been reported that WSP mice undergoing acute withdrawal from ethanol (testing at six hours after a 24-hour ethanol exposure) were more sensitive to NMDA-induced enhancement of handling-induced seizures and to the anticonvulsant effects of dizocilpine (Crabbe et al., 1993) than control WSP mice, supporting a role of NMDA receptors in ethanol withdrawal seizures. However, there was no difference between WSP and WSR mice when certain other effects of dizocilpine (e.g., loss of righting reflex and open field activity) were tested, even after acute ethanol withdrawal, and in some cases (reduction of seizure threshold for ECS), WSR mice were more sensitive than WSP mice to dizocilpine (Crabbe et al., 1994). WSR mice were also reported to be innately more sensitive to NMDA-induced seizures (in contrast to NMDA enhancement of handling-induced seizures during ethanol withdrawal) than WSP mice (Kosobud and Crabbe, 1993). These data suggest that "genetic variation which has increased sensitivity to [handling-induced convulsions] during [ethanol] withdrawal has tended to decrease sensitivity to convulsions directly elicited by NMDA infusions in naive [WSP] mice ..." (Kosobud and Crabbe,

1993). However, the results with WSP and WSR mice are overall consistent with the hypothesis that the NMDA receptor is one of the key systems that mediate the occurrence of ethanol withdrawal seizures (others include the $GABA_A$ receptor and voltage-sensitive Ca^{2+} channels [see Tabakoff and Hoffman, 1992]).

B. INVOLVEMENT OF NMDA RECEPTOR UP-REGULATION IN ETHANOL-INDUCED EXCITOTOXICITY

The "up-regulation" of NMDA receptors induced by chronic ethanol exposure may have consequences other than ethanol withdrawal seizures. It is generally believed that the NMDA receptor plays an important role in glutamate-induced excitotoxic damage in CNS neurons (Choi 1988; 1992), and this damage could be enhanced after chronic ethanol treatment. Rats that had been chronically exposed to ethanol by inhalation were found to be more sensitive than controls to the effects of intrahippocampal NMDA injection (i.e., decrease in hippocampal glutamate decarboxylase activity and mortality) (Davidson et al., 1993). The lethal effect of NMDA was suggested to be a reflection of enhanced NMDA receptor function in the ethanol-withdrawn animals (Davidson et al., 1993). In a more direct study, glutamate-induced neurotoxicity, mediated by the NMDA receptor, was significantly increased in cerebellar granule cells that had been exposed chronically to ethanol, then withdrawn from ethanol, and exposed to glutamate (Iorio et al., 1993). NMDA receptor function and the excitotoxic effect of NMDA were also enhanced in cerebral cortical cells that had been chronically exposed to and withdrawn from ethanol (Chandler et al., 1993a; Ahern et al., 1994). It is likely that the enhanced sensitivity to neurotoxic damage is a consequence of ethanol withdrawal, since it has been shown that ethanol, while present in the cellular milieu, inhibits NMDA receptor function and blocks NMDA receptor-mediated neurotoxicity (Chandler et al., 1993b; Takadera et al., 1990). Therefore, the presence of ethanol in the brain of an animal would be expected to protect against excitotoxicity, even in the presence of an increased number of NMDA receptors. However, once ethanol is eliminated, the increased susceptibility to glutamate excitotoxicity becomes apparent.

The suggestion that brain damage produced by chronic ethanol exposure and withdrawal may be in part due to increased NMDA receptor function underlines the contention (Tabakoff, 1989) that ethanol withdrawal hyperexcitability should be treated with drugs that are antagonists at the NMDA receptor. Work with NMDA receptor antagonists is indicating specific therapeutic approaches for ameliorating ethanol withdrawal seizures and preventing neuronal damage (Grant et al., 1992). Recent work has substantiated the efficacy of NMDA receptor antagonists that act at various sites on the receptor, as well as the ganglioside GM_1, which is believed to act at a step beyond the receptor itself, in attenuating glutamate neurotoxicity in ethanol-exposed cerebellar granule cells (Hoffman et al., in press). Heaton et al. (1994) also recently demonstrated ganglioside GM_1 protection against "direct" ethanol neurotoxicity in cultures of septal and hippocampal neurons from fetal rat and dorsal root ganglion neurons from embryonic chick, although they did not assess the role of glutamate in cell death under their conditions.

C. NMDA RECEPTORS AND WITHDRAWAL INDUCED "KINDLING"

There is substantial evidence to indicate that ethanol withdrawal symptoms are exacerbated following multiple episodes of withdrawal (i.e., that withdrawal seizures result from a "kindling" phenomenon [Becker and Hale, 1993]). Most alcoholics undergo withdrawal numerous times throughout their lives, and it is likely that multiple episodes of up-regulation of NMDA receptor function play a key role not only in the worsening seizures but also in the well-described neuronal damage in alcoholics (Charness, 1993) (see Chapter 20). A recent study of 72 alcoholics, in fact, demonstrated that cortical and ventricular atrophy were significantly greater in those individuals with a previous history of ethanol withdrawal episodes, and a greater number of previous episodes was associated with significantly greater ventricular size. Of all variables tested, only previous withdrawal episodes were predictive of cerebral atrophy

(Daryanani et al., 1994). Thus, the treatment of ethanol withdrawal seizures with NMDA receptor antagonists may well prevent the development of neuronal damage in alcoholics.

Another consequence of NMDA receptor up-regulation after chronic ethanol exposure may arise from the fact that activation of NMDA receptors in certain cell types has been shown to increase the expression of the proto-oncogene, c-*fos* (Szekely et al., 1989; Cole et al., 1989). This gene encodes a DNA binding protein that, in conjunction with the product of another proto-oncogene, c-*jun,* can serve as a transcription factor that regulates the expression of other genes (Verma, 1986; Halazonetis et al., 1988). Activation of c-*fos* can therefore lead to *long-term* alterations in CNS function, and it has been suggested, for example, that this proto-oncogene, which can be activated in the adult CNS by neurotransmitters, may be involved in adaptive processes such as learning or memory (Greenberg et al., 1986). In mice that were undergoing ethanol withdrawal and displayed seizure activity, the levels of mRNA for c-*fos* were greatly increased in several brain areas (Dave et al., 1990). This increase in c-*fos* expression was apparent at the time that seizures occurred, and mRNA levels returned to control values by 24 hours after a seizure (Dave et al., 1990). It has recently been reported that, in rats, the increase in c-*fos* expression during ethanol withdrawal can be blocked by administration of an NMDA receptor antagonist (Morgan et al., 1992). It is possible that the repeated up-regulation of NMDA receptors, which in turn results in increased c-*fos* expression in certain brain areas, could contribute to the long-term changes in CNS function that underlie the "kindling" phenomenon of ethanol withdrawal, discussed above.

Overall, the data clearly show that the function of the NMDA receptor is very sensitive to acute inhibition by ethanol. This sensitivity may translate into acute effects of ethanol on learning or memory processes. After chronic ethanol treatment, the number of NMDA receptors in brain or cultured cells appears to increase, and, in animals, this change seems to be intimately involved in the generation or expression of ethanol withdrawal seizures and may also mediate long-term neuroadaptive changes and neurotoxicity induced by chronic ethanol exposure and withdrawal.

V. CHRONIC EFFECTS OF ETHANOL ON NON-NMDA GLUTAMATE RECEPTORS

Although it appears that under some conditions, non-NMDA receptor function may be as sensitive to acute ethanol inhibition as NMDA receptor function, to date only a few studies have evaluated non-NMDA receptor function following chronic ethanol treatment. In cerebellar granule cells exposed chronically to ethanol, no change in response to kainate (measured as an increase in intracellular Ca^{++}) was observed (Iorio et al., 1993). Similarly, in *Xenopus* oocytes expressing rat hippocampal mRNA, there was no difference in kainate-stimulated currents between control oocytes and those that had been exposed to 100 m*M* ethanol for 1–5 days. However, acute ethanol inhibition of the response to low concentrations of kainate was reduced after chronic ethanol treatment (i.e., tolerance to ethanol developed); no such change was observed in the maximum currents stimulated by kainate or NMDA (Dildy-Mayfield and Harris, 1992a). "Acute tolerance" to ethanol inhibition of NMDA-stimulated EPSPs in hippocampal slices was, however, reported to develop over 15 minutes of ethanol exposure (Grover et al., 1994), and further work will be necessary to elucidate the nature and mechanisms of the development of resistance to the effect of ethanol on glutamate receptor function.

VI. THE NMDA RECEPTOR AS A MODULATOR OF ETHANOL TOLERANCE

The NMDA receptor has been suggested to modulate tolerance to various effects of ethanol (Khanna et al., 1992a,b; Szabó et al., 1994), although the function of the receptor itself may or may not become "tolerant" to the effects of ethanol (Iorio et al., 1992; Grover et al., 1994).

In several studies, dizocilpine or ketamine (a dissociative anesthetic that binds to the PCP site within the ion channel), when administered together with ethanol, were demonstrated to block the development of ethanol tolerance (Khanna et al., 1992a,b; Szabó et al., 1994), while an *agonist* at the glycine site on the NMDA receptor was found to *enhance* the development of ethanol tolerance (Khanna et al., 1993a). Treatment of rats with the nitric oxide synthase inhibitor, L-nitro-arginine, also prevented the development of ethanol tolerance (Khanna et al., 1993b), and the authors interpreted these data to reflect an interference with nitric oxide facilitation of NMDA receptor-mediated transmission. An interaction of the NMDA receptor and serotonin systems in modulating ethanol tolerance has also been demonstrated (Khanna et al., 1994a). Although tolerance may be simply defined as an acquired resistance to the effects of ethanol, in reality it is a more complex phenomenon that can include metabolic (pharmacokinetic) and functional (pharmacodynamic) components (Tabakoff et al., 1982). In addition, tolerance can be conditioned or associative, meaning that tolerance generated under certain paradigms reflects a conditioned response to the cues associated with ethanol administration (Poulos and Cappell, 1991). It appears that conditioned tolerance is most susceptible to blockade by NMDA receptor antagonists. When animals were given ethanol using a paradigm in which conditioned tolerance was prominent (repeated ethanol injections in a particular environment), this tolerance was blocked by dizocilpine (Khanna et al., 1992b; Szabó et al., 1994). When a paradigm (chronic ingestion of a liquid diet containing ethanol) was used in which nonassociative tolerance was generated, development of tolerance was not blocked even by higher doses of dizocilpine (Szabó et al., 1994). Similarly, tolerance develops more rapidly when animals are given ethanol injections prior to being tested for motor function ("intoxicated practice"), compared to animals given the same dose *after* testing. Ketamine inhibited the development of tolerance to ethanol in rats given ethanol before testing but not in those that received ethanol after testing (Khanna et al., 1994b). That is, ketamine attenuated tolerance when practice or learning was involved but not when tolerance was due only to exposure to ethanol. Therefore, it seems that the effect of NMDA receptor antagonists may not be on ethanol tolerance *per se* but may reflect the role of the NMDA receptor in the processes of learning (conditioning) and/or memory.

VII. CONCLUSION

The consideration of ionotropic glutamate receptors as "receptive elements" for ethanol's actions is well supported by the experimental evidence. Particularly prominent is the information regarding ethanol's acute inhibitory effects on the NMDA subtype of glutamate receptor. The acute inhibitory effects of ethanol on the NMDA receptors can certainly account for ethanol's disruption of long term potentiation and can provide a rational explanation of ethanol-induced memory loss (i.e., blackouts). Other aspects of acute ethanol intoxication are not as well linked to ethanol's actions at the various glutamate receptors, although there is evidence, cited above, that ethanol's anxiolytic actions and even some of the motor impairment seen with ethanol may involve ethanol's effects at the glutamate (NMDA?) receptors.

The data on the chronic effects of ethanol on the NMDA receptors has profound implications in the management of ethanol withdrawal. There is good evidence that medications acting specifically to suppress the activity of the ethanol-upregulated NMDA receptors can control a broad spectrum of the signs (including seizures) of ethanol withdrawal-induced CNS hyperexcitability. More important may be the demonstrations that up-regulation of NMDA receptors during chronic ingestion of ethanol enhances the prospect of glutamate-induced excitoxicity in the brain of the alcohol-dependent individual. These data certainly suggest that ethanol withdrawal may have ramifications beyond the display of overt signs of CNS hyperexcitability and suggest that treatment of ethanol withdrawal with NMDA receptor-active pharmacologic agents may be important for protecting the brain of the ethanol withdrawing subject.

The work on the glutamate receptor systems is also providing insights into mechanisms by which certain types of tolerance to ethanol develop, and the interactions between NMDA, serotonergic, and dopaminergic systems in brain may contribute to an explanation of the mechanism of the reinforcing effects of ethanol. The explosion of studies over the last five to six years on ethanol's actions on glutamate receptors has certainly provided revolutionary thinking and progress in understanding the ways in which ethanol acts in the CNS and the ways in which the damaging actions of ethanol can be controlled.

ACKNOWLEDGMENTS

Portions of this work also appear in other recent reviews of ethanol's actions by Tabakoff, B. et al. (in press) and Tabakoff, B. and Hoffman, P.L. (in press). This work was supported in part by the Banbury Foundation and the National Institute on Alcohol Abuse and Alcoholism.

REFERENCES

Abel, E.L., (1980), The fetal alcohol syndrome: behavioral teratology. *Psychol. Bull.,* 87: 29–50.

Ahern, K. von B., Lustig, H.S., and Greenberg, D.A. (1994), Enhancement of NMDA toxicity and calcium responses by chronic exposure of cultured cortical neurons to ethanol. *Neurosci. Lett.,* 165: 211–214.

Aizenman, E., Lipton, S.A., and Loring, R.H. (1989), Selective modulation of NMDA responses by reduction and oxidation. *Neuron,* 2: 1257–1263.

Balster, R.L. (1987), The behavioral pharmacology of phencyclidine. In: *Psychopharmacology: The Third Generation of Progress.* H.Y. Meltzer, Ed., pp. 1573–1579. Raven Press, New York.

Balster, R.L., Wiley, J.L., Tokarz, M.E., and Tabakoff, B. (1993), Effects of ethanol and NMDA antagonists on operant behavior in ethanol withdrawal seizure-prone and -resistant mice. *Behav. Pharmacol.,* 4: 107–113.

Becker, H.C. and Hale, R.L. (1993), Repeated episodes of ethanol withdrawal potentiate the severity of subsequent withdrawal seizures: an animal model of alcohol withdrawal "kindling". *Alcoholism: Clin. Exp. Res.,* 17: 94–98.

Becquet, D., Faudon, M., and Hery, F. (1990), *In vivo* evidence for an inhibitory glutamatergic control of serotonin release in the cat caudate nucleus: involvement of GABA neurons. *Brain Res.,* 519: 82–88.

Bhave, S.V., Snell, L.D., Tabakoff, B., and Hoffman, P.L. (1994), Effect of ethanol on N-methyl-D-aspartate receptor function in cultured cerebral cortical cells. *Alcoholism: Clin. Exp. Res.,* 18: 444.

Blandina, P., Goldfarb, J., Craddock-Royal, B., and Green, J.P. (1989), Release of endogenous dopamine by stimulation of 5-hydroxytryptamine$_3$ receptors in rat striatum. *J. Pharmacol. Exp. Ther.,* 251: 803.

Blitzer, R.D., Gil, O., and Landau, E.M. (1990), Long-term potentiation in rat hippocampus is inhibited by low concentrations of ethanol. *Brain Res.,* 537: 203–208.

Brown, L.M., Leslie, S.W., and Gonzales, R.A. (1991), The effects of chronic ethanol exposure on N-methyl-D-aspartate-stimulated overflow of [^{3}H]catecholamines from rat brain. *Brain Res.,* 547: 289–294.

Cerne, R., Jiang, M., and Randic, M. (1992), Cyclic adenosine 3′,5′-monophosphate potentiates excitatory amino acid and synaptic responses of rat spinal dorsal horn neurons. *Brain Res.,* 596: 111–123.

Chandler, L.J., Sumners, C., and Crews, F.T. (1991), Ethanol inhibits NMDA-stimulated excitoxicity. *Alcoholism: Clin. Exp. Ther.,* 15: 323.

Chandler, L.J., Newsom, H., Sumners, C., and Crews, F.T. (1993a), Chronic ethanol exposure potentiates NMDA excitotoxicity in cerebral cortical neurons. *J. Neurochem.,* 60: 1578–1581.

Chandler, L.J., Sumners, C., and Crews, F.T. (1993b), Ethanol inhibits NMDA receptor-mediated excitotoxicity in rat primary neuronal cultures. *Alcoholism: Clin. Exp. Res.,* 17: 54–60.

Chandler, L.J., Guzman, N.J., Sumners, C., and Crews, F.T. (1994), Magnesium and zinc potentiate ethanol inhibition of *N*-methyl-D-aspartate-stimulated nitric oxide synthase in cortical neurons. *J. Pharmacol. Exp. Ther.,* 271: 67–75.

Charness, M.E. (1993), Brain lesions in alcoholics. *Alcoholism: Clin. Exp. Res.,* 17: 2–11.

Chen, L. and Huang, L.-Y.M. (1991), Sustained potentiation of NMDA receptor-mediated glutamate responses through activation of protein kinase C by a mu opioid. *Neuron,* 7: 319–326.

Choi, D.W. (1988), Calcium-mediated neurotoxicity: relationship to specific channel types and role in ischemic damage. *Trends Neurosci.,* 11: 465–469.

Choi, D.W. (1992), Excitotoxic cell death. *J. Neurobiol.,* 23: 1261–1276.

Cole, A.J., Saffern, D.W., Baraban, J.M., and Worley, P.F. (1989), Rapid increase of an immediate early gene messenger RNA in hippocampal neurons by synaptic NMDA receptor activation. *Nature,* 340: 474–476.

Collingridge, G.L. (1985), Long term potentiation in the hippocampus: mechanisms of initiation and modulation by neurotransmitters. *Trends Pharmacol. Sci.,* 6: 407–411.

Collingridge, G.L. and Lester, R.A.J. (1989), Excitatory amino acid receptors in the vertebrate central nervous system. *Pharmacol. Rev.,* 40: 143–210.

Courtney, M.J. and Nicholls, D.G. (1992), Interactions between phospholipase C-coupled and N-methyl-D-aspartate receptors in cultured cerebellar granule cells: protein kinase C mediated inhibition of N-methyl-D-aspartate responses. *J. Neurochem.,* 59: 983–992.

Crabbe, J., Young, E.R., and Dorow, J. (1994), Effects of dizocilpine in withdrawal seizure-prone (WSP) and withdrawal seizure-resistant (WSR) mice. *Pharmacol. Biochem. Behav.,* 47: 443–450.

Crabbe, J.C., Merrill, C.M., and Belknap, J.K. (1993), Effect of acute alcohol withdrawal on sensitivity to pro- and anti-convulsant treatments in WSP mice. *Alcoholism: Clin. Exp. Res.,* 17:1233–1239.

Daryanani, H.E., Santolaria, F.J., Reimers, E.G., Jorge, J.A., Lopez, N.B., Hernandez, F.M., Riera, M., and Rodriguez, E.R. (1994), Alcoholic withdrawal syndrome and seizures. *Alc. Alcoholism,* 29: 323–328.

Dave, J.R., Tabakoff, B., and Hoffman, P.L. (1990), Ethanol withdrawal seizures produce increased c-*fos* mRNA in mouse brain. *Mol. Pharmacol.,* 37: 367–371.

Davidson, M.D., Wilce, P., and Shanley, B.C. (1993), Increased sensitivity of the hippocampus in ethanol-dependent rats to toxic effect of N-methyl-D-aspartic acid *in vivo*. *Brain Res.,* 606: 5–9.

Dawson, V.L., Dawson, T.M., London, E.D., Bredt, D.S., and Snyder, S.H. (1991), Nitric oxide mediates glutamate neurotoxicity in primary cortical cultures. *Proc. Natl. Acad. Sci. U.S.A.,* 88: 6368–6371.

DeMontis, G., Devoto, P., Giorgi, G., Taliamonte, A., and Gessa, G.L. (1991), Ethanol, at micromolar concentrations, increases the affinity of [^{3}H]MK-801 binding in rat brain. *Eur. J. Pharmacol.,* 199: 139–140.

Dildy-Mayfield, J.E. and Leslie, S.W. (1989), Ethanol inhibits NMDA-induced increases in free intracellular Ca^{2+} in dissociated brain cells. *Brain Res.,* 499: 383–387.

Dildy-Mayfield, J.E. and Leslie, S.W. (1991), Mechanism of inhibition of N-methyl-D-aspartate-stimulated increases in free intracellular Ca^{2+} concentration by ethanol. *J. Neurochem.,* 56: 1536–1543.

Dildy-Mayfield, J.E. and Harris, R.A. (1992a), Comparison of ethanol sensitivity of rat brain kainate, DL-α-amino-3-hydroxy-5-methyl-4-isoxalone proprionic acid and N-methyl-D-aspartate receptors expressed in *Xenopus* oocytes. *J. Pharmacol. Exp. Ther.,* 262: 487–494.

Dildy-Mayfield, J.E. and Harris, R.A. (1992b), Acute and chronic ethanol exposure alters the function of hippocampal kainate receptors expressed in *Xenopus* oocytes. *J. Neurochem.,* 58: 1569–1572.

Dildy-Mayfield, J.E. and Harris, R.A. (1994a), Activation of protein kinase C inhibits kainate-induced currents in oocytes expressing glutamate receptor subunits. *J. Neurochem.,* 62: 1639–1642.

Dildy-Mayfield, J.E. and Harris, R.A. (1994b), Ethanol inhibition of AMPA/kainate receptor function in *Xenopus* oocytes: role of calcium and protein kinase C. *Alcoholism: Clin. Exp. Ther.,* 18: 445.

Dingledine, R. (1991), New wave of non-NMDA excitatory amino acid receptors. *TIPS,* 12: 360–362.

Durand, G.M., Bennett, M.V., and Zukin, R.S. (1993), Splice variants of the N-methyl-D-aspartate receptor NR1 identify domains involved in regulation by polyamines and protein kinase C. *Proc. Natl. Acad. Sci. U.S.A.,* 90: 6731–6735.

Frye, G.D. (1991), Interaction of ethanol and L-glutamate in the guinea pig ileum myenteric plexus. *Eur. J. Pharmacol.,* 192: 1–7.

Gessa, G.L., Muntoni, F., Collu, M., Vargiu, L., and Mereu, G. (1985), Low doses of ethanol activate dopaminergic neurons in the ventral tegmental area. *Brain Res.,* 348: 201–203.

Gonzales, R.A. and Woodward, J.J. (1990), Ethanol inhibits N-methyl-D-aspartate-stimulated [^{3}H]norepinephrine release from rat cortical slices. *J. Pharmacol. Exp. Ther.,* 253: 1138–1144.

Gonzales, R.A., Westbrook, S.L., and Bridges, L.T. (1991), Alcohol-induced inhibition of N-methyl-D-aspartate-evoked release of [^{3}H]norepinephrine from brain is related to lipophilicity. *Neuropharmacology,* 30: 441–446.

Göthert, M. and Fink, K. (1989), Inhibition of N-methyl-D-aspartate (NMDA)- and L-glutamate-induced noradrenaline and acetylcholine release in the rat brain by ethanol. *Naunyn-Schmiedebergs Arch. Pharmacol.,* 340: 516–521.

Göthert, M. and Fink, K. (1991), Stimulation of noradrenaline release in the cerebral cortex via presynaptic N-methyl-D-aspartate (NMDA) receptors and their pharmacological characterization. *J. Neural Transm. (Suppl).,* 34: 121–127.

Grant, K.A., Valverius, P., Hudspith, M., and Tabakoff, B. (1990), Ethanol withdrawal seizures and the NMDA receptor complex. *Eur. J. Pharmacol.,* 176: 289–296.

Grant, K.A., Knisely, J.S., Tabakoff, B., Barrett, J.E., and Balster, R.L. (1991), Ethanol-like discriminative stimulus effects of non-competitive N-methyl-D-aspartate antagonists. *Behav. Pharmacol.,* 2: 87–95.

Grant, K.A., Snell, L.D., Rogawski, M.A., Thurkauf, A., and Tabakoff, B. (1992), Comparison of the effects of the uncompetitive N-methyl-D-aspartate antagonist (±)-5-aminocarbonyl-10,11-dihydro-5H-dibenzo[a,d]cyclohepten-5,10-imine (ADCI) with its structural analogs dizocilpine (MK-801) and carbamazepine on ethanol withdrawal seizures. *J. Pharmacol. Exp. Ther.,* 260: 1017–1022.

Greenberg, M.E., Ziff, E.B., and Greene, L.A. (1986), Stimulation of neuronal acetylcholine receptors induces rapid gene transcription. *Science,* 234: 80–83.

Greengard, P., Jen, J., Nairn, A.C., and Stevens, C.F. (1991), Enhancement of the glutamate response by cAMP-dependent protein kinase in hippocampal neurons. *Science,* 253: 1135–1138.

Grover, C.A., Frye, G.D., and Griffith, W.H. (1994), Acute tolerance to ethanol inhibition of NMDA-mediated EPSPs in the CA1 region of the rat hippocampus. *Brain Res.,* 642: 70–76.

Gulya, K., Grant, K.A., Valverius, P., Hoffman, P.L., and Tabakoff, B. (1991), Brain regional specificity and time course of changes in the NMDA receptor-ionophore complex during ethanol withdrawal. *Brain Res.,* 547: 129–134.

Halazonetis, T.D., Georgopoulos, K., Greenberg, M.E., and Leder, P. (1988), c-*Jun* dimerizes with itself and c-*fos*, forming complexes of different DNA binding affinities. *Cell,* 55: 917–924.

Heaton, M.B., Paiva, M., Swanson, D.J., and Walker, D.W. (1994), Ethanol neurotoxicity *in vitro:* effects of GM1 ganglioside and protein synthesis inhibition. *Brain Res.,* 654: 336–342.

Heinemann, S., Bettler, B., Boulter, J., Deneris, E., Gasic, G., Hartley, M., Hollmann, M., Hughes, T.E., O'Shea-Greenfield, A., and Rogers, S. (1991), The glutamate receptors: Genes, structure and expression. In: *Neurotransmitter Regulation of Gene Transcription,* Vol. 7., E. Costa and T.H. Joh, Eds., pp. 143–165, Thieme Medical Publishers, New York.

Hoffman, P.L., Rabe, C.S., Moses, F., and Tabakoff, B. (1989a), N-methyl-D-aspartate receptors and ethanol: Inhibition of calcium flux and cyclic GMP production. *J. Neurochem.,* 52: 1937–1940.

Hoffman, P.L., Moses, F., and Tabakoff, B. (1989b), Selective inhibition by ethanol of glutamate-stimulated cyclic GMP production in primary cultures of cerebellar granule cells. *Neuropharmacology,* 28: 1239–1243.

Hoffman, P.L., Iorio, K.R., Snell, L.D., and Tabakoff, B. (1995), Attenuation of glutamate-induced neurotoxicity in chronically ethanol-exposed cerebellar granule cells by NMDA receptor antagonists and ganglioside GM_1. *Alcoholism: Clin. Exp. Res.,* 19:721–726.

Hollmann, M., Boulter, J., Maron, C., Beasley, L., Sullivan, J., Pecht, G., and Heinemann, S. (1993), Zinc potentiates agonist-induced currents at certain splice variants of the NMDA receptor. *Neuron,* 10: 943–954.

Hollmann, M. and Heinemann, S. (1994), Cloned glutamate receptors. *Annu. Rev. Neurosci.,* 17: 31–108.

Ikeda, K., Nagasawa, M., Mori, H., Araki, K., Sakimura, K., Watanabe, M., Inoue, Y., and Mishina, M. (1992), Cloning and expression of the ε4 subunit of the NMDA receptor channel. *FEBS Lett.,* 313: 34–38.

Imperato, A. and Di Chiara, G. (1986), Preferential stimulation of dopamine release in the nucleus accumbens of freely moving rats by ethanol. *J. Pharmacol. Exp. Ther.,* 239: 219–228.

Imperato, A., Scrocco, M.G., Bacchi, S., and Angelucci, L. (1990), NMDA receptors and *in vivo* dopamine release in the nucleus accumbens and caudatus. *Eur. J. Pharmacol.,* 187: 555–556.

Iorio, K.R., Reinlib, L., Tabakoff, B., and Hoffman, P.L. (1992), Chronic exposure of cerebellar granule cells to ethanol results in increased NMDA receptor function. *Mol. Pharmacol.,* 41: 1142–1148.

Iorio, K.R., Tabakoff, B., and Hoffman, P.L. (1993), Glutamate-induced neurotoxicity is increased in cerebellar granule cells exposed chronically to ethanol. *Eur. J. Pharmacol.,* 248: 209–212.

Ishii, T., Moriyoshi, K., Sugihara, H., Sakurada, K., Kadotani, H., Yokoi, M., Akazawa, C., Shigemoto, R., Mizuno, N., Masu, M., and Nakanishi, S. (1993), Molecular characterization of the family of the N-methyl-D-aspartate receptor subunits. *J. Biol. Chem.,* 268: 2836–2843.

Isokawa, M. and Levesque, M.F. (1991), Increased NMDA responses and dendritic degeneration in human epileptic hippocampal neurons in slices. *Neurosci. Lett.,* 132: 212–216.

Keller, B.U., Hollmann, M., Heinemann, S., and Konnerth, A. (1992), Calcium influx through subunits GluR1/GluR3 of kainate/AMPA receptor channels is regulated by cAMP dependent protein kinase. *EMBO J.,* 11: 891–896.

Khanna, J.M., Kalant, H., Shah, G., and Chau, A. (1992a), Effect of (+)MK-801 and ketamine on rapid tolerance to ethanol. *Brain Res. Bull.,* 28: 311–314.

Khanna, J.M., Weiner, J., Kalant, H., Chau, A., and Shah, G. (1992b), Ketamine blocks chronic but not acute tolerance to ethanol. *Pharm. Biochem. Behav.,* 42: 347–350.

Khanna, J.M., Kalant, H., Shah, G., and Chau, A. (1993a), Effect of D-cycloserine on rapid tolerance to ethanol. *Pharm. Biochem. Behav.,* 45: 983–986.

Khanna, J.M., Morato, G.S., Shah, G., Chau, A., and Kalant, H. (1993b), Inhibition of nitric oxide synthesis impairs rapid tolerance to ethanol. *Brain Res. Bull.,* 32: 43–47.

Khanna, J.M., Kalant, H., Chau, A., Shah, G., and Morato, G.S. (1994a), Interaction between *N*-methyl-D-aspartate (NMDA) and serotonin (5-HT) on ethanol tolerance. *Brain Res. Bull.,* 35: 31–35.

Khanna, J.M., Morato, G.S., Chau, A., Shah, G., and Kalant, H. (1994b), Effect of NMDA antagonists on rapid and chronic tolerance to ethanol: importance of intoxicated practice. *Pharm. Biochem. Behav.,* 48: 755–763.

Kitamura, Y., Miyazaki, A., Yamanaka, Y., and Nomura, Y. (1993), Stimulatory effects of protein kinase C and calmodulin kinase II on N-methyl-D-aspartate receptor/channels in the postsynaptic density of rat brain. *J. Neurochem.,* 61: 100–109.

Kleckner, N.W. and Dingledine, R. (1988), Requirement for glycine in activation of NMDA receptors expressed in *Xenopus* oocytes. *Science,* 241: 835–837.

Koltchine, V., Anantharam, V., Wilson, A., Bayley, H., and Treistman, S.N. (1993), Homomeric assemblies of NMDAR1 splice variants are sensitive to ethanol. *Neurosci. Lett.,* 152: 13–16.

Kosobud, A.E. and Crabbe, J.C. (1993), Sensitivity to N-methyl-D-aspartic acid-induced convulsions is genetically associated with resistance to ethanol withdrawal seizures. *Brain Res.,* 610: 176–179.

Kumar, K.N., Tilakaratne, N., Johnson, P.S., Allen, A.E., and Michaelis, E.K. (1991), Cloning of cDNA for the glutamate-binding subunit of an NMDA receptor complex. *Nature,* 354: 70–73.

Kuner, T., Schoepfer, R., and Korpi, E.R. (1993), Ethanol inhibits glutamate-induced currents in heteromeric NMDA receptor subtypes. *Mol. Neurosci.,* 5: 297–300.

Kutsuwada, T., Kashiwabuchi, N., Mori, H., Sakimura, K., Kushiya, E., Araki, K., Meguro, H., Masaki, H., Kumanishi, T., Arakawa, M., and Mishina, M. (1992), Molecular diversity of the NMDA receptor channel. *Nature,* 358: 36–41.

Liljequist, S. (1991), The competitive NMDA receptor antagonist, CGP 39551, inhibits ethanol withdrawal seizures. *Eur. J. Pharmacol.,* 192: 197–198.

Lima-Landman, M.T. and Albuquerque, E.X. (1989), Ethanol potentiates and blocks NMDA-activated single-channel currents in rat hippocampal pyramidal cells. *FEBS Lett.,* 247: 61–67.

Lister, R.G., Eckhardt, M.J., and Weingartner, H. (1987), Ethanol intoxication and memory. Recent developments and new directions. In: *Recent Developments in Alcoholism.* Vol. 5., M. Galanter, Ed., pp. 111–126, Plenum Press, New York.

Löscher, W., Annies, R., and Hönack, D. (1991), The N-methyl-D-aspartate receptor antagonist MK-801 induces increases in dopamine and serotonin metabolism in several brain regions of rats. *Neurosci. Lett.,* 128: 191–194.

Lovinger, D.M., White, G., and Weight, F.F. (1989), Ethanol inhibits NMDA-activated ion current in hippocampal neurons. *Science,* 243: 1721–1724.

Lovinger, D.M., White, G., and Weight, F.F. (1990), NMDA receptor-mediated synaptic excitation selectively inhibited by ethanol in hippocampal slice from adult rat. *J. Neurosci.,* 10: 1372–1379.

Lustig, H.S., Chan, J., and Greenberg, D.A. (1992), Ethanol inhibits excitotoxicity in cerebral cortical cultures. *Neurosci. Lett.,* 135: 259–261.

Lynch, D.R., Anegawa, N.J., Verdoorn, T., and Pritchett, D.B. (1994), N-methyl-D-aspartate receptors: different subunit requirements for binding of glutamate antagonists, glycine antagonists, and channel-blocking agents. *Mol. Pharmacol.,* 45: 540–545.

Makram, H. and Segal, M. (1992), Activation of protein kinase C suppresses response to NMDA in rat CA1 hippocampal neurones. *J. Physiol.,* 457: 491–501.

Martin, D., Morrisett, R.A., Bian, X.-P., Wilson, W.A., and Swartzwelder, H.S. (1991), Ethanol inhibition of NMDA mediated depolarizations is increased in the presence of Mg^{2+}. *Brain Res.,* 546: 227–234.

Masood, K., Wu, C., Brauneis, U., and Weight, F.F. (1994), Differential ethanol sensitivity of recombinant N-methyl-D-aspartate receptor subunits. *Mol. Pharmacol.,* 45: 324–329.

Matsumoto, I., Davidson, M., and Wilce, P.A. (1993), Polyamine-enhanced NMDA receptor activity: effect of ethanol. *Eur. J. Pharmacol.,* 247: 289–294.

McBain, C.J. and Mayer, M.L. (1994), N-methyl-D-aspartic acid receptor structure and function. *Physiol. Rev.,* 74: 723–760.

McDonald, J.W. and Johnston, M.V. (1990), Physiological and pathophysiological roles of excitatory amino acids during central nervous system development. *Brain Res. Rev.,* 15: 41–70.

McDonald, J.W., Johnston, M.V., and Young, A.B. (1990), Differential ontogenic development of three receptors comprising the NMDA receptor/channel complex in the rat hippocampus. *Exp. Neurol.,* 110: 237–247.

McGlade-McCulloh, E., Yamamoto, H., Tan, S.-E., Brickey, D.A., and Soderling, T.R. (1993), Phosphorylation and regulation of glutamate receptors by calcium/calmodulin-dependent protein kinase II. *Nature,* 362: 640–642.

Michaelis, E.K., Mulvaney, M.J., and Freed, W.J. (1978), Effects of acute and chronic ethanol intake on synaptosomal glutamate binding activity. *Biochem. Pharmacol.,* 27: 1685–1691.

Michaelis, E.K., Freed, W.K., Galton, N., Foye, J., Michaelis, M.L., Phillips, I., and Kleinman, J.E. (1990), Glutamate receptor changes in brain synaptic membranes from human alcoholics. *Neurochem. Res.,* 15: 1055–1063.

Monyer, H., Sprengel, R., Schoepfer, R., Herb, A., Higuchi, M., Lomeli, H., Burnashev, N., Sakmann, B., and Seeburg, P.H. (1992), Heteromeric NMDA receptors: molecular and functional distinction of subtypes. *Science,* 256: 1217–1221.

Morgan, P.F., Nadi, N.S., Karanian, J., and Linnoila, M. (1992), Mapping rat brain structures activated during ethanol withdrawal. Role of glutamate and NMDA receptors. *Eur. J. Pharmacol. — Mol. Pharmacol.,* 225: 217–223.

Mori, H., Yamakura, T., Masaki, H., and Mishina, M. (1993), Involvement of the carboxyl-terminal region in modulation by TPA of the NMDA receptor channel. *Mol. Neurosci.,* 4: 519–522.

Moriyoshi, K., Masu, M., Ishii, T., Shigemoto, R., Mizuno, N., and Nakaniski, S. (1991), Molecular cloning and characterization of the rat NMDA receptor. *Nature,* 354: 31–37.

Morris, R.G., Anderson, E., Lynch, G.S., and Baudry, M. (1986), Selective impairment of learning and blockade of long-term potentiation by an N-methyl-D-aspartate receptor antagonist, AP5. *Nature,* 319: 774–776.

Morrisett, R.A., Rezvani, A.H., Overstreet, D., Janowsky, D.S., Wilson, W.A., and Swartzwelder, H.A. (1990), MK-801 potently inhibits alcohol withdrawal seizures in rats. *Eur. J. Pharmacol.,* 176: 103–105.

Morrisett, R.A., Martin, D., Oetting, T.A., Lewis, D.V., Wilson, W.A., and Swartzwelder, H.S. (1991), Ethanol and magnesium ions inhibit N-methyl-D-aspartate-mediated synaptic potentials in an interactive manner. *Neuropharmacology,* 30: 1173–1178.

Moss, S.J., Blackstone, C.D., and Huganir, R.L. (1993), Phosphorylation of recombinant non-NMDA glutamate receptors on serine and tyrosine residues. *Neurochem. Res.,* 18: 105–110.

Murphy, N.P., Cordier, J., Glowinski, J., and Premont, J. (1994), Is protein kinase C activity required for the N-methyl-D-aspartate-evoked rise in cytosolic Ca^{2+} in mouse striatal neurons? *Eur. J. Neurosci.,* 6: 854–860.

Nakanishi, S. (1992), Molecular diversity of glutamate receptors and implications for brain function. *Science,* 258: 597–603.

Nie, Z., Yuan, X., Madamba, S.G., and Siggins, G.R. (1993), Ethanol decreases glutamatergic synaptic transmission in rat nucleus accumbens *in vitro*: naloxone reversal. *J. Pharmacol. Exp. Ther.,* 266: 1705–1712.

Perez-Velazquez, J.L. and Zhang, L. (1994), *In vitro* hypoxia induces expression of the NR2C subunit of the NMDA receptor in rat cortex and hippocampus. *J. Neurochem.,* 63: 1171–1173.

Phillips, T.J., Feller, D.J., and Crabbe, J.C. (1989), Selected mouse lines, alcohol and behavior. *Experientia,* 45: 805–827.

Poulos, C.X. and Cappell, H. (1991), Homeostatic theory of drug tolerance: a general model of physiological adaptation. *Psychol. Rev.,* 98: 390–408.

Rabe, C.S. and Tabakoff, B. (1990), Glycine site directed agonists reverse ethanol's actions at the NMDA receptor. *Mol. Pharmacol.,* 38: 753–757.

Raymond, L.A., Blackstone, C.D., and Huganir, R.L. (1993), Phosphorylation and modulation of recombinant GluR6 glutamate receptors by cAMP-dependent protein kinase. *Nature,* 361: 637–641.

Reynolds, I.J., Rush, E.A., and Aizenman, E. (1990), Reduction of NMDA receptors with dithiothreitol increases [^{3}H]-MK-801 binding and NMDA-induced Ca^{2+} fluxes. *Br. J. Pharmacol.,* 101: 178–182.

Sanger, D.J. (1993), Substitution by NMDA antagonists and other drugs in rats trained to discriminate ethanol. *Behav. Pharmacol.,* 4: 523–528.

Sanna, E., Serra, M., Cossu, A., Colombo, G., Follesa, P., Cuccheddu, T., Concas, A., and Biggio, G. (1993), Chronic ethanol intoxiation induces differential effects on $GABA_A$ and NMDA receptor function in the rat brain. *Alcoholism: Clin. Exp. Res.,* 17: 115–123.

Schechter, M.D., Meehan, S.M., Gordon, T.L., and McBurney, D.M. (1993), The NMDA receptor antagonist MK-801 produces ethanol-like discrimination in the rat. *Alcohol,* 10: 197–201.

Sigel, E., Baur, R., and Malherbe, P. (1994), Protein kinase C transiently activates heteromeric N-methyl-D-aspartate receptor channels independent of the phosphorylatable C-terminal splice domain and of consensus phosphorylation sites. *J. Biol. Chem.,* 269: 8204–8208.

Simson, P.E., Criswell, H.E., Johnson, K.B., Hicks, R.E., and Breese, G.R. (1991), Ethanol inhibits NMDA-evoked electrophysiological activity *in vivo*. *J. Pharmacol. Exp. Ther.,* 257: 225–231.

Simson, P.E., Criswell, H.E., and Breese, G.R. (1993), Inhibition of NMDA-evoked electrophysiological activity by ethanol in selected brain regions: evidence for ethanol-sensitive and ethanol-insensitive NMDA-evoked responses. *Brain Res.,* 607: 9–16.

Smirnova, T., Stinnakre, J., and Mallet, J. (1993a), Characterization of a presynaptic glutamate receptor. *Science,* 262: 430–433.

Smirnova, T., Laroche, S., Errington, M.L., Hicks, A.A., Bliss, T.V.P., and Mallet, J. (1993b), Transsynaptic expression of a presynaptic glutamate receptor during hippocampal long-term potentiation. *Science,* 262: 433–436.

Snell, L.D., Tabakoff, B., and Hoffman, P.L. (1993), Radioligand binding to the N-methyl-D-aspartate receptor/ ionophore complex: Alterations by ethanol *in vitro* and by chronic *in vivo* ethanol ingestion. *Brain Res.,* 602: 91–98.

Snell, L.D., Iorio, K.R., Tabakoff, B., and Hoffman, P.L. (1994a), Protein kinase C activation attenuates N-methyl-D-aspartate-induced increases in intracellular calcium in cerebellar granule cells. *J. Neurochem.,* 62: 1783–1789.

Snell, L.D., Tabakoff, B., and Hoffman, P.L. (1994b), Involvement of protein kinase C in ethanol-induced inhibition of NMDA receptor function in cerebellar granule cells. *Alcoholism: Clin. Exp. Res.,* 18: 81–85.

Sugihara, H., Moriyoshi, K., Ishii, T., Masu, M., and Nakanishi, S. (1992), Structures and properties of seven isoforms of the NMDA receptor generated by alternative splicing. *Biochem. Biophys. Res. Commun.,* 185: 826–832.

Szabó, G., Tabakoff, B., and Hoffman, P.L. (1994), The NMDA receptor antagonist dizocilpine differentially affects environment-dependent and environment-independent ethanol tolerance. *Psychopharmacology,* 113: 511–517.

Szekely, A.M., Barbaccia, J.L., Alho, H., and Costa, E. (1989), In primary cultures of cerebellar granule cells the activation of N-methyl-D-aspartate-sensitive glutamate receptors induces c-*fos* mRNA expression. *Mol. Pharmacol.,* 35: 401–408.

Tabakoff, B. and Boggan, W.O. (1974), Effects of ethanol on serotonin metabolism in brain. *J. Neurochem.,* 22: 759–764.

Tabakoff, B., Jaffe, R.C., and Ritzmann, R.F. (1978), Corticosterone concentrations in mice during ethanol drinking and withdrawal. *J. Pharm. Pharmac.,* 30: 371–374.

Tabakoff, B., Melchior, C.L., and Hoffman, P.L. (1982), Commentary on ethanol tolerance. *Alcoholism: Clin. Exp. Res.,* 6: 252–259.

Tabakoff, B. and Rothstein, J.D. (1983), Biology of tolerance and dependence. In: *Medical and Social Aspects of Alcohol Abuse.* B. Tabakoff, P.B. Sutker, and C.L. Randall, Eds., pp. 187–220, Plenum Press, New York.

Tabakoff, B. and Hoffman, P.L. (1987), Biochemical pharmacology of alcohol. In: *Psychopharmacology — The Third Generation of Progress,* H.Y. Meltzer, Ed., pp. 1521–1526, Raven Press, New York.

Tabakoff, B. (1989), Treatment of alcoholism. *N. Engl. J. Med.,* 231: 400.

Tabakoff, B. and Hoffman, P.L. (1992), Alcohol: Neurobiology. In: *Substance Abuse. A Comprehensive Textbook.* J.H. Lowenstein, P. Ruiz, and R.B. Millman, Eds., pp. 152–185, Williams and Wilkins, Baltimore.

Tabakoff, B., Hellevuo, K., and Hoffman, P.L. Alcohol. In: *Handbook of Experimental Pharmacology. Pharmacological Aspects of Drug Dependence — Toward an Integrated Neurobehavioral Approach.* C.R. Schuster, S.W. Gust, and M.J. Kuhar, Eds., Springer-Verlag, Heidelberg, (in press).

Tabakoff, B. and Hoffman, P.L. Effect of alcohol on neurotransmitters and their receptors and enzymes. In: *Alcohol and Alcoholism,* Vol 2. H. Begleiter and B. Kissin, Eds., Oxford University Press, New York, (in press).

Takadera, T., Suzuki, R., and Mohri, T. (1990), Protection by ethanol of cortical neurons from N-methyl-D-aspartate-induced neurotoxicity is associated with blocking calcium influx. *Brain Res.,* 537: 109–115.

Tan, S.-E., Wenthold, R.J., and Soderling, T.R. (1994), Phosphorylation of AMPA-type glutamate receptors by calcium/calmodulin-dependent protein kinase II and protein kinase C in cultured hippocampal neurons. *J. Neurosci.,* 14(3): 1123–1129.

Tingley, W.G., Roche, K.W., Thompson, A.K., and Huganir, R.L. (1993), Regulation of NMDA receptor phosphorylation by alternative splicing of the COOH-terminal domain. *Nature,* 364: 70–73.

Tremwel, M.F., Anderson, K.J., and Hunter, B.E. (1994), Stability of [^{3}H]MK-801 binding sites following chronic ethanol consumption. *Alcoholism: Clin. Exp. Res.,* 18: 1004–1008.

Trevisan, L., Fitzgerald, L.W., Brose, N., Gasic, G.P., Heinemann, S.F., Duman, R.S., and Nestler, E.J. (1994), Chronic ingestion of ethanol up-regulates NMDAR1 receptor subunit immunoreactivity in rat hippocampus. *J. Neurochem.,* 62: 1635–1638.

Urushihara, H., Tohda, M., and Nomura, Y. (1992), Selective potentiation of N-methyl-D-aspartate-induced current by protein kinase C in *Xenopus* oocytes injected with rat brain RNA. *J. Biol. Chem.,* 267: 11697–11700.

Valverius, P., Crabbe, J.C., Hoffman, P.L., and Tabakoff, B. (1990), NMDA receptors in mice bred to be prone or resistant to ethanol withdrawal seizures. *Eur. J. Pharmacol.,* 184: 185–189.

Verma, I.M. (1986), Proto-oncogene *fos*: a multifaceted gene. *Trends Genet.,* 2: 93–96.

Walker, D.W. and Hunter, B.E. (1978), Short-term memory impairment following chronic alcohol consumption in rats. *Neuropsychology,* 16: 545–553.

Walker, D.W., Barnes, D.E., Zornetzer, S.F., Hunter, B.E., and Kubanis, P. (1980), Neuronal loss in hippocampus induced by prolonged ethanol consumption in rats. *Science,* 209: 711–713.

Wang, L.-Y., Salter, M.W., and MacDonald, J.F. (1991), Regulation of kainate receptors by cAMP-dependent protein kinase and phosphatases. *Science,* 253: 1132–1135.

Wang, L.-Y., Taverna, F.A., Huang, X.-P., MacDonald, J.F., and Hampson, D.R. (1993), Phosphorylation and modulation of a kainate receptor (GluR6) by cAMP-dependent protein kinase. *Science,* 259: 1173–1175.

Weight, F.F., Lovinger, D.M., White, G., and Peoples, R.W. (1991), Alcohol and anesthetic actions on excitatory amino acid-activated ion channels. *Ann. N.Y. Acad. Sci.,* 625: 97–107.

Woodward, J.J. and Gonzales, R.A. (1990), Ethanol inhibition of N-methyl-D-aspartate stimulated endogenous dopamine release from rat striatal slices: reversal by glycine. *J. Neurochem.,* 54: 712–715.

Woodward, J.J. (1994a), The effects of thiol reduction and oxidation on the inhibition of NMDA-stimulated neurotransmitter release by ethanol. *Neuropharmacology,* 33: 635–640.

Woodward, J.J. (1994b), A comparison of the effects of ethanol and the competitive glycine antagonist 7-chlorokynurenic acid on N-methyl-D-aspartic acid-induced neurotransmitter release from rat hippocampal slices. *J. Neurochem.,* 62: 987–991.

Wozniak, K.M., Pert, A., and Linnoila, M. (1990), Antagonism of 5-HT_3 receptors attenuates the effects of ethanol on extracellular dopamine. *Eur. J. Pharmacol.,* 187: 287–289.

Wozniak, K.M., Pert, A., Mele, A., and Linnoila, M. (1991), Focal application of alcohols elevates extracellular dopamine in rat brain: a microdialysis study. *Brain Res.,* 540: 31–40.

Yamakura, T., Mori, H., Shimoji, K., and Mishina, M. (1993), Phosphorylation of the carboxyl-terminal domain of the $\zeta 1$ subunit is not responsible for potentiation by TPA of the NMDA receptor channel. *Biochem. Biophys. Res. Commun.,* 196: 1537–1544.

Yamazaki, M., Mori, H., Araki, K., Mori, K.J., and Mishina, M. (1992), Cloning, expression and modulation of a mouse NMDA receptor subunit. *FEBS Lett.,* 300: 39–45.

Yoneda, Y., Han, D., and Ogita, K. (1994), Preferential induction by stress of the N-methyl-D-aspartate recognition domain in discrete structure of rat brain. *J. Neurochem.,* 63: 1863–1871.

Chapter 6

THE NICOTINIC CHOLINERGIC RECEPTOR AS A POTENTIAL SITE OF ETHANOL ACTION

Allan C. Collins

TABLE OF CONTENTS

I. INTRODUCTION

The membrane has long been considered as the major site of action for ethanol and related compounds primarily because early "structure-activity" studies of anesthetics, including alcohols, indicated that lipid solubility is a reliable predictor of anesthetic potency (Meyer, 1899; Overton, 1901). These findings led to the postulate that disruption of membrane lipid structure (fluidization) somehow results in depressant actions. This postulate is no longer in fashion, partly because fluidization of the bulk membrane lipids and anesthesia can be separated (Buck et al., 1989), and has been replaced by the suggestion that ethanol exerts its behavioral effects by modifying the activities of membrane-associated proteins (Goldstein, 1984; Franks and Lieb, 1987; Harris et al., 1987; Suzdak et al., 1988). Special attention has been paid to the effects of ethanol on membrane-associated proteins such as ligand- and voltage-gated ion channels. Curiously, the nicotinic acetylcholine (nAChR), which is the prototype of the ligand-gated ion channels, has received only limited attention.

0-8493-8389-7/95/$0.00+$.50

This review will attempt to:

1. Provide a synopsis of the interactions between alcohol and tobacco use in humans. These findings argue that ethanol effects on the nAChR should be studied. The major findings from the literature will be summarized, and testable explanations for the interactions will be offered.
2. Provide an overview of the structure, function, and regulation of "peripheral" type nAChRs, focusing on the receptor found in electric organ of *Torpedo*. The focus of this discussion will be the data that indicate that the receptor is a dynamic structure, the activities of which are modulated by the lipids that surround it.
3. Briefly summarize the brain nAChRs and discuss the similarities and differences between peripheral and brain nAChRs.
4. Provide an in depth summary and critique of the ethanol-nAChR literature, focusing on the studies that have used *Torpedo* nAChRs.

Papers have been selected from the literature that help make what I believe are important points. Seminal findings were not always cited if a more recent paper helps make the point. My hope is that this chapter will not be a dry review of the literature. Instead, I hope it will stimulate thought and discussion which will result in research which will help explain why alcoholics are almost invariably heavy smokers.

II. ALCOHOL AND TOBACCO USE

The observation that alcohol use, particularly excessive alcohol use, is almost invariably associated with tobacco use was first made by Cartwright et al. (1959) who noted that individuals who consume alcohol regularly are more likely to be cigarette smokers. The data available 10 years ago (reviewed by Istvan and Matarazzo, 1984) clearly demonstrated that smokers consume more alcohol than nonsmokers, and, among smokers, heavy smokers consume more alcohol than light smokers. Even though the prevalence of smoking has decreased in the general population since the early 1970s, smoking has not decreased markedly in alcoholics (Kozlowski et al., 1986; Bobo, 1989; Hughes, 1993, 1994). Current data (Hughes, 1994) indicate that about 70% of alcoholics are heavy smokers (>29 cigarettes per day) compared with 10% of the general population. One of every seven heavy smokers meets the criteria for active alcoholism whereas among never-smokers and ex-smokers only one of 20 meets these criteria.

Keenan et al. (1990) examined several aspects of smoking (number, years, nicotine yield, puff number, and duration) in four groups of smokers: nondrinkers, regular (low dose) drinkers, abstinent alcoholics, and current alcoholics. Active drinkers (alcoholic and nonalcoholic) smoked more than their control group, and both current and abstinent alcoholics smoked more than did either of the nonalcoholic groups. Both Keenan et al. (1990) and Hughes (1993) noted that smokers who were active alcoholics, or had been alcoholics, started smoking at an earlier age than did nonalcoholic smokers and continued to smoke more even if they discontinued drinking.

Controlled (laboratory) studies have consistently shown that alcohol administration increases smoking in male alcoholics (Mello and Mendelson, 1972; Griffiths et al., 1976; Henningfield et al., 1983, 1984), but there is some controversy whether alcohol ingestion increases tobacco use in nonalcoholics. Henningfield et al. (1984) failed to detect a consistent effect of alcohol ingestion on smoking in a short term study (90 min), but Mello et al. (1980) did detect covariation between alcohol and tobacco consumption in a 30-d study of hospitalized male subjects. More recently Mello et al. (1987) reported that tobacco use increased in harmony with alcohol intake in nonalcoholic women. Thus, an effect might not be seen in nonalcoholics during a short term experiment but, when 24 h/d data are obtained, it seems as though even nonalcoholics exhibit increases in tobacco use when alcohol is consumed.

There seems to be solid agreement that alcohol and tobacco are frequently used together, particularly by alcoholics, but potential explanations have not been provided. One possibility is alcohol and nicotine might increase one another's reinforcing effects. Since both drugs elicit dopamine release in the nucleus accumbens (Di Chiara and Imperato, 1985; Imperato et al., 1986), it may be that additive or synergistic effects are seen on dopamine release when the alcohol and tobacco are consumed together which could lead to an enhanced reinforcing effect. Another possible explanation for increases in tobacco use when alcohol is consumed is that alcohol might reduce one or more of the toxic actions of nicotine. Several studies from our group (Burch et al., 1988; Collins et al., 1988, 1993; de Fiebre and Collins, 1993; Luo et al., 1994) using mice have demonstrated that cross-tolerance develops between ethanol and nicotine using locomotor activity, body temperature, and heart rate as the drug responses. Since smokers seem to reduce their tobacco intake when toxic responses to nicotine develop (Kozlowski and Herman, 1984), alcohol-induced decreases in the toxic actions of nicotine could facilitate an increase in tobacco use.

III. PERIPHERAL-TYPE nAChRs

The interactions between ethanol and nicotine might be explained if ethanol alters brain nAChR function. Gaining an understanding of how ethanol might affect the nAChR should be facilitated by understanding the structure and regulation of function of the nAChR. Studies of the nAChR date back well over 100 years to Claude Bernard's studies of the sites of action of curare (Bernard, 1857) and Langley's studies of the actions of nicotine and curare at the neuromuscular junction (Langley, 1907–08). Indeed, the early work done with the nAChR was critical in the development of the receptor concept. Terms such as agonist, antagonist, and desensitization were developed largely on the basis of studies done with the nAChR.

Much of the early work on nAChRs focused on the skeletal muscle receptor, but our knowledge of the structure and regulation of function of the nAChRs has been facilitated, in no small way, by studying the nAChRs found in the electric organ of electric fish such as *Torpedo* and *Electrophorus*. These organs have extraordinarily high concentrations of nAChRs which made it easier to isolate and purify the receptor as well as to study function. The knowledge obtained from studies of the electric organ nAChR has been of vital importance in advancing our knowledge of the skeletal muscle nAChR, which, it turns out, is very similar to the *Torpedo* nAChR. It should not be a surprise, as a result, that our most comprehensive knowledge of ethanol effects on nAChR function has been obtained in studies that utilized the electric organ and/or skeletal muscle. A major challenge is to ascertain which, if any, of the effects that ethanol has on electric organ and skeletal muscle nAChRs apply to neuronal nAChRs.

A. STRUCTURE OF PERIPHERAL-TYPE nAChRs

The nAChR in the electric organ, like the skeletal muscle nAChR, is largely postsynaptic and is normally activated by presynaptically released acetylcholine (ACh). The receptor was first isolated from electric organ (Nachmansohn, 1959) and is made up of four subunits organized as a heterologous pentamer $2\alpha\beta\gamma\delta$ (Devillers-Thiéry et al., 1979; Raftery et al., 1980; Popot et al., 1981). Reconstitution experiments in frog oocytes (Numa, 1989) using mRNA for the four species of mRNA demonstrated that the $2\alpha\beta\gamma\delta$ oligomer contains all of the elements required for the physiological response to ACh, i.e., increased conductance of, primarily, monovalent cations. Continued debate exists concerning the precise arrangement of the five receptor subunits with the exception that it is very clear that the two α subunits are not immediately adjacent to one another, and it is likely that the γ subunit separates the two α subunits (Bon et al., 1982, 1984; Karlin, 1983, 1993; Stroud et al., 1990).

Based on primary amino acid sequence data obtained from the N-terminal ends of purified subunits (Devillers-Thiéry et al., 1979; Raftery et al., 1980), the genes coding for the four subunits were cloned and sequenced (reviewed in Numa, 1989). The four subunits are very similar and have virtually identical hydrophobicity profiles which suggest a division of each of the chains into a large (210–220 amino acids) hydrophilic N-terminal domain, a 70 amino acid sequence that is highly hydrophobic which may make up three (M1, M2, and M3) transmembrane α-helical sequences, a second variable length hydrophilic domain that may be an intracellular cytoplasmic loop, a second hydrophobic segment that may make up another transmembrane sequence (M4), and a short (~20 amino acids) C-terminal hydrophilic sequence. The most commonly accepted model (Claudio et al., 1983; Di Paolo et al., 1989) postulates that the large N-terminal domain is extracellular, the small hydrophilic domain is intracellular, and the four hydrophobic sequences (M1–M4) traverse the membrane as α-helices resulting in an extracellular C-terminal end.

B. AGONIST BINDING SITES

Until relatively recently it was thought that the α subunits were totally responsible for agonist binding. Early on it was established that each receptor molecule has two ACh binding sites found per receptor molecule (Reynolds and Karlin, 1978). One binding site is probably located on each α subunit. Binding to these two sites shows positive cooperativity (Weber and Changeux, 1974a–c; Neubig and Cohen, 1980; Pedersen and Cohen, 1990a). When binding occurs at one site the affinity of the second site is increased. It does not seem likely that the differences in binding arise because of differences in α subunit primary sequence, since both electric fish (Klarsfeld et al., 1984) and mouse (Merlie et al., 1983) have only one gene that encodes for the α protein. This suggests that the positive cooperativity seen in binding to the nAChR occurs as a result of allosteric interactions between the subunits of the receptor complex.

The binding of high affinity chemically reactive analogs of nicotinic ligands has been used successfully to identify amino acids that are involved in agonist binding. Initial studies demonstrated that the cysteine residues found at positions 192 and 193 of the *Torpedo* α subunit are critical components of the binding site (Kao et al., 1984). Consistent with this argument, the site directed mutagenesis of either of these amino acids results in receptors that are not responsive to ACh (Mishina et al., 1985). Other amino acids are also labeled by high affinity agonist alkylating agents and fall into three distinct regions, or loops, of the α subunit (Dennis, et al., 1986). They include Tyr 193 (Loop A), Trp 149 (Loop B), and Tyr 190 (Loop C). Trp 86, Tyr 151, and Tyr 198 (Dennis et al., 1986; Galzi et al., 1990) also bind to the affinity ligands but with low affinity. Other amino acids in the hydrophilic N-terminal domain have also been implicated in binding (see Changeux et al., 1992), but those listed above seem to be most critical.

Binding apparently also involves other subunits of the complex as demonstrated by the observations that: UV irradiation of [^{3}H]-α-bungarotoxin binding yields incorporation of radioactivity into the γ and δ subunits as well as the α subunit (Oswald and Changeux, 1982); carbamylcholine sensitive incorporation of the affinity ligand, N, N-(dimethylamine) benzenediazonium fluoroborate (DDF) occurs in the γ subunit (Langenbuch-Cachat et al., 1988), and [^{3}H]-d-tubocurarine photo affinity labels the α and γ subunits when it binds to its high affinity site and the α and δ subunits when it binds to its low affinity site (Pedersen and Cohen, 1990b). Similar results have been obtained with mouse muscle nAChR (Kurosaki et al., 1987; Blount and Merlie, 1989), and Sine and Claudio (1991) have argued that the γ and δ subunits affect the cooperativity of ligand binding. Changeux et al. (1992) suggest that the binding site spans the boundaries between the α-γ and α-δ subunits which would help explain nonequivalence of the two binding sites as well as cooperativity in binding.

C. ION FLUX-CHANNEL OPENING

The usual consequence of agonist binding is enhanced ion flux through the ion channel that is formed by the nAChR (Kasai et al., 1970; Kasai and Changeux, 1971; Popot et al., 1974). Changeux et al. (1992) argue that agonist binding does not induce conformational changes that result in opening of the ion channel. Rather, the channel spontaneously opens (Jackson, 1984, 1988), and agonist binding changes the equilibrium between closed and open states toward that of the open state. The precise changes involved in channel opening are unknown, but the two most readily accepted are the global twist model (Unwin and Zampighi, 1980; Furois-Corbin and Pullman, 1988) which postulates that the five subunits rigidly tilt to an angle tangential to the axis of the molecule resulting in a change in diameter of the central pore, and the bloom model (Changeux, 1990) which postulates that the top of the molecule found above the membrane opens up like an umbrella, revealing the ion channel. These, and other models that have been proposed, try to accommodate the data which suggest that the nAChR is dynamic and structure, somehow, correlates to function.

D. DESENSITIZATION

Katz and Thesleff (1957) were the first to characterize nAChR desensitization which consists of a reversible decline in agonist-induced ion flux that develops during persistent exposure to agonists. Desensitization is seen with both *Torpedo* and muscle nAChRs and probably involves at least two processes, one that occurs in the millisecond range and another that develops in tens of seconds (Sine and Taylor, 1979; Aoshima et al., 1981; Feltz and Trautmann, 1982; Udgaonkar and Hess, 1987; Forman and Miller, 1988). Since desensitization is seen in receptors that have been purified and reconstituted in lipid bilayers (McNamee and Ochoa, 1982), it seems highly likely that desensitization is a property of the receptor itself and is not due to a change in one or more related processes. However, the rates of desensitization, and recovery from desensitization, are affected by a number of factors including phosphorylation (Huganir and Greengard, 1990; Mulle et al., 1988).

Studies of ion fluxes and agonist binding using either radiolabeled ACh (Boyd and Cohen, 1980) or fluorescent dansyl C6 choline (Heidmann and Changeux, 1979, 1980) have shown that agonist activation of ion flux is a unimodal process with a K_D of ~50 μM, while binding is bimodal with two K_D values (~10 nM and 1 μM). The popular notion is that the high affinity binding occurs to pre-existent desensitized receptors. Low affinity binding occurs to the ground state or activatable form of the receptor. Preincubation of the nAChR with agonists results in a decreased ion flux response to agonists as well as loss of the low affinity binding site (Boyd and Cohen, 1980). This change in affinity is also detectable using α-bungarotoxin as a ligand. Preincubation with agonists results in a decreased initial rate of α-bungarotoxin binding to the nAChRs as compared with nAChRs that have not been preincubated with agonist. This change in affinity can be elicited by concentrations of agonist that do not activate the receptor and mirrors inactivation of the receptor.

The binding of affinity ligands such as DDF (Galzi et al., 1991) to the γ and δ subunits is altered under conditions where the receptor is stabilized in the desensitized state by the allosteric effector, meproadifen. Changes in binding to amino acids found in the A, B, and C loops of the α subunit are also seen under these circumstances (Galzi et al., 1991). These findings argue that desensitization changes the quaternary structure of the receptor oligomer as well as the tertiary structure of (at least) the α subunit.

E. INTERACTIONS WITH MEMBRANE LIPIDS

The currently accepted model of nAChR structure argues that each of the subunits has four α-helical transmembrane sequences (M1–M4). Studies have attempted to identify which of the putative transmembrane sequences is in maximal contact with the membrane lipids. The M1, M3, and M4 transmembrane sequences of all four of the receptor subunits are labeled by hydrophobic fluorescent probes (Middlemas and Raftery, 1983; Blanton and Cohen, 1992), but the M4 sequence is labeled best. The periodicity and distribution of labeling argues that the M4 region is α-helical and presents a broad face to the membrane lipid.

Even though M4 is far from agonist binding sites, it is important in regulating receptor function as revealed by point mutation studies of Cys 418. Photo affinity labeling studies indicate that the Cys 418 residue on the α subunit is located at the lipid-protein interface (Giraudat et al., 1985; Pradier et al., 1989; Li et al., 1990). Site-directed mutation of α Cys 418 to α Trp 418 results, even though this site is not directly involved in agonist binding, in a marked increase in open time of the ion channel when expressed in *Xenopus* oocytes. Mutation of α Cys 418 to Trp 418 also increases the rate of receptor desensitization and decreases the EC_{50} for acetylcholine (Lee et al., 1994). These effects, as the reader will see, are virtually identical to those elicited by ethanol. In contrast, mutation of the homologous Cys (Cys 451) found in the M4 region of the γ subunit decreased ion flux (Li et al., 1992). Thus, changes in protein structure at the putative lipid-protein interface can dramatically affect receptor function.

F. REGULATION BY LIPIDS

Perhaps the first evidence that indicated membrane lipids modulate receptor activation and desensitization came in isolation-purification-reconstitution studies (see, for examples: Heidmann et al., 1980; Jones et al., 1988). Detergent extracted nAChRs do not exhibit agonist-induced changes in α-bungarotoxin binding nor do agonists stimulate ion flux, but, if extraction procedures are used that do not strip the protein of its lipids, the allosteric properties of the receptor are maintained (Heidmann et al., 1980). Jones et al. (1988) argued that a single shell of lipids around the perimeter of the receptor (annular lipids) are absolutely required to yield a receptor that is both activated and desensitized by agonists.

Reconstitution experiments have also demonstrated that the ion permeability of the receptor is dramatically affected by lipid composition of the membrane. The receptor seems to require negatively charged lipids, such as phosphatidic acid, and a steroid, such as cholesterol, for maximal activity (Criado et al., 1982; Ochoa et al., 1983; Fong and McNamee, 1986, 1987).

Phosphatidic acid activates the nAChR and also seems to regulate receptor structure. Butler and McNamee (1993) used Fourier-transform infrared resonance spectroscopy to study the structure of *Torpedo* nAChR in reconstituted dioleoylphosphatidylcholine membranes. Note that nAChRs reconstituted in these membranes do not exhibit agonist-induced changes in binding and do not flux ions. The addition of phosphatidic acid to the reconstituted system activated the receptor and produced significant increases in the β structure of the receptor. Spin label studies have shown that phosphatidic acid has a high affinity for the receptor (Ellena et al., 1983) and the addition of nAChR to phosphatidic acid-containing lipid bilayers changes the pKa of the phosphatidic acid (Bhushan and McNamee, 1993). Therefore, it seems very likely that the negatively charged head group of the phosphatidic acid interacts with the nAChR.

Cholesterol enhances both the conductance and cooperativity of the agonist-activated ion channel when the receptor is reconstituted into lipid bilayers (Nelson et al., 1980; Schindler, 1982), but these effects are modulated by the other lipids present in the bilayer (Dalziel et al., 1980; Criado et al., 1982; Fong and McNamee, 1986). Although it is clear that cholesterol modulates receptor properties, it is not clear how this occurs. Cholesterol does not compete effectively with phospholipids for annular binding sites (Jones and McNamee, 1988).

Cholesterol also modulates receptor structure. Butler and McNamee (1993) detected significant increases in α-helix content induced by cholesterol. Criado et al. (1982) thought that cholesterol is absolutely required for allosteric state transitions, but several lipids and lipid mixtures can facilitate the allosteric transitions even in the absence of cholesterol (Fong and McNamee, 1986). Thus, it does not seem likely that the cholesterol binding sites have high structural requirements. Rather, it seems likely that the cholesterol affects the microenvironment in a way that modulates the ability of the receptor to undergo conformational changes.

G. MEMBRANE FLUIDITY AND THE nAChR

Lipid substitution often is accompanied by changes in bulk membrane fluidity, and several studies have assessed the relationship between lipid fluidity and nAChR function. Fong and McNamee (1986) studied *Torpedo* nAChRs that were reconstituted into artificial lipid membranes containing different lipid compositions including phosphatidylcholine, cholesterol, phosphatidic acid, phosphatidylethanolamine, asolectin, neutral lipid depleted asolectin, native *Torpedo* membrane lipids, and cholesterol-depleted native lipids. Phosphatidylcholines with different configurations of fatty chains were also used. Altering lipid combinations of the lipid bilayers altered the transition from low to high affinity binding as well as ion flux. None of the lipid modifications affected equilibrium carbamylcholine binding, but clear effects on the interconversion between low and high affinity binding were obtained. Similar, but not quite so dramatic, effects were seen on agonist-induced ion flux. In those systems where the affinity state transition in binding was seen, only those lipids containing both cholesterol and negatively charged phospholipids would show an agonist-induced increase in ion flux. Comparisons between binding/ion flux and order parameter of the membrane revealed a relatively small range of membrane fluidity that would support optimal receptor function. The conclusion drawn from these findings is that lipids affect receptor desensitization.

More recently, Sunshine and McNamee (1992) assessed the effects of three different lipid environments (with and without neutral lipids) on affinity state transition and ion flux. All three of the negatively charged lipid mixtures tested (phosphatidylcholine/phosphatidylserine, phosphatidylcholine/phosphatidic acid, and phosphatidylcholine/cardiolipin) gave fully functional ion channels as long as neutral lipids (e.g., cholestanol, α-tocopherol, squalene, and cholesterol) were added. With regard to the neutral lipids tested, no structural requirements were evident in that all of the lipids tested could fully replace cholesterol, but both the negatively charged lipids and neutral lipids markedly affected the rate of receptor desensitization.

In a more recent series of experiments (Sunshine and McNamee, 1994), *Torpedo* receptors were purified and reconstituted into lipid bilayers made up of sphingomyelin, phosphatidylcholines with different degrees of unsaturation, or different neutral lipids. The responses (ion flux) to agonist of nAChR-containing vesicles varied with degree of unsaturation of the phosphatidylcholines. Membranes containing fully saturated phosphatidylcholine (along with dioleoylphosphatidic acid and cholesterol — 60:20:20) were much less responsive to agonist than were membranes containing a phosphatidylcholine with oleic acid (18:1) residues. As the unsaturation of the C-18 fatty acids increased, however, the ion flux response decreased. Indeed, 18:3 fatty acid-containing phosphatidylcholines were almost nonresponsive to agonist. These findings are consistent with the earlier studies of Fong and McNamee (1986) and suggest that bulk membrane fluidity may be important in regulating receptor response. However, when bulk membrane fluidity was altered by manipulating the concentrations of sphingomyelin, agonist response did not change. Sunshine and McNamee (1994) argue that these results indicate that lipid type is far more important in regulating receptor function than is bulk membrane fluidity.

H. SUMMARY

The evidence that the electric organ nAChRs are dynamic structures that have multiple conformations which covary with changes in function (ground state, activated, and desensitized) is overwhelming. Apparently the receptor complex normally oscillates between a closed and open conformer when in the ground state and agonists serve to stabilize the receptor in the open channel, activated state. However, if activation persists, the receptor once again changes its conformation which results in increased affinity of the receptor for agonists and a loss of function (desensitization). All of these properties seem to be modulated by the membrane lipids, particularly those found in close association with the nAChR.

IV. BRAIN nAChRs

Up until the late 1970s, little was known about brain nAChRs. A binding site for α-bungarotoxin had been characterized in brain and chromatography done with α-bungarotoxin affinity columns had been used successfully to isolate and partially purify brain nAChRs, but it was not clear whether this binding site served as a neurotransmitter receptor. In the early 1980s, a second nAChR was identified that bound radiolabeled nicotine with high affinity (Romano and Goldstein, 1980). Early results showed that α-bungarotoxin and nicotine binding sites are distributed differently in brain (Marks and Collins, 1982) which suggested that they represent different brain nAChRs. In the last decade, molecular methods have been used to identify multiple brain nAChR subtypes (reviewed in Sargent, 1993; Lindstrom, 1994). At last count, molecular methods have successfully identified a large number of α subunits that are found in brain (α2 … α9) and several β subunits (β2–β4). (The muscle subunits are now designated α1, β1, etc.) The large number of subunits predicts incredible complexity, but it may not be that complex; four of the α subunits (α2, α5, α6, and α9) are found in only a few small brain regions (Marks et al., 1992), and α8 has been found only in chick brain (Schoepfer et al., 1990). The dominant subtypes of neuronal nAChR, as measured by *in situ* hybridization of mRNA, should include the α4, α7, and/or β2 gene products (Goldman et al., 1986; Wada et al., 1989; Marks et al., 1992).

Relatively little is known about the structure of the brain nAChRs. Hydrophobicity profiles of the subunits suggest that a four transmembrane protein should be formed. However, we do not have accurate knowledge of subunit composition in most cases. Receptors that seem to be responsible for the high affinity binding of nicotine seem to be made up of two α4 subunits and three β2 subunits (Whiting and Lindstrom, 1988; Flores et al., 1992). The other major subtype expressed in brain, α7, seems to code for a subunit of the protein that binds α-bungarotoxin with high affinity (Schoepfer et al., 1990). It is probable, given expression of mRNA, that selected brain regions such as striatum also contain α3/β4/α5-containing receptors (Marks et al., 1992) similar to those found in autonomic ganglia (Conroy et al., 1992).

The hydrophobicity profiles of each of the α genes predicts the protein will have four transmembrane domains found in the same approximate position as the *Torpedo* and muscle nAChRs. The overall amino acid homology between rat brain α nAChRs and rat skeletal muscle nAChR subunits is 40–55%, but the homology approaches 100% in the putative M1–M3 transmembrane domains (Sargent, 1993). The structure of the probable extracellular domains of the brain nAChR α subunits should be similar to that of the muscle-type receptor. For example, each of the α subunit genes encodes for a protein with two cysteines separated by 13 residues which is equivalent to Cys 128 and Cys 142 in the α1 subunit. Similarly, all of the neuronal α subunits have two adjacent cysteines that align with Cys 192–193 of the α1 subunit, and all, except α5, have a Tyr two amino acids upstream from these two cysteines. The only major difference in gross structure is that the cytoplasmic loop between M3 and M4 is much longer in the brain nAChRs. This is especially true for the α4-encoded protein.

There are several dissimilarities in function between peripheral-type and brain nAChRs that have emerged. Included among these are:

1. Unlike muscle type receptors, most neuronal nAChRs are not highly selective for the cations that will be transported through these ion channels (Mulle and Changeux, 1990; Fieber and Adams, 1991). Some neuronal nicotinic receptors, when expressed in oocytes, have calcium permeabilities that are greater than NMDA receptors (Vernino et al., 1992; Séguéla et al., 1993).
2. Open-channel conductances vary markedly among different brain nAChRs (Ifune and Steinbach, 1990, 1991), and the burst durations vary markedly depending upon which subunit composition of the brain nAChRs. In general, the burst durations of brain

nAChRs are longer than those recorded from muscle fibers (Kuba et al., 1989; Moss et al., 1989).
3. Agonist potencies vary among neuronal nAchRs (Luetje and Patrick, 1991; Mulle et al., 1991) and generally differ from those obtained with skeletal muscle.

Another major difference is desensitization rate. All of the brain nAChRs seem to desensitize following exposure to agonists and can be produced by concentrations of agonists that do not increase ion flux (Grady et al., 1994; Marks et al., 1994). Brain nAChRs containing the protein encoded by the α7 gene desensitize very rapidly — the rate resembles the fast desensitization of *Torpedo* and muscle nAChR (Ballivet et al., 1988; Bertrand et al., 1990; Couturier et al., 1990; Zhang et al., 1994). Receptors made up of α3 and α4 subunits desensitize more slowly than do muscle-type receptors (Connolly et al., 1992; Grady et al., 1994; Marks et al., 1994).

The primary amino acid sequences of the various brain α and β nAChR subunits predict that the receptors should have structures that resemble the peripheral-type nAChRs. It is clear, however, that the brain receptors differ functionally, but our knowledge of these differences is very preliminary. We do not know, for example, whether lipids modulate the function of the brain nAChRs and, as will be outlined below, virtually nothing is known about potential effects of ethanol on brain nAChR function.

V. ETHANOL AND THE *TORPEDO* nAChR

The evidence that ethanol affects *Torpedo* nAChRs is growing steadily. As will be outlined below, ethanol seems to affect several properties of the receptors, but a vital concern is that high concentrations of ethanol are required to detect measurable effects. This result could mean that the nAChR is not affected by ethanol at pharmacologically relevant doses. However, virtually all of the *Torpedo* nAChR studies, particularly those studies that monitored receptor function, were done at 4°C. This temperature was presumably chosen because at higher temperatures the rate of ion flux is so fast that accurate measurements are very difficult to obtain, but the gains made by decreasing temperature to 4°C may be offset by loss of sensitivity to alcohol. Thus, a major problem that must be resolved is whether ethanol affects nAChR function at reasonable concentrations.

A. EFFECTS OF ETHANOL ON BINDING

Many studies have shown that agonists will inhibit the rate of α-bungarotoxin binding to *Torpedo* membranes, but, if the membranes are preincubated with agonist, the inhibition of α-bungarotoxin binding is enhanced. The standard interpretation of this effect is: continued exposure to agonist has resulted in a conformational change that results in higher affinity of the receptor for the agonist. The greater affinity of the receptor for agonist results in greater inhibition, by agonist, of toxin binding. This change in affinity is affected by ethanol and other anesthetic agents (Young et al., 1978), and preincubation with ethanol, butanol, or octanol results in alterations in the association of α-bungarotoxin that resemble those induced by preincubation with agonist (Young and Sigman, 1981), which led to the suggestion that alcohols change the ratio of desensitized to ground state receptors and stabilize the receptor in its desensitized form.

Ethanol also affects [^{3}H]-ACh binding to *Torpedo* membranes, but this effect is seen only at subsaturating concentrations. El-Fakahany et al. (1983) reported that ethanol, propanol, and butanol (0.01–1 *M*) did not affect equilibrium binding of [^{3}H]-ACh (1 μ*M*) to the *Torpedo* receptor, whereas Firestone et al. (1986) reported that ethanol, butanol, hexanol, and octanol elicited dose-dependent increases in [^{3}H]-ACh binding. The latter studies used a concentration of ligand that produced half maximal receptor occupancy, whereas the former used a supramaximal concentration. If alcohols increase the receptor's affinity for agonists, enhanced binding should be seen only at concentrations that do not saturate the receptor. Consistent with this prediction, Firestone et al. (1986) detected decreases in K_D (increased affinity) of the receptor for ACh following the addition of alcohols.

Ethanol also affects the rate of association of [^{3}H]-ACh. Firestone et al. (1994) examined the effects of 18 general anesthetics, including eight n-alkanols on initial rates of [^{3}H]-ACh binding. According to their data, which agree with virtually all other reports, only 15–20% of native *Torpedo* nAChRs are desensitized in the absence of agonist. Treatment of the membranes with all of the anesthetics produced dose-dependent increases in the fraction of receptors that show fast association (high affinity [^{3}H]-ACh binding). The ethanol concentration that produced half-maximal increases in high affinity binding is 570 m*M*. The dose-response curves were quite steep, and the Hill coefficient for ethanol was 3.2. This effect occurs quickly; a concentration of butanol that produced a half-maximal increase in high affinity binding at 20°C resulted in a near maximal change in affinity within three minutes after the alcohol was added to the membrane preparation.

Propanol and butanol also evoke a concentration-dependent increase in high affinity [^{3}H]-ACh binding (Boyd and Cohen, 1984). This high affinity conformation appears to be identical to that stabilized by agonist as judged by the rates of reisomerization to ground state and dissociation of [^{3}H]-ACh from the receptor complex.

Additional binding data that suggest ethanol and other alcohols modulate the conformational changes involved in desensitization was obtained by El-Fakahany et al. (1983) in a study that examined the binding of several compounds, such as perhydrohistrionicotoxin (HTX), which binds in, or near, the receptor's ion channel. Ethanol did not alter equilibrium binding of HTX, but initial rates of binding were enhanced in the absence of carbamylcholine which is consistent with an effect of ethanol on open channel time.

B. ETHANOL EFFECTS ON *TORPEDO* nAChR FUNCTION

The best evidence for ethanol effects on *Torpedo* nAChR function has come from an elegant and comprehensive series of studies emanating from Keith Miller's laboratory. One of the first of these (Forman et al., 1989) examined the effects of ethanol on ACh- and carbamylcholine-induced $^{86}Rb^+$ efflux from *Torpedo* vesicles. These studies were done at 4°C presumably because at higher temperatures the ion flux occurs so quickly that effects of drugs on rate of efflux are very difficult to detect. The number of active receptor-channel complexes was reduced by pretreating the vesicle preparation with α-bungarotoxin. This is necessary for detecting changes in agonist potency and efficacy; fully 50–60% of the receptors must be blocked before reduced response to full agonists such as ACh or carbamylcholine are observed.

Forman et al. (1989) examined the effects of ethanol (0–3 *M*) on agonist-induced ion flux in experiments where ethanol and agonist were added simultaneously. When $^{86}Rb^+$ efflux was measured over a 10 s time period, ethanol had no effect on maximal efflux, but the K_A for carbamylcholine-induced increase in flux was reduced by ethanol. The control value (120 μ*M*) was significantly reduced by 300 m*M* ethanol to 52 μ*M*. At 3 *M* ethanol the K_A was nearly 200-fold less (0.67 μ*M*) than the control value. This finding is consistent with increases in affinity detected by binding studies. Virtually identical results were obtained using a quench flow method that examined flux for the first 30 ms after agonist administration. When lower concentrations of ethanol (50–200 m*M*) were tested using submaximal concentrations of agonist, significant enhancement (~20%) was seen even at 50 m*M* ethanol. These effects of ethanol were readily reversed after ethanol was removed. Subsequent studies by Miller's group (Wood et al., 1991; Tonner et al., 1992) have replicated the finding that ethanol enhances the action of agonists at the nAChR.

Studies using partial agonists have provided added insight into the actions of ethanol on the nAChR (Tonner et al., 1992; Wu et al., 1994). Nicotine has very low efficacy at the *Torpedo* nAChR, but, upon addition of ethanol, readily detectable nicotine-induced increases in $^{86}Rb^+$

efflux are seen. The dose-response curves for nicotine are bell-shaped (the nicotine dose-dependent increase in ion flux is reversed at higher concentrations). Ethanol evoked a concentration dependent decrease in K_A. An inhibition constant (K_B) can also be calculated from the negative slope of nicotine's bell-shaped dose-response curve. Ethanol produced a small dose-dependent increase in the K_B value, leading to a greater K_B/K_A ratio. At high concentrations of ethanol, nicotine will elicit an increase in maximal $^{86}Rb^+$ efflux that is about one-third of the full agonist, carbamylcholine. If initial rates of efflux are determined, ethanol makes nicotine and carbamylcholine equieffective.

Wu et al. (1994) obtained virtually identical results when suberyldicholine was used as the partial agonist. Suberyldicholine also generates bell-shaped dose-response curves, but it has somewhat greater efficacy than does nicotine. Ethanol enhanced the actions of suberyldicholine, but lower concentrations of ethanol were required to enhance the actions of this agonist than were required with nicotine. Maximal suberyldicholine-induced increases in ion flux were enhanced four-fold by 0.25 *M* ethanol, whereas 0.5 *M* ethanol was without effect in similar experiments that used nicotine as the agonist (Tonner et al., 1992). Thus, the data obtained with two partial agonists (nicotine and suberyldicholine) agree with the data obtained for full agonists; ethanol increases the apparent affinity of the nAChR for agonists, leading to enhanced ion flux at low concentrations of full agonists and substantial increases in the efficacy and potency of partial agonists.

Wu et al. (1994) argue that ethanol increases the fraction of occupied receptors that actually open, but its main action is to stabilize the channel in its open state. Since it has been argued that agonist binding stabilizes the receptor in its open channel state (Changeux et al., 1992), the observations that ethanol increases binding affinity and response to agonists as mediated by increased duration of channel opening seem entirely consistent and predict, all other things being equal, that ethanol might potentiate the actions of nicotine.

C. ETHANOL EFFECTS ON DESENSITIZATION

Receptor activation is almost invariably accompanied by a desensitization process which serves to attenuate ion flux and reduce the response of the receptor to subsequent application of agonist. Desensitization of the *Torpedo* nAChR is characterized by two processes, one that occurs very quickly and a second, slower process. Several studies from Miller's laboratory (Forman et al., 1989; Fraser et al., 1990; Wu and Miller, 1994) indicate that ethanol increases the rates of both of these desensitization processes. In the first of these studies, Forman et al. (1989) studied fast desensitization by preincubating *Torpedo* vesicles with ACh (1 μ*M*–10 m*M*) for very short time periods (0.1–0.5 s). Response to agonist was measured using a quench flow method. Ethanol (0.5 *M*) enhanced fast desensitization by increasing maximal rate of desensitization by approximately 50%, and the ACh concentration-desensitization rate curve was shifted to the left by ethanol.

Wu and Miller (1994) subsequently demonstrated that preincubation of *Torpedo* vesicles with various concentrations of ACh changed the maximal rate of fast desensitization. A 1 *M* solution of ethanol doubled the rate of desensitization and decreased the ACh concentration required to elicit 50% desensitization by more than 10-fold. Ethanol did not, however, alter the Hill coefficient. Wu and Miller (1994) concluded that ethanol alters the rate of desensitization of all states of the receptor (open, closed, liganded, and unliganded) as well as increasing the stability of the open channel form of the receptor. As a consequence of this dual action, ethanol's net effect will depend on the concentration of agonist as well as its duration of action. The greatest effect of lower concentrations of ethanol, Wu and Miller (1994) argue, will be observed at low agonist concentrations.

Slow desensitization is also enhanced by ethanol (Forman et al., 1989). In these experiments, Forman et al. (1989) preincubated *Torpedo* vesicles with subactivating (50 n*M*) concentrations of carbamylcholine and ethanol (0.25–1.0 M) for 1–30 min before activation with 5 m*M* carbamylcholine. $^{86}Rb^+$ flux was measured over a 10 s time period. Pretreatment with nonactivating concentrations of carbamylcholine resulted in decreases in ion flux evoked by activating concentrations. The desensitization rate constant increased by 50% in the presence of 0.25 *M* ethanol.

D. ETHANOL'S SITE(S) OF ACTION

The data argue that ethanol affects a minimum of three processes [agonist affinity (increase), initial ion flux (increase), and desensitization rate (increase)] which are also affected by lipids. This suggests that ethanol might modulate the receptor in much the same way as lipids do, i.e., the site of action of ethanol on the nAChR may be the "lipid" site. If ethanol is acting like lipids in modulating the nAChR, the question is, is there a specific alcohol/lipid binding site associated with the receptor, or do both alcohol and lipid work by altering bulk membrane lipid properties?

As noted previously, substitution of one lipid for another often results in an altered nAChR function. The lipid substitution also often results in altered membrane fluidity, and it was argued (Fong and McNamee, 1986) that an optimal fluidity is required for proper receptor function. In a later study, however, Sunshine and McNamee (1994) observed that the optimal fluidity varied depending on what lipids were incorporated into the bilayer and argued that bulk lipid fluidity is not the critical factor that regulates nAChR function. Since ethanol also modulates bulk membrane fluidity, several studies have tried to assess whether ethanol's effects on receptor function correlate with membrane fluidity changes. An early study that used a series of alcohols and other anesthetics (Miller et al., 1987) determined that the ability of alcohols to desensitize the nAChR correlates well with ability to disorder membranes using a C-12 nitroxide labeled stearic acid probe. This probe should report on effects in the inner core of the lipid bilayer. Miller et al. (1987) observed that ethanol would increase the fluidity of the membrane as reported by this probe, but a concentration that produced 50% increase in the desensitized form of the receptor produced only a very small (~4%) increase in bulk membrane fluidity which led to the conclusion that changes in bulk membrane fluidity do not cause desensitization. This was reinforced by a more recent study (Firestone et al., 1994) which used probes that measured fluidity at several levels of the membrane. While bulk fluidity changes evoked by the anesthetics correlated with desensitization, in all cases 50% desensitization of the receptor was evoked by concentrations that produced only marginal changes in fluidity, i.e., the anesthetics were much more potent at desensitizing the receptor than they were at producing fluidity changes. Firestone et al. (1994) argue that, because changes in cholesterol content of the membrane or temperature do not cause desensitization even though they cause changes in bulk fluidity that match or exceed those caused by anesthetics, changes in bulk membrane fluidity are not related to changes in the receptor.

While it does not seem likely that bulk lipid fluidity changes are involved in ethanol-modulated changes in nAChR function, the jury is still out on whether disruption of specific lipid-protein interactions are involved. Two studies using spin label probes (Fraser et al., 1990; Abadji et al., 1994) have obtained evidence that argues that the receptor is surrounded by a layer of lipids, the annular lipids, that have restricted mobility. The mobility of these lipids is modified by anesthetics, but the concentration dependence is different than that seen for bulk lipid. Fraser et al. (1990) reported that concentration dependent changes in annular lipid fluidity correspond more closely than do bulk lipid changes in fluidity with concentration dependent changes in desensitization when hexanol is used as the test compound. This finding argues that specific lipid-protein interactions might be disrupted by alcohols and is certainly consistent with the argument made by Sunshine and McNamee (1994) that specific lipid-protein interactions are more important in regulating nAChR function than is bulk lipid fluidity.

Abadji et al. (1994) analyzed the effects of ethanol on lipid-protein interactions using probes that monitor the fluidity of lipids in the inner core of the membrane. Their methods were capable of detecting changes in fluidity in a mobile pool of lipids and in the annular lipids. Ethanol up to 0.9 *M* did not alter the mobility of the annular lipids which argues that ethanol's actions on the nAChR are not mediated by changes in the lipid-protein interface near the center of the bilayer. It is unknown whether lipid-protein interactions at the hydrophilic edge of the bilayer are affected by ethanol.

It is also possible that ethanol may act by binding at one or more sites on one or more of the nAChR subunits. Abadji et al. (1994) labeled disulfides of reduced nAChRs with a maleimide spin label (MSL). High concentrations of ethanol (1.6 *M*) produced changes in the spectra obtained from the MSL probe, and Abadji et al. (1994) concluded that this result suggests either a direct effect of ethanol on the receptor or it (ethanol) disrupts interactions between the polar head groups of the lipids and the receptor protein.

E. STRUCTURE-ACTIVITY STUDIES

Analyses of the consequences of structural change on activity have often been very useful in identifying sites and mechanisms of drug action. As might be expected, the actions of ethanol on the nAChR have been compared with the actions of other alcohols and some surprising results have been obtained. The n-alkanols have been tested for their effects on the kinetics of binding to the *Torpedo* nAChR and, consistent with the observation that the potency of alkanols or anesthetics correlates highly with lipid solubility (Seeman, 1972), potency correlates highly with effects on desensitization. However, when function was assessed, the correlation between lipid solubility and activity breaks down. Wood et al. (1991) observed that longer chain alcohols (≥C5) will inhibit agonist-induced increases in ion flux. Methanol (3 *M*) and ethanol (2 *M*), in contrast, did not exhibit this property. These same concentrations of methanol and ethanol, however, shifted the dose-response curves for ACh-induced increases in ion flux to the left (increased agonist affinity) whereas longer chain alcohols decreased both the potency and efficacy of ACh. These results suggest that short and long chain alcohols may act at different sites associated with the receptor. This conclusion is supported by a study of the actions of a series of cycloalkanemethanols $(C_nH_{(2n-1)})$ CH_2OH (Wood et al., 1993). These alcohols would also inhibit or activate the receptor, but the properties were clearly separable, suggesting that the two effects may be mediated by physically distinct sites. Most recently Wood et al. (1995) presented data that argues that long chain alkanols inhibit the receptor by binding to a site that is distinct from, but allosterically coupled to, the agonist self-inhibition site.

F. SUMMARY

Ethanol seems to increase agonist affinity and enhances agonist-induced ion flux. It also increases the rate of nAChR desensitization and stabilizes the receptor in the high affinity desensitized state. These actions should oppose one another and suggest that alcohol might enhance receptor function under some circumstances and inhibit it under other circumstances. Given that desensitization is longer lasting, it seems probable that the dominant action would be inhibition, particularly when alcohol is consumed.

VI. ETHANOL AND THE MUSCLE nAChR

Several studies have examined the effects of ethanol on skeletal muscle nAChRs, and the results obtained are quite similar to those obtained with the *Torpedo* nAChR, i.e., ethanol can activate the receptor, alter affinity, and enhance desensitization rate.

One of the first studies of ethanol effects on the muscle nAChR (Gage et al., 1975) examined the effects of several n-alkanols (ethanol to hexanol) on toad neuromuscular junction miniature endplate potentials. Ethanol, and all of the other alcohols, prolonged the decay phase of the miniature endplate currents. Ethanol also affected iontophoretically applied, ACh-induced changes in frog neuromuscular junction current flow (Bradley et al., 1980). Ethanol treatment produced a dose dependent increase in peak current; a doubling was seen with 0.2% ethanol, and dose-response curves for ACh-induced increases in current flow were shifted to the left by ethanol. Similar results were obtained by Linder et al. (1984) who studied the effects of ethanol on mouse diaphragm miniature endplate currents; ethanol enhanced endplate current, prolonged channel lifetime, and increased the apparent affinity of ACh for its receptors. These results are consistent with the suggestion that ethanol increases the affinity of the nAChR for ACh.

Not all alcohols affect the muscle receptor in the same way. Bradley et al. (1984) studied the effects of ethanol, propanol, butanol, hexanol, and octanol on frog endplate currents. Consistent with their earlier observations, the addition of ethanol (20–200 m*M*) enhanced, in a concentration dependent way, ACh-induced increases in endplate currents. Ethanol also shifted the dose-response curves for ACh-induced increases in peak current to the left, and the ACh dissociation constant was decreased. This effect was quickly reversed upon removal of ethanol. Similar effects were obtained following the addition of propanol. In contrast, butanol, hexanol, and octanol inhibited ACh-induced increases in endplate current. Thus, short chain alcohols seem to enhance ACh effects whereas longer chain alcohols block the receptor. Long chain n-alkanols (pentanol to octanol) also reduce the mean duration of ACh-induced channel openings in rat myotubes (Murrell et al., 1991). Slightly different results were obtained by Liu et al. (1994) in a study of the effects of butanol and pentanol on ACh-induced changes in single channel activity. This study used clonal BC3H-1 cells, a muscle-derived cell line. The n-alkanols in this system increased the frequency of bursts of nAChR channels induced by both ACh and the partial agonist, decamethonium, possibly by increasing channel opening rate.

VII. ETHANOL AND THE BRAIN nAChR

The data obtained with *Torpedo* and muscle nAChRs argue that ethanol may increase receptor function under certain circumstances and inhibit under others (via accelerating desensitization). Studies dealing with the effects of ethanol on brain nAChR function are virtually nonexistent, but a study by Mancillas et al. (1986) suggests that such studies would be productive. These investigators reported that systemic administration of ethanol (0.75 and 1.5 g/kg) produced an increase in excitatory response to ACh in hippocampus; CA1 cells were more reliably affected than were CA3 cells. Responses to glutamate and GABA, in contrast, were not affected by ethanol. Unfortunately, Mancillas et al. (1986) did not attempt to ascertain whether the ACh-induced changes in firing rate are due to actions of muscarinic or nicotinic nAChRs, but these results certainly argue that such studies should be done.

Two preliminary studies (de Fiebre and Meyer, 1995; Collins et al., 1995) have examined effects of ethanol on brain nAChR function. de Fiebre and Meyer (1995) studied the effects of ethanol using *Xenopus* oocytes that had been injected with mRNA for several different brain nAChR subtypes. Oocytes that had been injected with α7 mRNA were sensitive to ethanol (ethanol inhibited agonist-induced depolarization) whereas oocytes injected with α4 and β2 mRNA were unaffected by ethanol.

Studies from my laboratory using synaptosomes obtained from mouse brain also suggest that ethanol may affect some nAChRs whereas others may be unaffected. Nicotine, and other nicotinic agonists, will stimulate the release of dopamine from synaptosomes obtained from striatum. Based on expression of mRNA (Marks et al., 1992) and sensitivity to agonists and antagonists (Grady et al., 1992), it seems likely that the dopamine release process is modulated by a receptor that contains the α3 subunit. We (Collins et al., 1995) have observed that ethanol (100–500 m*M*) will produce a concentration-dependent decrease in agonist-induced dopamine release. In contrast, similar concentrations of ethanol had no effect on nicotine-induced increases in $^{86}Rb^+$ efflux from synaptosomes obtained from thalamus. Our data (Marks et al., 1993, 1994) argue that $^{86}Rb^+$ efflux from thalamus is produced by stimulation of an α4/β2-containing nAChR. Thus, our preliminary results agree with those of de Fiebre and Meyer in that the α4/β2 variant of the nAChR may be relatively insensitive to ethanol, whereas α3- and α7-containing receptors may be inhibited by concentrations of ethanol that are physiologically relevant.

VIII. CONCLUSION

The evidence that ethanol and tobacco are used together and may even promote one another's use is almost undeniably clear. It is also clear that ethanol will alter the function of muscle type nAChRs; under some circumstances enhancement of function might be seen, but it seems most probable that the dominant action will be inhibition because of ethanol-induced increases in desensitization rate and stabilization of the receptor in the desensitized form. Most of the effects of ethanol require large concentrations, but almost all of the studies have been done at temperatures that are below the transition temperature of mammalian membrane lipids. Given that lipids modulate nAChR function, it may very well be that the potency of ethanol will vary with temperature. Perhaps *Torpedo* and muscle nAChRs will be affected by physiologically relevant concentrations of ethanol at higher temperatures.

Many of the studies done with *Torpedo* and muscle nAChRs are done in the hope that they may provide insight into how ethanol might affect brain nAChRs. All of the brain nAChRs share structured similarities with the *Torpedo* and muscle nAChRs, but differences in function have been identified. It will be important to determine whether the similarities include similar response to ethanol and whether the differences include sensitivity to low concentrations of ethanol. Moreover, it may be that different subtypes have differential sensitivity to ethanol (preliminary data from our laboratory and that of Chris de Fiebre suggest this). Resolving these issues could provide the key to understanding why alcohol use promotes tobacco use, and perhaps vice versa.

ACKNOWLEDGMENTS

Studies emanating from the author's laboratory were supported by grants from the National Institute on Alcohol Abuse and Alcoholism (AA-06391) and the National Institute on Drug Abuse (DA-03194). The author is supported, in part, by a Research Scientist Award from the National Institute on Drug Abuse (DA-00197). Special thanks are due to Dawn Caillouet for assistance in preparation of the manuscript.

REFERENCES

Abadji, V., Raines, D. E., Dalton, L. A., and Miller, K. W., 1994: Lipid-protein interactions and protein dynamics in vesicles containing the nicotinic acetylcholine receptor: a study with ethanol. *Biochim. Biophys. Acta,* 1194: 25–34.

Aoshima, H., Cash, D. J., and Hess, G. P., 1981: Mechanism of inactivation (desensitization) of acetylcholine receptor. Investigations by fast reaction techniques with membrane vesicles. *Biochemistry,* 20: 3467–3474.

Ballivet, M., Nef, P., Couturier, S., Rungger, D., Bader, C. R., Bertrand, D., and Cooper, E., 1988: Electrophysiology of a chick neuronal nicotinic acetylcholine receptor expressed in *Xenopus* oocytes after cDNA injection. *Neuron,* 1: 847–852.

Bernard, M. C., 1857: Leçon sur les effects des substances toxiques et médicamenteuses. Paris: Baillière, 238–306.

Bertrand, D., Ballivet, M., and Rungger, D., 1990: Activation and blocking of neuronal nicotinic acetylcholine receptor reconstituted in *Xenopus* oocytes. *Proc. Natl. Acad. Sci. U.S.A.,* 87: 1993–1997.

Bhushan, A. and McNamee, M. G., 1993: Correlation of phospholipid structure with functional effects on the nicotinic acetylcholine receptor: a modulatory role for phosphatidic acid. *Biophys. J.,* 64: 716–723.

Blanton, M. P. and Cohen, J. B., 1992: Mapping the lipid-exposed regions in the *Torpedo californica* nicotinic acetylcholine receptor. *Biochemistry,* 31: 3738–3750.

Blount, P. and Merlie, J. P., 1989: Molecular basis of the two nonequivalent ligand binding sites of the muscle nicotinic acetylcholine receptor. *Neuron,* 3: 349–357.

Bobo, J. K., 1989: Nicotine dependence and alcoholism epidemiology and treatment. *J. Psychoactive Drugs,* 21: 323–329.

Bon, F., Lebrun, E., Gomel, J., Van Rappenbusch, R., Cartaud, J., Popot, J. L., and Changeux, J.-P., 1982: Orientation relative de deux oligomères constituant la forme lourde du récepteur de l'acétylcholine chez la *Torpille marbrée. C. R. Acad. Sci. Paris,* 295: 199–203.

Bon, F., Lebrun, E., Gomel, J., Van Rappenbusch, R., Cartaud, J., Popot, J. L., and Changeux, J.-P., 1984: Image analysis of the heavy form of the acetylcholine receptor from *Torpedo marmorata. J. Mol. Biol.,* 176: 205–237.

Boyd, N. D. and Cohen, J. B., 1980: Kinetics of binding of [^{3}H]acetylcholine and [^{3}H]carbamoylcholine to *Torpedo* postsynaptic membranes: slow conformational transitions of the cholinergic receptor. *Biochemistry,* 19: 5344–5358.

Boyd, N. D. and Cohen, J. B., 1984: Desensitization of membrane-bound *Torpedo* acetylcholine receptor by amine noncompetitive antagonists and aliphatic alcohols: studies of [^{3}H]acetylcholine binding and $^{22}Na^+$ ion fluxes. *Biochemistry,* 23: 4023–4033.

Bradley, R. J., Peper, K., and Sterz, R., 1980: Postsynaptic effects of ethanol at the frog neuromuscular junction. *Nature,* 284: 60–62.

Bradley, R. J., Sterz, R., and Peper, K., 1984: The effects of alcohols and diols at the nicotinic acetylcholine receptor of the neuromuscular junction. *Brain Res.,* 295: 101–112.

Buck, K. J., Allan, A. M., and Harris, R. A., 1989: Fluidization of brain membranes by A_2C does not produce anesthesia and does not augment muscimol-stimulated $^{36}Cl^-$ influx. *Eur. J. Pharmacol.,* 160: 359–367.

Burch, J. B., De Fiebre, C. M., Marks, M. J., and Collins, A. C., 1988: Chronic ethanol or nicotine treatment results in partial cross-tolerance between these agents. *Psychopharmacology,* 95: 452–458.

Butler, D. H. and McNamee, M. G., 1993: FTIR analysis of nicotinic acetylcholine receptor secondary structure in reconstituted membranes. *Biochim. Biophys. Acta,* 1150: 17–24.

Cartwright, A., Martin, F. M., and Thomson, J. G., 1959: Distribution and development of smoking habits. *Lancet,* 2: 725–727.

Changeux, J.-P., 1990: Functional architecture and dynamics of the nicotinic acetylcholine receptor: an allosteric ligand-gated ion channel. *Fidia Research Foundation Neuroscience Award Lectures,* vol. 4 (eds. J.-P. Changeux, R. R. Llinas, D. Purves, and F. E. Bloom), pp. 21–168; Raven Press Ltd., New York.

Changeux, J.-P., Galzi, J.-L., Devillers-Thiéry, A., and Bertrand, D., 1992: The functional architecture of the acetylcholine nicotinic receptor explored by affinity labelling and site-directed mutagenesis. *Q. Rev. Biophys.,* 25: 395–432.

Claudio, T., Ballivet, M., Patrick, J., and Heinemann, S., 1983: Nucleotide and deduced amino acid sequences of *Torpedo californica* acetylcholine receptor gamma-subunit. *Proc. Natl. Acad. Sci. U.S.A.,* 80: 1111–1115.

Collins, A. C., Burch, J. B., De Fiebre, C. M., And Marks, M. J., 1988: Tolerance to and cross tolerance between ethanol and nicotine. *Pharmacol. Biochem. Behav.,* 29: 365–373.

Collins, A. C., Romm, E., Selvaag, S., and Marks, M. J., 1993: A comparison of the effects of chronic nicotine infusion on tolerance to nicotine and cross-tolerance to ethanol in Long-Sleep and Short-Sleep mice. *J. Pharmacol. Exp. Ther.,* 266: 1390–1397.

Collins, A. C., Grady, S. R., and Marks, M. J., 1995: Ethanol selectively modulates brain nicotinic receptor function. *Alcoholism: Clin. Exp. Res.,* 19: 502, P87A.

Connolly, J., Boulter, J., and Heinemann, S. F., 1992: α4-2β2 and other nicotinic acetylcholine receptor subtypes as targets of psychoactive and addictive drugs. *Br. J. Pharmacol.,* 105: 657–666.

Conroy, W. G., Vernallis, A. B., and Berg, D. K., 1992: The α5 gene product assembles with multiple acetylcholine receptor subunits to form distinctive receptor subtypes in brain. *Neuron,* 9: 679–691.

Couturier, S., Bertrand, D., Matter, J.-M., Hernandez, M.-C., Bertrand, S., Millar, N., Valera, S. Barkas, T., and Ballivet, M., 1990: a neuronal nicotinic acetylcholine receptor subunit (α7) is developmentally regulated and forms a homo-oligomeric channel blocked by α-Btx. *Neuron,* 5: 847–856.

Criado, M., Eibl, H., and Barrantes, F. J., 1982: Effects of lipids on acetylcholine receptor. Essential need of cholesterol for maintenance of agonist-induced state transitions in lipid vesicles. *Biochemistry,* 21: 3622–3629.

Dalziel, A. W., Rollins, E. S., and McNamee, M. G., 1980: The effect of cholesterol on agonist-induced flux in reconstituted acetylcholine receptor vesicles. *FEBS Lett.,* 122: 193–196.

De Fiebre, C. M. and Collins, A. C., 1993: A comparison of the development of tolerance to ethanol and cross-tolerance to nicotine following chronic ethanol treatment in Long-Sleep and Short-Sleep mice. *J. Pharmacol. Exp. Ther.,* 266: 1398–1406.

De Fiebre, C. M. and Meyer, E., 1995: Ethanol affects the function of nicotinic receptor subtypes expressed in *Xenopus* oocytes. *Alcoholism Clin. Exp. Res.,* 19: 504, P87A.

Dennis, M., Giraudat, J., Kotzyba-Hibert, F., Goeldner, M., Hirth, C., Chang, J. Y., and Changeux, J.-P. 1986: A photoaffinity ligand of the acetylcholine binding site predominantly labels the region 179–207 of the α-subunit on native acetylcholine receptor from *Torpedo marmorata. FEBS Lett.,* 207: 243–249.

Devillers-Thiéry, A., Changeux, J.-P., Paroutaud, P., and Strosberg, A. D., 1979: The amino-terminal sequence of the 40000 molecular weight subunit of the acetylcholine receptor protein from *Torpedo marmorata. FEBS Lett.,* 104: 99–105.

Di Chiara, G. and Imperato, A., 1985: Ethanol preferentially stimulates dopamine release in the nucleus accumbens of freely moving rats. *Eur. J. Pharmacol.,* 115: 131.

DiPaola, M. Czajkowski, C., and Karlin, A., 1989: The sidedness of the COOH terminus of the acetylcholine receptor δ-subunit. *J. Biol. Chem.,* 264: 1–7.

El-Fakahany, E. F., Miller, E. R., Abbassy, M. A., Eldefrawi, A. T., and Eldefrawi, M. E., 1983: Alcohol modulation of drug binding to the channel sites of the nicotinic acetylcholine receptor. *J. Pharmacol. Exp. Ther.,* 224: 289–296.

Ellena, J. F., Blazing, M. A., and McNamee, M. G., 1983: Lipid-protein interactions in reconstituted membranes containing acetylcholine receptor. *Biochemistry,* 22: 5523.

Feltz, A. and Trautmann, A., 1982: Desensitization at the frog neuromuscular junction: a biphasic process. *J. Physiol.,* 322: 257–272.

Fieber, L. A. and Adams, D. J., 1991: Acetylcholine-evoked currents in cultured neurones dissociated from rat parasympathetic cardiac ganglia. *J. Physiol.,* 434: 215–237.

Firestone, L. L., Alifimoff, J. K., and Miller, K. W., 1994: Does general anesthetic-induced desensitization of the *Torpedo* acetylcholine receptor correlate with lipid disordering? *Mol. Pharmacol.,* 46: 508–515.

Firestone, L. L., Sauter, J.-F., Braswell, L. M., and Miller, K. W., 1986: Actions of general anesthetics on acetylcholine receptor-rich membranes from *Torpedo californica. Anesthesiology,* 64: 694–702.

Flores, C. M., Rogers, S. W., Pabreza, L. A., Wolfe, B. B., and Kellar, K. J., 1992: A subtype of nicotinic cholinergic receptor in rat brain is composed of α4 and β2 subunits and is up-regulated by chronic nicotine treatment. *Mol. Pharmacol.,* 41: 31–37.

Fong, T. M. and McNamee, M. G., 1986: Correlation between acetylcholine receptor function and structural properties of membranes. *Biochemistry,* 25: 830–840.

Fong, T. M. and McNamee, M. G., 1987: Stabilization of acetylcholine receptor secondary structure by cholesterol and negatively charged phospholipids in membranes. *Biochemistry,* 26: 3871–3880.

Forman, S. A. and Miller, K. W., 1988: High acetylcholine concentrations cause rapid inactivation before fast desensitization in nicotinic acetylcholine receptors from *Torpedo. Biophys. J.,* 54: 149–158.

Forman, S. A., Righi, D. L., and Miller, K. W., 1989: Ethanol increases agonist affinity for nicotinic receptors from *Torpedo. Biochim. Biophys. Acta,* 987: 95–103.

Franks, N. P. and Lieb, W. R., 1987: What is the molecular nature of general anesthetic target sites? *Trends Pharmacol. Sci.,* 8: 169.

Fraser, D. M., Louro, S. R., Horváth, L. I., Miller, K. W., and Watts, A., 1990: A study of the effect of general anesthetics on lipid-protein interactions in acetylcholine receptor enriched membranes from *Torpedo nobiliana* using nitroxide spin-labels. *Biochemistry,* 29: 2664–2669.

Furois-Corbin, S. and Pullman, A., 1988: Theoretical study of potential ion-channels formed by bundles of α-helices. Partial modelling of the acetylcholine receptor channel. In *Transport through Membranes: Carriers, Channels and Pumps* (Eds. A. Pullman et al.), pp. 337–357; Kluwer Academic Publishers, Austin, TX.

Gage, P. W., McBurney, R. N., and Schneider, G. T., 1975: Effects of some aliphatic alcohols on the conductance change caused by a quantum of acetylcholine at the toad end-plate. *J. Physiol.,* 244: 409–429.

Galzi, J.-L., Revah, F., Black, D., Goeldner, M., Hirth, C., and Changeux, J.-P., 1990: Identification of a novel amino acid α-Tyr_{93} within the active site of the acetylcholine receptor by photoaffinity labeling: additional evidence for a three-loop model of the acetylcholine binding site. *J. Biol. Chem.,* 265: 10430–10437.

Galzi, J.-L., Revah, F., Bouet, F., Menez, A., Goeldner, M., Hirth, C., and Changeux, J.-P., 1991: Allosteric transitions of the acetylcholine receptor probed at the amino acid level with a photolabile cholinergic ligand. *Proc. Natl. Acad. Sci. U.S.A.,* 88: 5051–5055.

Giraudat, J., Montecucco, C., Bisson, R., and Changeux, J., 1985: Transmembrane topology of the acetylcholine receptor probed with photoreactive phospholipids. *Biochemistry,* 24: 3121–3127.

Goldman, D., Simmons, D., Swanson, L. W., Patrick, J., and Heinemann, S., 1986: Mapping of brain areas expressing RNA homologous to two different acetylcholine receptor α-subunit cDNAs. *Neurobiology,* 83: 4076–4080.

Goldstein, D. B., 1984: The effects of drugs on membrane fluidity, *Annu. Rev. Pharmacol. Toxicol.,* 24: 43.

Grady, S. R., Marks, M. J., and Collins, A. C., 1994: Desensitization of nicotine-stimulated [^{3}H]dopamine release from mouse striatal synaptosomes. *J. Neurochem.,* 62: 1390–1398.

Grady, S., Marks, M. J., Wonnacott, S., and Collins, A. C., 1992: Characterization of nicotinic receptor mediated [^{3}H]dopamine release from synaptosomes prepared from mouse striatum. *J. Neurochem.,* 59: 848–856.

Griffiths, R. R., Bigelow, G. E., and Liebson I., 1976: Facilitation of human tobacco self-administration by ethanol: a behavioral analysis. *J. Exp. Anal. Behav.,* 25: 279–292.

Harris, R. A., Burnett, R., McQuilkin, S., McClard, A., and Simon, F. R., 1987: Effects of ethanol on membrane order: fluorescence studies. *Ann. N. Y. Acad. Sci.,* 492: 125.

Heidmann, T. and Changeux, J.-P., 1979: Fast kinetic studies on the interaction of a fluorescent agonist with the membrane-bound acetylcholine receptor from *Torpedo marmorata. Eur. J. Biochem.,* 94: 281–296.

Heidmann, T. and Changeux, J.-P., 1980: Interaction of a fluorescent agonist with the membrane-bound acetylcholine receptor from *Torpedo marmorata* in the millisecond time range: resolution of an 'intermediate' conformational transition and evidence for positive cooperative effects. *Biochem. Biophys. Res. Commun.,* 97: 889–896.

Heidmann, T., Sobel, A., Popot, J.-L., and Changeux, J.-P., 1980: Reconstitution of a functional acetylcholine receptor. *Eur. J. Biochem.,* 110: 35–55.

Henningfield, J. E., Chait, L. D., and Griffiths, R. R., 1983: Cigarette smoking and subjective response in alcoholics: effects of pentobarbital. *Clin. Pharmacol. Ther.,* 33: 806–812.

Henningfield, J. E., Chait, L. D., and Griffiths, R. R., 1984: Effects of ethanol on cigarette smoking by volunteers without histories of alcoholism. *Psychopharmacology,* 82: 1–5.

Huganir, R. L. and Greengard, P., 1990: Regulation of neurotransmitter receptor desensitization by protein phosphorylation. *Neuron,* 5: 555–567.

Hughes, J. R., 1993: Treatment of smoking cessation in smokers with past alcohol/drug problems. *J. Subst. Abuse Treat.,* 10: 181–187.

Hughes, J. R., 1994: Smoking and alcoholism. In: *Behavioral Approaches to Addiction* (Eds. J. L. Cos and D. K. Hatsukami) pp. 1–3; Belle Mead, NJ: Cahners Healthcare Communications.

Ifune, C. K. and Steinbach, J. H., 1990: Rectification of acetylcholine-elicited currents in PC12 pheochromocytoma cells. *Proc. Natl. Acad. Sci. U.S.A.,* 87: 4794–4798.

Ifune, C. K. and Steinbach, J. H., 1991: Voltage-dependent block by magnesium of neuronal nicotinic acetylcholine receptor channels in rat pheochromocytoma cells. *J. Physiol.,* 443: 683–701.

Imperato, A., Mulas, A., and Di Chiara, G., 1986: Nicotine preferentially stimulates dopamine release in the limbic system of freely moving rats. *Eur. J. Pharmacol.,* 132: 337–338.

Istvan, J. and Matarazzo, J. D., 1984: Tobacco, alcohol and caffeine use: a review of their interrelationships. *Psychol. Bull.,* 95: 301.

Jackson, M. B., 1984: Spontaneous openings of the acetylcholine receptor channel. *Proc. Natl. Acad. Sci. U.S.A.,* 81: 3901–3904.

Jackson, M. B., 1988: Dependence of acetylcholine receptor channel kinetics on agonist concentration in cultured mouse muscle fibres. *J. Physiol.,* 397: 555–583.

Jones, O. T., Eubanks, J. H., Earnest, J. P., and McNamee, M. G., 1988: A minimum number of lipids are required to support the functional properties of the nicotinic acetylcholine receptor. *Biochemistry,* 27: 3733–3742.

Jones, O. T. and McNamee, M. G., 1988: Annular and nonannular binding sites for cholesterol associated with the nicotinic acetylcholine receptor. *Biochemistry,* 27: 2364–2374.

Kao, P. N., Dwork, A. J., Kaldany, R. R. J., Silver, M. L., Wideman, J., Stein, S., and Karlin, A., 1984: Identification of the alpha-subunit half-cystine specifically labeled by an affinity reagent for the acetylcholine receptor binding site. *J. Biol. Chem.,* 259: 11662–11665.

Karlin, A., 1983: Anatomy of a receptor. *Neurosci. Commun.,* I: 111–123.

Karlin, A., 1993: Structure of nicotinic acetylcholine receptors. *Curr. Opin. Neurobiol.,* 3: 299–309.

Kasai, M., Podleski, T. R., and Changeux, J.-P., 1970: Some structural properties of excitable membranes labeled by fluorescent probes. *FEBS Lett.,* 7: 13–19.

Kasai, M. and Changeux, J.-P., 1971: *In vitro* excitation of purified membrane fragments by cholinergic agonists. I. Pharmacological properties of the excitable membrane fragments. II. The permeability change caused by cholinergic agonists. III. Comparison of the dose response curves to decamethonium with the corresponding binding curves of decamethonium to the cholinergic receptor. *J. Membr. Biol.,* 6: 1–80.

Katz, B. and Thesleff, S., 1957: A study of the 'desensitization' produced by acetylcholine at the motor end-plate. *J. Physiol.,* 138: 63–80.

Keenan, R. M., Hatsukami, D. K., Pickens, R. W., Gust, S. W., and Strelow, L. J., 1990: The relationship between chronic ethanol exposure and cigarette smoking in the laboratory and the natural environment. *Psychopharmacology,* 100: 77–83.

Klarsfeld, A., Devillers-Thiéry, A., Giraudat, J., and Changeux, J.-P., 1984: A single gene codes for the nicotinic acetylcholine receptor alpha-subunit in *Torpedo marmorata*: structural and developmental implications. *EMBO J.,* 3: 35–41.

Kozlowski, L. T. and Herman, C. P., 1984: The interaction of psychosocial and biological determinants of tobacco use: more on the boundary model. *J. Appl. Soc. Psychol.,* 14: 244–256.

Kozlowski, L. T., Jelinek, L. C., and Pope, M. A., 1986: Cigarette smoking among alcohol abusers: a continuing and neglected problem. *Can. J. Public Health,* 77: 205–207.

Kuba, K. Tanaka, E., Kumamoto, E., and Minota, S., 1989: Patch clamp experiments on nicotinic acetylcholine receptor-ion channels in bullfrog sympathetic ganglion cells. *Pflugers Arch.,* 414: 105–112.

Kurosaki, T., Fukuda, K., Konno, T., Mori, Y., Tanaka, K. I., Mishina, M., and Numa, S., 1987: Functional properties of nicotinic acetylcholine receptor subunits expressed in various combinations. *FEBS Lett.,* 214: 253–258.

Langenbuch-Cachat, J., Bon, C., Goeldner, M., Hirth, C., and Changeux, J.-P., 1988: Photoaffinity labeling by aryldiazonium derivatives of *Torpedo marmorata* acetylcholine receptor. *Biochemistry,* 27: 2337–2345.

Langley, J. N., 1907–08: On the contraction of muscle, chiefly in relation to the presence of "receptive" substances. *J. Physiol.,* 36: 347–384.

Lee, Y.-H., Li, L., Lasalde, J., Rojas, L., McNamee, M., Ortiz-Miranda, S. I., and Pappone, P., 1994: Mutations in the M4 domain of *Torpedo californica* acetylcholine receptor dramatically alter ion channel function. *Biophys. J.*, 66: 646–653.

Li, L., Schuchard, M., Palma, A., Pradier, L., and McNamee, M. G., 1990: Functional role of the cysteine 451 thiol group in the M4 helix of the γ subunit of *Torpedo californica* acetylcholine receptor. *Biochemistry,* 29: 5428–5436.

Li, L., Lee, Y.-H., Pappone, P., Palma, A., and McNamee, M. G., 1992: Site-specific mutations of nicotinic acetylcholine receptor at the lipid-protein interface dramatically alter ion channel gating. *Biophys. J.*, 62: 61–63.

Linder, T. M., Pennefather, P., and Quastel, D. M. J., 1984: The time course of miniature endplate currents and its modification by receptor blockade and ethanol. *J. Gen. Physiol.*, 83: 435–468.

Lindstrom, J. M., 1994: Nicotinic acetylcholine receptors. In: *Handbook of Receptors and Channels: Ligand- and Voltage-gated Ion Channels.* (Ed. R. A. North) pp. 153–175; Boca Raton, FL: CRC Press.

Liu, Y., Dilger, J. P., and Vidal, A. M., 1994: Effects of alcohols and volatile anesthetics on the activation of nicotinic acetylcholine receptor channels. *Mol. Pharmacol.*, 45: 1235–1241.

Luetje, C. W. and Patrick, J., 1991: Both α- and β-subunits contribute to the agonist sensitivity of neuronal nicotinic acetylcholine receptors. *J. Neurosci.*, 11: 837–845.

Luo, Y. Marks, M. J., and Collins, A. C., 1994: Genotype regulates the development of tolerance to ethanol and cross-tolerance to nicotine. *Alcohol,* 11: 167–176.

Mancillas, J. R., Siggins, G. R., and Bloom, F. E., 1986: Systemic ethanol: selective enhancement of responses to acetylcholine and somatostatin in hippocampus. *Science,* 231: 161–163.

Marks, M. J. and Collins, A. C., 1982: Characterization of nicotine binding in mouse brain and comparison with the binding of alpha-bungarotoxin and quiliculidinyl benzilate. *Mol. Pharmacol.*, 22: 554–564.

Marks, M. J., Farnham, D. A., Grady, S. R., and Collins, A. C., 1993: Nicotinic receptor function determined by stimulation of rubidium efflux from mouse brain synaptosomes. *J. Pharmacol. Exp. Ther.*, 264: 542–552.

Marks, M. J., Grady, S. R., Yang, J.-M., Lippiello, P. M., and Collins, A. C., 1994: Desensitization of nicotine-stimulated $^{86}Rb^{+}$ efflux from mouse brain synaptosomes. *J. Neurochem.*, 63: 2125–2135.

Marks, M. J., Pauly, J. R., Gross, S. D., Deneris, E. S., Hermans-Borgmeyer, I., Heinemann, S. F., and Collins, A. C., 1992: Nicotine binding and nicotinic receptor subunit RNA after chronic nicotine treatment. *J. Neurosci.*, 12: 2765–2784.

McNamee, M. G. and Ochoa, E. L. M., 1982: Reconstitution of acetylcholine receptor function in model membranes. *Neuroscience,* 7: 2305–2319.

Mello, N. K. and Mendelson, J. H., 1972: Drinking patterns during work-contingent and noncontingent alcohol acquisition. *Psychosom. Med.*, 34: 139–164.

Mello, N. K., Mendelson, J. H., and Palmieri, S. L., 1987: Cigarette smoking by women: interactions with alcohol use. *Psychopharmacology,* 93: 8–15.

Mello, N. K., Mendelson, J. H., Sellers, M. L., and Kuehnle, J. C., 1980: Effect of alcohol and marihuana on tobacco smoking. *Clin. Pharmacol. Ther.*, 27: 202–209.

Merlie, J. P., Sebbane, R., Gardner, S., and Lindstrom, J., 1983: cDNA clone for the alpha-subunit of the acetylcholine receptor from the mouse muscle cell line BC_3H-I. *Proc. Natl. Acad. Sci. U.S.A.*, 80: 3845–3849.

Meyer, H., 1899: Welche eigenschaft der anasthetica bedingt ihre narkotische wirkung. *Arch. Exp. Pathol. Pharmakol.*, 42: 109.

Middlemas, D. S. and Raftery, M. A., 1983: Exposure of acetylcholine receptor to the lipid bilayer. *Biochem. Biophys. Res. Commun.*, 115: 1075–1082.

Miller, K. W., Firestone, L. L., and Forman, S. A., 1987: General anesthetic and specific effects of ethanol on acetylcholine receptors. *Ann. N. Y. Acad. Sci.*, 492: 71–87.

Mishina, M., Tobimatsu, T., Imoto, K., Tanaka, K., Fujita, Y. Fukuda, K., Kurasaki, M., Takahashi, H., Morimoto, Y., Hirose, T., Inayama, S., Takahashi, T., Kuno, M., and Numa, S., 1985: Location of functional regions of acetylcholine receptor alpha-subunit by site-directed mutagenesis. *Nature,* 313: 364–369.

Moss, B. L., Schuetze, S. M., and Role, L. W., 1989: Functional properties and developmental regulation of nicotinic acetylcholine receptors on embryonic chicken sympathetic neurons. *Neuron,* 3: 597–607.

Mulle, C., Benoit, P., Pinset, C. Roa, M., and Changeux, J.-P., 1988: Calcitonin gene-related peptide enhances the rate of desensitization of the nicotinic acetylcholine receptor in cultured mouse muscle cells. *Proc. Natl. Acad. Sci. U.S.A.*, 85: 5728–5732.

Mulle, C. and Changeux, J.-P., 1990: A novel type of nicotinic receptor in the rat central nervous system characterized by patch-clamp techniques. *J. Neurosci.*, 10: 169–175.

Mulle, C., Vidal, C., Benoit, P., and Changeux, J.-P., 1991: Existence of different subtypes of nicotinic acetylcholine receptors in the rat habenulo-interpeduncular system. *J. Neurosci.*, 11: 2588–2597.

Murrell, R. D., Braun, M. S., and Haydon, D. A., 1991: Actions of *n*-alcohols on nicotinic acetylcholine receptor channels in cultured rat myotubes. *J. Physiol.*, 437: 431–448.

Nachmansohn, D., 1959: *Chemical and Molecular Basis of Nerve Activity.* pp. 235; New York: Academic Press.

Nelson, N., Anholt, R., Lindstrom, J., and Montal, M., 1980: Reconstitution of purified acetylcholine receptor with functional ion channels in planar lipid bilayers. *Proc. Natl. Acad. Sci. U.S.A.*, 77: 3057–3061.

Neubig, R. R. and Cohen, J. B., 1980: Permeability control by cholinergic receptors in *Torpedo* post synaptic membranes: agonist dose response relations measured at second and millisecond times. *Biochemistry,* 19, 2770–2779.

Numa, S., 1989: A molecular view of neurotransmitter receptors and ionic channels. *Harvey Lect.,* Series 83, pp. 121–165.

Ochoa, E. L. M., Dalziel, A. W., and McNamee, M. G., 1983: Reconstitution of acetylcholine receptor function in lipid vesicles of defined composition. *Biochim. Biophys. Acta,* 727: 151.

Oswald, R. E. and Changeux, J.-P., 1982: Crosslinking of alpha-bungarotoxin to the acetylcholine receptor from *Torpedo marmorata* by ultraviolet light irradiation. *FEBS Lett.,* 139: 225–229.

Overton, E., 1901: *Studien über die Narkose* (Fischer, Jena).

Pedersen, S. E. and Cohen, J. B., 1990a: d-tubocurarine binding sites are located at α-γ and α-δ subunit interfaces of the nicotinic acetylcholine receptor. *Proc. Natl. Acad. Sci. U.S.A.,* 87: 2785–2789.

Pedersen, S. E. and Cohen, J. B., 1990b: [^{3}H]meproadifen mustard reacts with glu-262 of the nicotinic acetylcholine receptor (AChR) α-subunit. *Biophys. J.,* 57, 126a.

Popot, J. L., Sugiyama, H., and Changeux, J.-P., 1974: Démonstration de la désensibilisation pharmacologique du récepteur de l'acetylcholine *in vitro* avec des fragments de membrane excitable de *Torpille. C. R. Acad. Sci. Paris,* 279: 1721–1724.

Popot, J. L., Cartaud, J., and Changeux, J.-P., 1981: Reconstitution of a functional acetylcholine receptor: incorporation into artificial lipid vesicles and pharmacology of the agonist-controlled permeability changes. *Eur. J. Biochem.,* 118: 203–214.

Pradier, L., Yee, A. S., and McNamee, M. G., 1989: Use of chemical modifications and site-directed mutagenesis to probe the functional role of thiol groups on the γ subunit of *Torpedo californica* acetylcholine receptor. *Biochemistry,* 28: 6562–6571.

Raftery, M. A., Hukapiller, M., Strader, C. D., and Hood, L. E., 1980: Acetylcholine receptor complex of homologous subunits. *Science,* 208: 1454–1457.

Reynolds, J. A. and Karlin, A., 1978: Molecular weight in detergent solution of acetylcholine receptor from *Torpedo californica. Biochemistry,* 17: 2035–2038.

Romano, C. and Goldstein, A. M., 1980: Stereospecific nicotine receptors in rat brain. *Science,* 210: 647–649.

Sargent, P. B., 1993: The diversity of neuronal nicotinic acetylcholine receptors. In: *Annu. Rev. Neurosci.,* Vol. 16. (Eds: W. M. Cowan, E. M. Shooter, C. F. Stevens, and R. F. Thompson) pp. 403–443; Palo Alto, CA: Annual Reviews Inc.

Schindler, H., 1982: Reconstitution of acetylcholine receptor in planar bilayers. *Neurosci. Res. Prog. Bull.,* 20: 295–301.

Schoepfer, R., Conroy, W. G., Whiting, P., Gore, M., and Lindstrom, J., 1990: Brain α-bungarotoxin binding protein cDNAs and MAbs reveal subtypes of this branch of the ligand-gated ion channel gene superfamily. *Neuron,* 5: 35–48.

Seeman, P., 1972: The membrane actions of anesthetics and tranquilizers. *Pharmacol. Rev.,* 24: 583.

Séguéla, P., Wadiche, J., Dineley-Miller, K., Dani, J. A., and Patrick, J. W., 1993: Molecular cloning, functional properties and distribution of rat brain α7: a nicotinic cation channel highly permeable to calcium. *J. Neurosci.,* 13: 596–604.

Sine, S. M. and Claudio, T., 1991: γ- and δ-subunits regulate the affinity and the cooperativity of ligand binding to the acetylcholine receptor. *J. Biol. Chem.,* 266: 19369–19377.

Sine, S. and Taylor, P., 1979: Functional consequences of agonist-mediated state transitions in the cholinergic receptor. *J. Biol. Chem.,* 254: 3315–3325.

Stroud, R. M., McCarthy, M. P., and Shuster, M., 1990: Nicotinic acetylcholine receptor superfamily of ligand-gated ion channels. *Biochemistry,* 29: 11010–11023.

Sunshine, C. and McNamee, M. G., 1992: Lipid modulation of nicotinic acetylcholine receptor function: the role of neutral and negatively charged lipids. *Biochim. Biophys. Acta,* 1108: 240–246.

Sunshine, C. and McNamee, M. G., 1994: Lipid modulation of nicotinic acetylcholine receptor function: the role of membrane lipid composition and fluidity. *Biochim. Biophys. Acta,* 1191: 59–64.

Suzdak, P. D., Schwartz, R. D., Skolnick, P., and Paul, S. M., 1988: Alcohols stimulate gamma-aminobutyric acid receptor-mediated chloride uptake in brain vesicles: correlation with intoxication potency. *Brain Res.,* 444: 340.

Tonner, P. H., Wood, S. C., and Miller, K. W., 1992: Can nicotine self-inhibition account for its low efficacy at the nicotinic acetylcholine receptor from *Torpedo? Mol. Pharmacol.,* 42: 890–897.

Udgaonkar, J. B. and Hess, G. P., 1987: Chemical kinetic measurements of a mammalian acetylcholine receptor by a fast-reaction technique. *Proc. Natl. Acad. Sci. U.S.A.,* 84: 8758–8762.

Unwin, P. N. T. and Zampighi, G., 1980: Structure of the junction between communicating cells. *Nature,* 283: 545–549.

Vernino, S., Amador, M., Luetje, C. W., Patrick, J., and Dani, J. A., 1992: Calcium modulation and high calcium permeability of neuronal nicotinic acetylcholine receptors. *Neuron,* 8: 127–134.

Wada, E., Wada, K., Boulter, J., Deneris, E., Heinemann, S., Patrick, J., and Swanson, L. W., 1989: Distribution of alpha2, alpha3, alpha4, and beta2 neuronal nicotinic receptor subunit mRNAs in the central nervous system: a hybridization histochemical study in the rat. *J. Comp. Neurol.,* 284: 314–335.

Weber, M. and Changeux, J.-P., 1974a: Binding of *Naja nigricollis* ^{3}H-alpha-toxin to membrane fragments from *Electrophorus* and *Torpedo* electric organs. I. Binding of the tritiated alpha-neurotoxin in the absence of effector. *Mol. Pharmacol.,* 10: 1–14.

Weber, M. and Changeux, J.-P., 1974b: id. 2. Effect of the cholinergic agonists and antagonists on the binding of the tritiated α-neurotoxin. *Mol. Pharmacol.,* 10: 13–34.

Weber, M. and Changeux, J.-P., 1974c: id. 3. Effect of local anaesthetics on the binding of the tritiated α-neurotoxin. *Mol. Pharmacol.,* 10: 35–40.

Whiting, P. J. and Lindstrom, J. M., 1988: Characterization of bovine and human neuronal nicotinic acetylcholine receptors using monoclonal antibodies. *J. Neurosci.,* 8: 3395–3404.

Wood, S. C., Forman, S. A., and Miller, K. W., 1991: Short chain and long chain alkanols have different sites of action on nicotinic acetylcholine receptor channels from *Torpedo. Mol. Pharmacol.,* 39: 332–338.

Wood, S. C., Hill, W. A., and Miller, K. W., 1993: Cycloalkanemethanols discriminate between volume- and length-dependent loss of activity of alkanols at the *Torpedo* nicotinic acetylcholine receptor. *Mol. Pharmacol.,* 44: 1219–1226.

Wood, S. C., Tonner, P. H., De Armendi, A. J., Bugge, B., and Miller, K. W., 1995: Channel inhibition by alkanols occurs at a binding site on the nicotinic acetylcholine receptor. *Mol. Pharmacol.,* 47: 121–130.

Wu, G. and Miller, K. W., 1994: Ethanol enhances agonist-induced fast desensitization in nicotinic acetylcholine receptors. *Biochemistry,* 33: 9085–9091.

Wu, G., Tonner, P. H., and Miller, K. W., 1994: Ethanol stabilizes the open channel state of the *Torpedo* nicotinic acetylcholine receptor. *Mol. Pharmacol.,* 45: 102–108.

Young, A. P., Brown, F. F., Halsey, M. J., and Sigman, D. S., 1978: Volatile anesthetic facilitation of *in vitro* desensitization of membrane-bound acetylcholine receptor from *Torpedo californica. Proc. Natl. Acad. Sci. U.S.A.,* 75: 4563–4567.

Young, A. P. and Sigman, D. S., 1981: Allosteric effects of volatile anesthetics on the membrane-bound acetylcholine receptor protein. *Mol. Pharmacol.,* 20: 498–505.

Zhang, Z.-W., Vijayaraghavan, S., and Berg, D. K., 1994: Neuronal acetylcholine receptors that bind α-bungarotoxin with high affinity function as ligand-gated ion channels. *Neuron,* 12: 167–177.

Chapter 7

ETHANOL AND THE PHOSPHORYLATION OF LIGAND-GATED ION CHANNELS

Kathryn T. Eggeman and Michael D. Browning

TABLE OF CONTENTS

I. INTRODUCTION

The physiological effects of acute ethanol exposure are profound and diverse. Despite intense research efforts for the past several decades, we still have no clear idea of the molecular mechanism(s) by which ethanol exerts its various effects. Protein phosphorylation is one process that has gained recent attention as a possible target of ethanol. Protein phosphorylation is widely recognized as the primary mechanism for posttranslational regulation of protein function. Thus a parsimonious explanation of the diversity of ethanol's effects would be that ethanol modulates a variety of different molecular processes not by interacting directly with the proteins that mediate these processes but rather that ethanol acts by influencing phosphorylation. Such an explanation does not require that ethanol have the ability to interact in a specific manner with all the various proteins it has been shown to affect. Rather ethanol could affect a varied group of proteins simply by modulating the single process, phosphorylation, which regulates the activity of these various proteins. Considerable support for this view has come from studies of the effects of phosphorylation and ethanol on the ligand-gated ion channels. Such studies will be the principle focus of this review. As will become apparent, we are still far from either affirming or rejecting the hypothesis that the acute effects of ethanol could be produced, in whole or in part, via modulation of phosphorylation. Nevertheless, the data at hand do warrant discussion. However, before discussing

0-8493-8389-7/95/$0.00+$.50

ethanol and protein phosphorylation, a brief overview of the protein phosphorylation process is warranted.

II. PROTEIN PHOSPHORYLATION

Protein phosphorylation is mediated by protein kinases which catalyze the transfer of the terminal phosphate from ATP to the protein. The transfer of the charged phosphate group to the protein modifies the structure and hence the activity of the protein. The phosphate can be removed by a protein phosphatase which restores the protein to its original confirmation and activation state. Historically most interest in phosphorylation research has focused on protein kinases, however, recent work on phosphatases has demonstrated the critical importance of these enzymes to the phosphorylation process. The vast majority of all phosphorylation is on serine and threonine residues in the target protein. However, it has become clear that tyrosine phosphorylation is also a critically important form of protein phosphorylation. There are a large number of serine/threonine protein kinases. Most of these kinases are activated by various second messengers such as cAMP and calcium while much less is known about the mechanism(s) of activation of the tyrosine kinases. One possible mechanism by which ethanol might influence phosphorylation would be by modulation of the production of such second messengers. A discussion of such mechanisms can be found in a recent review by Tabakoff and Hoffman (in press). The focus of the present review will be on the effects of ethanol on events downstream from second messenger, namely on the protein kinases and on specific phosphoproteins with particular emphasis on the family of ligand-gated ion channels.

III. ETHANOL EFFECTS ON PROTEIN KINASES

As a first step in evaluating the possibility that ethanol might affect protein kinases we tested the effects of various concentrations of ethanol on kinase activity in an *in vitro* assay with purified kinases and purified substrates (Machu et al., 1991). We tested three prominent protein kinases: cAMP-dependent protein kinase (PKA), Ca^{2+}/Calmodulin-dependent protein kinase II (CAM kinase II), and Ca^{2+} and phosphatidylserine-dependent protein kinase (PKC). Each of these kinases phosphorylate a wide variety of different substrates, and thus if ethanol were to modulate one of these kinases it would provide thereby a mechanism for ethanol effects on a number of different proteins. However, we saw no effect of ethanol at concentrations up to 200 m*M* on these kinases (Figure 7.1). It is important to note that we used saturating concentrations of activators of PKC and CAM kinase II in this study. Thus it still possible that ethanol might alter the activity of these kinases when the concentration of such factors (e.g., Ca^{2+}, calmodulin, and phospholipids) was limiting. Such data argue that if ethanol does affect phosphorylation *in vivo* or *in situ* in intact preparations, it is not likely to be due to direct effects on PKA, CAM kinase II, or PKC. Similar results were obtained by Deitrich et al. (1989) who examined the effects of ethanol on PKC activity in rat mouse brain cytosol and found no effects of acute ethanol (100 m*M*). However, these authors did observe an inhibition of PKC by 100 m*M* ethanol in a membrane fraction prepared from mouse brain raising thereby the possibility alluded to above that ethanol might have effects on protein kinase activity *in situ*. Slater et al. (1993) report that ethanol inhibited PKC but only when extremely high doses were used. These authors indicated that the extent of inhibition depended on the composition of the lipid vesicles used to activate PKC. In contrast to these negative results for acute effects of ethanol on protein kinase activity *in vitro,* there is good evidence that chronic ethanol treatment *in vivo* leads to alterations in protein kinase activity. Thus chronic ethanol exposure has been reported to affect PKC (Messing et al., 1991), PKA (Rius et al., 1986), and Ca^{2+}/calmodulin-dependent protein phosphorylation (Shanley et al., 1985). However, such chronic effects are beyond the scope of this review.

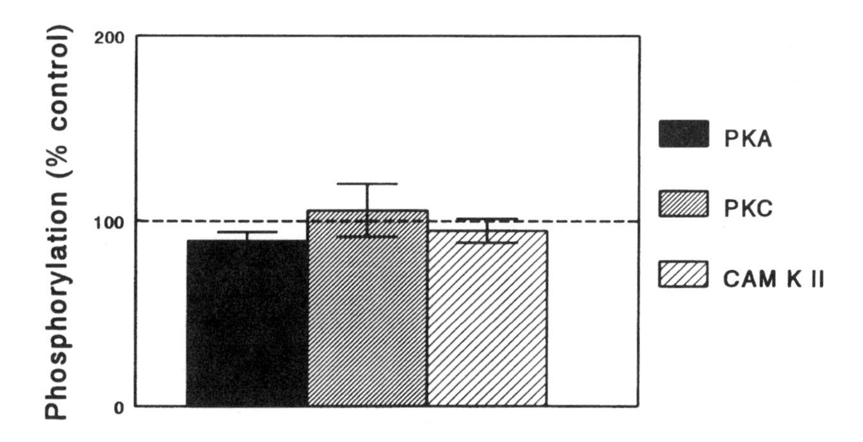

FIGURE 7.1. Effects of high concentrations of ethanol on the ability of purified protein kinase to phosphorylate known substrates. The substrates used are as follows; Histone IIa for PKA, Histone IIIs for PKC, and Synapsin 1 for CAM Kinase II. Data taken from Machu et al., 1991.

IV. LIGAND-GATED ION CHANNELS

Ethanol has been shown to have profound effects on a number of ligand-gated ion channels. The receptors that have been the focus of the most intense research in this area are the gamma aminobutyric acid-A ($GABA_A$-R) and the N-methyl-D-Aspartate (NMDA-R) and kainate/AMPA subtypes of the glutamate receptors. In addition, considerable recent attention has been focused on the 5-hydroxy-tryptamine (5-HT) or serotonin receptor. As might be expected given the focus of this review, all of these receptors are also regulated by phosphorylation. A working hypothesis of our group has been that the effects of ethanol on these receptors could be due, at least in part, to modulation of receptor phosphorylation by ethanol. The evidence linking the effects of ethanol to protein phosphorylation is most compelling for the $GABA_A$-R. Accordingly this receptor will be discussed first.

A. $GABA_A$-RECEPTORS

γ-Amino-butyric acid (GABA) is the major inhibitory neurotransmitter in the CNS. There are two major classes of receptors for GABA. The $GABA_A$ receptor (benzodiazapene receptor) is a ligand-gated chloride ion channel, and the $GABA_B$ receptor is a G-protein coupled receptor. GABA receptor function and dysfunction has been implicated in several neurological and psychiatric disorders. Not surprisingly, the GABA receptors are the targets for several clinically important drugs. The $GABA_A$-R is made up of several subunits, α (1–6), β (1–4), γ(1–4), δ, and ρ (1–2) (Schofield et al., 1987; Burt and Kamatchi, 1991; Harvey et al., 1993). *In situ* hybridization histochemistry studies have determined that the α_1, β_2, and γ_2 transcripts are the most abundant and widely distributed in the rat CNS, with the remaining subunit mRNAs having more restricted expression patterns (Persohn et al., 1992). Functional receptors can be formed by combining only α and β subunits, but the γ subunit appears to be required to impart benzodiazepene sensitivity to recombinant GABA receptors (DeLorey and Olsen, 1992).

1. Ethanol and the $GABA_A$-R

Ethanol potentiation of GABA induced chloride channel activity has been reported in several systems. In biochemical studies, 15–50 m*M* ethanol has been shown to potentiate the uptake of $^{36}Cl^-$ into synaptoneurosomes or microsacs (Allan and Harris, 1986; Harris et al., 1988; Suzdak et al., 1986). In electrophysiological studies, GABA responses were shown to be potentiated in mouse hippocampal neurons in culture or slices by 1–50 m*M* ethanol (Reynolds et al., 1992; Aguayo 1990,1991; Weiner et al., 1994). In chick (Celentano et al., 1988) and mouse (Mehta and Ticku, 1988) spinal cord neurons, 5–100 m*M* ethanol also increased responses to GABA. In addition, the responses of dorsal root ganglion (DRG) neurons in primary culture to GABA are potentiated by relatively high ethanol concentrations (30–300 m*M*) in a dose dependent manner (Nakahiro et al., 1991; Nishio and Narahashi, 1990). Ethanol has also been shown to enhance the initial peak current and shorten the decay time course of GABA induced current by 66% in DRG neurons (Nakahiro et al., 1991). However, potentiation of GABA responses by ethanol has not always been detected. Ethanol has been reported to have no effect on GABA responses in spinal cord (Gruol, 1982), in hippocampal neurons (Mancillas et al., 1986; Siggins et al., 1987a, 1987b), and in freshly isolated dorsal root ganglion neurons (White et al., 1990). The differences may simply be due to the fact that the ability of ethanol to potentiate $GABA_A$ responses is brain-region dependent (Proctor et al., 1992a, 1992b). This difference may be due to different subunit composition of the $GABA_A$ receptor in these regions. In addition, a clue might be found in the work of Lin et al. (1991) who reported ethanol potentiation of GABA induced depression of Purkinje cell firing *only* when norepinephrine (NE) or isoproterenol (ISO) were present. As both NE and ISO produce their effects, at least in part, through activation of cAMP-dependent protein phosphorylation, it is possible that ethanol potentiation might depend on the phosphorylation state and/or the phosphorylatability of the $GABA_A$-R. For example, phosphorylation, as will be discussed below, has been generally reported to inhibit $GABA_A$-R function. Ethanol could therefore potentiate $GABA_A$-R function by inhibiting or reversing the phosphorylation of the receptor. Alternatively ethanol might act selectively on a phosphorylated receptor and thus potentiate only those receptors that were phosphorylated.

2. Phosphorylation and the $GABA_A$-R

There have been a number of reports indicating that the $GABA_A$-R is a phosphoprotein. cAMP-dependent protein kinase (PKA) has been shown to phosphorylate the purified $GABA_A$-R (Kirkness et al., 1989; Browning et al., 1990) as has protein kinase C (PKC) (Browning et al., 1990). Serine 409 of the β_1 subunit has been shown to be phosphorylated *in vitro* by PKA and PKC (Moss et al., 1992a) as well as cGMP-dependent protein kinase (PKG) and Cam KII (McDonald and Moss, 1994). Cam KII also phosphorylates the β_1 subunit on serine 384 (McDonald and Moss, 1994). The γ_{2S} and γ_{2L} subunits can be phosphorylated by PKC on serine 327, and the γ_{2L} subunit has an additional PKC site at serine 343 (Moss et al., 1992a; Kellenberger et al., 1992). PKA increased the phosphorylation of several microsac proteins including a polypeptide specifically immunoprecipitated by an antibody specific for the α_1 subunit of the $GABA_A$-R (Leidenheimer et al., 1991). In addition, Sweetnam et al. (1988) demonstrated that a kinase present in a partially purified preparation of the $GABA_A$-R phosphorylated the receptor. However, this kinase has not yet been identified. The physiological effects of phosphorylation of the receptor have been studied by a number of investigators. PKA has been clearly shown to inhibit $GABA_A$-R function by Porter et al. (1990) and Moss et al. (1991, 1992b). Porter et al. (1990) demonstrated that introduction of the catalytic subunit of PKA into cultured mouse spinal neurons led to a marked reduction in $GABA_A$-R mediated currents. Moss et al. (1991) expressed the $\alpha_1\beta_1$ subunits of the $GABA_A$-R in human embryonic kidney (HEK) 293 cells and demonstrated that intracellular perfusion with cAMP inhibited GABA responses. Moreover, Moss et al. (1991) were also able to show that site directed mutagenesis of serine 409 in the beta subunit to an alanine completely abolished the modulation of GABA responses by cAMP. Thus PKA phosphorylation inhibits $GABA_A$-R function. The effects of PKC phosphorylation on the $GABA_A$-R have been less well established. There is widespread agreement that activators of PKC produce an inhibition of GABA-activated currents in either *Xenopus* oocytes expressing whole brain mRNA (Sigel and Baur, 1988; Moran and Dascal, 1989), in *Xenopus* oocytes expressing $\alpha_1\beta_1\gamma_{2L}$ subunits (Leidenheimer et al., 1992), or in HEK 293 cells expressing various recombinant $GABA_A$-Rs (Krishek et al., 1994). For example, peak amplitudes of GABA activated currents in HEK 293 cells expressing recombinant $GABA_A$-Rs were reduced up to 45% over 10–15 minutes following intracellular application of PdBu (Krishek et al., 1994). In contrast to these reports, Lin et al. (1994) demonstrated that intracellular dialysis of catalytically active PKC potentiated $GABA_A$-R currents in L929 cells expressing $\alpha_1\beta_1\gamma_2$ subunits of the receptor. It is PKC phosphorylation of the $GABA_A$-R that has drawn most attention in studies of ethanol's effects on the $GABA_A$-R. This is due in large part to studies of an alternatively spliced variant of the $\gamma 2$ subunit of the $GABA_A$-R.

The $\gamma 2$ subunit has two alternative splice products, short and long, differing by an eight amino acid insert between the putative third and fourth membrane spanning regions of the molecule. These eight amino acids contain a consensus sequence site for PKC (Whiting et al., 1990; Kofuji et al., 1991; Wafford et al., 1991). Whiting et al. (1990) presented evidence that the γ_{2L} subunit of the $GABA_A$-R could be phosphorylated by PKC; however, these authors did not provide data on the stoichiometry of this phosphorylation. Subsequently Moss et al. (1992a) demonstrated that purified fusion proteins of the major intracellular domain of the γ_{2L} and the γ_{2S} subunits of the receptor were rapidly and stoichiometrically phosphorylated by PKC but not PKA. Moreover, these authors demonstrated with site directed mutagenesis that PKC phosphorylated serine 327 in the γ_{2S} subunit and serine 343 in the γ_{2L} subunit. Recently, Machu et al. (1993) have demonstrated that Ca^{2+}/calmodulin-dependent protein kinase II can phosphorylate a 20 amino acid peptide corresponding to the region of γ_{2L} which contains serine 343.

3. Phosphorylation and Ethanol's Effect on the $GABA_A$-R

The phosphorylation of the γ_{2L} subunit appears to play a critical role in the potentiating effect of alcohol on the $GABA_A$-R. Wafford et al (1991) examined the effects of ethanol on the $GABA_A$-R which had been expressed in *Xenopus* oocytes using either brain mRNA or cRNA. These authors demonstrated that ethanol potentiation of the $GABA_A$-R was prevented only by antisense oligonucleotides to γ_{2L}. Moreover, expression of $\alpha_1\beta_1\gamma_{2S}$ or the $\alpha_1\beta_1\gamma_{2L}$ subunits cRNA combinations yielded functional $GABA_A$-Rs, but only the combination containing the γ_{2L} subunit was affected by ethanol. Additional evidence suggesting that *phosphorylatability* of the γ_{2L} subunit is required for ethanol's effect on the $GABA_A$-R comes from a recent report from Wafford et al. (1992). These authors demonstrated, using point mutations, that the serine residue which is the putative site for PKC in the eight amino acid insert of γ_{2L} is required for ethanol's effect on the $GABA_A$-R. Such data raised the possibility that ethanol may modulate the $GABA_A$-R by influencing the PKC phosphorylation of the γ_{2L} subunit of receptor. Additional support for a role for PKC in ethanol's effects on the $GABA_A$-R has come from Weiner et al. (1994). The authors demonstrated that ethanol (10–50 m*M*) potentiated pharmacologically isolated $GABA_A$-R-mediated inhibitory synaptic currents recorded from rat hippocampal CA1 neurons. These effects were not seen when the cells were intracellularly perfused with PKC_{19-31} which is a specific inhibitor of PKC. However, not all evidence supports a role for PKC and the γ_{2L} subunit of the receptor in ethanol's effects.

No effect of 20 m*M* ethanol on GABA-induced current was seen in *Xenopus* oocytes expressing $\alpha_1\beta_1\gamma_{2S}$ or $\alpha_1\beta_1\gamma_{2L}$ subunit combinations (Sigel et al., 1993). These authors reported that 100 m*M* ethanol did produce a "very small (<20%)" stimulation of GABA induced currents in both subunit combinations. Similarly, no effect of 10–100 m*M* ethanol was observed on GABA activated channels in HEK 293 cells expressing the $\alpha_1\beta_2\gamma_{2S}$, $\alpha_6\beta_2\gamma_{2S}$, or $\alpha_6\beta_2\gamma_{2L}$ subunit combinations (Marszalec et al., 1994). Ethanol was effective in increasing the rate of desensitization produced by higher (300 μm) GABA concentrations in the $\alpha_6\beta_2\gamma_{2S}$ but not the $\alpha_1\beta_2\gamma_{2S}$ subunit combination. This effect on desensitization rate was not altered by the protein kinase inhibitor H-7 (Marszalec et al., 1994). Studies of the effects of ethanol and GABA iontophoretically applied to different brain regions found several areas where ethanol did enhance the subsequent activation of GABA induced channel currents (Criswell et al., 1993); however, these areas did not coincide with brain regions expressing solely the γ_{2L} subunit of the $GABA_A$-R as shown by PCR analysis of mRNA content.

B. NMDA RECEPTORS

Ion channels activated by glutamate are typically divided into two classes. Those that are sensitive to N-methyl-D-aspartate (NMDA) are designated NMDA receptors (NMDAR) while those activated by kainate and α-amino-3-hydroxy-5-methyl-4-isoxalone propionic acid (AMPA) are known as kainate/AMPA receptors (K/AMPA-R). NMDA receptors have been the subject of intense scrutiny as they appear to be involved in long-term potentiation (LTP), a form of synaptic plasticity linked to learning. As ethanol clearly inhibits this receptor, it has been hypothesized that this inhibition might be responsible, at least in part, for the amnestic effects of acute ethanol exposure. These receptors have also been linked to neuronal development, and they have been implicated in several disorders of the central nervous system including epilepsy and ischemic neuronal cell death. NMDA receptors like the $GABA_A$-R described above are also regulated by protein phosphorylation. Unfortunately, there are few studies which have directly addressed the linkage between phosphorylation and the effects of ethanol on the NMDAR. Before discussing the effects of ethanol on the NMDAR, some background on the subunit composition of the receptor is needed.

The rat NMDAR1 (NR1) was the first to be cloned (Moriyoshi et al., 1991). The NR1 protein can form NMDA activated channels when expressed in *Xenopus* oocytes by itself, but channels with more physiological characteristics are produced when the NR1 subunit is combined with one of the NMDAR2 (NR2) subunits (Monyer et al., 1992; Ishii et al., 1993). The brain regional expression of two of the NMDA receptor subunits have been studied using immunocytochemistry with subunit specific antibodies. Brose et al. (1993) demonstrated that NR1 was widely distributed in the CNS with prominent staining in specific layers of cerebral cortex, in the hippocampus and dentate gyrus, and in the cerebellum. A number of authors have studied the expression on mRNA for the various subunits. Unfortunately, expression of mRNA is an unreliable predictor of protein expression. Messenger RNA for the four NR2 subunits is found at varying levels in different brain regions. NR2A message is seen in the olfactory bulb, anterior olfactory nuclei, cortex, hippocampus, and cerebellum. NR2B message is detected in the olfactory bulb, cortex, and hippocampus. NR2C message has been detected only in the cerebellum (Monyer et al., 1992; Ishii et al., 1993). NR2D message is expressed in the olfactory bulb, ventral palladium, hypothalamus, substantia nigra, superior colliculus, cerebellum, and deep cerebellar nuclei (Ishii et al., 1993).

Proteins with similar properties and significant sequence homology to the NMDA receptor subunits described above have also been cloned from mouse. The mouse ζ1 subunit is equivalent to the NR1 (Yamazaki et al., 1992). The mouse ε1 (Meguro et al., 1992), ε2, ε3 (Kutsuwada et al., 1992), and ε4 (Ikeda et al., 1992) are all similar in sequence and distribution to the NR2A, B, C, and D from rat, respectively. For clarity we will use the terminology for the rat receptor in all subsequent discussions. Several of these subunits contain consensus phosphorylation sites in the putative intracellular region of the protein which will be discussed in more detail shortly. In contrast to the clones described above, a second putative type of NMDA receptor has been cloned which shows little homology with the NMDA or non-NMDA type glutamate receptors (Kumar et al., 1991). The sequence of this protein indicates that it contains three potential sites for phosphorylation by Cam KII; however, this protein has not yet been functionally expressed so the effects of phosphorylation or ethanol on this protein are not known.

1. Ethanol and the NMDA-R

There is virtually unanimous agreement that ethanol inhibits the NMDAR. NMDA activated ion currents in voltage clamped hippocampal neurons are reduced 61% by 50 m*M* ethanol (Lovinger et al., 1989). Ion currents activated by NMDA in voltage clamped sensory neurons are inhibited by ethanol with an IC_{50} equal to 10 m*M* (White et al., 1990). NMDA-stimulated Ca^{2+} uptake into cerebellar granule cells was reduced 30% by ethanol concentrations as low as 10 m*M*. Ethanol also inhibited NMDA stimulated, Ca^{2+} dependent cyclic GMP accumulation in cerebellar granule cells in a dose dependent manner (Hoffman et al., 1989). Gothert and Fink (1989) demonstrated that NMDA-stimulated release of tritiated noradrenaline from rat brain cortex was inhibited by ethanol in the absence of Mg^{2+} ions with an IC_{50} of 45 m*M*. There has been one report of ethanol potentiation of NMDA receptor function at low (2–9 m*M*) ethanol concentration (Lima-Landman and Albuquerque, 1989).

Ethanol sensitivity appears to be linked to the subunit composition of the NMDA receptor. Using the two electrode voltage clamp technique Masood et al. (1994) have shown that combinations of the mouse NMDA receptor subunits expressed in oocytes are differentially sensitive to ethanol. NMDA activated currents through channels composed of the subunit combinations NR1/NR2A or NR1/NR2B were significantly inhibited by 50 m*M* ethanol, while channels composed of NR1/NR2C subunits were significantly inhibited only by 100 m*M* ethanol. In contrast, homomeric NR1 channels were not significantly inhibited by ethanol at concentrations of 100 m*M*. An additional level of complexity in terms on NMDAR sensitivity to ethanol was provided by evidence indicating that splice variants of the NR1 subunit itself are differentially sensitive to ethanol (Koltchine et al., 1993) (See Chapter 5 for a detailed discussion of this issue).

2. Phosphorylation of the NMDA-R

Recent work has shown that the NR1 subunit can be phosphorylated by PKC. The C-terminal of the NR1 subunit can be found in two forms due to alternative splicing of the gene (Sugihara et al., 1992). One form contains several consensus phosphorylation sites for PKC. NMDA receptors expressed in HEK 293 cells are highly phosphorylated under basal conditions; however, the phosphorylation state can be increased by treatment with the phorbol ester TPA (Tingley et al., 1993). These authors also demonstrated that a purified fusion protein containing the C-terminal region of the NR1 protein could be phosphorylated *in vitro* by purified PKC. Site directed mutagenesis of the serine residues in this fusion protein abolished the ability of PKC to phosphorylate the protein. Similarly Yamakura et al. (1993) demonstrated that the carboxy terminal of the NR1 subunit of the mouse NMDA receptor (equivalent to the rat NR1) expressed as a fusion protein could be phosphorylated *in vitro* with PKC. However, recent evidence indicates that phosphorylation of the C-terminal portion of the NR1 is *not* required for modulation of the receptor by PKC. A mutant NR1 subunit was constructed such that it no longer contained serine and threonine residues in the carboxy terminus. Expression of the mutant NR1 subunit alone or in combination with the NR2A subunit in *Xenopus* oocytes formed glutamate activated channels that were potentiated by the PKC activator TPA, but the extent of channel enhancement was not different from wild type channels expressed in oocytes (Yamakura et al., 1993). Similar results were obtained with rat NMDA receptors expressed in oocytes (Sigel et al., 1994). A deletion mutant coding for the NMDAR1 protein was constructed lacking the C-terminal 37 amino acids ("NR1C"). This deleted region contains the four consensus phosphorylation sites for PKC previously described (Tingley et al., 1993). Glutamate activated channels formed in *Xenopus* oocytes by the NR1C–NR2A subunit combination were stimulated by the phorbol ester, 4-beta-phorbol 12-myristrate 13-acetate, twice as much as the glutamate activated channels formed by the wild type subunits (NR1–NR2A). Additional mutant NR1 and NR2A subunits were constructed to remove the remaining PKC consensus phosphorylation sites located between the putative third and fourth membrane spanning regions of the proteins. All channels formed in oocytes with the various mutant subunits responded to phorbol ester treatment similarly to the equivalent wild type recombinant receptors (Sigel et al., 1993). These studies suggest that phosphorylation of consensus PKC sites on C-terminal of the NR1 is not responsible for the potentiation of NMDA receptors by phorbol ester treatment. Interestingly we have recently obtained evidence that the carboxy terminus of NR2A expressed as a fusion protein can be phosphorylated *in vitro* by both CAM Kinase II and PKC (Lickteig and Browning, 1994). It is possible that phosphorylation of the NR2A subunit of the NMDAR may be responsible for the physiological effects of phosphorylation on NMDAR function.

The physiological effects of PKC phosphorylation on NMDAR function were examined by two groups (Chen and Huang, 1991, 1992; Wang et al., 1994) who introduced purified PKC into neurons using the whole cell patch technique. Both groups reported that introduction of PKC into the neurons led to potentiation of NMDA currents. Chen and Huang (1991) examined NMDA currents in spinal trigeminal neurons in thin medullary slices. They demonstrated that intracellular application of PKC potentiated NMDA currents in these neurons and this potentiation was antagonized by the specific peptide inhibitor of PKC, $PKCI_{19-31}$. In contrast, PKA had no effect on NMDA currents. Similarly, we reported (Wang et al., 1994) that introduction of the catalytic fragment of PKC potentiated NMDA currents in cultured hippocampal neurons. NMDA activated currents on spinal dorsal horn neurons were potentiated by intracellular application of the protein tyrosine kinase $pp60_{c\text{-}src}$ and inhibited by Genistein and Lavendustin A which are known to inhibit protein tyrosine kinases (Wang and Salter, 1994). In contrast to these studies using purified kinases, studies using activators of PKC such as phorbol esters have produced conflicting results. Kelso et al. (1992) and Urushihara et al. (1992) demonstrated that phorbol esters potentiated NMDA currents in *Xenopus* oocytes injected with whole brain mRNA. Similar potentiation of NMDA currents by PKC activators was also observed in oocytes expressing specific subunits of either mouse or rat NMDAR subunits (Yamakura et al., 1993; Sigel et al., 1994). Murphy et al. (1994) also saw phorbol ester induced potentiation of NMDA currents in cultured mouse striatal neurons. In contrast, two groups report that PKC activators inhibit NMDA currents in cultured cerebellar granule cells (Courtney and Nicholls, 1992; Snell et al., 1994a). These conflicting results on the effects of PKC on NMDAR function may reflect differences in NMDAR subunit expression in cerebellar granule cells. However, there is also a report that PKC activators inhibit NMDA currents in hippocampal neurons (Markram and Segal, 1992), data which conflict with the report of Wang et al. (1994) who saw potentiation of NMDA current by purified PKC in hippocampal neurons. As indicated above in the discussion of the GABA receptor, resolution of these conflicting reports will require direct measurement of the phosphorylation state of the NMDAR.

Protein phosphatases have also been shown to modulate NMDA receptor activity. Addition of purified protein phosphatase 1 (PP1) or 2A (PP2A) to inside-out patches from hippocampal neurons decreased the channel open probability (Wang et al., 1994). In addition, application of calyculin A (a phosphatase inhibitor specific for PP1 and PP2A) to perforated hippocampal patches increased the channel open probability by up to 204% without affecting the single channel conductance level. Lieberman and Mody (1994) report that purified calcineurin (also known as protein phosphatase 2B), a calcium and calmodulin dependent protein phosphatase, shortens NMDA activated channel openings in acutely dissociated rat dentate gyrus granule cells. These authors also demonstrated that the phosphatase inhibitor, okadaic acid, prolongs NMDA activated channel openings. Okadaic acid had no effect when Ca^{2+} is prevented from entering the cells. Similarly, FK-506 which inhibits calcineurin also prolonged NMDA currents. Finally, Wang and Salter (1994) demonstrated that exposure of cultured dorsal horn neurons to the protein tyrosine phosphatase inhibitor orthovanadate potentiated NMDA activated currents 1.5–2 fold. Thus four phosphatases with differing substrate specificities decrease NMDA currents. It is not yet clear whether such data indicate that there are numerous phosphorylation sites regulating NMDAR function or whether these phosphatases and/or phosphatase inhibitors can all influence a single site under conditions used in these studies. Resolution of this issue will await direct studies of the phosphorylation of the NMDAR *in situ*. Nevertheless, all of these studies are consistent with previous reports (see above) suggesting that phosphorylation enhances NMDAR function (Chen and Huang, 1991, 1992; Wang et al., 1994).

3. Ethanol and Phosphorylation of the NMDA-R

Only one recent report has linked the ability of ethanol to inhibit NMDA receptor function to activation of PKC (Snell et al., 1994b). These authors studied cultured cerebellar granule cells and demonstrated that in the presence of 0.1 μ*M* glycine, ethanol caused a concentration-dependent inhibition of the Ca^{2+} influx stimulated by 100 μ*M* NMDA, with an IC_{50} of 12 m*M*. Similarly, PMA also produced an inhibition of NMDA stimulated calcium influx. The inhibitory effect of ethanol could be reversed by treatment of the cells not only with a high concentration of glycine but also with the protein kinase inhibitor, staurosporine, or with the more specific PKC inhibitor, calphostin. Because the effects of PMA and ethanol were not additive, one may postulate that these two compounds act via the same pathway. Taken together, these results suggest that ethanol may activate PKC to phosphorylate the NMDA receptor (or an associated protein) at a site that inhibits the ability of glycine to act as a coagonist. Clearly more studies are required to settle this issue, particularly since in most reports PKC has been shown to potentiate the NMDAR.

C. NON-NMDA RECEPTORS

The glutamate-activated ion channels (GluR) that are insensitive to NMDA but are activated by kainate and/or AMPA are also important in CNS functioning. This class of receptors can be further subdivided into two classes. The first class is the kainate/AMPA receptor which is activated by both kainate and AMPA, while the second class is sensitive only to kainate (Hollmann and Heinemann, 1994). A number of different GluR proteins have been identified with GluR1–4, GluR5–7, and KA1–2 representing different families of GluRs. GluR1–4 have been shown to constitute the kainate/AMPA receptors, while the GluR5–7 and KA1–2 are thought to constitute receptors sensitive only to kainate. In the functional studies of these receptors described below, it is important to keep in mind that typically kainate is used to activate these receptors. This is due to the fact that kainate/AMPA receptors rapidly desensitize to AMPA application. The use of kainate does not permit one to unequivocally determine which of the various GluR family of receptors has been activated as all respond to kainate. However, the receptors that are sensitive only to kainate desensitize rapidly and thus in most cases, kainate stimulated responses are likely to be due to the kainate/AMPA receptor family of GluRs. There is also another class of GluRs which is not coupled to ion channels and is known as metabotropic glutamate receptors. This class of receptors has been the subject of a number of recent reviews (Nakanishi, 1994; Schoepp, 1993; Schoepp and Conn, 1993) and will not be discussed further here.

A number of groups have reported that the function of the GluRs is potentiated during the maintenance phase of LTP. Thus while the NMDAR is critical for inducing LTP, the GluRs appear to play a critical role in the long-term components of LTP. As will be discussed below, ethanol also appears to inhibit these GluRs although some report higher doses of ethanol are needed to inhibit the GluR than that required to inhibit the NMDAR. A difference in the doses of ethanol required for inhibiting receptors involved in induction versus maintenance provide a basis, at least in part, for the amnestic effects of different doses of ethanol. Moreover, GluRs also appear to be involved in development, epileptic seizures, and acute neurotoxicity.

1. GluRs and Ethanol

In their original report of ethanol's effect on the NMDAR-mediated currents in hippocampal neurons, Lovinger et al. (1989) also noted an inhibitory effect of ethanol on kainate-stimulated currents. However, ethanol was less potent in inhibiting these responses than in inhibiting NMDA currents. Similar results were obtained by Hoffman et al. (1989) who examined calcium uptake in cerebellar granule cells. These authors reported IC_{50}s for ethanol of 41 m*M* and 118 m*M* for the NMDA- and kainate-stimulated responses, respectively. However, in a recent study of GluRs expressed in *Xenopus* oocytes, moderate doses of ethanol (50 m*M*) were shown to inhibit NMDAR and GluRs to a similar degree (Dildy-Mayfield and Harris, 1992). These authors also found that ethanol inhibition of GluR's function was significantly influenced by the concentration of *kainate* used to elicit these responses. Responses induced by 12.5 μm kainate were inhibited 45% by 50 m*M* ethanol, whereas responses induced by 400 μm kainate were only inhibited 15% by 50 m*M* ethanol. Similar results have recently been obtained by Snell et al. (1994b) who reported that 100 m*M* ethanol significantly inhibited the increase in intracellular Ca^{2+} stimulated by 50 μ*M* kainate but had no effect on the response to 100 μ*M* kainate in cerebellar granule cells. Lovinger et al. (1993) demonstrated that expression of different combinations of cloned glutamate receptors (GluR1, GluR2, GluR4, and all "flip") in HEK 293 cells produced channels that were activated by 1 m*M* kainate. These channels were inhibited by ethanol as follows: GluR1 alone, IC_{50} 119 m*M*; GluR1+4, IC_{50} 165 m*M*; GluR2+4, IC_{50} 139m*M*; and GluR4 alone, IC_{50} 133 m*M*. When primary cultures of neocortical neurons were used in the same experimental system, the IC_{50} for ethanol inhibition of kainate-induced current increased to 300–400 m*M*. Thus recombinant GluRs appear to differ in some respects from GluRs expressed endogenously in cultured cortical neurons. In summary, it appears clear that ethanol acts to inhibit GluR currents although some uncertainty exists concerning the relative sensitivity of the GluRs and the NMDAR to the effects of ethanol (for a more detailed discussion of these issues, see Chapter 5).

2. Phosphorylation of the GluRs

Three groups have demonstrated that introduction of protein kinases (PKA, PKC, or CAM kinase II) into cultured hippocampal neurons potentiates kainate activated currents. Wang et al. (1991) and Greengard et al. (1991) demonstrated that the catalytic fragment of PKA potentiated kainate currents and these effects were blocked by the peptide inhibitor of PKA. This inhibitor also inhibited control responses to glutamate indicating that endogenous basal PKA activity modulates these currents. The large conductance, NMDA type channels were not affected by activators of PKA (Greengard et al., 1991). Wang et al. (1994) subsequently demonstrated that PKC potentiated kainate currents in hippocampal neurons. Interestingly, the modulation of kainate currents by PKC was dependent on the concentration of agonist applied. Thus potentiation was seen at kainate concentrations above those which caused 50% of the maximal response, and currents were depressed by PKC at lower agonist concentrations. The specific peptide inhibitor of PKC blocked the potentiating effect of added PKC and also produced a significant inhibition of the kainate currents suggesting that endogenous PKC also modulates these currents. These data contrast with work by Chen and Huang (1992) who used purified PKC and that of Kelso et al. (1992), Yamazaki et al. (1992), and Urushihara et al. (1992) who used PKC activators and saw no effects on kainate currents. However, all these groups used concentrations of kainate substantially lower than that shown by Wang et al. (1994) to be required for PKC potentiation of kainate currents. More recently, Dildy-Mayfield and Harris (1994) have shown that PKC activators *inhibit* kainate currents in oocytes expressing specific GluR subunits. It is difficult to reconcile this result with the work of Wang et al. (1994), as Dildy-Mayfield and Harris (1994) used relatively high concentrations of kainate (400 μ*M*). However, it is possible, as suggested in the work of Lovinger (1993), that recombinant GluRs differ in some respects from GluRs expressed endogenously in cultured hippocampal neurons. Lastly, McGlade-McCulloh et al. (1993) have recently demonstrated that activated CAM kinase II produced a 3–4 fold enhancement of kainate currents in cultured hippocampal neurons. In summary, it is clear that GluR function can be modulated by protein phosphorylation; however, addition of kinases or kinase activators is not sufficient to determine whether the modulation is due to direct phosphorylation of the receptor.

Two groups have recently provided evidence that the GluRs can be directly phosphorylated. Tan et al. (1994) demonstrated that activators of PKC produce a 227% increase in the phosphorylation of GluRs in cultured hippocampal neurons. Moreover these authors also demonstrated that treatments that activate CAM kinase II and other calcium-dependent kinases also produced significant increases in GluR phosphorylation. Stimulation of the cells with forskolin to activate PKA had no effect on GluR phosphorylation. These data contrast with the work of Moss et al. (1993) who studied recombinant GluRs in human embryonic kidney 293 cells. Moss et al. (1993) detected a high basal level of phosphorylation of GluRs but saw no effect of PKC activation on this phosphorylation.

From the literature cited above, it is clear that GluRs are modulated by phosphorylation and that in most cases the effect of phosphorylation is to potentiate GluR function. This effect is, once again, in contrast to the effect of ethanol which appears to inhibit GluR function. However, no studies to date have been done to show that effects of ethanol and phosphorylation on the GluRs interact with each other.

D. ACETYLCHOLINE RECEPTOR

The acetylcholine receptor (AChR) at the neuromuscular junction is the best characterized of the ligand-gated ion channels both in terms of its structure and in terms of its regulation by phosphorylation. The effects of phosphorylation of the receptor have been the subject of a number of recent reviews (Huganir, 1991; Swope et al., 1992), and these effects will only be briefly summarized here. Three different kinases that are found in postsynaptic membranes of the *Torpedo* electroplax have been shown to phosphorylate specific subunits of the receptor. These kinases include PKA, PKC, and a protein tyrosine kinase (Huganir and Greengard, 1983; Huganir, 1987; Safran et al., 1987; Huganir et al., 1984). The effects of these phosphorylations have been directly examined in reconstitution experiments using liposomes. PKA phosphorylation produces an 8-fold increase in the desensitization of the receptor (Huganir et al., 1986). Similarly there was a marked correlation between the extent of tyrosine phosphorylation and the desensitization of the receptor (Hopfield et al., 1988). The effects of PKC phosphorylation has been examined in muscle cells using phorbol esters to activate PKC (Eusebi et al., 1985). Such treatments produce a decreased sensitivity of the receptor to ACh and an increased rate of desensitization. In addition to these studies suggesting that phosphorylation inhibits AChR function by enhancing desensitization, there are several recent reports suggesting that tyrosine phosphorylation may play a role in the aggregation of the receptor (Wallace et al., 1991; Wallace, 1994).

Ethanol has been shown to increase ion flux through ACh activated ion channels in several systems. In neuromuscular junctions of frogs, ethanol increases the amplitude and duration of miniature end-plate potentials (Sterz et al., 1986). Similar results have been found in mouse where it appears that prolongation of channel open time by ethanol is accompanied by increased affinity of the AChRs for acetylcholine (Linder et al., 1984). In Torpedo, physiological concentrations of ethanol increase $^{86}Rb^{+}$ flux from AChR rich vesicles by 15–35% only when submaximal concentrations of agonist (carbamylcholine) were used (Forman et al., 1989). Treating AChR rich vesicles with higher concentrations of ethanol (1.5 *M*) increases the efficacy of the partial agonist, nicotine, to levels comparable with the full agonist carbamylcholine (Tonner et al., 1992). Kinetic analysis of $^{86}Rb^{+}$ flux data suggests that the primary action of ethanol on nAChRs is to increase the fraction of occupied receptors that open, or stabilize the open channel state (Wu et al., 1994). Although there has been one report that ethanol inhibits excitatory postsynaptic current duration in crustacean neuromuscular junctions (Adams et al., 1977), the majority of evidence indicates that ethanol increases ion flux through ACh activated ion channels. There have been no studies to date that indicate that ethanol affects the phosphorylation state of the nAChR.

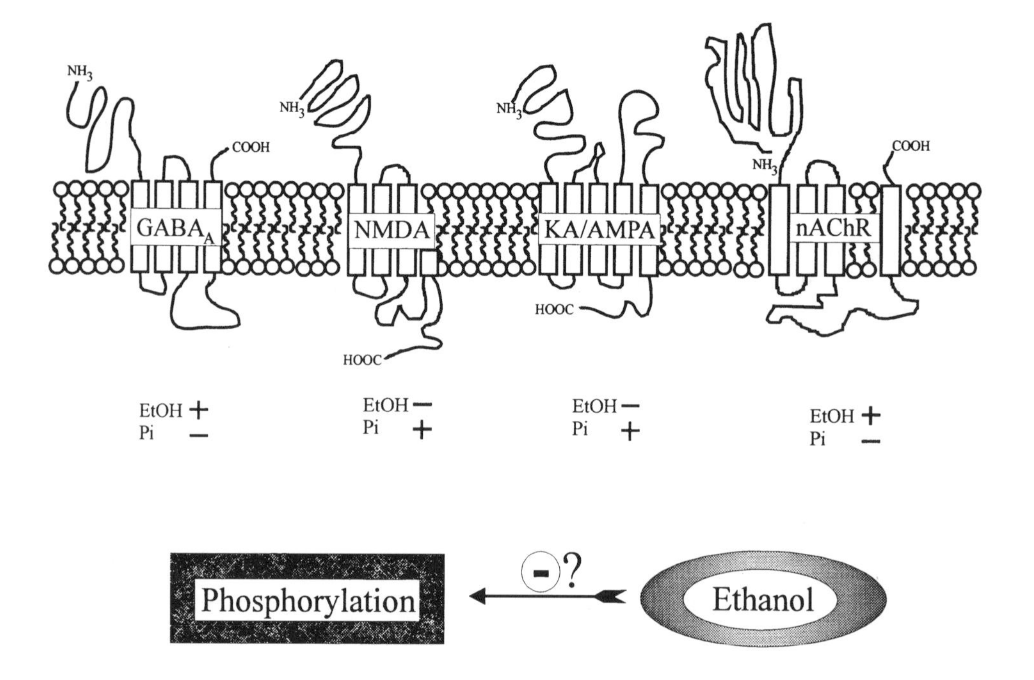

FIGURE 7.2. Schematic representation of the effects of ethanol and phosphorylation on ligand-gated ion channels. In general, the effect of ethanol on these receptors is opposite the effect of increasing the phosphorylation state of the receptors. This may indicate that ethanol has a negative effect on various aspects of the phosphorylation process.

E. SEROTONIN RECEPTOR

The ligand-gated ion channel activated by serotonin has not been extensively studied in terms of the effects of ethanol on this channel. Lovinger and colleagues have shown that ethanol (25–100 m*M*) potentiates currents mediated by $5HT_3$ receptors in NCB-20 neuroblastoma cells or in freshly isolated Nodose ganglion neurons (Lovinger, 1991a, 1991b; Lovinger and White, 1991). However, there have been no studies to date which examine the phosphorylation of this receptor.

V. SUMMARY

As is clearly apparent from the forgoing review, there is no clear consensus concerning the hypothesis that ethanol modulates the function of ligand-gated ion channels by affecting the phosphorylation of these receptors. This is in large part due to the paucity of data which have directly examined the relationship between phosphorylation and ethanol. However, as shown in Figure 7.2 there is general agreement that the effects of ethanol on these receptors are usually the opposite of the effects of phosphorylation. Thus ethanol typically inhibits NMDAR and K/AMPAR while phosphorylation is predominantly reported to potentiate these receptors. Moreover, ethanol has generally been reported to potentiate $GABA_A$-R and nAChR, while phosphorylation is typically reported to inhibit these receptors. Such data support for the hypothesis depicted in Figure 7.2, namely that ethanol modulates the function of these receptors by inhibiting phosphorylation. However, such a view is still quite controversial. To resolve this issue we need a good deal of additional data in two areas:

1. We need to resolve the controversies concerning the functional effects that ethanol and phosphorylation have on these receptors.
2. We need more experiments which directly examine the effects of ethanol on phosphorylation of these receptors.

Future studies should help resolve these issues which are of critical importance to our understanding of the molecular mechanisms of ethanol's acute effects.

REFERENCES

Adams, D.J., Gage, P.W., and Hamill, O.P. (1977) Ethanol reduces excitatory postsynaptic current duration at a crustacean neuromuscular junction. *Nature,* 266, 739–741.

Aguayo, L.G. (1990) Ethanol potentiates the $GABA_A$-activated Cl^- current in mouse hippocampal and cortical neurons. *Eur. J. Pharmacol.,* 187, 127–130.

Aguayo, L.G. (1991) Demonstration that ethanol potentiates the $GABA_A$-activated Cl^- current in central mammalian neurons. *Alcohol Alcohol Suppl.,* 1, 187–190.

Allan, A.M. and Harris, R.A. (1986) Gamma-aminobutyric acid and alcohol actions: neurochemical studies of long sleep and short sleep mice. *Life Sci.,* 39, 2005–2015.

Brose, N., Gasic, G.P., Vetter, D.E., Sullivan, J.M., and Heinemann, S.F. (1993) Protein chemical characterization and immunocytochemical localization of the NMDA receptor subunit NMDA R1. *J. Biol. Chem.,* 268, 22663–22671.

Browning, M.D., Bureau, M., Dudek, E.M., and Olsen, R.W. (1990) Protein kinase C and cAMP-dependent protein kinase phosphorylate the β subunit of the purified γ-aminobutyric acid receptor. *Proc. Natl. Acad. Sci. U.S.A.,* 87, 1315–1317.

Burt, D.R. and Kamatchi, G.L. (1991) $GABA_A$ receptor subtypes: from pharmacology to molecular biology. *FASEB J.,* 5, 2916–2923.

Celentano, J.J., Gibbs, T.T., and Farb, D.H. (1988) Ethanol potentiates GABA- and glycine-induced chloride currents in chick spinal cord neurons. *Brain Res.,* 455, 377–380.

Chen, L. and Huang, L.-Y.M. (1991) Sustained potentiation of NMDA receptor-mediated glutamate responses through activation of protein kinase C by a μ opioid. *Neuron,* 7, 319–326.

Chen, L. and Huang, L.-Y.M. (1992) Protein kinase C reduces Mg^{++} block of NMDA-receptor channels as a mechanism of modulation. *Nature,* 356, 521–523.

Courtney, M.J. and Nicholls, D.G. (1992) Interactions between phospholipase C-coupled and *N*-methyl-D-aspartate receptors in cultured cerebellar granule cells: protein kinase C mediated inhibition of *N*-methyl-D-aspartate responses. *J. Neurochem.,* 59, 983–992.

Criswell, H.E., Simson, P.E., Duncan, G.E., McCown, T.J., Herbert, J.S., Morrow, A.L., and Breese, G.R. (1993) Molecular basis for regionally specific action of ethanol on gamma-aminobutyric $acid_A$ receptors: generalization to other ligand-gated ion channels. *J. Pharmacol. Exp. Ther.,* 267, 522–537.

Deitrich, R.A., Bludeau, P.A., and Baker, R.C. (1989) Investigations of the role of protein kinase C in the acute sedative effects of ethanol. *Alcoholism: Clin. Exp. Res.,* 13, 737–745.

DeLorey, T.M. and Olsen, R.W. (1992) Gamma-aminobutyric acid A receptor structure and function. *J. Biol. Chem.,* 267, 16747–16750.

Dildy-Mayfield, J.E. and Harris, R.A. (1992) Acute and chronic ethanol exposure alters the function of hippocampal kainate receptors expressed in *Xenopus* oocytes. *J. Neurochem.,* 58, 1569–1572.

Dildy-Mayfield, J.E. and Harris, R.A. (1994) Activation of protein kinase C inhibits kainate-induced currents in oocytes expressing glutamate receptor subunits. *J. Neurochem.,* 62, 1639–1642.

Eusebi, F., Molinaro, M., and Zani, B.M. (1985) Agents that activate protein kinase C reduce acetylcholine sensitivity in cultured myotubes. *J. Cell Biol.,* 100, 1339–1342.

Forman, S.A., Righi, D.L., and Miller, K.W. (1989) Ethanol increases agonist affinity for nicotinic receptors from Torpedo. *Biochim. Biophys. Acta,* 987, 95–103.

Gothert, M. and Fink, K. (1989) Inhibition of N-methyl-D-aspartate (NMDA)- and L-glutamate-induced noradrenaline and acetylcholine release in the rat brain by ethanol. *Naunyn Schmiedebergs Arch Pharmacol.,* 340, 516–521.

Greengard, P., Jen, J., Nairn, A.C., and Stevens, C.F. (1991) Enhancement of the glutamate response by cAMP-dependent protein kinase in hippocampal neurons. *Science,* 253, 1135–1138.

Gruol, D.L. (1982) Ethanol alters synaptic activity in cultured spinal cord neurons. *Brain Res.,* 243, 25–33.

Harris, R.A., Allan, A.M., Daniell, L.C., and Nixon, C. (1988) Antagonism of ethanol and pentobarbital actions by benzodiazepine inverse agonists: neurochemical studies. *J. Pharmacol. Exp. Ther.,* 247, 1012–1017.

Harvey, R.J., Kim, H.-C., and Darlison, M.G. (1993) Molecular cloning reveals the existence of a fourth gamma subunit of the vertebrate brain $GABA_A$ receptor. *FEBS Lett.,* 331, 211–216.

Hoffman, P.L., Rabe, C.S., Moses, F., and Tabakoff, B. (1989) N-methyl-D-aspartate receptors and ethanol: inhibition of calcium flux and cyclic GMP production. *J. Neurochem.,* 52, 1937–1940.

Hollmann, M. and Heinemann, S. (1994) Cloned glutamate receptors. *Annu. Rev. Neurosci.*, 17, 31–108.

Hopfield, J.F., Tank, D.W., Greengard, P., and Huganir, R.L. (1988) Functional modulation of the nicotinic acetylcholine receptor by tyrosine phosphorylation. *Nature,* 336, 677–680.

Huganir, R.L. (1987) Phosphorylation of purified ion channel proteins. In: Kaczmarek, L. and Levitan, I., (Eds.) *Neuromodulation*, pp. 86–99. New York: Raven.

Huganir, R.L. (1991) Regulation of the nicotinic acetylcholine receptor by serine and tyrosine protein kinases. *Adv. Exp. Med Biol.*, 287, 279–294.

Huganir, R.L., Delcour, A.H., Greengard, P., and Hess, G.P. (1986) Phosphorylation of the nicotinic acetylcholine receptor regulates its rate of desensitization. *Nature,* 321, 774–776.

Huganir, R.L. and Greengard, P. (1983) cAMP-dependent protein kinase phosphorylates the nicotinic acetylcholine receptor. *Proc. Natl. Acad. Sci. U.S.A.*, 80, 1130–1134.

Huganir, R.L., Miles, K., and Greengard, P. (1984) Phosphorylation of the nicotinic acetylcholine receptor by an endogenous tyrosine-specific protein kinase. *Proc. Natl. Acad. Sci. U.S.A.*, 81, 6968–6972.

Ikeda, K., Nagasawa, M., Mori, H., Araki, K., Sakimura, K., Watanabe, M., Inoue, Y., and Mishina, M. (1992) Cloning and expression of the ε4 subunit of the NMDA receptor channel. *FEBS Lett.*, 313, 34–38.

Ishii, T., Moriyoshi, K., Sugihara, H., Sakurada, K., Kadotani, H., Yokoi, M., Akazawa, C., Shigemoto, R., Mizuno, N., Masu, M., and Nakanishi, S. (1993) Molecular characterization of the family of the *N*-methyl-D-aspartate receptor subunits. *J. Biol. Chem.*, 268, 2836–2843.

Kellenberger, S., Malherbe, P., and Sigel, E. (1992) Function of the $\alpha_1\beta_2\gamma_{2S}$ gamma-aminobutyric acid type A receptor is modulated by protein kinase C via multiple phosphorylation sites. *J. Biol. Chem.*, 267, 25660–25663.

Kelso, S.R., Nelson, T.E., and Leonard, J.P. (1992) Protein kinase C-mediated enhancement of NMDA currents by metabotropic glutamate receptors in *Xenopus* oocytes. *J. Physiol.*, 449, 705–718.

Kirkness, E.F., Bovenkerk, C.F., Ueda, T., and Turner, A.J. (1989) Phosphorylation of gamma-aminobutyrate (GABA)/benzodiazepine receptors by cyclic AMP-dependent protein kinase. *Biochem. J.*, 259, 613–616.

Kofuji, P., Wang, J.B., Moss, S.J., Huganir, R.L., and Burt, D.R. (1991) Generation of two forms of the gamma-aminobutyric acid_A receptor gamma_2-subunit in mice by alternative splicing. *J. Neurochem.*, 56, 713–715.

Koltchine, V., Anantharam, V., Wilson, A., Bayley, H., and Treistman, S.N. (1993) Homomeric assemblies of NMDAR1 splice variants are sensitive to ethanol. *Neurosci. Lett.*, 152, 13–16.

Krishek, B.J., Xie, X., Blackstone, C., Huganir, R.L., Moss, S.J., and Smart, T.G. (1994) Regulation of $GABA_A$ receptor function by protein kinase C phosphorylation. *Neuron,* 12, 1081–1095.

Kumar, K.N., Tilakaratne, N., Johnson, P.S., Allen, A.E., and Michaelis, E.K. (1991) Cloning of cDNA for the glutamate-binding subunit of an NMDA receptor complex. *Nature,* 354, 70–73.

Kutsuwada, T., Kashiwabuchi, N., Mori, H., Sakimura, K., Kushiya, E., Araki, K., Meguro, H., Masaki, H., Kumanishi, T., Arakawa, M., and Mishina, M. (1992) Molecular diversity of the NMDA receptor channel. *Nature,* 358, 36–40.

Leidenheimer, N.J., Machu, T.K., Endo, S., Olsen, R.W., Harris, R.A., and Browning, M.D. (1991) Cyclic AMP-dependent protein kinase decreases gamma-aminobutyric acid A receptor-mediated $^{36}Cl^-$ uptake by brain microsacs. *J. Neurochem.*, 57, 722–725.

Leidenheimer, N.J., McQuilkin, S.J., Hahner, L.D., Whiting, P., and Harris, R.A. (1992) Activation of protein kinase C selectively inhibits the gamma-aminobutyric acid A receptor: role of desensitization. *Mol. Pharmacol.*, 41, 1116–1123.

Lickteig, R.L. and Browning, M.D. (1994) Calmodulin kinase II and PKC phosphorylate a fusion protein derived from the C-terminal region of the NMDAR2A. *Society for Neuroscience Abstract,* 20, 312.4.

Lieberman, D.N. and Mody, I. (1994) Regulation of NMDA channel function by endogenous Ca^{2+}-dependent phosphatase. *Nature,* 369, 235–239.

Lima-Landman, M.T. and Albuquerque, E.X. (1989) Ethanol potentiates and blocks NMDA-activated single-channel currents in rat hippocampal pyramidal cells. *FEBS Lett.*, 247, 61–67.

Lin, A.M., Freund, R.K., and Palmer, M.R. (1991) Ethanol potentiation of GABA-induced electrophysiological responses in cerebellum: requirement for catecholamine modulation. *Neurosci. Lett.*, 122, 154–158.

Lin, Y.-F., Browning, M.D., Dudek, E.M., and Macdonald, R.L. (1994) Ca^{2+}/Phospholipid-dependent protein kinase (PKC) enhances recombinant bovine $\alpha_1\beta_1\gamma_{2L}$ $GABA_A$ receptor whole-cell currents expressed in L929 Fibroblasts. *Neuron,* 131, 1421–1431.

Linder, T.M., Pennefather, P., and Quastel, D.M. (1984) The time course of miniature endplate currents and its modification by receptor blockade and ethanol. *J. Gen. Physiol.*, 83, 435–468.

Liu, Y., Dilger, J.P., and Vidal, A.M. (1994) Effects of alcohols and volatile anesthetics on the activation of nicotinic acetylcholine receptor channels. *Mol. Pharmacol.*, 45, 1235–1241.

Lovinger, D.M. (1991a) Ethanol potentiates ion current mediated by 5-HT_3 receptors on neuroblastoma cells and isolated neurons. *Alcohol Alcohol. Suppl.*, 1, 181–185.

Lovinger, D.M. (1991b) Ethanol potentiation of 5-HT_3 receptor-mediated ion current in NCB-20 neuroblastoma cells. *Neurosci. Lett.*, 122, 57–60.

Lovinger, D.M. (1993) High ethanol sensitivity of recombinant AMPA-type glutamate receptors expressed in mammalian cells. *Neurosci. Lett.*, 159, 83–87.

Lovinger, D.M. and White, G. (1991) Ethanol potentiation of 5-hydroxytryptamine3 receptor-mediated ion current in neuroblastoma cells and isolated adult mammalian neurons. *Mol. Pharmacol.*, 40, 263–270.

Lovinger, D.M., White, G., and Weight, F.F. (1989) Ethanol inhibits NMDA-activated ion current in hippocampal neurons. *Science*, 243, 1721–1724.

Machu, T.K., Firestone, J.A., and Browning, M.D. (1993) Ca^{2+}/calmodulin-dependent protein kinase II and protein kinase C phosphorylate a synthetic peptide corresponding to a sequence that is specific for the γ_{2L} subunit of the $GABA_A$ receptor. *J. Neurochem.*, 61, 375–377.

Machu, T.K., Olsen, R.W., and Browning, M.D. (1991) Ethanol has no effect on cAMP-dependent protein kinase-, protein kinase C-, or Ca^{2+}-calmodulin-dependent protein kinase II-stimulated phosphorylation of highly purified substrates *in vitro*. *Alcoholism: Clin. Exp. Res.*, 15, 1040–1044.

Mancillas, J.R., Siggins, G.R., and Bloom, F.E. (1986) Systemic ethanol: selective enhancement of responses to acetylcholine and somatostatin in hippocampus. *Science*, 231, 161–163.

Markram, H. and Segal, M. (1992) Activation of protein kinase C suppresses responses to NMDA in rat CA1 hippocampal neurones. *J. Physiol.*, 457, 491–501.

Marszalec, W., Kurata, Y., Hamilton, B.J., Carter, D.B., and Narahashi, T. (1994) Selective effects of alcohols on gamma-aminobutyric acid A receptor subunits expressed in human embryonic kidney cells. *J. Pharmacol. Exp. Ther.*, 269, 157–163.

Masood, K., Wu, C., Brauneis, U., and Weight, F.F. (1994) Differential ethanol sensitivity of recombinant *N*-methyl-D-aspartate receptor subunits. *Mol. Pharmacol.*, 45, 324–329.

McDonald, B.J. and Moss, S.J. (1994) Differential phosphorylation of intracellular domains of γ-aminobutyric acid type A receptor subunits by calcium/calmodulin type 2-dependent protein kinase and cGMP-dependent protein kinase. *J. Biol. Chem.*, 269, 18111–18117.

McGlade-McCulloh, E., Yamamoto, H., Tan, S.-E., Brickey, D.A., and Soderling, T.R. (1993) Phosphorylation and regulation of glutamate receptors by calcium/calmodulin-dependent protein kinase II. *Nature*, 362, 640–642.

Meguro, H., Mori, H., Araki, K., Kushiya, E., Kutsuwada, T., Yamazaki, M., Kumanishi, T., Arakawa, M., Sakimura, K., and Mishina, M. (1992) Functional characterization of a heteromeric NMDA receptor channel expressed from cloned cDNAs. *Nature*, 357, 70–74.

Mehta, A.K. and Ticku, M.K. (1988) Ethanol potentiation of GABAergic transmission in cultured spinal cord neurons involves gamma-aminobutyric acid A-gated chloride channels. *J. Pharmacol. Exp. Ther.*, 246, 558–564.

Messing, R.O., Petersen, P.J., and Henrich, C.J. (1991) Chronic ethanol exposure increases levels of protein kinase C delta and epsilon and protein kinase C-mediated phosphorylation in cultured neural cells. *J. Biol. Chem.*, 266, 23428–23432.

Monyer, H., Sprengel, R., Schoepfer, R., Herb, A., Higuchi, M., Lomeli, H., Burnashev, N., Sakmann, B., and Seeburg, P.H. (1992) Heteromeric NMDA receptors: molecular and functional distinction of subtypes. *Science*, 256, 1217–1221.

Moran, O. and Dascal, N. (1989) Protein kinase C modulates neurotransmitter responses in *Xenopus* oocytes injected with rat brain RNA. *Mol. Brain Res.*, 5, 193–202.

Moriyoshi, K., Masu, M., Ishii, T., Shigemoto, R., Mizuno, N., and Nakanishi, S. (1991) Molecular cloning and characterization of the rat NMDA receptor. *Nature*, 354, 31–37.

Moss, S.J., Blackstone, C.D., and Huganir, R.L. (1993) Phosphorylation of recombinant non-NMDA glutamate receptors on serine and tyrosine residues. *Neurochem. Res.*, 18, 105–110.

Moss, S.J., Doherty, C.A., and Huganir, R.L. (1992a) Identification of the cAMP-dependent protein kinase and protein kinase C phosphorylation sites within the major intracellular domains of the $\beta 1$, γ_{2S} and γ_{2L} subunits of the $GABA_A$ receptor. *J. Biol. Chem.*, 267, 14470–14476.

Moss, S.J., Ravindran, A., Mei, L., Wang, J.B., Kofuji, P., Huganir, R.L., and Burt, D.R. (1991) Characterization of recombinant $GABA_A$ receptors produced in transfected cells from murine α_1, β_1, and γ_2 subunit cDNAs. *Neurosci. Lett.*, 123, 265–268.

Moss, S.J., Smart, T.G., Blackstone, C.D., and Huganir, R.L. (1992b) Functional modulation of $GABA_A$ receptors by cAMP-dependent protein phosphorylation. *Science*, 257, 661–665.

Murphy, N.P., Cordier, J., Glowinski, J., and Prémont, J. (1994) Is protein kinase C activity required for the *N*-methyl-D-aspartate-evoked rise in cytosolic Ca^{2+} in mouse striatal neurons. *Eur. J. Neurosci.*, 6, 854–860.

Nakahiro, M., Arakawa, O., and Narahashi, T. (1991) Modulation of gamma-aminobutyric acid receptor-channel complex by alcohols. *J. Pharmacol. Exp. Ther.*, 259, 235–240.

Nakanishi, S. (1994) Metabotropic glutamate receptors: synaptic transmission, modulation, and plasticity. *Neuron*, 13, 1031–1037.

Nishio, M. and Narahashi, T. (1990) Ethanol enhancement of GABA-activated chloride current in rat dorsal root ganglion neurons. *Brain Res.*, 518, 283–286.

Persohn, E., Malherbe, P., and Richards, J.G. (1992) Comparative molecular neuroanatomy of cloned $GABA_A$ receptor subunits in the rat CNS. *J. Comp. Neurol.*, 326, 193–216.

Porter, N.M., Twyman, R.E., Uhler, M.D., and Macdonald, R.L. (1990) Cyclic AMP-dependent protein kinase decreases $GABA_A$ receptor current in mouse spinal neurons. *Neuron*, 5, 789–796.

Proctor, W.R., Allan, A.M., and Dunwiddie, T.V. (1992a) Brain region-dependent sensitivity of $GABA_A$ receptor-mediated responses to modulation by ethanol. *Alcoholism: Clin. Exp. Res.*, 16, 480–489.

Proctor, W.R., Soldo, B.L., Allan, A.M., and Dunwiddie, T.V. (1992b) Ethanol enhances synaptically evoked $GABA_A$ receptor-mediated responses in cerebral cortical neurons in rat brain slices. *Brain Res.*, 595, 220–227.

Reynolds, J.N., Prasad, A., and MacDonald, J.F. (1992) Ethanol modulation of GABA receptor-activated Cl^- currents in neurons of the chick, rat and mouse central nervous system. *Eur. J. Pharmacol.*, 224, 173–181.

Rius, R.A., Govoni, S., Battaini, F., and Trabucchi, M. (1986) Cyclic AMP-dependent protein phosphorylation is reduced in rat striatum after chronic ethanol treatment. *Brain Res.*, 365, 355–359.

Safran, A., Sagi-Eisenberg, R., Neumann, D., and Fuchs, S. (1987) Phosphorylation of the acetylcholine receptor by protein kinase C and identification of the phosphorylation site within the γ subunit. *J. Biol. Chem.*, 262, 10506–10510.

Schoepp, D.D. (1993) The biochemical pharmacology of metabotropic glutamate receptors. *Biochem. Soc. Trans.*, 21, 97–102.

Schoepp, D.D. and Conn, P.J. (1993) Metabotropic glutamate receptors in brain function and pathology. *Trends Pharmacol. Sci.*, 14, 13–20.

Schofield, P.R., Darlison, M.G., Fujita, N., Burt, D.R., Stephenson, F.A., Rodriguez, H., Rhee, L.M., Ramachandran, J., Reale, V., Glencorse, T.A., Seeburg, P.H., and Barnard, E.A. (1987) Sequence and functional expression of the $GABA_A$ receptor shows a ligand-gated receptor super-family. *Nature*, 328, 221–227.

Shanley, B., Gurd, J., and Kalant, H. (1985) Ethanol tolerance and enhanced calcium/calmodulin-dependent phosphorylation of synaptic membrane proteins. *Neurosci. Lett.*, 58, 55–59.

Sigel, E. and Baur, R. (1988) Activation of protein kinase C differentially modulates neuronal Na^+, Ca^{2+}, and gamma-aminobutyrate type A channels. *Proc. Natl. Acad. Sci. U.S.A.*, 85, 6192–6196.

Sigel, E., Baur, R., and Malherbe, P. (1993) Recombinant $GABA_A$ receptor function and ethanol. *FEBS Lett.*, 324, 140–142.

Sigel, E., Baur, R., and Malherbe, P. (1994) Protein kinase C transiently activates heteromeric *N*-methyl-D-aspartate receptor channels independent of the phosphorylatable C-terminal splice domain and of consensus phosphorylation sites. *J. Biol. Chem.*, 269, 8204–8208.

Siggins, G.R., Bloom, F.E., French, E.D., Madamba, S.G., Mancillas, J., Pittman, Q.J., and Rogers, J. (1987a) Electrophysiology of ethanol on central neurons. *Ann N. Y. Acad. Sci.*, 492, 350–366.

Siggins, G.R., Pittman, Q.J., and French, E.D. (1987b) Effects of ethanol on CA1 and CA3 pyramidal cells in the hippocampal slice preparation: an intracellular study. *Brain Res.*, 414, 22–34.

Slater, S.J., Cox, K.J.A., Lombardi, J.V., Ho, C., Kelly, M.B., Rubin, E., and Stubbs, C.D. (1993) Inhibition of protein kinase C by alcohols and anaesthetics. *Nature*, 364, 82–84.

Snell, L.D., Iorio, K.R., Tabakoff, B., and Hoffman, P.L. (1994a) Protein kinase C activation attenuates N-methyl-D-aspartate-induced increases in intracellular calcium in cerebellar granule cells. *J. Neurochem.*, 62, 1–7.

Snell, L.D., Tabakoff, B., and Hoffman, P.L. (1994b) Involvement of protein kinase C in ethanol-induced inhibition of NMDA receptor function in cerebellar granule cells. *Alcoholism*, 18, 81–85.

Sterz, R., Peper, K., Simon, J., Ebert, J.P., Edge, M., Pagala, M., and Bradley, R.J. (1986) Agonist and blocking effects of choline at the neuromuscular junction. *Brain Res.*, 385, 99–114.

Sugihara, H., Moriyoshi, K., Ishii, T., Masu, M., and Nakanishi, S. (1992) Structures and properties of seven isoforms of the NMDA receptor generated by alternative splicing. *Biochem. Biophys. Res. Commun.*, 185, 826–836.

Suzdak, P.D., Schwartz, R.D., Skolnick, P., and Paul, S.M. (1986) Ethanol stimulates gamma-aminobutyric acid receptor-mediated chloride transport in rat brain synaptoneurosomes. *Proc. Natl. Acad. Sci. U.S.A.*, 83, 4071–4075.

Sweetnam, P.M., Lloyd, J., Gallombardo, P., Malison, R.T., Gallager, D.W., Tallman, J.F., and Nestler, E.J. (1988) Phosphorylation of the $GABA_A$/benzodiazepine receptor alpha subunit by a receptor-associated protein kinase. *J. Neurochem.*, 51, 1274–1284.

Swope, S.L., Moss, S.J., Blackstone, C.D., and Huganir, R.L. (1992) Phosphorylation of ligand-gated ion channels: a possible mode of synaptic plasticity. *FASEB*, 6, 2514–2523.

Tabakoff, B. and Hoffman, P.L. (in press) Effect of alcohol on neurotransmitters and their receptors and enzymes. In: *Alcohol and Alcoholism*, volume 2, Oxford: Oxford University Press.

Tan, S.-E., Wenthold, R.J., and Soderling, T.R. (1994) Phosphorylation of AMPA-type glutamate receptors by calcium/calmodulin-dependent protein kinase II and protein kinase C in cultured hippocampal neurons. *J. Neurosci.*, 14, 1123–1129.

Tingley, W.G., Roche, K.W., Thompson, A.K., and Huganir, R.L. (1993) Regulation of NMDA receptor phosphorylation by alternative splicing of the C-terminal domain. *Nature*, 364, 70–73.

Tonner, P.H., Wood, S.C., and Miller, K.W. (1992) Can nicotine self-inhibition account for its low efficacy at the nicotinic acetylcholine receptor from Torpedo? *Mol. Pharmacol.*, 42, 890–897.

Urushihara, H., Tohda, M., and Nomura, Y. (1992) Selective potentiation of *N*-methyl-D-aspartate-induced current by protein kinase C in *Xenopus* oocytes injected with rat brain RNA. *J. Biol. Chem.*, 267, 11697–11700.

Wafford, K.A., Burnett, D.M., Leidenheimer, N.J., Burt, D.R., Wang, J.B., Kofuji, P., Dunwiddie, T.V., Harris, R.A., and Sikela, J.M. (1991) Ethanol sensitivity of the $GABA_A$ receptor expressed in xenopus oocytes requires 8 amino acids contained in the γ_{2L} subunit. *Neuron*, 7, 27–33.

Wafford, K.A. and Whiting, P.J. (1992) Ethanol potentiation of $GABA_A$ receptors requires phosphorylation of the alternatively spliced variant of the gamma 2 subunit. *FEBS Lett.,* 313, 113–117.

Wallace, B.G. (1994) Staurosporine inhibits agrin-induced acetylcholine receptor phosphorylation and aggregation. *J. Cell Biol.,* 125, 661–668.

Wallace, B.G., Qu, Z., and Huganir, R.L. (1991) Agrin induces phosphorylation of the nicotinic acetylcholine receptor. *Neuron,* 6, 869–878.

Wang, L., Salter, M.W., and MacDonald, J.F. (1991) Regulation of kainate receptors by cAMP-dependent protein kinase and phosphatases. *Science,* 253, 1132–1135.

Wang, L.-Y., Dudek, E.M., Browning, M.D., and MacDonald, J.F. (1994) Modulation of AMPA/kainate receptors in cultured murine hippocampal neurones by protein kinase C. *J. Physiol.,* 475, 431–437.

Wang, L.-Y., Orser, B.A., Brautigan, D.L., and MacDonald, J.F. (1994) Regulation of NMDA receptors in cultured hippocampal neurons by protein phosphatases 1 and 2A. *Nature,* 369, 230–232.

Wang, Y.T. and Salter, M.W. (1994) Regulation of NMDA receptors by tyrosine kinases and phosphatases. *Nature,* 369, 233–235.

Weiner, J.L., Zhang, L., and Carlen, P.L. (1994) Potentiation of $GABA_A$-mediated synaptic current by ethanol in hippocampal CA1 neurons: possible role of protein kinase C. *J. Pharmacol. Exp. Ther.,* 268, 1388–1395.

White, G., Lovinger, D.M., and Weight, F.F. (1990) Ethanol inhibits NMDA-activated current but does not alter GABA-activated current in an isolated adult mammalian neuron. *Brain Res.,* 507, 332–336.

Whiting, P., McKernan, R.M., and Iversen, L.L. (1990) Another mechanism for creating diversity in gamma-aminobutyrate type A receptors: RNA splicing directs expression of two forms of γ_2 subunit, one of which contains a protein kinase C phosphorylation site. *Proc. Natl. Acad. Sci. U.S.A.,* 87, 9966–9970.

Wu, G., Tonner, P.H., and Miller, K.W. (1994) Ethanol stabilizes the open channel state of the *Torpedo* nicotinic acetylcholine receptor. *Mol. Pharmacol.,* 45, 102–108.

Yamakura, T., Mori, H., Shimoji, K., and Mishina, M. (1993) Phosphorylation of the carboxyl-terminal domain of the zeta-1 subunit is not responsible for potentiation by TPA of the NMDA receptor channel. *Biochem. Biophys. Res. Commun.,* 196, 1537–1544.

Yamazaki, M., Mori, H., Araki, K., Mori, K.J., and Mishina, M. (1992) Cloning, expression and modulation of a mouse NMDA receptor subunit. *FEBS Lett.,* 300, 39–45.

Chapter 8

ETHANOL AND GUANINE NUCLEOTIDE BINDING PROTEINS

Francis Y.G. Lee, Sr. and Gary S. Wand

TABLE OF CONTENTS

I. INTRODUCTION

Guanine nucleotide binding proteins (GTP-binding proteins) are a superfamily of proteins that regulate a diverse array of cellular activities. They are important in such functions as directing transmembrane signals, inducing protein synthesis, guiding vesicular transport, and controlling cellular growth and differentiation.

This chapter will focus on the family of membrane signal transducing GTP-binding proteins called G proteins. The membrane bound G proteins regulate the activity of several enzymes and ion channels, including cGMP-specific phosphodiesterase, hormone sensitive adenylyl cyclase, phospholipase C, and at least six sodium-, potassium-, and calcium-ion channels (Brown and Birnbaumer, 1988; Brown, 1990; Yatani et al., 1988a; Yatani et al., 1988b). Some receptors may couple with more than one G protein, allowing a finite number of receptors to elicit a myriad of cell responses in different tissues.

G proteins are trimeric in structure, composed of an α chain and a smaller tightly coupled $\beta\gamma$ dimer. Binding and hydrolysis of GTP by GTP-binding proteins is the mechanism by which their activity is regulated. In this scenario, a ligand-stimulated receptor induces the heterotrimeric G protein to dissociate into its α and $\beta\gamma$ forms allowing the α subunit to bind GTP, and in so doing, to liberate a previously bound GDP. Bound GTP causes a conformational change in the GTP-binding protein that increases its affinity for effector proteins (e.g., adenylyl cyclase). The activity of the GTP-binding protein persists until an intrinsic GTPase hydrolyses GTP to GDP.

Seventeen unique α genes have been identified, coding for proteins which range in size from 350 to 395 amino acids. Among all identified α chains, there is 40% homology; among members within subfamilies, there is between 60–90% homology (Simon et al., 1991). The α chains have been subdivided into four subfamilies according to their amino acid homology: $G_s\alpha$ and G_{olf}; $G_i\alpha$ (Sugimoto et al., 1985; Fong et al., 1986; Gao et al., 1987), $G_o\alpha$, $G_z\alpha$, transducins 1 and 2, and gustducin; $G_q\alpha$, $G_{11}\alpha$, $G_{14}\alpha$, and $G_{15}\alpha$; and $G_{12}\alpha$ and $G_{13}\alpha$ (Simon et al., 1991). Some α subunits, such as $G_s\alpha$, $G_i\alpha$, and $G_q\alpha$, have a ubiquitous tissue distribution (Birnbaumer et al., 1990). Other α subunits, such as the transducins $T\alpha1$ and $T\alpha2$ (Lerea et al.,

0-8493-8389-7/95/$0.00+$.50

1986; Lerea et al., 1989), G_{olf} (Jones and Reed, 1987), and gustducin (McLaughlin et al., 1992) are quite tissue specific. The α subunit discriminates between different receptor and effector molecules, conferring functional specificity to the G protein. Additionally, the α subunits of many G proteins are posttranslationally modified. The amino-terminal glycines of $G_i\alpha$, $G_o\alpha$, $G_z\alpha$, and $G_t\alpha$ undergo N-myristoylation, which probably enhances the α subunits' affinity for βγ dimers. Fatty acids modify the N-terminal region of transducin; the functional significance of which is unclear at the present time (Kokame et al., 1992). And, certain G_i proteins can be phosphorylated (Bushfield et al., 1990a), presumably inactivating the protein.

The βγ dimer is tightly associated with the α chain and is important in allowing the G protein to interact with a receptor molecule. Four different β subunits (Sugimoto et al., 1985; Fong et al., 1986; Gao et al., 1987; Evans et al., 1987; Levine et al., 1990a; Fung et al., 1992; von Weizsacker et al., 1992; Gautam et al., 1990) and six γ subunits (Simon et al., 1991; Robishaw et al., 1989; Cali et al., 1992; Fisher and Aronson, 1992; Birnbaumer, 1990) have been identified. Dimerization of β and γ subunits is quite specific (Schmidt and Neer, 1991; Schmidt et al., 1992). Furthermore, βγ dimers differ in their ability to couple α subunits to receptors and to modulate effectors. The effects of βγ dimers on adenylyl cyclase also depends upon the type of adenylyl cyclase present. There is evidence that type-I adenylyl cyclase is inhibited by βγ dimers, whereas type-II adenylyl cyclase is stimulated by βγ dimers (Tang and Gilman, 1991). Further investigations should elucidate the unclearly defined and complex role of the βγ subunit in signal transduction.

Inherited as well as acquired alterations in G protein function play a role in several clinical disorders. Activating mutations of the gene encoding the $G_s\alpha$ protein (GNAS1) have been identified in tissues from individuals with McCune-Albright syndrome, a disorder characterized phenotypically by a mosaic distribution of bone lesions, cutaneous hyperpigmentation, and secretory endocrine neoplasia (Schwindinger et al., 1992; Weinstein et al., 1991). Inherited mutations in GNAS1 that decrease $G_s\alpha$ expression or function have been identified in individuals with Albright hereditary osteodystrophy (AHO) or pseudohypoparathyroidism. AHO is characterized by resistance of target tissues to the biological actions of parathyroid hormone (Patten et al., 1990). Sporadic human pituitary (Vallar et al., 1987) and thyroid adenomas (Landis et al., 1989) have been identified in which GNAS1 mutations inactivate the intrinsic GTPase activity of $G_s\alpha$, leading to constitutive activation of adenylyl cyclase activity. For the pituitary adenomas, the mutations result in neoplasia as well as in hypersecretion of hormone.

The bacterial toxins of *Vibrio cholera* and *Bordetella pertussis* covalently modify certain α chains. Cholera toxin will block the GTPase activity of $G_s\alpha$, resulting in enhanced activation of $G_s\alpha$ and in the clinical manifestation of secretory diarrhea. In the case of pertussis toxin, certain $G_i\alpha$ subunits, which inhibit adenylyl cyclase, are modified by the toxin, resulting in increased cyclic AMP (cAMP) accumulation. Clinically, this explains the protracted respiratory course of whooping cough and presumably several other manifestations of pertussis, such as hypoglycemia and histamine sensitivity.

In addition to mutations and to toxins, altered expression of G proteins can occur under a variety of circumstances:

1. hypo- or hyperthyroidism (Levine et al., 1990b)
2. glucocorticoid treatment (Chang and Bourne, 1987)
3. insulin deficiency (Bushfield et al., 1990a)
4. heart failure (Insel and Ransnas, 1988; Longabaugh et al., 1988; Feldman et al., 1990; Katoh et al., 1990)
5. certain psychiatric illnesses (Okada et al., 1990; Young et al., 1991; Schreiber et al., 1991)
6. exposure to certain antidepressant medications (Lesch et al, 1991)

7. exposure to lithium (Avissar et al., 1988; Mork et al., 1990; Kawamoto et al., 1991; Ellis and Lenox, 1991; Manji et al., 1991; Colin et al., 1991)
8. exposure to opiates (Parolaro et al., 1990; Attali and Vogel, 1989; Beitner et al., 1989; Ammer et al., 1991)
9. exposure to cocaine (Nestler et al., 1990; Nestler, 1992)
10. exposure to ethanol (Wand and Levine, 1991; Wand et al., 1993; Waltman et al., 1993; Charness et al., 1988; Mochly-Rosen et al., 1988).

Ethanol is one of several drugs that alters the expression of G proteins. Ethanol may either increase or decrease adenylyl cyclase activity, depending upon in part whether the exposure is acute or chronic (Stenstrom and Richelson, 1982; Saito et al., 1985; Rabe et al., 1990; Rabin and Molinoff, 1983; Chung et al., 1989). Data from studies of neural cells in culture, of animal models, and of humans will be reviewed below.

II. *IN VITRO* STUDIES

Early studies demonstrated that acute exposure to ethanol increases adenylyl cyclase activity and intracellular cAMP accumulation in many tissues (see review by Hoffman and Tabakoff, 1990). For example, brief exposure to low concentrations of ethanol (e.g., 20 m*M*) potentiates ligand-stimulated intracellular cAMP accumulation in intact lymphocytes, platelets, granulocytes, alveolar macrophage, as well as in a variety of cell lines and tissue preparations (Hoffman and Tabakoff, 1990). In cell preparations of adipocytes, pancreatic tissue, and parotid gland tissue, ethanol activates adenylyl cyclase activity more potently in the presence of guanine nucleotides (Tabakoff et al., 1988a), implicating $G_s\alpha$ as a mediator of some of the cellular responses of ethanol on cAMP accumulation.

Studies of central nervous system tissue preparations have corroborated the initial hypothesis that G proteins mediate some of the cellular actions of ethanol and that $G_s\alpha$ may be of particular importance. For example, in mouse striatal membranes, physiologically relevant concentrations of ethanol will increase adenylyl cyclase membrane activity, provided that guanine nucleotides are present in the reaction (Luthin and Tabakoff, 1984). Additionally, adenylyl cyclase activity in cerebral cortical membranes has been shown to increase to a greater extent when stimulated with guanine nucleotides and isoproterenol in the presence of ethanol than in its absence (Saito et al., 1985). Overall, these studies support the hypothesis that ethanol alters the rate of activation (dissociation) of G_s and increases $G_s\alpha$ binding to guanine nucleotides. However, these studies do not preclude the possibility that ethanol can directly interact with adenylyl cyclase. In fact, studies with forskolin, a diterpene activator of adenylyl cyclase, support the likelihood of this interaction. Ethanol has been shown to inhibit forskolin-stimulated adenylyl cyclase activity in several tissues, including bovine corpora lutea (Huang et al., 1982), rat heart (Robberecht et al., 1983), and murine neuroblastoma N1E-115 cells (Stenstrom et al., 1985).

Repeated exposure to ethanol promotes a cellular adaptation resulting in ethanol having reduced ability to potentiate ligand-stimulated cAMP accumulation. This homeostatic adaptation to ethanol is often referred to as cellular tolerance and in certain settings resembles heterologous desensitization. For example, using NG108–15 cells, a hybrid line of mouse neuroblastoma cells and rat glioma cells, investigators have shown that a brief exposure to 50 m*M* ethanol potentiates ligand-stimulated cAMP accumulation. However, after prolonged ethanol exposure (e.g., 48 hours), there is a significant reduction in the ability of ethanol to potentiate ligand-stimulated cAMP accumulation within these cells (Gordon et al., 1986). This ethanol-induced reduction in cAMP accumulation coincides with reduced membrane $G_s\alpha$ content but no change in $G_i\alpha$ expression (Mochly-Rosen et al., 1988). These findings suggest that the ethanol-induced fall in membrane $G_s\alpha$ content is responsible for this form of desensitization or tolerance. Inhibitory effects of ethanol on the adenosine transporter as demonstrated in NG108–15 cells (Gordon et al., 1986) and S49 lymphoma cells (Nagy et al., 1990) probably play a role in this response as well. More recently, a similar effect of chronic ethanol exposure on $G_s\alpha$ and cAMP accumulation was demonstrated in the pheochromocytoma cell line, PC 12 cells (Rabin, 1990; Rabin, 1993). In the case of the PC 12 cells, the effects of ethanol on an adenosine transporter does not seem to be involved (Rabin, 1993). Interestingly, in other cell lines, the reduction in cAMP accumulation induced by ethanol involves inhibitory rather than stimulatory G proteins. For example, in the murine neuroblastoma N1E-115 cells, a cell line related to NG108, chronic ethanol exposure also decreases cAMP responses to several ligands; however, this change in cAMP accumulation is associated with an *increase* in the inhibitory protein $G_i\alpha$, rather than a *decrease* in the stimulatory protein $G_s\alpha$. Longer incubations of N1E-15 with ethanol ultimately results in decreased expression of $G_s\alpha$ as well (Charness et al., 1988).

It remains unclear how chronic ethanol exposure results in diminished membrane $G_s\alpha$ content or in enhanced $G_i\alpha$ content. Speculative mechanisms include:

1. regulation at the transcriptional level, perhaps induced by chronic activation of protein kinase A (PKA)
2. alterations in the translocation of G proteins from membrane to cytosol
3. altered kinetics of degradation.

The molecular consequences of chronic ethanol exposure are complex, involving more participants than merely ethanol and G proteins. Chronic ethanol exposure may have effects on certain forms of adenylyl cyclase. This is underscored by observations regarding how chronic ethanol exposure inhibits forskolin-stimulated (Hoffman and Tabakoff, 1990), and manganese-stimulated adenylyl cyclase activity (Wand and Levine, 1991). Furthermore, chronic ethanol exposure can perturb the function of protein kinase A (Rabin et al., 1992) and protein kinase C (Hoek et al., 1988). This may have significant functional consequences since kinase-induced phosphorylation of certain $G_i\alpha$ proteins (Bushfield et al., 1990b) and of adenylyl cyclases (Yoshimasa et al., 1987) may alter protein function without necessarily changing membrane protein content.

III. ANIMAL MODELS

We have examined the effects of chronic ethanol exposure on G proteins and adenylyl cyclase signalling in the pituitary and central nervous system of the selectively bred Long-Sleep (LS) and Short-Sleep (SS) lines of mice. Transcription of the proopiomelanocortin gene (POMC, the prohormone for ACTH and β-endorphin), as well as ACTH biosynthesis in the anterior pituitary, is dependent upon activation of adenylyl cyclase and the generation of cAMP. POMC production is stimulated by acute ethanol exposure in both LS and SS lines of mice (Wand, 1989). However, LS and SS lines of mice differ in their ability to sustain POMC production during repeated ethanol exposure. Hormonal tolerance develops in LS mice, manifested by a reduction in the ability of ethanol to stimulate POMC mRNA levels and ACTH synthesis following chronic ethanol exposure. In the LS line of mice, this form of hormonal tolerance is associated with decreased membrane adenylyl cyclase activity. The ethanol-induced diminution in membrane adenylyl cyclase activity is coincident with a fall in membrane $G_s\alpha$ content but without a change in the content of $G_i\alpha$. Since POMC biosynthesis is regulated by adenylyl cyclase signal transduction and adenylyl cyclase signal transduction is in part regulated by $G_s\alpha$ levels, the observations suggest that ethanol-induced regulation of $G_s\alpha$ results in reduced adenylyl cyclase signal transduction and hormonal tolerance within the POMC system. Moreover, the SS line which did not develop hormonal tolerance during chronic ethanol administration did not have any alterations in adenylyl cyclase activity or G protein expression (Wand and Levine, 1991). These observations in selectively bred lines of mice support the notion that certain forms of tolerance may in part be genetically determined.

Is the reduction in membrane $G_s\alpha$ content in the anterior pituitary of LS mice a direct result of ethanol (i.e., similar to the effects of ethanol on NG-108 cells described above)? Or, is this G protein modulation the indirect result of a primary effect of ethanol elsewhere? Both *in vivo* and *in vitro* studies show that ethanol stimulates the release of the major hypothalamic ACTH releasing factor, corticotropin-releasing factor (CRH) (Rivier et al., 1984; Zgombick and Erwin, 1988; Redei et al., 1988). It is plausible that the ethanol-induced fall in anterior pituitary membrane $G_s\alpha$ is really a consequence of chronic CRH release induced by repeated ethanol exposure. To test this hypothesis, we exposed primary cultures of anterior pituicytes, as well as Att20 cells (a mouse anterior pituitary ACTH-producing cell line), with physiologically relevant concentrations of CRH for 12 hour intervals. This treatment resulted in a 50% fall in membrane $G_s\alpha$ content as well as in membrane adenylyl cyclase activity, mimicking the effects of ethanol just described (unpublished results from Wand lab). In contrast, 12 hours of glucocorticoid exposure actually increased membrane adenylyl cyclase activity and content of $G_s\alpha$, supporting what we have previously described in several *in vivo* stress paradigms (Wolfgang et al., 1994; Morrill et al., 1993).

The effects of ethanol on adenylyl cyclase activity and G protein expression in LS and SS mice is not limited to the pituitary. Chronic ethanol exposure also reduces adenylyl cyclase activity in membranes derived from cerebellum and pons in LS and SS mice (Wand et al., 1993). However, in this case, the reduction in adenylyl cyclase activity is coincident with enhanced expression of $G_i\alpha(1)$ and $G_i\alpha(2)$ compared to levels in control mice. Other G protein levels (e.g., $G_s\alpha$, $G_o\alpha$, and $G\beta$) were not altered by the ethanol treatments. The reduction in adenylyl cyclase activity was reversed after pretreating membranes with pertussis toxin, implying a causal relationship between the ethanol-induced increase in membrane G_i content and the reduction in membrane adenylyl cyclase activity.

These *in vivo* studies of ethanol exposure on pituitary and brain tissue corroborate what has been shown in the *in vitro* models described above and again suggest that there are several mechanisms by which chronic ethanol can decrease membrane adenylyl cyclase activity that results in a form of heterologous desensitization. In the pituitary, desensitization to ethanol was accompanied by decreased membrane $G_s\alpha$ content as well as diminished membrane adenylyl cyclase activity. In contrast, in the cerebellum and pons, desensitization was accompanied by inhibition of adenylyl cyclase activity via enhanced expression of $G_i\alpha(1)$ and $G_i\alpha(2)$ (Wand et al., 1993).

It must be emphasized that the *in vivo* studies described above were conducted by delivering ethanol as a daily intraperitoneal (i.p.) injection. As opposed to oral ethanol administration where blood ethanol levels (BELs) are elevated during the nocturnal feeding time and often throughout the day (thereby preventing any significant emergence of withdrawal), in the i.p. design, elevations in BELs are confined to a 3–4 hour time interval. After several days of repeated i.p. ethanol administration, mice begin to display symptoms of withdrawal (e.g., hyperkinetic and aggressive behavior) for a part of each day. Withdrawal is a significant stressor resulting in increased sympathetic tone, glucocorticoid release, and release of other biologically active humoral factors. It is plausible that alterations in G proteins observed with the i.p. design result from these turbulent periods of withdrawal and that these alterations are not the consequences of a direct effect of ethanol on G protein expression. Work with membranes derived from human alcoholic subjects, abstinent alcoholic subjects, and nonalcoholics suggests this may be the case (see below). This may also explain why some studies utilizing oral ethanol paradigms do not show significant changes in G protein content (Druse et al., 1994).

IV. HUMAN STUDIES

A. BLOOD CELLS ISOLATED FROM ALCOHOLIC AND NONALCOHOLIC SUBJECTS

What evidence is there that expression and function of the protein coupled adenylyl cyclase system is altered in human subjects with alcoholism? Lymphocytes and platelets isolated from alcoholics show reduced accumulation of intracellular cAMP and reduced adenylyl cyclase activity when incubated in the presence of several stimulators (Mochly-Rosen et al., 1988; Nagy et al., 1988; Diamond et al., 1987; Tabakoff et al., 1988b; Waltman et al., 1993). Although freshly isolated lymphocytes from alcoholic subjects have reduced cAMP levels, after several generations in alcohol-free culture media, lymphocytes from alcoholics appear more sensitive to factors that stimulate cAMP production than lymphocytes from nonalcoholic controls (Nagy et al., 1988). It is likely that the reduced cAMP levels found in freshly isolated lymphocytes from alcoholics are the result of prior alcohol exposure, and it is possible that the alcohol-free environment allows the return of $G_s\alpha$ expression to higher steady state levels, as seen in our nonalcoholic family history positive (FHP) men (see below).

To investigate G protein changes that might be associated with reduced adenylyl cyclase activity in alcoholics, we studied adenylyl cyclase activity in peripheral lymphocytes isolated from abstinent alcoholic men, actively drinking alcoholic men, and nonalcoholic men (Waltman et al., 1993). Immunoblot analysis indicated that lymphocyte membranes from abstinent alcoholics (7–35 days abstinent) contained 3-fold greater levels of $G_i\alpha(2)$ protein than did membranes from control subjects. Lymphocytes from actively drinking alcoholics had levels of $G_s\alpha$ and $G_i\alpha(2)$ protein that were nonsignificantly elevated compared to control subjects. The enhanced $G_i\alpha(2)$ expression in lymphocytes from abstinent alcoholics was associated with decreased basal, PGE_1-, GTPγS-, and forskolin-stimulated adenylyl cyclase activity compared to activity in control subjects and actively drinking alcoholics. These results confirm the previous observation that lymphocyte adenylyl cyclase activity is reduced during early abstinence from alcohol and raises the question of whether alterations in the inhibitory G protein, $G_i\alpha(2)$, account for this change. These observations are similar to findings described in N1E-115 cells (Charness et al., 1988) and in membranes from pons and cerebellum of ethanol-treated mice (Wand et al., 1993). Enhanced expression of $G_i\alpha(2)$ tends to normalize during prolonged abstinence (our unpublished observation) suggesting that the elevation of $G_i\alpha(2)$ during drinking and early abstinence is a *reversible biochemical event* triggered by heavy drinking and/or withdrawal. However, only prospective studies can confirm this speculation.

B. BLOOD CELLS ISOLATED FROM HIGH RISK AND LOW RISK NONALCOHOLIC SUBJECTS

Although the etiologies of alcoholism most certainly include inherited factors that create a physiological vulnerability to addiction, such mechanisms remain unclear. This ambiguity, in part, results from the fact that long-term use of alcohol obscures factors that play a role in the etiology of alcoholism (Newlin and Thomson, 1990). For example, comparing the adenylyl cyclase signal transduction system of alcoholics and nonalcoholics, as described above, may show us some of the effects of alcohol and/or withdrawal on this signal transduction system, but this does not necessarily elucidate how the adenylyl cyclase system in certain individuals may be particularly vulnerable to ethanol (e.g., tolerance, activation, and addiction). One approach to studying causation is to investigate the offspring of alcoholics who are at increased risk for alcoholism but who are nonalcoholics at the time of testing. Characteristics of the membrane adenylyl cyclase system of at-risk individuals may reveal factors that promote development of the disorder. At the same time, the analysis of the signalling system will not have been obscured by the long-term consequences of alcohol abuse.

We chose to investigate adenylyl cyclase signal transduction for this type of design because of several lines of evidence implicating adenylyl cyclase signalling as a candidate system for study in relationship to genetics of differential tissue responsiveness to ethanol. These observations include:

1. differential responsiveness of G proteins to ethanol among various cell lines (Charness et al., 1988)
2. differential responsiveness of adenylyl cyclase signalling between certain lines of selectively bred mice (Wand and Levine, 1991)
3. the observation that long term abstinent alcoholics have lower platelet membrane adenylyl activity compared to nonalcoholics (Tabakoff et al., 1988b; Tabakoff and Hoffman, 1989)
4. the observation that the magnitude of fluoride-stimulated adenylyl cyclase activity in platelet membranes is an inherited trait (Devor et al., 1991).

To this end, we evaluated membrane G protein content in erythrocytes and lymphocytes isolated from 26 FHP nonalcoholic and 26 family history negative (FHN) nonalcoholic subjects (Wand et al., 1994). Subjects were classified as FHP if their fathers met criteria for alcohol dependence; as FHN if there was no family history of alcoholism in any first or second degree relatives. Immunoblot analysis indicated that levels of erythrocyte membrane $G_s\alpha$ in FHP subjects were 1.7-fold greater than levels in FHN. To confirm the results of the immunoblot analysis, $G_s\alpha$ was quantitated by cholera toxin-dependent ADP ribosylation. Levels of erythrocyte ^{32}P-ADP ribose $G_s\alpha$ from FHP subjects were 2-fold greater than levels in FHN subjects. $G_s\alpha$ levels did not correlate with age or alcohol consumption. By contrast to differences in $G_s\alpha$, immunoblot analysis showed similar levels of $G_i\alpha(2)$ and $G_i\alpha(3)$ in erythrocyte membranes from both groups. To assess whether enhanced expression of $G_s\alpha$ in FHP men was unique to the RBC or generalized to other tissue, similar analyses were conducted on lymphocytes isolated from 16 FHP and 19 FHN subjects. For each subject, there was a significant correlation between erythrocyte membrane $G_s\alpha$ and lymphocyte membrane $G_s\alpha$ ($r = 0.6$, $P < 0.0001$). Similar to the erythrocytes, enhanced expression of $G_s\alpha$ was also documented in lymphocyte membranes from FHP subjects compared to FHN subjects. Expression of other G proteins did not show group differences. As only a subset FHP subjects will develop problems with alcohol over a lifetime, it is not surprising that some of our FHP subjects had $G_s\alpha$ levels which were similar to those of low-risk men.

It remains uncertain if membrane $G_s\alpha$ content is a primary marker for alcoholism (i.e., one causal determinant of phenotype) or a secondary marker, bearing no causal relationship to the alcoholism phenotype, but genetically linked to some other causal factors. Regardless of whether $G_s\alpha$ could be a primary or secondary marker, it is important to determine if enhanced expression of $G_s\alpha$ is a reliable predictor of individuals who will develop alcoholism.

If, in fact, alcoholics and their family members have low adenylyl cyclase activity as described in the platelet studies above, enhanced expression of $G_s\alpha$ may be compensatory for and balancing the effects of an adenylyl cyclase with reduced ability to generate cAMP. In this scenario, membrane adenylyl cyclase activity would be low or unchanged, depending on the degree of $G_s\alpha$ compensation. However, if alcoholics and their family members do not have an adenylyl cyclase with reduced activity, it is also plausible that enhanced expression of $G_s\alpha$ results in a cell more capable of generating cAMP following stimulation with ligand and ethanol. We speculate that genetic differences in G protein expression allows for differential amplification of receptor-stimulated cAMP production following ethanol exposure and that the resulting differences in cAMP levels allow for differences in the magnitude of certain forms of ethanol induced activation and tolerance. This may in part explain the observation of differential sensitivity among individuals.

V. SUMMARY

G proteins discriminate and amplify signals originating from the cell's external environment and destined for transmission into the cytosolic milieu. These proteins play a ubiquitous role in transmembrane signal transduction, highlighted by the rapidly expanding list of receptors and effector molecules that are coupled through G proteins. Altered expression or altered forms of these important proteins can lead to a variety of metabolic disturbances. Certain aspects of differential tissue sensitivity to alcohol as well as certain forms of tolerance to ethanol may be mediated, in part, through the interactions of ethanol and these important proteins.

REFERENCES

Ammer, H., Nice, L., Lang, J., and Schulz, R. (1991) Regulation of G proteins by chronic opiate and clonidine treatment in the guinea pig myenteric plexus. *J. Pharmacol. Exp. Ther.,* 258:790–796.

Attali, B. and Vogel, Z. (1989) Long-term opiate exposure leads to reduction of the alpha i-1 subunit of GTP-binding proteins. *J. Neurochem.,* 53:1636–1639.

Avissar, S., Schreiber, G., Danon, A., and Belmaker, R.H. (1988) Lithium inhibits adrenergic and cholinergic increases in GTP binding in rat cortex. *Nature,* 331(6155):440–442.

Beitner, D.B., Duman, R.S., and Nestler, E.J. (1989) A novel action of morphine in the rat locus coeruleus: persistent decrease in adenylate cyclase. *Molec. Pharmacol.,* 35 (5):559–564.

Birnbaumer, L. (1990) G proteins in signal transduction. *Rev. Pharmacol. Toxicol.,* 30:675–705.

Birnbaumer, L., Abramowitz, J., and Brown, A.M. (1990) Receptor-effector coupling by G proteins. *Biochem. Biophys. Acta,* 1031(2):163–224.

Brown, A.M. and Birnbaumer, L. (1988) Direct G protein gating of ion channels. Am. *J. Physiol.,* 254(3 Pt 2):H401–H410.

Brown, D.A. (1990) G proteins and potassium currents in neurons. *Annu. Rev. Physiol.,* 52:215–242.

Bushfield, M., Griffiths, S.L., Murphy, G.J., Knowler, J.T., Milligan, G., Parker, P.J., Mollner, S., and Houslay, M.D. (1990a) Diabetes-induced alterations in the expression, functioning and phosphorylation state of the inhibitory guanine nucleotide regulatory protein G_i2 in hepatocytes. *Biochem. J.,* 271(2):365–3372.

Bushfield, M., Murphy, G.J., Lavan, B.E., Parker, P.J., Hruby, V.J., Milligan, G., and Houslay, M.D. (1990b) Hormonal regulation of $G_i2\alpha$-subunit phosphorylation in intact hepatocytes. *Biochem. J.,* 268(2):449–457.

Cali, J.J., Balcueva, E.A., Rybalkin, I., and Robishaw, J.D. (1992) Selective tissue distribution of G protein gamma subunits, including a new form of the gamma subunits identified by cDNA cloning. *J. Biol. Chem.,* 267(33):24023–24027.

Chang, F. and Bourne, H.R. (1987) Dexamethasone increases adenylyl cyclase activity and expression of the alpha-subunit of G_s in GH3 cells. *Endocrinology,* 121(5):1711–1715.

Charness, M.E., Querimit, L.A., and Henteleff, M. (1988) Ethanol differentially regulates G proteins in neural cells. *Biochem. Biophys. Res. Commun.,* 155(1):138–143.

Chung, C.T., Tamarkin, L., Hoffman, P.L., and Tabakoff, B. (1989) Ethanol enhancement of isoproterenol-stimulated melatonin and cAMP release from cultured pineal glands. *J. Pharmacol. Exp. Ther.,* 249(1):16–22.

Colin, S.F., Chang, H.C., Mollner, S., Pfeuffer, T., Reed, R.R., Duman, R.S., and Nestler, E.J. (1991) Chronic lithium regulates the expression of adenylate cyclase and G_i protein alpha subunit in rat cerebral cortex. *Proc. Natl. Acad. Sci. U.S.A.,* 88(23):10634–10637.

Devor, E.J., Cloninger, C.R., Hoffman, P.L., and Tabakoff, B. (1991) A genetic study of platelet adenylate cyclase activity: evidence for a single major locus effect in fluoride-stimulated activity. *Am. J. Hum. Genet.,* 49(2):372–377.

Diamond, I., Wrubel, B., Estrin, W., and Gordon, A. (1987) Basal and adenosine receptor-stimulated levels of cAMP are reduced in lymphocytes from alcoholic patients. *Proc. Natl. Acad. Sci. U.S.A.,* 84(5):1413–1416.

Druse, M.J., Tajuddin, N.F., Eshed, M., and Gillespie, R. (1994) Maternal ethanol consumption: effects on G proteins and second messengers in brain regions of offspring. *Alcoholism: Clin. Exp. Res.,* 18(1):47–52.

Ellis, J. and Lenox, R.H. (1990) Chronic lithium treatment prevents atropine-induced supersensitivity of the muscarinic phospoinsitide response in rat hippocampus. *Biol. Psychiatry,* 28(7):609–619.

Evans, T., Fawzi, A., Fraser, E.D., Brown, M.L., and Northup, J.K. (1987) Purification of a beta 35 form of the beta gamma complex common to G proteins from human placental membranes. *J. Biol. Chem.,* 262(1):176–181.

Feldman, A.M., Tena, R.G., Kessler, P.D., Weisman, H.F., Schulman, S.P., Blumenthal, R.S., Jackson, D.G., and Vandop, C. (1990) Diminished beta-adrenergic receptor responsiveness and cardiac dilation in hearts of myopathic Syrian hamsters. *Circulation,* 81(4):1341–1352.

Fisher, K.J. and Aronson, N.N., Jr. (1992) Characterization of the cDNA and genomic sequence of a G protein gamma subunit (gamma 5). *Mol. Cell. Biol.,* 12(4):1585–1591.

Fong, H.K.W., Hurley, J.B., Hopkins, R.S., Miake-Lye, R., Johnson, M.S., Doolittle, R.F., and Simon, M.I. (1986) Repetitive segmental structure of the transducin beta subunit: homology with the CDC4 gene and identification of related mRNAs. *Proc. Natl. Acad. Sci. U.S.A.,* 83(7): 2162–2166.

Fung, B.K., Lieberman, B.S., and Lee, R.H. (1992) A third form of the G protein beta subunit. 2. Purification and biochemical properties. *J. Biol. Chem.,* 267(34):24782–24788.

Gao, B., Gilman, A.G., and Robishaw, J.D. (1987) A second form of the beta subunit of signal-transducing G proteins. *Proc. Natl. Acad. Sci. U.S.A.,* 84(17):6122–6125.

Gautam, N., Northup, J., Tamir, H., and Simon, M.I. (1990) G protein diversity is increased by associations with a variety of gamma subunits. *Proc. Natl. Acad. Sci. U.S.A.,* 87(20):7973–7977.

Gordon, A.S., Collier, K., and Diamond, I. (1986) Ethanol regulation of adenosine receptor-stimulated cAMP levels in a clonal neural cell line: an *in vitro* model of cellular tolerance to ethanol. *Proc. Natl. Acad. Sci. U.S.A.,* 83(7):2105–2108.

Hoek, J.B., Taraschi, T.F., and Rubin, E.R. (1988) Functional implications of the interaction of ethanol with biologic membranes: actions of ethanol on hormonal signal transduction systems. *Semin. Liver Dis.,* 8(1):36–46.

Hoffman, P.L. and Tabakoff, B. (1990) Ethanol and guanine nucleotide binding proteins: a selective interaction. *FASEB J.,* 4(9):2612–2622.

Huang, R.D., Smith, M.F., and Zahler, W.L. (1982) Inhibition of forskolin-activated adenylate cyclase by ethanol and other solvents. *J. Cyclic Nucleotide Res.,* 8(6):385–395.

Insel, P.A. and Ransnas, L.A. (1988) G proteins and cardiovascular disease. *Circulation,* 78(6):1511–1513.

Jones, D.T. and Reed, R.R. (1987) Molecular cloning of five GTP-binding protein cDNA species from rat olfactory neuroepithelium. *J. Biol. Chem.,* 262:14241–14249.

Katoh, Y.K., Komuro, I., Takaku, F., Yamaguchi, H., and Yazaki, Y. (1990) Messenger RNA levels of guanine nucleotide-binding proteins are reduced in the ventricle of cardiomyopathic hamsters. *Circulation Res.,* 67(1):235–239.

Kawamoto, H., Watanabe, Y., Imaizumi, T., Iwasaki, T., and Yoshida, H. (1991) Effects of lithium ion on ADP ribosylation of inhibitory GTP-binding protein by pertussis toxin, islet-activating protein. *Eur. J. Pharmacol.,* 206(1):33–37.

Kokame, K., Fukada, Y., Yoshizawa, T., Takao, T., and Shimonishi, Y. (1992) Lipid modification at the N terminus of photoreceptor G-protein alpha-subunit. *Nature,* 359(6397):749–752.

Landis, C.A., Masters, S.B., Spada, A., Pace, A.M., et al. (1989) GTPase inhibiting mutations activate the alpha chain of G_s and stimulate adenylyl cyclase in human pituitary tumors. *Nature,* 340(6236):692–696.

Lerea, C.L., Bunt-Milam, A.H., and Hurley, J.B. (1989) Alpha transducin is present in blue-, green-, and red-sensitive cone photoreceptors in the human retina. *Neuron,* 3(3):367–376.

Lerea, C.L., Somers, D.E., Hurley, J.B., Klock, I.B., and Bunt-Milam, A.H. (1986) Identification of specific transducin alpha subunits in retinal rod and cone photoreceptors. *Science,* 234(4772):77–80.

Lesch, K.P., Aulakh, C.S., Tolliver, T.J., Hill, J.L., and Murphy, D.L. (1991) Regulation of G proteins by chronic antidepressant drug treatment in rat brain: tricyclics but not clorgyline increase $G_o\alpha$-subunits. *Eur. J. Pharmacol.,* 207(4):361–364.

Levine, M.A., Feldman, A.M., Robishaw, J.D., Ladenson, P.W., Ahn, T.G., Moroney, J.F., and Smallwood, P.M. (1990b) Influence of thyroid hormone status on expression of genes encoding G protein subunits in the rat heart. *J. Biol. Chem.,* 265(6):3553–3560.

Levine, M.A., Smallwood, P.M., Moen, P.T., Jr., Helman, L.J., and Ahn, T.G. (1990a) Molecular cloning of beta 3 subunit, a third form of the G protein beta-subunit polypeptide. *Proc. Natl. Acad. Sci. U.S.A.,* 87(6):2329–2333.

Longabaugh, J.P., Vatner, D.E., Vatner, S.F., and Homcy, C.J. (1988) Decreased stimulatory guanosine triphosphate binding protein in dogs with pressure-overload left ventricular failure. *J. Clin. Invest.,* 81(2):420–424.

Luthin, G.R. and Tabakoff, B. (1984) Activation of adenylate cyclase by alcoholics requires the nucleotide-binding protein. *J. Pharmacol. Exp. Ther.,* 228(3):579–587.

Manji, H.K., Bitran, J.A., Masana, M.I., Chen, G.A., Hsiao, J.K., Risby, E.D., Rudorfer, M.V., and Potter, W.Z. (1991) Signal transduction modulation by lithium: cell culture, cerebral microdialysis and human studies. *Psychopharmacol. Bull.,* 27(3):199–208.

McLaughlin, S.K., McKinnon, P.J., and Margolskee, R.F. (1992) Gustducin is a taste-cell-specific G protein closely related to the transducins. *Nature,* 357(6379):563–569.

Mochly-Rosen, D., Chang, F.H., Cheever, L., Kim, M., et al. (1988) Chronic ethanol causes heterologous desensitization of receptors by reducing alpha-s messenger RNA. *Nature,* 333(6176):848–850.

Mork, A., Klysner, R., and Geisler, A. (1990) Effects of treatment with a lithium-imipramine combination on components of adenylate cyclase in the cerebral cortex of the rat. *Neuropharmacology,* 29(3):261–267.

Morrill, A.C., Wolfgang, D., Levine, M.A., and Wand, G.S. (1993) Stress alters adenylyl cyclase activity in the pituitary and frontal cortex of the rat. *Life Sci.,* 53(23):1719–1727.

Nagy, L.E., Diamond, I., Casso, D.J., Franklin, C., and Gordon, A.S. (1990) Ethanol increases extracellular adenosine by inhibiting adenosine uptake via the nucleotide transporter. *J. Biol. Chem.,* 265(4):1946–1951.

Nagy, L.E., Diamond, I., and Gordon, A. (1988) Cultured lymphocytes from alcoholic subjects have altered cAMP signal transduction. *Proc. Natl. Acad. Sci. U.S.A.,* 85(181:6973–6976.

Nestler, E.J. (1992) Molecular mechanisms of drug addiction. *J. Neurosci.,* 12(7):2439–2450.

Nestler, E.J., Terwilliger, R.Z., Walker, J.R., Sevarino, K.A., and Duman, R.S. (1990) Chronic cocaine treatment decreases levels of the G protein subunits $G_i\alpha$ and $G_o\alpha$ in discrete regions of rat brain. *J. Neurochem.,* 55(3):1079–1082.

Newlin, D.B. and Thomson, J.B. (1990) Alcohol challenge with sons of alcoholics: a critical review and analysis. *Psychol. Bull.,* 108(3):383–402.

Okada, F., Crow, T.J., and Roberts, G.W. (1990) G proteins (G_i, G_o) in the basal ganglia of control and schizophrenic brain. *J. Neural Transm.,* 79(3):227–234.

Parolaro, D., Patrini, G., Giagnoni, G., Massi, P., Groppetti, A., and Parenti, M. (1990) Pertussis toxin inhibits morphine analgesia and prevents opiate dependence. *Pharmacol. Biochem. Behav.,* 35(1):137–141.

Patten, J.L., Johns, D.R., Valle, D., Eil, C., Gruppuso, P.A., Steele, G., Smallwood, P.M., and Levine, M.A. (1990) Mutation in the gene encoding the stimulatory G protein of adenylate cyclase in Albright's hereditary osteodystrophy. *N. Engl. J. Med.,* 322(20):1412–1419.

Rabe, C.S., Giri, P.R., Hoffman, P.L., and Tabakoff, B. (1990) Effect of ethanol on cyclic AMP levels in intact PC12 cells. *Biochem. Pharmacol.,* 40(3):565–571.

Rabin, R.A. (1990) Chronic ethanol exposure of PC 12 cells alters adenylate cyclase activity and intracellular cyclic AMP content. *J. Pharmacol. Exp. Ther.,* (252(3):1021–1027.

Rabin, R.A. (1993) Ethanol-induced desensitization of adenylate cyclase: role of the adenosine receptor and GTP-binding proteins. *J. Pharmacol. Exp. Ther.,* 264(2):977–983.

Rabin, R.A., Edelman, A.M., and Wagner, J.A. (1992) Activation of protein kinase A is necessary but not sufficient for ethanol-induced desensitization of cyclic AMP production. *J. Pharmacol. Exp. Ther.,* 262(1):257–262.

Rabin, R.A. and Molinoff, P.B. (1983) Multiple sites of action of ethanol on adenylate cyclase. *J. Pharmacol. Exp. Ther.,* 227(3):551–556.

Redei, E., Branch, E.J., Gholami, S., Lin, E.Y., and Taylor, A.N. (1988) Effects of ethanol on CRF release *in vitro. Endocrinology,* 123(6):2736–2743.

Rivier, C., Bruhn, T., and Vale, W. (1984) Effect of ethanol on the hypothalamic-pituitary-adrenal axis in the rat: role of corticotropin-releasing factor (CRF). *J. Pharmacol. Exp. Ther.,* 229(1):127–131.

Robberecht, P., Waelbroeck, M., Chatelain, P., Camus, J., and Christopher, J. (1983) Inhibition of forskolin-stimulated cardiac adenylate cyclase by short-chain alcohols. *FEBS Lett.,* 154:205–208.

Robishaw, J.D., Kalman, V.K., Moomaw, C.R., and Slaughter, C.A. (1989) Existence of two gamma subunits of the G proteins in brain. *J. Biol. Chem.,* 264(27):15758–15761.

Saito, T., Lee, J.M., and Tabakoff, B. (1985) Ethanol's effects on cortical adenylate cyclase activity. *J. Neurochem.,* 44(4):1037–1044.

Schmidt, C.J. and Neer, E.J. (1991) *In vitro* synthesis of G protein beta gamma dimers. *J. Biol. Chem.,* 266(7):4538–4544.

Schmidt, C.J., Thomas, T.C., Levine, M.A., and Neer, E.J. (1992) Specificity of G protein beta and gamma subunit interactions. *J. Biol. Chem.,* 267(20):13807–13810.

Schreiber, G., Avissar, S., Danon, A., and Belmaker, R.H. (1991) Hyperfunctional G proteins in mononuclear leukocytes of patients with mania. *Biol. Psychiatry,* 29(3):273–280.

Schwindinger, W.F., Francomano, C.A., and Levine, M.A. (1992) Identification of a mutation in the gene encoding the alpha subunit of the stimulatory G protein of adenylyl cyclase in McCune-Albright syndrome. *Proc. Natl. Acad. Sci. U.S.A.,* 89(11):5152–5156.

Simon, M.I., Strathmann, M.P., and Gautam, N. (1991) Diversity of G proteins in signal transduction. *Science,* 252(5007):802–808.

Stenstrom, S. and Richelson, E. (1982) Acute effect of ethanol on prostaglandin E1-mediated cAMP formation by a murine neuroblastoma clone. *J. Pharmacol. Exp. Ther.,* 221(2):334–341.

Stenstrom, S., Seppala, M., Pfenning, M., and Richelson, E. (1985) Inhibition by ethanol of forskolin-stimulated adenylate cyclase in a murine neuroblastoma clone (N1E-115). *Biochem. Pharmacol.,* 34(20):3655–3659.

Sugimoto, K., Nukada, T., Tanabe, T., Takahashi, H., Noda, M., Minamino, N., Kandawa, K., Matsuo, H., Hirose, T., Inayama, S., and Numa, S. (1985) Primary structure of the beta-subunit of bovine transducin deduced from the cDNA sequence. *FEBS Lett.,* 191(2):235–240.

Tabakoff, B. and Hoffman, P.L. (1989) Genetics and biological markers of risk for alcoholism. In *Genetic Aspects of Alcoholism* (Kiianmaa, K., Tabakoff, B., and Saito, T., Eds.) 127–142, The Finnish Foundation for Alcohol Studies, Helsinki.

Tabakoff, B., Hoffman, P.L., Lee, J.M., Saito, T., et al. (1988b) Differences in platelet enzyme activity between alcoholics and nonalcoholics. *N. Engl. J. Med.,* 318(3):134–139.

Tabakoff, B., Hoffman, P.L., and McLaughlin, A. (1988a) Is ethanol a discriminating substance? *Semin. Liver Dis.,* 8(1):26–35.

Tang, W.J. and Gilman, A.G. (1991) Type-specific regulation of adenylyl cyclase by G protein beta gamma subunits. *Science,* 254(5037):1500–1503.

Vallar, L., Spada, A., and Giannattasio, G. (1987) Altered G_s and adenylate cyclase activity in human GH-secreting pituitary adenomas. *Nature,* 330(6148):566–568.

von Weizsacker, E., Strathmann, M.P., and Simon, M.I. (1992) Diversity among the beta subunits of heterotrimeric GTP-binding proteins: characterization of a novel beta-subunit cDNA. *Biochem. Biophys. Res. Commun.,* 183(1):350–356.

Waltman, C., Levine, M.A., McCaul, M.E., Svikis, D.S., and Wand, G.S. (1993) Enhanced expression of the inhibitory protein $G_i2\alpha$ and decreased activity of adenylyl cyclase in lymphocytes of abstinent alcoholics. *Alcoholism: Clin. Exp. Res.,* 17(2):315–320.

Wand, G.S. (1989) Ethanol differentially regulates proadrenocorticotropin/endorphin production and corticosterone secretion in LS and SS lines of mice. *Endocrinology,* 124(1):518–526.

Wand, G.S., Diehl, A.M., Levine, M.A., Wolfgang, D., and Samy, S. (1993) Chronic ethanol treatment increases expression of inhibitory G-proteins and reduces adenylylcyclase activity in the central nervous system of two lines of ethanol-sensitive mice. *J. Biol. Chem.,* 268(4):2595–2601.

Wand, G.S. and Levine, M.A. (1991) Hormonal tolerance to ethanol is associated with decreased expression of the GTP-binding protein, $G_s\alpha$, and adenylyl cyclase activity in ethanol-treated LS mice. *Alcoholism,* 15(4):705–710.

Wand, G.S., Waltman, C., Martin, C.S., McCaul, M.E., Levine, M.A., and Wolfgang, D. (1994) Differential expression of guanosine triphosphate binding proteins in men at high and low risk for the future development of alcoholism. *J. Clin. Invest.,* 94:1004–1011.

Weinstein, L.S., Shenker, A., Gejman, P.V., Merino, M.J., Friedman, E., and Spiegel, A.M. (1991) Activating mutations of the stimulatory G protein in the McCune-Albright syndrome. *N. Engl. J. Med.,* 325(24):1688–1695.

Wolfgang, D., Chen, I., and Wand, G.S. (1994) Effects of restraint stress on components of adenylyl cyclase signal transduction in the rat hippocampus. *Neuropsychopharmacology,* 11(3):187–193.

Yatani, A., Hamm, H., Codina, J., Mazzoni, M.R., Birnbaumer, L., and Brown, A.M. (1988a) A monoclonal antibody to the alpha subunit of GK blocks muscarinic activation of atrial K+ channels. *Science,* 241(4867):828–831.

Yatani, A., Mattera, R., Codina, J., Graf, R., Okabe, K., Padrell, E., Iyengar, R., Brown, A.M., and Birnbaumer, L. (1988) The G protein-gated atrial K+ channel is stimulated by three distinct $G_i\alpha$-subunits. *Nature,* 336(6200):680–682.

Yoshimasa, T., Sibley, D.R., Bouvier, M., Lefkowitz, R.J., and Caron, M.G. (1987) Cross-talk between cellular signalling pathways suggested by phorbolester-induced adenylate cyclase phosphorylation. *Nature,* 327(6117):67–70.

Young, L.T., Li, P.P., Kish, S.J., Siu, K.P., and Warsh, J.J. (1991) Postmortem cerebral cortex $G_s\alpha$-subunit levels are elevated in bipolar affective disorder. *Brain Res.,* 553(2):323–326.

Zgombick, J.M. and Erwin, V.G. (1988) Ethanol differentially enhances adrenocortical response in LS and SS mice. *Alcohol,* 5(4):287–294.

Chapter 9

ACUTE AND CHRONIC EFFECTS OF ETHANOL ON THE BRAIN: INTERACTIONS OF ETHANOL WITH ADENOSINE, ADENOSINE TRANSPORTERS, AND ADENOSINE RECEPTORS

Thomas V. Dunwiddie

TABLE OF CONTENTS

I. INTRODUCTION

A. THE EFFECTS OF PURINERGIC AGENTS ON THE BRAIN

Adenosine is an important inhibitory modulator of the activity of the central nervous system. At the cellular level, adenosine has a variety of actions including activation and inhibition of adenylyl cyclase, hyperpolarization of neurons, inhibition of transmitter release, and inhibitory modulation of Ca^{2+} channels; taken together, these effects produce a profound suppression of the electrical activity of the brain (Dunwiddie, 1985). Behavioral studies of the effects of stable analogs of adenosine support this basic idea, in that agonists induce a high degree of sedation (Dunwiddie and Worth, 1982; Katims et al., 1983; Nikodijevic et al., 1991), whereas adenosine receptor antagonists such as caffeine, theophylline, and a variety of more potent and selective antagonists generally elicit increases in spontaneous motor activity (Snyder et al., 1981). Despite the profound and widespread actions of adenosine, most current evidence does not support the idea that adenosine is a neurotransmitter per se, but rather a neuromodulator that is an important regulator of neuronal activity (see Dunwiddie, 1985 for review). There is no convincing evidence for the existence of "purinergic" neurons that release adenosine as their major transmitter either in brain or in the peripheral nervous system. However, there appears to be a generalized release of adenosine from brain tissue that is adequate to activate extracellular adenosine receptors in brain even under normal conditions. In addition to the normal low concentrations of extracellular adenosine, it has been hypothesized that in the brain as well as in other tissues such as the heart, adenosine serves as a kind of "retaliatory metabolite" (Newby, 1984). Under conditions of metabolic stress (ischemia, seizures, etc.), increased adenosine that is formed by the breakdown of ATP is released from many types of cells to inhibit the activity of neighboring cells. Considerable evidence suggests that whatever the source, adenosine is found in the extracellular space of normal brain in sufficient concentrations to exert a tonic inhibitory effect on neural activity; the increase in behavioral and electrophysiological activity that is observed upon the administration of adenosine antagonists such as caffeine or theophylline (Dunwiddie et al., 1981; Haas and Greene, 1988; Motley and Collins, 1983; Snyder et al., 1981) is thought to reflect the antagonism of the effects of this endogenous adenosine.

The profound sedative effects of adenosine analogs has suggested that purinergic receptors in brain might be activated in some way as a result of ethanol consumption, and this might contribute to the pharmacological effects of ethanol. It seems unlikely that this type of mechanism could be completely responsible for the effects of ethanol, because adenosine receptor antagonists are unable to reverse ethanol intoxication; but it is certainly possible that purinergic receptor activation contributes to the pharmacological response to this drug.

B. POSSIBLE MECHANISMS UNDERLYING ADENOSINE/ETHANOL INTERACTIONS

There are at least three specific mechanisms that can explain how ethanol could lead to increased activation of adenosine receptors in brain, involving changes in adenosine formation, adenosine uptake, and direct effects on adenosine receptors. The first of these mechanisms is based on the fact that blood acetate concentrations rise to the 1–2 m*M* range as a result of ethanol metabolism, even with relatively low blood concentrations of ethanol (Carmichael et al., 1991). When acetate is incorporated into acetyl-CoA, significant amounts of 5′-AMP are formed, and the action of 5′-nucleotidases leads to the subsequent formation of adenosine. The second potential site of adenosine/ethanol interaction involves adenosine transport systems. Adenosine can cross cell membranes either by active transport or by facilitated diffusion, a carrier-mediated process in which adenosine follows its concentration gradient across the membrane (Geiger and Fyda, 1991). There are two major facilitatory transporters, one of which is potently inhibited by nitrobenzylthioinosine (NBTI) and the other by dipyridamole (DIPY). These transporters can mediate both uptake as well as release of adenosine, depending upon the intra- vs. extracellular adenosine concentrations; pharmacological inhibition of these transporters can block transport in either direction but in most cases leads to increases in the interstitial concentration of adenosine. Work from a number of laboratories has indicated that even relatively low concentrations of ethanol can significantly inhibit adenosine transport, which in turn could increase extracellular brain concentrations of adenosine.

A third potential site of adenosine/ethanol interaction is the adenosine receptor itself, or more likely, the G-protein coupled effector mechanisms through which adenosine receptors act. There is considerable evidence to suggest that ethanol can facilitate the receptor mediated activation of adenylyl cyclase by various hormones and neurotransmitters (Rabin and Molinoff, 1981; Rabin, 1990; Luthin and Tabakoff, 1984; Hoffman and Tabakoff, 1990). The adenosine A2 receptor was originally defined based upon positive coupling to adenylyl cyclase and subdivided into A2a and A2b when it became clear that certain A2a selective agonists (e.g., CGS 21680) could discriminate between A2 receptor subtypes. However, it is also possible that it is coupled to activation of phospholipase C (PLC), as it does when expressed in *Xenopus* oocytes (Yakel et al., 1993). The A1 and the A3 receptors inhibit adenylyl cyclase but may also be linked to other second messenger systems as well (e.g., increased phospholipid

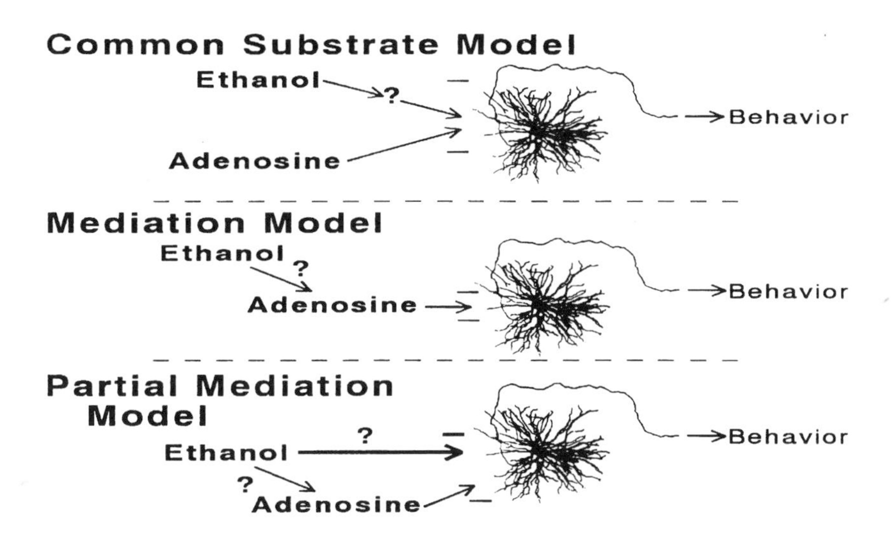

FIGURE 9.1. Models of ethanol/adenosine interactions in the brain. The Common Substrate Model hypothesizes that adenosine and ethanol produce similar changes in a common brain region or group of neurons but do so through independent and unrelated mechanisms. The Mediation Models hypothesize that ethanol either directly increases adenosine concentrations in brain or perhaps facilitates ligand-mediated activation of adenosine receptors, so that the effects of both agents at some point involve increased activity of adenosine receptors. The Partial Mediation Model proposes that this is only one effect of ethanol and that other unrelated pathways also contribute to the ethanol response.

turnover; Ramkumar et al., 1994). Because ethanol enhances basal and receptor stimulated adenylyl cyclase activity and disrupts the inhibitory regulation of cyclase (Bauche et al., 1987; Gordon et al., 1986; Hynie et al., 1980; Luthin and Tabakoff, 1984; Rabe et al., 1990; Rabin and Molinoff, 1981), G-proteins linked to adenylyl cyclase have often been considered targets for ethanol action. Because the A2a and A2b receptors activate and A1 and A3 receptors inhibit adenylyl cyclase, this provides yet another point at which ethanol could modulate effects mediated via adenosine receptors.

Thus, these three mechanisms — increases in adenosine formation, inhibition of adenosine transport, and interactions with GTP binding protein-coupled adenosine receptors — provide a number of potential ways in which ethanol could inhibit the activity of the nervous system through mechanisms that would involve adenosine receptors. In this chapter we will first examine the general (primarily behavioral) evidence that implicates purinergic systems in responses to ethanol and then evaluate the more specific mechanistic studies that have suggested particular ways in which these effects could arise.

II. EVIDENCE FOR ADENOSINE INVOLVEMENT IN ETHANOL RESPONSES IN MAN

Ethanol, adenosine, and adenosine receptor antagonists all are behaviorally active, and there are a number of similarities between ethanol and adenosine receptor agonists, which have a generally depressant effect on the nervous system. Adenosine receptor agonists can produce a complete cessation of motor activity (Dunwiddie and Worth, 1982) but do not cause unconsciousness or loss of righting reflex in rodents, such as occurs with ethanol. There are at least three ways in which one can conceptualize the actions of adenosine and ethanol, as presented schematically in Figure 9.1. The Common Substrate Model suggests that ethanol and adenosine may act on the same brain regions or neural substrates to produce similar types of behaviors, but mechanistically there is no interaction or common mechanism linking the two agents. The Mediation Model suggests that adenosine directly mediates the effects of ethanol and that all ethanol responses are expressed as a result of activation of adenosine receptors. The third model, the Partial Mediation Model, hypothesizes a role for adenosine in mediating some but not all of the effects of ethanol. Although it is somewhat difficult to differentiate these models experimentally, this is a critical step in interpreting the results of behavioral experiments such as are discussed below. In particular, a number of studies which have been proposed to demonstrate a role for adenosine in ethanol responses are really more consistent with the Common Substrate Model than with either of the Mediation models.

If adenosine mediates a significant component of the physiological response to ethanol, then one would predict that a) there should be similarities between the actions of the two agents and that b) antagonists of the actions of adenosine should have a similar effect upon the actions of ethanol. The data bearing upon these issues provides some support for this hypothesis, although it is not compelling. As far as the actions of purinergic *agonists* on mental state or behavior are concerned, there do not appear to be any reports of the objective or subjective effects of adenosine receptor agonists in man. This issue has been addressed in nonhuman studies (see Section III.A), but it should be pointed out that purinergic agonists have profound effects on various physiological parameters (heart rate, blood pressure, etc.), that make the interpretation of such studies difficult. As far as antagonists are concerned, both conventional wisdom and experimental studies have suggested that caffeine can antagonize acute intoxication to some extent, but it is difficult to determine whether this reflects a specific interaction between ethanol and adenosine. For example, although it has been reported that ethanol slows reaction time on several types of tasks and that this effect can be at least partially reversed by caffeine (Franks et al., 1975; Hasenfratz et al., 1993; but see Oborne and Rogers, 1983), caffeine alone is able to reduce reaction time in normal controls. Thus, it is unclear whether caffeine is antagonizing a specific component of the response to ethanol that is mediated via adenosine receptors or whether the response to caffeine is merely offsetting an ethanol response which is in the opposing direction. Furthermore, there are a number of reports where methylxanthines clearly do not reverse specific responses to ethanol (Nuotto et al., 1982; Alkana et al., 1977). Thus, although one can conclude from these experiments that not *all* of the effects of ethanol are mediated via adenosine (i.e., the Mediation Model can be ruled out, because not all ethanol responses can be blocked by adenosine receptor antagonists), they do not clearly differentiate between Common Substrate and the Partial Mediation models.

III. BEHAVIORAL STUDIES IN ANIMALS

A. ADENOSINE RECEPTOR AGONIST/ANTAGONIST INTERACTIONS WITH ETHANOL RESPONSES

Behavioral studies in animals have lent support to the hypothesis that some components of the behavioral response to ethanol might be mediated via adenosine and adenosine receptors. One of the first papers to address this issue was that of Dar et al. (1983), who reported that theophylline could decrease the duration of alcohol-induced loss of righting reflex (LORR, or "sleep" time) and that dipyridamole, which is an adenosine uptake inhibitor that increases brain concentrations of adenosine, could increase the duration of LORR. The motor incoordinating effects of ethanol were also partially reversed by adenosine antagonists such as theophylline and augmented by dipyridamole (Dar et al., 1983; Dar, 1990). Subsequent studies have provided some additional support for these initial conclusions. One study of LORR in CD-1 mice found no effect of the adenosine antagonists caffeine and isobutylmethylxanthine (IBMX) on sleep time (Dar et al., 1987), but confirmed the decrease originally observed with theophylline (Dar et al., 1983). Smolen and Smolen (1991) observed that adenosine receptor agonists increased the duration of LORR, theophylline reduced the duration, and caffeine had significant effects in both directions that were dependent upon the line of animals studied. Caffeine and IBMX were also observed to *increase* ethanol induced ataxia, and theophylline was inactive in this regard (Dar et al., 1987). Because all three of these drugs are effective adenosine antagonists and yet had differing effects on these two response measures, this suggested that the effects of theophylline were perhaps unrelated to adenosine antagonism.

These studies, as with the human studies described above, underscore the difficulty in making conclusions based upon antagonistic interactions between adenosine antagonists and ethanol. Response measures such as LORR and motor incoordination are obviously unaffected by agents with excitatory actions (such as adenosine receptor antagonists), but can affect these responses only when tested in conjunction with CNS depressants such as ethanol. The doses of caffeine and theophylline which interacted with ethanol responses in these studies were quite high (between 10–65× the threshold concentrations for adenosine receptor-mediated behavioral responses in mice; Katims et al., 1983; Snyder et al., 1981), so even though these concentrations by themselves had no effect on LORR or motor incoordination, this was a reflection of the nature of the response measure not an indication of a lack of pharmacological activity in those concentrations. Thus, the antagonism that is reported between ethanol and adenosine receptor antagonists might simply reflect opposing drug actions that are completely unrelated at the mechanistic level but which would be observed with any CNS depressant drug and adenosine antagonists.

While these studies provide evidence that is suggestive of an interaction between ethanol and adenosine, the evidence is not compelling. At this point, at least two types of experimental strategies might bolster support for this hypothesis. First, if the effects of adenosine receptor antagonists were specific to ethanol (as compared with other CNS depressants), this would reinforce the putative role of adenosine receptors in ethanol mediated depression. For example, if the duration of LORR in response to barbiturates was unaffected by theophylline, while the same dose of theophylline reduced ethanol sleep time, this would support the idea that adenosine is specifically involved in ethanol responses. A complementary approach would be to test the effectiveness of a variety of CNS excitants with differing mechanisms of action in antagonizing ethanol-induced LORR; if theophylline could be shown to be unique in this regard, this again would point to a specific adenosine-related interaction.

Yet another way to evaluate the role of adenosine receptors in ethanol action is to characterize drug interactions using behaviors that do not suffer from the same ceiling/floor effects as LORR and ataxia. One such behavior is spontaneous locomotor activity, which is affected by both purinergic drugs and ethanol. Although studies of this type have found instances where adenosine receptor antagonists such as caffeine disrupt some of the actions of ethanol (Hilakivi et al., 1989), no behaviors have been found where antagonists produce a predictable, dose-dependent inhibition of a response to ethanol that is clearly an antagonism (i.e., a drug interaction) and not simply the result of the addition of offsetting responses. Furthermore, with certain kinds of behaviors, methylxanthines actually accentuate the effects of ethanol, rather than antagonizing them (Kuribara, 1993). As with the human literature, the difficulty may be that the appropriate behaviors have not yet been tested, but there are no clear indications as to what these behaviors might be.

A final approach that has also been employed to establish a possible mediator role for adenosine has been to determine whether adenosine receptors are involved in the discriminative stimulus properties of ethanol. In this type of paradigm, animals are trained to identify particular drugs (in this case ethanol) by emitting an appropriate behavioral response, and the ability of various other drugs to substitute for the training drug is then tested. In one such study, i.p. administration of adenosine analogs did not generalize to the ethanol cue (Michaelis et al., 1987). However, the observation that these analogs show poor penetration of the blood-brain barrier following i.p. administration (Brodie et al., 1987) make these results difficult to interpret. In addition, it is also possible that the lack of generalization reflects the fact that activation of adenosine receptors is simply not the most salient aspect of the ethanol cue.

B. STUDIES OF ADENOSINE RECEPTOR SENSITIVITY IN SELECTED LINES OF MICE

Another approach to determining whether adenosine plays a role in responses to ethanol has been to determine whether mice that have been selectively bred for ethanol sensitivity (using the LORR as the measure of ethanol sensitivity) show a parallel difference in adenosine sensitivity. If endogenous adenosine modulates the ethanol disruption of the righting reflex, then breeding protocols that select for or against ethanol sensitivity should select for or against genes that determine adenosine sensitivity as well. One study of purinergic modulation of ethanol-induced sleep time found exactly that, i.e., that adenosine receptor agonists increased sleep time, antagonists decreased sleep time, and both of these interactions were more pronounced in an ethanol sensitive line of animals (Smolen and Smolen, 1991). In addition, most of these effects could not be accounted for by changes in ethanol metabolism, because they were observed not only in sleep time but also in terms of blood or brain ethanol concentration upon regaining the righting reflex. In addition, studies from our laboratory have showed that the long sleep (LS) or ethanol sensitive line of mice and the short sleep (SS) or ethanol insensitive line of mice, show a nearly 10-fold difference in sensitivity to the behavioral actions of adenosine receptor agonists and to the excitatory actions of the adenosine antagonist theophylline (Proctor and Dunwiddie, 1984; Fredholm et al., 1985) when tested in the *absence* of ethanol. Thus, these effects represent a difference in adenosine receptor mediated responses per se and not just a difference in the modulation of ethanol responses by adenosine.

C. ETHANOL SENSITIVITY DIFFERENCES IN ANIMALS CHRONICALLY TREATED WITH PURINERGIC AGONISTS AND ANTAGONISTS

If behavioral responses to ethanol can be modulated by purinergic drugs, one would hypothesize that treatments that modify adenosine sensitivity should be able to alter ethanol sensitivity as well. Dar and colleagues have tested this hypothesis in several ways. First, they demonstrated that chronic treatment with methylxanthines (adenosine receptor antagonists) increased the incoordinating effects of acute ethanol administration (Dar and Wooles, 1986). Furthermore, these changes were only observed in those instances where the chronic drug treatment caused an upregulation of adenosine receptors. A complementary study showed that chronic administration of adenosine receptor agonists reduced the incoordinating effects of ethanol (Dar and Clark, 1992). Furthermore, chronic ethanol treatment reduced the ability of adenosine agonists to facilitate ethanol-induced incoordination. Thus, there was a bidirectional cross tolerance between the incoordinating effects of adenosine agonists and ethanol. However, neither chronic ethanol nor chronic adenosine agonists produced any change in the number of adenosine A1 receptors, suggesting that either adenosine receptors were still present but had desensitized or that the locus of this effect was at some point beyond the receptor level.

D. BEHAVIORAL EFFECTS MEDIATED VIA ACETATE/ADENOSINE

One specific mechanism by which adenosine could serve as a mediator of ethanol actions is via the formation of acetate from ethanol. During ethanol metabolism, millimolar concentrations of acetate are formed and circulate through the blood. The subsequent metabolic conversion of acetate to acetyl CoA results in the formation of 5′-AMP, which can then be converted to adenosine by the action of 5′-nucleotidase. It has been demonstrated that this mechanism generates measurable concentrations of adenosine (Clark and Dar, 1989; Carmichael et al., 1988), and in some systems (e.g., the ethanol-induced vasodilation in the portal vasculature), adenosine formed by this mechanism can account for the entire physiological response (Carmichael et al., 1988; Orrego et al., 1988). Several studies have suggested that a similar process might occur in the brain as well, with acetate formed from hepatic metabolism entering the brain and raising endogenous adenosine levels. Carmichael et al. (1991) have demonstrated that both acetate and ethanol impair motor coordination and potentiate the effects of general anesthetics; in both cases, the adenosine receptor antagonist 8-phenyltheophylline completely reversed the effects of acetate but only partially reversed the effects of ethanol. A noteworthy aspect of these studies is that because the metabolic pathways for acetate production are saturated by relatively low doses of ethanol (approximately 0.5 g/kg of ethanol in rat), acetate levels reach their peak concentration of about 1 m*M* with low doses of ethanol and do not rise further with additional ethanol (Carmichael et al., 1991). In sum, these data provide strong support for the idea that acetate has biological activity and that these effects are mediated primarily via adenosine receptors. These experiments clearly support the "Partial Mediation" model of ethanol/adenosine interactions, where low-dose effects of ethanol could be mediated via adenosine and high-dose effects via some other unspecified mechanism. Those responses that are seen only with high concentrations of ethanol (e.g., LORR) cannot be mediated solely by adenosine but must involve other cellular mechanisms of ethanol action as well.

IV. BIOCHEMICAL STUDIES

A. THE EFFECT OF ETHANOL ON ADENOSINE UPTAKE IN THE BRAIN

The preceding studies involving ethanol-derived acetate have suggested one mechanism by which extracellular adenosine concentrations could be increased by ethanol. An entirely different kind of mechanism has been suggested by biochemical studies indicating that ethanol can inhibit the uptake of adenosine into cells. Adenosine uptake is mediated primarily by two major facilitatory transporters, the NBTI-sensitive transporter and the DIPY-sensitive transporter. Work from several laboratories has indicated that even relatively low concentrations of ethanol (10 m*M*) can significantly inhibit adenosine transport (Clark and Dar, 1989; Gordon et al., 1990; Nagy et al., 1990), and a recent study has shown that this effect is restricted to the NBTI-sensitive transporter (Krauss et al., 1993). The inhibition of adenosine uptake by ethanol that has been described in brain is not particularly profound; for example, Clark and Dar (1989) showed that the maximal inhibition that could be achieved in cerebellar synaptosomes was about 15% and that this occurred at an ethanol concentration of about 25 m*M*. Two factors might explain the relatively small magnitude of this effect. First, even 200 m*M* ethanol only inhibits the NBTI sensitive transporter by approximately 50% (Nagy et al., 1990). Second, there are fewer NBTI-sensitive transporters than DIPY-sensitive transporters in brain (Bisserbe et al., 1985; Bisserbe et al., 1986). The relatively small magnitude of the ethanol inhibition may explain the occasional reports which suggest that ethanol does not elevate extracellular brain adenosine levels at all (Clark and Dar, 1988; Phillis et al., 1980; Phillis et al., 1992). We have demonstrated using electrophysiological techniques that in hippocampus, where there are relatively few NBTI-sensitive transporters, inhibition of uptake by NBTI does not appear to increase extracellular adenosine (Dunwiddie and Diao, 1994), whereas dipyridamole produces a marked increase in extracellular adenosine. In the olfactory cortex, where the proportion of NBTI-sensitive sites is higher (Bisserbe et al., 1985; Geiger and Nagy, 1984), NBTI produces electrophysiological responses that are consistent with elevations in extracellular adenosine (Sanderson and Scholfield, 1986). On this basis, we hypothesize that ethanol inhibition of adenosine uptake might be functionally significant only in brain regions that have relatively high levels of NBTI-sensitive transport relative to DIPY-sensitive transport (olfactory cortex, locus coeruleus, and superior colliculus), whereas in brain regions that have low ratios (hippocampus and cerebellum), this particular interaction may be functionally unimportant.

B. STUDIES OF THE EFFECT OF ETHANOL ON ADENOSINE UPTAKE IN CULTURED CELLS

Although the functional importance of inhibition of the NBTI-sensitive transporter in brain is somewhat unclear, extensive studies of the effects of ethanol and adenosine on cultured cells have been quite informative in establishing the nature of these interactions (Gordon et al., 1990; Krauss et al., 1993). These studies, which have been conducted primarily on the NG108–15 neuroblastoma-glioma, S49 lymphoma, and related cell lines, have led to the development of a model that incorporates mechanisms to account for both alcohol tolerance and dependence. The basic observations that have been made are that ethanol potentiates the increases in cAMP that occur with activation of the A2 adenosine receptor (Gordon et al., 1986), perhaps via a G protein interaction, and inhibits the function of the NBTI sensitive adenosine transporter (Krauss et al., 1993). An interesting aspect of this inhibition is that unlike all the other known inhibitors of transporter function, ethanol only inhibits adenosine uptake and does not reduce adenosine efflux (Nagy et al., 1990). With chronic ethanol treatment, the inhibition of uptake results in increases in extracellular concentrations of adenosine, which then activates A2 receptors and produces increases in intracellular cAMP. With prolonged treatment with ethanol, the corresponding persistent increase in cAMP concentrations leads to a heterologous desensitization of adenylyl cyclase-coupled receptor systems (Gordon et al., 1986); similar phenomena have been described in brain or cultured neurons as well (Saito et al., 1987; Rabin, 1990). This heterologous desensitization appears to have two primary consequences. First, with chronic ethanol treatment, the continued presence of ethanol is required to see normal sensitivity to adenosine (Gordon et al., 1986). In other words, following chronic treatment, the increase in cAMP induced by ethanol+adenosine is the same as that induced by adenosine alone in the control condition. In the absence of ethanol, activation of adenosine receptors produces an increase in cAMP that is approximately 50% of normal. This effect has been proposed as a mechanism of cellular dependence, i.e., a condition in which the continued presence of ethanol is required to maintain a normal physiological state.

A second effect occurs as a result of a regulation of adenosine transporter activity by protein kinase A (PKA). Phosphorylation of the transporter does not appear to change its function but has a permissive effect on the modulation of function by ethanol. Thus, the activity of the dephosphorylated state of the transporter appears to be relatively unaffected by ethanol. In cells that have been chronically treated with ethanol, ethanol is no longer able to inhibit transporter activity, presumably because the reduction in cAMP levels associated with chronic ethanol results in a reduced level of phosphorylation of the transporter. This is a cellular analog of ethanol tolerance, i.e., a reduced responsiveness to previously effective concentrations of a drug.

A final aspect of these studies that bears mention concerns differences in adenosine receptor mediated responses in lymphocytes from alcoholics. Both basal and adenosine stimulated cAMP formation was found to be reduced by approximately 75% in alcoholics when compared with appropriate controls (Diamond et al., 1987). Furthermore, the ability of ethanol to enhance stimulated cAMP production was blunted as well, and these changes were not observed in lymphocytes from abstinent alcoholics (Gordon et al., 1990). Thus, the reduced sensitivity of adenylyl cyclase coupled receptors appeared to be a consequence of acute ethanol exposure in the alcoholics; whether adenosine itself mediates this response *in vivo* as it appears to *in vitro* is not yet clear.

C. IMPLICATIONS OF ETHANOL-INDUCED CHANGES IN ADENYLYL CYCLASE ACTIVITY FOR THE BRAIN

Although the consequences of these types of changes in terms of neuronal function have not been explored, it seems highly likely that if parallel changes are taking place in the brain then they will contribute significantly to the neuropharmacological response to alcohol. As but one example, adenosine is thought to exert a tonic inhibitory influence over the activity of the brain and has been proposed as an endogenous anticonvulsant agent (Dunwiddie, 1980; Dunwiddie, 1985). The changes induced by chronic ethanol, which would include heterologous desensitization of A2 receptors that are coupled to adenylyl cyclase, might contribute to the ethanol withdrawal syndrome. It has been proposed that ethanol withdrawal seizures might result from the uncoupling of these adenosine receptors and the loss of the tonic inhibitory effect of adenosine (Diamond et al., 1991); however, the anticonvulsant effects of adenosine appear to be mediated primarily via A1 receptors (which *inhibit* adenylyl cyclase), rather than the A2 receptors that activate adenylyl cyclase. Nevertheless, persistent increases in adenosine resulting from inhibition of adenosine uptake during ethanol administration might uncouple A1 receptors as well through an as yet unknown mechanism to produce a proconvulsant effect.

An important caveat with respect to these experiments concerns whether effects such as have been described in cultured cells occur in brain. Unlike lymphocytes, NBTI does not appear to block the majority of adenosine transport in brain, so it is difficult to determine the functional consequences of inhibition of this transporter. Furthermore, other cell lines that maintain some neuronal characteristics (e.g., PC12 cells) do not show the same regulation of transporter and cyclase function as do the NG108 cells. Rabin et al. (1993) have shown that in PC12 cells, chronic ethanol induces a similar increase in extracellular adenosine and a similar desensitization of cAMP production, but there is no relationship between the two, i.e., desensitization is still observed under conditions where extracellular adenosine does not accumulate. Thus, although the results obtained with lymphocytes, NG108 cells, and S49 cells seem fairly unequivocal, their applicability to other systems in not yet clear.

D. EFFECTS OF ETHANOL ON ADENOSINE RELEASE

Several reports have suggested that ethanol may be able to enhance the release of adenosine from brain through an unknown type of mechanism. In an initial study, Clark and Dar (1989) indirectly characterized adenosine actions by studying the purinergic modulation of glutamate release from cerebellar synaptosomes. They observed that adenosine receptor agonists inhibited glutamate release, whereas antagonists enhanced release, presumably via a block of the effects of endogenous adenosine. These observations are consistent with the well-established tonic inhibitory modulation of excitatory glutamatergic transmission by adenosine (Dunwiddie, 1985). However, 25 m*M* ethanol *antagonized* the effect of theophylline; since the effect of theophylline is thought to depend upon endogenous adenosine, the simplest interpretation of this result would be that ethanol *inhibits* the basal release of adenosine. However, were this the case, one would expect ethanol by itself to enhance glutamate release, and this was not observed. At this point, the most that can be concluded from these indirect experiments is that ethanol probably has multiple actions on this system and that a modulation of adenosine release (or uptake) by ethanol is one such action.

Another study from this same group directly characterized the net efflux of endogenous adenosine from cerebellar synaptosomes (Clark and Dar, 1989). These studies demonstrated a dose-dependent enhancement of adenosine overflow by ethanol concentrations ranging from 25–100 m*M* but did not distinguish between a change in release per se or a possible inhibition of the subsequent uptake of adenosine. As has been reported previously in other systems, the adenosine transport inhibitor dilazep *inhibited* endogenous adenosine release, which is consistent with a bidirectional inhibition of transport, i.e., both efflux as well as influx are inhibited. However, ethanol clearly does not resemble other transport inhibitors in this respect, since it produced only increases in release. These results could be explained by the mechanism outlined previously with respect to S49 cells (Nagy et al., 1990), where ethanol selectively affected uptake but not the release of adenosine. As a final note, however, there are also indications that ethanol may directly stimulate the release of adenosine and/or adenine nucleotides. Cahill et al. (1993) reported that ethanol (approximately 175 m*M*) could increase the net efflux of both adenosine and adenine nucleotides from spinal cord synaptosomes; while the increase in adenosine efflux could be explained on the basis of uptake inhibition, this is not the case for adenine nucleotides, which are not substrates for the transporter.

E. ETHANOL EFFECTS ON ADENOSINE RECEPTORS AND EFFECTOR MECHANISMS

Multiple subtypes of adenosine receptors exist, all of which belong to the G-protein coupled family of membrane receptors. The A1 and A3 receptors are both coupled in an inhibitory fashion to adenylyl cyclase, and the multiple subtypes of A2 receptors all activate adenylyl cyclase. In addition, the A1 and A3 receptors also appear to be coupled to a wide variety of other effector mechanisms such as activation of K^+ channels, inhibition of Ca^{2+} channels, and other effectors which differ based upon the particular system under study. Because GTP binding proteins are a major target of ethanol action, responses mediated via adenosine receptors can clearly be affected by ethanol via this mechanism. An early study by Hynie et al. (1980) demonstrated an ethanol enhancement of adenosine-stimulated cAMP production in lymphocytes, but this effect was clearly not specific for adenosine, because cAMP responses to isoproterenol and prostaglandin E_2 were potentiated as well. Because this effect is not specific to adenosine systems per se, it will not be discussed further here (see Chapter 8 for a discussion of G proteins and ethanol; see also Hoffman and Tabakoff, 1990, for a review). Nevertheless, a potentiation of the acute effects of A2 receptor action by ethanol and a subsensitivity of such systems induced by chronic ethanol appear to be significant insofar as interactions between ethanol and adenosine are concerned.

Given that adenosine receptors (perhaps via G-protein interactions) are a significant target for ethanol action, one might hypothesize that differences in adenosine receptors could be a significant factor in determining ethanol sensitivity, at least for those behaviors that involve an adenosine link. The behavioral studies outlined previously provide some encouraging support for this hypothesis. However, while an early study of adenosine receptors in the LS and SS lines of mice suggested that there might be differences between these lines in terms of A1 but not A2 receptors (Fredholm et al., 1985), subsequent studies have failed to find significant differences in either A1 or A2a receptors (Fredholm, unpublished; Smolen and Smolen, 1993; Smolen et al., 1993). Because of this, it is somewhat difficult to explain why behavioral responses to purinergic agonists and antagonists differ in these two lines of mice, and yet there are no differences at the receptor level. One possible explanation for these observations is outlined in Figure 9.2.

A related issue concerns possible changes in adenosine receptors that occur as a result of either acute or chronic ethanol treatment. Clark and Dar (1988) have reported that acute ethanol administration (1.5 g/kg, *in vivo*) produces a 40% increase of A1 receptors as determined by ligand binding *in vitro*. However, this effect did not appear to be mediated via an activation of adenosine receptors; in fact, if animals were pretreated with theophylline and ethanol, there was an even larger increase in adenosine receptors when compared to the theophylline controls. Thus, it seems likely that this may reflect an interaction between ethanol and GTP-binding proteins, that indirectly causes changes in post-mortem adenosine receptor binding. The effects of *chronic* ethanol treatment on adenosine receptors have not been fully explored. One study suggested that chronic ethanol had no effect on cerebellar A1 receptors (Dar and Clark, 1992), and another report has suggested that A2a receptors in the striatum are down-regulated by chronic ethanol (Smolen and Smolen, 1993). This latter effect was only observed in the ethanol sensitive LS line of mice and not in the SS line. Although these differences may implicate changes in purinergic systems in the response to chronic ethanol, at this point it is premature to speculate on the mechanisms that might mediate these effects.

V. ELECTROPHYSIOLOGICAL STUDIES OF ADENOSINE/ETHANOL INTERACTIONS

The putative link between acetate formed from ethanol metabolism and adenosine has been tested both in intact animals and in brain slices by using electrophysiological measures of adenosine-sensitive responses. Phillis et al. (1992) characterized ethanol and acetate effects on

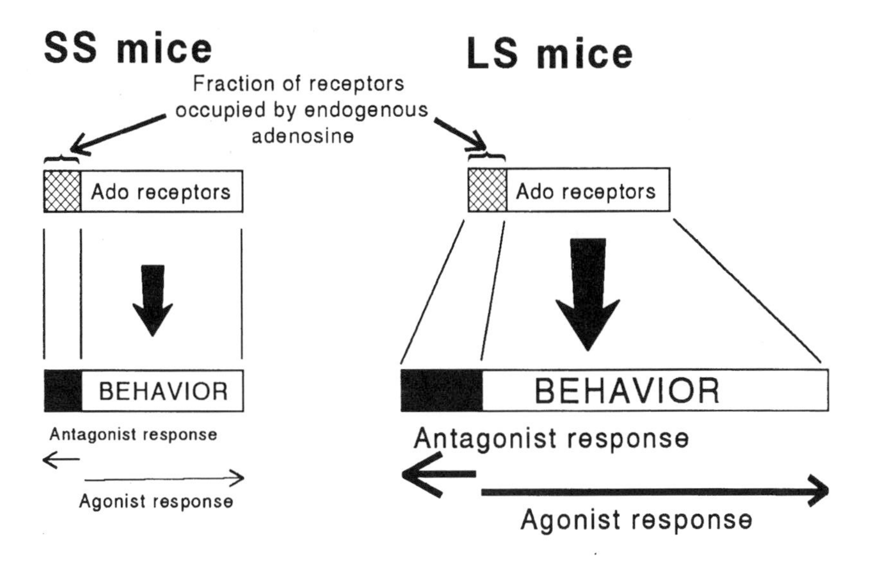

FIGURE 9.2. Putative mechanism to explain behavioral differences in LS and SS mice. If LS and SS mice have the same numbers of adenosine receptors and the same levels of endogenous adenosine, there might still be differences in behavioral responses to purinergic drugs if the neural mechanisms downstream from the adenosine receptor differ. In the case of the LS mice, either occupation of adenosine receptors by an agonist or displacement of endogenous adenosine by an antagonist would translate into a larger behavioral response than would be observed in the SS mice. In this model, the critical differences between the mice are in the neural substrates that are involved in the elaboration of the behavioral response and not in purinergic systems per se.

the spontaneous firing of cerebral cortical neurons in anesthetized rats. In this study, intraperitoneal injections of ethanol inhibited cortical neurons and prolonged the depression in firing induced by local iontophoretic application of adenosine. Although local application of acetate inhibited firing, this response was insensitive to the adenosine receptor antagonist 8-p-sulfophenyltheophylline. These authors concluded that the ability of ethanol to potentiate adenosine responses was most likely the result of an inhibition of adenosine uptake and that their evidence did not support the hypothesis that acetate could lead to increased adenosine formation in the brain. One potential consideration that should be raised concerns the local application of acetate; at this point little is known about the dynamics of adenosine formation through this mechanism, and it is possible that local administration of acetate does not generate sufficient adenosine to affect cellular activity. In this context, it would have been interesting to see whether parenteral administration of acetate would inhibit cortical neurons via a purinergic mechanism.

Some additional support for a role of acetate in ethanol action has been provided by a study conducted in the dentate gyrus in brain slices by Cullen and Carlen (1992). Ethanol, acetate, and adenosine were found to produce a variety of effects that could be antagonized by 8-phenyltheophylline, which included a hyperpolarization of the resting membrane potential and an enhancement of a calcium-dependent afterhyperpolarization (AHP) following depolarizing current injection. However, several aspects of this study deserve mention. First, acetate had no significant effect on the amplitude of evoked EPSPs, which is probably the best characterized and most robust response to adenosine in this region. This would suggest that bath superfusion with acetate leads to the formation of adenosine in a restricted region surrounding the cell body, a region that does not include the dendrites of these cells where excitatory inputs terminate. Cullen and Carlen (1992) suggest that the lack of effects on EPSPs might be because A_2 rather than A_1 receptors are activated under their conditions, but there is certainly nothing about the pharmacological properties of adenosine itself that would make it selective for either type of receptor. The primary effect of acetate appeared to be on the AHP, which is a complicated response whose amplitude is affected by the resting membrane potential, cAMP levels, resting intracellular Ca^{2+} concentration, the influx of Ca^{2+} through voltage-dependent Ca^{2+} channels on the cell body, and the intra- and extracellular K^+ concentrations. Furthermore, adenosine itself has biphasic effects on this response, increasing it at low concentrations (Haas and Greene, 1984) and decreasing it at higher concentrations (Proctor and Dunwiddie, unpublished). Because of the complexity of this response, it is often not possible to determine which of these regulatory processes is affected by acetate/ethanol/adenosine, but it is clearly a response which can be readily perturbed by a variety of cellular events. Finally, this effect of acetate appears to be a regionally specific response, since we have found acetate to have no significant effect on the resting membrane potential, input resistance, or intracellular EPSP amplitude in the CA1 region of hippocampus, nor did it affect field EPSPs in either CA1 or dentate gyrus (Brundege and Dunwiddie, 1995).

At this point, it is difficult to evaluate the relative role of adenosine receptor mediated actions in the central effects of acetate based upon electrophysiological studies. It is possible that some of the differences outlined above between different laboratories may relate to differences in the strain of animals tested, the conditions under which the slices are maintained, or to the brain region tested. If the effects of acetate in the dentate gyrus can be confirmed, they suggest that the acetate/adenosine response may have a very specific localization and that systemic ethanol or acetate will not necessarily result in a similar increase in activation of all adenosine receptors in brain.

VI. CONCLUSIONS

The studies summarized above certainly suggest that some component of the acute response to ethanol may be mediated via an increased activation of adenosine receptors in the brain. However, the relative roles of the multiple putative mechanisms that could underlie this interaction are unclear. Nevertheless, a number of tentative conclusions based upon this evidence might be proposed:

A purinergic mechanism alone cannot account for the behavioral and cellular actions of ethanol. Although adenosine receptor antagonists appear to reverse some of the effects of ethanol, there are many responses that are unaffected by antagonists, and hence must be mediated through other mechanisms.

Some ethanol responses may be completely mediated via a purinergic mechanism. The behavioral responses to ethanol and adenosine receptor agonists show many similarities, and antagonists appear in some instances to antagonize ethanol responses. At this point what is needed is perhaps a better understanding of the behavioral pharmacology of adenosine receptor agonists; with this as a foundation, it may then be possible to refine behavioral paradigms to isolate adenosine-specific responses to ethanol and to demonstrate in a more unequivocal fashion that such responses can be completely blocked by adenosine antagonists.

Several candidate mechanisms have been defined at the cellular level through which ethanol can lead to the activation of adenosine receptors. The ethanol mediated inhibition of adenosine uptake and facilitation of adenosine receptor activation (probably through an interaction with GTP binding proteins) have been clearly established at the cellular level. At this point, what is needed are functional studies on more intact systems, such as brain slices or intact animals, to establish the relative importance of these effects in ethanol actions when compared to the numerous alternative candidate mechanisms. Circumstantial evidence suggests that they may be important, but hard evidence is in most cases lacking.

A final point that should be raised is that it is important to consider all of the potential interactions of ethanol with adenosine systems *in toto*, rather than individually, because there might be important interactions that are not apparent when studied in isolation. For example, the acetate-mediated increases in adenosine and the ethanol inhibition of adenosine uptake may by themselves produce only marginally significant changes in physiology or behavior when studied alone. However, if the combined effects of acetate and ethanol were studied on isolated preparations such as cultured cells or brain slices, they might have synergistic actions that could not be inferred based only on their effects in isolation.

REFERENCES

Alkana, R.L., Parker, E.S., Cohen, H.B., Birch, H., and Noble, E.P.: Reversal of ethanol intoxication in humans: an assessment of the efficacy of l-dopa, aminophylline, and ephedrine. *Psychopharmacology,* 55: 203–212, 1977.

Bauche, F., Bourdeaux-Jaubert, A.M., Giudicelli, Y., and Nordmann, R.: Ethanol alters the adenosine receptor-Ni-mediated adenylate cyclase inhibitory response in rat brain cortex *in vitro. FEBS Lett.,* 219: 296–300, 1987.

Bisserbe, J.C., Deckert, J., and Marangos, P.: Autoradiographic localization of adenosine uptake sites in guinea pig brain using [3H]dipyridamole. *Neurosci. Lett.,* 66: 341–345, 1986.

Bisserbe, J.C., Patel, J., and Marangos, P.J.: Autoradiographic localization of adenosine uptake sites in rat brain using [3H]nitrobenzylthioinosine. *J. Neurosci.,* 5: 544–550, 1985.

Brodie, M.S., Lee, K.S., Fredholm, B.B., Stahle, L., and Dunwiddie, T.V.: Central versus peripheral mediation of responses to adenosine receptor agonists: evidence against a central mode of action. *Brain Res.,* 415: 323–330, 1987.

Brundege, J.M. and Dunwiddie, T.V.: The role of acetate as a potential mediator of the effects of ethanol in the brain. *Neurosci. Lett.,* 186: 214–218, 1995.

Cahill, C.M., White, T.D., and Sawynok, J.: Influence of calcium on the release of endogenous adenosine from spinal cord synaptosomes. *Life Sci.,* 53: 487–496, 1993.

Carmichael, F.J., Israel, Y., Crawford, M., Minhas, K., Saldivia, V., Sandrin, S., Campisi, P., and Orrego, H.: Central nervous system effects of acetate: contribution to the central effects of ethanol. *J. Pharmacol. Exp. Ther.,* 259: 403–408, 1991.

Carmichael, F.J., Saldivia, V., Varghese, G.A., Israel, Y., and Orrego, H.: Ethanol-induced increase in portal blood flow: role of acetate and A1- and A2-adenosine receptors. *Am. J. Physiol.,* 255: G417–G423, 1988.

Clark, M. and Dar, M.S.: The effects of various methods of sacrifice and of ethanol on adenosine levels in selected areas of rat brain. *J. Neurosci. Meth.,* 25: 243–249, 1988.

Clark, M. and Dar, M.S.: Mediation of acute ethanol-induced motor disturbances by cerebellar adenosine in rats. *Pharmacol. Biochem. Behav.,* 30: 155–161, 1988.

Clark, M. and Dar, M.S.: Effect of acute ethanol on uptake of [3H]adenosine by rat cerebellar synaptosomes. *Alcoholism: Clin. Exp. Res.,* 13: 371–377, 1989.

Clark, M. and Dar, M.S.: Release of endogenous glutamate from rat cerebellar synaptosomes: interactions with adenosine and ethanol. *Life Sci.,* 44: 1625–1635, 1989.

Clark, M. and Dar, M.S.: Effect of acute ethanol on release of endogenous adenosine from rat cerebellar synaptosomes. *J. Neurochem.,* 52: 1859–1865, 1989.

Cullen, N. and Carlen, P.L.: Electrophysiological action of acetate, a metabolite of ethanol, on hippocampal dentate granule neurons: interaction with adenosine. *Brain Res.,* 588: 49–57, 1992.

Dar, M.S.: Central adenosinergic system involvement in ethanol-induced motor incoordination in mice. *J. Pharmacol. Exp. Ther.,* 255: 1202–1209, 1990.

Dar, M.S. and Clark, M.: Tolerance to adenosine's accentuation of ethanol-induced motor incoordination in ethanol-tolerant mice. *Alcoholism: Clin. Exp. Res.,* 16: 1138–1146, 1992.

Dar, M.S., Jones, M., Close, G., Mustafa, S.J., and Wooles, W.R.: Behavioral interactions of ethanol and methylxanthines. *Psychopharmacology,* 91: 1–4, 1987.

Dar, M.S., Mustafa, S.J., and Wooles, W.R.: Possible role of adenosine in the effects of ethanol. *Life Sci.,* 33: 1363–1374, 1983.

Dar, M.S. and Wooles, W.R.: Effect of chronically administered methylxanthines on ethanol-induced motor incoordination in mice. *Life Sci.,* 39: 1429–1437, 1986.

Diamond, I., Nagy, L., Mochly-Rosen, D., and Gordon, A.: The role of adenosine and adenosine transport in ethanol-induced cellular tolerance and dependence. Possible biologic and genetic markers of alcoholism. *Ann. N. Y. Acad. Sci.,* 625: 473–487, 1991.

Diamond, I., Wrubel, B., Estrin, W., and Gordon, A.: Basal and adenosine receptor-stimulated levels of cAMP are reduced in lymphocytes from alcoholic patients. *Proc. Natl. Acad. Sci. U.S.A.,* 84: 1413–1416, 1987.

Dunwiddie, T.V.: Endogenously released adenosine regulates excitability in the *in vitro* hippocampus. *Epilepsia,* 21: 541–548, 1980.

Dunwiddie, T.V.: The physiological role of adenosine in the central nervous system. *Int. Rev. Neurobiol.,* 27: 63–139, 1985.

Dunwiddie, T.V. and Diao, L.H.: Extracellular adenosine concentrations in hippocampal brain slices and the tonic inhibitory modulation of evoked excitatory responses. *J. Pharmacol. Exp. Ther.,* 268: 537–545, 1994.

Dunwiddie, T.V., Hoffer, B.J., and Fredholm, B.B.: Alkylxanthines elevate hippocampal excitability: evidence for a role of endogenous adenosine. *Naunyn-Schmiedebergs Arch. Pharmacol.,* 316: 326–330, 1981.

Dunwiddie, T.V. and Worth, T.S.: Sedative and anticonvulsant effects of adenosine analogs in mouse and rat. *J. Pharmacol. Exp. Ther.,* 220: 70–76, 1982.

Franks, H.M., Hagedorn, H., Hensley, V.R., Hensley, W.J., and Starmer, G.A.: The effect of caffeine on human performance, alone and in combination with ethanol. *Psychopharmacologia,* 45: 177–181, 1975.

Fredholm, B.B., Zahniser, N.R., Weiner, G.R., Proctor, W.R., and Dunwiddie, T.V.: Behavioural sensitivity to PIA in selectively bred mice is related to a number of A1 adenosine receptors but not to cyclic AMP accumulation in brain slices. *Eur. J. Pharmacol.,* 111: 133–136, 1985.

Geiger, J.D. and Fyda, D.M.: Adenosine transport in nervous system tissues. In *Adenosine in the Nervous System,* Ed. by T. W. Stone, pp. 1–20, Academic Press, London, 1991.

Geiger, J.D. and Nagy, J.I.: Heterogeneous distribution of adenosine transport sites labelled by [3H]Nitrobenzylthioinosine in rat brain: An autoradiographic and membrane binding study. *Brain Res. Bull.,* 13: 657–666, 1984.

Gordon, A.S., Collier, K., and Diamond, I.: Ethanol regulation of adenosine receptor-stimulated cAMP levels in a clonal neural cell line: an *in vitro* model of cellular tolerance to ethanol. *Proc. Natl. Acad. Sci. U.S.A.,* 83: 2105–2108, 1986.

Gordon, A.S., Nagy, L., Mochly-Rosen, D., and Diamond, I.: Chronic ethanol-induced heterologous desensitization is mediated by changes in adenosine transport. *Biochem. Soc. Symp.,* 56: 117–136, 1990.

Haas, H.L. and Greene, R.W.: Adenosine enhances afterhyperpolarization and accommodation in hippocampal pyramidal cells. *Pflugers Arch.,* 402: 244–247, 1984.

Haas, H.L. and Greene, R.W.: Endogenous adenosine inhibits hippocampal CA1 neurones: further evidence from extra- and intracellular recording. *Naunyn-Schmiedebergs Arch. Pharmacol.,* 337: 561–565, 1988.

Hasenfratz, M., Bunge, A., Dal Pra, G., and Battig, K.: Antagonistic effects of caffeine and alcohol on mental performance parameters. *Pharmacol. Biochem. Behav.,* 46: 463–465, 1993.

Hilakivi, L.A., Durcan, M.J., and Lister, R.G.: Effects of caffeine on social behavior, exploration and locomotor activity: interactions with ethanol. *Life Sci.,* 44: 543–553, 1989.

Hoffman, P.L. and Tabakoff, B.: Ethanol and guanine nucleotide binding proteins: a selective interaction. *FASEB J.,* 4: 2612–2622, 1990.

Hynie, S., Lanefelt, F., and Fredholm, B.B.: Effects of ethanol on human lymphocyte levels of cyclic AMP *in vitro:* potentiation of the response to isoproterenol, prostaglandin E2 or adenosine stimulation. *Acta Pharmacol. Toxicol.,* 47: 58–65, 1980.

Katims, J.J., Annau, Z., and Snyder, S.H.: Interactions in the behavioral effects of methylxanthines and adenosine derivatives. *J. Pharmacol. Exp. Ther.,* 227: 167–173, 1983.

Krauss, S.W., Ghirnikar, R.B., Diamond, I., and Gordon, A.S.: Inhibition of adenosine uptake by ethanol is specific for one class of nucleoside transporters. *Mol. Pharmacol.,* 44: 1021–1026, 1993.

Kuribara, H.: Enhancement of the behavioral toxicity induced by combined administration of ethanol with methylxanthines: evaluation by discrete avoidance in mice. *J. Toxicol. Sci.,* 18: 95–101, 1993.

Luthin, G.R. and Tabakoff, B.: Activation of adenylate cyclase by alcohol requires the nucleotide-binding protein. *J. Pharmacol. Exp. Ther.,* 228: 579–587, 1984.

Michaelis, R.C., Holohean, A.M., Hunter, G.A., and Holloway, F.A.: Endogenous adenosine-receptive systems do not mediate the discriminative stimulus properties of ethanol. *Alcohol Drug Res.,* 7: 175–185, 1987.

Motley, S.J. and Collins, G.G.S.: Endogenous adenosine inhibits excitatory transmission in the rat olfactory cortex slice. *Neuropharmacology,* 22: 1081–1086, 1983.

Nagy, L.E., Diamond, I., Casso, D.J., Franklin, C., and Gordon, A.S.: Ethanol increases extracellular adenosine by inhibiting adenosine uptake via the nucleoside transporter. *J. Biol. Chem.,* 265: 1946–1951, 1990.

Newby, A.C.: Adenosine and the concept of "retaliatory metabolites." *Trends Biochem.,* 9: 42–44, 1984.

Nikodijevic, O., Sarges, R., Daly, J.W., and Jacobson, K.A.: Behavioral effects of A1- and A2-selective adenosine agonists and antagonists: evidence for synergism and antagonism. *J. Pharmacol. Exp. Ther.,* 259: 286–294, 1991.

Nuotto, E., Mattila, M.J., Seppala, T., and Konno, K.: Coffee and caffeine and alcohol effects on psychomotor function. *Clin. Pharmacol. Ther.,* 31: 68–76, 1982.

Oborne, D.J. and Rogers, Y.: Interactions of alcohol and caffeine on human reaction time. *Aviat. Space. Environ. Med.,* 54: 528–534, 1983.

Orrego, H., Carmichael, F.J., Saldivia, V., Giles, H.G., Sandrin, S., and Israel, Y.: Ethanol-induced increase in portal blood flow: role of adenosine. *Am. J. Physiol.,* 254: G495–G501, 1988.

Phillis, J.W., O'Regan, M.H., and Perkins, L.M.: Actions of ethanol and acetate on rat cortical neurons: ethanol/adenosine interactions. *Alcohol,* 9: 541–546, 1992.

Phillis, J.W., Ziang, Z.G., and Chelack, B.J.: Effects of ethanol on acetylcholine and adenosine efflux from in vivo rat cerebral cortex. *J. Pharm. Pharmacol.,* 32: 871–872, 1980.

Proctor, W.R. and Dunwiddie, T.V.: Behavioral sensitivity to purinergic drugs parallels ethanol sensitivity in selectively bred mice. *Science,* 224: 519–521, 1984.

Rabe, C.S., Giri, P.R., Hoffman, P.L., and Tabakoff, B.: Effect of ethanol on cyclic AMP levels in intact PC12 cells. *Biochem. Pharmacol.,* 40: 565–571, 1990.

Rabin, R.A.: Direct effects of chronic ethanol exposure on beta-adrenergic and adenosine-sensitive adenylate cyclase activities and cyclic AMP content in primary cerebellar cultures. *J. Neurochem.,* 55: 122–128, 1990.

Rabin, R.A., Fiorella, D., and Van Wylen, D.G.: Role of extracellular adenosine in ethanol-induced desensitization of cyclic AMP production. *J. Neurochem.,* 60: 1012–1017, 1993.

Rabin, R.A. and Molinoff, P.B.: Activation of adenylate cyclase by ethanol in mouse striatal tissue. *J. Pharmacol. Exp. Ther.,* 216: 129–134, 1981.

Ramkumar, V., Stiles, G.L., Beaven, M.A., and Ali, H.: The A3 adenosine receptor is the unqiue adenosine receptor which facilitates release of allergic mediators in mast cells. *J. Biol. Chem.,* 268: 16887–16890, 1994.

Saito, T., Lee, J.M., Hoffman, P.L., and Tabakoff, B.: Effects of chronic ethanol treatment on the beta-adrenergic receptor-coupled adenylate cyclase system of mouse cerebral cortex. *J. Neurochem.,* 48: 1817–1822, 1987.

Sanderson, G. and Scholfield, C.N.: Effects of adenosine uptake blockers and adenosine on evoked potentials of guinea-pig olfactory cortex. *Pflugers Arch.,* 406: 25–30, 1986.

Smolen, T.N. and Smolen, A.: Purinergic modulation of ethanol-induced sleep time in long-sleep and short-sleep mice. *Alcohol,* 8: 123–130, 1991.

Smolen, T.N. and Smolen, A.: Down-regulation of adenosine A2 receptors in long-sleep mice following chronic ethanol administration. *Alcoholism: Clin. Exp. Res.,* 17: 498, 1993 (Abstract).

Smolen, T.N., Smolen, A., and Han, P.C.: Upregulation of adenosine A1 receptors following chronic purinergic agonist and antagonist administration in mice. *FASEB J.,* 7: A255, 1993 (Abstract).

Snyder, S.H., Katims, J.J., Annau, Z., Bruns, R.F., and Daly, J.W.: Adenosine receptors and behavioral actions of methylxanthines. *Proc. Natl. Acad. Sci. U.S.A.,* 78: 3260–3264, 1981.

Yakel, J.L., Warren, R.A., Reppert, S.M., and North, R.A.: Functional expression of adenosine A2b receptor in Xenopus oocytes. *Mol. Pharmacol.,* 43: 277–280, 1993.

Chapter 10

NEUROTENSIN: A POTENTIAL MEDIATOR OF ETHANOL ACTIONS

V. Gene Erwin

TABLE OF CONTENTS

I. INTRODUCTION

This chapter focuses on the neurobiology and pharmacology of neurotensin (NT) with specific emphasis on interactions of this neuropeptide with ethanol-related behaviors. Pharmacological studies have revealed many similarities in CNS effects of NT and ethanol leading to hypotheses that some of the acute and chronic effects of ethanol may be mediated or modulated by NT or NT-mediated processes. It is recognized that genetic factors contribute to the development of alcoholism, and selective breeding studies in rats and mice have demonstrated genetic influences on ethanol drinking as well as behavioral responses to ethanol including loss of righting response, locomotor activity, hypothermia, and acquisition of tolerance. Recent studies in inbred strains of mice have indicated genetic-based differences in NT levels and NT receptor densities in specific brain regions. These observations support the hypothesis that innate differences in ethanol sensitivity are mediated, in part, by genetic differences in NTergic processes. Thus, comparisons of NTergic systems in rodent models have been useful in investigations of the role of NT in actions of ethanol.

II. NEUROBIOLOGY OF NEUROTENSIN AND NEUROTENSIN RECEPTORS

This summary of the neurobiology and pharmacology of neurotensin and its receptors is intended to provide sufficient background to enable the reader to understand interactions of ethanol with NTergic systems.

0-8493-8389-7/95/$0.00+$.50

A. NEUROTENSIN

Since the isolation of NT (p-Glu-Leu-Tyr-Glu-Asn-Lys-Pro-Arg-Arg-Pro-Tyr-Ile-Leu-OH) from hypothalamus and the demonstration of its potent vascular effects (Carraway and Leeman, 1973), convincing evidence has been obtained supporting a neuromodulator or neurotransmitter role of this tridecapeptide in the CNS. Neurotensin-immunoreactivity (NT-ir), is widely distributed in mammalian brain (Emson et al., 1982; Cooper et al., 1981) with highest levels in the hypothalamus (HYP), ventral midbrain (VMB), and nucleus accumbens (NA). It is often colocalized with dopamine (Emson et al., 1982; Jennes et al., 1982; Bean et al., 1989), and electrical stimulation of the median forebrain bundle, *in vivo,* causes corelease of NT-ir and dopamine in the medial prefrontal cortex (MPFC) (Bean and Roth, 1991; Bean et al., 1989). Release of the peptide from brain slices is Ca^{++}-dependent (Iversen et al., 1978). Following its release, NT is rapidly inactivated by endopeptidases including 24.15 and 24.11 (Checler et al., 1986).

The gene for preproNT/neuromedin N (preproNT) has been cloned from rat, dog, cow, and human (Tanaka et al., 1990; Gerhard et al., 1989). In rats, levels of preproNT mRNA in specific brain regions, except for frontal cortex, correspond reasonably well with levels of the peptide (Tanaka et al., 1990). This gene has been mapped to human chromosome 12 (Gerhard et al., 1989) but to our knowledge has not been mapped in the mouse genome. Evidence has been obtained (Dobner et al., 1992) for a cooperative regulation of preproNT gene expression by transcription factors including c-*fos*, CREB (cAMP responsive element binding protein), and glucocorticoids. In PC12 cells there is a striking enhancement of expression of the preproNT gene when cells are incubated with nerve growth factor, forskolin, and dexamethasone. In brain tissues, drugs that elevate levels of NT-ir, e.g., haloperidol, produce increases in c-*fos* mRNA prior to increases in preproNT mRNA (Merchant and Dorsa, 1993). Interestingly, ethanol administration alters most, if not all of these factors that regulate preproNT expression. For example, it is well known that ethanol causes a rapid increase in plasma glucocorticoids in mice (Zgombick and Erwin, 1987). In cells and brain tissues, ethanol stimulates cAMP production via dopamine-coupled adenylyl cyclase (Rabe et al., 1990; Rabin and Molinoff, 1981). Ethanol increases dopamine overflow in nucleus accumbens and striatum (Lai et al., 1979; Imperato and Dichiara, 1986), and evidence has been reported indicating that chronic ethanol exposure and withdrawal alters c-*fos* mRNA levels (Dave et al., 1989). As discussed below, ethanol administration has been shown to alter NT-ir levels. Thus, it is possible that ethanol-induced changes in NT-ir levels are mediated by effects of ethanol on these systems, particularly in brain regions containing dopamine terminals.

B. NEUROTENSIN-DOPAMINE INTERACTIONS

Pharmacological and biochemical studies have established that NT modulates central dopaminergic function (Nemeroff and Cain, 1985). Administration of NT into the cerebral ventricles (icv) or nucleus accumbens (NA) elicits neuroleptic-like effects and blocks locomotor activation produced by cocaine and amphetamine (Nemeroff and Cain, 1985; Robledo et al., 1993). When administered into the medial prefrontal cortex (MPFC), NT blocks inhibitory effects of dopamine on motor neuron firing (Shi and Bunney, 1992). Recent studies have shown that, in the MPFC, NT blocks locomotor inhibition induced by GBR 12909, a selective dopamine transporter blocker (Erwin et al., 1994c). Other investigators (Kalivas and Taylor, 1985) have shown that intraVTA injections of NT increase dopamine turnover in the NA, an effect associated with enhanced locomotor activity. A number of investigations with dopaminergic drugs show that chronic administration of indirect dopamine agonists, 3,4-methylenedioxymethamphetamine and cocaine, or D2 dopamine receptor antagonists, e.g., haloperidol and sulpiride, produce increases in both NT mRNA and the peptide in the striatum (Merchant et al., 1992; Merchant et al., 1988). These increases were blocked by the dopamine, D1, antagonist, SCH 23390 indicating that dopamine, via D1 receptors, regulates levels of expression of NT-ir and NTR. Since, as noted above, ethanol also elicits an increase in dopamine overflow it is possible that effects of ethanol on expression of NT-ir or NTR might be mediated, in part, by D1 receptor activation. These observations raise an important question of whether chronic ethanol administration alters expression of preproNT and/or the NTR gene.

C. NEUROTENSIN RECEPTORS AND SECOND MESSENGERS

Evidence indicates that the pharmacological effects of NT are mediated by NTR in the brain. Neurotensin receptors have been characterized in brains of several species (Kitabgi et al., 1987; Mazella et al., 1983), including human. Two receptor subtypes, a high-affinity (Kd = 0.2 n*M*) and a low-affinity (Kd = 3.0 n*M*) subtype, have been characterized from rat and mouse brains; in adult human brain the high-affinity receptor is predominant. A recent report indicates three distinct NT binding proteins of 50, 60, and 100 kd are present in cells cultured from embryonic mouse brains (Chabry et al., 1993). High-affinity receptors are located on membranes of cell bodies, fibers, and terminals of dopaminergic neurons (Dana et al., 1989; Young and Kuhar, 1979) and on cholinergic neurons of the basal forebrain (Szigethy and Beaudet, 1987). In N1E-115 neuroblastoma cells, NT receptors are coupled to both cyclic GMP formation and hydrolysis of inositol phospholipid (Kanba and Richelson, 1987), and evidence in brain slices indicates that NT receptors are coupled to production of inositol 1,4,5-trisphosphate (Goedert et al., 1984; Erwin and Radcliffe, 1993). Studies on cells in culture have demonstrated NT-induced increases in intracellular Ca^{++} ion (Bozou et al., 1989; Snider et al., 1986).

Several studies (Faggin et al., 1990; Hetier et al., 1988) have shown that NT enhances K^+-stimulated or evoked release of dopamine from slices of rat striatum, nucleus accumbens, and prefrontal cortex. This effect is mediated by a direct action on dopamine terminals. Other studies have demonstrated that NT enhances K^+-stimulated release of acetylcholine and gamma amino butyric acid (GABA) from brain slices (Lapchak et al., 1991; O'Connor et al., 1992). Evidence is presented in these studies indicating that NT stimulation of neurotransmitter release is the result of a direct action of NT on GABAergic or cholinergic cell bodies. More recent electrophysiological studies indicate that NT excites cholinergic cells of the basal forebrain by pertussis toxin-insensitive G protein (Farkas et al., 1994). The direct action of NT was to reduce inwardly rectifying K^+ conductance. It is of interest that the neurochemical effects of NT in cells or slices, including increased intracellular Ca^{++}, enhanced neurotransimitter release, and electrophysiological actions are rapidly desensitized (Faggin and Cubeddu, 1990; Hermans et al., 1994). This rapid desensitization has been attributed to a rapid internalization of the high-affinity NT receptor (DiPaola et al., 1993; Mazella et al., 1992). It has been suggested that rapid receptor desensitization and internalization may be responsible for acquisition of acute tolerance to pharmacological effects of centrally administered NT (Erwin et al., 1995; Erwin et al., 1994a).

Two non-peptide compounds, levocabastine and SR 48692, have been shown to be relatively selective for the low- and high-affinity NTRs, respectively. Levocabastine has a Kd of 1 μ*M* for the low affinity NTR, and SR 48692 has a Kd value of approximately 1 n*M* for the high-affinity NTR; SR 48692 has a Kd of about 10 n*M* for the low-affinity NTR (Gully et al., 1993). Neither levocabastine nor SR 48692 exhibits potent pharmacological effects when administered peripherally; however, this route of administration may not be ideal because both are relatively large organic molecules, are negatively charged at physiological pH, and may not readily cross the blood/brain barrier. Unfortunately, little is known at present regarding the pharmacokinetics of these compounds. Recent reports show that SR 48692, administered peripherally, reduces NT-induced circling following intra-striatal injections (Poncelet et al., 1994) and inhibits NT (injected into nucleus accumbens) reversal of amphetamine-induced activation. *In vitro*, SR 48692 appears to inhibit NT-induced [3H]-dopamine release from brain slices (Brouard et al., 1994), and it is reported to prevent retrograde axonal transport of NT, presumably by blocking internalization of the high-affinity NTR (Steinberg et al., 1994).

In rats, SR 48692 produced anxiolytic-like effects and attenuated the locomotor inhibitory effects of ethanol (Amin et al., 1994); similar results have been obtained in LS mice where 80 μg/kg SR 48692 attenuates locomotor inhibitory effects of ethanol at 1.5 and 2.0 g/kg (Erwin, unpublished data).

High-affinity NTR has been purified from bovine (Mills et al., 1988) and mouse (Mazella et al., 1989) brain, and the gene has been cloned from rat (Tanaka et al., 1990) and human (Vita et al., 1993; Watson et al., 1993) brain. The cloned receptors consist of 424 (rat) and 418 (human) amino acids with seven putative transmembrane domains belonging to the family of G-protein coupled receptors. The genes have been expressed in *Xenopus* oocytes, fibroblasts, CHO-K1, and COS cells, and resulting receptors are of the levocabastine insensitive, high affinity type. Expressed receptors are coupled to ion channels and phosphatidylinositol hydrolysis (Watson et al., 1993; Vita et al., 1993; Chabry et al., 1994). The NTR gene has been localized to the mouse chromosome 2 (H region) and human chromosome 20 (long arm) (Laurent et al., 1994). Other studies (Watson et al., 1993) identified a polymorphism in the coding region of the human NTR gene which results in a single amino difference in the fourth transmembrane region of the receptor. This polymorphism apparently does not cause functional differences between the two receptor forms (Watson et al., 1993).

III. NEUROTENSIN AND ETHANOL INTERACTIONS

A. PHARMACOLOGICAL INTERACTIONS OF ETHANOL AND NT

Pharmacological effects of centrally administered NT include hypothermia (Nemeroff et al., 1977), altered locomotor activity (Kalivas et al., 1982), analgesia (Clineschmidt et al., 1979), muscle relaxation (Osbahr et al., 1979), and neuroendocrine effects (Rowe et al., 1992). Interestingly, acute ethanol administration produces many of the behavioral and pharmacological effects elicited by NT (see review by Deitrich et al., 1989), including effects on dopaminergic systems. Similarities in actions of centrally administered NT and ethanol, i.p., are summarized in Table 1.

TABLE 1
Comparisons of Actions of Ethanol and Centrally Administered Neurotensin

Ethanol	Neurotensin	Interaction
Hypothermia	Hypothermia (icv)	Positive
Locomotor Inhibition (>2.5 g/kg)	Locomotor Inhibition (icv)	Positive
Locomotor Activation (<2.0 g/kg)	Locomotor Activation (intraVTA)	+ or –
Analgesia	Analgesia (icv)	Unknown
Self Administration (oral)	Self Administration (intraVTA)	Unknown
Tolerance	Tolerance	Crosstoler
Dopamine Turnover (mesolimbic)	Dopamine Turnover (mesolimbic)	Unknown
Hyperglycemia	Hyperglycemia (icv)	Unknown
HPA Axis Stimulation	HPA Axis Stimulation	Unknown
Anesthesia	No Anesthesia	Positive

From Erwin, V.G. et al. (1994b), *Ann. N. Y. Acad. Sci.*, 739: 185–196.

Effects of NT on locomotor activity are brain regions specific. For example, NT injected icv or into the NA produces locomotor inhibition and blocks amphetamine-induced locomotor activation, whereas injections into the VTA produces locomotor activation (Kalivas et al., 1982). Ethanol administered in low doses in mice (1.0 to 2.0 g/kg) generally produces locomotor activation, and higher doses up to 3.0 g/kg usually produce locomotor inhibition.

These effects of ethanol are genotype dependent. Early studies observed that central administration of NT lengthened duration of loss of righting response and potentiated hypothermia following hypnotic doses of ethanol in mice, suggesting that NT may mediate some of the actions of ethanol (Frye et al., 1981; Luttinger et al., 1981). Similar results have been reported in rats (Widdowson, 1987). Subsequently, several genotype-dependent interactions between ethanol and NT have been reported (Erwin and Su, 1989; Erwin and Jones, 1989; Erwin et al., 1987) using LS and SS lines of mice which were selectively bred for differences in ethanol sensitivity (McClearn and Kakihana, 1983). NT, icv, increased hypnotic and hypothermic effects of ethanol to a greater extent in SS than in LS mice (Erwin et al., 1987).

Dose-response relationships indicate that NT administered into the VTA produces greater locomotor activation in SS than in LS mice (Erwin and Jones, 1989). These differences in sensitivity were attributed to higher NTR densities in ventral midbrain of SS compared with LS mice. Injections of NT into the NA or lateral ventricles inhibited locomotor activating effects of low doses of ethanol (Erwin and Su, 1989). The effects of NT on ethanol-induced behaviors were specific as indicated by:

1. dose-response and structure-activity relationships
2. lack of efficacy of other peptides
3. absence of interaction with other anesthetics.

These observations supported the hypothesis that some of the actions of ethanol are modulated by NT and NT receptor-coupled processes.

B. EFFECTS OF ETHANOL ON NT LEVELS

Studies have been conducted on the time-course and dose-response relationships for effects of acute ethanol administration on NT-ir in LS and SS mice (Erwin et al., 1990). At doses between 0.5 and 2.5 g/kg, ethanol produced a decrease in NT-ir in hypothalamus of LS but not SS mice. None of the ethanol doses altered NT levels in the ventral midbrain, and ethanol-induced decreases in NT levels in the NA and CP were only significant at 2 g/kg. Results show that ethanol significantly decreased hypothalamic NT-ir levels in LS mice within five minutes and that the effect was observed up to four hours; longer times have not been studied. In both LS and SS mice, hypnotic doses of ethanol (4 g/kg) produced significant increases in NT-ir levels in hypothalamus, nucleus accumbens, and caudate-putamen. In other studies (Erwin et al., unpublished), it was shown that following hypnotic doses of ethanol in HAS and LAS rats selectively bred for differences in ethanol sensitivity, NT-ir levels were increased in these brain regions. In other studies (Erwin and Radcliffe, unpublished) the effects of 2.0 g/kg ethanol on NT-ir levels in brain regions from 24 LS × SS RI strains of mice were determined. Thirty minutes after ethanol, NT-ir levels in the HYP and NA were significantly decreased in about 25% of the strains and were unchanged or increased in 75% of the strains.

Chronic ethanol administration produces a time-dependent acquisition of tolerance to locomotor inhibitory effects of ethanol in LS mice and sensitization to locomotor activating effects of ethanol in SS mice (Erwin et al., 1992b). Interestingly, this treatment produced a significant increase in NT-ir levels in nucleus accumbens of LS and SS mice (Erwin et al., 1992a). Increases corresponded, over the time course of chronic ethanol consumption and withdrawal, with tolerance to ethanol-induced locomotor inhibition in LS mice. It was hypothesized that ethanol enhances NT release (and dopamine release) from mesolimbic dopaminergic neurons where it is colocalized with dopamine and that chronic increases in synaptic dopamine might lead to increased NT-ir. Also, chronic increases in synaptic NT might cause a down-regulation of NT receptors.

C. EFFECTS OF ETHANOL ON NT RECEPTORS

Several studies have been devoted to characterization of NT receptor subtypes in brain regions from LS, SS, and LS × SS RI strains of mice (Campbell et al., 1991; Erwin and Jones, 1989). Saturation isotherms and Scatchard analyses of NT binding were biphasic indicating two specific receptor subtypes with Kd values of about 0.3 and 4 n*M* for the high- and low-affinity receptors, respectively. NT binding to the low-affinity receptor was completely inhibited by levocabastine, a putative histamine antagonist. Competitive binding experiments with NT fragments showed that the C-terminal amino acids are essential for specific binding to either the low- or high-affinity receptor. No differences were found in Kd values for NT binding in various brain regions from either LS or SS mice. Both low- and high-affinity binding capacities differed significantly between brain regions, and low-affinity receptor densities were higher in SS than in LS in areas of the cortex, nucleus accumbens, and ventral midbrain. High-affinity receptor densities were higher in VMB from SS than LS. Neither acute ethanol administration nor ethanol, *in vitro*, appears to alter NT binding to brain membranes.

In other experiments we demonstrated that chronic ethanol treatment (Campbell and Erwin, 1993), which produced functional tolerance to ethanol-induced locomotor inhibition and hypothermia in LS mice, caused a significant decrease in NT binding capacity for both high- and low-affinity receptors in specific brain regions. This apparent down-regulation of NT receptors was maximal after two weeks of chronic ethanol consumption and returned to control values at two to three weeks following withdrawal. Interestingly, the time course for changes in NT receptor density paralleled the time course for acquisition and decay of tolerance to ethanol in LS mice. Response of SS mice following chronic ethanol administration differed from LS in that, at the challenge dose of 2.5 g/kg ethanol, SS mice were more activated than control animals, indicating either sensitization to locomotor activating effects of ethanol or acquisition of tolerance to the inhibitory effects had occurred allowing for greater activity. NT receptors were down-regulated in both LS and SS mice following chronic ethanol. More recent studies have demonstrated acquisition of cross tolerance between ethanol and NT (Erwin et al., 1995). In these studies, chronic icv administration of NT produced a decrease in NT receptor density in the nucleus accumbens that was associated with acquisition of tolerance to ethanol-induced locomotor inhibition in LS mice.

Recently, studies on the effects of NT injections into VTA on dopamine metabolism in the NA and CP were reported. Since it has been shown that intraVTA administration of NT produces an increase in dopamine turnover in the NA and CP, it was of interest to determine whether NT, injected into theVTA, might differentially increase dopamine metabolism in the NA and CP of LS and SS mice. Results from these studies (Erwin et al., 1994a) show that NT produces a dose-dependent increase in the ipsilateral NA and CP but not the contralateral regions of LS mice; stimulation of dopamine metabolism was greater in NA than CP, but within-region LS and SS were not significantly different. As noted above, we have shown that chronic ethanol decreases NT receptors in the ventral midbrain with a time-course similar to development of tolerance to ethanol-induced locomotor inhibition in LS mice. Therefore, it was of interest to determine whether effects of centrally administered NT on dopamine metabolism was altered by chronic ethanol administration in LS mice. Studies show that NT-enhanced dopamine metabolism in NA and CP is significantly inhibited following chronic ethanol treatment (Erwin et al., 1994a). This is direct evidence of a neurochemical tolerance which accompanies the behavioral tolerance; that is, animals, made tolerant to hypothermic and locomotor inhibitory effects of ethanol, are tolerant to behavioral and neurochemical effects of centrally administered NT.

D. EFFECTS OF ETHANOL ON NT RECEPTOR-MEDIATED PROCESSES IN SPECIFIC BRAIN REGIONS OF LS AND SS MICE

As noted above NT stimulates phosphatidylinositol (PI) hydrolysis and enhances dopamine release in slices from specific brain regions. The effect of ethanol on these processes have been examined in the LS and SS mice (Erwin and Radcliffe, 1993; Duncan and Erwin, 1991). The results showed that NT-stimulation of PI hydrolysis was brain region specific with significant between line (LS > SS) differences in the nucleus accumbens and entorhinal cortex. There were no line differences for carbachol-stimulated PI hydrolysis. Ethanol enhanced NT- but not carbachol-stimulated PI hydrolysis in nucleus accumbens slices from LS and SS brains. These results were surprising because NT receptor densities are somewhat higher in brain regions from SS than LS. However, in both lines of mice NT-stimulated PI hydrolysis was correlated with NT receptor density (across brain regions) with the slope of the correlation being greater in the LS than SS mice. These results were cautiously interpreted as indicating there may be some differences between LS and SS mice in the coupling of NT receptors to PI hydrolysis. Other investigators (Smith, 1991) observed little effect of ethanol on phospholipid hydrolysis in N1E-115 cells except at high concentrations where inhibition was noted.

In another series of studies the effects of ethanol on NT enhancement of K^+-stimulated dopamine (dopamine) release was measured in caudate-putamen (CP) and nucleus accumbens (NA) slices from LS and SS mice (Erwin and Radcliffe, 1993). Slices from LS or SS (CP or NA) did not differ in submaximal (25 m*M*) K^+-stimulated dopamine release. Interestingly, NT enhanced K^+-stimulated release from CP but not NA slices. Augmentation of dopamine release was NT concentration-dependent and tetrodotoxin insensitive, implicating a role of presynaptic NT receptors located on nigrostriatal dopaminergic terminals. Ethanol (100 m*M*) had no effect on K^+-stimulated dopamine release, but when applied concomitantly with NT, it blocked NT enhancement of dopamine release from CP slices of LS but not SS mice. Since ethanol has no effect on NT receptor, in either LS or SS brain membranes, these results and those on PI hydrolysis suggest that some of the interactions between ethanol and NT may be at the level of NT receptor coupling to second messenger systems. This interaction appears to be greater in LS than SS brain slices. Caution is needed in relating these *in vitro* differences in specific brain regions to behavioral effects of NT or ethanol, *in vivo*. This is particularily true since it is well-known that NT has markedly different effects depending on the central site of administration.

IV. GENETIC CORRELATIONS AMONG NEUROTENSIN AND ETHANOL ACTIONS

A. GENETIC CORRELATIONS AMONG ETHANOL ACTIONS AND NTergic PROCESSES

An exhaustive and detailed set of experiments have been conducted to determine ethanol-related behaviors, NT-ir levels, and densities of NT receptor subtypes in specific brain regions from 24 LS × SS RI strains of mice (Erwin and Jones, 1993; Erwin et al., 1993; Erwin et al., 1994a). Analyses of these data has provided important information regarding the genetic associations among measures of NTergic processes and ethanol behaviors. Analyses showed highly significant, $p < 0.01$, genetic correlations between NT-ir levels in the hypothalamus ($r = 0.70$) and ventral midbrain ($r = 0.40$) and hypnotic sensitivity (sleep time and blood ethanol concentration at regaining righting reflex). Moreover, significant, $p < 0.05$, correlations were found between high-affinity NT receptors densities in the frontal cortex (FC) and locomotor ($r = 0.47$) or hypnotic ($r = 0.50$) effects of ethanol. These results indicate common genes that regulate NTergic processes and ethanol sensitivity. Current studies are evaluating the chromosomal location of those genes using the relatively new approach of quantitative trait locus

(QTL) analyses. Results of preliminary studies indicate a number of common QTLs regulating NT-ir levels and/or NTR densities and hypnotic, locomotor, and hypothermic sensitivities to ethanol.

These data show common QTLs for NT-ir levels and/or NTR densities and locomotor and/or hypnotic effects of ethanol on chromosomes 1, 2, 3, 4, 6, 7, 8, 10, 17, and 18. The QTLs for hypnotic sensitivity and locomotor effects of ethanol do not overlap completely with those reported for B × D RI strains of mice (Crabbe et al., 1994), but there are a number of regions where there is overlap. Also, there are regions where QTLs for ethanol preference have been reported, in the B × D RI strains, that are in common with QTLs for the neurotensin measures, e.g., on chromosomes 1, 2, 4, 7, and 17. Candidate genes with loci within flanking markers of the significant common QTLs include but are not limited to: the NMDA receptor, dopamine beta-hydroxylase, and the high-affinity neurotensin receptor on chr. 2; inhibitory G proteins and Nras oncogene on chr. 3; NMDA 2b receptor and neuropeptide Y on chr. 6; protein kinase c, gamma, and beta on chr. 4; tyrosine kinase on chr. 8; $GABA_A$ receptor subunits alpha and gamma on chr. 11; proopiomelanocortin, alpha on chr. 12; dopamine D1 receptor on chr. 13; and adrenergic receptor, beta 2 and calmodulin kinase II, alpha subunit on chr. 18. Clearly, some of these correlations are fortuitous because of type II errors given the limited degrees of freedom with only 24 RI strains. Thus, it is essential to confirm significant QTLs using a large number of F2 generation mice.

V. SUMMARY

A number of observations support the hypotheses that ethanol acts via NTergic systems in the brain and that individual differences in ethanol sensitivity are mediated, in part, by NTergic processes. These include:

1. Central administration of NT produces genotype-dependent changes in ethanol sensitivity.
2. Acute and chronic ethanol administration alters NT levels in specific brain regions in a genotype-dependent manner.
3. Chronic ethanol administration decreases brain NTR densities in parallel with acquisition of tolerance.
4. There are significant genetic correlations between NTR densities and ethanol sensitivity and between endogenous NT-ir levels and ethanol sensitivity in LS × SS RI strains.
5. Common QTLs for NT-ir levels and hypnotic sensitivity and for NTR levels and locomotor activation have been identified in LS × SS RI strains.
6. Genotype-dependent interactions have been found between NT and ethanol on dopamine release and phosphatidyl inositol hydrolysis in brain slices.
7. Cross-tolerance has been observed between ethanol- and NT-induced locomotor inhibition and hypothermia.

Potential sites of action of ethanol on NTergic systems include:

1. NT levels and turnover (synthesis, release, and degradation) in neurons projecting to the frontal cortex, striatum, NA, and in the hypothalamus
2. NT receptors with potential alteration in Kd or Bmax values
3. receptor-coupled second-messenger production and subsequent processes.

Acute and chronic actions of ethanol could be altered by genetic differences in one or more of these neurotensinergic systems in discrete brain regions. For example, NT receptors in the NA, FC, and VTA are undoubtedly associated with motor behaviors, and hypothalamic receptors may be associated with thermoregulation.

REFERENCES

Amin, R., Pohorecky, L. A., and Benjamin, D., (1994), The neurotensin antagonist SR 48692 attenuates the locomotor depressant effects of ethanol. *Alcoholism: Clin. Exp. Res.* (Abstract).

Bean, A. J., During, M. J., and Roth, R. H., (1989), Stimulation-induced release of coexistent transmitters in the prefrontal cortex: an *in vitro* microdialysis study of dopamine and neurotensin release. *J. Neurochem.*, 53: 655–657.

Bean, A. J. and Roth, R. H., (1991), Extracellular dopamine and neurotensin in rat prefrontal cortex *in vivo:* effects of median forebrain bundle stimulation frequency, stimulation pattern, and dopamine autoreceptors. *J. Neurosci.*, 11: 2694–2702.

Bozou, J. C., Rochet, N., Magnaldo, I., Vincent, J. P., and Kitabgi, P., (1989), Neurotensin stimulates inositol trisphosphate-mediated calcium mobilization but not protein kinase C activation in HT29 cells. Involvement of a G-protein. *Biochem. J.*, 264: 871–878.

Brouard, A., Heaulme, M., Leyris, R., Pelaprat, D., Gully, D., Kitabgi, P., Le Fur, G., and Rostene, W., (1994), SR 48692 inhibits neurotensin-induced [3H]dopamine release in rat striatal slices and mesencephalic cultures. *Eur. J. Pharmacol.*, 253: 289–291.

Campbell, A. D., Jones, B. C., and Erwin, V. G., (1991), Regional characterization of brain neurotensin receptor subtypes in LS and SS mice. *Alcohol Clin. Exp Res.*, 15: 1011–1017.

Campbell, A. D. and Erwin, V. G., (1993), Chronic ethanol administration downregulates neurotensin receptors in long- and short-sleep mice. *Pharmacol. Biochem. Behav.*, 45: 95–106.

Carraway, R. and Leeman, S. E., (1973), The isolation of a new hypertensive peptide, neurotensin, from bovine hypothalmus. *J. Biol. Chem.*, 248: 6854–6861.

Chabry, J., Gaudriault, G., Vincent, J. P., and Mazella, J., (1993), Implication of various forms of neurotensin receptors in the mechanism of internalization of neurotensin in cerebral neurons. *J. Biol. Chem.*, 268: 17138–17144.

Chabry, J., Labbe-Jullie, C., Gully, D., Kitabgi, P., Vincent, J. P., and Mazella, J., (1994), Stable expression of the cloned rat brain neurotensin receptor into fibroblasts: binding properties, photoaffinity labeling, transduction mechanisms, and internalizations. *J. Neurochem.*, 63: 19–27.

Checler, F., Vincent, J. P., and Kitabgi, P., (1986), Purification and characterization of a novel neurotensin-degrading peptidase from rat brain synaptic membranes. *J. Biol. Chem.*, 261: 11274–11281.

Clineschmidt, B. V., McGuffin, J. C., and Bunting, P. B., (1979), Neurotensin: antinocisponsive action in rodents. *Eur. J. Pharmacol.*, 54: 129–139.

Cooper, P. E., Fernstrom, M. H., Rorstad, O. P., Leeman, S. E., and Martin, J. B., (1981), The regional distribution of somatostatin, substance P and neurotensin in human brain. *Brain Res.*, 218: 219–232.

Crabbe, J. C., Buck, K.J., and Belknap, J.K., (1994), Genetic animal models of alcohol and drug abuse. *Science*, 264:1715–1723.

Dana, C., Vial, M., Leonard, K., Beauregard, A., Kitabgi, P., Vincent, J. P., Rostene, W., and Beaudet, A., (1989), Electron microscopic localization of neurotensin binding sites in the midbrain tegmentum of the rat, I. ventral tegmental area and the interfascicular nucleus. *J. Neurosci.*, 9: 2247–2257.

Dave, J. R., Tabakoff, B., and Hoffman, P. L., (1989), Ethanol withdrawal seizures produce increased c-fos mRNA in mouse brain. *Mol. Pharmacol.*, 37: 367–371.

Deitrich, R. A., Dunwiddie, T. V., Harris, R. A., and Erwin, V. G., (1989), Mechanisms of action of ethanol: initial central nervous system actions. *Pharmacol. Rev.*, 41: 489–537.

DiPaola, E. D., Cusack, B., Yamada, M., and Richelson, E., (1993), Desensitization and down-regulation of neurotensin receptors in murine neuroblastoma clone N1E-115 by [D-lYS8] neurotensin (8–13). *J. Pharmacol. Exp. Ther.*, 264: 1–5.

Dobner, P. R., Kislauskis, E., and Bullock, B. P., (1992), Cooperative regulation of neurotensin/neuromedin N gene expression in PC12 cells involves AP-1 transcription factors. *Ann. N. Y. Acad. Sci.*, 668: 17–29.

Duncan, C. C. and Erwin, V. G., (1991), Neurotensin modulateds K+-stimulated dopamine release from the caudate-putamen but not the nucleus accumbens of mice selectively bred for differential hypnotic sensitivity to ethanol. *Alcohol*, 9: 23–29.

Emson, P. C., Goedert, M., Benton, H., St.-Pierre, S., and Rioux, F., (1982), The regional distribution and chromatographic characterization of neurotensin-like immunoreactivity in the rat. *Adv. Biochem. Psychopharmacol.*, 33: 477–485.

Erwin, V. G., Korte, A., and Marty, M., (1987), Neurotensin selectively alters ethanol-induced anesthesia in LS/Ibg and SS/Ibg lines of mice. *Brain Res.*, 400: 80–90.

Erwin, V. G., Jones, B. C., and Radcliffe, R. A., (1990), Low doses of ethanol reduce neurotensin levels in discrete brain regions from Ls/Ibg and SS/Ibg mice. *Alcoholism: Clin. Exp. Res.*, 14: 42–47.

Erwin, V. G., Campbell, A. D., and Radcliffe, R. A., (1992a), Effects of chronic ethanol administration on neurotensinergic processes: correlations with tolerance in LS and SS mice. *Ann. N. Y. Acad. Sci.*, 654: 441–443.

Erwin, V. G., Radcliffe, R. A., and Jones, B. C., (1992b), Chronic ethanol consumption produces genotype-dependent tolerance to ethanol in LS/Ibg and SS/Ibg mice. *Pharmacol. Biochem. Behav.*, 41: 275–281.

Erwin, V. G., Radcliffe, R., Hinkle, B., and Jones, B. C., (1993), Genetic-based differences in neurotensin levels and receptors in brains of LL × SS mice. *Peptides*, 14: 821–828.

Erwin, V. G., Campbell, A. D., Myers, R., and Womer, D. E., (1995), Cross tolerance between ethanol and neurotensin in mice selectively bred for differences in ethanol sensitivity. *Pharmacol. Biochem. Behav.*, 51: 891–899.

Erwin, V. G., Jones, B. C., Myers, R., and Womer, D. E., (1994a), Effects of acute and chronic ethanol administration on neurotensic systems. *Ann. N. Y. Acad. Sci.*, (in press).

Erwin, V. G., Jones, B. C., and Radcliffe, R. A., (1994b), Neuotensin receptor densities correlate with ethanol induced locomotor activation in LS, SS and LS × SS RI strains of mice. *Alcoholism: Clin. Exp. Res.*, 14: 480.

Erwin, V. G. and Jones, B. C., (1989), Comparison of neurotensin levels, receptors and actions in LS/Ibg and SS/Ibg mice. *Peptides,* 10: 435–440.

Erwin, V. G. and Jones, B. C., (1993), Genetic correlations among ethanol-related behaviors and neurotensin receptors in long sleep (LS) × short sleep (SS) recombinant inbred strains of mice. *Behav. Genet.*, 23: 191–196.

Erwin, V. G. and Radcliffe, R. A., (1993), Characterization of neurotensin-stimulated phosphoinositide hydrolysis in brain regions of long-sleep and short-sleep mice. *Brain Res.*, 629: 59–66.

Erwin, V. G. and Su, N. C., (1989), Neurotensin and ethanol interactions on hypothermia and locomotor activity in LS and SS mice. *Alcoholism: Clin. Exp. Res.*, 13: 91–94.

Faggin, B., Zubieta, J. K., Rezvani, A. H., and Cubeddu, L. X., (1990), Neurotensin-induced dopamine release *in vivo* and *in vitro* from substantia nigra and nucleus caudate. *J. Pharmacol. Exp. Ther.*, 252: 817–825.

Faggin, B. M. and Cubeddu, L. X., (1990), Rapid desensitization of dopamine release induced by neurotensin and neurotensin fragments. *J. Pharmacol. Exp. Ther.*, 253: 812–818.

Farkas, R. H., Nakajima, S., and Nakajima, Y., (1994), Neurotensin excites basal forebrain cholinergic neurons: ionic and signal-transduction mechanisms. *Proc. Natl. Acad. Sci. U.S.A.*, 91: 2853–2857.

Frye, G. D., Luttinger, D., Nemeroff, C. B., Vogel, R. A., Prange, A. J., Jr., and Breese, G. R., (1981), Modification of the actions of ethanol by centrally active peptides. *Peptides 2 Suppl.*, 1: 99–106.

Gerhard, G. S., Dobner, D. S., and Bruns, G. A. P., (1989), Localization of the neurotensin gene to human chromosome 12. *Cytogenet. Cell. Genet.*, 51: 1003–1006.

Goedert, M., Pinnock, R. D., Downes, C. P., Mantyh, P. W., and Emson, P. C., (1984), Neurotensin stimulates inositol phospholipid hydrolysis in rat brain slices. *Brain Res.*, 323: 193–197.

Gully, D., Canton, M., Boigegrain, R., Jeanjean, F., Molinard, J. -C., Poncelet, M., Gueudet, C., Heaulme, M., Leyris, R., Brouard, A., Pelaprat, D., Labbe-Jullie, C., Mazella, J., Soubrie, P., Maffrand, J. -P., Rostene, W., Kitabgi, P., and Le Fur, G., (1993), Biological and pharmacological profile of a potent and selective nonpeptide antagonist of the neurotensin receptor. *Proc. Natl. Acad. Sci. U.S.A.*, 90: 65–69.

Hermans, E., Gailly, P., Octave, J. N., and Maloteaux, J. M., (1994), Rapid desensitization of agonist-induced calcium mobilization in transfected PC12 cells expressing the rat neurotensin receptor. *Biochem. Biophys. Res. Commun.*, 198: 400–407.

Hetier, L., Boireau, A., Dubedat, P., and Blanchard, J. C., (1988), Neurotensin effects on evoked release of dopamine in slices from striatum, nucleus accumbens, and prefrontal cortex in rat. *Naunyn-Schmiedebergs Arch. Pharmacol.*, 337: 13–17.

Imperato, A. and Dichiara, G., (1986), Preferential stimulation of dopamine release in the nucleus accumbens of freely moving rats by ethanol. *J. Pharmacol. Exp. Ther.*, 239: 219–228.

Iversen, L. L., Iversen, S. D., Bloom, F., Douglas, C., Brown, M., and Vale, W., (1978), Calcium-dependent release of somatostatin and neurotensin from rat brain *in vitro. Nature,* 273: 161–163.

Jennes, L., Stumpf, W. E., and Kalivas, P. W., (1982), Neurotensin: topographical distribution in rat brain by immunohistochemistry. *J. Comp. Neurol.*, 210: 211–224.

Kalivas, P. W., Nemeroff, C. B., and Prange, A. J., Jr., (1982), Neuroanatomical site specific modulation of spontaneous motor activity by neurotensin activity by neurotensin. *Eur. J. Pharmacol.*, 78: 471–474.

Kalivas, P. W., and Taylor, S., (1985), Behavioral and neurochemical effect of daily injection with neurotensin into the ventral tegmental area. *Brain Res.*, 358: 70–76.

Kanba, K. S. and Richelson, E., (1987), Comparison of the stimulation of inositol phospholipid hydrolysis and of cyclic GMP formation by neurotensin, some of its analogs, and neuromedin N in neuroblastoma clone N1E-115. *Biochem. Pharmacol.*, 36: 869–874.

Kitabgi, P., Rostene, W., Dussaillant, M., Schotte, A., Laduron, P. M., and Vincent, J. P., (1987), Two populations of neurotensin binding sites in murine brain: discrimination by the antihistamine levocabastine reveals markedly different radioautographic distribution. *Eur. J. Pharmacol.*, 140: 285–293.

Lai, H., Makans, W. L., Horita, A., and Leung, H., (1979), Effects of ethanol on turnover and function of striatal dopamine. *Psychopharmacology,* 61: 1–9.

Lapchak, P. A., Araujo, D. M., Quirion, R., and Beaudet, A., (1991), Neurotensin regulation of endogenous acetylcholine release from rat striatal slices is independent of dopaminergic tone. *J. Neurochem.*, 56: 651–657.

Laurent, P., Clerc, P., Mattei, M.-G., Forgez, P., Dumont, S., Ferrara, P., Caput, D., and Rostene, W., (1994), Chromosomal localization of mouse and human neurotensin receptor genes. *Mammalian Genome,* 5: 303–306.

Luttinger, D., Nemeroff, C. B., Mason, G. A., Frye, G. D., Breese, G. R., and Prange, A. J., Jr., (1981), Enhancement of ethanol-induced sedation and hypothermia by centrally administered neurotensin, beta-endorphorphin and bombesin. *Neuropharmacology,* 20: 305–309.

Mazella, J., Poustis, C., Labbe, C., Checler, F., Kitabgi, P., Granier, C., Van Rietschoten, J., and Vincent, J. P., (1983), Monoiodo-[Trp11]neurotensin, a highly radioactive ligand of neurotensin receptors. Preparation, biological activity, and binding properties to rat brain synaptic membranes. *J. Biol. Chem.,* 258: 3476–3781.

Mazella, J., Chabry, J., Zsurger, N., and Vincent, J. P., (1989), Purification of the neurotensin receptor from mouse brain by affinity chromatography. *J. Biol. Chem.,* 264: 5559–5563.

Mazella, J., Chabry, J., Leonard, K., Kitabgi, P., Vincent, J. P., and Beaudet, A., (1992), Internalization of neurotensin in cultured neurons. *Ann. N. Y. Acad. Sci.,* 668: 356–358.

McClearn, G. E. and Kakihana, R., (1983), Selective breeding for ethanol sensitivity in mice. *Behav. Genet.,* 3: 409.

Merchant, K. M., Letter, A. A., Gibb, J. W., and Hanson, G. R., (1988), Changes in the limbic neurotensin systems induced by dopaminergic drugs. *Eur. J. Pharmacol.,* 153: 1–9.

Merchant, K. M., Dobie, D. J., and Dorsa, D. M., (1992), Expression of the proneurotensin gene in the rat brain and its regulation by antipsychotic drugs. *Ann. N. Y. Acad. Sci.,* 668: 54–69.

Merchant, K. M. and Dorsa, D. M., (1993), Differential induction of neurotensin and c-fos gene expression by typical versus atypical antipsychotics. *Proc. Natl. Acad. Sci. U.S.A.,* 90: 3447–3451.

Mills, A., Demoliou-Mason, C. D., and Barnard, E. A., (1988), Purification of the neurotensin receptor from bovine brain. *J. Biol. Chem.,* 263: 13–16.

Nemeroff, C. B., Bissette, G., Prange, A. J., Jr., Loosen, P. T., Barlow, T. S., and Lipton, M. A., (1977), Neurotensin: central nervous system effects of a hypothalamic peptide. *Brain Res.,* 128: 485–496.

Nemeroff, C. B. and Cain, S. T., (1985), Neurotensin-dopamine interactions in the CNS. *Trends Pharmacol. Sci.,* 6: 201–205.

O'connor, W. T., Tanganelli, S., Ungerstedt, U., and Fuxe, K., (1992), The effects of neurotensin on GABA and acetylcholine release in the dorsal striatum of the rat: an *in vivo* microdialysis study. *Brain Res.,* 573: 209–216.

Osbahr, A. J., Nemeroff, C. B., Manberg, P. J., and Prange, A. J., Jr. (1979), Centrally administered neurotensin: activity in the Julou-Courvoisier muscle relaxation test in mice. *Eur. J. Pharmacol.,* 54: 299–302.

Poncelet, M., Gueudet, C., Gully, D., Soubrie, P., and Le Fur, G., (1994), Turning behavior induced by intrastriatal injection of neurotensin in mice: sensitivity to non-peptide neurotensin antagonists. *Naunyn-Schmiedebergs Arch. Pharmacol.,* 349: 57–60.

Rabe, C. S., Giri, P. R., Hoffman, P. L., and Tabakoff, B., (1990), Effects of ethanol on Cyclic AMP levels in intact PC12 cells. *Biochem. Pharmacol.,* 40: 565–571.

Rabin, R. A. and Molinoff, P. B., (1981), Activation of adenylate cyclase by ethanol in mouse striatal tissue. *J. Pharmacol. Exp. Ther.,* 216: 129–134.

Robledo, P., Maldonado, R., and Koob, G. F., (1993), Neurotensin injected into the nucleus accumbens blocks the psychostimulant effects of cocaine but does not attenuate cocaine self-administration in the rat. *Brain Res.,* 622: 105–120.

Rowe, W., Viau, V., Meaney, M. J., and Quirion, R., (1992), Central administration of neurotensin stimulates hypothalamic-pituitary-adrenal activity. The paraventricular CRF neuron as a critical site of action. *Ann. N. Y. Acad. Sci.,* 668: 365–367.

Shi, W. X. and Bunney, B. S., (1992), Actions of neurotensin: a review of the electrophysiological studies. *Ann. N. Y. Acad. Sci.,* 668: 129–145.

Smith, T. L., (1991), Selective effects of acute and chronic ethanol exposure on neuropeptide and guanine nucleotide stimulated phospholipase C activity in intact N1E-115 neuroblastoma. *J. Pharmacol. Exp. Ther.,* 258: 410–415.

Snider, R. M., Forray, C., Pfenning, M., and Richelson, E., (1986), Neurotensin stimulates inositol phospholipid metabolism and calcium mobilization in murine neuroblastoma clone N1E-115. *J. Neurochem.,* 47: 1214–1218.

Steinberg, R., Bougault, I., Souilhac, J., Gully, D., Le Fur, G., and Soubrie, P., (1994), Blockade of neurotensin receptors by the antagonist SR 48692 partially prevents retrograde axonal transport of neurotensin in rat nigrostriatal system. *Neurosci. Lett.,* 166: 106–108.

Szigethy, E. and Beaudet, A. (1987), Selective association of neurotensin receptors with cholinergic neurons in the rat forebrain. *Neurosci. Lett.,* 83: 47–52.

Tanaka, K., Masu, M., and Nakanishi, S., (1990), Structure and functional expression of the cloned rat neurotensin receptor. *Neuron,* 4: 847–854.

Vita, N., Laurent, P., Lefort, S., Chalon, P., Dumont, X., Kaghad, M., Gully, D., Le Fur, G., Ferrara, P., and Caput, D., (1993), Cloning and expression of a complementary DNA encoding a high affinity human neurotensin receptor. *FEBS Lett.,* 317: 139–142.

Watson, M., Isackson, P. J., Makker, M., Yamada, M. S., Yamada, M., Cusack, B., and Richelson, E., (1993), Identification of a polymorphism in the human neurotensin receptor gene. *Mayo Clin. Proc.,* 68: 1043–1048.

Widdowson, P. S., (1987), The effect of neurotensin, TRH and the delta-opioid receptor antagonist ICI 174864 on alcohol-induced narcosis in rats. *Brain Res.,* 424: 281–289.

Young, W. S. and Kuhar, M. J., (1979), Neurotensin receptors: autoradiographic localization in rat CNS. *Eur. J. Pharmacol.,* 59: 161–163.

Zgombick, J. M. and Erwin, V. G., (1987), Central mechanisms of ethanol-induced adrenocortical response in selectively bred lines of mice. *Neuroendocrinology,* 46: 324–332.

Chapter 11

ELECTROPHYSIOLOGICAL INTERACTIONS OF ETHANOL WITH CENTRAL NEUROPEPTIDE SYSTEMS: OPIOIDS AND SOMATOSTATINS

George Robert Siggins, Zhiguo Nie, Paul Schweitzer, Samuel Madamba, and Steven J. Henriksen

TABLE OF CONTENTS

0-8493-8389-7/95/$0.00+$.50

I. INTRODUCTION: WHY NEUROPEPTIDES?

Past studies showing a strong correlation between the chain length or partition coefficient of a series of alcohols and their behavioral or biochemical potency suggested that alcohol's effects were mediated by an alteration of the ordering or fluidity of the lipids in cellular membranes (see Deitrich et al., 1989; Goldstein et al., 1982). However, more recent work in several disciplines (Franks and Lieb, 1985; Li et al., 1994a; Li et al., 1994b; Stubbs and Rubin, 1993) strongly implicates cellular proteins as a major substrate of alcohol and anesthetic action. Key among these proteins are the membrane-spanning receptors for neurotransmitters, associated GTP-binding proteins, and enzymatic second messenger transduction systems. As it is now clear that peptides such as the various somatostatins (SS) and opioids can act on proteinacious effector systems in the same way as many other, nonpeptide neurotransmitters do (e.g., see Evans et al., 1992; Siggins and Gruol, 1986; Siggins and Zieglgänsberger, 1990), there is the presumption of potential interactions between peptide receptors and ethanol effects.

Unfortunately, the isolation and expression of receptors for neuropeptides has not yet reached the stage of development obtained to date for the transmitter-gated channels. Therefore, we must use the more inferential methods available to electrophysiologists and pharmacologists to ferret out plausible interactions between ethanol and peptide receptors. For these reasons, in this chapter we will focus on two such interactions that seem the most convincing (between ethanol and somatostatin receptors and between ethanol and the opioid system) as revealed by electrophysiological methods and attempt to predict other potential instances of ethanol and peptide interactions. We will also point up similarities, superficial or otherwise, between the actions of ethanol and peptides on the excitability of neurons.

II. METHODS USED

A. *IN VIVO*

Most of the early studies of ethanol actions on CNS neurons involved *in vivo* studies. Such studies have the advantage of testing the effect of systemic or local ethanol effects on neurons in a fairly normal external milieu and with intact neuronal interconnections. Investigations of the neuropharmacology of ethanol have traditionally employed two types of *in vivo* preparations. First, most studies have been undertaken in acutely anesthetized subjects (mostly rodents), where brain areas under study have been investigated using traditional neurophysiological techniques including pathway stimulation and transmitter/drug interactions employing electrophoresis (for review, see Bloom and Siggins, 1987; Bloom et al., 1984; Siggins et al., 1982; Siggins and Bloom, 1981; Siggins et al., 1987). Second, a smaller number of studies of ethanol's action on the central nervous system have employed the freely moving, unanesthetized preparation. Such studies avoid the possible confounding interaction of ethanol with the anesthetic. In addition, these studies have investigated not only the neuropharmacology of systemic and locally applied ethanol but also the actions of ethanol on various quantifiable behaviors including self-administration of ethanol. Our own studies have employed both forms of *in vivo* preparations.

B. *IN VITRO*

In the late 1970s, the development of stable, *in vitro* slice preparations ushered in a new era of routine use of drug superfusion and intracellular recording methods that allowed the application to neurons of defined drug concentrations and determination of cellular and ionic mechanisms responsible for drug action. These methods have lent themselves admirably to alcohol studies of single CNS neurons, albeit with a potentially altered external milieu and disrupted afferent connections. In our studies we have used single-electrode voltage- and current-clamp recording techniques in submerged slices from both rat hippocampus and rat nucleus accumbens (NAcc). In addition, in some studies of the hippocampus, we have examined ethanol effects on extracellular recordings of field potentials evoked by pathway stimulation.

III. OPIOID SYSTEMS AND ETHANOL

A. HIPPOCAMPUS

Opioid peptides, and especially the enkephalins and dynorphins, have been localized by immunohistochemical methods to several circuits in the CA1, CA3, and dentate hippocampal regions (Gall et al., 1981; McGinty et al., 1983). In general, in all regions there are scattered interneurons containing enkephalin-like immunoreactivity, and the dentate gyrus receives a profuse enkephalinergic input from the entorhinal cortex via perforant path fibers. In addition, the dentate granule cells contain both enkephalin and dynorphin in processes that project to CA3 neurons via mossy fibers (Gall et al., 1981; McGinty et al., 1983) and perhaps as recurrent collaterals (or as dentritic release sites) back to perforant path terminals (Terman et al., 1994). The latter dynorphin pathway may serve to reduce excitatory transmitter release from perforant path endings.

1. *In Vivo* Studies

Our original *in vivo* studies of hippocampal and other brain neurons using local application of ethanol by electroosmosis showed some similarities between the actions of ethanol and opiates. In general, μ and δ receptor-preferring opiates inhibit the firing frequency of neurons in other brain regions, yet excite most hippocampal pyramidal and dentate granule neurons by inhibiting inhibitory interneurons (the process of disinhibition) (Mayer et al., 1994; Nicoll et al., 1980; Siggins and Gruol, 1986; Siggins and Zieglgänsberger, 1981; Zieglgänsberger et al., 1979). Interestingly, ethanol applied locally from micropipettes had similar excitatory actions on the majority of hippocampal pyramidal neurons tested *in vivo* (Berger et al., 1982; Siggins et al., 1982). However, the mechanism(s) of action of this excitatory effect was difficult to discern from these *in vivo* studies.

More recent investigations of systemically administered ethanol on hippocampal neuronal responsiveness have indicated complex actions of ethanol involving both intrinsic and extrinsic hippocampal circuits. Acute intoxicating doses of ethanol producing blood alcohol levels of 120–200 mg% increased paired-pulse inhibition in the dentate gyrus while decreasing spontaneous discharge of individually recorded granule cells, suggesting that ethanol increases recurrent inhibitory processes (Wiesner and Henriksen, 1987). In these studies ethanol had no effect on population excitatory postsynaptic potential (pEPSP) amplitudes or slopes. On the other hand, ethanol decreased population spike amplitudes, increased paired-pulse inhibition, and markedly increased post field-potential evoked interneuron discharges. Recent studies have indicated that some of these effects of systemically administered ethanol on hippocampal physiology are, in part, mediated by indirect actions on other major hippocampal afferent structures, including the ventral tegmental area and the lateral septal nucleus (Criado et al., 1994b; Givens and Breese, 1990; Steffensen et al., 1993).

2. Hippocampal Slices *In Vitro*

a. Opiates

The effects of opiate agonists on CA1 pyramidal neurons in the slice preparation generally helped prove the disinhibition hypothesis of μ and δ receptor action in CA1 neurons developed from *in vivo* studies (Zieglgänsberger et al., 1979). Thus, most intracellular studies of CA1 hippocampal pyramidal neurons *in vitro* have shown primarily a reduction of the size and duration of inhibitory postsynaptic potentials (IPSPs) and often also of excitatory postsynaptic potentials (EPSPs; Siggins and Zieglgänsberger, 1981), with activation of μ or δ receptors (Nicoll et al., 1980; Siggins and Zieglgänsberger, 1981). This action is thought to arise either from:

1. a hyperpolarization of inhibitory interneurons (Madison and Nicoll, 1988).
2. reduction of release of inhibitory transmitter (GABA) by an action at the terminals of inhibitory interneurons (Lambert et al., 1991).
3. the postsynaptic modulation of glutamate receptor activation (Zieglgänsberger et al., 1979).

However, early electrophysiological studies on mossy fiber targets — the CA3 neurons — often found mixed effects of opiates on the excitability of these neurons (Gruol et al., 1983; Henriksen et al., 1982; Iwama et al., 1986). In our laboratory, we reasoned that these mixed effects might involve voltage-dependent actions; we thus examined the effect of several opiates on CA3 pyramidal neurons using single-electrode voltage-clamp recording in the rat hippocampal slice (Moore et al., 1994). In most of the neurons, the voltage-dependent K^+ current known as the M-current (I_M; see Figure 11.1) was uniquely sensitive to the opioid peptides, with the direction of the response dependent upon the opiate type and concentration. Thus, opiates selective for κ receptors (e.g., U-50,488H or nanomolar concentrations of dynorphin A or B) significantly augmented I_M. (Note that the dynorphins are known to co-exist with glutamate in mossy fiber afferents to CA3 pyramidal neurons.) By contrast, several opiates known to act on the δ receptor subtypes (e.g., D-Ala^2, D-Leu^5 enkephalin: DADL; [D-Pen^{2,5}]-enkephalin: DPDPE) reduced the I_M. Neither the selective μ-receptor agonist [D-Ala^2, NMe-Phe^4, Gly-ol]-Enkephalin (DAMGO) nor the nonopioid fragment of dynorphin, des-Tyr-dynorphin, consistently altered I_M. Dynorphin did not measurably affect the Q current, a conductance known to contribute to inward rectification in hippocampal pyramidal neurons. The opposing effects of the dynorphins (both A and B) and DADL on I_M were both antagonized by naloxone, and the dynorphin-induced augmentations of I_M were usually reversed by the κ receptor antagonist norbinaltorphimine. These results suggested to us that the opiates have opposing or reciprocal effects on the same voltage-dependent K^+ channel type (the M-channel) in the rat CA3 pyramidal neuron, with the direction of the response depending on which receptor subtype is activated. Such data not only help explain the mixed effects of opiates seen in previous studies, but also suggested a postsynaptic function for the endogenous opiates contained in the CA3 mossy fibers.

b. Ethanol

In some respects the actions of ethanol we find *in vitro* mimic those we saw *in vivo* and the effects of opiates tested *in vitro*. Although one group (Carlen et al., 1982) has reported ethanol-evoked hyperpolarizations and enhancements of IPSPs and afterhyperpolarizations (AHPs), in CA1 and dentate, early studies from our laboratory (Siggins et al., 1987) showed mixed or little effect of ethanol on membrane properties at rest but clear reductions of IPSPs and EPSPs in CA1 CA3 and reductions of AHPs in CA3 neurons (there was no change in CA1 AHPs). Such ethanol reduction of IPSPs and AHPs would have overall excitatory actions on the neurons; indeed increases in spike discharge were often quite pronounced, especially in CA3 hippocampal pyramidal neurons (Siggins et al., 1987).

Previous *in vivo* studies had shown that systemic ethanol enhanced hippocampal neuronal responses to iontophoretically applied acetylcholine and somatostatin (Mancillas et al., 1986a; see below) while having little or no effect on responses to other transmitters such as GABA, norepinephrine, or serotonin. Since our laboratory also had reported that somatostatin and acetylcholine reciprocally regulate the M-current (Moore et al., 1988), we tested ethanol superfusion on this current in rat hippocampal pyramidal neurons *in vitro* using intracellular single electrode voltage-clamp methods. We found that ethanol in low

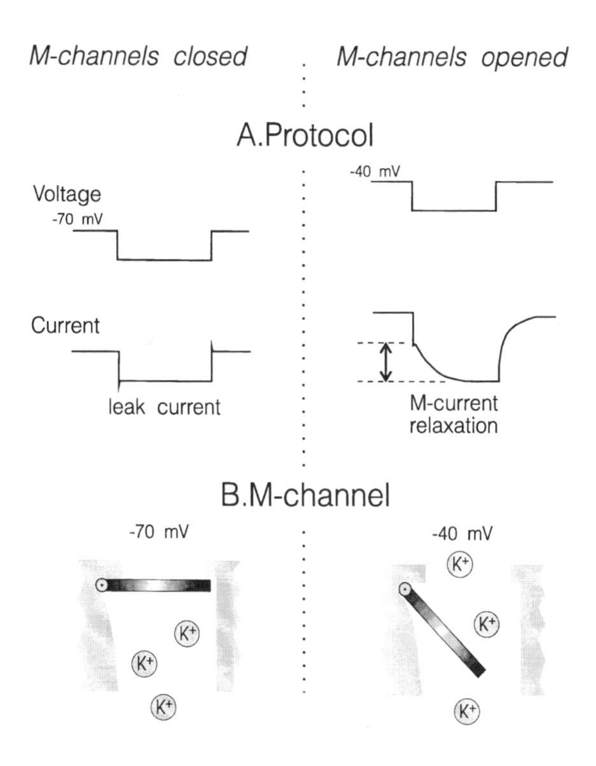

FIGURE 11.1. Schematic of the voltage-dependent properties of the M-channel and a simple voltage-clamp protocol for recording the M-current. A. Protocol: The left panels show the expected current response to a small (5–10 mV) hyperpolarizing voltage command if there were no open M-channels in the membrane (for example, at holding potentials of –70 or more hyperpolarized). Here, only an abrupt, non-voltage-dependent 'ohmic' drop (due to the so-called 'leak' current) is evoked, giving rise to a rectangularly shaped current response. However, if the membrane is depolarized by clamping it at around –50 mV or more depolarized and the same-sized voltage command is delivered (right panel), the abrupt ohmic drop is followed by an inward current relaxation that is actually due to the slow (time- and voltage-dependent) closing of many M-channels that were persistently open at the more depolarized potential (i.e., a reduction of an outward K^+ current). A measure of the M-conductance at the holding potential can be obtained by measuring the current difference (the 'relaxation') between the end of the ohmic drop and the 'steady-state' current toward the end of the hyperpolarizing command. Duration of voltage command: about 1 s. B. Schematic representation of the closed (left) and open (right) states of the M channels during the two conditions shown in A.

concentrations (22–44 m*M*), like muscarinic agonists, greatly reduced I_M amplitude (Moore et al., 1990). These changes (always at depolarized membrane potentials optimal for I_M) were accompanied by an inward (excitatory) baseline current with a conductance decrease. However, other than a small inward current in some cells, there was little or no consistent ethanol effect at resting membrane potentials. In addition, ethanol, like the opioids and SS, had no consistent effect on I_Q or other inwardly rectifying K^+ conductance. Since atropine (the muscarinic antagonist) and tetrodotoxin (used to block transmitter release) did not alter the ethanol effect on I_M, the site of ethanol action was considered most likely distal to the muscarinic receptor, such as at the M-channel itself or at some signal transduction site (e.g., G-proteins). This ethanol reduction of the M-current provides a mechanism for the excitatory effects of ethanol seen in some central neurons like hippocampal pyramidal neurons (see also the section on Somatostatins and ethanol) and is comparable to the I_M reduction by δ receptors in CA3 pyramidal neurons.

B. NUCLEUS ACCUMBENS

The brain mechanisms responsible for human addiction to drugs have been the subject of extensive research, yet are still little understood. Behavioral studies demonstrate that the NAcc may be a key area supporting self-administration of heroin (Koob and Bloom, 1988; Vaccarino et al., 1985) and possibly also alcohol dependence (DiChiara and Imperato, 1991). Morphological studies show a profuse network of opioid peptide-containing fibers in the NAcc (Wamsley et al., 1980), as well as several different kinds of opiate-receptor subtypes (Herkenham et al., 1984; Mansour et al., 1988; Quirion et al., 1983). Thus, opiate addiction may involve a direct action of the opiate in this brain region. Functionally, this area is also of interest because it may regulate extrapyramidal motor function and motivation for motor activity (Mogenson, 1987).

There are many similarities between the effects of ethanol and opiates in the NAcc. *In vivo*, both drugs primarily depress cell firing, and *in vitro*, both have been shown to reduce synaptic transmission, with little effect on cell membrane properties (Nie et al., 1993; Yuan et al., 1991).

1. *In Vivo* Ethanol Studies

Studies in our laboratory have demonstrated that systemic administration of ethanol (0.5–2 g/kg i.p.) reduces the firing rate of both spontaneous and glutamate-activated NAcc neurons in both acutely anesthetized and unrestrained rats (Criado et al., 1994a). In contrast, ethanol does not affect the recruitment of fimbria-activated accumbens neurons. To further characterize the effects of ethanol on limbic-accumbens inputs, we have studied the sensitivity of amygdala-activated NAcc neurons to intoxicating doses of ethanol. Systemic administration of ethanol (1.2–1.4 g/kg, i.p.) significantly reduced the firing rate of spontaneous NAcc neurons in both electrophysiological preparations (Criado et al., 1994a). Similarly, ethanol inhibited the occurrence of amygdala-activated single units in NAcc neurons.

The functional significance of these inhibitory effects of ethanol on spontaneous, evoked, and glutamate-activated NAcc neurons is unknown, but many drugs that have reinforcing properties (e.g., heroin and cocaine) elicit similar inhibitory responses on NAcc neurons *in vivo*. In view of the possible role of the NAcc in mediating the reinforcing properties of ethanol, we also studied the effects of systemic administration of naloxone (5 mg/kg, i.p.) on ethanol-dependent inhibition of spontaneous and fimbria-evoked firing of accumbens neurons in acute anesthetized and unanesthetized rats. However, naloxone did not reverse these inhibitory effects. These findings suggest that heterogeneous inputs to the NAcc exhibit a differential neuropharmacological profile in response to ethanol challenge and that, *in vivo*, the fimbria pathway does not appear to utilize opioids as an important mediator in ethanol's action in this structure.

2. Nucleus Accumbens Slices *In Vitro*

Because of the proposed role of the NAcc in the rewarding effects of opiates, our laboratories (Yuan et al., 1991; Nie et al., in preparation) studied the effect of a variety of opioids on membrane properties and responses to synaptic stimulation in a slice preparation of the NAcc using intracellular recording. Electrical stimulation of the region just dorsal to the olfactory tubercle evoked EPSPs that were abolished at normal resting potentials by 6-cyano-7-nitroquinoxaline-2,3-dione (CNQX, the AMPA/kainate glutamate receptor antagonist), suggesting their mediation by non-N-methyl-D-aspartate type glutamate receptors. At more depolarized potentials (–70 to –55 mV), CNQX-resistant EPSPs were revealed that were blocked by the N-methyl-D-aspartate (NMDA) receptor antagonist DL-2-amino-5-phosphonovalerate (APV). Superfusion of several types of opioid peptides (selective for μ, δ, or κ receptors) over a wide range of concentrations did not affect membrane potential or input resistance of these NAcc core neurons but significantly reduced both depolarizing and hyperpolarizing synaptic potentials. Naloxone superfusion significantly reversed the depressant effects of the μ and δ receptor agonists (but not those of the κ agonist) on synaptic transmission, suggesting involvement of opiate receptors. The most potent effect was exerted by the μ-receptor agonist DAMGO, which reduced synaptic potentials at concentrations as low as 30 n*M*. Superfusion of 1 μ*M* DAMGO also reduced the depolarizing responses to iontophoretic or pressure application of glutamate from micropipettes, in about half of all NAcc cells tested, an effect also reversed by naloxone. These results suggest that the predominant effect of opiates in NAcc is to reduce glutamatergic synaptic transmission, an effect that could be exerted at either pre- or postsynaptic sites.

a. Naloxone-Dependent Ethanol Effects

In the NAcc slice, the similarities between opiate and alcohol actions are pronounced. To investigate ethanol effects on NAcc neuronal properties, our laboratory (Nie et al., 1993) initially used intracellular voltage recording in this slice preparation. In these early studies, ethanol 11–66 m*M* had little reproducible effect on membrane potential or input slope resistance but reduced the amplitude of EPSPs evoked by stimulation of the peritubercle region. Ethanol 22, 44, and 66 m*M* all significantly decreased the EPSPs evoked by half-maximal stimulation to 80%, 60%, and 68% of control, respectively. Superfusion of 11 m*M* ethanol had no effect. Since ethanol still decreased EPSP size in the presence of 30 μ*M* bicuculline superfused to block $GABA_A$ergic IPSPs, such IPSPs cannot be involved in this effect. Because ethanol actions mimicked those of opiates in reducing EPSPs without effect on membrane potentials in the NAcc, we administered the opiate antagonist naloxone together with ethanol. Naloxone 1–2 μ*M* significantly reversed the 44 m*M* ethanol-evoked reduction of EPSP amplitude. Thus, our early data suggested that a major effect of intoxicating concentrations of ethanol (like opioids) in NAcc is to reduce glutamatergic synaptic transmission. These results showed a clear similarity between mechanisms of ethanol and opiate effects in NAcc and suggested that ethanol may act via some opiate mechanism, either indirectly by releasing endogenous opioids or by directly activating opiate receptors.

b. Naloxone-Independent Effects

In more recent *in vitro* studies of NAcc neurons, our laboratory (Nie et al., 1994) has begun characterizing the receptors involved in the evoked EPSPs and examining ethanol's action on these specific receptors, using both voltage- and current-clamp recording. As noted above, at depolarized membrane potentials NAcc core neurons have CNQX-resistant EPSPs that are blocked by the NMDA receptor antagonist APV. In CNQX, ethanol 66 m*M* dramatically decreased these NMDA components of the EPSPs. Application (either via pipette or superfusion) of exogenous NMDA or non-NMDA glutamate agonists (kainate, (R,S)-α-amino-3-hydroxy-5-methylisoxazole-4-propionic acid [AMPA] or quisqualate [QUIS]) evoked reversible depolarizations or inward currents. Ethanol 11–200 m*M* decreased the NMDA currents (evoked at depolarized resting potentials) significantly and dose-dependently, with an IC_{50} of 28 m*M* ethanol. Ethanol 5 m*M* had no effect. In contrast to the ethanol reduction of EPSPs at resting potentials, naloxone 1–2 μ*M* had no effect on the ethanol inhibition of NMDA responses. Higher ethanol concentrations (44–66 m*M*) also slightly but significantly reduced kainate-induced currents (again without a naloxone effect) but not AMPA or QUIS currents; however, even higher ethanol concentrations (100 m*M*) also reduced AMPA currents.

We interpreted these data to suggest that NAcc core neurons express both NMDA and non-NMDA glutamate receptors. Because low ethanol concentrations reduced the EPSPs at normal resting potentials but not responses to non-NMDA glutamate agonists, ethanol probably acts both pre- and postsynaptically in NAcc; by an opioid-dependent reduction of glutamate release, and by postsynaptically reducing NMDA and kainate currents, respectively.

3. Confounds and Implications for Alcoholism

With respect to the involvement of opioid systems in ethanol actions in the accumbens, the more recent *in vitro* data may help to explain the negative naloxone data obtained from the *in vivo* studies. First, the excitatory pathways activated (hippocampal-subicular to accumbens) in the *in vivo* studies are clearly quite different from the 'peri-tubercle' input tested *in vitro.* Second, it is obvious from the *in vitro* studies of postsynaptic responses to exogenously applied glutamate agonists that there are sites of ethanol action that clearly do not involve opioid systems. It is possible that the pathways activated in the *in vivo* studies activate one or more of these sites of ethanol action, such as postsynaptic NMDA or kainate receptors.

By virtue of the likely role NAcc plays in alcoholism, these multiple ethanol actions could represent major determinants in ethanol's intoxicating and reinforcing properties. A better perception of the role these sites play in alcoholism will require further studies of these systems in ethanol-dependent or self-administering animals. If such studies bear up the early findings with acutely applied ethanol, they could provide a rational basis for future strategies for the treatment or prevention of alcoholism. For example, combination therapies using both opiate and glutamate (NMDA or kainate) antagonists might be considered. Interestingly, at least two groups (O'Malley et al., 1992; Volpicelli et al., 1992) recently reported clinical success using naltrexone administration combined with traditional therapies for alcoholism.

IV. SOMATOSTATINS AND ETHANOL

A. HIPPOCAMPUS

The hippocampal formation has been a favorite subject for several decades of research on both somatostatin and ethanol, studied separately. In most brain regions, including hippocampus, somatostatin is among the top 2–3 neuropeptides in total abundance (Crawley, 1985). The profuse somatostatin-containing fibers in hippocampus derive from both intrinsic interneurons (e.g., basket-like cells in the CA1 stratum oriens and in the CA3–CA4 hylar region) and from extrinsic afferents (see Bakst et al., 1986; Joëls et al., 1990). These fibers appear to contact hippocampal pyramidal neurons (Joëls et al., 1990) in CA1 and may also innervate interneurons as well. In studies since the early 1980s it has become apparent that ethanol is able to interact with the hippocampal somatostatin system.

1. *In Vivo* Studies

Early studies in our laboratory of rat hippocampus *in vivo*, using extracellular recording and local application of somatostatin from micropipettes by iontophoresis (electroosmosis) or pressure, showed that somatostatin had primarily depressant actions on spontaneous activity of CA1 and CA3 hippocampal pyramidal neurons (Mancillas et al., 1986b). However, despite these direct inhibitory effects, somatostatin application enhanced the excitatory effects of acetylcholine, operating through muscarinic receptors in these CA1 and CA3 neurons (Mancillas et al., 1986b). Interestingly, as described above, *in vivo* studies using equivalent methods had also shown that systemic ethanol could enhance both the inhibitory effects of somatostatin and the excitatory effects of acetylcholine (Mancillas et al., 1986a). Because such *in vivo* studies do not easily reveal the mechanisms of these multiple drug and transmitter interactions, and because there were so many known sites of action of these various players, our laboratory turned to *in vitro* studies of hippocampal slices to pursue these sites and mechanisms.

2. *In Vitro* Studies on Hippocampal Slices

a. *The M-Current*

Our earliest studies on hippocampal slices showed that somatostatin superfusion most often slightly hyperpolarized hippocampal pyramidal neurons, in association with a small increase in input conductance (Pittman and Siggins, 1981). Since the M-current had been shown previously to be a major site of acetylcholine action in hippocampus (I_M is decreased by acetylcholine), we were excited to find, in subsequent voltage-clamp studies in our laboratory, that somatostatin markedly *increased* the I_M relaxation amplitude in most CA1 pyramidal neurons (Moore et al., 1988). Therefore, as noted above (see Opioid Systems and Ethanol) we tested ethanol superfusion on this current in rat CA1 pyramidal neurons *in vitro*, again using voltage-clamp methods, and found ethanol reduction of the I_M amplitude at depolarized membrane potentials (Moore et al., 1990). Surprisingly, at 44 m*M*, ethanol also antagonized augmentation of the I_M by somatostatin. However, 22 m*M* ethanol did not appear to alter the somatostatin augmentation of the M-current; thus somatostatin *in vitro* may provide a 'protective' action against the I_M-diminishing effect of ethanol.

As atropine did not alter the ethanol effect on the I_M (see above), and thus the site of ethanol action is not likely due to alteration of muscarinic receptors, further studies will be needed to verify the M-channel as the common site of interaction between ethanol, somatostatin, and acetylcholine. Ethanol reduction of the M-current, by summation of like effects, may account for the potentiation of acetylcholine responses seen *in vivo* and *in vitro* (see below).

b. Leak K^+ Current

However, the limited expression of I_M at resting membrane potential (RMP) and a recent study from our laboratory on the I_M augmentation by somatostatin (Schweitzer et al., 1993) suggest the involvement of another conductance in the inhibitory effect of somatostatin. This conductance may be the principal mechanism for the somatostatin-induced hyperpolarization at RMP. We (Schweitzer, Madamba, Piomelli, and Siggins, in preparation) are currently characterizing this second effect of somatostatin on CA1 hippocampal pyramidal neurons in the slice preparation. Preliminary results indicate that this current is voltage-independent and does not show rectification. This conductance is likely to be a K^+ leak current ($I_{K(L)}$). Thus, it would appear that SS inhibits HPNs by activating two different K^+ conductances: a voltage-independent $I_{K(L)}$ and the voltage-dependent I_M.

With regard to the possible sites of ethanol interaction, it is of some interest that both of these two conductances are major ones inactivated by muscarinic agonists. Further studies will be required to determine if ethanol in intoxicating concentrations has any effect on the leak K^+ conductance and if this channel is a site responsible for the augmenting effects of ethanol on somatostatin and acetylcholine action.

3. More on Acetylcholine and Muscarinic Receptors

As noted previously, studies from our laboratory showed that:

1. systemic ethanol can enhance excitatory responses of rat hippocampal neurons to iontophoretically applied somatostatin and acetylcholine *in vivo*.
2. ethanol, like acetylcholine, reduces the M-current in the rat hippocampal slice preparation.

Therefore, recently we (Madamba et al., 1995) have used extracellular and intracellular recording techniques in this slice preparation to further examine the mechanisms underlying this ethanol-acetylcholine interaction.

Superfusion or local application (by pipette) of low concentrations of acetylcholine, carbachol (CCh), or muscarine reduced the amplitudes of CAl field potentials evoked by stimulation of the stratum radiatum (SR). As this effect was blocked by 1 μ*M* atropine, muscarinic receptors are likely to be involved. Combined intracellular and extracellular single unit recordings showed that the cholinergic depressions of field potentials were usually correlated with:

1. depolarization of pyramidal neurons.
2. increase in spike firing rate.
3. reduction of the amplitude of postsynaptic potentials.
4. reduction of late afterhyperpolarizations (AHPs) following current-evoked spike trains.

Superfusion of ethanol alone in concentrations (11–22 m*M*) that would yield just-intoxicating blood alcohol levels in humans usually had little effect on the SR evoked field potentials. However, when superfused together with acetylcholine, CCh, or muscarine, these same ethanol concentrations enhanced (by 10–90%) the depressions of evoked field potentials elicited by the cholinergic agonists.

Release of endogenous acetylcholine evoked by trains of stimuli applied to several loci in the hippocampal slice elicits a prolonged depolarization (the slow excitatory postsynaptic potential: sEPSP) in both CA1 and CA3 neurons. Ethanol (22–44 m*M*) superfusion dramatically enhanced both the amplitude and duration of these sEPSPs, recorded intracellularly in CA1 and CA3 neurons. The sEPSPs also were enhanced by eserine and blocked by atropine, verifying that ethanol's interaction is on a muscarinic receptor-mediated effect. These results confirmed earlier reports of presynaptic inhibitory and postsynaptic excitatory actions of acetylcholine in the hippocampus and suggest that ethanol in low concentrations can enhance the effects of endogenous muscarinic synaptic transmission as well as responses to exogenous muscarinic agonists in hippocampus *in vitro.*

However, we were interested in the role somatostatin might play in these acetylcholine and ethanol actions and interactions and how somatostatin might enhance acetylcholine effects. As already noted, in CA1 hippocampal pyramidal neurons somatostatin increases the M-current, whereas CCh decreases it. Our laboratory (Schweitzer et al., 1994) has further investigated the interaction of these two substances on I_M using voltage-clamp recordings in the rat hippocampal slice preparation and a computerized method of I_M quantification. Superfusion of CCh 1 μ*M* largely decreased the M-current in hippocampal pyramidal neurons to 25–45% of control. In attempting to reverse the CCh inhibition of I_M by adding somatostatin 1 μ*M* together with CCh, somatostatin had no effect on I_M, which remained inhibited. Reversing the order of agonist application gave different results: when somatostatin was superfused first, the M-current was increased to 160% of control. Then adding CCh 1 μ*M* to the somatostatin-augmented I_M returned its amplitude back to 85–95% of control. As CCh did not decrease I_M below control in the presence of somatostatin as compared to CCh alone, somatostatin thus provides a *protective effect*. However, such a protective effect was not seen with a high concentration (30 μ*M*) of CCh. Thus, in rat hippocampus CCh prevails over somatostatin for the control of the M-current. However, somatostatin can protect some of the M-channel pool against low (1 μ*M*) but not high (30 μ*M*) concentrations of CCh. Since the CCh concentration and sequence of superfusion determines the end-effect, the interaction between these two agonists could take place at the second messenger level (see Figure 11.2 and below). These complex interactions are interesting with regard to the seemingly parallel effect of somatostatin in 'protecting' the M-current against low but not high ethanol concentrations (see "Somatostatins and Ethanol; *In Vitro* Studies on Hippocampal Slices; the M-Current" above).

4. Second Messenger Considerations

There is considerable information on the second messengers likely to regulate or mediate the responses to activation of muscarinic and somatostatin receptors. Previous data from several laboratories suggest that the reduction of the M-current by muscarinic agonists is mediated by the intracellular generation of some product of phospholipase C, either diacylglycerol in peripheral neurons (Brown et al., 1989) or inositol triphosphate in hippocampal neurons (Dutar and Nicoll, 1988). Data from our laboratory suggests that the mechanism of action of somatostatin in hippocampus occurs through the activation of the phospholipase A_2 (PLA_2) pathway and the release of arachidonic acid and its metabolites. The somatostatin-evoked I_M increase probably is mediated by the 5-lipoxygenase metabolite leukotriene C_4 (Schweitzer et al., 1990; Schweitzer et al., 1993), whereas the voltage-independent 'leak' K^+ current elicited by somatostatin after block of the 5-lipoxygenase pathway probably involves arachidonic acid itself (Schweitzer et al., 1993).

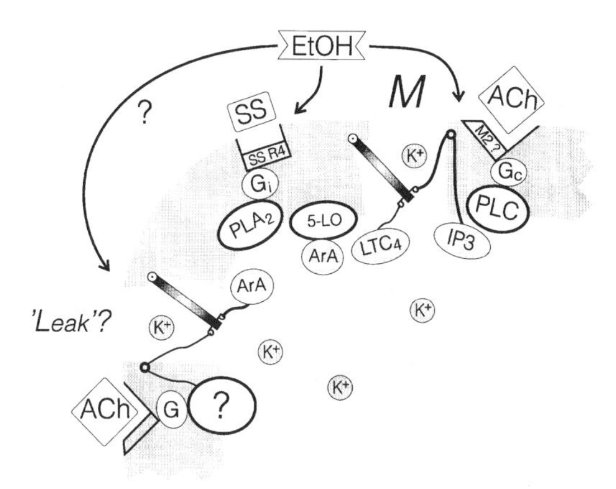

FIGURE 11.2. Schematic of the two K^+ channels acted upon in hippocampal neurons by ethanol (EtOH), somatostatin (SS), and acetylcholine (ACh) and of the postulated receptor-channel linkages governing interactions between these agents. ArA = arachidonic acid; IP3 = inositol triphosphate; PLC = phospholipase C; PLA_2 = phospholipase A_2; 'Leak' = 'leak' channel; M = M-channel; SSR4 = somatostatin receptor; m2 = muscarinic receptor; G = GTP binding protein; ? = unknown; 5-LO = 5-lipoxygenase.

Specifically, our laboratory (Schweitzer et al., 1990; Schweitzer et al., 1993) used voltage-clamp methods in the hippocampal slice preparation to study these signal transduction mechanisms of somatostatin effects. As noted above, somatostatin hyperpolarizes hippocampal pyramidal neurons in part by augmenting the M-current and in part by augmenting a conductance likely to be a 'leak' K^+ current. Both of these somatostatin effects are abolished by the PLA_2 inhibitors 4-bromophenacyl bromide and quinacrine. Furthermore, arachidonic acid mimics both effects of somatostatin on hippocampal pyramidal neurons. The effects of arachidonic acid and somatostatin on I_M are totally blocked by the general lipoxygenase inhibitor nordihydroguaiaretic acid but not by the cyclooxygenase inhibitor indomethacin (Schweitzer et al., 1990; Schweitzer et al., 1993). Prostaglandins E_2, $F_{2\alpha}$, and I_2 did not increase the M-current. The specific 5-lipoxygenase inhibitors 5,6-methanoleukotriene A_4 methylester and 5,6-dehydroarachidonic acid both blocked the I_M-augmenting action of either somatostatin or arachidonic acid. Leukotriene C_4 (but not leukotriene B_4) increased I_M to the same extent as did arachidonic acid and somatostatin. These and other data have been taken to implicate some 5-lipoxygenase metabolite (probably leukotriene C_4) as a mediator for the M-current augmenting effect of somatostatin.

In addition, when the M-current effect is blocked by lipoxygenase inhibitors, both somatostatin and arachidonic acid elicit another outward current likely to be carried by K^+, suggesting a direct role for arachidonic acid itself distinct from the I_M effect. Somatostatin did not alter significantly Ca^{++}-dependent action potentials or inward high threshold Ca^{++} currents, and conversely, the two somatostatin effects were not altered by Ca^{++} channel blockers. The combined results of these studies suggest that somatostatin hyperpolarizes hippocampal neurons by two mechanisms, both mediated through the arachidonic acid system. However, one mechanism (the M-current) involves a metabolite of arachidonic acid and is most effective at slightly depolarized potentials, whereas the other (probably a 'leak' K^+ current) may involve arachidonic acid itself and be more effective at membrane potentials near rest. Figure 11.2 schematizes the complicated interactions of these first and second messenger systems with the relevant ion channels.

The importance of these data on second messenger systems will become more apparent when we are able to express the relevant somatostatin and muscarinic receptors together in *Xenopus* oocytes or cell lines for analysis of the regulatory elements (e.g., post-translational factors) required for their interaction and for sensitivity to ethanol (see below). Unfortunately, the likely involvement of G-proteins or their subunits in these receptors and in ethanol complicates these sorts of molecular studies. For example, a recent study by Sanna et al. (1994) has shown ethanol inhibition of M_1 muscarinic receptors expressed in *Xenopus* oocytes; however, the origin of the G-proteins involved in these muscarinic responses is not clear. They may have been either native to the oocytes or newly synthesized. Whether such an inhibitory ethanol interaction could occur with M_1 receptors coupled to the native transduction machinery of normal brain neurons is unknown.

V. MOLECULAR BIOLOGICAL CONSIDERATIONS

A. G-PROTEIN LINKED RECEPTORS WITH SEVEN MEMBRANE-SPANNING REGIONS

Recent cloning studies within the last three years has verified that transmitter receptors for both the opiate and the somatostatin families of peptides belong to the larger family of receptors likely to consist of seven-putative membrane spanning (hydrophobic) domains, with intracellular loops containing consensus sites for linkage of GTP binding proteins and for phosphorylation of the receptor. Thus, one would expect both the somatostatin and opiate receptors to utilize G-proteins to produce their respective responses (i.e., ion channels changes via direct action of the G-protein or by some other second messenger) and to be regulated in their activity by phosphorylation mechanisms such as provided by the various protein kinases and phosphatases (see Dascal et al., 1993; Evans et al., 1992; Kieffer et al., 1992; Kong et al., 1993). Actions of ethanol on GTP binding proteins have been reported (Candura et al., 1992; Hoek et al., 1987; Hoffman and Tabakoff, 1990; Rubin and Hoek, 1988); thus, the ethanol interactions with responses to opioid and somatostatin peptides could involve an effect of ethanol either on the G-protein itself or on the G-protein linkage site at the receptor (e.g., see Sanna et al., 1994). However, we cannot rule out possible ethanol interactions with phophorylation sites or on the conformation of the receptors or relevant ion channels themselves.

Interestingly, the opiate receptors cloned and sequenced to date fall into the three classical subtypes, μ, δ, and κ (Chen et al., 1993a; Chen et al., 1993b; Evans et al., 1992; Fukuda et al., 1993; Kieffer et al., 1992; Li et al., 1993; Meng et al., 1993; Minami et al., 1993; Nishi et al., 1993; Reisine and Bell, 1993; Thompson et al., 1993; Wang et al., 1993; Yasuda et al., 1993), plus new subtypes recently sequenced that are distinct from, but with relatively high homology to, the classic subtypes (Bunzow et al., 1994; Fukuda et al., 1994; Mollereau et al., 1994). In addition, at least five different somatostatin receptor subtypes (with considerable sequence homologies between them) have been characterized and localized in brain and peripheral tissues (Demchyshyn et al., 1993; Kaupmann et al., 1993; Raynor et al., 1993a; Raynor et al., 1993b; Reisine et al., 1993; Xu et al., 1993; Yamada et al., 1993). It is of relevance to the interaction of ethanol with somatostatin effects that Bito et al. (1994) have recently described a somatostatin receptor highly expressed in rat hippocampus that is capable of generating arachidonic acid. This receptor is likely to be involved in some of the somatostatin actions on the two K^+ channel types described above.

B. HOMOLOGOUS PEPTIDE RECEPTORS

Significant homologies (from 35–60%) exist between the opiate and somatostatin receptors (e.g., see Evans et al., 1992; Yasuda et al., 1993), as well as between the subtypes of each family of receptors. In addition, there are known pharmacological cross reactivities between drugs acting on the opiate and somatostatin receptor types; for example, the potent and selective μ receptor antagonist [Cys2,Tyr3,Orn5,Pen7]somatostatinamide (CTOP) is actually a somatostatin analogue (Piguet and North, 1993). In addition, several other somatostatin agonists have been shown to have antagonist or mixed agonist-antagonist actions on opiate receptors (Betoin et al., 1994; Chiu et al., 1994). For these reasons we find it of some interest that κ opiate receptors and somatostatin receptors both enhance the same K^+ current type (M-current) that is blocked by ethanol in hippocampal pyramidal neurons; indeed, these two peptide types are the only ones known to increase rather than decrease the M-current in neurons. It seems logical, therefore, that these two peptide receptor types would be among the first to show an ethanol interaction; conversely, M-channels, which appear to be directly closed by just-intoxicating ethanol concentrations, may be a major neuronal site of some aspect of ethanol intoxication.

VI. NEUROPEPTIDES FOR FUTURE STUDIES

A. CRF

Because of the clear ethanol-sensitivity of the M-channel, its presence and regulation by peptides might be used to predict other peptide receptors affected by low doses of alcohol. High on such a list would be the receptor for corticotropin releasing factor (CRF); CRF has been found to act much like muscarinic agonists in several neuron types, including hippocampal pyramidal neurons (Aldenhoff, 1983), cerebellar Purkinje cells (Fox and Gruol, 1993), and a pituitary cell line (Mollard et al., 1987). In the first two cell types, CRF (like muscarinic agonists) reduces the AHP following spike trains, and in the cell line it also reduces an M-like current. Such effects could underlie the known interactions of ethanol with CRF-related components of the hypothalamic-pituitary-adrenocortical axis (Baldwin et al., 1991; Ehlers and Chaplin, 1987; Ehlers et al., 1992; Rassnick et al., 1993; Redei et al., 1988; Rivier et al., 1990; Thatcher and Koob, 1986). An interesting related finding is the lower CRF levels and greater CRF responsivity, including CRF-induced EEG activation, in alcohol-preferring rat strains (Ehlers et al., 1992; George et al., 1990).

As far as we are aware, there are no cellular electrophysiological studies of possible interactions between ethanol and CRF receptors. However, there are preliminary reports (Ehlers, 1990; Wall et al., 1989) that systemic ethanol can inhibit the excitatory effects of icv CRF on rat EEG and event-related potentials (e.g., the P300 wave). The implications here are important in the light of behavioral studies showing a similar interaction between ethanol, ethanol withdrawal, and the postulated stress-related role of CRF (Baldwin et al., 1991; Rassnick et al., 1993; Thatcher and Koob, 1986).

B. OTHER PEPTIDES

Angiotensin II is another peptide that acts like muscarinic agonists in reducing the M-current, in sympathetic ganglia neurons (see Brown, 1988). *Bradykinin* also decreases the M-current in sympathetic ganglia neurons and in cell lines, as does *luteinizing hormone releasing hormone* (LHRH). Thus it would be interesting to test the effects of ethanol alone and on these peptide actions in these cell types.

Neurotensin is another peptide worth considering for future electrophysiological study. First, there are significant differences in both neurotensin levels and neurotensin receptors in discrete brain regions of mice selected for their brain sensitivity to ethanol (Erwin and Jones, 1993; Erwin et al., 1993): the long-sleep (LS) and short sleep (SS) mice. Second, neurtotensin, like muscarinic agonists, enhances phophoinositide hydrolysis; interestingly, this effect is more pronounced in LS than SS mice and is enhanced by ethanol (Erwin and Radcliffe, 1993). Furthermore, both acute and chronic ethanol treatments alter neurotensin levels and binding in brain (Campbell and Erwin, 1992; Campbell and Erwin, 1993; Erwin et al., 1992a; Erwin et al., 1992b; Erwin et al., 1990). These studies suggest the possibility of parallels to the opioid, somatostatin, and cholinergic systems in terms of possible ethanol interactions with the ion channels regulated by these systems (see Chapter 10).

VII. CONCLUSIONS

Unlike the transmitter-gated superfamily of receptors (Deitrich et al., 1989; Dildy-Mayfield and Harris, 1992; Harris and Allan, 1989; Koltchine et al., 1993; Li et al., 1993; Lovinger and White, 1991; Lovinger et al., 1989; Masood et al., 1994), the interaction of ethanol with the putative seven-membrane spanning G-protein linked receptors is not so well-understood at the molecular level (but see Sanna et al., 1994). This arises in part because of the complex involvement of separate, multiple transduction molecules — the various G-proteins, second messengers, kinases, phosphatases, and ion channels. It is obvious that, based on the suggestive but indirect physiological and pharmacological evidence obtained to date, more research on these peptide receptors could be fruitful in deciphering loci of potentially important ethanol effects. Especially interesting would be tests of ethanol effects on peptide receptors co-expressed with various neuronal G-proteins and appropriate ion channels in *Xenopus* oocytes or cell lines, first to determine if such isolated receptor-channel transduction systems can be affected by ethanol, and second to allow assessment of the ethanol-sensitive site(s) on the peptide receptor-complex (for example, via site-directed mutagenesis) or ion channel. The ethanol study by Sanna et al. (1994) of 5HT and muscarinic receptors expressed in oocytes might serves as model in this regard. However, isolation and cloning of the pertinent peptide regulated ion channels (e.g., M-channels) and better clarification of the relevant G-proteins and regulatory elements for each peptide receptor will be a prerequisite for a meaningful analysis of such ethanol interactions.

Another area of potentially fruitful research would be the study of models of chronic ethanol (and opiate) treatment. For example, it will be of considerable interest to know if chronic ethanol causes compensatory changes in the peptide receptors or their transduction mechanisms, as might be expected from their alteration by acute ethanol. Also, one would like to know if such changes play a role in ethanol tolerance or dependence, as recently suggested, for example, for NMDA receptors (Chandler et al., 1993; Grant et al., 1990; Khanna et al., 1993; Lijequist, 1991; Wu et al., 1993). If so, drugs interacting with the peptide receptors or their associated transduction systems could provide new therapeutic avenues for the treatment of alcoholism, as exemplified by the successful use of naltrexone in reducing remission in abstaining alcoholics (O'Malley et al., 1992; Volpicelli et al., 1992).

REFERENCES

Aldenhoff, J. B., Gruol, D.L., Rivier, J., Vale, W., and Siggins, G.R.: Corticotropin releasing factor decreases post-burst hyperpolarizations and excites hippocampal pyramidal neurons *in vitro*. *Science,* 221: 875–877, 1983.

Bakst, I., Avendano, C., Morrison, J., and Amaral, D.: An experimental analysis of the origins of somatostatin-like immunoreactivity in the dentate gyrus of the rat. *J. Neurosci.,* 6: 1452–1462, 1986.

Baldwin, H. A., Rassnick, S., Rivier, J., Koob, G. F., and Britton, K. T.: CRF antagonist reverses the "anxiogenic" response to ethanol withdrawal in the rat. *Psychopharmacology,* 103(2): 227–232, 1991.

Berger, T., French, E. D., Siggins, G. R., and Bloom, F. E.: Ethanol and some tetrahydroisoquinolines alter the discharge of rat hippocampal neurons *in vivo* when applied by microelectro-osmosis or pressure: relationship to opiate action. *Pharmacol. Biochem. Behav.,* 17: 813–821, 1982.

Betoin, F., Ardid, D., Herbet, A., Aumaitre, O., Kemeny, J. L., Duchene-Marullaz, P., Lavarenne, J., and Eschalier, A.: Evidence for a central long-lasting antinociceptive effect of vapreotide, an analog of somatostatin, involving an opioidergic mechanism. *J. Pharmacol. Exp. Ther.,* 269(1): 7–14, 1994.

Bito, H., Mori, M., Sakanaka, C., Takano, T., Honda, Z., Gotoh, Y., Nishida, E., and Shimizu, T.: Functional coupling of SSTR4, a major hippocampal somatostatin receptor, to adenylate cyclase inhibition, arachidonate release, and activation of the mitogen-activated protein kinase cascade. *J. Bio. Chem.,* 269: 12722–12730, 1994.

Bloom, F. E. and Siggins, G. R.: Electrophysiological action of ethanol at the cellular level. *Alcohol,* 4: 331–337, 1987.

Bloom, F. E., Siggins, G. R., Foote, S. L., Gruol, D., Aston-Jones, G., Rogers, J., Pittman, Q., and Staunton, D.: Noradrenergic involvement in the cellular actions of ethanol, in *Catecholamines, Neurology and Neurobiology.* (E. Usdin, Ed.), Alan R. Liss, New York, 159–168, 1984.

Brown, D., Marrion, N., and Smart, T.: On the transduction mechanism for muscarinic-induced inhibition of M-current in cultured rat sympathetic neurones. *J. Physiol.,* 413: 469–488, 1989.

Brown, D. A.: M-currents: an update. *Trends Neurosci.,* 11: 294–299, 1988.

Bunzow, J. R., Saez, C., Mortrud, M., Bouvier, C., Williams, J. T., Low, M., and Grandy, D. K.: Molecular cloning and tissue distribution of a putative member of the rat opioid receptor gene family that is not a μ, δ or κ opioid receptor type. *FEBS Lett.,* 347: 284–288, 1994.

Campbell, A. D. and Erwin, V. G.: Changes in neurotensin receptor binding in mice after chronic ethanol consumption. *Ann. N.Y. Acad. Sci.,* 668: 314–6, 1992.

Campbell, A. D. and Erwin, V. G.: Chronic ethanol administration downregulates neurotensin receptors in long- and short-sleep mice. *Pharmacol. Biochem. Behav.,* 45(1): 95–106, 1993.

Candura, S. M., Manzo, L., and Costa, L. G.: Inhibition of muscarinic receptor- and G protein-dependent phosphoinositide metabolism in cerebrocortical membranes from neonatal rat by ethanol. *Neurotoxicology,* 13: 281–288, 1992.

Carlen, P. L., Gurevich, N., and Durand, D.: Ethanol in low doses augments calcium mediated mechanisms measured intracellularly in hippocampal neurons. *Science,* 215: 306–309, 1982.

Chandler, L. J., Newsom, H., Sumners, C., and Crews, F.: Chronic ethanol exposure potentiates NMDA excitotoxicity in cerebral cortical neurons. *J. Neurochem.,* 60(4): 1578–1581, 1993.

Chen, Y., Mestek, A., Liu, J., Hurley, J. A., and Yu, L.: Molecular cloning and functional expression of a mu-opioid receptor from rat brain. *Mol. Pharmacol.,* 44(1): 8–12, 1993a.

Chen, Y., Mestek, A., Liu, J., and Yu, L.: Molecular cloning of a rat kappa opioid receptor reveals sequence similarities to the mu and delta opioid receptors. *Biochem. J.,* 295(3): 625–628, 1993b.

Chiu, T. H., Yeh, M. H., and Chen, M. F.: Actions of a long-acting somatostatin analog SMS201–995 (sandostatin) on rat locus coeruleus neurons. *Life Sci.,* 54(18): 1313–1320, 1994.

Crawley, J.: Comparative distribution of cholecystokinin and other neuropeptides. *Ann. N.Y. Acad. Sci.,* 448: 1–8, 1985.

Criado, J. R., Lee, R.-S., Berg, G., and Henriksen, S. J.: Sensitivity of nucleus accumbens neurons *in vivo* to intoxicating doses of ethanol. *Alcoholism: Clin. Exp. Res.,* 19: 164–169, 1994a.

Criado, J. R., Steffensen, S. C., and Henriksen, S. J.: Ethanol acts via the ventral tegmental area to influence hippocampal physiology. *Synapse,* 17: 84–91, 1994b.

Dascal, N., Schreibmayer, W., Lim, N. F., Wang, W., Chavkin, C., DiMagno, L., Labarca, C., Kieffer, B. L., Gaveriaux, R. C., Trollinger, D., Lester, H. A., and Davidson, N.: Atrial G protein-activated K+ channel: expression cloning and molecular properties. *Proc. Natl. Acad. Sci. U.S.A.,* 90(21): 10235–10239, 1993.

Deitrich, R. A., Dunwiddie, T. V., Harris, R. A., and Erwin, V. G.: Mechanism of action of ethanol: initial central nervous system actions. *Pharmacol. Rev.,* 41: 489–537, 1989.

Demchyshyn, L. L., Srikant, C. B., Sunahara, R. K., Kent, G., Seeman, P., Van, T. H., Panetta, R., Patel, Y. C., and Niznik, H. B.: Cloning and expression of a human somatostatin-14-selective receptor variant (somatostatin receptor 4) located on chromosome 20. *Mol. Pharmacol.,* 43(6): 894–901, 1993.

DiChiara, G. and Imperato, A.: Preferential stimulation of dopamine release in the nucleus accumbens by opiates, alcohol, and barbiturates: studies with transcerebral dialysis in freely moving rats. *Ann. N.Y. Acad. Sci.,* 473: 367–381, 1986.

Dildy-Mayfield, J. E. and Harris, R. A.: Comparison of ethanol sensitivity of rat brain kainate, DL-a-Amino-3-Hydroxy-5-Methyl-4-lsoxalone proprionic acid and N-Methyl-D-Aspartate receptors expressed in Xenopus oocytes. *J. Pharmacol. Exp. Ther.,* 262: 487–493, 1992.

Dutar, P. and Nicoll, R. A.: Classification of muscarinic responses in hippocampus in terms of receptor subtype. *J. Neurosci.,* 8: 4214–4224, 1988.

Ehlers, C., CRF effects on EEG activity: implications for the modulation of normal and abnormal brain states, in *Corticotropin-releasing factor: Basic and clinical studies of a neuropeptide* (E. DeSouza and C. Nemeroff, Eds.), CRC Press, Boca Raton, 234–251, 1990.

Ehlers, C. L. and Chaplin, R. I.: Chronic ethanol exposure potentiates the locomotor-activating effects of corticotropin-releasing factor (CRF) in rats. *Regul. Pept.,* 19(5–6): 345–353, 1987.

Ehlers, C. L., Chaplin, R. I., Wall, T. L., Lumeng, L., Li, T. K., Owens, M. J., and Nemeroff, C. B.: Corticotropin releasing factor (CRF): studies in alcohol preferring and non-preferring rats. *Psychopharmacology,* 106(3): 359–364, 1992.

Erwin, V. G., Campbell, A. D., and Jones, B. C.: Interactions of ethanol with neurotensinergic processes. *Ann. N.Y. Acad. Sci.,* 668: 277–286, 1992a.

Erwin, V. G., Campbell, A. D., and Radcliffe, R.: Effects of chronic ethanol administration on neurotensinergic processes: correlations with tolerance in LS and SS mice. *Ann. N.Y. Acad. Sci.,* 654: 441–443, 1992b.

Erwin, V. G. and Jones, B. C.: Genetic correlations among ethanol-related behaviors and neurotensin receptors in long sleep (LS) × short sleep (SS) recombinant inbred strains of mice. [Review]. *Behav. Genet.,* 23(2): 191–196, 1993.

Erwin, V. G., Jones, B. C., and Radcliffe, R.: Low doses of ethanol reduce neurotensin levels in discrete brain regions from LS/Ibg and SS/Ibg mice. *Alcoholism,* 14(1): 42–47, 1990.

Erwin, V. G., Radcliffe, R., Hinkle, B., and Jones, B. C.: Genetic-based differences in neurotensin levels and receptors in brains of LS × SS mice. *Peptides,* 14(4): 821–828, 1993.

Erwin, V. G. and Radcliffe, R. A.: Characterization of neurotensin-stimulated phosphoinositide hydrolysis in brain regions of long sleep and short sleep mice. *Brain Res.,* 629(1): 59–66, 1993.

Evans, C. J., Keith, D. J., Morrison, H., Magendzo, K., and Edwards, R. H.: Cloning of a delta opioid receptor by functional expression. *Science,* 258(5090): 1952–1955, 1992.

Fox, E. and Gruol, D.: Corticotropin-releasing factor suppresses the afterhyperpolarization in cerebellar purkinje neurons. *NSL,* 149: 103–107, 1993.

Franks, N. and Lieb, W.: Mapping of general anesthetic target sites provides a molecular basis for cut-off effects. *Nature,* 316: 349–351, 1985.

Fukuda, K., Kato, S., Mori, K., Nishi, M., and Takeshima, H.: Primary structures and expression from cDNAs of rat opioid receptor delta- and mu-subtypes. *FEBS Lett.,* 327(3): 311–314, 1993.

Fukuda, K., Kato, S., Mori, K., Nishi, M., Takeshima, H., Iwabe, N., Miyata, T., Houtani, T., and Sugimoto, T.: cDNA cloning and regional distribution of a novel member of the opioid receptor family. *FEBS Lett.,* 343(1): 42–46, 1994.

Gall, C. M., Brecha, H. J., Karten, H., and Chang, K.: Localization of enkephalin-like immunoreactivity to identified axonal and neuronal populations of the rat hippocampus. *J. Comp. Neurol.,* 198: 335–350, 1981.

George, S. R., Fan, T., Roldan, L., and Naranjo, C. A.: Corticotropin-releasing factor is altered in brains of animals with high preference for ethanol. *Alcoholism: Clin. Exp. Res.,* 14(3): 425–429, 1990.

Givens, B. S. and Breese, G. R.: Electrophysiological evidence that ethanol alters function of medial septal area without affecting later septal function. *J. Pharmacol. Exp. Therap.,* 253(1): 95–103, 1990.

Goldstein, D. B., Chin, J. H., and Lyon, R. C.: Disordering of spin-labeled mouse brain membranes. Correlation with genetically determined ethanol sensitivity of mice. *Proc. Natl. Acad. Sci. U.S.A.,* 79: 4231–4233, 1982.

Grant, K. A., Valverius, P., Hudspith, M., and Tabakoff, B.: Ethanol withdrawal seizures and the NMDA receptor complex. *Eur. J. Pharmacol.,* 176: 289–296, 1990.

Gruol, D. L., Chavkin, C., Valentino, R. J., and Siggins, G. R.: Dynorphin-A alters the excitability of pyramidal neurons of the rat hippocampus *in vitro. Life Sci.,* 1: 533–536, 1983.

Harris, A. R. and Allan, A. M.: Alcohol intoxication: ion channels and genetics. *FASEB J.,* 3: 1689–1695, 1989.

Henriksen, S. J., Chouvet, G., and Bloom, F. E.: *In vivo* cellular responses to electrophoretically applied dynorphin in the rat hippocampus. *Life Sci.,* 31: 1785–1788, 1982.

Herkenham, M., Edley, S. M., and Stuart, J.: Cell clusters in the nucleus accumbens of the rat, and the mosaic relationship of opiate receptors, acetylcholinesterase and subcortical afferent terminations. *Neuroscience,* 11: 561–593, 1984.

Hoek, J. B., Thomas, A. P., Rubin, R., and Rubin, E.: Ethanol-induced mobilization of calcium by activation of phosphoinositide specific phopholipase C in intact hepatocytes. *J. Biol. Chem.,* 262: 682–691, 1987.

Hoffman, P. L. and Tabakoff, B.: Ethanol and guanine nucleotide binding proteins: a selective interaction. *FASEB J.,* 4: 2612–2622, 1990.

Iwama, T., Ishihara, K., Satoh, M., and Takagi, H.: Different effects of dynorphin A on *in vitro* guinea pig hippocampal CA3 pyramidal cells with various degrees of paired-pulse facilitation. *Neurosci. Lett.,* 63: 190–194, 1986.

Joëls, M., Madamba, S. G., Moore, S. D., Morrison, J. H., and Siggins, G. R.: Somatostatin immunohistochemistry of hippocampal slices with lucifer yellow stained pyramidal neurons responding to somatostatin. *Reg. Peptides,* 28: 215–221, 1990.

Kaupmann, K., Bruns, C., Hoyer, D., Seuwen, K., and Luebbert, H.: Distribution and second messenger coupling of four somatostatin receptor subtypes expressed in brain. *FEBS Lett.,* 331(1–2): 53–59, 1993.

Khanna, J. M., Shah, G., Weiner, J., Wu, P.H., and Kalant, H.: Effect of NMDA receptor antagonists on rapid tolerance to ethanol. *Eur. J. Pharmacol.,* 230: 23–31, 1993.

Kieffer, B. L., Befort, K., Gaveriaux, R. C., and Hirth, C. G.: The delta-opioid receptor: isolation of a cDNA by expression cloning and pharmacological characterization. *Proc. Natl. Acad. Sci. U.S.A.,* 89(24): 12048–12052, 1992.

Koltchine, V., Anantharam, V., Wilson, A., Bayley, H., and Treistman, S. N.: Homomeric assemblies of NMDAR1 splice variants are sensitive to ethanol. *Neurosci. Lett.,* 152: 13–16, 1993.

Kong, H., Raynor, K., Yasuda, K., Bell, G. I., and Reisine, T.: Mutation of an aspartate at residue 89 in somatostatin receptor subtype 2 prevents Na+ regulation of agonist binding but does not alter receptor-G protein association. *Mol. Pharmacol.,* 44(2): 380–384, 1993.

Koob, G. F. and Bloom, F. E.: Cellular and molecular mechanisms of drug dependence. *Science,* 242: 715–722, 1988.

Lambert, N. A., Harrison, N. L., and Teyler, T. J.: Evidence for μ opiate receptors on inhibitory terminals in area CA1 of rat hippocampus. *Neurosci. Lett.,* 124: 101–104, 1991.

Li, C., Aguayo, L., Peoples, R., and Weight, F.: Ethanol inhibits a neuronal ATP-gated ion channel. *Mol. Pharmacol.,* 44: 871–875, 1993.

Li, C., Peoples, R., and Weight, F.: Alcohol action on a neuronal membrane receptor: evidence for a direct interaction with the receptor protein. *Proc. Natl. Acad. Sci. U.S.A.*, 91: 8200–8204, 1994a.

Li, C., Peoples, R., and Weight, F.: A cutoff in the potency of alcohols for inhibiting a neurotransmitter receptor. *Proc. Natl. Acad. Sci. U.S.A.*, in press, 1994b.

Li, S., Zhu, J., Chen, C., Chen, Y. W., Deriel, J. K., Ashby, B., and Liu-Chen, L.,Y.: Molecular cloning and expression of a rat kappa opioid receptor. *Biochem. J.*, 295(3): 629–633, 1993.

Lijequist, S.: The competitive NMDA receptor antagonist, CGP 39551, inhibits ethanol withdrawal seizures. *Eur. J. Pharmacol.*, 192: 197–198, 1991.

Lovinger, D. M., and White, G.: Ethanol potentiation of 5-HT3 receptor mediated ion current in neuroblastoma cells and adult mammalian neurons. *Mol. Pharmacol.*, 40(2): 263–270, 1991.

Lovinger, D. M., White, G., and Weight, F. F.: Ethanol inhibits NMDA-activated ion currents in hippocampal neurons. *Science,* 243: 1721–1724, 1989.

Madamba, S., Hsu, M., Schweitzer, P., and Siggins, G.R.: Ethanol enhances muscarinic cholinergic neurotransmission in rat hippocampus *in vitro. Brain Res.*, 685: 21–32, 1995.

Madison, D. V. and Nicoll, R. A.: Enkephalin hyperpolarizes interneurones in the rat hippocampus. *J. Physiol.*, 398: 123–130, 1988.

Mancillas, J. R., Siggins, G. R., and Bloom, F. E.: Ethanol selectively enhances responses to acetylcholine and somatostatin in the rat hippocampus. *Science,* 231: 161–163, 1986a.

Mancillas, J. R., Siggins, G. R., and Bloom, F. E.: Somatostatin selectively enhances acetylcholine-induced excitations in rat hippocampus and cortex. *Proc. Natl. Acad. Sci. U.S.A.*, 83: 7518–7521, 1986b.

Mansour, A., Kachaturian, H., Lewis, M. E., Akil, H., and Watson, S. J.: Anatomy of CNS opioid receptors. *TINS,* 11: 308–314, 1988.

Masood, K., Wu, C., Brauneis, U., and Weight, F.: Differential ethanol sensitivity of recombinant N-methyl-D-aspartate receptor subunits. *Mol. Pharmacol.*, 45: 324–329, 1994.

Mayer, J. H., Steffensen, S. C., and Henriksen, S. J.: Opioid effects on cellular excitability in the dentate gyrus. *Neuropharmacology,* 33: 963–975, 1994.

McGinty, J. F., Henriksen, S. J., Goldstein, A., Terenius, L., and Bloom, F. E.: Dynorphin is contained within hippocampal mossy fibers: immunochemical alterations after kainic acid administration and colchicine-induced neurotoxicity. *Proc. Natl. Acad. Sci. U.S.A.*, 80: 589–593, 1983.

Meng, F., Xie, G. X., Thompson, R. C., Mansour, A., Goldstein, A., Watson, S. J., and Akil, H.: Cloning and pharmacological characterization of a rat kappa opioid receptor. *Proc. Natl. Acad. Sci. U.S.A.*, 90(21): 9954–9958, 1993.

Minami, M., Toya, T., Katao, Y., Maekawa, K., Nakamura, S., Onogi, T., Kaneko, S., and Satoh, M.: Cloning and expression of a cDNA for the rat kappa-opioid receptor. *FEBS Lett.*, 329(3): 291–295, 1993.

Mogenson, G. J.: Limbic-motor integration. *Prog. Physiol. Psychol.*, 12: 117–170, 1987.

Mollard, P., Vacher, P., Guerin, J., Rogawski, M., and Dufy, B.: Electrical properties of cultured human adrenocortropin-secreting adenoma cells: effects of high K+ corticotropin-releasing factor, and angiotensin II. *Endocrinology,* 121: 395, 1987.

Mollereau, C., Parmentier, M., Mailleux, P., Butour, J. L., Moisand, C., Chalon, P., Caput, D., Vassart, G., and Meunier, J. C.: ORL1, a novel member of the opioid receptor family. *FEBS Lett.*, 341(1): 33–38, 1994.

Moore, S. D., Madamba, S. G., Joels, M., and Siggins, G. R.: Somatostatin augments the M-current in hippocampal neurons. *Science,* 239: 278–280, 1988.

Moore, S. D., Madamba, S. G., and Siggins, G. R.: Ethanol diminishes a voltage-dependent K^+ current, the M-current, in CA1 hippocampal pyramidal neurons *in vitro. Brain Res.*, 516: 222–228, 1990.

Moore, S. D., Madamba, S. G., Schweitzer, P., and Siggins, G. R.: Voltage-dependent effects of opioid peptides on hippocampal CA3 pyramidal neurons *in vitro. J. Neurosci.*, 14(2): 809–820, 1994.

Nicoll, R. A., Alger, B. E., and Yahr, C. E.: Enkephalin blocks inhibitory pathways in the vertebrate CNS. *Nature,* 287: 22–25, 1980.

Nie, Z., Madamba, S. G., and Siggins, G. R.: Ethanol inhibits glutamatergic neurotransmission in nucleus accumbens neurons by multiple mechanisms. *J. Pharmacol. Exp. Therap.*, 271: 1566–1573, 1994.

Nie, Z., Yuan, X., Madamba, S. G., and Siggins, G. R.: Ethanol decreases glutamatergic synaptic transmission in rat nucleus accumbens *in vitro*: naloxone reversal. *J. Pharmacol. Exp. Ther.*, 266(3): 1705–1712, 1993.

Nishi, M., Takeshima, H., Fukuda, K., Kato, S., and Mori, K.: cDNA cloning and pharmacological characterization of an opioid receptor with high affinities for kappa-subtype-selective ligands. *FEBS Lett.*, 330(1): 77–80, 1993.

O'Malley, S. S., Jaffe, A., Chang, G., Schottenfeld, R. S., Meyer, R. E., and Rounsaville, B.: Naltrexone and coping skills therapy for alcohol dependence: a controlled study. *Arch. Gen. Psychiatry,* 49: 881–887, 1992.

Piguet, P. and North, R. A.: Opioid actions at mu and delta receptors in the rat dentate gyrus *in vitro. J. Pharmacol. Exp. Ther.*, 266(2): 1139–1146, 1993.

Pittman, Q. J. and Siggins, G. R.: Somatostatin hyperpolarizes hippocampal pyramidal cells *in vitro. Brain Res.*, 221: 402–408, 1981.

Quirion, R., Zajac, J. M., Morgat, J. L., and Roques, B. P.: Autoradiographic distribution of mu and delta opiate receptors in rat brain using highly selective ligands. *Life Sci.*, 33 Suppl. 1: 227–230, 1983.

Rassnick, S., Heinrichs, S. C., Britton, K. T., and Koob, G. F.: Microinjection of a corticotropin-releasing factor antagonist into the central nucleus of the amygdala reverses anxiogenic-like effects of ethanol withdrawal. *Brain Res.*, 605(1): 25–32, 1993.

Raynor, K., Murphy, W. A., Coy, D. H., Taylor, J. E., Moreau, J. P., Yasuda, K., Bell, G. I., and Reisine, T.: Cloned somatostatin receptors: identification of subtype-selective peptides and demonstration of high affinity binding of linear peptides. *Mol. Pharmacol.*, 43(6): 838–844, 1993a.

Raynor, K., O'Carroll, A. M., Kong, H., Yasuda, K., Mahan, L. C., Bell, G. I., and Reisine, T.: Characterization of cloned somatostatin receptors SSTR4 and SSTR5. *Mol. Pharmacol.*, 44(2): 385–392, 1993b.

Redei, E., Branch, B. J., Gholami, S., Lin, E. Y., and Taylor, A. N.: Effects of ethanol on CRF release *in vitro*. *Endocrinology*, 123(6): 2736–2743, 1988.

Reisine, T. and Bell, G. I.: Molecular biology of opioid receptors. *Trends Neurosci.*, 16(12): 506–510, 1993.

Reisine, T., Kong, H., Raynor, K., Yano, H., Takeda, J., Yasuda, K., and Bell, G. I.: Splice variant of the somatostatin receptor 2 subtype, somatostatin receptor 2B, couples to adenylyl cyclase. *Mol. Pharmacol.*, 44(5): 1016–1020, 1993.

Rivier, C., Imaki, T., and Vale, W.: Prolonged exposure to alcohol: effect on CRF mRNA levels, and CRF- and stress-induced ACTH secretion in the rat. *Brain Res.*, 520(1–2): 1–5, 1990.

Rubin, R. and Hoek, J. B.: Alcohol-induced stimulation of phospholipase C in human platelets requires G protein activation. *Biochem. J.*, 254: 147–153, 1988.

Sanna, E., Dildy-Mayfield, J. E., and Harris, R. A.: Ethanol inhibits the function of 5-hydroxytryptamine type 1c and muscarinic M_1 G protein-linked receptors in *Xenopus* oocytes expressing brain mRNA: role of protein kinase C. *Mol. Pharmacol.*, 45: 1004–1012, 1994.

Schweitzer, P., Madamba, M., and Siggins, G. R.: Arachidonic acid metabolites as mediators of somatostatin-induced increase of neuronal M-current. *Nature*, 346: 464–467, 1990.

Schweitzer, P., Madamba, S., Champagnat, J., and Siggins, G. R.: Somatostatin inhibition of hippocampal CA1 pyramidal neurons: mediation by arachidonic acid and its metabolites. *J. Neurosci.*, 13: 2033–2049, 1993.

Schweitzer, P., Madamba, S., and Siggins, G. R.: Interaction of somatostatin and carbachol on hippocampal M-currents. *Soc. Neurosci. Abstr.*, 20: 1555, 1994.

Siggins, G. R., Berger, T., French, E. D., Shier, T., and Bloom, F. E.: Ethanol, salsolinol and tetrahydropapaveroline alter the discharge of neurons in several brain regions: comparisons to opioid effects, in *Beta Carbolines and Tetrahydroisoquinolines* (F. E. Bloom, J. Barchus, M. Sandler, and E. Usdin, Eds.), Alan R. Liss, New York, 275–288, 1982.

Siggins, G. R. and Bloom, F. E.: Alcohol-related electrophysiology. *Pharmacol. Biochem. Behav.*, 13: 203–211, 1981.

Siggins, G. R., Bloom, F. E., French, E. D., Madamba, S. G., Mancillas, J. R., Pittman, Q. J., and Rogers, J.: Electrophysiology of ethanol on central neurons. *Ann. N.Y. Acad. Sci.*, 492: 350–366, 1987.

Siggins, G. R. and Gruol, D. L.: Mechanisms of transmitter action in the vertebrate central nervous system, in *Handbook of Physiology. The Nervous System IV* (F. E. Bloom, Ed.), The American Physiological Society, Bethesda, MD, 1–114, 1986.

Siggins, G. R., Pittman, Q., and French, E.: Effects of ethanol on CA1 and CA3 pyramidal cells in the hippocampal slice preparation: an intracellular study. *Brain Res.*, 414: 22–34, 1987.

Siggins, G. R. and Zieglgänsberger, W.: Morphine and opioid peptides reduced inhibitory synaptic potentials in hippocampal pyramidal cells *in vitro* without alteration of membrane potential. *Proc. Natl. Acad. Sci. U.S.A.*, 78: 5235–5239, 1981.

Siggins, G. R. and Zieglgänsberger, W.: How neuropeptides alter neuronal excitability, In *Proceedings 16th CINP Congress Neuropsychopharmacology* (W. E. J. Bunney, H. Hippius, G. Laackmann, and M. Schmauss, Eds.), Springer-Verlag, Berlin, 488–505, 1990.

Steffensen, S. C., Yeckel, M. F., Miller, D. R., and Henriksen, S. J.: Ethanol-induced suppression of hippocampal long-term potentiation is blocked by lesions of the septohippocampal nucleus. *Alcoholism: Clin. Exp. Res.*, 17(3): 655–659, 1993.

Stubbs, C. and Rubin, E.: Molecular mechanisms of ethanol and anesthetic actions: lipid- and protein-based theories, in *Alcohol, Cell Membranes, and Signal Transduction in Brain* (C. Alling et al., Eds.), Plenum Press, New York, 1–11, 1993.

Terman, G. W., Wagner, J. J., and Chavkin, C.: Kappa opioids inhibit induction of long-term potentiation in the dentate gyrus of the guinea pig hippocampus. *J. Neurosci.*, 14(8): 4740–4747, 1994.

Thatcher, B. K. and Koob, G. F.: Alcohol reverses the proconflict effect of corticotropin-releasing factor. *Regul. Pept.*, 16(3–4): 315–320, 1986.

Thompson, R. C., Mansour, A., Akil, H., and Watson, S. J.: Cloning and pharmacological characterization of a rat Mu opioid receptor. *Neuron*, 11(5): 903–913, 1993.

Vaccarino, F. J., Bloom, F. E., and Koob, G. F.: Blockade of nucleus accumbens opiate receptors attenuates the intravenous heroin reward in the rat. *Psychopharmacology*, 86: 37–42, 1985.

Volpicelli, J. R., Alterman, A. I., Hayashida, M., and O'Brien, C. P.: Naltrexone in the treatment of alcohol dependence. *Arch. Gen. Psychiatry*, 49(Nov.): 876–880, 1992.

Wall, T., Chaplin, R., and Ehlers, C.: Does CRF contribute to the cognitive changes seen in depression. *Soc. Neurosci. Abstr.,* 15: 1069, 1989.

Wamsley, J. K., Young, W. S., and Kuhar, M. J.: Immunohistochemical localization of enkephalin in rat forebrain. *Brain Res.,* 190: 153–174, 1980.

Wang, J. B., Imai, Y., Eppler, C. M., Gregor, P., Spivak, C. E., and Uhl, G. R.: Mu opiate receptor: cDNA cloning and expression. *Proc. Natl. Acad. Sci. U.S.A.,* 90(21): 10230–10234, 1993.

Wiesner, J. B. and Henriksen, S. J.: Ethanol enhances recurrent inhibition in the dentate gyrus of the hippocampus. *Neurosci. Lett.,* 79: 169–173, 1987.

Wu, P. H., Mihic, S. J., Liu, J., Le, A. D., and Kalant, H.: Blockade of chronic tolerance to ethanol by the NMDA antagonist,(+)-MK-801. *Eur. J. Pharmacol.,* 231: 157–164, 1993.

Xu, Y., Song, J., Bruno, J. F., and Berelowitz, M.: Molecular cloning and sequencing of a human somatostatin receptor, hSSTR4. *Biochem. Biophys. Res. Commun.,* 193(2): 648–652, 1993.

Yamada, Y., Kagimoto, S., Kubota, A., Yasuda, K., Masuda, K., Someya, Y., Ihara, Y., Li, Q., Imura, H., Seino, S., and Seino, Y.: Cloning, functional expression and pharmacological characterization of a fourth (hSSTR4) and a fifth (hSSTR5) human somatostatin receptor subtype. *Biochem. Biophys. Res. Commun.,* 195(2): 844–852, 1993.

Yasuda, K., Raynor, K., Kong, H., Breder, C. D., Takeda, J., Reisine, T., and Bell, G. I.: Cloning and functional comparison of kappa and Delta opioid receptors from mouse brain. *Proc. Natl. Acad. Sci. U.S.A.,* 90(14): 6736–6740, 1993.

Yuan, X., Madamba, S. G., and Siggins, G. R.: Opioid peptides reduce synaptic transmission in the nucleus accumbens. *Neurosci. Lett.,* 134: 223–228, 1991.

Zieglgänsberger, W., French, E. D., Siggins, G. R., and Bloom, F. E.: Opioid peptides may excite hippocampal pyramidal neurons by inhibiting adjacent inhibitory interneurons. *Science,* 205: 415–417, 1979.

Chapter 12

THE INTERACTIONS OF THYROTROPIN-RELEASING HORMONE WITH ETHANOL

Thomas J. McCown and George R. Breese

TABLE OF CONTENTS

I. INTRODUCTION

Thyrotropin-releasing hormone (TRH) is a tripeptide (L-pyroglutamyl-L-histidyl-L-prolineamide) that was discovered in the hypothalamus by two groups of investigators, one led by Schally (Boler et al., 1969) and the other by Guillemin (Burgus et al., 1970). The intense search to identify this hypothalamic peptide arose from the widely held belief that "hypothalamic factors" regulated the release of hormones from the anterior pituitary. Identification of TRH not only validated this widely held belief, but created the field of neuroendocrinology. As the scope of studies on TRH expanded, it became clear that TRH also had actions upon the central nervous system (CNS) that were distinct from its neuroendocrine role (Plotnikoff et al., 1972). Now it is widely accepted that numerous actions of TRH are mediated in specific brain areas distinct from the hypothalamus. Studies on TRH's interactions with sedatives, especially pentobarbital and ethanol, proved to be a major contribution to this advance.

II. CNS EFFECTS AND LOCALIZATION OF TRH

In normal rats, the administration of TRH causes an increase in respiratory rate, an increase in locomotor behaviors, the appearance of "wet dog shakes," piloerection, tremor, and a decrease in both food and water ingestion (Myers, et al., 1977; Vogel et al., 1979). Certainly, TRH administration has neuroendocrine consequences, but many of these behaviors depend upon site-specific actions within the CNS. For example, TRH changes respiratory function when microinjected into the nucleus tractus solitarius, raphe obscurus, or interpeduncular nucleus (McCown et al., 1986a; Hedner et al., 1987), but TRH microinjection into the nucleus accumbens increases locomotor activity (Miyamoto et al., 1979) (see Table 12.1). In contrast to these stimulant actions, TRH also exerts effects upon the CNS indicative of anxiolytic, sedative agents, such as the benzodiazepines. Using a behavioral screen for anxiolytic activity, Vogel et al. (1980) found that, like chlordiazepoxide, TRH increased punished operant behavior without changing unpunished responding. Similar results have been obtained in other species and with different conflict tasks (Barrett and

0-8493-8389-7/95/$0.00+$.50

TABLE 12.1

TRH Action	Brain Site	Dose
Sedative antagonism	medial septum	0.03 μg[1]
	lateral posterior hypothalamus	2.0 μg[2]
Locomotor activity increases	nucleus accumbens	1 μg[3]
	lateral ventricle	1 μg[4]
Analgesia	periaquiductal grey	10 μg[5]
Respiratory changes		
rate increases	interpeduncular nucleus	0.1 μg[6]
inspiratory changes	raphe obscurus/nucleus tractus solitarius	0.1 μg[7]

[1] McCown et al., 1986a; [2] Yamamura et al., 1991, [3] Miyamoto et al., 1979; [4] Katsuura et al., 1984; [5] Webster et al., 1984; [6] Hedner et al., 1987; [7] McCown et al., 1986.

Brady, 1985; Criswell and Breese, 1989). Also like benzodiazepines, TRH administration has been shown to attenuate seizure activity, both in animals and humans (Minabe et al., 1987; Matsumoto et al., 1987; Sakai et al., 1991; Renming et al., 1992), and to potentiate phenobarbital attenuation of maximal electroshock seizures (Nemeroff et al., 1975). These studies illustrate that TRH exhibits site specific actions within the CNS and that the nature of TRH action on the CNS is mixed.

One source of these diverse TRH actions derives from a wide distribution of TRH receptors and TRH-containing cell bodies throughout the brain. Using immunohistochemical techniques, TRH-like immunoreactivity has been found in the hypothalamus, olfactory bulbs, nucleus accumbens, caudate nucleus, piriform and entorhinal cortices, hippocampus, amygdala, periaqueductal central grey, raphe nuclei, and dorsal horn of the spinal cord (Brownstein et al., 1974; Winokur and Utiger, 1974). Even though this list is only partial (see Hökfelt et al., 1989 for review), it illustrates TRH localization from the forebrain to the brainstem. Likewise, TRH receptor localization extends throughout the brain. For example, the olfactory bulbs, nucleus accumbens, amygdala, medial septum, hippocampus, entorhinal cortex, interpeduncular nucleus, and spinal trigeminal nucleus all contain TRH receptors (Taylor and Burt, 1982; Mantyh and Hunt, 1985; Sharif, 1989). In the face of this diverse distribution, an additional complexity exists. In some brain areas, TRH is co-localized with classical neurotransmitters, such as serotonin, and other neuropeptides, such as substance P (Johansson et al., 1981). Given this wide sphere of influence, it is not surprising that TRH alters a number of different behaviors.

Finally, studies on TRH's molecular mechanisms further illustrate the diversity of TRH's actions on the CNS. In the brain, the TRH receptor is G-protein linked and exhibits a great degree of homology across different species (Martin et al., 1986; Yamada et al., 1993). Activation of the TRH receptor stimulates polyphosphoinositide hydrolysis, a property common to many agonists that mobilize calcium (Imai and Gershengorn, 1985). This phosphoinositol hydrolysis does depend upon GTP, but not a cholera or pertussis toxin sensitive G-proteins (Martin et al., 1986). Presumably, this TRH second-messenger induction leads to the subsequent alterations in calcium flux both into and within the cell (Gershengorn and Thaw, 1985). As a result of these calcium fluxes, changes occur in both potassium flux across the neural membrane (Toledo-Aral et al., 1993) and sodium channel properties (Lopez-Barneo et al., 1990). However, recent studies have identified two isoforms of the TRH receptor, a long and a short (de la Pena et al., 1992). The long isoform appears to be linked to changes in inositol phosphates, but not TRH-induced changes in adenyl cyclase activity (Kim et al., 1994). These studies support the conclusion that two distinct TRH receptors exist in the brain. Clearly, TRH receptor activation produces a wide range of cellular changes that could account for the diversity of TRH action in the CNS.

III. TRH INTERACTIONS WITH ETHANOL-INDUCED SEDATION

Up to this point, the one action of TRH not discussed proves to be the most dramatic one: centrally administered TRH can antagonize sedation induced by a wide range of anesthetic agents. Initially, Prange et al. (1974) found that TRH would reverse pentobarbital-induced sedation, while Breese et al. (1974) found that TRH dramatically shortened ethanol-induced sleep time. These investigators also reported that TRH antagonized the hypothermia produced by pentobarbital or ethanol. A key component to these early studies was the demonstration that TRH's analeptic action was independent of its neuroendocrine effects or any pharmacokinetic interaction (Breese et al., 1975). TRH antagonized drug-induced sedation through a direct action on the brain. Subsequently, this analeptic action of TRH has been demonstrated not only in rodents, but in rabbits, dogs, and monkeys (Kraemer et al., 1976; Horita et al., 1976; Hernandez et al., 1987). However, unlike other ethanol antagonists (Suzdak et al., 1986; Lister and Nutt, 1987), TRH did not increase seizure susceptibility, even though it antagonized ethanol's anticonvulsant actions (Breese, et al., 1974; McCown and Breese, 1989). Thus, TRH effectively antagonized ethanol-induced sedation, hypothermia, and anticonvulsant actions, but did not cause the hyperexcitability that is characteristic of most sedative antagonists. Given this unique profile of action on the CNS, two obvious questions were these: where and how did TRH exert this antagonism within the CNS?

The analeptic action of TRH proved to be highly localized within the brain, whereas other actions were more diffuse in localization. Using CNS site microinjections, Kalivas and Horita (1980) found that the medial septum was exquisitely sensitive to TRH's analeptic actions on pentobarbital. Other areas of the brain did respond to TRH's analeptic effect, but these areas were at least fivefold less sensitive compared to the medial septum. For example, the TRH ED_{50} was 0.095 n*M* for the medial septum, whereas the next most sensitive area to TRH analepsis was the interpeduncular nucleus with an ED_{50} of 0.509 n*M* (Kalivas and Horita, 1980). In contrast, the ability of TRH to antagonize hypothermia or cause "wet dog shakes" was not nearly as well localized. A number of brain areas were capable of mediating these TRH actions with the same potency (Kalivas and Horita, 1980). More recently, other investigators found that TRH and a TRH analogue, TA-0910, antagonized pentobarbital sedation, but not in the medial septum (Yamamura et al., 1991). In these studies, the posterior hypothalamus and midbrain reticular nucleus were most sensitive to this action of TRH. Although these two studies have a number of methodological differences, at present, there is no obvious explanation for the conflicting results on the medial septum. However, when the analeptic actions of TRH were evaluated for ethanol-induced sedation, the medial septum proved to be a central site of action. In rats, septal microinjection of TRH reversed the sedative actions of both low and high doses of ethanol (Breese, et al., 1984; McCown et al., 1986b). This septal TRH effect depended upon specific interactions of the peptide, because co-infusion of TRH antibody fragments prevented TRH's analeptic action. Furthermore, the net effect of TRH on septal neurons appeared to be an increase in neural activity, because electrical stimulation of the medial septum also antagonized ethanol-induced sedation (McCown et al., 1986b). In contrast, microinjection of TRH into brain areas, where TRH alters locomotor activity or respiratory function, had no effect on ethanol-induced sedation (McCown et al., 1986b). These studies proved that TRH did antagonize ethanol-induced sedation from a specific brain area. Table 12.1 summarizes some site-specific actions of TRH and the relative potency of these actions.

If TRH administration antagonizes most sedatives, then one might expect sedatives to alter endogenous TRH function. When this question was considered, it was found that sedatives do alter TRH function within the brain. For example, Lighton et al. (1986) treated rats with a TRH antibody fragment over 14 days and then evaluated the effects of a TRH analogue on pentobarbital sleep time. Intraseptal administration of this TRH analogue proved somewhat more potent in antagonizing pentobarbital sedation. More impressively, the TRH antibody infusion alone caused the animals to sleep 2.5 times longer than the controls. Since chronic impairment of TRH function changed the course of sedation, it seemed probable that TRH activity in the CNS maintained an appropriate state of alertness. Similar conclusions can be drawn from studies with ethanol. During ethanol-induced sedation, the concentrations of septal TRH are not different from those of alert untreated controls. However, at the time that these animals regain their righting reflex, medial septal TRH concentrations are significantly elevated (Morzorati and Kubek, 1993). This association of increased endogenous TRH with a change in the level of sedation shows that ethanol administration alters endogenous TRH function.

IV. POTENTIAL MECHANISMS OF TRH ANALEPSIS

Over the years, many studies have attempted to delineate the neurochemical basis for TRH's analeptic action, but to date, definitive answers are lacking. Numerous TRH-neurotransmitter interactions have been demonstrated, but in many cases, the functional significance is not readily apparent. For example, when applied to the same hypothalamic or septal neuron, TRH or norepinephrine produced similar electrophysiological consequences (Winokur and Beckman, 1978). Also, using long-sleep (LS) and short-sleep (SS) mice, French et al. (1993) found that TRH reduced ethanol sleep time and increased tyrosine hydroxylase activity in the LS mice while having neither of these actions in SS mice. Certainly, these studies implicate catecholamine mechanisms in TRH's analeptic actions; yet in another study (Cott et al., 1976), whole-brain catecholamine reduction with α-methyltyrosine did not alter TRH's analeptic action. A more direct association has been established between TRH and acetylcholine. Using mice, early studies reported that muscarininc cholinergic blockade prevented the analeptic action of TRH (Breese, et al., 1975; Cott et al., 1976). This TRH-cholinergic link was further elaborated using a number of *in vivo* and *in vitro* techniques (see Yarbrough, 1983 for review). However, there was one complication. In rats, peripheral atropine or scopolamine administration did not alter TRH's analeptic action (Santori et al., 1981; Miyamoto et al., 1982). From another perspective, there is a major cholinergic projection from the septum to the hippocampus, and TRH alters brain activity localized to the hippocampus (Kalivas et al., 1980). It seemed reasonable that this pathway mediated the analeptic actions, and in support of this hypothesis, lesions of the major septohippocampal projection did abolish the analeptic effects of TRH (Kalivas et al., 1981). However, lesions of the dorsal hippocampus did not alter TRH analepsis (Kalivas et al., 1981). Thus, one origin of TRH analepsis is the medial septum, but where this septal information projects is not presently known. These investigations of cholinergic links to TRH analepsis illustrate an important point. There is no doubt that TRH antagonizes sedative actions across a number of species, and there is good evidence that TRH interacts with cholinergic mechanisms in the brain. However, there appears to be species differences in the importance of cholinergic function to TRH analepsis.

Another proposed mechanism of TRH action is the GABA/benzodiazepine receptor site (Cott and Engel, 1977). For example, the GABA antagonist, bicuculline, or the chloride channel antagonist, RO 15–4513, both can reverse ethanol-induced sedation (Breese et al., 1984; Suzdak et al., 1986), and benzodiazepines can displace TRH receptor binding (Rinehart et al., 1986). As attractive as this association might be, other findings temper support for a TRH-GABA association. First, direct antagonism of GABA receptors, or the associated chloride channel, creates a proconvulsant state that can progress to seizure activity. TRH exhibits no such action. Second, like TRH, medial septal bicuculline reverses ethanol impairment, but unlike TRH, bicuculline causes a dramatic increase in body temperature (Breese et al., 1984). Thus, in the medial septum, there are clear differences between the effects of TRH and GABA antagonism. Finally, if the interaction were through the benzodiazepine modulatory site of the GABA-chloride channel complex, then benzodiazepine antagonists should antagonize ethanol-induced sedation. They do not exert such an action. Thus, comparisons of GABA antagonists and TRH show some similarities, but also marked differences. How TRH antagonizes ethanol's sedative effects remains to be determined.

V. TRH INTERACTIONS WITH ETHANOL PREFERENCE AND CONSUMPTION IN THE RAT

Another area of TRH/ethanol interactions involves TRH contributions to ethanol preference and consumption in the rat. It has been demonstrated that ethanol exposure alters the hypothalamic-pituitary-thyroid (HPT) axis (Loosen et al., 1979; Singh et al., 1979), but the peripheral components of this axis appear unrelated to ethanol preference. In rats, neither thyroidectomy nor T4 augmentation changed ethanol preference (Noonan et al., 1993). In contrast, when studies focused upon TRH, the results proved quite different. Rezvani et al. (1992) found that a TRH analogue, TA-0910, significantly reduced ethanol intake and increased water intake in an ethanol preferring strain of rats (P rats). This TRH analogue did not alter caloric intake or ethanol pharmacokinetics, so it appears that the preference interaction was mediated in the CNS. In a subsequent set of investigations, Mason et al. (1994) found that after as few as three TA-0910 treatments, tolerance developed to the ethanol preference effects of TA-0910. Also, these authors reported that cross tolerance developed to the ethanol preference effects of the dopamine agonist, bromocriptine. This development of cross tolerance between TA-0910 and bromocriptine implicates dopaminergic mechanisms in the action of the TRH analogue. However, it must be emphasized that a number of neurotransmitter systems have been associated with ethanol preference in rats, and TRH clearly influences these neurotransmitter systems, particularly serotonin and enkephalins/endorphins. Therefore, there could be a number of mechanisms by which TRH modulates ethanol preference in rats.

TABLE 12.2

TRH/Ethanol Interactions in Rats

1. Antagonizes high-dose sedation
2. Antagonizes low-dose coordination impairment
3. Antagonizes ethanol's anticonvulsant effects
4. Reduces ethanol preference

VI. CLINICAL APPLICATION OF BASIC RESEARCH FINDINGS

The basic research findings provide an attractive basis for TRH application in humans, but unfortunately, clinical studies have produced a number of conflicting results. For example, Linnoila et al. (1981) reported that in humans a 10 μg i.v. dose of TRH actually worsened measures of inebriation after a 1.5 g/kg dose of ethanol. In contrast, Knutsen et al.(1989) administered five oral doses of TRH (20 mg/kg each) to human subjects over a 30-h period prior to the administration of a 1.5 g/kg dose of ethanol. This TRH pretreatment significantly improved the reaction test, as well as tests of inebriation. Thus, using a different route of administration and a higher dose of TRH, these investigators found that TRH antagonized some of ethanol's sedative actions. However, more recently, Garbutt et al. (1991) found that a 500 μg i.v. dose of TRH did not alter impairments caused by a moderate dose of ethanol (0.8 g/kg). When the dose of TRH was increased, some subjects showed a reduction in ethanol impairments, but many subjects did not (Garbutt, personal communication). At the doses used above, Garbutt et al. (1989) have reported that TRH significantly reduced benzodiazepine impairments, so clearly, the doses of TRH used in these studies have effects upon the central nervous system. Two explanations could account for these mixed clinical reports. First, the dose of TRH may need to be increased further in order to obtain the analeptic action. Second, there may be a species difference, such that the analeptic action of TRH is far less effective in humans as compared to rats. Clearly, further research will be needed to resolve these issues

VII. CONCLUSIONS

From the basic research studies, it is clear that TRH potently antagonizes ethanol-induced depression of the CNS. In addition, a TRH analogue decreases ethanol preference in an ethanol-preferring strain of rat. Some of the TRH-ethanol interactions have been localized to specific brain areas, particularly the medial septum. However, translation of the analeptic actions to the clinical setting has not been straightforward. Also, the rapid development of tolerance to effects of a TRH analogue on ethanol preference could limit the usefulness of this drug for long-term administration. In spite of these limitations, TRH-ethanol interactions still provide unique opportunities to unravel the actions of ethanol upon the central nervous system.

REFERENCES

Barrett, J.E. and Brady, L.S., Neuropeptide modulation of the behavioral effects of drugs. *Prog. Clin. Biol. Res.,* 192:37–50, 1985.

Boler J., Enzmann, F., Folkers, K., Bowers, C.Y., and Schally, A.V., The identity of chemical and hormone properties of the thyrotropin releasing hormone and pyroglutamyl-histidyl-prolineamide. *Biochem. Biophys. Commun.,* 37:505–510, 1969.

Breese, G.R., Cott, J.M., Cooper, B.R., Prange, A.J., Jr., and Lipton, M.A., Antagonism of ethanol narcosis by thyrotropin-releasing hormone. *Life Sci.,* 14:1053–1063, 1974.

Breese, G.R., Cott, J.M., Cooper, B.R., Prange, A.J., Jr., Lipton, M.A., and Plotnikoff, N.P., Effects of thyrotropin-releasing hormone (TRH) on the actions of pentobarbital and other centrally acting drugs. *J. Pharmacol. Exp. Ther.,* 193:11–22, 1975.

Breese, G.R., Frye, G.D., McCown, T.J., and Mueller, R.A., Comparison of the CNS effects induced by TRH and bicuculline after microinjection into medial septum, substantia nigra and inferior colliculus: absence of support for a GABA antagonist action for TRH. *Pharmacol. Biochem. Behav.,* 21:145–149, 1984.

Brownstein, M.J., Palkovits, M., Saavedra, J.M., Bassiri, R.M., and Utiger, R.D., Thyrotropin-releasing hormone in specific nuclei of rat brain. *Science,* 185:267–269, 1974.

Burgus, R., Dunn, T.F., Desiderio, D.M., Ward, D.N., Vale, W., and Guillemin, R., Biological activity of synthetic polypeptide derivatives related to the structure of hypothalamic TRF. *Endocrinology,* 86:573–582, 1970.

Cott, J.M., Breese, G.R., Cooper, B.R., Barlow, T.S., and Prange, A.J., Jr., Investigations into the mechanism of reductions of ethanol sleep time by thyrotropin-releasing hormone (TRH). *J. Pharmacol. Exp. Ther.,* 196:594–604, 1976.

Cott, J. and Engel, J., Antagonism of the activity of thyrotropin-releasing hormone by agents which enhance GABA transmission. *Psychopharmacol.,* 52:145–149, 1977.

Criswell, H.E. and Breese, G.R., A conflict procedure not requiring deprivation: evidence that chronic ethanol treatment induces tolerance to the anticonflict action of ethanol and chlordiazepoxide. *Alcoholism: Clin. Exp. Res.,* 13:680–685, 1989.

French, T.A., Masserano, J.M., and Weiner, N., Influence of thyrotropin-releasing hormone and catecholaminergic interactions on CNS ethanol sensitivity. *Alcoholism: Clin. Exp. Res.,* 17:99–106, 1993.

Garbutt, J.C., Gillette, G.M., and Hicks, R.E., TRH effects on chlordiazepoxide sedation. *Biol. Psychiatr.,* 25:983–985, 1989.

Garbutt, J.C., Hicks, R.E., Clayton, C.J., Andrews, R.T., and Mason, G.A., Behavioral and endocrine interactions between thyrotropin-releasing hormone and ethanol in normal human subjects. *Alcoholism: Clin. Exp. Res.,* 15:1045–1049, 1991.

Gershengorn, M.C. and Thaw, C., Thyrotropin-releasing hormone (TRH) stimulates biphasic elevation of cytoplasmic free calcium in GH 3 cell. Further evidence that TRH mobilizes cellular and extracellular Ca^{2+}. *Endocrinology,* 116:591–596, 1985.

Hedner, J., McCown, T.J., Mueller, R.A., Hedner, T., Jonason, J., and Breese, G.R., Respiratory stimulant effects by TRH into the mesencephalic region in the rat. *Acta Physiol. Scand.,* 130:69–75, 1987.

Hernandez, D.E., Meyer, R.E., Irving, P.E., and Crane, S.W., Reversal of pentobarbital-induced narcosis by thyrotropin-releasing hormone (TRH) in dogs. *Pharmacol. Res. Comm.,* 19:567–577, 1987.

Hökfelt, T., Tsuruo, Y., Ulfhake, B., Culheim, S., Arvidsson, U., Foster, G.A., Schultzberg, M., Schalling, M., Arborelius, L., Freedman, J., Psot, C., and Visser, T., Distribution of TRH-like immunoreactivity with special reference to coexistence with other neuroactive compounds. *Ann NY Acad. Sci.,* 306:145–155, 1989.

Horita, A., Carino, M.A., and Chesnet, R.M., Influence of thyrotropin releasing hormone (TRH) on drug-induced narcosis and hypothermia in rabbits. *Psychopharmacol.,* 49:57–62, 1976.

Imai, A. and Gershengorn, M.C., Evidence for tight coupling of thyrotropin-releasing hormone receptors to stimulated inositol triphosphate formation in rat pituitary cells. *J. Biol. Chem.,* 260:10536–10540, 1985.

Johansson, O, Hökfelt, T., Pernow, B., Jeffcoate, S.L., White, N., Steinbusch, H.W.M., Berhofstad, A.A.J., Emson, P.D., and Spindel, E., Immunohistochemical support for three putative transmitters in one neuron: coexistence of 5-hydroxytryptamine, substance P- and thyrotropin-releasing hormone-like immunoreactivity in medullary neurons projecting to the spinal cord. *J. Neurosci.,* 6:1857–1882, 1981.

Kalivas, P.W. and Horita, A., Thyrotropin-releasing hormone: neurogenesis of actions in the pentobarbital narcotized rat. *J. Pharmacol. Exp. Ther.,* 212:203–210, 1980.

Kalivas, P.W., Halpern, L.M., and Horita, A., Synchronization of hippocampal and cortical electroencephalogram by thyrotropin-releasing hormone. *Exp. Neurol.,* 69:627–638, 1980.

Kalivas, P.W., Simasko, S.M., and Horita, A., Effect of septohippocampal lesions on thyrotropin-releasing hormone antagonism of pentobarbital narcosis. *Brain Res.,* 222:253–265, 1981.

Katsuura, G., Yoshikawa, K., Itoh, S., and Hsiao, S., Behavioral effects of thyrotropin releasing hormone in frontal decorticated rats. *Peptides,* 5:899–903, 1984.

Kim, G., Carr, I.C., Anderson, L.A., Zababnik, J., Eidne, K.A., and Milligan, G., The long isoform of the rat thyrotropin-releasing hormone receptor down-regulates G_q proteins. *J. Biol. Chem.,* 269:19933–19940, 1994.

Knutsen, H., Dolva, L.O., Skrede, S., Bjorklund, R., and Morland, J., Thyrotropin-releasing hormone antagonism of ethanol inebriation. *Alcoholism: Clin. Exp. Res.,* 13:365–370, 1989.

Kraemer, G.W., Mueller, R.A., Breese, G.R., Prange, A.J., Jr., Lewis, J.K., Morrison, H., and McKinnery, W.T., Thyrotropin-releasing hormone: antagonism of pentobarbital narcosis in the monkey. *Pharmacol. Biochem. Behav.,* 4:709–712, 1976.

Lighton, C., Bennett, G.W., and Marsden, C.A., Chronic immunization of endogenous thyrotropin-releasing hormone (TRH) in brain alters the behavioral response to pentobarbital and a TRH analogue. *Brain Res.,* 378:385–389, 1986.

Linnoila, M., Mattila, M.J., Karhunen, P., Nuotto, E., and Seppala, T., Failure of TRH and ORG 2766 hexapeptide to counteract alcoholic inebriation in man. *Eur. J. Clin. Pharmacol.,* 21:27–32, 1989.

Lister, R.G. and Nutt, D.J., Is RO 15–4513 a specific alcohol antagonist? *Trends Neurosci.,* 10:223–225, 1987.

Loosen, P.T., Prange, A.J., Jr., and Wilson, I.C., TRH (Protirelin) in depressed alcoholic men. *Arch. Gen. Psychiatr.,* 36:540–547, 1979.

Lopez-Barneo, J, Castellano, A., and Toledo-Aral, J., Thyrotropin-releasing-hormone (TRH) and its physiological metabolite TRH-OH inhibit Na^+ channel activity in mammalian septal neurons. *Proc. Natl. Acad. Sci. U.S.A.,* 87:8150–8154, 1990.

Mantyh, P.W. and Hunt, S.P., Thyrotropin-releasing hormone (TRH) receptors: localization by light microscopic autoradiography in rat brain using [^{3}H][3-Me-His^2]TRH as the radioligand. J. Neuroscience, 5:551–561, 1985.

Martin, T.F., Bajjalieh, S.M., Lucas, D.O., and Kowalchyk, J.A., Thyrotropin-releasing hormone stimulation of polyphosphoinositide hydrolysis in GH3 cell membranes is GTP dependent but insensitive to cholera or pertussis toxin. *J. Biol. Chem.,* 261:10041–10049, 1986.

Mason, G.A., Rezvani, A.H., Grady, D.R., and Garbutt, J.C., The sub-chronic effects of the TRH analog TA-0910 and bromocriptine on alcohol preference in alcohol preferring (P) rats: development of tolerance and cross tolerance. *Alcoholism: Clin. Exp. Res.,* 18:1196–1201, 1994.

Matsumoto, A., Kumagai, T., Takeuchi, T., Miyazaki, S., and Watanabe, K., Clinical effects of thyrotropin-releasing hormone for severe epilepsy in childhood. *Epilepsia,* 28:49–55, 1987.

McCown, T.J., Hedner, J.A., Towle, A.C., Breese, G.R., and Mueller, R.A., Brainstem localization of a thyrotropin-releasing hormone-induced change in respiratory function. *Brain Res.,* 373:189–196, 1986a.

McCown, T.J., Moray, L.J., Kizer, J.S., and Breese, G.R., Interactions between TRH and ethanol in the medial septum. *Pharmacol. Biochem. Behav.,* 24:1269–1274, 1986b.

McCown, T.J. and Breese, G.R., Mechanistic and functional divergence between thyrotropin-releasing hormone and RO 15–4513 interactions with ethanol. *Alcoholism: Clin. Exp. Res.,* 13:660–663, 1989.

Minabe, Y., Tanii, Y., Tsunoda, M., and Kurachi, M., Acute effect of TRH, flunarizine, lithium and zotepine on amygdaloid kindled seizures induced with low-frequency stimulation. *Jpn. J. Psychiatr. Neurol.,* 41:685–691, 1987.

Miyamoto, M., Narumi, S., Nagai, Y., Shima, T., and Nagawa, Y., Thyrotropin-releasing hormone: hyperactivity and mesolimbic dopamine system in rats. *Jpn. J. Pharmacol.,* 29:335–347, 1979.

Miyamoto, M, Nagai, Y., Narumi, S., Saji, Y., and Nagawa, Y., TRH and its novel analog (DN-1417): antipentobarbital action and involvement of cholinergic mechanisms. *Pharmacol. Biochem. Behav.,* 17:797–806, 1982.

Morzorati, S. and Kubek, M.J., Septal TRH in alcohol-naive P and NP rats and following alcohol challenge. *Brain Res. Bull.,* 31:301–304, 1993.

Myers, R.D., Metcalf, G., and Rice, J.C., Identification by microinjection of TRH-sensitive sites in the cat's brain stem that mediate respiratory, temperature and other autonomic changes. *Brain Res.,* 126:105–115, 1977.

Nemeroff, C.B., Prange, A.J., Bissette, G., Breese, G.R., and Lipton, M.A., Thyrotropin-releasing hormone (TRH) and its beta-alanine analogue: potentiation of the anticonvulsant potency of phenobarbital in mice. *Psychopharmacol. Commun.,* 1:305–317, 1975.

Noonan, L.R., Walker, C.H., Li, L., Garbutt, J.C., Prange, A.J., Jr., and Mason, G.A., Effects of thyroid state on preference for and sensitivity to ethanol in Fischer-344 rats. *Prog. Neuro-Psychopharmacol. Biol. Psychiat.,* 17:475–486, 1993.

de la Pena, P., Delgado, L.M., del Camino, D., and Barros, F., Two isoforms of the thyrotropin-releasing hormone receptor generated by alternative splicing have indistinguishable functional properties. *J. Biol. Chem.,* 267:25703–25708, 1992.

Plotnikoff, N.P., Prange, A.J., Jr., Breese, G.R., Anderson, M.S., and Wilson, I.C., Thyrotropin-releasing hormone: enhancement of DOPA activity by a hypothalamic hormone. *Science,* 178:417–418, 1972.

Prange, A.J., Jr., Breese, G.R., Martin, B.R., Cooper, B.R., Wilson, I.C., and Plotnikoff, N.P., Thyrotropin-releasing hormone: antagonism of pentobarbital in rodents. *Life Sci.,* 14:447–455, 1974.

Renming, X., Ishihara, K., Sasa, M., Ujihara, H., Momiyama, T., Fyjita, Y., Todo, N., Serikawa, T., Yamada, J., and Takaori, S., Antiepileptic effects of CNK-602A, a novel thyrotropin-releasing hormone analog, on absence-like and tonic seizures of spontaneously epileptic rats. *Eur. J. Pharmacol.,* 17:185–192, 1992.

Rezvani, A.H., Garbutt, J.C., Garges, P.L., Janowsky, D.S., and Mason, G.A., Attenuation of alcohol preference in alcohol-preferring rats by a novel TRH analogue, TA-0910. *Alcoholism: Clin. Exp. Res.,* 16:326–330, 1992.

Rhinehart, R.K., Barbaz, B., Iyengar, S., Ambrose, F., Steel, D.J., Neale, R.F., Petrack, B., Bittiger, H., Wood, P.L., and Williams, M., Benzodiazepine interactions with central thyroid-releasing hormone binding sites: characterization and physiological significance. *J. Pharmacol. Exp. Ther.,* 238:178–185, 1986.

Sakai, S., Baba, H., Sato, M., and Wada, J.A., Effect of DN-1417 on photosensitivity and cortically kindled seizure in Senegalese baboons, Papio Papio. *Epilepsia,* 32:16–21, 1991.

Santori, E.M., Schmidt, D.E., Kalivas, P.W., and Horita, A., Failure of muscarinic blockade to antagonize analepsis induced by thyrotropin-releasing hormone and MK-771 in the rat. *Psychopharmacol.,* 74:13–16, 1981.

Sharif, N.A., Quantitative autoradiography of TRH receptors in discrete brain regions of different mammalian species. *Ann NY Acad. Sci.,* 553:147–175, 1989.

Singh, S.P., Patel, D.g., Kabie, M.A., and Premachandra, B.N., Serum T4, T3 and reverse T3 in ethanol fed rats. *Life Sci.,* 25:889–894, 1979.

Suzdak, P.D., Paul, S.M., and Crawley, N.J., Effects of RO 15–4513 and other benzodiazepine inverse agonists on alcohol-induced intoxication in the rat. *Science,* 234:1243–1247, 1986.

Taylor, R.L. and Burt, D.R., Species differences in the brain regional distribution of receptor binding for thyrotropin-releasing hormone. *J. Neurochem.,* 38:1649–1656, 1982.

Toledo-Aral, J., Castellano, A., Urena, J., and Lopez-Barneo, J., Dual modulation of K^+ currents and cytosolic Ca^{2+} by the peptide TRH and its derivatives in guinea-pig septal neurones. *J. Physiol. (Lond),* 472:327–340, 1993.

Vogel, R.A., Cooper, B.R., Barlow, T.S., Prange, A.J., Jr., Mueller, R.A., and Breese, G.R., Effects of thyrotropin-releasing hormone on locomotor activity, operant performance and ingestive behavior. *J. Pharmacol. Exp. Ther.,* 208:161–168, 1979.

Vogel, R.A., Frye, G.D., Wilson, J.H., Kuhn, C.M., Mailman, R.B., Mueller, R.A., and Breese, G.R., Attenuation of the effect of punishment by thyrotropin-releasing hormone: comparisons with chlordiazepoxide. *J. Pharmacol. Exp Ther.,* 212:153–161, 1980.

Webster, V.A.D., Griffiths, E.C., and Slater, P., Antinociceptive effects of thyrotropin-releasing hormone and its analogues in the rat periaqueductal grey region. *Neurosci. Lett.,* 42:67–70, 1983.

Winokur, A. and Utiger, R.D., Thyrotropin-releasing hormone: regional distribution in rat brain. *Science,* 185:265–267, 1974.

Winokur, A. and Beckman, A.L., Effects of thyrotropin releasing hormone, norepinephrine and acetylcholine on the activity of neurons in the hypothalamus, septum and cerebral cortex of the rat. *Brain Res.,* 150:205–209, 1978.

Yamada, M., Monden, T., Satoh, T., Satoh, N., Murakami, M., Iriuchihima, T., Kakegawa, T., and Mori, M., Pituitary adenomas of patients with acromegaly express thyrotropin-releasing hormone receptor messenger RNA: cloning and functional expression of the human thyrotropin-releasing hormone receptor gene. *Biochem. Biophys. Res. Comm.,* 195:737–745, 1993.

Yamamura, M., Kinoshita, K., Nakagawa, H., and Ishida, R., Pharmacological study of TA-0910, a new thyrotropin-releasing hormone (TRH) analog (III): inhibition of pentobarbital anesthesia. *Jpn. J. Pharmacol.,* 55:69–80, 1991.

Yarbrough, G.G., Thyrotropin releasing hormone and CNS cholinergic neurons. *Life Sci.,* 33:111–118, 1983.

Chapter 13

NEUROBEHAVIORAL REGULATION OF ETHANOL INTAKE

Herman H. Samson and Clyde W. Hodge

TABLE OF CONTENTS

I. INTRODUCTION

The excessive drinking of beverages containing alcohol (most often defined by social norms [Babor, 1986]) has been reported for centuries. Countless theories and hypotheses for these incidents of abuse have been formulated (Littieri et al., 1980), many of which include the contribution of organismic (genetic and physiological), environmental, social, and cultural factors. Central to most of these theories of excessive alcohol consumption is the concept that some individuals can apparently "control" their consumption of alcohol, whereas others cannot (i.e., loss of control).

In the study of regulatory behavior of biologically required substances (i.e., nutrients and water) the concept of control has played a central role (Stricker, 1990). The failure to regulate food and fluid intakes can lead to life-threatening consequences. In this chapter, we have utilized the concepts of neurobehavioral regulatory processes of feeding and drinking as a

0-8493-8389-7/95/$0.00+$.50

framework upon which to examine mechanisms involved in the "control" (regulation) of alcohol consumption. By understanding how alcohol consumption is regulated under nonexcessive "normal" conditions, alterations of these mechanisms that result in excessive consumption potentially can be elucidated.

The field of behavioral biology has for many years encompassed studies whose goal is to understand the processes that control ingestion of food and water, maintenance of body temperature, and other survival requirements (see Stellar, 1990 for a historical review). A common theme for this work emanated from the classical studies of Cannon (1932), whose use of the concept of homeostasis in the maintenance of the *internal milieu* formed the theoretical frame work for much of the research on the physiological regulation process for decades. This work has validated the usefulness of understanding how the body regulates "normal" homeostatic control, in order to diagnose and develop effective treatments of irregular or "disease" states.

As a general principle, hemostasis represents the organism as defending some internal "set point" through behavioral and/or physiological processes. In the case of "abnormal" states, this set point is thought to be altered in a manner disadvantageous to the organism. For example, in the case of nutritional status, deviation from some required set point normally results in increased activity by the organism to obtain and ingest the required nutrient in order to restore the balance. However, if the set point is altered, the amount of nutrient required to reach the new set point could be excessive, leading to overeating and obesity (for a more complete review of the concept of set point and its heuristic role in regulatory behavior, see Stricker, 1990).

As part of the study of the regulatory behavior of eating and drinking, a sequence of behavioral actions related to set point maintenance are observed in the consumption of a substance. Different authors have described these processes in different terms, but the overall consensus supports the notion that multiple factors control both the initial seeking behavior and the resulting consummatory response regulating the amount ingested. In his initial studies of these behaviors, Craig (1918) alluded to these two distinct processes as the *appetitive* and *consummatory* phases. Within each phase, different controlling variables were considered important in regulating behavior. During the appetitive phase, an increase in activity was postulated as critical for procurement of the "needed" substance (i.e., food or water), along with appropriate responses to environmental stimuli relevant to their procurement. Thus, during this phase, past experience was considered to play a central role in the behavior observed. In the consummatory phase the procured substance was ingested. During this phase, the behavior was seen as predominately reflexive, with minimal alteration occurring from past experience.

Two central questions related to these response classes have been suggested as keys to understanding the regulatory process. First, what initiates the onset of the appetitive seeking behavior? Specifically, what factors (i.e., discriminative stimuli, both internal and external; conditioned reinforcers, etc.) regulate the intervals between consummatory responses. Second, once consumption has begun, what controls the amount of a substance consumed? That is, what, besides a limited supply of consumables, functions to determine when a consummatory response will end. Clearly, total ingestion of a substance is regulated by both the frequency of consummatory events and by their magnitude. By understanding the variables related to these two processes, the factors regulating total intake can be defined (for a more detailed discussion of these issues, see Campfield & Smith, 1990).

An additional factor has to be considered when applying this theoretical framework to alcohol consumption. A variety of foods available to organisms, which provide important and usable nutrients, do not always have palatable tastes (i.e., chili peppers). Thus, learning about "tastes" appears necessary as an interface between the appetitive responses and the consummatory responses, in order to make unpalatable foods and fluids acceptable and even sought out by the animal (for a review of these issues, see Rozin and Schulkin, 1990). We refer to the development of a "taste" for alcohol as the process of initiation. It should be pointed out that a "taste" does not necessarily imply excessive consumption in this context, only that an initially less palatable substance can become more palatable.

II. REGULATORY PROCESSES RELATED TO ETHANOL INTAKE

The application of the regulatory framework provides a valuable heuristic model for studying the control of ethanol drinking. A variety of different procedures have been used to examine the consumption of alcohol but, for the most part, these models do not make detailed measurements of drinking in terms of time (i.e., reporting two-bottle home cage ethanol intakes for 24 h does not provide information about the frequency and magnitude of individual drinking episodes). As a result, the data gathered from many studies related to the pharmacological manipulation of drinking cannot address the effects these manipulations have upon regulatory processes. This chapter concentrates only on those studies in which the measures of drinking allow for the detailed analysis of the patterns of consumption. Also, the studies discussed are limited primarily to those using rats, as they are the most prevalent ones employing this methodology.

A variety of animal studies have shown that under conditions of unlimited access to ethanol, intakes can be observed to occur in discrete drinking bouts separated by various intervals during the day (Gill et al., 1988; Samson et al., 1991). We will use the term drinking bout to describe the ongoing appetitive-consummatory behavior during ethanol consumption periods. When ethanol access is limited each day, then bout size increases as a function of the time limit, whereas total daily intakes decrease (Files et al., 1994; Marcucella and Munro, 1987). Using this type of consumption pattern analysis, ethanol drinking bouts can be examined as other ingestive behaviors have been (Stricker, 1990). In many cases, human consumption of alcohol can be seen to fit into several aspects of this framework, with the intervals between drinking occasions (i.e., drinking frequency = bout number) and the quantity drunk on each occasion (amount consumed/occasion = bout size) both being used as part of the clinical determination of alcohol abuse and alcoholism (i.e., the quantity/frequency measure index, Heather et al., 1993).

A. APPETITIVE AND CONSUMMATORY PROCESSES

The consumption of ethanol can be considered as resulting from the regulation of feeding, drinking, and/or drug-related processes. Therefore, when determining ethanol intake patterns, it is important to determine the temporal relation of ethanol drinking to the consumption of other substances available in the environment. By examining ethanol consumption in this manner, it is possible to make inferences as to what properties of ethanol (i.e., nutrient, fluid, or pharmacological) are influencing intake. Unfortunately, not many studies have examined this relationship. Although these issues will not be addressed in this chapter, the application of a regulatory process framework to ethanol consumption cannot be done in the absence of understanding that ethanol intake interacts with both food and fluid ingestion (Samson et al., 1991; Files et al., 1993). For example, some authors (Dole, 1986; Gentry and Dole, 1987) have suggested that in alcohol-preferring mice, ethanol consumption is basically a replacement for water consumption. These authors suggest that alcohol drinking in this context is primarily related to prandial water drinking resulting from feeding, and that these "ethanol-preferring" mice do not drink ethanol for its pharmacological effects. Clearly, when using animal models to determine processes regulating ethanol consumption, understanding the complex factors involved in the ingestion of other nutrients is also critical.

Rat studies using operant methodology suggest that in some cases ethanol consumption is not primarily related to food or fluid properties (Samson et al., 1991; Files et al., 1993). Using computer assisted analysis of ingestion patterns, our laboratory has examined the nature of

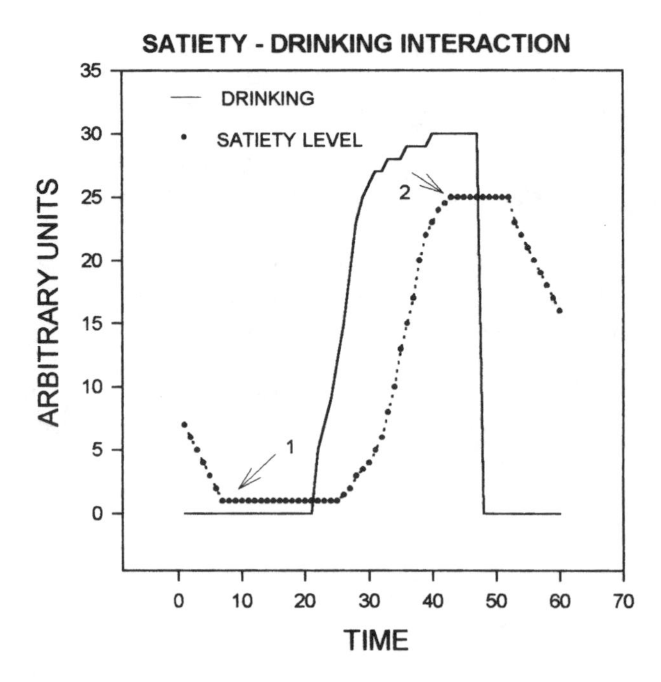

FIGURE 13.1. Hypothetical relationship between "satiety" and alcohol drinking plotted as a function of time in a drinking bout. The onset of satiety is inferred from the decrease in drinking rate. The arrow at 1 depicts the minimal influence of satiety, while the arrow at 2 indicates maximal satiety activation.

ethanol, water, and food consumption in detail (Samson et al., 1991). These studies indicate that ethanol is consumed in discrete bouts, with bout frequency and amount dependent on a variety of factors (both environmental and behavioral). For example, with rats previously initiated to drink ethanol in limited ethanol access periods, ethanol drinking in the continuous access situation occurs in relation to feeding (prandial drinking) in about 40% to 50% of bout occurrences, whereas the remaining bouts are not as clearly related to feeding bouts (nonprandial drinking). Using this procedure, the periods between ethanol drinking bouts (the inter-bout interval) can be used as a partial measure of the appetitive processes involved in alcohol-seeking behaviors.

Regulation of an ethanol drinking bout can be considered to be an interaction between the appetitive behavioral processes, which direct behavior towards obtaining and consuming ethanol (alcohol-seeking behavior), and the actual consummatory processes which terminate the appetitive behaviors as a result of ingestion and satiety (see Figure 13.1). While conceptually, the appetitive processes occurring between ethanol drinking bouts can be considered as a partial measure of alcohol-seeking behavior, the timing of this interval is also influenced by satiety mechanisms. Furthermore, there are appetitive processes controlling the amount of alcohol consumed in a bout, particularly during the initial consumption phase of drinking. Therefore, the determination of specific ethanol seeking requires a detailed analysis of the total sequences of behavior occurring in relation to the environment. Whereas it could be speculated that the concept of "ethanol craving" would be included in the processes that are important for control over the appetitive behaviors, the actual understanding of a role of craving is difficult to determine experimentally in these situations.

The discrete ethanol consummatory bout is controlled by a complex interaction between continued appetitive processes, initial ingestive processes related to the substance (i.e., taste and contextual cues related to the actual ethanol presentation), and satiety processes that will eventually terminate consumption (see Figure 13.1). Although it is beyond the scope of this chapter to detail the various theories that have been postulated in regulatory behavior to account for the control of the size of a consummatory bout, central to most of these theories are a variety of feedback processes, including the release of specific gut peptide hormones (Gibbs et al., 1993). It is unclear how these food-related satiety mechanisms might be involved in regulating an ethanol bout, but one of these gut hormones, CCK, has been shown to impact the amount of ethanol ingested in a limited access situation (Kulkosky and Glazner, 1988; Kulkosky et al., 1993).

B. BEHAVIORAL PROCESSES

A variety of behavioral processes related to alcohol consumption can be conceptualized as operating within the appetitive-consummatory framework. It is hoped that by indicating how these behavioral concepts interface with the aforementioned concepts of appetitive and consummatory processes, a useful heuristic synthesis of these approaches can be accomplished.

When considering appetitive behavior, the role of environmental discriminative stimuli (i.e., stimuli that predict or signal reinforcer availability) is obviously critical. The presence or absence of these stimuli can guide appetitive behaviors and have been studied under the concepts of stimulus control for many years by operant psychologists. Some environmental stimuli, which become closely associated with the delivery of a reinforcer, can become conditioned reinforcers, resulting in the ability to maintain appetitive response classes. It should be noted that internal stimuli (i.e., hunger cues) can also function as discriminative stimuli, directing the nature of appetitive activities. Internal and external stimuli interactively control responding. For example, the appearance of stimuli in the environment that are related to food availability do not always result in the onset of food-seeking appetitive behaviors. Only in combination with internal stimuli do these external stimuli provide the appropriate cues for the maintenance of appetitive activity.

The stimuli that control consummatory activities are equally complex. The consumable substance is associated with both external and internal stimuli (i.e., the sight, smell, and taste of food, as well as the physiological actions resulting from ingestion). Many of the external stimuli are important in maintaining consummatory activities, especially when the physiological actions of the substance may be delayed due to factors of absorption and distribution. In particular for ethanol consumption, the physiological actions are clearly delayed. Thus, the taste and smell (and in some cases the sight) of alcohol may become conditioned reinforcers that maintain the consummatory behavior prior to the onset of ethanol's pharmacological actions.

An additional consideration for both appetitive and consummatory processes in ethanol consumption is the actual amounts available to be consumed at each occasion. Although in a "natural" setting an organism may be able to obtain a large quantity of a substance to consume on a single occasion, in most laboratory studies of operant self-administration, the amount of ethanol immediately available for consumption at each opportunity is usually limited to a small amount. These experimental protocols, which measure response rates and patterns as an indication of ethanol reinforcement, require the animal to continue to respond in some specified manner (i.e., press a lever) in order to gain continued access to small quantities (between 0.1 and 0.5 ml) of ethanol at each presentation. This experimental condition may or may not be similar to what is perceived as the human ethanol consummatory response condition. When the response cost in these operant situations is small for each presentation of alcohol (i.e., the number of lever presses required is low), the situation may not be greatly different from that of drinking when larger quantities of the substance are continuously available (i.e., a rat drinking from a water bottle tube or a human sipping from a glass or bottle). Similar consumption patterns have been noted in these operant situations for both rats (Samson, 1986) and humans in an experimental bar-type environment (Samson & Fromme, 1984). In the operant situations with rats, we have found that the total amount of ethanol consumed in a 23-h ethanol access operant situation is reduced compared to the home-cage two-bottle situation, when as few as four lever presses (a fixed ratio 4 schedule of reinforcement) are required for each 0.1 ml dipper presentation. However, when a single lever press is required for dipper presentation, intakes in the operant situation are more similar to the home-cage condition. Therefore, when using operant self-administration paradigms to examine appetitive-consummatory behaviors, the nature of the appetitive or consummatory processes being measured is related to the schedule of reinforcement and amount of reinforcer presented. If response cost/reinforcer becomes too high, the resultant bout pattern may not closely model most human drinking situations. Although studies in operant situations using higher response costs are informative as to the reinforcing function of ethanol, they remove an important component of the human consummatory response situation, and thus may be less informative regarding drinking regulatory processes.

In these operant situations, it would appear that an appetitive process and a consummatory process are co-occurring during the drinking bout. This occurs even when reinforcement schedules requiring only a few responses and presenting small amounts of ethanol are employed. In these situations, lever pressing during a bout can most likely be considered an appetitive behavior, whereas licking each dipper of fluid presented must be seen as a consummatory response. However, some of the lever pressing/dipper licking behavior sequences can become almost stereotypic in appearance in rats that have had an extended period of experience in this situation. In these cases, it is unclear to what extent the lever pressing remains as an appetitive behavior or becomes part of the complex consummatory response.

It has been suggested that a drug's ability to function as a reinforcer is the key to understanding abuse liability (Katz, 1989). That alcohol can function as a reinforcer has been demonstrated in a variety of situations (Meisch, 1977; Samson, 1987) and will not be reviewed in this chapter. It should be pointed out however, that almost all of the operant studies concerned with evaluating ethanol's reinforcing capabilities have been done using limited access procedures, with only a few studies addressing this issue in continuous access situations. A point that needs to be stressed, when using a regulatory behavioral analysis of ethanol drinking, is that ethanol's capacity to function as a reinforcer (i.e., ethanol's ability to maintain behavior that provides continued access to self-administer more ethanol) will vary in relationship to the overall appetitive-consummatory cycle regulating intake. Thus, understanding the Central Nervous System (CNS) mechanisms related to the reinforcing ability of ethanol are by nature also a function of the procedures used to measure the regulation of self-administration.

III. CENTRAL NERVOUS SYSTEM PROCESS IN ETHANOL INTAKE

In the last few years, specific brain systems through which various abused substances appear to influence drug-taking behavior have begun to be clarified (Koob, 1992; Liebman & Cooper, 1989). These include a variety of CNS structures that have been defined to some extent by specific neurotransmitter-receptor systems and the ability of a particular drug to interact with those systems. It is often suggested that the drug, because of the dose and time course of action resultant from self-administration, is more potent than the "natural" ligand. Consequently, a drug can overshadow other stimuli with less ability to activate the system, resulting in drug choice over other reinforcing stimuli.

The dopamine (DA) neurotransmitter systems have been proposed to be primarily involved for the stimulant drugs like amphetamine and cocaine that act at the DA receptors and re-uptake sites, whereas the endogenous opiate neurotransmitter systems are potentially more related to analgesic drugs like heroin and morphine (Koob, 1992). To some degree, the anatomical distributions of these receptor systems in the brain can be used to suggest underlying CNS pathways for the actions of each drug class (Koob and Goeders, 1989; Dworkin and Smith, 1987). However, because of the overlap of these transmitter-receptor classes in most brain areas known to be related to psychoactive drug actions, the conceptual isolation of one brain system based solely on neurotransmitter type from another brain system has become an untenable explanation for drug reward. Rather, brain systems based on behavioral functions of reinforcement/reward have become more popular in the past few years. The most prominent of these systems is the mesolimbic "reward" system (Bozarth, 1991). This system has been postulated to underlie all reward/reinforcement functions regardless of the receptors upon which a given substance might act. As a basic tenant of this concept, any drug that can access and alter activity in this brain system will have some ability to function as a reinforcer/reward. A second CNS system, which has long been known to be related to the control of a variety of consummatory behaviors, is the hypothalamic regulatory system. Although it has received less study in relation to drug self-administration, this system has been proposed to be involved in the control of appetitive-consummatory behaviors (Booth, 1990) and warrants consideration in relation to alcohol self-administration. There are undoubtedly other "systems" that can be considered as important to the reinforcing/rewarding actions of ethanol, but for the purposes of this chapter, only these two are considered.

A. THE MESOLIMBIC REWARD SYSTEM

Several recent reviews have covered in detail the current hypotheses related to the mesolimbic reward system's function (Koob, 1992a,b; Le Moal & Simon, 1991). This chapter does not cover these in detail, but briefly reviews the anatomy and function of this system and then discusses the role that it may have in the regulation of ethanol self-administration.

1. Neuroanatomy

Several important studies on the anatomy of the mesolimbic system have been published in the last few years (Berendse et al., 1992; Domesick, 1988; Kalivas & Samson, 1992; Le Moal & Simon, 1991; McDonald, 1991a; McDonald, 1991b; Pennartz et al., 1994; Pirot et al., 1992; Salamone, 1992; Willner & Scheel-Kruger, 1991). In general, these studies have centered on the projections of the DA cell bodies located in the ventral tegmental area (VTA) to various regions located in the "limbic" system, including the nucleus accumbens, amygdala, olfactory bulb, frontal cortex, and hippocampus. However, it is clear that there are a variety of feed-forward and feedback pathways in the mesolimbic system related to a variety of neurotransmitter systems that regulate the activity of the nucleus accumbens and its GABAergic output (Berendse & Groenewegen, 1990; Pennartz et al., 1994). For example, the DA projections to the frontal cortex and the nucleus accumbens arise from the same areas in the VTA. This frontal cortical area projects, via a glutamatergic pathway, back to the nucleus accumbens, which receives the same VTA DA input (Groenewegen et al., 1990). Modulation of this excitatory output by DA in the cortex occurring concurrently with DA effects in the nucleus accumbens could lead to a variety of changes in nucleus accumbens functions. Given the number of potential transmitters within this circuit that can act as points of entry for various drugs, it seems likely that all psychoactive agents, both those generated by "normal bodily function" (i.e., corticosterone from stress) and those self-administered, will influence system activity in some way. Understanding how this mesolimbic system might be influenced by ethanol and/or possibly subserve ethanol reinforcement has been examined in the past few years (see below).

2. Function

The mesolimbic DA system (Kalivas & Nemeroff, 1988; Willner & Scheel-Kruger, 1991) has been conceptually hypothesized to be related to the processing of reward and reinforcement in a general sense (Dworkin & Smith, 1987; Kalivas & Samson, 1992; Koob, 1992a; Koob, 1992b, Koob & Goeders 1989). This includes both positive and negative stimuli (see Salamone, 1992 for recent review). This system has been shown to have interactions with both memory and motivational functions via its relationship with several limbic structures thought to be involved in these processes (i.e., prefrontal cortex, hippocampus, amygdala). However, much of the research regarding the role of this system in drug and alcohol reinforcement, has focused only on the nucleus accumbens and/or the VTA (see reviews in Kalivas & Samson, 1992). Although the meso-accumbens DA pathway appears critical for some reinforcement/reward stimuli (i.e., stimulant class drugs), it fails to account for several important findings with other drug reinforcers (Koob, 1992a; Le Moal & Simon, 1991). For example, 6-OHDA lesions of the nucleus accumbens effectively block IV amphetamine self-administration, but have minimal effects upon IV morphine self-administration (Koob & Goeders, 1989). Although these lesions can affect electrical brain stimulation reward, this only occurs for certain placements of the stimulating electrodes, with little or no effect on hypothalamic stimulation sites (Phillips & Fibiger, 1989). Recent data also indicate that 6-OHDA lesions of nucleus accumbens do not alter established oral ethanol self-administration maintained on an FR 1 schedule (Rassnick et al., 1993), which suggests that the VTA-nucleus accumbens DA segment of the mesolimbic system is not the only system involved in ethanol reinforcement. As noted previously, the mesolimbic DA system is complicated, both within each component area and between the various components. For example, known neurotransmitters in the nucleus accumbens include DA, serotonin, norepinephrine, GABA, glutamate, substance P, and acetylcholine (Dworkin & Smith, 1987). A similar list is applicable to the VTA. Clearly, if DA system function is altered in the nucleus accumbens or VTA, other neurotransmitter systems within these structures could still play important roles related to the reinforcing actions of various stimuli. Thus, following DA terminal destruction by 6-OHDA lesions in the nucleus accumbens, other neurotransmitter systems could remain involved in a variety of physiological regulatory processes and compensate for the lack of DA input. Consequently, the activity of the nucleus accumbens in some reinforcement processes following DA lesions could remain functional, independent of the meso-accumbens DA input.

The following sections describe evidence from microinjection studies that elucidates, in part, the role of this mesolimbic system in ethanol reinforcement. The research has almost exclusively focused on the VTA-nucleus accumbens pathway, which limits our understanding of how the more complex mesolimbic system may function to regulate ethanol consumption.

B. NEUROBEHAVIORAL PHARMACOLOGY OF ETHANOL REINFORCEMENT

In order to determine the effects of manipulations of the mesolimbic DA system on ethanol reinforcement, the need for a quantifiable, consistent model of daily ethanol self-administration via the oral route is required. As discussed earlier, there are a variety of procedures that have been used to initiate oral ethanol self-administration, but those which provide for the analysis of drinking patterns are more useful in understanding the effects of manipulations of this neural system. It is important to emphasize the importance of the response (self-administration) patterns, as alterations in these patterns can have a dramatic effect on the resultant blood-ethanol profile related to the consumption bout, even when similar quantities of ethanol are drunk in limited-acess time periods.

1. Quantification of Ethanol Reinforced Bouts

When rats are trained to lever press according to fixed ratio schedules of ethanol (10%) reinforcement, a complex but consistent response pattern develops (Samson, 1986). During limited-access conditions, responding begins at a high rate shortly after ethanol becomes available. After approximately 5 to 10 min, the response rate slows and then terminates abruptly. We have previously described this response pattern in terms of three components: onset, maintenance, and termination of responding (Samson et al., 1992). Understanding the variables that control ethanol reinforcement (and potentially a part of limited-access bout regulation) involves elucidating neurochemical and neurophysiological factors that determine these three features of an ethanol bout.

Figure 13.2 shows the typical pattern of responding seen with ethanol reinforcement in this operant situation. Responding is characterized by an initial high rate that continues for about

Variables:

T_0	=	Start of Session
T_1	=	Onset of Responding
$T_{1/2}$	=	Half-time
T_2	=	Termination of Responding
T_2-T_1	=	T total
R	=	Total Responses
T_1-T_0	=	Response Latency

$$\frac{R_{1/2}}{T_{1/2}} = T_{1/2} \text{ Response Rate}$$

$$\frac{R}{T_2} = \text{Total Response Rate}$$

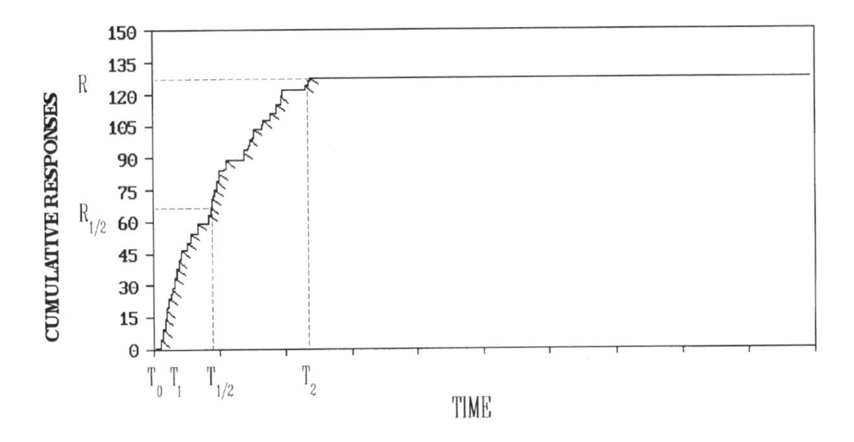

FIGURE 13.2. Cumulative response graph showing the typical pattern of lever-press responding when 10% ethanol is presented as the reinforcer on an FR 4 schedule (bottom) and measures that can be taken to quantitatively describe the temporal distribution of responses (top). Downward diagonal pips indicate presentation of 0.1 ml ethanol.

10 to 15 min and then terminates abruptly with few responses occurring thereafter (Samson, 1986). Response latency, the delay from the beginning of ethanol availability to the onset of responding, can be conceptualized as an index of initial appetitive processes. Under control conditions, discriminative stimuli set the occasion for the operant lever press and collaterally engage neurological (i.e., appetitive) processes that correlate with the onset of responding. Decreasing or increasing response latency may indicate enhancement or disruption of appetitive brain mechanisms, respectively.

During the time of the first one-half of the bout ($T_{1/2}$) (what we have previously called the early maintenance phase, [Samson et al., 1992]), ethanol functionally reinforces lever pressing. However, neurochemical effects directly related to the presence of ethanol in the CNS, such as DA release in the nucleus accumbens, do not consistently occur for approximately 5 to 15 min after the onset of drug ingestion or administration (Imperato & DiChiara, 1986; Weiss et al., 1992). Thus, behavior during this initial portion of the ethanol self-administration bout most likely involves control by conditioned reinforcers that have been paired with ethanol's CNS pharmacological effects on prior consummatory occasions. One function of "initiation procedures," such as sucrose-substitution, may be to establish control by conditioned reinforcers so that responding is maintained during this delay. Thus, $T_{1/2}$ and the response rate during this time ($Rate_{1/2}$) represent, to some extent, indices of behavior occurring during a combined appetitive and consummatory phase prior to any CNS pharmacological effect of ethanol.

After the initial burst of responses, the response rate decreases and then terminates. A measure of the time course of satiety can be obtained by comparing the half-time ($T_{1/2}$), total

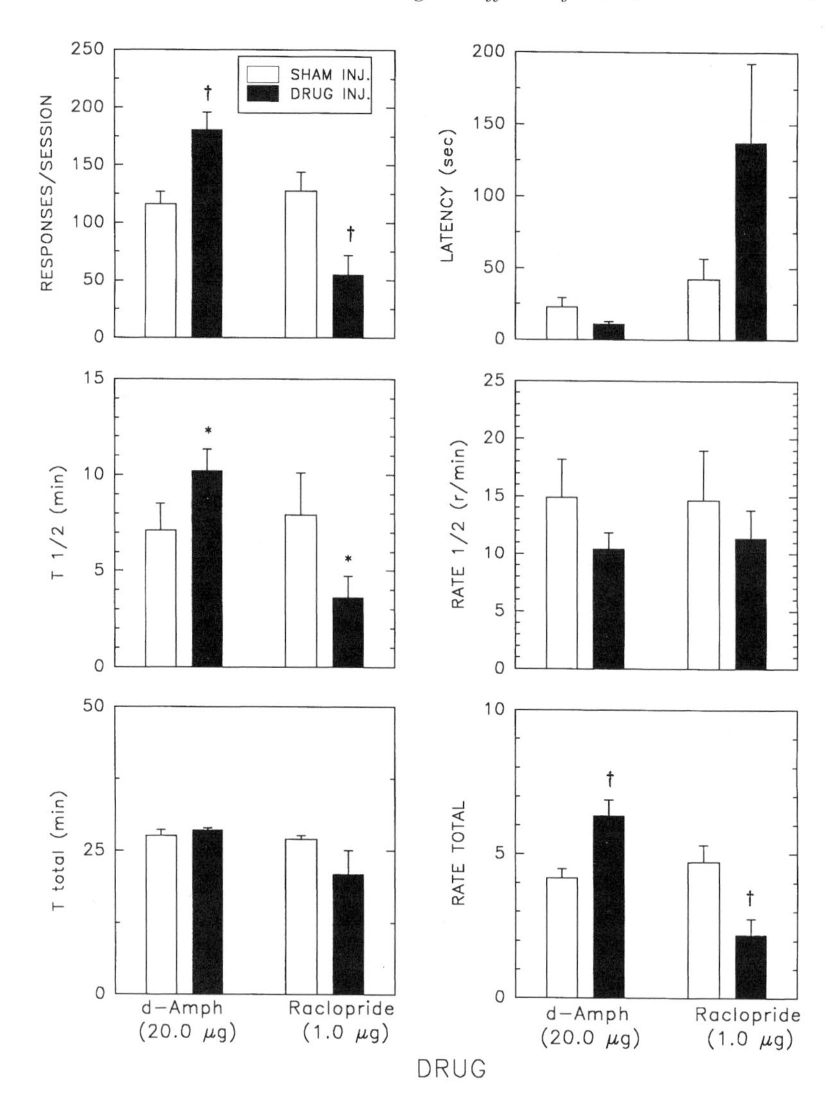

FIGURE 13.3. Response measures derived from the analysis of data such as those shown in Figure 13.2. Each measure is plotted from sham injection and drug injection sessions for an effective dose of d-amphetamine (20 µg) and raclopride (1.0/ig) injected in nucleus accumbens prior to a 30-min. ethanol reinforcement session. Error bars represent SEM, * = $p < 0.05$, = $p < 0.01$, paired t-test.

bout duration (T_{total}), and $Rate_{1/2}$ with total rate ($Rate_{total}$) measures. For instance, a constant response rate that terminates abruptly, as shown in Figure 13.2, would result in equal temporal measures (i.e., $2T_{1/2} = T_{total}$). This indicates a rapid onset of satiety. If satiety is disrupted, the relative difference between these two measures may increase as the termination of responding occurs later in time (i.e., $4\,T_{1/2} = T_{total}$). Measuring response rates and time during the first half of the responses and the total responses can also reveal differential changes in appetitive and consummatory processes. For example, a drug that alters satiety, but not appetitive processes, may increase $Rate_{total}$ without changing $Rate_{1/2}$. In addition, response rate measures during each of the components of the bout can be used as descriptive measures of ethanol's reinforcement function.

The following section reviews data from brain microinjection studies illustrating how these descriptive measures can indicate neurochemical and neurophysiological processes that regulate the onset, maintenance, and termination of an ethanol reinforcement bout.

2. Mesolimbic System and Ethanol Reinforcement

a. Nucleus Accumbens

Microinjection of the nonspecific DA agonist, d-amphetamine (4.0 to 20.0 μg), in a dose-related manner increased the number of ethanol (10% v/v) reinforced responses in 30-min and 60-min sessions by producing a continuous steady response rate, but only at the highest dose (Hodge et al., 1992; Samson et al., 1992). Alternatively, the D2-like antagonist, raclopride, terminated early an otherwise normal response pattern (Hodge et al., 1992; Samson et al, 1992).

Figure 13.3 shows a summary of six descriptive measures of ethanol-reinforced response patterns (as indicated on Figure 13.2) following injection in the nucleus accumbens of effective doses of the nonspecific agonist, d-amphetamine, and the D2-like antagonist, raclopride. Neither the agonist nor the antagonist significantly altered response latency, suggesting that DA activity in the nucleus accumbens is not primarily involved in the initial (i.e., response onset) appetitive or seeking phase of ethanol intake. This implies that other neurochemical/neurophysiological systems related to the prevailing discriminative stimulus conditions are important in the onset of responding. Although increased DA release from the nucleus accumbens has been reported in some ethanol-experienced rats during a presession waiting period (Weiss et al., 1992), these microinjection data would suggest that, whereas early DA activity may be part of initial appetitive behavioral control, it is not necessary for bout onset.

The differential effect on $T_{1/2}$ produced by the agonist and antagonist suggests that DA activity in the nucleus accumbens may be involved in the maintenance of consummatory responding, or the onset of satiety. That is, the changes in $T_{1/2}$ produced by both compounds, without a corresponding change in $Rate_{1/2}$, indicates that DA activity in the nucleus accumbens does not influence responding during the initial phase of an ethanol bout. However, the differential changes in $Rate_{total}$ with no change in T_{total}, shows that d-amphetamine increased, whereas raclopride decreased, the response rate in later portions of the bout. Thus, d-amphetamine extended ethanol reinforced responding in time, and raclopride shortened the temporal extent of the bout. This suggests that DA transmission in the nucleus accumbens is involved in the neurobehavioral processes that influence the offset of ethanol-reinforced responding (Hodge et al., 1992; Samson et al., 1992).

Figure 13.3 also shows that the total number of responses/session and the total response rate were differentially affected by the agonist and the antagonist. From a strictly behavioral standpoint, this suggests that DA activity in the nucleus accumbens may be positively correlated with ethanol's reinforcement function. However, this interpretation should be viewed with caution because only one concentration of ethanol was tested, and similar results have been obtained using sucrose as the reinforcer (Hodge et al., 1994).

Recent findings indicate that the microinjections of the D2-like agonist, quinpirole, in nucleus accumbens have a dose-related effect on ethanol reinforced lever pressing (unpublished observations). Nucleus accumbens injections of quinpirole (1.0 μg) increased total responses/session by approximately 50%, but higher doses (4.0 and 10.0 μg) each produced about a 50% reduction in total responses. The finding that quinpirole at low doses increases responding suggests that increases seen following d-amphetamine injections in the nucleus accumbens could be due, in part, to D2-like receptor activation. Consistent with this interpretation, quinpirole (1.0 μg) also increased $T_{1/2}$, which indicates a prolongation of responding. The D1-like agonist, SKF 38393 (0.03 to 3.0 μg), dose-relatedly decreased responding and did not change the temporal pattern of response bouts (unpublished observations). Additionally, the D1-like antagonist, SCH 233390 (.05 to 2.0 μg), terminated early an otherwise normal response bout in a manner similar to that seen with raclopride (unpublished observation). When taken together with the findings for raclopride, the data suggest that DA in the nucleus accumbens is involved with the maintenance and termination, but not the onset, of an ethanol bout. A similar decrease in ethanol reinforced responding has also been observed following systemic administration of DA receptor blockers (Pfeffer & Samson, 1985; Pfeffer & Samson, 1988) and during extinction procedures (Grant,1984). Importantly, however, DA blockade does not eliminate the initial components of an ethanol bout, suggesting that DA function via these receptors in the nucleus accumbens is not necessary for the onset of responding, and therefore may be more involved with the stimulus salience of ethanol once consummatory behavior has started, than with the control of appetitive ethanol-seeking behavior.

Additionally, nucleus accumbens administration of the mixed dopamine antagonist, fluphenazine (0.0 to 4.0 μg), and the competitive NMDA receptor antagonist, AP-5 (Rassnick et al., 1992), reduced, but did not eliminate, total ethanol-reinforced responses. This provides further evidence that the nucleus accumbens DA, and possibly excitatory amino acid processes, can regulate the maintenance and termination, but affect minimally the onset, of an ethanol-reinforced behavioral bout sequence.

b. Ventral Tegmental Area

Microinjection of the D2-like agonist, quinpirole, into the ventral tegmental area (VTA) decreased total responses/session maintained by ethanol reinforcement in a dose-related manner (Hodge et al., 1993a). The response pattern alteration was similar to the early termination effect seen following administration of the D2-like antagonist, raclopride, into the nucleus accumbens. Because quinpirole in the VTA inhibits DA cell body activity through a proposed DA feedback of the cell upon itself (White, 1991), the effect of quinpirole on meso-accumbens DA transmission should be similar to that of a blockade of DA neurotransmission in the terminal field by raclopride, as was observed.

Microinjection of the $GABA_A$ agonist muscimol into the VTA produced no overall significant changes in total responding, but resulted in a mixed effect on response pattern (see Figure 13.4, left). Following muscimol (30.0 ng), a prolonged response pattern, somewhat similar to that observed following DA agonist administration in the nucleus accumbens, occurred. At a 100.0 ng dose (data not shown), the effect on the response pattern was similar to that of quinpirole injected into the VTA or raclopride into the nucleus accumbens (i.e., early termination of the bout). GABA actions in the VTA are mixed (Harris et al., 1992). Low doses of GABA agonists stimulate DA cell activity, most likely by inhibiting inhibitory interneurons, whereas higher doses inhibit DA cell firing, probably by direct action on the DA cell. The observation of differing effects upon response patterns by muscimol may be a result of increased DA activity at low doses and suppressed activity at high doses.

When taken together with the nucleus accumbens findings, these data provide additional support for the hypothesis that activation and transmission within the meso-accumbens DA pathway is involved in the maintenance and termination of ethanol reinforced responding. Again, however, this transmission does not appear to influence appetitive processes involved in the onset of responding.

3. Mesolimbic System and Sucrose Reinforcement

One way of addressing the specificity of the effects observed on ethanol reinforcement is to test the effects of microinjections of the same compounds on responding reinforced by another substance such as sucrose. Since response rate can affect the action of a drug (McKearney, 1981), we have utilized both low and high concentrations of sucrose reinforcement in an attempt to match response patterns and rates to those observed with ethanol reinforcement (Hodge et al., 1994). The lowest concentrations of sucrose capable of maintaining consistent behavior in nondeprived animals (2.0 to 3.0% w/v) resulted in higher response rates and longer bout durations than those observed with ethanol reinforcement. Very high sucrose concentrations (75% w/v) produced response patterns more like those of ethanol reinforcement. As would be predicted from the behavioral pharmacology literature on rate

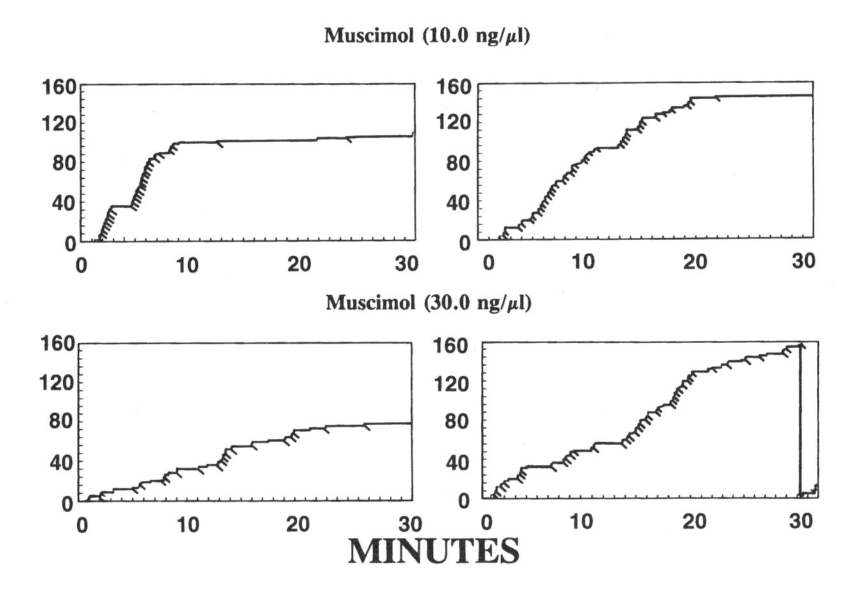

FIGURE 13.4. Cumulative response record showing differential sensitivity of ethanol- (left) and sucrose-reinforced (right) lever presses on an FR 4 schedule of reinforcement following microinjection of two doses of the $GABA_A$ agonist muscimol into the VTA. Downward diagonal pips indicate presentation of 0.1 mil ethanol (10%).

dependency (McKearney, 1981), different effects occurred following the microinjection of agonists and antagonists, depending upon the sucrose concentration used as the reinforcer.

a. Nucleus Accumbens

With low concentrations of sucrose presented as the reinforcer, microinjections of d-amphetamine in the nucleus accumbens (4.0 to 20.0 μg) resulted only in decreases in the total number of responses per session with no significant changes in the response pattern. With high sucrose concentrations (75% w/v) as well, in which baseline response patterns more closely resemble those observed with 10% ethanol reinforcement, no effects were observed with any dose of d-amphetamine on total session responding. This is in contrast to the increases in total ethanol-reinforced responding discussed above (see Section 2a). However, with this higher sucrose concentration, a change in response pattern similar to that observed with ethanol reinforcement occurred at the higher amphetamine doses and with 4 μg quinpirole (Hodge et al., 1994). Thus, when appropriate sucrose and ethanol concentrations are used to generate similar baseline response rates and patterns, similar changes in the response pattern are observed following manipulations of the mesolimbic DA system.

Raclopride injections into the nucleus accumbens produced no statistically significant decreases in total session responding with sucrose (75%) reinforcement (Hodge et al., 1994). However, an effect on the pattern of responding was initially similar to that observed with ethanol reinforcement, but many rats in this sucrose group resumed responding during the later parts of the session with a slowed and erratic pattern, an effect that was not observed when ethanol reinforcement was employed. This later session responding is most likely the reason why total responding was not significantly reduced in the sucrose reinforcement conditions. It is unlikely that this late session response resumption is related to an elimination of the drug or decreased drug action, and it most probably is a function of a differential reinforcer efficacy between ethanol and sucrose, even when each reinforcer produced similar baseline FR4 response patterns. This suggests that D2-like receptor activity may be involved to a lesser extent in the maintenance of sucrose- than of ethanol-reinforced responding.

b. Ventral Tegmental Area

Quinpirole microinjections in the VTA suppressed total responses per session with 75% sucrose reinforcement, but required a dose 100 times greater than that needed for suppression of ethanol-reinforced responding (Hodge et al., 1993a). The effect on the temporal distribution of responses was somewhat similar to that observed with raclopride in the nucleus accumbens in these reinforcement conditions. These observations support the overall hypothesis concerning the activation of this pathway for reinforcement in general and indicate that ethanol-reinforced responding is more sensitive to the blockade of D2-like transmission in this pathway than is responding reinforced by sucrose. Because both 10% ethanol and 75% sucrose reinforcement generated behavioral baselines with equivalent response rates and patterns, the shift to the right of the quinpirole dose-response curve using sucrose reinforcement conditions suggests differential regulation by the mesolimbic DA system of responding controlled by these two reinforcers.

Further support for this hypothesis comes from a similar finding that muscimol microinjections in the VTA increased sucrose (75%), but not ethanol reinforced responding (see Figure 13.4, right). Muscimol (10.0 ng and 30.0 ng) resulted in prolonged sucrose-reinforced response patterns similar to those observed with ethanol reinforcement after DA agonists were injected in nucleus accumbens. Thus, VTA administration of doses of muscimol prolonged ethanol- and sucrose-reinforced response bouts, but at different doses, which further supports the hypothesis that nucleus accumbens DA activity is differentially involved in the maintenance and termination of ethanol and sucrose reinforcement bouts.

4. Conditioned Reinforcement and the Control of Ethanol Consumption

As the previous section indicates, DA activity in the nucleus accumbens does not appear to be primarily involved in regulating the onset of ethanol-reinforced responding. However, it is of particular interest in the study of the neurobehavioral regulation of ethanol intake to understand the variables that control the onset and early maintenance of ethanol-reinforced responding prior to the onset of ethanol's pharmacological effects. Accordingly, recent findings have implicated involvement of the nucleus accumbens in conditioned reinforcement (Bakshi & Kelley, 1991a; Bakshi & Kelley, 1991b; Taylor & Robbins, 1984). Stimuli paired with the presentation of primary reinforcers can maintain responding for a period of time in the absence of the primary reinforcer (i.e., a red light paired with cocaine self-administration can then maintain lever pressing when presented without concurrent cocaine administration [see Goldberg, 1976]). Conditioned reinforcers are behaviorally important for maintaining responding in situations in which the availability of a primary reinforcer occurs infrequently or has a delayed onset of action. Thus, environmental stimuli can maintain procurement behavior required to obtain the primary reinforcer over a longer time span, increasing the probability of primary reinforcer attainment. As indicated earlier, the time between alcohol ingestion and rise in blood ethanol to levels that can impact CNS functioning represents a delay in what is hypothesized to be the "primary" reinforcing quality of ethanol. This delay effect is even more critical if ethanol consumption is done in small incremental amounts spread out over minutes to hours, as occurs in many operant drinking situations. If some CNS physiological impact (defined as the primary reinforcing event for the sake of this discussion) is necessary to maintain drinking behavior, this delay of reinforcement would make sustained drinking unlikely. However, stimuli that occur in a drinking situation, and which have been paired with ethanol reinforcement, can become conditioned reinforcers and maintain drinking behavior between the onset of a drinking sequence and the delayed pharmacological effects of ethanol.

The role of the nucleus accumbens in conditioned reinforcement has been demonstrated by several different studies in the last few years. Taylor and Robbins (1984) showed enhanced behavioral control by conditioned stimuli following the microinjection of d-amphetamine into the nucleus accumbens. Kelley and colleagues (Bakshi & Kelley, 1991a, 1991b; Cunningham & Kelley, 1992a, 1992b; Kelley & Throne, 1992) have further elucidated this effect, examining the relation of amphetamine activation to other receptor systems. They have implicated both the opiate (Cunningham & Kelley, 1992a, 1992b) and the NMDA system (Kelley & Throne, 1992) in this enhancement effect. Cador et al. (1989) demonstrated that lesions of the amygdala reduced responding for a conditioned reinforcer, even following amphetamine microinjections into the nucleus accumbens, which suggests that the amygdala can play a role in conditioned reinforcement independent of the nucleus accumbens. Wadenberg et al. (1990) also demonstrated that sulpiride injected into the ventral striatum (actual injection sites indicated shell portions of the nucleus accumbens) resulted in a suppression of a conditioned avoidance response, which indicated the inability of the conditioned stimuli to control the avoidance behavior. Gratton and colleagues (Gratton et al., 1992; Gratton and Wise, 1994) demonstrated a significant increase in extracellular DA in the nucleus accumbens following the presentation of a light that had been paired with the IV self-administration of cocaine. This DA increase was observed after a single paired trial in some rats. Weiss et al. (1992) and Weiss & Koob (1991) have reported an increase in extracellular DA levels in the nucleus accumbens preceding the onset of a cocaine self-administration session, and in some cases, for alcohol self-administration. Thus, a critical function of the mesolimbic system via the nucleus accumbens may be to integrate motor activity via stimulus control or conditioned reinforcement. This integration can provide for behavior maintenance over an increased time span required for primary reinforcers to either occur or have some direct effect on the organism.

The output of the nucleus accumbens through the ventral pallidum has been hypothesized to be an "enabling" switch, determining which motor behavior pattern will be emphasized at any given moment (Cador et al., 1991; Le Moal & Simon, 1991; Mogenson and Yim, 1991; Oades et al., 1985; Phillips & Fibiger, 1991; Salamone, 1991, 1992; Scheel-Kruger & Willner, 1991). These authors have suggested that the integration of information from both the frontal cortex and limbic structures, acting through the nucleus accumbens, enables and/or disables motor actions. This enabling function is thought to be regulated in part by motivational states encompassing both positive and negative reinforcing consequences. Rather than subserving a strictly motor function, the nucleus accumbens is postulated to direct motor actions based on sensory and motivational input. For ethanol reinforcement, these inputs presumably arise from afferent projections into the nucleus accumbens, which are influenced by the organism's history of ethanol reinforcement and serve to select response parameters that have a high probability of resulting in reinforcement.

Since nucleus accumbens administration of DA agonists and antagonists or NMDA antagonists do not disrupt the onset of ethanol reinforced responding, it would appear that neither DA nor glutamate is involved in the behavioral control by discriminative stimuli that set the occasion for the beginning of an ethanol bout. However, nucleus accumbens injections of D1-like (unpublished observations) and D2-like antagonists (Samson et al., 1992, 1993) decrease ethanol-reinforced responding by terminating early an otherwise normal response bout. This supports the hypothesis that DA transmission in nucleus accumbens is involved in the conditioned reinforcement processes that control responding prior to the onset of the central pharmacological effects of ethanol. The hypothesis that sensory and/or motivational input to the nucleus accumbens, other than VTA DA, may be involved in ethanol appetitive or seeking behavior has not been tested by these studies. CNS systems, known to be involved in appetitive processes with other reinforcers such as food, may be involved in determining the onset of responding reinforced by ethanol. The following section reviews data on one such system, the hypothalamus, and suggests that it might be involved in the regulation of ethanol reinforcement.

5. Potential Involvement of Hypothalamic Systems in Ethanol-Seeking Behavior

Research on feeding and energy regulation has focused for many years on the region of the hypothalamus (see Blundell, 1986, 1991 for reviews). The historically traditional view of specific feeding and satiety centers has more recently been incorporated into a systems view based on broader knowledge of neuroanatomy, neurochemistry, and behavior associated with the hypothalamic system (Stellar, 1990). The hypothalamus is involved in the regulation of numerous functions critical for the survival of the organism and is known to influence complex behavioral phenomena related to reinforcement/reward processes. This section briefly reviews some of the relevant research on the hypothalamic neuroanatomy, neurophysiology, and neurochemistry of regulatory function, bearing in mind that similar processes could be involved in the onset, maintenance, and termination of ethanol self-administration bouts.

a. Neuroanatomy

The hypothalamus represents a critical link in a complex neuroregulatory system. Lesions of the lateral hypothalamus (LH) result in the disruption of numerous homeostatic functions including feeding, drinking, body weight maintenance, sensorimotor effectiveness, and motivated behavior. Neurons in the LH connect via the medial forebrain bundle (MFB) to the cerebral cortex, septum, prefrontal cortex, cerebellum, hippocampus, and amygdala (see Bernardis & Bellinger, 1993 for a review). Within the hypothalamus, the dorsomedial nucleus (DMH) receives afferent connections from both the LH and the ventromedial hypothalamus (VMH), which places the DMH in a nodal point between the other two areas. Thus, the DMH may modulate the output function of the LH and VMH, which are reciprocally connected to each other. Additionally, a majority of the connections in the hypothalamus run from the LH and VMH through the DMH to the paraventricular nucleus (PVN) (Luiten et al., 1987). Output from the PVN may represent a critical link in the influence that hypothalamic mechanisms exert on the function of other brain regions and behavior.

b. Electrophysiology

The firing of neurons in the LH have been shown to be associated with the taste and sight of some foods in hungry monkeys (Rolls et al., 1980). Other neurons in the LH and substantia innominata respond at the highest rates at the sight of preferred food and not at all at the sight of nonfood objects (Rolls et al., 1976). The same neurons did not respond when the monkeys contacted food in a dark environment. Accordingly, LH neurons that respond to the sight or taste of food in hungry monkeys, do not respond if the monkeys are well-fed or as satiety occurs (Burton et al., 1976). This suggests that hypothalamic processes are modulated by stimulus conditions that a) set the occasion for the onset of a meal (i.e., discriminative stimuli associated with the availability of food when hungry) and b) physiological variables that may trigger the termination of eating (i.e., circulating glucose/insulin levels). Importantly, LH neurons that cease to respond to the sight of food after satiety will still respond to the presentation of a novel food substance (Rolls, 1981). These data suggest that some LH neurons fire only when current environmental and organismic stimulus conditions are both present, and this activity is predictive of the consumption of the substance.

Numerous questions arise regarding whether similar LH mechanisms are involved in the regulation of ethanol intake. Unfortunately, no data from explorations of this possible connection were found in the literature. Whether the firing of LH neurons is positively correlated with the onset of ethanol reinforcement bouts and negatively correlated with the termination of responding has not been studied. If LH neurons fire selectively during the onset of responding, would preloading with ethanol decrease or eliminate this activity? Preloading before an operant ethanol self-administration session reduces responding in a dose-related manner in our laboratory (unpublished observation). Do LH neurons respond differentially to the sight of food or ethanol? Clearly, these questions require further study to determine the potential hypothalamic involvement in the regulation of ethanol appetitive processes.

c. Neurochemistry

Neurochemical investigations suggest that catecholamine transmission in the LH may be involved in the inhibition of feeding. For example, injections of dopamine and norepinephrine into the LH have been shown to decrease feeding (Leibowitz, 1980). Beta-adrenergic and dopaminergic drugs, such as d-amphetamine, injected into the LH also reduce feeding, whereas destruction of the catecholamine terminals in the LH by 6-hydroxydopamine lesions blocks the anorectic effect of d-amphetamine (Leibowitz, 1980). Systemic administration of dopamine agonists decreases ethanol reinforced responding and intake in a manner suggestive of interference with onset and early maintenance phases of an ethanol bout (Pfeffer & Samson, 1985, 1988). This would indicate that increases in catecholamine levels in areas other than the nucleus accumbens may inhibit ethanol drinking in a manner similar to feeding. The potential antagonism of ethanol reinforcement by catecholamine agonists or enhancement by antagonists injected in the LH has not been examined, but could provide valuable information for our understanding of the relationship among the neurobehavioral factors that regulate food and ethanol consumption.

The PVN is richly innervated by terminals containing norepinephrine and has a dense population of α2-noradrenergic receptors (Cunningham & Sawchenko, 1988; Leibowiz et al., 1982). In contrast to the inhibitory role in the LH, monamine neurotransmission in the PVN has been shown to play a nutrient-specific role in the facilitation of feeding. For instance, injections of norepinephrine into the PVN selectively increased carbohydrate intake, decreased protein intake, and did not change fat consumption (Shor-Posner et al., 1985; Leibowitz et al., 1985a, 1985b). Similar increases in carbohydrate intake were found following peripheral administration of the α2 agonist, clonidine (Shor-Posner et al., 1988; McCabe et al., 1984). Increases in synaptic 5-HT in the PVN through local injection of 5-HT, 5-HT agonist, or reuptake blockers decrease food intake in food-deprived rats. Further, injections of 5-HT into the PVN inhibit α2-noradrenergic-induced eating, suggesting functionally opposing mechanisms for these two neurochemicals in satiety (see Leibowitz and Shor-Posner, 1986 for a review).

Although the specific role of α2-noradrenergic transmission in ethanol reinforcement is not known, the behavioral effects of ethanol are influenced by this receptor system. Ethanol-induced intoxication, as measured by exploration, loss of righting reflex, and an observer-rated intoxication score, was antagonized by the α2 antagonists, atipamezole and idazoxan (Lister et al., 1989). The ethanol-attenuating effects of the α2 antagonist, atipamezole, are thought to be due to central α2-adrenoceptors because similar results were obtained with RX 821002 (methoxy idazoxan), which has little or no affinity for imidazoline-binding sites (Durcan et al., 1991). Ethanol withdrawal symptoms are associated with a transient 24-h increase in plasma noradrenaline (Smith et al., 1990), suggesting that ethanol-seeking behavior that occurs during acute withdrawal may be related to the increased synaptic availability of noradrenaline. Mixed results have been obtained in studies using systemic administration of these agents on ethanol intake, which suggests that noradrenergic effects may show CNS regional specificity. In ethanol-preferring rats, peripherally administered clonidine reduced alcohol intake in a two-bottle drinking experiment (Opitz, 1990), but atipamezole transiently increased ethanol intake 24 h after implantation of osmotic minipumps and produced no change for 7 d thereafter (Korpi, 1990). No studies have tested the effects of centrally administered noradrenergic agonists or antagonists on ethanol reinforcement.

Brain 5-HT is thought to inhibit feeding by virtue of findings that peripheral administration of numerous compounds putatively acting at central 5-HT synapses decrease food intake (see Blundell, 1986 and Leibowitz & Shor-Posner, 1986 for reviews). This hypothesis is supported by findings that peripheral and PVN administration of 5-HT or fluoxetine resulted in a nutrient selective reduction in carbohydrate intake (Leibowitz et al., 1990; Weiss, et al., 1991). Suppressive effects on feeding behavior by central injection of 5-HT are not observed in other brain areas including the amygdala, nucleus accumbens, septum, and the nucleus reuniens that is dorsal to the PVN (Leibowitz et al., 1990). Further, systemic injection of fluoxetine resulted in a significant increase in extracellular 5-HT as measured by microdialysis, suggesting that the PVN may be a site of action of serotonergic inhibition of food intake (Paez & Leibowitz, 1993).

Alterations in synaptic availability of 5-HT influence ethanol intake. As with food intake, numerous studies have shown that increases in synaptic availability of serotonin (5-HT) decrease ethanol intake. For instance, precursor loading with 5-hydroxytryptophan (Zabik & Roache, 1983), administration of 5-HT agonists (McBride et al., 1990; Svenson et al., 1989), uptake inhibitors (Gill et al., 1988; Haraguchi et al., 1990; Murphy et al., 1985, 1988; Rockman et al., 1982), and high doses of the 5-HT_3 antagonist, tropisetron (Hodge et al., 1993b), decrease ethanol intake. Further support for the involvement of 5-HT receptors in ethanol drinking comes from the recent finding that systemic injections of low doses of 8-OH-DPAT increase ethanol intake in a limited-access situation (Tomkins et al., 1994). Low doses of 8-OH-DPAT have been shown to selectively activate 5-HT receptors in the dorsal and median raphae nuclei (Hutson et al., 1986), which results in a reduction of synthesis and release of 5-HT in terminal regions and a net decrease in 5-HT neurotransmission (Invernizzi et al, 1991; Hjorth & Sharp, 1991).

A good deal of attention has been paid to the role of the 5-HT_{1A} receptor subtype in ethanol intake because both the ethanol-preferring rats (P) and nonpreferring rats (NP) have been shown to have differential densities of $5HT_{1A}$ recognition sites in some brain areas (McBride et al., 1990). Importantly, P-rats have decreased binding of [^{3}H]8-OH DPAT in the dorsal raphe nucleus as compared to NP-rats (McBride et al., 1990). Accordingly, systemic administration of the selective 5-HT_{1A} agonist, 8-OH-DPAT, reduced ethanol consumption in P-rats in several models of drinking (Svenson et al., 1989; McBride et al., 1990; Kotowski & Dyr, 1992). Given the interaction between 5-HT and α2-noradrenergic systems in the regulation of feeding cited earlier, these data suggest similar mechanisms in the control of ethanol intake. Because P-rats have fewer 5-HT_{1A} recognition sites in the dorsal raphe nucleus, it is plausible that a net reduction in 5-HT_{1A} input to the PVN may disinhibit α2-noradrenergic neurons, which results in a net increase in ethanol (or food) intake. However, whether 5-HT and noradrenergic transmission in the PVN show a relationship in regulating ethanol satiety as they do for food remains to be tested.

In summary, hypothalamic nuclei form a complex regulatory system that is involved in consummatory behavior. Neuroanatomical data indicate that the majority of outputs from the LH and VMH run through the DMH to the PVN. Hypothalamic nuclei respond to discriminative stimulus conditions (i.e., the sight of food) that may regulate the onset of eating. This suggests that similar mechanisms may be involved in controlling the onset of ethanol intake and/or ethanol reinforced responding. Outputs from the PVN may influence other systems involved in the regulation of ethanol intake such as the nucleus accumbens. However, whether the PVN is important to ethanol intake remains to be tested.

IV. SUMMARY AND CONCLUSIONS

Over the last 15 years, the understanding of the neurobiology of ethanol consumption has seen important advances. These are due in part to the development of better animal models of ethanol self-administration, progress in our general understanding of brain processes involved in drug self-administration, and the marked advances in the field of neuroscience. It now seems clear that processes that underlie the reinforcement/reward actions for most reinforcing stimuli are also important for ethanol reinforcement. The mesolimbic pathways involved in directing an organism's behavior for "natural" reinforcers play important roles in the self-administration of drugs and alcohol. From the data reviewed in this chapter, it would appear that the meso-accumbens DA pathways are important for both the maintenance and termination of ethanol consumption once it has begun. It would seem probable that other brain processes are involved in the initial seeking behaviors and the onset of a drinking bout. It is suggested that studies of the involvement of hypothalamic mechanisms, known to be involved in the regulation of feeding and drinking behaviors, could provide additional important data on the regulation of ethanol consumption.

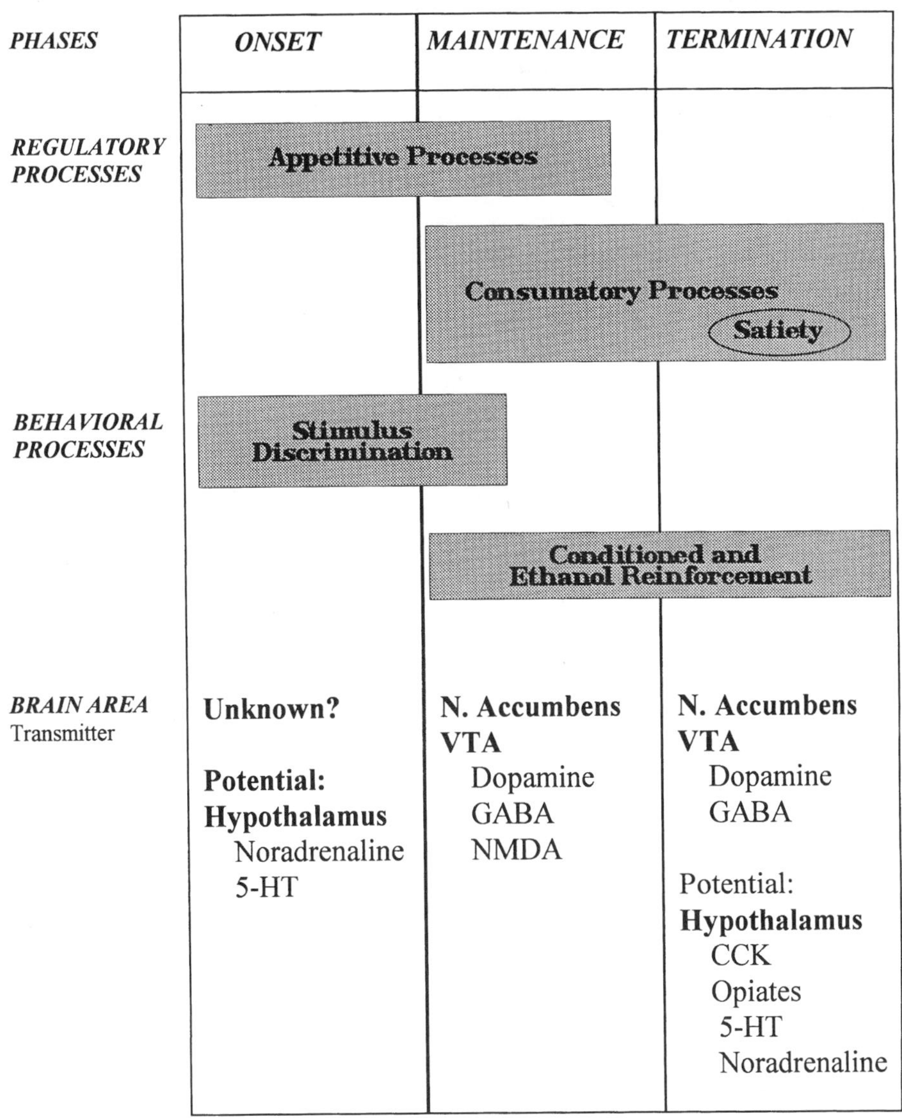

FIGURE 13.5. Involvement of Regulatory Processes, Behavioral Processes, Brain Areas, and Transmitters in the onset, maintenance, and termination of ethanol reinforcement bouts as shown by data from central administration studies.

Using a regulatory behavior framework, the phases of an alcohol bout can be conceptualized at several levels of analysis (see Figure 13.5). Using our previous terminology of onset, maintenance, and termination for a limited access ethanol bout (Samson et al., 1992), we conceptualize appetitive regulatory processes as occurring during the onset and maintenance phases, whereas consummatory processes are involved during the maintenance and termination processes. The termination of responding clearly includes the process of satiety. There is a great deal of literature about how these processes are involved in the regulation of feeding and drinking, and the application of this information to the study of alcohol bout regulation would appear to be of heuristic value in the study of ethanol drinking control.

These regulatory processes can also be described in terms of behavioral processes that provide a different level of analysis of the bout phases. The development of stimulus discrimination (stimulus control) as a result of initiation is conceived of as being involved primarily in the appetitive regulatory process during the onset and maintenance phases, with conditioned and ethanol reinforcement processes being important for the appetitive and consummatory processes occurring during the maintenance and termination phases. Clearly, there is overlap within all three levels of analysis during a drinking bout, with discriminative stimuli, conditioned reinforcers, and ethanol reinforcers all interacting at each phase of the drinking episode.

When neural substrate involvement is examined for each phase of ethanol bout regulation, the data suggest that for the onset phase, it is unknown what areas and transmitters are important (see Figure 13.5). It is suggested that potential areas of interest, based on their involvement with other appetitive processes, would be the hypothalamus and the transmitters, noradrenaline and serotonin. It is clear that dopamine actions within the nucleus accumbens play a limited role in the onset phases, but might be important in other areas of the mesolimbic system (i.e., prefrontal cortex, amygdala, and hippocampus). During the maintenance phase, both appetitive and consummatory processes are involved, and it appears that the mesoaccumbens DA pathway plays an important function in maintaining the complex behavior occurring during this phase. This brain system appears important for the termination phase of the bout because termination can be altered by dopamine agonist microinjection into the nucleus accumbens. The actions of a variety of satiety processes that involve other brain areas most likely play important roles in the termination of the bout, but these processes have only been explored on a limited basis.

In humans, a variety of factors come into play in the regulation of ethanol consumption at various times. Many of these factors, such as social norms, etc. are difficult, if not impossible, to replicate in rodent and even primate models. However, many of the factors related to the regulation of drinking elucidated in animal models have been shown to have predictive importance in humans. Thus, continued efforts to understand the neurobiology of alcohol regulation can provide important information that could lead to better pharmacological adjuncts for the treatment of alcohol abuse and alcoholism.

ACKNOWLEDGMENTS

Preparation of this manuscript was supported by grants from the National Institute on Alcohol Abuse and Alcoholism (ROl AA 07404, R01 AA 06845, and K05 AA 00142 to HHS and RO1 AA09981 to CWH).

REFERENCES

Babor, T.F. (Ed.), *Alcohol and Culture: Comparative Perspectives from Europe and America.* New York Academy of Sciences, New York, v472, 1986.

Bakshi, V.P. and Kelley, A.E., Dopaminergic regulation of feeding behavior: I. Differential effects of haloperidol microinfusion into three striatal subregions. *Psychobiology,* 19:223–232, 1991a.

Bakshi, V.P. and Kelley, A.E., Dopaminergic regulation of feeding behavior: II. Differential effects of amphetamine microinfusion into three striatal subregions. *Psychobiology,* 19:233–242, 1991b.

Berendse, H.W. and Groenewegen, H.J., Organization of the thalamostriatal projections in the rat, with special emphasis on the ventral striatum. *J. Comp. Neurol.,* 229:187–228, 1990.

Berendse, H.W., Groenewegen, H.J., and Lohman, A.H.M., Compartmental distribution of ventral striatal neurons projecting to the mesencephalon in the rat. *J. Neurosci.,* 12:2079–2103, 1992.

Bernardis, L.L. and Bellinger, L.L., The lateral hypothalamus area revisited: neuroanatomy, body weight regulation, neuroendocrinology and metabolism. *Neurosci. Biobehav. Rev.,* 17:141–193, 1993.

Blundell, J.E., Serotonin manipulations and the structure of feeding behaviour. *Appetite,* 7:39–56, 1986.

Blundell, J.E., Pharmacology of appetite control. *Trends Pharmacol. Sci.,* 12: 147–157, April 1991.

Booth, D.A., Ingestive behavior and neural processes. In E.M. Stricker (Ed) *Handbook of Behavioral Neurobiology: Neurobiology of Food and Fluid Intake.* New York: Plenum Press, pp. 465–488, 1990.

Bozarth, M.A., The mesolimbic dopamine system as a model reward system. In P. Willner and J. Scheel-Kruger (Eds.), *The Mesolimbic Dopamine System: From Motivation to Action.* New York: John Wiley & Sons, pp. 301–330, 1991.

Burton, M.J., Rolls, E.T., and Mora, F., Effects of hunger on the responses of neurons in the lateral hypothalamus to the sight and taste of food. *Exp. Neurol.,* 51:668–677, 1976.

Cador, M., Robbins, T.W., and Everitt, B.J., Involvement of the amygdala in stimulus-reward: interaction with the ventral striatum. *Neuroscience,* 30:77–86, 1989.

Cador, M., Robbins, T.W., Everitt, B.J., Simon, H., Le Moal, M., and Stinus, L., Limbic-striatal interactions in reward-related processes: modulation by the dopaminergic system. In P. Willner and J. Scheel-Kruger (Eds.) *The Mesolimbic Dopamine System: From Motivation to Action.* New York: John Wiley & Sons, pp. 225–250, 1991.

Campfield, L. A. and Smith, F.J., Systemic factors in the control of food intake: evidence for patterns as signals. In E.M. Stricker (Ed), *Handbook of Behavioral Neurobiology: Neurobiology of Food and Fluid Intake.* New York: Plenum Press, pp. 183–206, 1990.

Cannon, W.B., *The wisdom of the body.* New York: Norton, 1932.

Craig, W., Appetites and aversions as constituents of instincts. *Biol. Bull.,* 34:91–107, 1918.

Cunningham, E.T. and Sawchenko, P.E., Anatomical specificity of noradrenergic inputs to the paraventricular and supraoptic nuclei of the rat hypothalamus. *J. Comp. Neurol.,* 274:60–76, 1988.

Cunningham S.T. and Kelley A.E., Opiate infusion into nucleus accumbens: contrasting effects on motor activity and responding for conditioned reward. *Brain Res.,* 588:104–114, 1992a.

Cunningham S.T. and Kelley A.E., Evidence for opiate-dopamine cross-sensitization in nucleus accumbens: studies of conditioned reward. *Brain Res. Bull.,* 29:675–680, 1992b.

Dole, V.P., On the relevance of animal models to alcoholism in humans. *Alcoholism: Clin. Exp. Res.,* 10:361–363, 1986.

Domesick, V.B., Neuroanatomical organization of dopamine neurons in the ventral tegmental area. In P.W. Kalivas, and C.B. Nemeroff (Eds.), The mesocorticolimbic dopamine system. *Ann. NY Acad. Sci.,* 537, 10–26, 1988.

Durcan, M.J., Lister, R.G., and Linnoila, M., Evidence for central alpha-2 adrenoceptors, not imidazoline binding sites, mediating the ethanol-attenuating properties of alpha-2 adrenoceptor antagonists. *J. Pharmacol. Exp. Ther.,* 258:576–582, 1991.

Dworkin, S.I. and Smith, J.E., Neurobiological aspects of drug-seeking behaviors. In T. Thompson, P.B. Dews, and J.E. Barrett (Eds.), *Neurobehav. Pharmacol.,* Vol 6, 1–43, 1987.

Files, F.J., Andrews, C.M., Lewis, R.S., and Samson, H.H., Effects of ethanol concentration and fixed-ratio requirement on ethanol self-administration by P rats in a continuous access situation. *Alcoholism: Clin. Exp. Res.,* 17:61–68, 1993.

Files, F.J., Lewis, R.S., and Samson, H.H., Effects of continuous versus limited access to ethanol on ethanol self-administration. *Alcohol,* 11:523–532, 1994.

Gentry, R.T. and Dole, V.P., Why does a sucrose choice reduce the consumption of alcohol in C57BL/6J mice? *Life Sci.,* 40:2191–2194, 1987.

Gibbs, J., Smith, G.P., and Greenberg, D., Cholecystokinin: A neuroendocrine key to feeding behavior. In J. Schulkin (Ed), *Hormonally induced changes in mind and brain.* Orlando: Academic Press, 1993, 51–69.

Gill, K., Amit, Z., and Koe, B.K., Treatment with steraline, a new serotonin uptake inhibitor, reduces voluntary ethanol consumption in rats. *Alcohol,* 5:349–354, 1988.

Goldberg, S.R., Stimuli associated with drug injections as events that control behavior. *Pharmacol. Rev.,* 27:325–340, 1976.

Grant, K.A., An experimental analysis of oral ethanol self-administration in the free feeding rat. Unpublished dissertation, University of Washington, Seattle, 1984.

Gratton, A. and Wise, R.A., Drug- and behavior-associated changes in dopamine-related electrochemical signals during intravenous cocaine self-administration in rats. *J. Neurosci.,* 14:4130–4146, 1994.

Gratton, A., Wise, R.A., and Kiyatkin, E., Chronoamperometric measurements of dopamine levels in rat nucleus accumbens during cocaine self-administration. Society for Neuroscience Abstracts, 22nd Annual Meeting, Anaheim, California, p. 1076, 1992.

Groenewegen, H.J., Berendse, H.W., Wolters, J.G., and Lohman, A.H.M., The anatomical relationship of the prefrontal cortex with the striatopallidal system, the thalamus and the amygdala: evidence for a parallel organization. In H.B.M. Uylings, C.G. Van Eden, J.P.C. De Bruin, M.A. Corner, and M.G.P. Feenstra (Eds.), *Progress in Brain Research,* Vol 85, Elsevier Science Publishers, Amsterdam, 1990, 95–118.

Haraguchi, M., Samson, H.H., and Tolliver, G.A., Reduction in oral ethanol self-administration in the rat by the 5-HT uptake blocker fluoxetine. *Pharmacol. Biochem. Behav.,* 35:259–262, 1990.

Harris R.A., Brodie, M.S., and Dunwiddie, T.V., Possible substrates of ethanol reinforcement: GABA and dopamine. In P.W. Kalivas and H.H. Samson, (Eds.), The neurobiology of drug and alcohol addiction. *Ann. NY Acad. Sci.,* 654:61–69, 1992.

Heather, N., Tebbutt, J.S., Mattick, R.P., and Zamir, R., Development of a scale for measuring impaired control over alcohol consumption: a preliminary report. *J. Stud. Alcohol,* 54:700–709, 1993.

Hodge, C.W., Haraguchi, M., and Samson, H.H., Microinjections of dopamine agonists in nucleus accumbens increase ethanol reinforced responding. *Pharmacol. Biochem. Behav.,* 43:249–254, 1992.

Hodge, C.W., Haraguchi, M., Erickson, H.L., and Samson, H.H., Microinjections of quinpirole in the ventral tegmentum decrease ethanol reinforced responding. *Alcoholism: Clin. Exp. Res.,* 17:370–375, 1993a.

Hodge, C.W., Samson, H.H., Lewis, R.S., and Erickson, H.L., Specific decreases in ethanol- but not water-reinforced responding produced by the 5-HT3 antagonist ICS 205–930. *Alcohol,* 10:191–196, 1993b.

Hodge, C.W., Samson, H.H., Tolliver, G.T., and Haraguchi, M., Effects of intra-accumbens injections of dopamine agonists and antagonists on sucrose- and sucrose-ethanol reinforced responding. *Pharmacol. Biochem. Behav.,* 48:141–150, 1994.

Hjorth, S. and Sharp, T., Effects of the 5-HT1 A receptor agonist 8-OH-DPAT on the release of 5-HT in dorsal and median raphae-innervated rat brain regions as measured by *in vivo* microdialysis. *Life Sci.,* 48:1779–1786, 1991.

Hutson, P.H., Dourish, C.T., and Curzon, G., Neurochemical and behavioural evidence for mediation of the hyperphagic action of 8-OH-DPAT by 5-HT cell body autoreceptors. *Eur. J. Pharmacol.,* 1 29:347–352, 1986.

Imperato, A. and DiChiara, G., Preferential stimulation of dopamine release in the nucleus accumbens of freely moving rats by ethanol. *J. Pharmacol. Exp. Ther.,* 238:219–228, 1986.

Invernizzi, R., Carli, M., DiClemente, A., and Samanin, R., Administration of 8-hydroxy-2-(Di-N-propylamino)-tetralin in raphe nuclei dorsalis and medianus reduces serotonin synthesis in the rat brain: differences in potency and regional sensitivity. *J. Neurochem.,* 56:243–247, 1991.

Kalivas, P.W. and Nemeroff, C.B. (Eds.), The mesocorticolimbic dopamine system. *Ann. NY Acad. Sci.,* 537, 1988.

Kalivas, P.W. and Samson, H.H. (Eds.), The neurobiology of drug and alcohol addiction. *Ann. NY Acad. Sci.,* 654, 1992.

Katz, J.L., Drugs as reinforcers: pharmacological and behavioral factors. In J.M. Liebman and S.J. Cooper (Eds.), *The Neuropharmacological Basis of Reward.* Oxford: Oxford Press, 164–213, 1989.

Kelley, A.E. and Throne, L.C., NMDA receptors mediate the behavioral effects of amphetamine infused into the nucleus accumbens. *Brain Res. Bull.,* 29:247–254, 1992.

Koob, G.F., Neural Mechanisms of Drug Reinforcement. In P.W. Kalivas and H.H. Samson (Eds.), The Neurobiology of Drug and Alcohol Addiction, *Ann. NY Acad. Sci.,* 654:171–191, 1992a.

Koob, G.F., Drugs of abuse: anatomy, pharmacology and function of reward pathways. *TIPS,* 13:177–184, 1992b.

Koob, G.F. and Goeders, N. E., Neuroanatomical substrates of drug self-administration. In J.M. Liebman and S.J. Cooper (Eds.) *The Neuropharmacological Basis of Reward.* Oxford: Oxford Press, 214–263, 1989.

Korpi, E.R., Effects of alpha 2-adrenergic drugs on the alcohol consumption of alcohol-preferring rats. *Pharmacol. Toxicol.,* 66:283–286, 1990.

Kotowski, W. and Dyr, W., Effects of 5-HT1 A receptor agonists on ethanol preference in the rat. *Alcohol,* 9:283–286, 1992.

Kulkosky, P.J. and Glazner, G.W., Dose-additive inhibition of intake of ethanol by cholecystokinin and bombesin. *Alcoholism: Clin. Exp. Res.,* 1 2:277–281, 1988.

Kulkosky, P.J., Clayborne, Y.J., and Sandoval, S.L., CCK and bombesin inhibit ethanol and food intake in rats selectively bred for ethanol sensitivity. *Alcoholism: Clin. Exp. Res.,* 17:545–551, 1993.

Leibowitz, S.F., Neurochemical systems of the hypothalamus: control of feeding and drinking behavior and water-electrolyte excretion. In P.J. Morgane and J. Panksepp (Eds.), *Handbook of the hypothalamus. Vol 3a. Behavioral studies of the hypothalamus.* New York: Marcel Dekker, 229–237, 1980.

Leibowitz, S.F. and Shor-Posner, G., Brain serotonin and eating behavior. *Appetite,* 7:Suppl 1–14, 1986.

Leibowitz, S.F, Jhanwar-Uniyal, M., Dvorkin. B., and Makman, M.H., Distribution of α-adrenergic, β-adrenergic, and dopaminergic receptors in discrete hypothalamic areas of rat. *Brain Res.,* 233:97–114, 1982.

Leibowitz, S.F., Brown, O., Tretter, J.R., and Kirschgessner, A., Norepinephrine, clonidine, and tricyclic antidepressants selectively stimulate carbohydrate ingestion through noradrenergic system of the paraventricular nucleus. *Physiol. Pharmacol. Behav.,* 23:541–550, 1985a.

Leibowitz, S.F., Weiss, G.F., Yee F., and Tretter, J.B., Noradrenergic innervation of the paraventricular nucleus: Specific role in control of carbohydrate ingestion. *Brain Res. Bull.,* 14:561–567, 1985b.

Lelbowitz, S.F., Weiss, G.F., and Suh, J.S., Medial hypothalamic nuclei mediate serotonin's inhibitory effect on feeding behavior. *Pharmacol. Biochem. Behav.,* 37:735–742, 1990.

Le Moal, M. and Simon, H., Mesocorticolimbic dopaminergic network: functional and regulatory roles. *Physiol. Rev.,* 71:155–234, 1991.

Liebman, J.M. and Cooper, S.J., *The Neuropharmacological Basis of Reward.* Oxford: Oxford Press, 1989.

Lister, R.G., Durcan, M.J., Nutt, D.J., and Linnoila, M., Attenuation of ethanol intoxication by alpha-2 adrenoceptor antagonists. *Life Sci.,* 44:111–119, 1989.

Littieri, D.J., Sayers, M., and Pearson H.W. (Eds.), Theories on Drug Abuse: Selected Contemporary Perspectives. NIDA Research Monograph 30, U.S. Govern. Printing House, DHHS Pub #(80–967), 1980.

Luiten, P.G.M., Ter Horst, G.J., and Steffens, A.B., The hypothalamus, intrinsic connections and outflow pathways to the endocrine system in the relation to the control of feeding and metabolism. *Prog. Neurobiol.,* 28:1–54, 1987.

Marcucella, H. and Munro, I., Ethanol consumption of free feeding animals during restricted ethanol access. *Alcohol Drug. Res.,* 7:405–414, 1987.

McBride, W.J., Murphy, J.M., Lumeng, L., and Li, T.-K., Serotonin, dopamine and GABA involvement in alcohol drinking of selectively bred rats. *Alcohol,* 7:199–205, 1990.

McCabe, J.T., Debellis, M., and Leibowitz, S.F., Clonidine-induced feeding: analysis of central sites of action and fiber projections mediating this response. *Brain Res.,* 390:85–104, 1984.

McDonald, A.J., Organization of amygdaloid projections to the prefrontal cortex and associated striatum in the rat. *Neuroscience,* 44:1–14, 1991a.

McDonald, A.J., Topographical organization of amygdaloid projections to the caudatoputamen, nucleus accumbens, and related striatal-like areas of the rat brain. *Neuroscience,* 44:15–33, 1991b.

McKearney, J.W., Rate-dependency: Scope and limitations in the explanation and analysis of the behavioral effects of drugs. In T. Thompson, P.B. Dews, and W.A. McKim (Eds.), *Advances in Behavioral Pharmacology* (vol 3), New York: Academic Press, 1981, 91–110.

Meisch, R.A., Ethanol self-administration: Infrahuman studies. In T. Thompson and P.B. Dews (Eds.), *Advances in Behavioral Pharmacology* (vol 1), New York: Academic Press, pp. 36–85; 1977.

Mogenson, G.J. and Yim C.C., Neuromodulatory functions of the mesolimbic dopamine system: electrophysiological and behavioral studies. In P. Willner and J. Scheel-Kruger (Eds.), *The Mesolimbic Dopamine System: From Motivation to Action.* New York: John Wiley & Sons, pp. 105–130, 1991.

Murphy, J.M., Waller, M.B., Gatto, G.J., McBride, W.J., Lumeng, L., and Li, T.-K., Monamine uptake inhibitors attenuate ethanol intake in alcohol-preferring (P) rats. *Alcohol,* 2:349–352, 1985.

Murphy, J.M., Waller, M.B., Gatto, G.J., McBride, W.J., Lumeng, L., and Li, T.-K., Effects of fluoxetine on the intragastric self-administration of ethanol in the alcohol preferring P line of rats. *Alcohol,* 5:283–286, 1988.

Oades R.D., Simon, H., and Le Moal, M., Modulation by dopaminergic neurons of neurobiological systems processing strategies in adaptive behaviors and learning. *Behav. Proc.,* 10:180–181, 1985.

Opitz, K., The effect of clonidine and related substances on voluntary ethanol consumption in rats. *Drug Alcohol Dep.,* 25:43–48, 1990.

Paez, X. and Leibowitz, S.F., Changes in extracellular PVN monamines and macronutrient intake after idazoxan or fluoxetine injection. *Pharmacol. Biochem. Behav.,* 46:933–941, 1993.

Pfeffer, A.O. and Samson, H.H., Oral ethanol reinforcement: interactive effects of amphetamine, pimozide and food-restriction. *Alcohol Drug Res.,* 6:37–48, 1985.

Pfeffer, A.O. and Samson, H.H., Haloperidol and apomorphine effects on ethanol reinforcement in free feeding rats. *Pharmacol. Biochem. Behav.,* 29:343–350, 1988.

Pennartz, C.M.A., Lopes Da Silva, F.H., and Groenewegen, H.J., The nucleus accumbens as a complex of functionally distinct neuronal ensembles: an integration of behavioral, electrophysiological and anatomical data. *Prog. Neurobiol.,* 42:719–740, 1994.

Phillips, A.G. and Fibiger, H.C., Neuroanatomical bases of intracranial self-stimulation: untangling the Gordian knot. In J.M. Liebman and S.J. Cooper (Eds.), *The Neuropharmacological Basis of Reward.* Oxford: Oxford Press, 66–105, 1989.

Phillips, A.G. and Fibiger, H.C., Dopamine and motivated behavior: Insights provided by *in vivo* analyses. In P. Willner and J Scheel-Kruger (Eds.), *The Mesolimbic Dopamine System: From Motivation to Action.* New York: John Wiley & Sons, pp. 199–224, 1991.

Pirot, S., Godbout, R., Mantz, J., Tassin, J.-P., Glowinski, J., and Thierry, A.-M., Inhibitory effects of ventral tegmental area stimulation on the activity of prefrontal cortical neurons: evidence for the involvement of both dopaminergic and gabaergic components. *Neuroscience,* 49:857–865, 1992.

Rassnick, S., Pulvirenti, L., and Koob, G.F., Oral ethanol self-administration in rats is reduced by the administration of dopamine and glutamate receptor antagonists into the nucleus accumbens. *Pychopharmacology,* 109:92–98, 1992.

Rassnick, S., Stinus, L., and Koob, G.F., The effects of 6-hydroxydopamine lesions of the nucleus accumbens and the mesolimbic dopamine system on oral self-administration of ethanol in the rat. *Brain Res.,* 623:16–24, 1993.

Rockman, G.E., Amit, Z., Brown, Z.W., Bourque, C., and Ogren, S.O., An investigation of the mechanisms of action of 5-hydroxytryptamine in the suppression of ethanol intake. *Neuropharmcology,* 21:341–347, 1982.

Rolls, E.T., Central nervous systems mechanisms related to feeding and appetite. *Br. Med. Bull.,* 37:131–134; 1981.

Rolls, E.T., Burton, M.J., and Mora, F., Hypothalamic neuronal responses associated with the sight of food. *Brain Res.,* 111:53–66; 1976.

Rolls, E.T., Burton, M.J., and Mora, F., Neurophysiological analysis of brain-stimulation reward in the monkey. *Brain Res.,* 194:339–357; 1980.

Rozin, P.N. and Schulkin, J., Food selection. In E.M. Stricker (Ed.), *Handbook of Behavioral Neurobiology: Neurobiology of Food and Fluid Intake.* New York: Plenum Press, pp. 297–328, 1990.

Salamone, J.D., Behavioral pharmacology of dopamine systems: a new synthesis. In P. Willner and J. Scheel-Kruger (Eds.), *The Mesolimbic Dopamine System: From Motivation to Action.* New York: John Wiley & Sons, pp. 599–613, 1991.

Salamone, J.D., Complex motor and sensorimotor functions of striatal and accumbens dopamine: Involvement in instrumental behavior processes. *Psychopharmacology,* 107:160–174, 1992.

Samson, H.H., Initiation of ethanol reinforcement using a sucrose-substitution procedure in food- and water-sated rats. *Alcoholism: Clin. Exp. Res.,* 10:436–442, 1986.

Samson, H.H., Initiation of ethanol-maintained behavior: A comparison of animal models and their implication to human drinking. In T. Thompson, P.B. Dews, and J.E. Barrett (Eds.), *Advances in Behavioral Pharmacology: Neurobehavioral Pharmacology* (vol 6). Hillsdale: Lawrence Erlbaum, 1987, 221–248.

Samson, H.H. and Fromme, K., Social drinking in a simulated tavern: an experimental analysis. *Drug Alcohol Dep.,* 14:141–163, 1984.

Samson, H.H., Tolliver, G.A., and Schwarz-Stevens, K., Ethanol self-administration in a nonrestricted access situation: effect of ethanol initiation. *Alcohol,* 8:43–53, 1991.

Samson, H.H., Tolliver, G.A., Haraguchi, M., and Hodge, C.W., Alcohol self-administration: Role of mesolimbic dopamine. In P.W. Kalivas and H.H. Samson (Eds.), The Neurobiology of Drug and Alcohol Addiction, *Ann. NY Acad. Sci.,* 654:242–253, 1992.

Samson, H.H., Hodge, C.W., Tolliver, G.A., and Haraguchi, M., Effects of dopamine agonists and antagonists on ethanol reinforced behavior: the involvement of the nucleus accumbens. *Brain Res. Bull.,* 30:133–141, 1993.

Scheel-Kruger, J. and Willner, P., The mesolimbic system: principles of operation. In P. Willner and J. Scheel-Kruger (Eds.), *The Mesolimbic Dopamine System: From Motivation to Action.* New York: John Wiley & Sons, pp. 560–597, 1991.

Shor-Posner, G., Grinker, J.A., Marinescu, C., and Leibowitz, S.F., Role of hypothalamic norepinephrine in control of meal patterns. *Physiol. Behav.,* 35:209–214, 1985.

Shor-Posner, G., Azar, A.P., Volpe, M., Grinker, J.A., and Leibowitz, S.F., Clonidine hyperphagia: neuroanatomic substrates and specific function. *Pharmacol. Biochem. Behav.,* 30:925–932, 1988.

Smith, A.J., Brent, P.J., Henry, D.A., and Foy, A., Plasma noradrenaline, platelet alpha 2-adrenoceptors, and functional scores during ethanol withdrawal. *Alcoholism,* 14:497–502, 1990.

Stellar, E., Brain and behavior. In E.M. Stricker (Ed.), *Handbook of Behavioral Neurobiology: Neurobiology of Food and Fluid Intake.* New York: Plenum Press, pp. 3–22, 1990.

Stricker, E.M., Homeostatic origins of ingestive behavior. E.M. Stricker (Ed.), *Handbook of Behavioral Neurobiology: Neurobiology of Food and Fluid Intake.* New York: Plenum Press, pp. 45–60, 1990.

Swenson, L., Engel, J., and Hard, E., Effects of the 5-HT receptor agonist 8-OH-DPAT on ethanol preference in the rat. *Alcohol,* 6:17–21, 1989.

Taylor J.R. and Robbins, T.W., Enhanced behavioral control by conditioned reinforcers following microinjections of d-amphetamine into the nucleus accumbens. *Psychopharmacology,* 90:390–397, 1984.

Tomkins, D.M., Higgins, G.A., and Sellers, E.M., Effects of low doses of the 5-HT1 A agonist, 8-hydroxy-2-(Di-N-propylamino)-tetralin (8-OH-DPAT), on ethanol preference in the rat. *Psychopharmacology,* 115: 173–179, 1994.

Wadenberg, M., Ericson, E., Magnusson, O., and Ahlenius, S.A., Suppression of conditioned avoidance behavior by the local application of (–)sulpiride into the ventral, but not the dorsal, striatum of the rat. *Biol. Psychiatry,* 28:297–307, 1990.

Weiss, F. and Koob, G.F., The neuropharmacology of ethanol self-administration. In R.E. Meyer, G.F. Koob, M.J. Lewis, and S.M. Paul (Eds.), *Neuropharmacology of ethanol: New approaches,* Boston: Birkhauser, pp. 125–162, 1991.

Weiss, F., Hurd, Y.L., Ungerstedt, M.A., Plotsky, P.M., and Koob, G.F., Neurochemical correlates of cocaine and ethanol self-administration. In Kalivas, P.W. and Samson, H.H. (Eds.), The neurobiology of drug and alcohol addiction. *Ann. NY Acad. Sci.,* 654, 220–241, 1992.

Weiss, G.F., Rogacki, N., Fueg, A., Buchen, D., Suh, J.S., Wong, D.T., and Leibowitz, S.F., Effect of hypothalamic and peripheral fluoxetine injection on natural patterns of macronutrient intake in the rat. *Psychopharmacology,* 105:467–476, 1991.

White, F.J., Neurotransmission in the mesoaccumbens dopamine system. In P. Willner and J. Scheel-Kruger (Eds.), *The Mesolimbic Dopamine System: From Motivation to Action.* New York: John Wiley & Sons, pp. 61–103, 1991.

Willner, P. and Scheel-Kruger, J. (Eds.), *The Mesolimbic Dopamine System: From Motivation to Action.* New York: John Wiley & Sons, 1991.

Zabik, J. and Roache, J.D., 5-hydroxytryptophan-induced conditioned taste aversion to ethanol in the rat. *Pharmacol. Biochem. Behav.,* 18:785–790; 1983.

Chapter 14

INITIAL EFFECTS OF ETHANOL ON THE CENTRAL NERVOUS SYSTEM

Laura J. Draski and Richard A. Deitrich

TABLE OF CONTENTS

I. INTRODUCTION

The sensitivity of animals, including humans, to the first dose of ethanol is the topic of this chapter. The variability of this response is both environmental and genetic, as are all responses of animals to ethanol. This aspect of the effect of ethanol has received increased attention since the finding that the initial sensitivity/acute tolerance to ethanol in young men accurately predicts their chances of becoming alcoholic in later life (Schuckit, 1994).

We discuss those neuronal systems that have been implicated in this sensitivity as well as the neuronal systems that are altered by the initial exposure to ethanol. Details of how ethanol alters these systems on a molecular basis may be found in the chapters devoted to discussion of each of these systems.

In organizing this section, we have chosen to proceed from those measures that require the smallest doses to those that require larger doses. Table 14.1 lists the measures as well as a range of doses utilized in these tests.

0-8493-8389-7/95/$0.00+$.50

TABLE 14.1
Measures of Initial Sensitivity

	Dose range g/kg
Increased activity	0.5–2.0
Decreased temperature	1.0–4.0
Roto-Rod	1.0–2.5
Moving belt	1.0–2.5
Inclined plane	1.5–2.5
Aerial righting reflex	1.5–4.0
Descending platform	1.5–2.5
Locomotor depression	1.5–4.0
Sleep time	2.5–5.0
Death	>4.0

II. METHODS

There is a large variety of measures of initial sensitivity to ethanol. These are largely behavioral measures or, in some cases, physiological responses. Likewise, there are a number of neurochemical systems whose function is altered by ethanol.

A. LOCOMOTOR ACTIVATION

The physiological and behavioral effects of ethanol and many sedative-hypnotics are known to be biphasic in nature, with subhypnotic doses of ethanol (typically less than 2 g/kg i.p.) often inducing locomotor activation or arousal in rodents, and with higher doses resulting in marked locomotor depression (Frye and Breese, 1981). In some instances, a single dose may result in both initial stimulation and subsequent supression of motor activity, depending upon the length of time following drug administration (Crabbe et al., 1982; Pohorecky, 1977). The suggestion that the activating segment of the dose response may be viewed as a model of the reinforcing and euphoric properties of ethanol and other drugs of abuse makes research in this area invaluable to understanding the etiology of alcohol and drug addiction (Alhenius et al., 1973; Di Chiara and Imperato, 1988; McAuliff and Gordon, 1974; Phillips et al., 1991; Waller et al., 1986; Wise and Bozarth, 1987).

While a variety of different techniques have been utilized to quantify the locomotor-stimulating effects of ethanol in rodents, perhaps the most commonly used is the open field test (Crabbe, 1986). Briefly, an animal is placed into an enclosed area equipped either with infrared photocells and receptors distributed equally around its perimeter (Crabbe et al., 1987b) or with its interior otherwise divided into grids. Activity is determined either automatically by the number of photocell beam interruptions caused by the animal's movements or manually by counting the number of sectors entered by the animal. This open field test has been used successfully to specify the effects of low doses of ethanol on a number of genetically defined rodent groups. For example, differences in ethanol-induced increases in activity of DBA/2 and BALB/c mice, and decreases in C57BL/6 mice, have been well-established (Crabbe, 1986; Crabbe et al., 1982; Tabakoff and Kiianmaa, 1982; Crabbe et al., 1980). Nineteen highly inbred strains of mice were examined for differences in open field activity by Crabbe (1986), with marked differences observed between the strains in ethanol neurosensitivity. A genetic analysis of ethanol-induced locomotor activation was conducted using 23 genotypes of mice, including diallele crosses of four of the inbred strains tested, with high variation occurring across genotypes and minimal dominance effects observed in the direction of greater activation in the crosses (Dudek et al., 1991).

Mice selectively bred for either high (long sleep = LS) or low (short sleep = SS) initial sensitivity to ethanol-induced narcosis (McClearn and Kakihana, 1981) also have been tested for differential sensitivity to the activating effects of ethanol in the open field. In these studies, LS mice have been reported to be less stimulated by subhypnotic doses of ethanol than SS mice (Dudek and Abbott, 1984; Dudek et al., 1984b; Sanders, 1976). When LS and SS mice were cross-bred to two inbred lines with highly divergent ethanol-influenced activity, partial dominance in the direction of locomotor activation was observed, suggesting that genes other than those responsible for hypnotic sensitivity in LS and SS mice may influence locomotor activation produced by low doses of ethanol (Phillips and Dudek, 1991). Indeed, additional studies confirm that the genetic influences responsible for ethanol narcosis and activation appear relatively independent (Dudek et al., 1991; Dudek and Phillips, 1990).

Selective breeding for sensitivity to the activating effects of ethanol in the open field has been initiated by Crabbe and his colleagues (Phillips et al., 1991; Crabbe et al., 1990). Although the separation of the lines has not been extremely convincing, in general, FAST mice demonstrate an increased sensitivity to ethanol-induced locomotor stimulation, whereas SLOW mice appear less susceptible to the locomotor stimulant effects of ethanol. The selection for ethanol stimulation also seems to generalize to other drugs of abuse, with FAST mice showing more sensitivity to the stimulant effects of methanol, t-butanol, n-propanol, pentobarbital, and phenobarbital than SLOW mice, supporting the belief that a common mechanism of action exists for the stimulant properties of alcohols and barbiturates (Phillips et al., 1992). FAST mice also appear to be more sensitive to ethanol-induced ataxia than SLOW mice are, but do not differ in hypothermic or hypnotic response to ethanol (Crabbe et al., 1990).

Other variations of an automated activity chamber have been employed with equal success. For example, Randall et al. (1975) demonstrated dose-related increases in ethanol responses of BALB/cJ mice and decreases in C57BL/6J mice using stabilimeter cages mounted on a central axle. When movement in the cage disrupts balance and causes the apparatus to be tilted longitudinally, a mercury-activating switch positioned at the end of the cage detects and counts each tilt. Ethanol significantly increased exploratory behavior of DBA/2 mice in a holeboard apparatus that measures the amount of time an animal spends dipping its head into a 3-cm diameter hole drilled into the floor. Surface photobeam interruptions were increased by ethanol as well (Lister, 1987).

Recent advances in implantable radio-telemetry devices in small laboratory animals have made these systems more attractive to investigators. A small radio transmitter is surgically implanted into the intraperitoneal cavity, and a signal is sent to a receiver located beneath the animal cage. The strength of the received signal varies as the animal moves about its cage and is converted to one or more digital pulses by the receiver each time the signal changes by a specified amount (Clement et al., 1989). The telemetry system can be modified with ease to collect a variety of other physiological and behavioral variables, such as core body temperature, heart rate, and blood pressure (Van Den Buuse, 1994; Brockway et al., 1991; Dilsaver et al., 1990; Clement et al., 1989). The ability of this system to obtain accurate and continuous measurements makes it a powerful tool for studying the temporal and correlated effects of drugs.

While increases in spontaneous motor activity following subhypnotic doses of ethanol are commonly observed in mice, ethanol-induced behavioral activation in rats has been observed less frequently (Mason et al., 1979; Hunt and Overstreet, 1977; Pohorecky, 1977). Both of the selectively bred alcohol-accepting (AA) and alcohol-nonaccepting (ANA) rat lines have demonstrated increases in activity following ethanol doses of 0.6–1.0 g/kg (Basma et al., 1990; Hilakivi et al., 1984). In another group of rats bred for alcohol preference (P) and nonpreference (NP), only the P rats exhibited potentiation of locomotor activity by ethanol doses of 0.12 and 0.25 g/kg (Waller et al., 1986). To the contrary, other investigators have observed either no change or locomotor inhibition in rats following low doses of ethanol (Masur et al., 1986; Frye and Breese, 1981; Duncan and Baez, 1981; Linakis and Cunningham, 1979).

One hypothesis is that rats may be more sensitive than mice to the acute depressant effects of ethanol which, in turn, may mask its stimulatory actions in rats (Breese et al., 1984). When ethanol-induced locomotor depression was antagonized by the coadministration of thyrotropin-releasing hormone, significant increases in activity were evident in rats (Breese et al., 1984). However, when Masur et al. (1986) attempted to make rats less sensitive to the depressant effects of ethanol by chronically treating them with ethanol, they still were unable to induce locomotor stimulation in the ethanol-tolerant group. It is possible that the interspecies variability in ethanol-induced locomotor activation reflects differences that are genetic in nature.

B. TEMPERATURE RESPONSE

This topic is covered in much greater detail in Chapter 17, but the discussion here is included for completeness and to explore the usefulness of temperature measurements as indications of the initial effects of ethanol.

Administration of ethanol to an animal causes that animal to become poikothermic, and if it is in an environment where the temperature is less than that of the animal, its temperature will decline. In rodents the usual way of detecting this fall in temperature is by the insertion of a rectal thermistor, which is either left in place or inserted at regular intervals. Alternatively, a small transmitter ("minimitter") can be implanted in the abdominal cavity. After the animal's recovery from the surgery, a signal is generated from the temperature in the abdominal cavity. This is detected by a receiver close by and provides a continuous record of the temperature of the animal (Dilsaver et al., 1992). Such an arrangement is also capable of detecting movement by the animal, and two sets of data can be generated in this way. The question of which method more accurately reflects the temperature regulation of the animal under conditions of ethanol treatment has been addressed by several groups. Melchior and Allen (1993) found that probing mice led to a decline in temperature after each probe. This could be a response to the stress of probing or a reflection of the disturbance of the animal's thermal surroundings. The ratio of body surface area to body weight is large in an animal as small as a mouse, and mice lose heat rapidly in an environmental temperature less than body temperature.

In studies with rats, which have a smaller surface area to body weight ratio, Dilsaver et al. (1992) found a hyperthermic response to handling and postulated that this was responsible for the observed differences between results with the probe and the minimitter system. They did not study the effects of ethanol, however.

An obvious question is whether the hypothermic response to ethanol is correlated with other initial responses to ethanol or controlled by mechanisms separate from those other responses. In a group of recombinant inbred strains (RIs) between short and long sleep mice (SSXLS), DeFries et al. (1989) found that the correlation between body temperature and sleep time at 60 minutes following a dose of ethanol was only –0.01, clearly nonsignificant. In these same RI strains, but at subsequent generations, Erwin et al. (1990) found that hypothermia was not correlated with sleep time, blood ethanol at regain of righting reflex, or locomotor activation. They estimated that hypothermic effects of ethanol were controlled by about 4 genes, independent of genes controlling the other measures.

Crabbe and his associates (1987a, 1990) have selectively bred mice for sensitivity (COLD mice) or insensitivity (HOT mice) to ethanol-induced hypothermia. They found a correlation between activation by low-dose ethanol and resistance to the hypothermic effects of ethanol (Phillips and Crabbe, 1991).

Crabbe et al. (1994) have taken advantage of another group of RI strains, the BXD strains, to develop quantatitative trait loci (QTLs) for genes that influence hypothermia and tolerance to hypothermia in ethanol-treated mice. It is of interest that different QTLs were found, depending upon the dose with which the animals were tested. This would indicate that different systems are called into play as the dose increases. This approach should eventually enable researchers to identify the genes and their products that are responsible for this and other behavioral effects of ethanol.

C. STATIONARY DOWEL AND ROTO-ROD

The fixed dowel was popularized by Goldstein and colleagues (1982). In this paradigm animals, mice or rats, are trained to balance on a wooden dowel fixed some distance above a foam pad for a specificied time (usually 30 seconds). They are then injected with a small dose of ethanol, and the time required for them to regain their ability to remain on the dowel is noted, and a reading of blood ethanol level is taken. This gives a measure of the initial sensitivity of the animals. They are then usually given a second dose of ethanol, and the time to recovery and the corresponding blood ethanol level are taken again. The difference in blood ethanol level at time 2 minus the blood ethanol level at time 1 is a measure of tolerance developed in the interval between time 1 and time 2.

The rotating rod (roto-rod) taxes the animal more severely. Animals can be trained to some criterion, running either forward or backward at a constant speed of rotation or at an increasing speed of rotation. Similar procedures as those used for the fixed dowel can be employed to determine initial sensitivity and acute or rapid tolerance.

Of course, it is possible to utilize either of these procedures with untrained animals; however, it is the general experience that the ability, especially on the rotating rod, is so variable in a group of untrained animals that the effect of ethanol would be hidden in this variability. A second issue concerns the amount of training and the way the data are handled. One procedure is to train the animals to criteria on one or more days, then on the experimental day ascertain if the animals are still capable of achieving the level set for the training. If they are not, two choices are available. One is to discard the animal and use only those that have retained the ability to meet the criteria. The second choice is to determine what the animal's ability is and use a change score or a percent impairment from that level.

Another issue involves the effect of intoxicated practice. The question is primarily of concern for studies of tolerance and asks whether the practice of animals while intoxicated, even if they are trained to criteria previously, affords more, less, or the same amount of tolerance as would be found if practice was not permitted (Holloway et al., 1992; Bitrán and Kalant, 1991, 1993; Khanna et al., 1994; Lê and Kalant, 1992; Lê, 1990; Mayfield et al., 1992). This issue is dealt with in detail in Chapter 15.

D. MOVING BELT

Like the roto-rod, the moving belt test consists of a continuously moving motor-driven treadmill. Animals are trained to walk on a nylon belt placed over an electrified grid. Stainless steel staples are inserted into the nylon webbing of the belt, such that if the animal steps or falls off of the belt, the electrical circuit between either the belt and the grid or the grid bars is closed and the animal receives a shock (Gibbins et al., 1968). Impairment is measured as the total time off of the belt during the testing interval (typically 2 minutes). This procedure has proven to be extremely sensitive in discriminating the effects of ethanol intoxication resulting from doses of 1.0–2.5 g/kg in rats (Gibbins et al., 1968), and the test is commonly used in evaluating the acquisition and loss of tolerance to ethanol (LeBlanc et al., 1969, 1975).

The same limitations regarding training and intoxicated practice on the rotating rod are also true for the moving belt test (Le and Kalant, 1992; Gallaher and Loomis, 1979). For example, rats maintained at a constant, elevated blood ethanol level demonstrated tolerance after 6 hours to the effects of ethanol on the moving belt when they were allowed to practice hourly while intoxicated versus rats which were not provided practice time (Gallaher and Loomis, 1979). More recent studies also suggest that the development of acute tolerance may be accelerated and confounded by intoxicated practice (Le and Kalant, 1992).

E. INCLINED PLANE

This procedure is one in which animals, usually rats, are placed on a slightly rough surfaced wooden platform. The platform is automatically raised slowly until the rat begins to slide. The angle at which this occurs is noted for each animal. After the administration of ethanol, this process is repeated. The difference in the angles at which the rat begins to slide is a measure of the initial intoxicating effect of the administered ethanol. This procedure has been used as the measure for the development of selected lines of rats that are alcohol tolerant (AT) or alcohol nontolerant (ANT) by the research group at ALKO in Helsinki (Kiianmaa et al., 1988; Kiianmaa and Hellevuo, 1990; Eriksson, 1990). The dose of ethanol used for selection was 2 gm/kg. Larger doses do not reveal any differences between the lines. The selection has been primarily in the direction of the ANT line, that is, for a greater decrease in the sliding angle. The value for the AT line probably is not significantly different from that of the starting population even after 30 generations of selection (Eriksson, 1990). Even this separation has been slow. It required nearly 25 generations before there was no overlap between the standard deviations of the AT and the ANT lines. This would indicate that either there are a great number of genes involved or the task is not very efficient in separating sensitive from insensitive animals or both.

A number of correlated measures have been carried out with these animals. The ANT animals consume less ethanol in a free choice situation. Ethanol narcosis at 3.5 and 4 gm/kg is greater in ANT than in AT rats. The blood ethanol concentrations at onset of the loss of righting reflex, however, was not different suggesting that the initial brain sensitivity is not different in these two lines of rats. This suggests that the difference between the lines in this test of hypnosis is in the development rate of acute tolerance (Eriksson, 1990). Similar results were obtained in young SS and LS mice by Keir and Deitrich (1990).

Ethanol hypothermia is greater in ANT rats at 4 gm/kg, but not significantly different at 2.5 or 3.5 gm/kg. We have already pointed out that in studies in mice, hypothermia and hypnotic effects are not genetically correlated.

F. AERIAL RIGHTING REFLEX

Early studies of rodent neuro-ontogeny by Fox (1965) and Hard and Larsson (1975) suggest that the ability of an animal to turn itself over and land on all four legs after being dropped from a position with its back down is a good measure of the responsiveness of the central nervous system. Subsequently, this task has been modified to demonstrate impairment of neural functioning and coordination by pharmaceuticals, such as ethanol (Leitch et al., 1977).

Two methods of testing the aerial righting reflex have emerged with equal success. In the first case, all animals are tested at a specified post-injection interval by grasping the back of the neck and the base of the tail of each animal and releasing it from an inverted position onto either a foam rubber pad, a suspended towel, or another shock-absorbing material. The initial distance from the animal's back to the surface corresponds to the minimum height required for control subjects to successfully right themselves prior to landing, typically 3 cm for mice and 5 cm for rats (Frye and Breese, 1981). Prior to drug administration, all animals should be screened under control conditions, and those failing to right themselves at these minimal distances should be excluded for the study (Vogel et al., 1981). If an animal is unable to right itself at this height during testing, it is retested immediately at successively higher increments (e.g., 6 cm for mice and 10 cm for rats) until a height is determined in which the animal can land upright in at least two out of three attempts. Using this method, impairment of motor coordination was observed following doses of 1.5 (Frye and Breese, 1981) and 2.0 (Mattucci-Schiavone and Ferko, 1986) g/kg of ethanol in rats, and following 4.0 g/kg ethanol in mice (Frye and Breese, 1981).

Alternatively, other investigators have injected rats with ethanol (2.5 g/kg) and tested them periodically at a constant height of 37 cm for their ability to right themselves (Grant et al., 1989; Spuhler and Deitrich, 1984; Waller et al., 1983). In this paradigm, the duration of time that an animal requires to regain its aerial righting reflex is measured as an index of impairment.

G. DESCENDING PLATFORM

The jumping apparatus used in this paradigm provides delineative measurements of sensitivity and acute tolerance to ethanol (Lumeng et al., 1982; Tullis et al., 1977). In this test, rats are first trained to escape from a shock grid by jumping onto an elevated platform. Most rats will easily learn to jump heights of 45–50 cm in order to escape shock. Following an injection of ethanol, however, locomotor-impaired rats placed onto the shock grid may be unable to accomplish this feat. In these instances, the safety platform is lowered from a starting height of 50 cm at a rate of 1 cm/sec until the rat successfully reaches the platform, and the degree of locomotor disturbance is inversely related to the height of the platform. Alternatively, the recovery of jumping ability may be compared by measuring the elapsed time from ethanol administration until the animal is able to reach the platform at a specified height (Waller et al., 1983).

Lines of rats selectively bred for differences in the degree of ethanol-induced locomotor inhibition also have been shown to differ in ethanol sensitivity as measured by the descending platform apparatus (Bass and Lester, 1981). In this study, lines of rats that were "most affected" (MA) by ethanol depression in an activity device showed greater decreases in the height jumped following ethanol doses of 1.25–2.25 g/kg than the "least affected" (LA) rats. P rats also have been shown to be more sensitive to doses of ethanol ranging from 1.5–2.5 g/kg than were NP rats (Waller et al., 1983; Lumeng et al., 1982). More specifically, this paradigm was used to demonstrate that P rats developed acute tolerance to the effects of ethanol at a faster rate and to a larger degree than NP rats (Waller et al., 1983). Rapidly acquired, or acute, tolerance to ethanol may occur in response to a single administration of ethanol and typically is demarcated by comparing the blood ethanol concentrations from the ascending and descending portions of the blood ethanol curve that correspond to the loss and recovery of a specific behavior or function. If the ability to perform a behavior is recovered at a higher blood ethanol concentration than when it was lost, then acute tolerance is said to be present. Since the development of acute tolerance to ethanol may contribute to overall observed differences in sensitivity to ethanol, it is valuable to identify and isolate this phenomenon as separate from initial ethanol sensitivity, which reflects primarily differences in ability during the ascending limb of the blood ethanol curve.

H. SPONTANEOUS LOCOMOTOR ACTIVITY DEPRESSION

Variations of the open field test previously described to assess the effects of activating doses of ethanol have been adapted to quantify the reduction of spontaneous locomotor activity that typically is observed following central nervous system depression induced by ethanol. Studies examining the shape of the biphasic ethanol dose response curve for locomotor activity in inbred and selectively bred lines of mice clearly reflect strong genetic influences (Dudek et al., 1991; Phillips and Dudek, 1991). However, while both stimulating and intoxicating doses of ethanol produce alterations in locomotor activity, it is important to note that the effects of locomotor activation and depression do not appear to be correlated. For example, selectively bred LS and SS mice have been reported to show a negative relationship between the degree of ethanol-induced stimulation and depression, with SS mice being extremely more resistant to locomotor depression yet considerably more sensitive to locomotor activation following subhypnotic doses of ethanol than LS mice (Dudek and Abbott, 1984; Dudek et al., 1984; Sanders, 1976). On the contrary, inbred MOLD/RkAbg mice demonstrate equally strong locomotor activation and depression in their ethanol activity curves, while C57BL/6Abg mice show no measurable locomotor activation and minimal depression of activity following subhypnotic doses of ethanol (Dudek and Phillips, 1990). Thus, while both ethanol-induced locomotor inhibition and stimulation appear to be under genetic control, the genetic mechanisms responsible for each phenotype are thought to be relatively independent.

One selective breeding study has been conducted using impairment of spontaneous locomotor activity following ethanol as its selection phenotype (Riley et al., 1977; Riley et al., 1976). Rats were tested on an enclosed stabilimeter platform following an intraperitoneal injection of 1.5 g/kg ethanol, and the animals that were most affected (MA) and least affected (LA) by subsequent locomotor depression were chosen as breeders for each generation. Significant bidirectional changes in ethanol sensitivity were evident after five generations of selection. Furthermore, this difference in sensitivity is measurable across ethanol doses ranging from 0.75 to 2.25 g/kg, all of which produce ethanol depression in these lines of rats (Worsham and Freed, 1977).

Besides demonstrating a greater reduction in activity following ethanol administration, MA rats also show greater sensitivity to ethanol than LA rats do on a variety of other behaviors. For example, MA rats are more ataxic on the moving belt (Mayer et al., 1982), cannot jump as high to escape foot shock in the descending platform test (Bass and Lester, 1981), lose their righting reflex for longer periods of time following hypnotic doses of ethanol (Mayer et al., 1982), and become more hypothermic following an injection of ethanol (Mayer et al., 1983) than do LA rats. A reversed order of sensitivity was noted for ethanol-induced impairment of swimming, with LA rats taking longer to finish a swim task than MA rats did following an injection of 1.75 g/kg ethanol (Bass and Lester, 1980). These results suggest that while the mechanisms of sensitivity governing the sedative and ataxic effects of ethanol are similar across a host of behaviors, task-dependent effects still may be evident.

I. LOSS AND RECOVERY OF RIGHTING REFLEX (SLEEP TIME)

One of the most commonly used assessments of initial ethanol sensitivity, loss of the righting reflex, or sleep time, is similar to the previously described test of the aerial righting reflex except that the animal is laid on its back following ethanol administration rather than being dropped from an elevated supine position. While commonly referred to as "sleep time," the actual measurement is the duration of time between the loss and regain of the righting reflex or righting response. An animal is identified as recovered when it is capable of righting itself several times (typically two or three) within a 30–60 second interval. Interestingly, despite its resemblance to the aerial righting test, the two behaviors were found to be uncorrelated when analyzed in eight inbred rat strains (Spuhler and Deitrich, 1984). These behaviors also demonstrate dissimilar ontological profiles, with the righting reflex appearing at about the fifth or sixth postnatal day in rats (Bignall, 1974), and the aerial reflex not appearing until approximately postnatal day 15 or 16 (Hard and Larsson, 1975), suggesting that they are controlled at different levels of neural organization.

This hypnotic response to ethanol was given the unfortunate nickname of "sleep time," but, of course, it has nothing to do with normally defined sleep. Neither can it appropriately be referred to as anesthetic time, since often animals will appear to be fully awake and alert yet unwilling or unable to right themselves from their back. In an effort to abate some of these potential confounds on the assessment of ethanol narcosis, some investigators have suggested that the animals be stimulated periodically if one is interested in identifying exclusively deeper levels of hypnosis than those commonly encompassed by the generalized description of loss of the righting reflex.

Rather than the duration of the loss of the righting response, one can confine the assessment of initial hypnotic sensitivity to ethanol solely to the amount of time between administration of ethanol and loss of the righting response. However, one problem with this technique is that typically the blood and brain ethanol levels are rising so rapidly following intraperitoneal ethanol injections that accurate assessments of neurosensitivity are difficult to attain. If time to the loss of the righting reflex is chosen as a primary indicator of ethanol sensitivity, a preferable measure of this early time point would be the determination of blood or brain ethanol levels following an intragastric dose of ethanol. In this case, the absorption of ethanol is slow enough to obtain a reasonable estimate of the initial sensitivity of the brain.

A final concern regarding the interpretation of sleep time data pertains to the specter of acute tolerance. In the case of sleep time, this measurement inevitably will be contaminated by the development of acute tolerance. Furthermore, the longer the sleep time following ethanol, the greater the amount of time that is available for development of acute tolerance (Keir and Deitrich, 1990).

Selection on the basis of the duration of ethanol "sleep time has led to the production of the well-chronicled long sleep and short sleep (LS/SS) lines of mice (McClearn and Kakihana, 1981). While the original dose designated in the selection protocol was 3.3 g/kg of ethanol, the doses used in distinguishing the LS and SS populations have been adjusted numerous times, with typical hypnotic doses for the lines now being 2.8–3 g/kg for LS mice and around 5 g/kg for SS mice. The use of differential dosing in guiding the application of selection pressure is by no means unique to this breeding paradigm. Particularly for a behavioral phenotype such as sleep time, a successful response to selection response rapidly diminishes the effectiveness of that dose in discerning the most and least affected animals. That is, a previously effective dose of ethanol soon would fail to intoxicate the least sensitive animals, and this floor effect would prevent the identification of animals conclusively demonstrating the lowest sensitivity. Likewise, the original selection dose eventually may become fatal for the most sensitive animals, rendering them incapable of propagation. Thus, in order to restore the phenotypic variation essential to the continuation of the selective process, it is not uncommon to individualize the testing doses for the high- and low-responding lines. Yet another way of looking at this procedure is that the selection is not for the degree of central nervous system effect produced by a given dose of ethanol, but a given level of central nervous system effect which may require differential dosing to achieve.

Given the likelihood that the genes responsible for greater and lesser susceptibility to ethanol may not be altogether inclusive, one benefit of this strategy is that the segregation of the disparate genotypes actually may be facilitated by the response to different doses. It is important to remember that the genes which are responsible for more sensitivity to ethanol are not necessarily the opposite alleles at the same locus that makes for greater resistance to ethanol. There may be different, although presumably overlapping, sets of genes responsible for the phenotypes in the two lines.

Since their inception, the LS and SS mice have been used in a multitude of behavioral, physiological, and biochemical studies designed to uncover the neurobiological and molecular bases of acute ethanol intoxication (for a review see Deitrich, 1993). The myriad of effects produced by an initial exposure to ethanol supports the contention that, minimally, several genes are responsible for ethanol susceptibility. Depending on the method of quantitative genetic analysis used in calculating the number of loci involved in the selection of the LS and SS phenotypes, estimates of either 3.32 (Howerton et al., 1984), 7.1 (DeFries et al., 1989), or 9 (Dudek and Abbott, 1984b) have been made. Heritability calculations of this same response have been reported to be 0.18 (McClearn and Kakihana, 1981), 0.30 (Dudek and Abbott, 1984), 0.42 (Masson et al., 1992), and 0.52 (Spuhler et al., 1982).

A similar program of selectively breeding rats for sensitivity to the anesthetic effects of ethanol (High Alcohol Sensitive, HAS, and Low Alcohol Sensitive, LAS) has been carried out as well with many of the same results as observed in SS and LS mice (Deitrich et al., 1994; Allan et al., 1991; De Fiebre et al., 1991; Draski et al., 1992). Doses utilized in the HAS and LAS selection tend to be somewhat lower than those utilized in the LS and SS selection, with HAS rats responding to ethanol doses between 2.3–2.8 g/kg, and LAS rats responding to doses between 3.5–4.5 g/kg. Overall realized heritability has been estimated to be 0.26, which is in agreement with estimates of additive genetic variation from other alcohol-related selection studies (Draski et al., 1992). One advantage of using the HAS and LAS selected rats is the availability of replicated selected lines and concurrent, nonselected control lines. If genetic correlational analyses are to be conducted using traits hypothesized to be associated with the selected phenotype, it is imperative that each selected replicate be tested for the traits as independent verification of the potential relationship.

J. DEATH

It may seem strange to include death in a discussion of the initial effects of ethanol; however, this is unfortunately too often the result of overdoses with ethanol or a result of the combination of other sedatives with ethanol in the human population. The endpoint is hardly ambiguous (even a biochemist can recognize this behavioral measure). However, there is one aspect that needs some explanation. This is the finding that following large doses of ethanol, mice or rats will often recover the righting response but be found dead some time later (Tsibulsky and Amit, 1993a). While this is clearly death not due to the initial effects of ethanol, how does one treat these events? It is likely that the deaths are due to the extreme hypothermia and/or the peritoneal damage caused by high doses of ethanol given intraperitoneally. Gastric lavage of such large doses would avoid this and more closely approximate the intake method seen in humans.

The other question is whether or not such experiments are warranted. Clearly, experiments where death is prevented by some experimental manipulation can be useful for untangling the mechanisms of ethanol's action. Unfortunately, enough deaths in the control population must be prevented by the experimental drug for the results to be statistically valid. Such experiments give useful information for the treatment of accidental or deliberate overdoses in humans.

It is surprising how high the blood levels of humans can be for those individuals who have died of unnatural causes. Many of these individuals have blood ethanol levels above 0.29% (Perola et al., 1994). While it is true that these individuals died, they must have been alive, if unconscious, in order for the ethanol to be absorbed. Although it is also likely that many of these individuals were chronic alcoholics, and thus would have developed tolerance to ethanol, in comparison to the normal intoxicating level of 0.05–0.1%, these values are quite high. The fact that lowered body temperature protects against lethality in mice (Finn et al., 1991) probably helps to explain the fact that the human who held the record for surviving the highest ethanol blood level was hypothermic (32°) when admitted with a blood ethanol level of 1127 mg/dl (Berild and Hasselbalch 1981). Another individual survived after a blood alcohol level of 1500 mg/dl (O'Neill et al., 1984).

Although there is little doubt that tolerance to the lethal effects of ethanol can develop (McGivern et al., 1993; Tsibulsky and Amit, 1993a, 1993b), the cited authors do not agree as to the role of Pavlovian or environmentally cued conditioning in the process.

Selective breeding for decreased susceptibility to the acute effects of ethanol in mice also resulted in an increase in the lethal dose (Baker et al., 1987).

III. NEUROTRANSMITTER SYSTEMS INVOLVED IN INITIAL SENSITIVITY

A. *IN VIVO* EVIDENCE

In vivo evidence for the involvement of any specific neurotransmitter system necessarily comes from behavioral experiments utilizing the interaction of ethanol with agonists and antagonists of the particular neuronal systems. The success or failure of such experiments depends upon the specificity of the test compound and the caution with which the experiments are carried out and interpreted. For example, nearly all "specific" agonists and antagonists for any given neurotransmitter system will have overlapping effects on other systems at higher doses.

1. GABA System

The GABA system probably affords the best evidence for the involvement of any single neurotransmitter system in the acute actions of ethanol.

a. Interaction with Agonists

Interactions can be additive or synergistic. A case can be made for the synergistic interaction of ethanol with other GABA agonists since there is evidence that they interact with the GABA receptor system at different sites (see Chapter 4).

The well-known problem of overdose and death from the combination of benzodiazepines and ethanol provides data with humans. This has been reviewed by Sellers and Busto (1982) as well as by Hollister (1990) more recently. These authors review the evidence that ethanol and benzodiazepines interact. However, they believe that the incidence of this interaction and the severity of the consequences have been overplayed. More recently several studies have confirmed the additive effect of benzodiazepines and ethanol (Dorian et al., 1985; Linnoila et al., 1990b). Regardless of the clinical significance of these observations, these studies provide evidence that a putative agonist for the GABA system and ethanol do interact in humans, thus fulfilling one of the criteria given earlier for involvement of a neurotransmitter system in the actions of ethanol.

A virtual plethora of studies with animals provides similar data. Inverse agonists generally counteract at least some effects of ethanol (see Table 14.2), whereas antagonists have a mixed bag of effects depending upon the compounds and the effect of ethanol measured (see Table 14.3). A large number of studies have been carried out with various agonists of the GABA system in combination with ethanol. In general, GABA agonists and ethanol are at least additive in their CNS depressant effects (see Table 14.4).

Table 14.5 gives the chemical names of at least some of the agents referred to in the previous table. Finally, Table 14.6 presents some of the genetic evidence for interactions between ethanol and the GABA system.

From Tables 14.2 to 14.4 it can be discerned that, in general, the criteria for involvement of the GABA system in the acute effects of ethanol are well documented at the behavioral level. Agonists potentiate the effects of ethanol and, in lines of rodents selectively bred for ethanol sensitivity, there is evidence that such rodents are also differentially sensitive to GABA agonists. The exceptions are somewhat puzzling, but perhaps offer an opportunity to guide molecular biological experiments into the species differences in the GABA receptor. The fact that SS/LS mice do not differ in their hypnotic response to pentobarbital whereas HAS/LAS rats do is surprising, but could indicate that there is genetic heterogeneity in the pentobarbital binding site in the HAS/LAS lines, whereas there is no such heterogeneity in the mouse lines. Also somewhat anomalous is the response of SS and LS mice to phenobarbital. The difference in hypnotic potency for phenobarbital is as large as it is for ethanol in these lines of mice, in spite of its being less water soluble. Marley et al. (1986) demonstrated a good correlation between the potencies of water-soluble anesthetics and their oil/water partition coefficient in LS and SS mice. Somewhat surprising as well is that propofol, a very lipid soluble anesthetic agent, shows a large differential potency between SS and LS mice, but not between HAS and LAS rats. Finally, the SS and LS mice do not show differential sensitivity to halothane, whereas the HAS and LAS rats do. However, both the SS/LS mice and the HAS/LAS rats show differential sensitivity to enflurane and isoflurane.

b. Inverse Agonists

The inverse GABA/benzodiazepine agonists, in general, have been shown to counteract at least low-dose effects of ethanol. They are less efficient in counteracting the high-dose effects, including sleep time and lethal effects of ethanol.

c. Antagonists

The antagonists, especially picrotoxin and bicuculline, also have been shown to counteract the actions of ethanol.

TABLE 14.2
GABA Active Drugs Interaction with Ethanol

Inverse Agonists

Compound	Effect	Reference
RO-19–4603	+ Intoxication	1
RO-15-4513	+ Intoxication	2
RO-15-4513	– Anesthesia, ataxia	3
RO-15-4513	+ Conflict	4
RO-15-4513	+ Discriminative stimulation	5
RO-15-4513	+ Amnesia	6
RO-15-4513	– Death	7
RO-15-4513	* Activity (depends on type and species)	8
RO-15-4513	– Hypothermia	9
RO-15-4513	+ EEG in rats	10
RO-15-4513	+ Anticonvulsant effect	11
RO-15-4513	+ Operant tasks	12
RO-15-4513	+ Anxiolytic	13
FG 7142	+ Ataxia	14
FG 7142	– Intoxication	15
FG 7142	+ Anticonvulsant effect	16
FG 7142	+ Anxiolytic	17
B CCE	– Intoxication	18
CGS-8216	– Intoxication	19
B CCM	+ Anxiolytic	20
RO-15-3505	+ Anxiolytic, Intoxication	21
RO-15-3505	– Anesthesia	22
RO-15-3505	– Anxiolytic	23
DM-9384	+ Amnesia	24

Note: A (+) indicates an interaction was observed. In this case the interaction was to counteract the effect of ethanol. A (–) indicates no effect, while a (*) indicates additivity with ethanol.

1. (Lister and Durcan, 1989) 2. (Lister and Durcan, 1989; Suzdak et al., 1988; Glowa et al., 1989; Marrosu et al., 1988; Bonetti et al., 1989; Nabeshima et al., 1988; Suzdak et al., 1986; Hoffman et al., 1987) 3. (Suzdak et al., 1988; Bonetti et al., 1989; Hellevuo and Korpi, 1988) 4. (Glowa et al., 1989; Suzdak et al., 1986) 5. (Rees and Balster, 1988) 6. (Nabeshima et al., 1988) 7. (Poling et al., 1988) 8. (Becker and Hale, 1989; June and Lewis, 1989) 9. (Hoffman et al., 1987; Syapin et al., 1987) 10. (Ehlers et al., 1990) 11. (Nutt and Lister, 1989) 12. (Koob et al., 1989) 13. (Lister, 1988) 14. (Marrosu et al., 1988) 15. (Lister and Durcan, 1989; Glowa et al., 1989) 16. (Nutt and Lister, 1989) 17. (Lister, 1988) 18. (Suzdak et al., 1988; Glowa et al., 1989) 19. (Suzdak et al., 1988) 20. (Belzung et al., 1988) 21. (Belzung et al., 1988; Bonetti et al., 1989) 22. (Bonetti et al., 1989) 23. (Lister, 1988) 24. (Nabeshima et al., 1990). In addition, see reviews (Lister, 1989; Yu and Ho, 1990).

2. Glutamate Receptors

Evidence for the involvement of glutamate receptors in the acute effects of ethanol is accumulating rapidly (see Tables 14.7 and 14.8).

The behavioral evidence for interactions of ethanol with glutamate receptors is not as extensive as that for GABA receptors. Nevertheless, this evidence is extensive, if somewhat limited. The evidence with N-methyl-D-asparate (NMDA) itself given intracerebralventrically (ICV) demonstrated no interaction in SS/LS mice. However, several investigators have found ethanol, pentobarbital, and halothane to have increased potency in the presence of polyamines, putative agonists for the NMDA receptor. Ethanol did not interact with NMDA in discriminative stimulus tests.

The experiments with antagonists, specifically MK-801 and ketamine, are much more numerous, if not more convincing. In nearly every case, these agents increase the potency of ethanol or other CNS depressants. In addition, rodents trained to discriminate MK-801 or

TABLE 14.3
GABA Antagonists

Compound	Effect	Reference
Picrotoxin	+ Rodent intoxication	1
Picrotoxin	Increased hypothermia	2
ZK-93426	– Anticonvulsant effect	3
RO-15-1788	– Human intoxication (1 mg)	4
RO-15-1788	+ Human intoxication (5 mg)	5
RO-15-1788	+ Human EEG (0.5 mg)	6
Bicuculline	Dec. mice intoxication	7
Bicuculline	Inc. LS mice intoxication	8
Bicuculline	Increased hypothermia	9
3-alpha-hydroxy-5-beta-pregnan-20-one sulphate	Increased hypothermia. No effect ataxia, anesthesia	10

Note: A (+) indicates an interaction was observed. In this case an interaction was to counteract the effects of ethanol. A (–) indicates that no effect was found.

1. (Liljequist and Engel, 1982; Martz et al., 1983; Dudek and Phillips, 1989) 2. (Liljequist and Engel, 1982) 3. (Nutt and Lister, 1989) 4. (Klotz et al., 1986; Clausen et al., 1990; Fluckiger et al., 1988) 5. (Martens et al., 1990) 6. (Klotz et al., 1986) 7. (Dudek and Phillips, 1989; Martz et al., 1983) 8. (Dudek and Phillips, 1989) 9. (Liljequist and Engel, 1982) 10. (Melchior and Allen, 1992).

TABLE 14.4
GABA Agonists

Compound	Effect	Reference
Mucimol	+ Rodent intoxication	1
Aminooxyacetic acid	+ Rodent intoxication	2
Aminooxyacetic acid	Activity inhibited	3
GABA	Activity inhibited	4
Barbiturates	Additive on discriminative stimulus	5
Alparazolam Diazepam	+ Human psychomotor and cognition	6
Adinazolam Diazepam	+ Human psychomotor and cognition	7
Baclophen	+ Rodent intoxication	8
Baclophen	Activity inhibited	9
Triazolam	Human Psychomotor impairment increased	10
THIP	+ Rodent intoxication	11
3-alpha-hydroxy-5-beta-pregnan-20-one	+ Ataxia, hypothermia, anesthesia	12
Dehydroepiandosterone	+ Rodent intoxication	13

1. (Dudek and Phillips, 1989; Liljequist and Engel, 1982) 2. (Dudek and Phillips, 1989; Biswas and Carlsson, 1978; Martz et al., 1983) 3. (Biswas and Carlsson, 1978; Cott et al., 1976) 4. (Biswas and Carlsson, 1978; Cott et al., 1976) 5. (Massey and Woolverton, 1994) 6. (Linnoila et al., 1990a) 7. (Linnoila et al., 1990b) 8. (Martz et al., 1983) 9. (Cott et al., 1976) 10. (Dorian et al., 1985) 11. (Martz et al., 1983) 12. (Melchior and Allen, 1992) 13. (Melchior and Ritzmann, 1992).

TABLE 14.5
Chemical and Generic Names

Compound	Chemical Name	Action (Name)
RO-15-4513	Ethyl, 8-azido-5.6-dihydro-5-methyl-6-oxo-4H-imidazol [1,5-a] [1,4]benzodiazepine-3-carboxylate	Inverse agonist
RO-15-1788	Ethyl 8-fluoro-5,6-dihydro-5-methyl-6-oxo-4H-imidazo [1,5a] [1,4]benzodiazepine-3-carboxylate	(Flumazenil) Antagonist
RO-15-3505	Ethyl 7-chloro-5,6-dihydro-5-methyl-6-oxo 4H imidazo [1,5a] [1,4]benzodiazepine 3 carboxylate	Inverse agonist
ZK-93426	Ethyl-5-isopropoxy-4-methyl-b-carboline-3-carboxylate	Antagonist
FG 7142	N-methyl-B-carboline-3-carboxamide	Inverse agonist
CGS 9896	2-(4-chlorophenyl)-2,5-dihydropyrazolo [4,3c] quinoline-3 (3H)-one	Partial Agonist
CGS-8216	2-phenylpryazolo[4,3-c]quinolin-3(5H)-one	Antagonist
B CCE	ethyl b-carboline-3-carboxylate	Inverse agonsit Antagonist ?
B CCM	methyl b-carboline-3-carboxylate	Inverse agonsit
Mucimol	5-aminomethyl-3-hydroxy-izoxalole	Agonist
THIP	4,5,6,7-tetrahydroisoxazolo[5,4,-c] pyridin-3-ol hydrate	
DMCM	methyl 6,7-dimethoxy-4-ethyl-b-carboline-3-carboxylate	Inverse agonist
TBPS	t-butylbicyclophosphothionate	(?)Antagonist
DM-9384	N-(2,6-dimethyl-phenyl)-2-(2-oxo-1-pyrrolidinyl) acetamide	?

ketamine from saline perceive ethanol as these drugs. Animals selectively bred for ethanol sensitivity are also differentially sensitive to NMDA antagonists. Several compounds such as pentobarbital and the gaseous anesthetics, thought to act at the GABA receptor, interact with NMDA antagonists in an additive way. Obviously the mechanism of action of these agents is extremely complicated, and interactions with several receptors or one receptor in ways that influence another are not only possible, but likely.

In conjunction with the *in vitro* evidence of interactions with the NMDA receptors as outlined by Tabakoff and Hoffman (Chapter 5), there is a strong case for a significant role for the NMDA receptor in the behavioral actions of ethanol.

3. Calcium Channels

The basic biochemistry, pharmacology, and action of ethanol on voltage-gated calcium channels is reviewed in Chapter 2. Here we present in tabular form the evidence that these actions on calcium channels by ethanol have some predictable behavioral consequences (see Tables 14.9 and 14.10).

4. Other Interactions

Several interactions between drugs commonly used to study receptor systems have been found. Thus pyrazole, a commonly used agent to inhibit alcohol dehydrogenase, has been found to block NMDA channels (Aracava et al., 1991). Likewise, nitrendipine, commonly used to block voltage-sensitive calcium channels, has been found to inhibit MK-801 (diclozapine) binding to mouse brain sections (Filloux et al., 1994). Finally, a number of steroids, especially pregnenolone sulfate, have been found to interact with the NMDA system (Maione et al., 1992; Bowlby, 1993; Wong and Moss, 1994; Mathis et al., 1994; Irwin et al., 1992). Most recently steroids have been found to enhance the effects of NMDA on calcium uptake into rat hippocampal neurons (Irwin et al., 1994). Pregnenolone sulfate and several structural analogs have been intensively studied in stimulation of the GABA system (Melchior and Ritzmann, 1994b; Melchior and Ritzmann, 1992, 1994a).

Results such as those cited above indicate one of several possibilities. The most likely is that the drugs studied are not sufficiently specific to give clean results with a given receptor

TABLE 14.6
Genetics

System	Drug	Result	Reference
SS/LS	Phenobarb. and many others	LS>SS sleep time. Depends on water solubility	1
SS/LS	Pentobarb.	SS and LS not different	2
SS/LS	Baclophen and THIP	LS>SS (Incoordination)	3
SS/LS	Picrotoxin	LS=SS dec sleep time	4
SS/LS	Benzodiazepines	LS>SS	5
SS/LS	Valproate	LS>SS	6
DS/DR	Ethanol	DR>DS	7
AT/ANT	Benzodiazepines	ANT>AT	8
AT/ANT	Barbital	ANT>AT	9
HAS/LAS	Pentobarb.	HAS>LAS sleep time. LAS>HAS brain levels at awakening	10
SS/LS	Halothane	SS and LS not different (ED_{50})	11
HAS/LAS	Halothane	LAS>HAS brain levels at awakening	12
SS/LS	Isoflurane	LS>SS	13
HAS/LAS	Isoflurane and Enflurane	LS>SS	14
SS/LS	Propofol	LS>>>SS	15
HAS/LAS	Propofol	LS>SS	Liu
SS/LS	Reserpine	LS dec but SS inc ETOH sleep time	16
SS/LS	Alpha methylparatyrosine	SS inc ETOH sleep time	17
SS/LS	Norepi, DA	LS>SS dec. ETOH sleep time	18
SS/LS	6-hydroxydopamine	Inc. ETOH sleep time SS>LS	19
SS/LS	Neurotensin	Inc. ETOH sleep time SS>LS	20
SS/LS	Neurotensin	SS>LS analgesic effect of neurotensin	21
SS/LS	Apomorphine	Inc. ETHO sleep time LS not SS	22
SS/LS	Gamma-butrylactone	LS>SS ETOH sleep time. Blocked by amphetamine	23
SS/LS	Propanolol	Dec. ETOH sleep time in both	24
SS/LS	Salsolinol	SS>LS activity. LS>SS sleep time	25
SS/LS	Nicotine	LS>SS behavioral battery	26
HAS/LAS	Nicotine	HAS>LAS behavioral battery	27
SS/LS	d-tubocurarine and hexamethonium	LS>SS dec ETOH sleep time	28
SS/LS	Taurine	LS>SS taurine inc ETOH sleep time	29
SS/LS	Adenosine agonists	LS>SS escape from elevated tower and ETOH sleep time	30
SS/LS	TRH	LS>SS dec. ETOH sleep time	31
SS/LS	Indomethacin	Dec. ETOH sleep time in both time	32

1. (Melchior and Ritzmann, 1992; De Fiebre et al., 1992; Marley et al., 1986) 2. (Allan and Harris, 1989; Dudek et al., 1984b) 3. (Martz et al., 1983) 4. (Masserano and Weiner, 1982) 5. (McIntyre and Alpern, 1986; Stinchcomb et al., 1989) 6. (Pearlman and Goldstein, 1984) 7. (Phillips and Gallaher, 1992) 8. (Hellevuo et al., 1987) 9. (Hellevuo et al., 1987) 10. (Deitrich et al., 1994) 11. (Baker et al., 1980) 12. (Deitrich et al., 1994) 13. (Simpson et al., 1993) 14. (Deitrich et al., 1994) 15. (Simpson and Blednov, 1994) 16. (French et al., 1988; Masserano and Weiner, 1982) 17. (French et al., 1988) 18. (Masserano and Weiner, 1982) 19. (French et al., 1988; Erwin and Cornell, 1986) 20. (Erwin, 1986) 21. (Erwin and Jones, 1989) 22. (Dudek et al., 1984a; Erwin et al., 1987) 23. (Dudek and Fanelli, 1980) 24. (Allan et al., 1984) 25. (Church et al., 1977; Church et al., 1976; Smolen and Collins, 1984; Smolen et al., 1984) 26. (De Fiebre et al., 1990; De Fiebre and Collins, 1992; De Fiebre et al., 1987) 27. (De Fiebre et al., 1991) 28. (Masserano and Weiner, 1982) 29. (Ferko and Bobyock, 1989) 30. (Fredholm et al., 1985; Proctor et al., 1985; Proctor and Dunwiddie, 1984; Smolen and Smolen, 1991) 31. (Masserano et al., 1989; French et al., 1990; French et al., 1993) 32. (George et al., 1985). Liu, unpublished data.

system and tend to interact with a number of receptors. The other possibility is that the various receptor systems do, in fact, have the capacity to interact with a number of endogenous and exogenous chemicals and that this is part and parcel of the normal homeostatic mechanisms in the brain.

TABLE 14.7
Interaction of Ethanol with Glutamate Receptors

Agonists

System	Drug	Result	Reference
Rat	NMDA	Rat hippocampal slices: ETOH inhib. LTP at same conc. as inhib. NMDA response.	1
Rat	NMDA	Discrimination. ETOH did not block NMDA, nor substitute completely for PCP or NPC 12626.	2
Mice	NMDA	SS and LS mice. No effect on ETOH LORR by 370 ng NMDA, ICV.	3
Mice	NMDA	Inc. CNS depression. Blocked by APV or bicuculline	4
Mice	spermine, spermidine	Inc. LRR duration with ETOH, pentobarb. Dec. MAC for halothane but not ether.	5

1. (Blitzer et al., 1990) 2. (Balster et al., 1992) 3. (Wilson et al., 1990) 4. (Ferko, 1992) 5. (Daniell, 1992).

TABLE 14.8
Antagonists/Genetics

System	Drug	Result	Reference
SS/LS	MK-801	LS<<SS MK-801 ED_{50} for LRR	Simpson
SS/LS	MK-801, APV	SS>LS Etoh inc. sleep time	1
HAS/LAS	MK-801	HAS<<LAS MK-801 ED_{50} for LRR	Liu
HAS/LAS	MK-801	HAS>>LAS Etoh inc. sleep time	Draski
HAS/LAS	Ketamine	HAS>LAS Ketamine sleep time	Liu
NMRI mice	MK-801 and CGP 39951	Inhibited ETOH stimulation due to inc. ETOH sedative effects	2
Rats	MK-801 and PCP	MK-801 and PCP discriminated as ETOH	3
Rats	MK-801	Rats discriminate MK-801 as ETOH	4
Rats	MK-801	ETOH dec. convulsions by NMDA and kainate. MK-801 inc. ETOH sleep time	5
Mice SSIA	MK-801	MK-801 dec. nonopioid component of ETOH analgesia	6
Rats	MK-801	MK-801 and ETOH not synergistic on EEG effects	7
Mice	CGS 19755	Inc. LRR duration due to ETOH, pentobarb., and halothane. Not ether.	8
Mice	MK-801, PCP, ketamine	Inc. LRR duration due to ETOH, pentobarb. halothane and ether	9
Mice	MK-801	Increased LRR duration due to ETOH	10
Mice	MK-801	MK-801 and ETOH synergistic anticonvulsants vs NMDA	11

1. (Wilson et al., 1990): 2. (Liljequist, 1991) 3. (Grant and Colombo, 1993) 4. (Schechter et al., 1993) 5. (Danysz et al., 1992) 6. (Mogil et al., 1993) 7. (Ehlers et al., 1992) 8. (Daniell, 1991) 9. (Daniell, 1990) 10. (Crabbe et al., 1994) 11. (Sharma et al., 1991). Draski, unpublished data; Simpson, unpublished data; Liu, unpublished data.

IV. CONCLUSION

The measurement of the initial effects of ethanol is achieved in a large number of ways utilizing a variety of doses. It is likely that as the dose is increased, more and more receptor systems are recruited. The correlation between biochemical and behavioral effects should be employed whenever possible, both with regard to the doses necessary and with respect to the genetics of the animals involved.

The evidence that ethanol interacts with the GABA, glutamate, and voltage-gated calcium channels is reasonably strong at the behavioral level. Other chapters present the evidence that there is an interaction at the biochemical level as well.

TABLE 14.9
Interaction of Ethanol with Calcium Channels

Calcium and Agonists

System	Compound	Result	Reference
Mice	calcium, ionophores	Inc. ETOH sleep time and ataxia in mice and rats	1
SS/LS mice	calcium, ionophore	BAC at LORR inc. in SS mice. Not halothane	2
SS/LS mice	calcium	Inc. ETOH sensitivity of cerebellar Purkinje cells in SS mice	3
Mice	BAY K 8644	Antagonized ETOH and argon at 1 mg/kg. Inc. ETOH and argon sleep time at 5 and 10 mg/kg. Antag. pentobarb.	4

1. (Erickson et al., 1978; Sutoo et al., 1985; Harris, 1979; Erickson et al., 1980) 2. (Morrow and Erwin, 1986; Morrow and Erwin, 1987) 3. (Palmer et al., 1987) 4. (Dolin et al., 1988).

TABLE 14.10
Calcium Antagonists

Mice, rats	Chelators	Dec. ETOH sleep time	1
Mice	Calmodulin antagonists	Antag. Ca^{++} induced inc. ETOH sleep time.	2
Rats	Nifedipine	Inc. ETOH hypnotic and hypothermia effects	3
Mice	verapamil, flunarizine, nitrendipine	Inc. anesthetic potency of ETOH and Pentobarb.	4
Mice	Nimodipine	Inc. ETOH rotorod ataxia and inc. hypothermia	5
Mice	Nifedipine	ETOH stimulation decreased. ETOH sedation increased	6
Humans	Nifedipine, verapamil	Physiological and mental impairment to ETOH. Not changed	7

1. (Erickson et al., 1980; Erickson et al., 1978; Harris, 1979) 2. (Sutoo et al., 1985) 3. (Pucilowski et al., 1989) 4. (Dolin and Little, 1986) 5. (Isaacson et al., 1985) 6. (Engel et al., 1988) 7. (Perez-Reyes et al., 1992).

REFERENCES

Alhenius, S., Carlsson, A., Engel, J., Svensson, T.H., and Sodersten, P. (1973). Antagonism by alpha-methyltyrosine of the ethanol-induced stimulation and euphoria in man. *Clin. Pharmacol. Ther.,* 14, 586–591.

Allan, A.M., Mayes, G.G., and Draski, L.J. (1991). Gamma-aminobutyric acid-activated chloride channels in rats selectively bred for differential acute sensitivity to alcohol. *Alcoholism: Clin. Exp. Res.,* 15, 212–218.

Allan, A.M., Horowitz, G.P., Isaacson, R.L., and Major, L.F. (1984). Adrenergic nervous system alteration and ethanol-induced narcosis in long-sleep and short-sleep mice. *Behav. Neural. Biol.,* 42, 134–139.

Allan, A.M. and Harris, R.A. (1989). Sensitivity to ethanol hypnosis and modulation of chloride channels does not cosegregate with pentobarbital sensitivity in HS mice. *Alcoholism: Clin. Exp. Res.,* 13, 428–434.

Aracava, Y., Froes-Ferrao, M.M., Pereira, E.F., and Albuquerque, E.X. (1991). Sensitivity of N-methyl-D-aspartate (NMDA) and nicotinic acetylcholine receptors to ethanol and pyrazole. *Ann. N.Y. Acad. Sci.,* 625, 451–472.

Baker, R., Melchior, C., and Deitrich, R. (1980). The effect of halothane on mice selectively bred for differential sensitivity to alcohol. *Pharmacol. Biochem. Behav.,* 12, 691–695.

Baker, R.C., Smolen, A., Smolen, T.N., and Deitrich, R.A. (1987). Relationship between acute ethanol-related responses in long-sleep and short-sleep mice. *Alcoholism: Clin. Exp. Res.,* 11, 574–578.

Balster, R.L., Grech, D.M., and Bobelis, D.J. (1992). Drug discrimination analysis of ethanol as an N-methyl-D-aspartate receptor antagonist. *Eur. J. Pharmacol.,* 222, 39–42.

Basma, A.N., Heikkila, R.E., Nicklas, W.J., Giovanni, A., and Geller, H.M. (1990). 1-Methyl-4-phenyl-1,2,3,6-tetrahydropyridine- and 1-methyl-4-(2′-ethylphenyl)-1,2,3,6-tetrahydropyridine-induced toxicity in PC12 cells: role of monoamine oxidase A. *J. Neurochem.,* 55, 870–877.

Becker, H.C. and Hale, R.L. (1989). Ethanol-induced locomotor stimulation in C57BL/6 mice following RO15-4513 administration. *Psychopharmacology,* 99, 333–336.

Belzung, C., Misslin, R., and Vogel, E. (1988). The benzodiazepine receptor inverse agonist β-CCM and RO 15-3505 both reverse the anxiolytic effects of ethanol in mice. *Life Sci.,* 42, 1765–1772.

Berlid, D. and Hasselbalch, H. (1981) *Lancet,* 363.

Biswas, B. and Carlsson, A. (1978). Effect of intraperitoneally administered GABA on the locomotor activity of mice. *Psychopharmacology,* 59, 91–94.

Bitrán, M. and Kalant, H. (1991). Learning factor in rapid tolerance to ethanol-induced motor impairment. *Pharmacol. Biochem. Behav.,* 39, 917–922.

Bitrán, M. and Kalant, H. (1993). Development of rapid tolerance to pentobarbital and cross-tolerance to ethanol on a motor performance test with intoxicated practice. *Pharmacol. Biochem. Behav.,* 44, 981–983.

Blitzer, R.D., Gil, O., and Landau, E.M. (1990). Long-term potentiation in rat hippocampus is inhibited by low concentrations of ethanol. *Brain Res.,* 537, 203–208.

Bonetti, E.P., Burkard, W.P., Gabl, M., Hunkeler, W., Lorez, H.-P., Martin, J.R., Moehler, H., Osterrieder, W., Pieri, L., Polc, P., Richards, J.G., Schaffner, R., Scherschlicht, R., Schoch, P., and Haefely, W.E. (1989). Ro 15-4513: partial inverse agonism at the BZR and interaction with ethanol. *Pharmacol. Biochem. Behav.,* 31, 733–749.

Bowlby, M.R. (1993). Pregnenolone sulfate potentiation of *N*-methyl-D-aspartate receptor channels in hippocampal neurons. *Mol. Pharmacol.,* 43, 813–819.

Breese, G.R., Coyle, S., Towle, A.C., Mueller, R.A., McCown, T.J., and Frye, G.D. (1984). Ethanol-induced locomotor stimulation in rats after thyrotropin-releasing hormone. *J. Pharmacol. Exp. Ther.,* 229, 731–737.

Brockway, B.P., Mills, P.A., and Azar, S.H. (1991). A new method for continuous chronic measurement and recording of blood pressure, heart rate and activity in the rat via radio-telemetry. *Clin. Exper. Hypertens. Theor. Pract.,* A13, 885–895.

Church, A.C., Fuller, J.L., and Dudek, B.C. (1976). Salsolinol differentially affects mice selected for sensitivity to alcohol. *Psychopharmacology,* 47, 49–52.

Church, A.C., Fuller, J.L., and Dudek, B.C. (1977). Behavioral effects of salsolinol and ethanol on mice selected for sensitivity to alcohol-induced sleep time. *Drug Alcohol Depend.,* 2, 443–452.

Clausen, T.G., Wolff, J., Carl, P., and Theilgaard, A. (1990). The effect of the benzodiazepine antagonist, flumazenil, on psychometric performance in acute ethanol intoxication in man. *Eur. J. Clin. Pharmacol.,* 38, 233–236.

Clement, J.G., Mills, P., and Brockway, B. (1989). Use of telemetry to record body temperature and activity in mice. *J. Pharmacol. Meth.,* 21, 129–140.

Cott, J., Carlsson, A., Engel, J., and Lindqvist, M. (1976). Suppression of ethanol-induced locomotor stimulation by GABA-like drugs. *Naunyn Schmiedebergs Arch. Pharmacol.,* 295, 203–209.

Crabbe, J., Young, E.R., and Dorow, J. (1994). Effects of dizocilpine in withdrawal seizure-prone (WSP) and withdrawal seizure-resistant (WSR) mice. *Pharmacol. Biochem. Behav.,* 47, 443–450.

Crabbe, J.C., Janowsky, J.S., Young, E.R., and Rigter, H. (1980). Strain-specific effects of ethanol on open-field activity in inbred mice. *Subst. Alcohol Actions Misuse,* 1, 537–543.

Crabbe, J.C. (1986). Genetic differences in locomotor activation in mice. *Pharmacol. Biochem. Behav.,* 25, 289–292.

Crabbe, J.C., Kosobud, A., Tam, B.R., Young, E.R., and Deutsch, C.M. (1987a). Genetic selection of mouse lines sensitive (cold) and resistant (hot) to acute ethanol hypothermia. *Alcohol. Drug. Res.,* 7, 163–174.

Crabbe, J.C., Young, E.R., Deutsch, C.M., Tam, B.R., and Kosobud, A. (1987b). Mice genetically selected for differences in open-field activity after ethanol. *Pharmacol. Biochem. Behav.,* 27, 577–581.

Crabbe, J.C., Belknap, J.K., Mitchell, S.R., and Cranshaw, L.I. (1994). Quantitative trait loci mapping of genes that influence the sensitivity and tolerance to ethanol-induced hypothermia in BXD recombinant inbred mice. *J. Pharmacol. Exp. Ther.,* 269, 184–192.

Crabbe, J.C., Johnson, N., Gray, D., Kosobud, A., and Young, E.R. (1982). Biphasic effects of ethanol on open-field activity: sensitivity and tolerance in C57BL/6N and DBA/2N mice. *J. Comp. Physiol. Psychol.,* 96, 440–451.

Crabbe, J.C., Jr., Feller, D.J., and Phillips, T.J. (1990). Selective Breeding for Two Measures of Sensitivity to Ethanol. In *Initial Sensitivity to Ethanol.* R.A. Deitrich and A.A. Pawlowski, Eds., Rockville, Maryland: National Institute on Alcohol Abuse and *Alcoholism: Clin. Exp. Res.,* pp. 123–154.

Daniell, L.C. (1990). The noncompetitive N-methyl-D-aspartate antagonists, MK-801, phencyclidine and ketamine, increase the potency of general anesthetics. *Pharmacol. Biochem. Behav.,* 36, 111–115.

Daniell, L.C. (1991). Effect of CGS 19755, a competitive N-Methyl-D-Aspartate antagonist, on general anesthetic potency. *Pharm. Biochem. Behav.,* 40, 767–769.

Daniell, L.C. (1992). Alteration of general anesthetic potency by agonists and antagonists of the polyamine binding site of the N-methyl-D-aspartate receptor. *J. Pharmacol. Exp. Ther.,* 261, 304–310.

Danysz, W., Dyr, W., Jankowska, E., Glazewski, S., and Kostowski, W. (1992). The involvement of NMDA receptors in acute and chronic effects of ethanol. *Alcoholism: Clin. Exp. Res.,* 16, 499–504.

De Fiebre, C.M., Medhurst, L.J., and Collins, A.C. (1987). Nicotine response and nicotinic receptors in long-sleep and short-sleep mice. *Alcoholism: Clin. Exp. Res.,* 4, 493–501.

De Fiebre, C.M., Marks, M.J., and Collins, A.C. (1990). Ethanol-nicotine interactions in long-sleep and short-sleep mice. *Alcoholism: Clin. Exp. Res.,* 7, 249–257.

De Fiebre, C.M., Romm, E., Collins, J.T., Draski, L.J., Deitrich, R.A., and Collins, A.C. (1991). Responses to cholinergic agonists of rats selectively bred for differential sensitivity to ethanol. *Alcoholism: Clin. Exp. Res.,* 15, 270–276.

De Fiebre, C.M. and Collins, A.C. (1992). Classical genetic analyses of responses to nicotine and ethanol in crosses derived from long- and short-sleep mice. *J. Pharmacol. Exp. Ther.,* 261, 173–180.

De Fiebre, N.C., Marley, R.J., Wehner, J.M., and Collins, A.C. (1992). Lipid solubility of sedative-hypnotic drugs influences hypothermic and hypnotic responses of long-sleep and short-sleep mice. *J. Pharmacol. Exp. Ther.,* 263, 232–240.

DeFries, J.C., Wilson, J.R., Erwin, V.G., and Petersen, D.R. (1989). LS × SS recombinant inbred strains of mice: initial characterization. *Alcoholism: Clin. Exp. Res.,* 13, 196–200.

Deitrich, R.A., Draski, L.J., and Baker, R.C. (1994). Effect of pentobarbital and gaseous anesthetics on rats selectively bred for ethanol sensitivity. *Pharmacol. Biochem. Behav.,* 47, 721–725.

Deitrich, R.A. (1993). Selective breeding for inital sensitivity to ethanol. *Behav. Genet.,* 23, 153–162.

Di Chiara, G. and Imperato, A. (1988). Drugs abused by humans preferentially increase synaptic dopamine concentrations in the mesolimbic system of freely moving rats. *Proc. Natl. Acad. Sci. U.S.A.,* 85, 5274–5278.

Dilsaver, S.C., Overstreet, D.H., and Peck, J.A. (1992). Measurement of temperature in the rat by rectal probe and telemetry yields compatible results. *Pharmacol. Biochem. Behav.,* 42, 549–552.

Dilsaver, S.C., Majchrzak, M.J., and Alessi, N.E. (1990). Telemetric measurement of core temperature in pharmacological research: validity and reliability. *Prog. Neuropsyc. Biol. Psychiat.,* 14, 591–596.

Dolin, S.J., Halsey, M.J., and Little, H.J. (1988). Effects of the calcium channel activator Bay K 8644 on general anaesthetic potency in mice. *Br. J. Pharmacol.,* 94, 413–422.

Dolin, S.J. and Little, H.J. (1986). Augmentation by calcium channel antagonists of general anaesthetic potency in mice. *Br. J. Pharmacol.,* 88, 909–914.

Dorian, P., Sellers, E.M., Kaplan, H.L., Hamilton, C., Greenblatt, D.J., and Abernethy, D. (1985). Triazolam and ethanol interaction: kinetic and dynamic consequences. *Clin. Pharmacol. Ther.,* 37, 558–562.

Draski, L.J., Spuhler, K.P., Erwin, V.G., Baker, R.C., and Deitrich, R.A. (1992). Selective breeding of rats differing in sensitivity to the effects of acute ethanol administration. *Alcoholism: Clin. Exp. Res.,* 16, 48–54.

Dudek, B.C., Abbott, M.E., Garg, A., and Phillips, T.J. (1984a). Apomorphine effects on behavioral response to ethanol in mice selectively bred for differential sensitivity to ethanol. *Pharmacol. Biochem. Behav.,* 20, 91–94.

Dudek, B.C., Abbott, M.E., and Phillips, T.J. (1984b). Stimulant and depressant properties of sedative-hypnotics in mice selectively bred for differential sensitivity to ethanol. *Psychopharmacology,* 82, 46–51.

Dudek, B.C. and Abbott, M.E. (1984c). A biometrical genetic analysis of ethanol response in selectively bred long-sleep and short-sleep mice. *Behav. Genet.,* 14, 1–19.

Dudek, B.C., Phillips, T.J., and Hahn, M.E. (1991). Genetic analyses of the biphasic nature of the alcohol dose-response curve. *Alcoholism: Clin. Exp. Res.,* 15, 262–269.

Dudek, B.C. and Abbott, M.E. (1984). The relationship between ethanol-induced locomotor activation and narcosis in long-sleep and short-sleep mice. *Alcoholism: Clin. Exp. Res.,* 8, 272–276.

Dudek, B.C. and Fanelli, R.J. (1980). Effects of gamma-butyrolactone, amphetamine, and haloperidol in mice differing in sensitivity to alcohol. *Psychopharmacology,* 68, 89–97.

Dudek, B.C. and Phillips, T.J. (1989). Genotype-dependent effects of GABAergic agents on sedative properties of ethanol. *Psychopharmacology,* 98, 518–523.

Dudek, B.C. and Phillips, T.J. (1990). Distinctions among sedative, disinhibitory, and ataxic properties of ethanol in inbred and selectively bred mice. *Psychopharmacology,* 101, 93–99.

Duncan, P.M. and Baez, A.M. (1981). The effect of ethanol on wheel running in rats. *Pharmacol. Biochem. Behav.,* 15, 819–821.

Ehlers, C.L., Chaplin, R.I., and Koob, G.F. (1990). EEG effects of Ro 15-4513 and FG 7142 alone and in combination with ethanol. *Pharmacol. Biochem. Behav.,* 36, 607–611.

Ehlers, C.L., Kaneko, W.M., Wall, T.L., and Chaplin, R.I. (1992). Effects of dizocilpine (MK-801) and ethanol on the EEG and event-related potentials (ERPs) in rats. *Neuropharmacology,* 31, 369–378.

Engel, J.A., Fahlke, C., Hulthe, P., Hard, E., Johannessen, K., Snape, B., and Svensson, L. (1988). Biochemical and behavioral evidence for an interaction between ethanol and calcium channel antagonists. *J. Neural. Transm.,* 74, 181–193.

Erickson, C.K., Tyler, T.D., and Harris, R.A. (1978). Ethanol: modification of acute intoxication by divalent cations. *Science,* 199, 1219–1221.

Erickson, C.K., Tyler, T.D., Beck, L.K., and Duensing, K.L. (1980). Calcium enhancement of alcohol and drug-induced sleeping time in mice and rats. *Pharmacol. Biochem. Behav.,* 12, 651–656.

Eriksson, C.K. and Peter, C.J. (1990). Finnish Selective Beeding Studies for Initial Sensitivity to Ethanol: Update 1988 on the AT and ANT Rat Lines. In *Initial Sensitivity to Alcohol.* R.A. Deitrich and A.A. Pawlowski, Eds., Rockville, Maryland: National Institute on Alcohol Abuse and *Alcoholism: Clin. Exp. Res.,* pp. 61–86.

Erwin, V.G. (1986). Effects of neurotensin on thermoregulation, locomotor activity, and ethanol induced sleep time in LS/Ibg and SS/Ibg mice. *Proc. West. Pharmacol. Soc.,* 29, 171–174.

Erwin, V.G., Korte, A., and Marty, M. (1987). Neurotensin selectively alters ethanol-induced anesthesia in LS/Ibg and SS/Ibg lines of mice. *Brain Res.,* 400, 80–90.

Erwin, V.G., Jones, B.C., and Radcliffe, R. (1990). Further characterization of LS × SS recombinant inbred strains of mice: activating and hypothermic effects of ethanol. *Alcoholism: Clin. Exp. Res.,* 14, 200–204.

Erwin, V.G. and Cornell, K. (1986). Effects of 6-hydroxydopamine on brain catecholamines and on acute actions of ethanol in LS/Ibg and SS/Ibg mice. *Alcoholism: Clin. Exp. Res.,* 10, 285–289.

Erwin, V.G. and Jones, B.C. (1989). Comparison of neurotensin levels, receptors and actions in LS/Ibg and SS/Ibg mice. *Peptides,* 10, 435–440.

Ferko, A.P. (1992). NMDA enhances the central depressant properties of ethanol in mice. *Pharmacol. Biochem. Behav.,* 43, 297–301.

Ferko, A.P. and Bobyock, E. (1989). Effect of taurine on ethnaol-induced sleep time in mice genetically bred for differences in ethanol sensitivity. *Pharmacol. Biochem. Behav.,* 31, 667–673.

Filloux, F.M., Fitts, R.C., Skeen, G.A., and White, H.S. (1994). The dihydropyridine nitrendipine inhibits [^{3}H]MK 801 binding to mouse brain sections. *Eur. J. Pharmacol. Mol. Pharmacol.,* 269, 325–330.

Finn, D.A., Syapin, P.J., Bejanian, M., Jones, B.L., and Alkana, R.L. (1991). Body temperature influences ethanol and ethanol/pentobarbital lethality in mice. *Alcohol,* 8, 39–41.

Fluckiger, A., Hartmann, D., Leishman, B., and Ziegler, W.H. (1988). Lack of effect of the benzodiazepine antagonist flumazenil (Ro 15-1788) on the performance of healthy subjects during experimentally induced ethanol intoxication. *Eur. J. Clin. Pharmacol.,* 34, 273–276.

Fox, M.W. (1965). Reflex-ontogeny and behavioral development of the mouse. *Anim. Behav.,* 13, 234–241.

Fredholm, B.B., Zahniser, N.R., Weiner, G.R., Proctor, W.R., and Dunwiddie, T.V. (1985). Behavioural sensitivity to PIA in selectively bred mice is related to a number of A1 adenosine receptors but not to cyclic AMP accumulation in brain slices. *Eur. J. Pharmacol.,* 111, 133–136.

French, T.A., Masserano, J.M., and Weiner, N. (1988). Further studies on the neurochemical mechanisms mediating differences in ethanol sensitivity in LS and SS mice. *Alcoholism: Clin. Exp. Res.,* 12, 215–223.

French, T.A., Masserano, J.M., and Weiner, N. (1990). The role of central and adrenal catecholaminergic systems in mediating TRH effects on ethanol sensitivity in LS and SS mice. *Alcoholism: Clin. Exp. Res.,* 14, 289.

French, T.A., Masserano, J.M., and Weiner, N. (1993). Influence of thyrotropin-releasing hormone and catecholaminergic interactions on CNS ethanol sensitivity. *Alcoholism: Clin. Exp. Res.,* 17, 99–106.

Frye, G.D. and Breese, G.R. (1981). An evaluation of the locomotor stimulating action of ethanol in rats and mice. *Psychopharmacology,* 75, 322–379.

Gallaher, E.J. and Loomis, T.A. (1979). The rapid onset of ethanol tolerance in Wistar rats following intensive practice on the moving-belt task. *Toxicol. Appl. Pharmacol.,* 48, 415–424.

Gallaher, E.W., Parsons, L.M., and Goldstein, D.B. (1982). The rapid onset of tolerance to ataxic effects of ethanol in mice. *Psychopharmacology,* 78, 67–70.

George, F.R., Ritz, M.C., and Collins, A.C. (1985). Indomethacin antagonism of ethanol-induced sleep time: sex and genotypic factors. *Psychopharmacology,* 85, 151–153.

Gibbins, R.J., Kalant, H., and Le Blanc, A.E. (1968). A technique for accurate measurement of moderate degrees of alcohol intoxication in small animals. *J. Pharmacol. Exp. Ther.,* 159, 236–242.

Glowa, J.R., Crawley, J., Suzdak, P.D., and Paul, S.M. (1989). Ethanol and the GABA receptor complex: studies with the partial inverse benzodiazepine receptor agonist Ro 15-4513. *Pharm. Biochem. Behav.,* 31, 767–772.

Grant, K.A. and Colombo, G. (1993). Discriminative stimulus effects of ethanol: effect of training dose on the substitution of N-methyl-D-aspartate antagonists. *J. Pharmacol. Exp. Ther.,* 264, 1241–1247.

Grant, K.A., Werner, R., Hoffman, P.L., and Tabakoff, B. (1989). Chronic tolerance to ethanol in the N:NIH rat. *Alcoholism: Clin. Exp. Res.,* 13, 402–406.

Hard, E. and Larsson, K. (1975). Development of air righting in rats. *Brain Behav. Evol.,* 11, 53–59.

Harris, R.A. (1979). Alteration of alcohol effects by calcium and other inorganic cations. *Pharmacol. Biochem. Behav.,* 10, 527–534.

Hellevuo, K., Kiianmaa, K., Juhakoski, A., and Kim, C. (1987). Intoxicating effects of lorazepam and barbital in rat lines selected for differential sensitivity to ethanol. *Psychopharmacology,* 91, 263–267.

Hellevuo, K. and Korpi, E.R. (1988). Failure of Ro 15-4513 to antagonize ethanol in rat lines selected for differential sensitivity to ethanol and in Wistar rats. *Pharmacol. Biochem. Behav.,* 30, 183–188.

Hilakivi, L., Eriksson, C.J.P., Sarviharju, M., and Sinclair, J.D. (1984). Revitalization of the AA and ANA rat lines: effects on some line characteristics. *Alcohol,* 1, 71–75.

Hoffman, P.L., Tabakoff, B., Szabo, G., Suzdak, P.D., and Paul, S.M. (1987). Effect of an imidazobenzodiazepine, Ro 15-4513, on the incoordination and hypothermia produced by ethanol and pentobarbital. *Life Sci.,* 41, 611–619.

Hollister, L.E. (1990). Interactions between alcohol and benzodiazepines. *Recent Dev. Alcohol,* 8, 233–239.

Holloway, F.A., Michaelis, R.C., Harland, R.D., Criado, J.R., and Gauvin, D.V. (1992). Tolerance to ethanol's effects on operant performance in rats: role of number and pattern of intoxicated practice opportunities. *Psychopharmacology,* 109, 112–120.

Howerton, T.C., Burch, J.B., O'Connor, M.F., Miner, L.L., and Collins, A.C. (1984). A genetic analysis of ethanol, pentobarbital, and methyprylon sleep-time response. *Alcoholism: Clin. Exp. Res.,* 8, 546–550.

Hunt, G.P. and Overstreet, D.H. (1977). Evidence for parallel development of tolerance to the hyperactivating and discoordinating effects of ethanol. *Psychopharmacology,* 55, 75–81.

Irwin, R.P., Maragakis, N.J., Rogawski, M.A., Purdy, R.H., Farb, D.H., and Paul, S.M. (1992). Pregnenolone sulfate augments NMDA receptor mediated increases in intracellular Ca^{2+} in cultured rat hippocampal neurons. *Neurosci. Lett.,* 141, 30–34.

Irwin, R.P., Lin, S.-Z., Rogawski, M.A., Purdy, R.H., and Paul, S.M. (1994). Steroid potentiation and inhibition of N-methyl-D-aspartate receptor-mediated intracellular Ca^{++} responses: structure-activity studies. *J. Pharmacol. Exp. Ther.,* 271, 677–682.

Isaacson, R.L., Molina, J.C., Draski, L.J., and Johnson, J.E. (1985). Nimodipine's interactions with other drugs: I. Ethanol. *Life Sci.,* 36, 2195–2199.

June, H.L. and Lewis, M.J. (1989). Ro 15-4513 enhances and attenuates motor stimulant effects of ethanol in rats. *Alcohol,* 6, 245–248.

Keir, W.J. and Deitrich, R.A. (1990). Development of central nervous system sensitivity to ethanol and pentobarbital in short- and long-sleep mice. *J. Pharmacol. Exp. Ther.,* 254, 831–835.

Khanna, J.M., Morato, G.S., Chau, A., Shah, G., and Kalant, H. (1994). Effect of NMDA antagonists on rapid and chronic tolerance to ethanol: importance of intoxicated practice. *Pharmacol. Biochem. Behav.,* 48, 755–763.

Kiianmaa, K., Hellevuo, K., and Korpi, E.R. (1988). The AT (alcohol tolerant) and ANT (alcohol nontolerant) rat lines selected for differences in sensitivity to ethanol: An overview. In *Biomedical and Social Aspects of Alcohol and Alcoholism,* K. Kuriyama, A. Takada, and H. Ishii, Eds., Elsevier Science Publishers, New York, pp. 415–418.

Kiianmaa, K. and Hellevuo, K. (1990). The alcohol tolerant and alcohol nontolerant rat lines selected for differential sensitivity to ethanol: a tool to study mechanisms of the actions of ethanol. *Ann. Med.,* 22, 283–287.

Klotz, U., Ziegler, G., Rosenkranz, B., and Mikus, G. (1986). Does the benzodiazepine antagonist Ro 15-1788 antagonize the action of ethanol? *Br. J. Clin. Pharmacol.,* 22, 513–520.

Koob, G.F., Percy, L., and Britton, K.T. (1989). The effects of Ro 15-4513 on the behavioral actions of ethanol in an operate reaction time task and a conflict test. *Pharm. Biochem. Behav.,* 31, 757–760.

Lê, A.D. (1990). Factors regulating ethanol tolerance. *Ann. Med.,* 22, 265–268.

Lê, A.D. and Kalant, H. (1992). Influence of intoxicated practice on the development of acute tolerance to the motor impairment effect of ethanol. *Psychopharmacology,* 106, 572–576.

Le Blanc, A.E., Kalant, H., and Gibbins, R.J. (1969). Acquisition and loss of tolerance to ethanol by the rat. *J. Pharmacol. Exp. Ther.,* 168, 244–250.

Le Blanc, A.E., Kalant, H., and Gibbins, R.J. (1975). Acute tolerance to ethanol in the rat. *Psychopharmacologia,* 41, 43–46.

Leitch, G.H., Backes, D.J., Siegman, F.S., and Guthrie, G.D. (1977). Possible role of GABA in the development of tolerance to alcohol. *Experientia,* 33, 496–497.

Liljequist, S. (1991). NMDA receptor antagonists inhibit ethanol-produced locomotor stimulation in NMRI mice. *Alcohol,* 8, 309–312.

Liljequist, S. and Engel, J. (1982). Effects of GABAergic agonists and antagonists on various ethanol-induced behavioral changes. *Psychopharmacology,* 78, 71–75.

Linakis, J.G. and Cunningham, C.L. (1979). Effects of concentration of ethanol injected intraperitoneally on taste aversion, body temperature and activity. *Psychopharmacology,* 64, 61–65.

Linnoila, M., Stapleton, J.M., Lister, R., Moss, H., Lane, E., Granger, A., and Eckardt, M.J. (1990a). Effects of single doses of alprazolam and diazepam, alone and in combination with ethanol, on psychomotor and cognitive performance and on autonomic nervous system reactivity in healthy volunteers. *Eur. J. Clin. Pharmacol.,* 39, 21–28.

Linnoila, M., Stapleton, J.M., Lister, R., Moss, H., Lane, E., Granger, A., Greenblatt, D.J., and Eckardt, M.J. (1990b). Effects of adinazolam and diazepam, alone and in combination with ethanol, on psychomotor and cognitive performance and on autonomic nervous system reactivity in healthy volunteers. *Eur. J. Clin. Pharmacol.,* 38, 371–377.

Lister, R.G. (1987). The effects of ethanol on exploration in DBA/2 and C57Bl/6 mice. *Alcohol,* 4, 17–19.

Lister, R.G. (1988). Interactions of three benzodiazepine receptor inverse agonists with ethanol in a plus-maze test of anxiety. *Pharm. Biochem. Behav.,* 30, 701–706.

Lister, R.G. (1989). Interactions of ethanol with benzodiazepine receptor ligands in tests of exploration, locomotion and anxiety. *Pharm. Biochem. Behav.,* 31, 761–765.

Lister, R.G. and Durcan, M.J. (1989). Antagonism of the intoxicating effects of ethanol by the potent benzodiazepine receptor ligand Ro 19–4603. *Brain Res.,* 482, 141–144.

Maione, S., Berrino, L., Vitagliano, S., Leyva, J., and Rossi, F. (1992). Pregnenolone sulfate increases the convulsant potency of N-methyl-D-aspartate in mice. *Eur. J. Pharmacol.,* 219, 477–479.

Marley, R.J., Miner, L.L., Wehner, J.M., and Collins, A.C. (1986). Differential effects of central nervous system depressants in long-sleep and short-sleep mice. *J. Pharmacol. Exp. Ther.,* 238, 1028–1033.

Marrosu, F., Mereu, G., Giorgi, O., and Corda, M.G. (1988). The benzodiazepine recognition site inverse agonists Ro 15-4513 and fg 7142 both antagonize the eeg effects of ethanol in the rat. *Life Sci.,* 43, 2151–2158.

Martens, F., Koppel, C., Ibe, K., Wagemann, A., and Tenczer, J. (1990). Clinical experience with the benzodiazepine antagonist flumazenil in suspected benzodiazepine or ethanol poisoning. *J. Toxicol. Clin. Toxicol.,* 28, 341–356.

Martz, A., Deitrich, R.A., and Harris, R.A. (1983). Behavioral evidence for the involvement of gamma-aminobutyric acid in the actions of ethanol. *Eur. J. Pharmacol.,* 89, 53–62.

Mason, S.T., Corcoran, M.E., and Fibiger, H.C. (1979). Noradrenergic processes involved in the locomotor effects of ethanol. *Eur. J. Pharmacol.,* 54, 383–387.

Masserano, J.M., Disbrow-Erickson, J., French, T.A., Zoeller, R.T., Zhao, H., and Weiner, N. (1989). LS and SS mice: models for the study of the role of TRH in ethanol sensitivity. *Ann. NY Acad. Sci.,* 553, 505–507.

Masserano, J.M. and Weiner, N. (1982). Investigations into the neurochemical mechanisms mediating differences in ethanol sensitivity in two lines of mice. *J. Pharmacol. Exp. Ther.,* 221, 404–409.

Massey, B.W. and Woolverton, W.L. (1994). Discriminative stimulus effects of combinations of pentobarbital and ethanol in rhesus monkeys. *Drug Alcohol Depend.,* 35, 37–43.

Masson, S., Desmoulin, F., Sciaky, M., and Cozzone, P.J. (1992). The effects of ethanol concentration on glycero-3-phosphate accumulation in the perfused rat liver—A reassessment of ethanol-induced inhibition of glycolysis using ^{31}P-NMR spectroscopy and HPLC. *Eur. J. Biochem.*, 205, 187–194.

Masur, J., Oliveira De Souza, M.L., and Zwicker, A.P. (1986). The excitatory effect of ethanol: absence in rats, no tolerance and increased sensitivity in mice. *Pharm. Biochem. Behav.,* 24, 1225–1228.

Mathis, C., Paul, S.M., and Crawley, J.N. (1994). The neurosteroid pregnenolone sulfate blocks NMDA antagonist-induced deficits in a passive avoidance memory task. *Psychopharmacology*, 116, 201–206.

Mattucci-Schiavone, L. and Ferko, A.P. (1986). An inhalation procedure to produce tolerance to the behavioral effects of ethanol. *Physiol. Behav.*, 36, 643–646.

Mayfield, R.D., Grant, M., Schallert, T., and Spirduso, W.W. (1992). Tolerance to the effects of ethanol on the speed and success of reaction time responding in the rat: effects of age and intoxicated practice. *Psychopharmacology*, 107, 78–82.

McAuliff, W.E. and Gordon, R.A. (1974). A test of Lindesmith's theory of addiction: the frequency of euphoria among long-term addicts. *Am. J. Sociol.*, 79, 795–840.

McClearn, G.E. and Kakihana, R. (1981). Selective breeding for ethanol sensitivity: Short-sleep and long-sleep mice. In *Development of animal models as pharmacogenetic tools.* G.E. McClearn, R.A. Deitrich, and V.G. Erwin, Eds. Rockville, Maryland: NIAAA Research Monograph, 20, 147–159.

McGivern, R.F., Melcer, T., and Melchior, C.L. (1993). Decreased tolerance to ethanol-induced hypothermia in long-term castrate male rats. *Pharmacol. Biochem. Behav.,* 46, 309–314.

McIntyre, T.D. and Alpern, H.P. (1986). Thiopental, phenobarbital, and chlordiazepoxide induce the same differences in narcotic reaction as ethanol in long-sleep and short-sleep selectively-bred mice. *Pharmacol. Biochem. Behav.,* 24, 895–898.

Melchior, C.L. and Allen, P.M. (1992). Interaction of pregnanolone and pregnenolone sulfate with ethanol and pentobarbital. *Pharmacol. Biochem. Behav.,* 42, 605–611.

Melchior, C.L. and Allen, P.M. (1993). Temperature in mice after ethanol: effect of probing and regain of righting reflex. *Alcohol,* 10, 17–20.

Melchior, C.L. and Ritzmann, R.F. (1992). Dehydroepiandrosterone enhances the hypnotic and hypothermic effects of ethanol and pentobarbital. *Pharmacol. Biochem. Behav.,* 43, 223–227.

Melchior, C.L. and Ritzmann, R.F. (1994a). Dehydroepiandrosterone is an anxiolytic in mice on the plus maze. *Pharmacol. Biochem. Behav.,* 47, 437–441.

Melchior, C.L. and Ritzmann, R.F. (1994b). Pregnanolone and pregnenolone sulfate, alone and with ethanol, in mice on the plus-maze. *Pharmacol. Biochem. Behav.,* 48, 893–897.

Mogil, J.S., Marek, P., Yirmiya, R., Balian, H., Sadowski, B., Taylor, A.N., and Liebeskind, J.C. (1993). Antagonism of the non-opioid component of ethanol-induced analgesia by the NMDA receptor antagonist MK-801. *Brain Res.,* 602, 126–130.

Morrow, E.L. and Erwin, V.G. (1986). Calcium influence on neuronal sensitivity to ethanol in selectively bred mouse lines. *Pharmacol. Biochem. Behav.,* 24, 949–954.

Morrow, E.L. and Erwin, V.G. (1987). Calcium influence on neurotensin and beta-endorphin enhancement of ethanol sensitivity in selectively bred mouse lines. *Alcohol Drug Res.,* 7, 225–232.

Nabeshima, T., Tohyama, K., and Kameyama, T. (1990). Effects of DM-9384, a pyrrolidone derivative, on alcohol- and chlordiazepoxide-induced amnesia in mice. *Pharmacol. Biochem. Behav.,* 36, 233–236.

Nabeshima, T., Tohyama, K., and Kameyama, T. (1988). Reversal of alcohol induced amnesia by the benzodiazepine inverse agonist Ro 15-4513. *Eur. J. Pharmacol.,* 155, 211–217.

Nutt, D.J. and Lister, R.G. (1989). Antagonizing the anticonvulsant effect of ethanol using drugs acting at the benzodiazepine/GABA receptor complex. *Pharm. Biochem. Behav.,* 31, 751–755.

Palmer, M.R., Morrow, E.L., and Erwin, V.G. (1987). Calcium differentially alters behavioral and electrophysiological responses to ethanol in selectively bred mouse lines. *Alcoholism: Clin. Exp. Res.,* 11, 457–463.

Pearlman, B.J. and Goldstein, D.B. (1984). Genetic influences on the central nervous system depressant and membrane-disordering actions of ethanol and sodium valproate. *Mol. Pharm.,* 26, 547–552.

Perez-Reyes, M., White, W.R., and Hicks, R.E. (1992). Interaction between ethanol and calcium channel blockers in humans. *Alcoholism: Clin. Exp. Res.,* 16, 769–775.

Perola, M., Vuori, E., and Penttilä, A. (1994). Abuse of alcohol in sudden out-of-hospital deaths in Finland. *Alcoholism: Clin. Exp. Res.,* 18, 255–260.

Phillips, T.J., Burkhart-Kasch, S., Terdal, E.S., and Crabbe, J.C. (1991). Response to selection for ethanol-induced locomotor activation: Genetic analyses and selection response characterization. *Psychopharmacology*, 103, 557–566.

Phillips, T.J. and Dudek, B.C. (1991). Locomotor activity responses to ethanol in selectively bred long- and short-sleep mice, two inbred mouse strains, and their F1 hybrids. *Alcoholism: Clin. Exp. Res.,* 15, 255–261.

Phillips, T.J. and Gallaher, E.J. (1992). Locomotor responses to benzodiazepines, barbiturates and ethanol in diazepam-sensitive (DS) and -resistant (DR) mice. *Psychopharmacology,* 107, 125–131.

Phillips, T.J., Burkhart-Kasch, S., Gwiazdon, C.C., and Crabbe, J.C. (1992). Acute sensitivity of FAST and SLOW mice to the effects of abused drugs in locomotor activity. *J. Pharmacol. Exp. Ther.,* 261, 525–533.

Phillips, T.J. and Crabbe, J.C., Jr. (1991). Behavioral studies of genetic differences in alcohol action. In *The Genetic Basis of Alcohol and Drug Actions.* J.C. Crabbe, Jr. and R.A. Harris, Eds. New York: Plenum Press, pp. 25–104.

Pohorecky, L.A. (1977). Biphasic action of ethanol. *Biobehav. Rev.,* 1, 231–240.

Poling, A., Schlinger, H., and Blakely, E. (1988). Failure of the partial inverse benzodiazepine agonist Ro 15-4513 to block the lethal effects of ethanol in rats. *Pharmacol. Biochem. Behav.,* 31, 945–947.

Proctor, W.R., Baker, R.C., and Dunwiddie, T.V. (1985). Differential CNS sensitivity to PIA and theophylline in long-sleep and short-sleep mice. *Alcohol,* 2, 287–291.

Proctor, W.R. and Dunwiddie, T.V. (1984). Behavioral sensitivity to purinergic drugs parallels ethanol sensitivity in selectively bred mice. *Science,* 224, 519–521.

Pucilowski, O., Krascik, P., Trzaskowska, E., and Kostowski, W. (1989). Different effect of diltiazem and nifedipine on some central actions of ethanol in the rat. *Alcohol,* 6, 165–168.

Randall, C.L., Carpenter, J.A., Lester, D., and Friedman, H.J. (1975). Ethanol-induced mouse strain differences in locomotor activity. *Pharmacol. Biochem. Behav.,* 3, 533–535.

Rees, D.C. and Balster, R.L. (1988). Attenuation of the discriminative stimulus properties of ethanol and oxazepam, but not of pentobarbital, by Ro 15-4513 in mice. *J. Pharmacol. Exp. Ther.,* 244, 592–598.

Sanders, B. (1976). Sensitivity to low doses of ethanol and pentobarbital in mice selected for sensitivity to hypnotic doses of ethanol. *J. Comp. Physiol. Psychol.,* 90, 394–398.

Schechter, M.D., Meehan, S.M., Gordon, T.L., and McBurney, D.M. (1993). The NMDA receptor antagonist MK-801 produces ethanol-like discrimination in the rat. *Alcohol,* 10, 197–201.

Schuckit, M.A. (1994). Low level of response to alcohol as a predictor of future alcoholism. *Am. J. Psychiatry,* 151, 184–189.

Sellers, E.M. and Busto, U. (1982). Benzodiazepines and ethanol: assessment of the effects and consequences of psychotropic drug interactions. *J. Clin. Psychopharm.,* 2, 249–262.

Sharma, A.C., Thorat, S.N., Nayar, U., and Kulkarni, S.K. (1991). Dizocilpine, ketamine and ethanol reverse NMDA-induced EEG changes and convulsions in rats and mice. *Indian J. Physiol. Pharmacol.,* 35, 111–116.

Simpson, V.J., Baker, R.C., and Timothy, S. (1993). Isoflurane but not halothane, demonstrates differential sleep times in long sleep and short sleep mice. *Anesthesiology,* 79, A387.

Simpson, V.J., and Blednov, Y. (1995). Propofol produces differences in behavior but not chloride channel function in selected lines of mice. In preparation.

Smolen, T.N., Howerton, T.C., and Collins, A.C. (1984). Effects of ethanol and salsolinol on catecholamine function in LS and SS mice. *Pharmacol. Biochem. Behav.,* 20, 125–131.

Smolen, T.N. and Collins, A.C. (1984). Behavioral effects of ethanol and salsolinol in mice selectively bred for acute sensitivity to ethanol. *Pharmacol. Biochem. Behav.,* 20, 281–287.

Smolen, T.N. and Smolen, A. (1991). Purinergic modulation of ethanol-induced sleep time in long-sleep and short-sleep mice. *Alcohol,* 8, 123–130.

Spuhler, K. and Deitrich, R.A. (1984). Correlative analysis of ethanol-related phenotypes in rat inbred strains. *Alcoholism: Clin. Exp. Res.,* 8, 480–484.

Spuhler, K., Hoffer, B., Weiner, N., and Palmer, M. (1982). Evidence for genetic correlation of hypnotic effects and cerebellar Purkinje neuron depression in response to ethanol in mice. *Pharmacol. Biochem. Behav.,* 17, 569–578.

Stinchcomb, A., Bowers, B.J., and Wehner, J.M. (1989). The effects of ethanol and Ro 15-4513 on elevated plus-maze and rotarod performance on long-sleep and short-sleep mice. *Alcohol,* 6, 369–376.

Sutoo, D'E., Akiyama, K., and Iimura, K. (1985). Effect of calmodulin antagonists on calcium and ethanol-induced sleeping time in mice. *Pharm. Biochem. Behav.,* 23, 627–631.

Suzdak, P.D., Glowa, J.R., Crawley, J.N., Schwartz, R.D., Skolnick, P., and Paul, S.M. (1986). A selective imidazobenzodiazepine antagonist of ethanol in the rat. *Science,* 234, 1243–1247.

Suzdak, P.D., Paul, S.M., and Crawley, J.N. (1988). Effects of Ro 15-4513 and other benzodiazepine receptor inverse agonists on alcohol-induced intoxication in the rat. *J. Pharmacol. Exp. Ther.,* 245, 880–886.

Syapin, P.J., Gee, K.W., and Alkana, R.L. (1987). Ro 15-4513 differentially affects ethanol-induced hypnosis and hypothermia. *Brain Res. Bull.,* 19, 603–605.

Tabakoff, B. and Kiianmaa, K. (1982). Does tolerance develop to the activating, as well as the depressant, effects of ethanol? *Pharmacol. Biochem. Behav.,* 17, 1073–1076.

Tsibulsky, V.L. and Amit, Z. (1993a). Tolerance to effects of high doses of ethanol: 1. Lethal effects in mice. *Pharmacol. Biochem. Behav.,* 45, 465–472.

Tsibulsky, V.L. and Amit, Z. (1993b). Role of environmental cues as Pavlovian-conditioned stimuli in enhancement of tolerance to ethanol effects: 1. Lethal effects in mice and rats. *Pharmacol. Biochem. Behav.,* 45, 473–479.

Van Den Buuse, M. (1994). Circadian rhythms of blood pressure, heart rate, and locomotor activity in spontaneously hypetensive rats as measured with radio-telemetry. *Physiol. Behav.,* 55, 783–787.

Vogel, R.A., Frye, G.D., Mann Koepke, K., Mailman, R.B., Mueller, R.A., and Breese, G.R. (1981). Differential effects of TRH, amphetamine, naloxone, and fenmetozole on ethanol actions: attenuation of the effects of punishment and impairment of aerial righting reflex. *Alcoholism: Clin. Exp. Res.*, 5, 386–392.

Waller, M.B., Murphy, J.M., McBride, W.J., Lumeng, L., and Li, T.-K. (1986). Effect of low dose ethanol on spontaneous motor activity in alcohol-preferring and -nonpreferring lines of rats. *Pharmacol. Biochem. Behav.,* 24, 617–623.

Wilson, W.R., Bosy, T.Z., and Ruth, J.A. (1990). NMDA agonists and antagonists alter the hypnotic response to ethanol in LS and SS mice. *Alcohol,* 7, 389–395.

Wise, R.A., and Bozarth, M.A. (1987). A psychomotor stimulant theory of addiction. *Psychol. Rev.,* 94, 469–472.

Wong, M. and Moss, R.L. (1994). Patch-clamp analysis of direct steroidal modulation of glutamate receptor-channels. *J. Neuroendocrinol.,* 6, 347–355.

Yu, S. and Ho, I.K. (1990). Effects of acute barbiturate administration, tolerance and dependence on brain GABA system: comparison to alcohol and benzodiazepines. *Alcohol,* 7, 261–272.

Chapter 15

ASPECTS OF ALCOHOL TOLERANCE IN HUMANS AND EXPERIMENTAL ANIMALS

A.D. Lê and J.M. Mayer

TABLE OF CONTENTS

I. INTRODUCTION

The term "alcohol* tolerance" has two different connotations: one is innate or initial tolerance, and the other is acquired tolerance. Innate or initial tolerance refers to variations in individual sensitivity to the acute effects of alcohol. On the other hand, acquired tolerance to alcohol is generally defined as a reduction in the magnitude of effects produced by a given

* The terms alcohol and ethanol will be used interchangeably.

0-8493-8389-7/95/$0.00+$.50

dose of alcohol following repeated or chronic exposure to this substance. Such tolerance may be due to a decrease in the sensitivity of the target tissue, or central nervous system (CNS), to the effects of alcohol (functional tolerance) and to an enhanced elimination of alcohol from the body (dispositional tolerance). Such a definition of tolerance may not be truly accurate, as it appears that chronic exposure is not essential for tolerance development. Indeed, tolerance has been shown to develop during a single exposure to alcohol, a phenomenon that was reported more than 70 years ago by Mellanby (1919) and commonly referred to as acute tolerance.

Contrary to earlier notions that tolerance reflects a physiological adaptation to the effects of a drug on the body, the systematic investigation of tolerance over the last three decades has shown that it is quite a complex phenomenon. It is now well accepted that tolerance is governed by a variety of neuronal systems and that the development and manifestation of tolerance can be influenced by a variety of nonpharmacological factors such as genetic, environmental, and learning processes (operant and associative). In this chapter, we concentrate on the acute and chronic functional tolerance to ethanol. The involvement of a metabolic component in the development of chronic tolerance will not be addressed in this chapter. For information in this area the reader is referred to some of the extensive reviews by Hawkins and Kalant (1972), Khanna and Israel (1980), and Lê and Khanna (1989). The area of innate tolerance, or initial sensitivity, and how acute tolerance might affect the measurement or assessment of initial sensitivity are covered extensively in Chapter 14 and, therefore, is not dealt with here. The relationship between initial sensitivity and acquired tolerance, however, is addressed.

We first examine the descriptive studies dealing with acute and chronic tolerance in animals and humans. This is followed by an examination of the various factors such as genetic, environmental, and learning processes that affect tolerance development. The possible relationship between acute and chronic tolerance, and the pharmacological and nonpharmacological aspects is also discussed.

II. ACUTE TOLERANCE

Acute tolerance was first described by Mellanby in 1919. He showed that the degree of alcohol-induced intoxication observed in dogs, at a given blood alcohol concentration (BAC), was much less severe on the descending than on the ascending portion of the BAC versus time curve. Initially, this report of acute tolerance was met with criticism for two reasons. First, as alcohol concentrations were determined in venous blood, the differences in the observed degrees of intoxication were thought to be related to the arterio-venous differences in the alcohol concentration during the absorptive (rising) phase. Second, as repeated testing was required, it was argued that the improvement in performance was due to a practice effect while the dogs were intoxicated, and thus acute tolerance was simply a learning artifact (Goldberg, 1943). Subsequent work by a number of investigators, employing a variety of experimental designs and techniques, has clearly shown that although these factors can affect the measurement of acute tolerance, the phenomenon does actually occur in humans and experimental animals. Evidence that confirms the existence of acute tolerance development and its nature in experimental animals and human subjects is reviewed below.

A. STUDIES WITH HUMAN SUBJECTS

1. Methodological Assessment of Acute Tolerance in Human

Over the years a number of experimental methods have been employed to examine the development of acute tolerance to alcohol in humans. Three basic experimental designs have been developed to evaluate acute tolerance, namely:

1. an assessment of the differences in the degree of intoxication at the same BAC on the ascending and descending portions of the BAC versus time curve (Within this approach, the extent of intoxication, assessed as the area under the curve during the ascending compared to the descending phase of the BAC-time curve, was also used [Martin and Moss, 1993])
2. a determination of the BAC at the onset of the impairment of a particular behavioral or physiological index and of the BAC at the offset of such impairment (Goldberg, 1943; Vogel-Sprott, 1979)
3. repeated measurements of the extent of intoxication while maintaining a steady-state BAC (Kaplan et al., 1985)

Each of these experimental designs has its own limitations. For example, the classical Mellanby measurement involves only two data points matched for the same BAC on the ascending and descending portions of the BAC-time curve. The magnitude of effects and the probability of detecting acute tolerance might be related to the time elapsed between ascending and descending measurement and must be of sufficient duration to allow tolerance to develop. The second approach does not require a differentiation between ascending and descending BACs. However, in order to measure the time and BACs at the onset and offset of a particular effect of alcohol, repeated measurements or assessments of intoxication are required, and such a procedure may result in an attenuation of the acute effects due to a practice effect (this issue will be discussed further in Section 2a). The third approach also suffers from the previously mentioned complication. In addition, this approach is rather labor intensive.

With the exception of the study by Mirsky et al. (1941), experiments on acute tolerance in humans have been restricted to low doses of alcohol that produce maximal BACs of no greater than 100 mg/dl. Most of the later studies on acute tolerance utilized breath testing for the determination of blood alcohol concentration. Such a procedure rules out most, if not all, of the complications likely to be associated with an arterio-venous lag during the early part of alcohol absorption, as breath alcohol more closely estimates arterial rather than venous BAC (Martin et al., 1984).

2. Evidence for Acute Tolerance to Alcohol

One of the earliest and most extensive investigations of acute tolerance came from the work by Goldberg (1943) that examined acute tolerance to a variety of ethanol-induced effects, ranging from effects on sensory and motor skills to results on cognitive tests. The BACs at the onset of intoxication for various functions varied between 31 and 65 mg/dl. However, the BACs at the offset of intoxication for these functions occurred at BACs ranging from 36 to 75 mg/dl. Acute tolerance was found to develop at different rates and extends to the different effects of alcohol, with cognitive skills returning to the prealcohol baseline sooner and at higher BACs compared to motor skills. This observation also raised the issue concerning the common assumption that intellectual processes are more sensitive to the disruptive effects of alcohol than are motor skills. It should be pointed out that Goldberg used capillary blood samples for determination of BACs and, therefore, provided a valid estimate of arterial BAC.

Other demonstrations of acute tolerance include those with respect to the impairing effects of alcohol on motor or psychomotor performance such as its effect on standing steadiness (Hurst and Bagley, 1972), reaction time (Young, 1970; Wilson et al., 1984; Nicholson et al., 1992), and pursuit rotor task (Vogel-Sprott, 1979; Wilson et al., 1984). Similarly, acute tolerance to the impairing effects of alcohol on a number of cognitive tasks such as anticipation time (Nicholson et al., 1992), recall (Jones, 1973), numerical coding of letter charts (Hurst and Bagley, 1972; Vogel-Sprott, 1979), and arithmetic (Ekman et al., 1964) have also been shown. In addition, acute tolerance to the subjective effects of alcohol has also been documented (Myrsten et al., 1975; Radlow and Hurst, 1985; Portans et al., 1989; Martin and Moss, 1993).

3. Factors Affecting Acute Tolerance

a. Test Dependency

The demonstration of acute tolerance will depend on the range of BAC produced and the test employed to assess tolerance. As a result, in some instances no tolerance or marginal tolerance was seen with respect to the acute effects of ethanol on automobile driving ability (Loomis and West, 1958) or visual masking task (Moskowitz and Burns, 1976). Vogel-Sprott (1979) demonstrated that under the same experimental conditions, tolerance to the impairing effect of ethanol on a coding task can be demonstrated, whereas the same subjects showed no evidence of acute tolerance on a pursuit motor task. The development of acute tolerance to the motor impairing effects of alcohol is also dependent on the test employed. For example, Wilson et al. (1984) showed that acute tolerance to the impairing effects of ethanol does develop and can be measured on the dowel balancing, hand steadiness, and reaction time, but not on pursuit motor task, rail walking, or body sway. Jones (1973) reported that although both short- and long-term memory are impaired by alcohol, acute tolerance was observed only for short-term memory impairment.

b. Effect of Repeated Testing

Goldberg (1943) raised the question of whether acute tolerance is, in fact, a learned adaptation following a subject's repeated performance while intoxicated. This issue has been examined quite extensively by Vogel-Sprott and co-workers, who used the BAC threshold for onset of and recovery from the impairing effect of alcohol on trachometer performance to evaluate tolerance development (Haubenreisser and Vogel-Sprott, 1987; Vogel-Sprott and Sdao-Jarvie, 1989). Acute tolerance was found to develop faster and at higher intensity in the group that received monetary or informative reward for maintaining their performance than in subjects that did not. This research indicates that when repeated testing is employed to assess acute tolerance, the consequences of task performance under such a condition play a critical role in tolerance development. Whether practice during intoxication, with or without rewarding consequences, can enhance tolerance development compared to a nonpractice situation remains to be examined.

c. Effect of Previous Exposure

The study by Goldberg (1943) provided some evidence for the differential development of acute tolerance in subjects with different histories of alcohol drinking. The BAC at which various signs of intoxication disappeared was found to be much higher in heavy drinkers than in moderate drinkers or abstainers. Similarly, Mirsky et al. (1941) found that alcoholics can develop a high degree of acute tolerance. Using gross clinical signs of intoxication as measures of the effects of ethanol and multiple doses of ethanol to maintain a constant BAC, these investigators reported that patients displayed signs of intoxication at BACs of approximately 200 mg/dl, but later became sober, even though the BAC was increased to approximately 300 mg/dl. Although these studies suggest that tolerance develops to a great extent in heavy drinkers, the data do not permit further conclusions, as no effort was made to experimentally control for the possible contribution of preexisting chronic tolerance in this group.

Most of the later studies (Banks et al., 1979; Portans et al., 1989) report that initially, light drinkers tend to be more affected by alcohol than heavy drinkers are, but the difference disappears sometime after peak BAC has been reached. Given that heavy and light drinkers have different initial responses to alcohol, it is difficult to draw any conclusions regarding the ability to develop acute tolerance among these subjects. There are, however, some indications that repeated exposure to alcohol can enhance the development of acute tolerance. For example, recent work by Bennet et al. (1993) examined the impairing effects on eye-hand coordination by the same dose of alcohol over five consecutive days. Chronic tolerance to this effect was observed in subjects receiving daily doses of 0.75 or 1 g/kg of alcohol. Acute tolerance, defined as differences in the level of impairment for the same BAC on the ascending and descending portions of BAC-time curve, was observed only by the 4th and 5th administration of alcohol in the lower-dose, but not in the higher-dose, group.

B. STUDIES WITH EXPERIMENTAL ANIMALS

1. Demonstration of Acute Tolerance

Probably the two most important studies demonstrating the existence of acute tolerance in experimental animals since the original report by Mellanby (1919) were carried out by Mirsky et al. (1941) and LeBlanc et al. (1975). The study by Mirsky et al. is important not only because of its chronology, but also because of its experimental and conceptual approach to the study of acute tolerance. Acute tolerance to alcohol was demonstrated in two different ways by Mirsky and co-workers. In the first instance, they demonstrated that various symptoms of intoxication, such as ear drop, nystagmus, and coma produced by the intravenous infusion of alcohol to hepatectomized animals, disappeared by 20–40 minutes after the beginning of such infusion. Raising the BAC to approximately 200 mg/dl reinstated such symptoms which, however, disappeared by 10–30 minutes after alcohol infusion, even though blood alcohol levels were about twice their initial value. In addition to these measurements of intoxication, evidence for this phenomenon was also provided by the electroencephalographic records. Acute tolerance was also demonstrated in a different experimental approach by these authors. They showed that a much higher BAC is required to produce the same degree of intoxication when ethanol is infused at a lower rate rather than a higher rate. In other words, time is an important variable for the manifestation of acute tolerance. The study by LeBlanc et al. (1975) contributes to the demonstration of acute tolerance to alcohol in two different ways. Theirs was the first study that employed brain rather than blood alcohol levels to assess acute tolerance. Second, animals were tested only once to assess intoxication, thereby eliminating the possible contribution of learning in such tolerance. In this study, acute tolerance was demonstrated by a parallel shift to the right of the regression line of motor impairment (response) versus brain ethanol concentration. The extent of the shift increased with the time elapsed since ethanol administration.

Various experimental designs which involve minor modifications of those employed by Mirsky et al. and LeBlanc et al. have been used to assess acute tolerance to ethanol (Lê et al., 1992b). Acute tolerance was defined experimentally by (1) a lower degree of intoxication at the same or higher blood ethanol levels following a single or two consecutive injections of ethanol, and (2) higher BACs or brain alcohol concentrations (BrAC) at the offset as compared to BAC or BrAC at the onset of intoxication (see Lê et al., 1992b for review). Acute tolerance to various effects of alcohol that involve a wide range of dosages has been demonstrated over the years by numerous investigators. Acute tolerance to the discriminative effect (Hiltunen and Jarbe, 1990), operant behavior (Hiltunen and Jarbe, 1992), motor impairment (Tullis et al., 1977; Gallaher et al., 1982; Campanelli et al., 1988; Lê and Kalant, 1992; Lê et al., 1992a), and hypnosis have all been demonstrated in experimental animals.

2. Factors Affecting Acute Tolerance

In contrast to human studies, experimental studies of acute tolerance to alcohol in animals have paid little attention to factors that might affect acute tolerance. The work has generally focused on demonstrating acute tolerance and the role of genetic factors in acute tolerance.

a. Effects of Test Systems and History of Alcohol Exposure

It can be said that factors affecting the development of acute tolerance to alcohol in humans have also been shown to influence acute tolerance to alcohol in animals in a similar manner. Acute tolerance has been shown to occur to many, but not all of the effects of ethanol in experimental animals. For example, recent work by Lê et al. (1992a) showed that under the same experimental conditions, acute tolerance to the motor impairing but not to the anticonvulsant effect of ethanol can be demonstrated. Practicing while intoxicated has also been shown to affect the development of acute tolerance. Lê and Kalant (1992) have shown that rats exposed to repeated testing under the influence of alcohol develop acute tolerance at a higher rate than those tested only once. Since the test employed in this study involved footshock, one cannot rule out the possibility that the consequences of performance (i.e., shock avoidance) might be a critical (rewarding) factor in influencing acute tolerance development rather than simply repeated testing as suggested by Vogel-Sprott (1979).

The effect of chronic treatment on the development of acute tolerance to ethanol has also been examined. Kalant et al. (1978) reported that although the maximum impairment induced by alcohol was essentially the same for naive rats and those withdrawn from chronic ethanol treatment for several weeks, acute tolerance to ethanol developed more rapidly in animals that were previously exposed to ethanol treatment. Recent work by Hiltunen and Jarbe (1992) showed that acute tolerance to the impairing effect of alcohol on operant performance can be demonstrated in rats chronically treated with alcohol. However, this can be demonstrated only when the test dose of alcohol employed is higher than that employed for chronic treatment. The development of acute tolerance is also dependent on the paradigm employed to assess tolerance development. Tabakoff and Ritzmann (1979) and Tabakoff et al. (1980) were unable to demonstrate acute tolerance in either long-sleep (LS) or short-sleep (SS) mice selectively bred for their differential sensitivity to ethanol, as no differences were seen in the BrAC at the onset and offset of loss of righting reflex (LORR). However, acute tolerance was apparent in both groups as mice that received higher doses of alcohol regained the righting reflex at higher BACs than those that received lower doses of ethanol (Keir and Deitrich, 1990).

b. Genetic Factors

Studies with various strains of rodents have shown that acute tolerance to ethanol can be influenced by genetic factors. Studies with inbred mice have shown that C57BL/6 mice develop more acute tolerance than do DBA/2N mice (Grieve and Littleton, 1979; Tabakoff and Ritzmann, 1979). C57BL/6 mice regain their righting reflex at a higher BrAC than DBA mice do, although BrACs at the loss of righting reflex (LORR) are the same between these two strains (Tabakoff and Ritzmann, 1979). SS mice have been shown to develop a greater degree of acute tolerance to ethanol than LS mice do (Keir and Deitrich, 1990). A study with both the alcohol tolerant (AT) and alcohol nontolerant (ANT) rats selectively bred for their differential response to the motor impairing effects of ethanol has also shown that acute tolerance to the motor impairing but not the hypothermic effect of ethanol is more pronounced in the AT rats, which are more resistant to such effects of ethanol (Lê and Kiianmaa, 1989). These studies together suggest that differences in the ability to develop acute tolerance might account in part for the differences in initial sensitivity to ethanol between these selected lines (Lê and Kiianmaa, 1989; Keir and Deitrich, 1990).

Differences in the ability to develop acute tolerance to ethanol have also been shown to occur in rats selectively bred for differences in alcohol consumption (Waller et al., 1983; Lê and Kiianmaa, 1988). Rats selected for high ethanol preference, such as the P and AA lines, have been shown to develop better acute tolerance than nonalcohol-preferring rats, such as the NP or ANA lines.

III. CHRONIC TOLERANCE

Chronic tolerance to alcohol, particularly in experimental animals, has been examined quite extensively over the last several decades. The literature on chronic tolerance to alcohol in animals has been the subject of many reviews. In this section the important features of chronic tolerance to alcohol in humans and experimental animals is described.

A. STUDIES WITH HUMAN SUBJECTS

The studies of tolerance to alcohol in humans can be classified into two different types:

1. a comparative examination of responses to alcohol by light and heavy drinkers or alcoholics
2. an examination of the development of experimentally-acquired tolerance under laboratory conditions.

Early evidence for chronic tolerance to alcohol was reported by Jetter (1938), who observed that at a BAC of 200 mg/dl, 83% of chronic alcoholics were clinically intoxicated, whereas 100% of occasional drinkers with the same BAC were intoxicated. The most systematic examination of tolerance to alcohol among different types of drinkers came from the study of Goldberg (1943), who examined the effects of challenge doses of alcohol on a battery of motor and cognitive tests in abstainers, moderate drinkers, and heavy drinkers. On tests of motor function, he found that the disappearance thresholds (BACs at which impairment disappeared) were 2 to 3 times higher in moderate and heavy drinkers, respectively, than in abstainers. In cognitive tests, the disappearance thresholds were about 1.5 and 2 times higher in moderate and heavy drinkers, respectively. In other words, the degree of tolerance is not only related to the history of drinking, but also depends on the tests employed to measure it.

Subsequent studies by a number of investigators also showed that heavy drinkers are less impaired by alcohol on motor tasks than light or social drinkers are (Goodwin et al., 1971; Moskowitz et al., 1974; Rosen and Lee, 1976; Mello and Mendelson, 1978). There is, however, considerable debate whether tolerance develops for the impairing effects of alcohol on memory or cognitive functions. Parker et al. (1974) reported that alcoholics and social drinkers are impaired in a similar manner by alcohol on tasks of digit-span and free-recall. Rosen and Lee (1976) showed that social drinkers display signs of gross intoxication at a BAC of 100 mg/dl, whereas such symptoms were absent in heavy drinkers or alcoholics. However, all groups were equally impaired on the cognitive performance tests. Since the basal performance on these tests was essentially similar among these experimental groups, it is difficult to suggest that the absence of tolerance on the cognitive tests in the alcoholic or heavy drinkers was a result of neuropsychologic damage associated with heavy drinking.

Laboratory studies showed that tolerance to the impairing effects of alcohol on motor performance can develop within a few sessions of exposure to low doses of alcohol (maximum BAC of about 80 mg/dl). Bennet et al. (1993) reported tolerance to the disruptive effect of ethanol on eye-hand coordination after 3 to 4 daily exposures to alcohol. However, the development of tolerance to impairment of motor performance is also dependent on the task and the experimental procedures employed. Tolerance to alcohol-induced impairment of performance on a trachometer, but not on a pursuit task, can develop after 3 to 4 sessions of exposure to alcohol (Beirness and Vogel-Sprott, 1984; Sdao-Jarvie and Vogel-Sprott, 1991; Zack and Vogel-Sprott, 1993). Tolerance to the latter developed only after subjects received feedback information regarding their performance.

Tolerance to impairment of cognitive functions has also been shown under experimental conditions. Vogel-Sprott (1979) reported that tolerance to impairment of cognitive tasks develops after four days of exposure to moderate doses of alcohol (maximum BAC 80 mg/dl), whereas impairment of a pursuit motor task is unaffected. In addition, Poulos et al. (1981) also showed that tolerance to the effects of alcohol on a memory task can be demonstrated after 10 consecutive daily exposures to 1.9 g/kg ethanol. Although a practice effect associated with repeated testing (Maylor and Rabbit, 1987) might account for some of the observed tolerance in these studies, it should be noted that acute tolerance to the effects of ethanol with respect to cognitive functions has also been reported by a number of investigators (Goldberg, 1943; Hurst and Bagley, 1972; Jones, 1973; Jones and Vega, 1972). Clearly, the questions of whether tolerance develops to the impairing effects of ethanol on cognitive functions and under what conditions it will manifest itself remain to be answered.

Most, if not all, of the experimental investigations of alcohol tolerance in humans have been limited to BACs no greater than 100 mg/dl. There are reports regarding high levels of alcohol tolerance in humans, and most deal with patients admitted to hospital emergency departments for alcohol-related, or other medical problems. Hammond et al. (1973) reported on a patient who was arrested for drunk driving and became comatose at a BAC of 780 mg/dl. Following treatment, she regained full consciousness at a BAC of 520 mg/dl. Similarly, Johnson et al. (1982) reported on a patient admitted to hospital emergency for abdominal pain, who had a BAC of 1510 mg/dl. She became sober after 12 hours or at a BAC > 900 mg/dl, assuming an elimination rate of 40 mg/dl per hour.

In a study involving 65 patients who were admitted to emergency for nonalcohol related illness and were diagnosed as nonintoxicated by the examining physician, the average BAC was 268 mg/dl (Urso et al., 1981). Among these patients, 19 males out of 47 had BAC ranging from 300–540 mg/dl. Perper et al. (1986) evaluated the clinical intoxication in alcoholics who voluntarily applied for admission to a hospital. They found normal speech and adequate verbal comprehension in 10 individuals who had BACs at or greater than 350 mg/dl. Similarly, Sullivan et al. (1987) also found no correlation between alcohol symptoms and BAC in 21 patients with an average plasma alcohol concentration of 300 mg/dl. These reports suggest that tolerance can develop to effects associated with high BACs. To what extent this phenomenon occurs in alcoholics or heavy drinkers remains to be examined.

B. STUDIES WITH EXPERIMENTAL ANIMALS

1. Features of Chronic Tolerance

Chronic tolerance to ethanol is commonly produced by exposing animals to ethanol for days or weeks by vapor inhalation, liquid diet, or by daily oral or intraperitoneal injections (see Lê et al., 1992b for recent review). To rule out the contribution of dispositional factors, the BAC or BrAC for control and ethanol-treated groups is usually compared. Alternatively, short-lasting physiological or behavioral indices may be employed to assess ethanol effects and tolerance, such that differences due to altered ethanol elimination would be insignificant. There is extensive literature on chronic functional tolerance to alcohol. This body of literature has been the subject of many extensive reviews over the last several years. In this section, a summary of important pharmacological features of tolerance is described.

a. Does Tolerance Develop to All Effects of Ethanol?

Tolerance has been shown to develop over a wide range of dosages to a variety of ethanol effects such as impairment of motor coordination, hypothermia, hypnosis, and lethality (see Lê et al., 1992b). The doses and duration of ethanol treatment are important factors that determine not only the rate and extent of tolerance development, but also the effect to which tolerance might develop. For example, the extent of tolerance to the hypothermic effect of ethanol is related to the treatment dose employed (Lê et al., 1984; Crabbe, 1994). A treatment regimen that produces tolerance to the motor impairment might not be sufficient to produce tolerance to the hypothermic or hypnotic effect of ethanol (Lê et al., 1989). The rate of tolerance development will differ for each of the effects of ethanol, often independently of the treatment dose or that required to elicit the effect. For example, tolerance to the hypothermic effect of ethanol develops at a much faster rate than that to the motor-impairing or hypnotic effects.

Few studies have examined the possible development of tolerance to the excitatory effects of ethanol. In contrast to the depressant effects of ethanol, the stimulatory effects of ethanol are difficult to detect and quantify. The stimulatory effect induced by low doses of ethanol as measured by hyperlocomotion is more readily seen in mice than in rats (Frye and Breese, 1981). In mice, the stimulation of locomotor activity induced by ethanol is also strain dependent, with the effect more pronounced in DBA mice but minimal or absent in other strains such as C57BL/6 or BALB (Tabakoff and Kiianmaa, 1980). The available data on the effects of chronic ethanol treatment on the stimulation of locomotor activity are conflicting since tolerance (Hunt and Overstreet, 1977), no change (Tabakoff and Kiianmaa, 1980), and sensitization (Crabbe et al., 1982; Cunningham and Noble, 1992) have all been reported. Because different strains of animals and different experimental procedures were employed in these studies, it is difficult to draw any definite conclusion regarding tolerance or sensitization to the stimulatory effect of ethanol.

b. Rapid Development of Tolerance

Although it is commonly believed that intensive treatment with ethanol for days or weeks is required to produce chronic tolerance to ethanol, single moderate to high doses of ethanol have also been shown to produce this phenomenon 24 hours after administration. Crabbe et al. (1979) were the first to demonstrate that tolerance to the hypothermic effect of ethanol can be induced by pretreatment, 24 hours prior to testing, with a single dose equivalent to the test dose. They referred to this phenomenon as "rapid tolerance," which has been subsequently demonstrated for the motor impairing and hypnotic effects of ethanol (Buck et al., 1991; Lê and Kiianmaa, 1988; Khanna et al., 1991). The extent of rapid tolerance is comparable to that produced following prolonged exposure to ethanol (Khanna et al., 1991).

Unlike some experimental designs used to study acute tolerance in which two consecutive injections of ethanol are administered, the second test dose in the rapid tolerance model is administered at a time when alcohol from the previous dose would have been completely eliminated. It is unlikely that such a phenomenon is confounded by acute tolerance unless one argues that a measure of acute tolerance has been retained. This is a possibility because rapid tolerance is demonstrable only at 24 hours, but not by 48 hours, after the first ethanol injection (Crabbe et al., 1979). The rapid dissipitation of this form of tolerance is also different from that induced by prolonged chronic treatment, which usually disappears by 3–7 days following discontinuation of ethanol treatment (Lê et al., 1992b).

c. Influence of Initial Sensitivity

There is a large variation in individual sensitivity to ethanol within and among different strains of rodents. Individual sensitivity to a particular effect of ethanol can also vary markedly depending on the circumstances in which ethanol is administered. It has been suggested that the degree of ethanol-induced functional disturbance or impairment is a critical variable in determining the development of chronic tolerance (Kalant et al., 1971). Crabbe et al. (1982) examined the relationship between initial sensitivity and the extent of tolerance development in 20 inbred strains of mice. They found a significant positive correlation between initial sensitivity to the hypothermic effect of ethanol and the subsequent tolerance development to this effect. Studies involving heterogeneous rat and mouse strains (San Marina et al., 1989; Khanna et al., 1989), inbred strains (Moore and Kakihana, 1978), rats or mice selectively bred for differences in sensitivity to ethanol (Crabbe et al., 1989; Crabbe, 1994; Khanna et al., 1985) have also supported such a relationship.

Since different lines of rodents were commonly used to examine the relationship between initial sensitivity and acquired tolerance, a genetic role has been implicated in controlling tolerance. Although there is no issue on whether genetics plays a role in determining tolerance development, genetics is not the only factor that determines the relationship between initial sensitivity and tolerance. For example, tolerance to the hypothermic effect of ethanol can develop at a higher rate when ethanol treatment is carried out at a low ambient temperature that enhances the hypothermic effect of ethanol. Conversely, tolerance develops at a lower rate when ethanol treatment is carried out at a high ambient temperature that reduces the hypothermic effect of ethanol (Lê et al., 1986).

d. Genetic Factors

Because initial sensitivity has been correlated with tolerance, it may be argued that genetic differences affect tolerance development indirectly by simply regulating the initial acute response to alcohol, but offer no control over other events involved in the process. A number of studies, however, have suggested a direct involvement of genetic factors in the regulation of tolerance development. The AT and ANT rats selectively bred for their differences to the motor-impairing effect of ethanol do not differ in their sensitivity to the hypothermic effect. Tolerance to ethanol-induced hypothermia, however, develops at a higher rate at the AT than in the ANT rats (Lê and Kiianmaa, 1990). Studies with other selected rat lines selectively bred for their differences in alcohol consumption also showed a differential rate of tolerance development between the high and low ethanol-preferring rats, even though these animals do not differ in their sensitivity to ethanol (Waller et al., 1983; Gatto et al., 1987; Lê and Kiianmaa 1988).

IV. REGULATION OF ALCOHOL TOLERANCE BY BEHAVIORAL PROCESSES

In the previous section, several pharmacological and biological factors affecting tolerance development were described. The development and manifestation of tolerance to alcohol, however, have been shown to be modified by a variety of nonpharmacological factors. In fact, the pronounced effect on tolerance by some learning or behavioral processes have led to the suggestion that tolerance might be a learned response rather than a biological adaptation. Opportunity to experience the drug effects or practice while under intoxication and Pavlovian conditioning have been shown to be critical behavioral factors that affect tolerance development. The involvement of these processes in the development of alcohol tolerance in humans and experimental animals is examined in this section.

A. PRACTICE UNDER INTOXICATION

The influence of practice during intoxication on tolerance development in humans was speculated on by Goldberg as early as 1943. The first experimental study to examine the role of task practice during intoxication on tolerance was conducted by Chen (1968). In this study, two groups of rats were trained to criterion performance on a maze. One group of rats received daily injections of 1.2 g/kg of ethanol before, whereas the other group received ethanol treatment after running the maze. On the fifth day, when tolerance was examined by administering alcohol to both groups before running the maze, tolerance to the impairing effect of ethanol on maze performance was found only in the "before" group. Since the groups had similar pharmacological history, Chen concluded that tolerance is a consequence of task practice during intoxication. Although the initial observation concerning the effects of task practice on tolerance made by Chen was confirmed by a number of investigators (Chen, 1972, 1979; LeBlanc et al., 1973, 1975; de Souza Moreira et al., 1981; Wenger et al., 1981), the role of the learning process was a subject of considerable debate. LeBlanc et al. (1973, 1975) suggested that task practice during intoxication only speeds the rate of tolerance development since tolerance was subsequently observed in the "after" group following an extended treatment period and the monitoring of performance at 4-day intervals. Wenger et al. (1981) demonstrated, however, that the observed tolerance in the "after" group in the LeBlanc et al. studies was due to repeated testing at 4-day intervals, because they did not observe any significant tolerance in an "after" group that was not repeatedly tested at 4-day intervals, even after 20 days of ethanol treatment. These studies support the notion of a role for task practice on tolerance development.

Because all of the above studies utilized a prior learned task to assess tolerance, the term "task practice" was coined to implicate a role for learning in tolerance. However, studies with other effects of ethanol such as hypothermia (Alkana et al., 1983), analgesia (Jorgensen and

Hole, 1984; Jorgensen et al., 1986), and anticonvulsion (Pinel et al., 1985) also revealed the influence of "experiencing the drug effect" on tolerance development. Employing a "before" and "after" design, Jorgensen and Hole (1984) and Jorgensen et al. (1986) have shown that tolerance develops to the analgesic effect of ethanol in a group of spinalized rats that received ethanol daily before, but not after, the tail-flick test. Similarly, the experience of hypothermia produced by ethanol has also been shown to be a prerequisite for tolerance to its hypothermic effect, because mice treated with ethanol at a high ambient temperature that antagonizes ethanol-induced hypothermia did not develop tolerance to this effect of ethanol. These observations question the "learning" interpretation of tolerance derived from those findings concerning the effects of task exposure during intoxication on tolerance. Kalant (1988) argued that since a cognitive process is not involved in the tail-flick response in spinalized animals, the term "learning" or "task practice" might not be appropriate. He proposed that "tolerance develops more readily in a neuronal pathway which is activated during the period of alcohol exposure than in one which is quiescent, regardless of whether or not cognition is involved" (Kalant 1988).

In fact, if one holds that the conditions (exposure or nonexposure to task during intoxication) might shift the sensitivity to the particular effects of ethanol, then such effects might be corrected by raising alcohol treatment doses. Results from studies that examined the development of tolerance to the motor impairment (Lê et al., 1989) as well as hypothermic (Lê et al., 1986) and analgesic (Jorgensen et al., 1986) effects of ethanol following treatment with various doses of alcohol tend to support this notion. When animals were treated with low doses of ethanol, tolerance was manifested for the motor impairment or analgesic effects of ethanol only when the animals were exposed to the test during intoxication. With higher doses of treatment (Lê et al., 1989; Jorgensen et al., 1986), tolerance to these effects was demonstrated, even though the animals were never exposed to testing while intoxicated. Alternatively, it is possible that the effect of intoxicated practice on tolerance as observed in the "before-after" design might be accounted for by Pavlovian conditioning (see below) because the environmental cues associated with ethanol administrations are consistent during ethanol treatment and testing for the "before," but not the "after," group (Hinson and Siegel, 1980).

Studies with human subjects dealing with the development of tolerance to alcohol following repeated testings during intoxication have been carried out primarily by Vogel-Sprott and co-workers (Beirnes and Vogel-Sprott, 1984; Vogel-Sprott, 1979; Zack and Vogel-Sprott, 1993). The emphasis of their studies, however, has been the role of expectancy, Pavlovian conditioning, and mental rehearsal on the development of tolerance. In these studies, subjects from all experimental groups were required to perform the tests while under alcohol intoxication. For these reasons, it is difficult to draw any conclusion regarding the effect of practice vs. nonpractice during intoxication on tolerance. Results form these studies, however, suggest that the nature or condition of practice rather than practice per se during intoxication is important to the development of alcohol tolerance. For example, tolerance to the impairment effect of ethanol on trachometer performance developed to a higher extent in subjects who received monetary reward for better performance or feedback information concerning their performance in a consistent manner during training sessions under the influence of alcohol than in those who did not or those who received feedback in a random manner.

B. PAVLOVIAN CONDITIONING

Siegel (1975, 1983) proposed that drug tolerance is a form of Pavlovian conditioning. In this model of drug tolerance, environmental cues associated with repeated drug administrations can serve as conditioned stimuli that elicit conditioned compensatory responses that are directly opposite to drug effects. The occurrence of a conditioned compensatory response in tolerance subjects can be elicited by administration of a placebo in association with cues predicting drug administration. To evaluate the role of conditioning on tolerance, experimental subjects usually receive several injections of alcohol in a distinctive environment and saline injections in a recognizably different environment. Manipulation of visual and/or auditory cues is usually employed to facilitate differentiation between the alcohol and saline environment. The doses of alcohol employed have been modest (1–2 g/kg) relative to those used to produce pharmacological tolerance.

Conditioned tolerance to various effects of ethanol such as hypothermia (Lê et al., 1979), narcosis (Melchior and Tabakoff, 1981), analgesia (Tiffany et al., 1987), and lethality (Melchior, 1990) has been demonstrated in experimental rodents. In humans, conditioned tolerance for tachycardia (Dafters and Anderson, 1982), impairment of cognition (Williams et al., 1981; Shapiro and Nathan, 1986; McCusker and Brown, 1990), motor performance (McCusker and Brown, 1990; Sdao-Jarvie and Vogel-Sprott, 1991), pulse rate and skin temperature (Newlin, 1985, 1986; McCusker and Brown, 1990) has been reported. A compensatory hyperthermic response induced by a saline challenge in an environment previously linked to alcohol administration was observed in rats tolerant to the hypothermic effect of alcohol (Lê et al., 1979; Crowell et al., 1981; Mansfield and Cunningham, 1980). A placebo challenge in an alcohol-linked environment has also been found to produce compensatory responses in humans (Staiger and White, 1988; Sdao-Jarvie and Vogel-Sprott, 1991. In addition to environmental cues, pharmacological effects induced by low doses of ethanol have also been shown to be capable of serving as conditioned stimuli for controlling tolerance to ethanol (Greeley et al., 1984).

It has been shown that the influence of conditioning on tolerance varies with the size of the daily drug dose used to produce tolerance. At low treatment doses, conditioning may be essential for the development of tolerance, whereas at high treatment doses it is not essential. For example, Lê et al. (1987) demonstrated that tolerance to the hypothermic effect of ethanol can be demonstrated only in an environment previously paired with ethanol administration in rats that received 11 training sessions with 2 g/kg of ethanol. In those that received the training dose of 4 g/kg, tolerance can be demonstrated in the presence or absence of cues associated with previous ethanol administration.

Another feature of conditioned tolerance is that it lasts much longer than tolerance acquired by purely pharmacological means. Tolerance to ethanol-induced hypothermia produced by treatment with ethanol in a form of liquid diet or by vapor inhalation has been shown to dissipate by 3–5 days after the termination of ethanol treatment (Ritzmann and Tabakoff, 1976; Goldstein and Zachelein, 1983; Lê et al., 1982). On the other hand, when ethanol tolerance is produced by a conditioning procedure, ethanol tolerance has been shown to persist for many weeks after the cessation of treatment. Another important feature of conditioned tolerance is that cross-tolerance to other drugs that produce similar effects can be readily seen. For example, cross-tolerance to the hypothermic effect of hydralazine was observed only in rats rendered tolerant to ethanol in a conditioning paradigm (Lê et al., 1987).

V. RELATIONSHIP AMONG VARIOUS FORMS OF TOLERANCE

A. ACUTE AND CHRONIC TOLERANCE

Acute and chronic tolerance have been though to be related in one of two ways: (1) acceleration of acute tolerance over repeated drug exposures or (2) slower dissipitation of acute tolerance with increasing persistence of tolerance from one exposure to the next (see Kalant et al., 1971; Littleton, 1980). The Pavlovian conditioning theory of tolerance also suggests that chronic tolerance is a consequence of an early elicitation of acute tolerance evoked by cues previously associated with alcohol administration (Siegel, 1978; Kalant and Lê, 1984). All of these hypotheses suggest that acute tolerance is a precursor of chronic tolerance.

The available experimental data dealing with this relationship, however, are correlational in nature. Basically, the ability to develop acute tolerance has been shown to correlate with that of chronic tolerance (San Marina et al., 1989). Chronic tolerance develops only to the effects of alcohol in which acute tolerance to such effects can be demonstrated (Vogel-Sprott, 1979). The most convincing data to support a relationship between acute and chronic tolerance are derived from the study by Kalant et al. (1978). When acute and chronic tolerance to ethanol were monitored frequently during the alcohol treatment period, tolerance was first manifested as an accentuation of acute tolerance, which was then followed by a reduction in the maximum effect. Experimental data inconsistent with such a relationship, however, have also been presented. For example, Tabakoff et al. (1980) reported that chronic, but not acute, tolerance develops to the hypnotic effect of ethanol in SS mice selectively bred for high resistance to such an effect of alcohol. Similarly, Lê et al. (1992) reported an absence of acute tolerance to the anticonvulsant effect of ethanol, whereas chronic tolerance to this effect has been demonstrated (Pinel et al., 1985). Clearly, further work in this area is required before any conclusions can be made regarding the relationship between these two forms of tolerance.

B. PHARMACOLOGICAL AND BEHAVIORAL TOLERANCE

It should be well accepted that alcohol tolerance can be influenced by pharmacological and behavioral factors. At issue, however, is whether these factors reflect separate or multiple mechanisms of tolerance, or whether tolerance is a unitary mechanism with behavioral factors that simply modulate tolerance development. Evidence to support a unitary mechanism can be advanced from experimental data which show that whether tolerance is induced by behavioral or pharmacological means or both, the same maximum level of tolerance is attained. For example, the same level of tolerance was observed in rats treated with high doses of ethanol in the presence or absence of conditions that provide maximum opportunity for intoxicated practice (LeBlanc et al., 1973) or conditioning (Lê et al., 1989). Kalant (1988) suggested that the degree of disturbance produced by alcohol is the common element or stimulus in producing tolerance, and that behavioral processes or behavioral manipulation modulate it.

There is, however, growing evidence to suggest that separate or different mechanisms of tolerance exist. First, tolerance induced by behavioral means, such as that of conditioned tolerance, lasts much longer than tolerance acquired by pharmacological means (Lê et al., 1989; Mansfield and Cunningham, 1980). Second, cross-tolerance to the hypothermic effects of pentobarbital and hydralazine or to the analgesic effect of clonidine is observed only in animals rendered tolerant to ethanol in a conditioning paradigm. Additional evidence for possible separate mechanisms of tolerance has been derived from recent work with the NMDA receptor system that has been implicated to play a key role in learning. Szabo et al. (1994) demonstrated that administration of dizocilpine, an NMDA receptor antagonist, blocks the development of a conditioned tolerance for alcohol, but does not affect tolerance induced by pharmacological means.

VI. SUMMARY

This review of the literature indicates that much progress has been made over the last several years toward the understanding of alcohol tolerance. Results from experimental studies with animals and human subjects are fairly consistent with one another. They clearly demonstrate the existence of various forms of tolerance based on the time course of development. Tolerance is clearly shown not to be governed only by pharmacological factors alone, but by a variety of factors ranging from genetic to behavioral processes. Although some progress has been made toward an understanding of how these various forms of tolerance or how behavioral and pharmacological factors interact with one another to regulate tolerance, much work is required in this area. Although not discussed in this chapter, there have been extensive investigations into the neurochemical or cellular mechanisms of tolerance, and how tolerance might contribute to the regulation of alcohol consumption. An understanding of how these various forms or types of tolerance are related, particularly behavioral and pharmacological tolerance, will be necessary for answering these questions.

REFERENCES

Alkana, R.L., Finn, D.A., and Malcolm, R.D. (1983). The importance of experience in the development of tolerance to ethanol hypothermia. *Life Sci.,* 32: 2685–2692.

Banks, W.P., Vogler, R.E., and Weissbach, T.A. (1979). Adaptation of ethanol intoxication. *Bull. Psychon. Soc.,* 14: 319–322.

Basile, A., Hoffer, B., and Dunwiddie, T. (1983). Differential sensitivity of cerebellar purkinje neurons to ethanol in selectively outbred lines of mice: maintenance *in vitro* independent of synaptic transmission. *Brain Res.,* 264: 69–78.

Beirness, D. and Vogel-Sprott, M. (1984). The development of alcohol tolerance: acute recovery as a predictor. *Psychopharmacology,* 84: 398–401.

Bennet, R.H., Cherek, D.R., and Spiga, R. (1993). Acute and chronic alcohol tolerance in humans: effects of dose and consecutive days of exposure. *Alcoholism: Clin. Exp. Res.,* 17: 740–745.

Buck, K.J., Heim, H., and Harris, R.A. (1991). Reversal of alcohol dependence and tolerance by a single administration of flumazepil. *J. Pharmacol. Exp. Ther.,* 257: 460–470.

Campanelli, C., Lê, A.D., Khanna, J.M., and Kalant, H. (1988). Effect of raphe lesions on the development of acute tolerance to ethanol and pentobarbital. *Psychopharmacology,* 96: 454–457.

Chen, C.S. (1968). A study of alcohol tolerance effect and an introduction of a new behavioral technique. *Psychopharmacologia,* 12: 433–440.

Chen, C.S. (1972). A further note on studies of acquired behavioral tolerance to alcohol. *Psychopharmacologia,* 27: 265–274.

Chen, C.S. (1979). Acquisition of behavioral tolerance to ethanol as a function of reinforced practice in rats. *Psychopharmacology,* 63: 285–288.

Chesher, G. and Greeley, J. (1992). Tolerance to the effects of alcohol. *Alcohol, Drugs and Driving,* 8(2): 93–105.

Crabbe, J.C. (1994). Tolerance to ethanol hypothermia in Hot and Cold mice. *Alcoholism: Clin. Exp. Res.,* 18: 42–46.

Crabbe, J.C., Feller, D., and Dorrow, J. (1989). Sensitivity and tolerance to ethanol-induced hypothermia in genetically selected mice. *J. Pharmacol. Exp. Ther.,* 49: 456–461.

Crabbe, J.C., Janowski, J.S., Young, R., Kosobud, A., Stack, J., and Rigter, H. (1982). Tolerance to ethanol hypothermia in inbred mice: genotypic correlations with behavioral responses. *Alcoholism: Clin. Exp. Res.,* 6: 446–458.

Crabbe, J.C., Johnson, N.A., Gray, D.K., Kosobud, A., and Young, R. (1982). Biphasic effects of ethanol on open-field activity: sensitivity and tolerance in C57BL/6N and DBA/2N mice. *J. Comp. Physiol. Psychol.,* 96: 440–451.

Crabbe, J.C., Rigter, H., Uijlen, J., and Strijbos, C. (1979). Rapid development of tolerance to the hypothermic effect of ethanol in mice. *J. Pharmacol. Exp. Ther.,* 208: 128–133.

Crowell, D.R., Hinson, R.E., and Siegel, S. (1981). The role of conditional drug responses in tolerance to the hypothermic effect of ethanol. *Psychopharmacology,* 73: 51–54.

Cunningham, C.L. and Noble, D. (1992). Conditioned activation induced by ethanol: role of sensitization and conditioned place preference. *Pharmacol. Biochem. Behav.,* 43: 307–313.

Dafters, R. and Anderson, G. (1982). Conditioned tolerance to the tachycardia effect of ethanol in humans. *Psychopharmacology,* 78: 365–367.

de Souza Moreira, L.F., Caprigliore, M.J., and Masur, J. (1981). Development and re-acquisition of tolerance to ethanol administered pre- and post-trials to rats. *Psychopharmacology,* 73: 165–167.

Ekman, G., Frankenhauser, M., Goldgerg, L., Bjerver, K., Jaerpe, G., and Myrsten, A.L. (1964). Subjective and objective effects of alcohol as functions of dosage and time. *Psychopharmacologia,* 6: 399–409.

Frye, G.D. and Breese, G.R. (1981). An evaluation of the locomotor stimulating action of ethanol in rats and mice. *Psychopharmacology,* 75: 372–379.

Gallaher, E.J., Parsons, L.M., and Goldstein, D.B. (1982). The rapid onset of tolerance to the ataxic effect of ethanol in mice. *Psychopharmacology,* 78: 67–70.

Gatto, G.J., Murphy, J.M., Waller, M.B., McBride, W.J., Lumeng, L., and Li, T.K. (1987). Persistence of tolerance to a single dose of ethanol in the selectively-bred alcohol-preferring P rat. *Pharmacol. Biochem. Behav.,* 28: 105–110.

Gilliam, D.M. (1989). Alcohol absorption rate affects hypothermic response in mice: evidence for rapid tolerance. *Alcohol,* 6: 357–362.

Goldberg, L. (1943). Quantitative studies on alcohol tolerance in man. *Acta. Physiol. Scand.* (suppl 16), 5: 1–126.

Goldstein, D.B. and Zaechelein, R. (1983). Time course of functional tolerance produced in mice by inhalation of ethanol. *J. Pharmacol. Exp. Ther.,* 227(1): 150–153.

Goodwin, D.W., Powell, B., and Stern, J. (1971). Behavioral tolerance to alcohol in moderate drinkers. *Am. J. Psychiat.,* 127: 1651–1658.

Greeley, J.D., Lê, A.D., Poulos, C.X., and Cappell, H. (1984). Alcohol is an effective cue in the condition control of tolerance to alcohol. *Psychopharmacology,* 83: 159–162.

Grieve, S.J. and Littleton, J.M. (1979). The rapid development of functional tolerance to ethanol by mice. *J. Pharm. Pharmacol.,* 31: 605–610.

Hammond, K.B., Rumack, B.H., and Rogerson, D.O. (1973). Blood ethanol: a report of unusually high levels in a living patient. *J. Amer. Med. Assoc.,* 226: 63–64.

Haubenreisser, T. and Vogel-Sprott, M. (1987). Reinforcement reduces behavioral impairment under an acute dose of alcohol. *Pharmacol. Biochem. Behav.,* 26: 29–33.

Hawkins, R.D. and Kalant, H. (1972). The metabolism of ethanol and its metabolic effects. *Pharmacol. Rev.,* 243: 65–157.

Hiltunen, A.J. and Järbe, T.U.C. (1990). Acute tolerance to ethanol using drug discrimination and open-field procedures in rats. *Psychopharmacology,* 102: 207–212.

Hiltunen, A.J. and Järbe, T.U.C. (1992). Acute and chronic ethanol tolerance: operant behaviour in naive and ethanol tolerant rats. *Psychopharmacology,* 107: 511–516.

Hinson, R. and Siegel, S. (1980). The contribution of Pavlovian conditioning to ethanol tolerance and dependence. In *Alcohol Tolerance and Dependence* (H. Rigter and J.C. Crabbe, Eds.), pp. 181–194. Amsterdam: Elsevier/ North-Holland Biomedical Press.

Hunt, G.P. and Overstreet, D.H. (1977). Evidence for parallel development of tolerance to the hyperactivating and discoordinating effects of ethanol. *Psychopharmacology,* 55: 75–81.

Hurst, P.M. and Bagley, S.K. (1972). Acute adaptation to the effects of alcohol. *Q. J. Stud. Alcohol,* 33: 358–378.

Jetter, W.W. (1938). Studies in alcohol. I. The diagnosis of acute alcohol intoxication by correlation of clinical and chemical findings. *Am. J. Med. Sci.,* 196: 475–487.

Johnson, R.A., Noll, E.C., and Rodney, W.W. (1982). Survival after a serum ethanol concentration of 1 1/2%. *Lancet,* p. ii:1394.

Jones, B.M. (1973). Memory impairment on the ascending and descending limbs of the blood alcohol curve. *J. Abnorm. Psychol.,* 82: 24–32.

Jones, B.M. and Vega, A. (1972). Cognitive performance measured on the ascending and descending limbs of the blood alcohol level curve. *Psychopharmacologia,* 23: 99–114.

Jorgensen, H.A., Fasmer, O.B., and Hole, K. (1986). Learned and pharmacologically-induced tolerance and cross-tolerance to morphine and clonidine. *Pharmacol. Biochem. Behav.,* 24: 1083–1088.

Jorgensen, H.A. and Hole, K. (1984). Learned tolerance to ethanol in the spinal cord. *Pharmacol. Biochem. Behav.,* 20: 789–792.

Kalant, H. (1988). Alcohol tolerance and withdrawal: an overview of current issue. *Aust. Drug Alc. Rev.,* 7: 27–34.

Kalant, H. and Lê, A.D. (1984). Effects of ethanol on thermoregulation. *Pharmacol. Ther.,* 23: 313–364.

Kalant, H., LeBlanc, A.E., and Gibbins, R.J. (1971). Tolerance to, and dependence on, some nonopiate psychotropic drugs. *Pharmacol. Rev.,* 23: 135–191.

Kalant, H., LeBlanc, A.E., Gibbins, R.J., and Wilson, A. (1978). Accelerated development of tolerance during repeated cycles of ethanol exposure. *Psychopharmacology,* 60: 59–65.

Kaplan, H., Sellers, E.M., Hamilton, C., Naranjo, C.A., and Dorian, P. (1985). Is there acute tolerance to alcohol at steady state? *J. Stud. Alcohol,* 46: 253–256.

Keir, W.J. and Deitrich, R.A. (1990). Development of central nervous system sensitivity to ethanol and pentobarbital in short- and long-sleep mice. *J. Exp. Pharmacol. Ther.,* 254: 831–835.

Khanna, J.M. and Israel, Y. (1980). Ethanol metabolism, In *Liver and Biliary Tract Physiology. I. International Review of Physiology,* Vol. 21, (N.B. Javitt, Ed.), pp. 275–315. Baltimore: University Park.

Khanna, J.M., Kalant, H., Shah, G., and Weiner, J. (1991). Rapid tolerance as an index of chronic tolerance. *Pharmacol. Biochem. Behav.,* 38: 427–432.

Khanna, J.M., Lê, A.D., LeBlanc, A.E., and Shah, G. (1985). Initial sensitivity versus acquired functional tolerance to ethanol in rats selectively bred for ethanol sensitivity. *Psychopharmacology,* 86: 302–306.

Khanna, J.M., San Marina, A., Kalant, H., and Lê, A.D. (1989). Relationship between initial sensitivity and chronic tolerance to ethanol and morphine in heterogeneous population of mice and rats. In *Genetic Aspects of Alcoholism,* Vol. 37 (K. Kiianmaa, B. Tabakoff, and T. Saito, Eds.), pp. 207–217. Helsinki, Finland: The Finnish Foundation for Alcohol Studies.

Lê, A.D. (1990). Factors regulating ethanol tolerance. *Ann. Med.,* 22: 265–268.

Lê, A.D. and Kalant, H. (1992). Influence of intoxicated practice on the development of acute tolerance to the motor-impairment effect of alcohol. *Psychopharmacology,* 106: 572–576.

Lê, A.D., Kalant, H., and Khanna, J.M. (1982). Interaction between desglycinamide-[ARG]vasopressin and serotonin on ethanol tolerance. *Eur. J. Pharmacol.,* 80: 337–345.

Lê, A.D., Kalant, H., and Khanna, J.M. (1986). Influence of ambient temperature on the development and maintenance of tolerance to ethanol-induced hypothermia. *Pharmacol. Biochem. Behav.,* 25: 667–672.

Lê, A.D., Kalant, H., and Khanna, J.M. (1989). Role of intoxicated practice in the development of ethanol tolerance. *Psychopharmacology,* 99: 366–370.

Lê, A.D. and Khanna, J.M. (1989). Dispositional mechanism in drug tolerance and sensitization. In *Psychoactive Drugs: Tolerance and Sensitization,* (A. Goudie and M.W. Emmett-Oglesby, Eds.), pp. 281–352, Clifton, New Jersey: Humana Press.

Lê, A.D., Khanna, J.M., and Kalant, H. (1984). Effects of treatment dose and test system on the development of ethanol tolerance and physical dependence. *Alcohol,* 1: 447–451.

Lê, A.D., Khanna, J.M., and Kalant, H. (1987). Role of Pavlovian conditioning in the development of tolerance and cross-tolerance to the hypothermic effect of ethanol and hydralazine. *Psychopharmacology,* 92: 210–214.
Lê, A.D. and Kiianmaa, K. (1988). Characteristics of ethanol tolerance in alcohol drinking (AA) and alcohol avoiding (ANA) rats. *Psychopharmacology,* 94: 479–483.
Lê, A.D. and Kiianmaa, K. (1989). Initial sensitivity and the development of acute and rapid tolerance to ethanol in the AT and ANT rats. In *Genetic Aspects of Alcoholism,* Vol. 37, (K. Kiianmaa, B. Tabakoff, and T. Saito, Eds.), pp. 147–155. Helsinki, Finland: The Finnish Foundation for Alcohol Studies.
Lê, A.D. and Kiianmaa, K. (1990). Role of initial sensitivity and genetic factors in the development of tolerance to ethanol in AT and ANT rats. *Psychopharmacology,* 94: 479–483.
Lê, A.D., Mana, M.J., Quan, B., and Kalant, H. (1992a). Differential development of acute tolerance to the motor impairment and anticonvulsant effects of ethanol. *Psychopharmacology,* 109: 107–111.
Lê, A.D., Mihic, J., and Wu, P. (1992b). Alcohol tolerance: Methodological and Experimental Issues. In *Animal Models of Drug Addiction, Neuromethods,* Vol. 24, (A. Boulton, G.B. Baker, and P.H. Wu, Eds.), pp. 95–124, Totawa, New Jersey: Humana Press.
Lê, A.D., Poulos, C.X., and Cappell, H. (1979). Conditioned tolerance to the hypothermic effect of ethyl alcohol. *Science,* 206: 1109–1110.
LeBlanc, A.E., Gibbins, R.J., and Kalant, H. (1973). Behaviorally augmentation of tolerance to ethanol in the rat. *Psychopharmacologia,* 30: 117–122.
LeBlanc, A.E., Gibbins, R.J., and Kalant, H. (1975). Generalization augmented tolerance to ethanol, and its relation to physical dependence. *Psychopharmacology,* 44: 241–246.
LeBlanc, A.E., Kalant, H., and Gibbins, R.J. (1975). Acute tolerance to ethanol in the rat. *Psychopharmacologia,* 41: 43–46.
LeBlanc, A.E., Kalant, H., and Gibbins, R.J. (1976). Acquisition and loss of behaviorally augmented tolerance to ethanol in the rat. *Psychopharmacology,* 48: 153–158.
Littleton, J.M. (1980). The assessement of rapid tolerance to alcohol. In *Alcohol Tolerance and Dependence,* (H. Rigter and J.C. Crabbe, Eds.), pp. 53–79. Amsterdam: Elsevier.
Loomis, T.A. and West, T.C. (1958). The influence of alcohol on automobile driving ability: an experimental study for evaluation of certain medicolegal aspects. *Q. J. Stud. Alc.,* 19: 30–46.
Mansfield, J.G. and Cunningham, C.L. (1980). Conditioning and extinction of tolerance to the hypothermic effect of ethanol in rats. *J. Comp. Physiol.,* 94: 962–969.
Martin, C.S. and Earleywine, M. (1990). Ascending and descending rates of change in blood alcohol concentrations and subjective intoxication ratings. *J. Sub. Abuse,* 2: 345–352.
Martin, C.S. and Moss, H.B. (1993). Measurement of acute tolerance to alcohol in human subjects. *Alcoholism: Clin. Exp. Res.,* 17(2): 211–216.
Martin, E., Moll, W., Schmid, P., and Detti, L. (1984). The pharmacokinetics of alcohol in human breath, venous and arterial blood after oral ingestion. *Eur. J. Clin. Pharmacol.,* 26: 619–626.
Maylor, E.A. and Rabbitt, P.M.A. (1987). Effects of practice and alcohol on performance of a perceptual motor task. *Q. J. Exp. Psychol.,* 39A: 777–795.
McCusker, C.G. and Brown, K. (1990). Alcohol-predictive cues enhance tolerance to the precipitate "craving" for alcohol in social drinkers. *J. Stud. Alcohol,* 51: 494–499.
Melchior, C.L. (1990). Conditioned tolerance provides protection against ethanol lethality. *Pharmacol. Biochem. Behav.,* 37: 205–206.
Melchior, C.L. and Tabakoff, B. (1981). Modification of environmentally cues tolerance to ethanol in mice. *J. Pharmacol. Exp. Ther.,* 219: 175–180.
Mellanby, E. (1919). Alcohol: its absorption into and disappearance from blood under different conditions. (Great Britain Medical Research Council, Special report series No 31) London: Her Majesty's Statistics Office.
Mello, K. and Mendelson, J.H. (1978). Alcohol and human behavior. In *Handbook of Psychopharmacology,* (L.L. Iversen, S.D. Iversen, and S. Snyder, Eds.), pp. 235–317. New York: Plenum Press.
Mirsky, I.A., Piker, P., Rosenbaum, M., and Lederer, H. (1941). "Adaptation" of the central nervous system to varying concentrations of alcohol in the blood. *Q. J. Stud. Alcohol.,* 2: 35–45.
Moore, J.A. and Kakihana, R. (1978). Ethanol-induced hypothermia in mice: influence of geneotype on development of tolerance. *Life Sci.,* 23: 2331–2337.
Moskowitz, H. and Burns, M. (1976). Effects of rate of drinking on human performance. *Q. J. Stud. Alcohol.,* 37: 598–605.
Moskowitz, H., Daily, J., and Henderson, R. (1974). Acute tolerance to behavioral impairment by alcohol in moderate and heavy drinkers, Washington, D.C.: Highway Safety Administration, U.S. Department of Transportation.
Myrsten, A.L., Hollstedt, C., and Homberg, L. (1975). Alcohol-induced changes in mood and activation in males and females as related to catecholamine excretion and blood alcohol level. *Scand. J. Psychol.,* 16: 303–310.
Newlin, D.B. (1985). Offsprings of alcoholics have enhanced antagonistic placebo response. *J. Stud. Alcohol.,* 45: 490–494.
Newlin, D.B. (1986). Conditioned compensatory response to alcohol placebo in human. *Psychopharmacology,* 88: 247–254.

Nicholson, M.E., Wang, M., Airhihenbuwa, C.O., Mahoney, B.S., Christina, R., and Maney, D.W. Variability in behavioral impairment in the rising and falling BAC curve. *J. Stud. Alcohol,* 53(4): 349–355.

Parker, E.S., Alkana, R.L., Birnbaum, I.M., Hartley, J.T., and Noble, E.P. (1974). Alcohol and the disruption of cognitive processes. *Arch. Gen. Psychiat.,* 31: 824–828.

Perper, J.A., Twerski, A., and Wienand, J.W. (1986). Tolerance at high blood alcohol concentrations: a study of 110 case and review of the literature. *J. Forens. Sci.,* 31: 212–221.

Pinel, J.P.J., Mana, M.J., and Renfrey, G. (1985). Contingent tolerance to the anticonvulsant effects of alcohol. *Alcohol,* 2: 495–499.

Portans, I., White, J.M., and Staiger, P.K. (1989). Acute tolerance to alcohol: changes in subjective effects among social drinkers. *Psychopharmacology,* 97: 365–369.

Poulos, C.X., Wolff, L., Zilm, D.H., Kaplan, H., and Cappell, H.D. (1981). Acquisition of tolerance to alcohol-induced memory deficits in humans. *Psychopharmacology,* 73: 176–179.

Radlow, R. and Hurst, P.M. (1985). Temporal relations between blood alcohol concentration and alcohol effect: an experiment with human subjects. *Psychopharmacology,* 85: 260–266.

Riley, E.P. and Lochry, E.A. (1977). Effects of initial tolerance on acquired tolerance to alcohol in two selectively bred rat strains. *Drug Alcohol Depend.,* 2: 485–494.

Ritzmann, R.F. and Tabakoff, B. (1976). Body temperature in mice: a quantitative measure of alcohol tolerance and dependence. *J. Pharmacol. Exp. Ther.,* 199: 158–170.

Rosen, L.J. and Lee, C.L. (1976). Acute and chronic effects of alcohol use on organization processes in memory. *J. Abn. Psychol.,* 85: 309–317.

San-Marina, A., Khanna, J.M., and Kalant, H. (1989). Relationship between initial sensitivity, acute tolerance and chronic tolerance to ethanol in a heterogeneous population of Swiss mice. *Psychopharmacology,* 99: 450–457.

Sdao-Jarvie, K. and Vogel-Sprott, M. (1991). Response expectancies affect the acquisition and display of behavioral tolerance to alcohol. *Alcohol,* 8: 491–498.

Shapiro, A.P. and Nathan, P.E. (1986). Human tolerance to alcohol: the role of Pavlovian conditioning process. *Psychopharmacology,* 88: 90–95.

Siegel, S. (1975). Evidence from rats that morphine tolerance is learned response. *J. Comp. Physiol. Psychol.,* 89: 498–506.

Siegel, S. (1983). Classical conditioning, drug tolerance, and drug dependence. In *Research Advances in Alcohol and Drugs Problems,* Vol. 7, (R.J. Smart, F.B. Glaser, Y. Israel, H. Kalant, R.E. Popham, and W. Schmidt, Eds.), pp. 207–246. New York: Plenum Press.

Staiger, P. and White, J. (1988). Conditioned alcohol-like and alcohol-opposite response in humans. *Psychopharmacology,* 95: 87–91.

Spuhler, K., Hoffer, B., Weiner, N., and Palmer, M. (1982). Evidence for genetic correlation of hypnotic effects and cerebellar Purkinje neuron depression in response to ethanol in mice. *Pharmacol. Biochem. Behav.,* 17: 569–578.

Sullivan, J.B., Hauptman, M., and Bronstein, A.C. (1987). Lack of observable intoxication in humans with high plasma alcohol concentrations. *J. Forens. Sci.,* 32: 1660–1665.

Szabo, G., Tabakoff, B., and Hoffman, P.L. (1994). The NMDA receptor antagonist dizocilpine differentially affects environment-dependent and environment-independent ethanol tolerance. *Psychopharmacology,* 113: 511–517.

Tabakoff, B. and Kiianmaa, K. (1980). Does tolerance develop to the activating, as well as the depressant, effects of ethanol? *Pharmacol. Biochem. Behav.,* 17: 1073–1076.

Tabakoff, B. and Ritzmann, R.F. (1979). Acute tolerance in inbred and selected lines of mice. *Drug Alcohol Depend.,* 4: 87–90.

Tabakoff, B., Ritzmann, R.J., Raju, T.S., and Deitrich, R.A. (1980). Characterization of acute and chronic tolerance in mice selected for inherent differences in sensitivity to ethanol. *Alcoholism: Clin. Exp. Res.,* 4: 70–73.

Tiffany, S.T., McCal, K.J., and Maude-Griffin, P.M. (1987). The contribution of classical conditioning to the antinociceptive effects of ethanol. *Psychopharmacology,* 92: 524–528.

Tullis, K.V., Sargent, W.Q., Simpson, J.R., and Beard, J.D. (1977). An animal model for the measurement of acute tolerance to ethanol. *Life Sci.,* 20: 875–880.

Urso, T., Gavaler, J.S., and Van Thiel, D.H. (1981). Blood ethanol levels in sober alcohol users seen in an emergency room. *Life Sci.,* 28: 1053–1056.

Vogel-Sprott, M. (1979). Acute recovery and tolerance to low doses of alcohol: differences in cognitive and motor skill performance. *Psychopharmacology,* 61: 287–291.

Vogel-Sprott, M. and Sdao-Jarvie, K. 91989). Learning alcohol tolerance: the contribution of response expectancies. *Psychopharmacology,* 98: 289–296.

Waller, M.B., Murphy, J.M., McBride, W.J., Lumeng, L., and Li, T.K. (1983). Effect of low dose ethanol on spontaneous motor activity in alcohol-preferring and -nonpreferring lines of rats. *Pharmacol. Biochem. Behav.,* 24: 617–623.

Wenger, J.R., Tiffany, T.M., Bombardier, C., Nichols, K., and Woods, S.C. (1981). Ethanol tolerance in rat is learned. *Science,* 213: 575–577.

Williams, R.M., Goldman, M.S., and Williams, D.L. (1981). Expectancy and pharmacological effects of alcohol on human cognitive and motor-performance: the compensation for alcohol effect. *J. Abnorm. Psychol.,* 90: 267–270.

Wilson, J.R., Erwin, V.G., McClearn, G.E., Plomin, R., Johnson, R.C., Ahern, F.M., and Cole, R.E. (1984). Effects of ethanol: II. Behavioral sensitivity and acute behavioral tolerance. *Alcoholism: Clin. Exp. Res.,* 8: 366–374.

Young, J.R. (1970). Blood alcohol concentration and reaction time. *Q. J. Stud. Alcohol,* 31: 823–831.

Zack, M. and Vogel-Sprott, M. (1993). Response outcomes affect the retention of behavior tolerance to alcohol: Information and incentive. *Psychopharmacology,* 113: 269–273.

Chapter 16

DEPENDENCE AND WITHDRAWAL

Pamela Metten and John C. Crabbe

TABLE OF CONTENTS

I. CENTRAL NERVOUS SYSTEM EXCITABILITY AND EXCITATION IN WITHDRAWAL

The complexity of alcoholism makes it a difficult entity to define. Embodied in most attempts is the concept of dependence, operationally defined as the state that manifests itself upon withdrawal from alcohol (Friedman, 1980; Kalant et al., 1971; Swift, 1994; but see Cappell and LeBlanc, 1979). Alcohol (ethanol) and other central nervous system depressants are now well-known to produce signs and symptoms during withdrawal that are opposite in direction to those induced by intoxication (Victor and Adams, 1953; Friedman, 1980; Jaffe, 1985; Kalant et al., 1971). Figure 16.1 shows the alteration over time in a hypothetical measure of ethanol effect during ethanol exposure and subsequent withdrawal. The zero line represents basal levels of the measure, which can be expected to vary among species or vary among different genotypes within a species. Points above the zero line represent stimulatory effects, whereas those below represent depressant effects. On most physiological/behavioral measures, an acute administration of ethanol produces short-term stimulation followed by a

0-8493-8389-7/95/$0.00+$.50

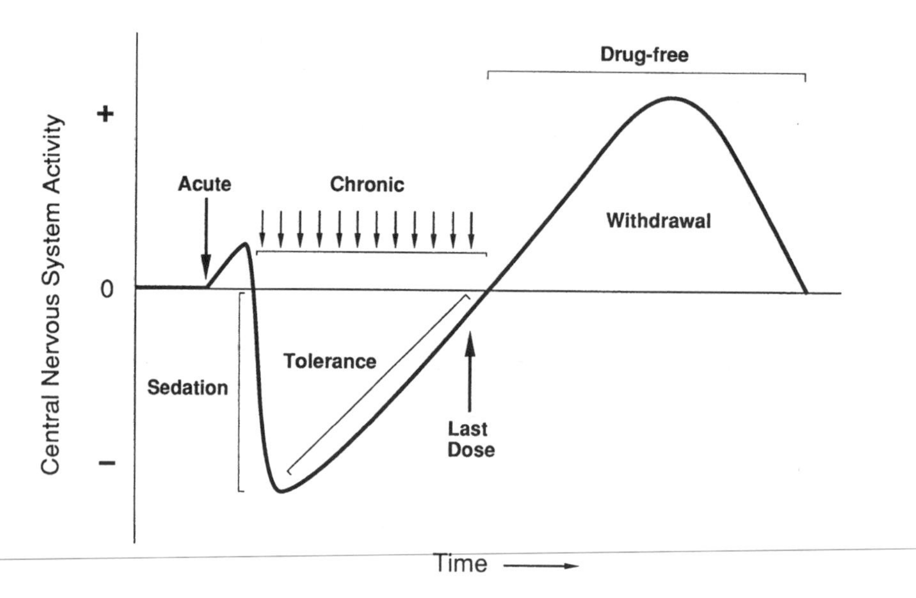

FIGURE 16.1. Hypothetical measure of ethanol effect on central nervous system excitability over time. See text for description.

depressive, or sedative, effect. Chronic ethanol administration prolongs and enhances the sedative effect, but this decays over time despite continued presence of ethanol, revealing a return toward basal levels of functioning. This reduction in initial effect defines tolerance. Dependence putatively occurs as the result of adaptation of the system in the presence of ethanol such that ethanol is required for quasi-normal function. Removal of ethanol through metabolism following cessation of treatment leads to a rebound stimulatory effect during withdrawal that is opposite in direction to that of the sedative effect. Cicero (1980) and others (e.g., Kalant et al., 1971; Kosobud and Crabbe, 1995) have argued that the measurement of dependence is problematic without a clear definition and that it is preferable to define and quantify withdrawal.

A. HUMAN WITHDRAWAL SIGNS AND THEIR TREATMENT

The model of withdrawal just presented in Figure 16.1 describes many responses to ethanol. For example, anxiety is lessened by ethanol consumption and is increased during the withdrawal episode (Isbell et al., 1955). Besides anxiety, the human ethanol withdrawal syndrome includes irritability, nausea, vomiting, insomnia, tremor, hyperthermia, hyperventilation, tachycardia, and neural hyperexcitation, manifestations of which include hallucinations, delusions, and grand mal seizures (Victor and Adams, 1953; Isbell et al., 1955; Jaffe and Ciraulo, 1985; Litten and Allen, 1991).

Pharmacologic treatments of human ethanol withdrawal symptoms have recently been reviewed (Litten and Allen, 1991). Briefly, benzodiazepines (anticonvulsants and anxiolytics), barbiturates (sedatives and anticonvulsants), β-adrenergic antagonists (like atenolol and propranolol), α2-adrenergic agonists (like clonidine), calcium channel antagonists (like caroverine), carbamazepine (tricyclic anticonvulsant), and haloperidol (sedative dopamine antagonist) have all been used recently with varying degrees of success.

B. WITHDRAWAL SIGNS IN ANIMALS

Ethanol withdrawal signs have been described in chimpanzees (Pieper et al., 1972), monkeys (Ellis and Pick, 1970), dogs (Essig and Lam, 1968), cats (Guerrero-Figueroa et al., 1970), rats (Majchrowicz, 1975), and mice (Freund, 1969; Goldstein, 1972a). All of these species, like the human, have been shown to display tremor and potentially fatal convulsions following withdrawal from ethanol (Friedman, 1980; Kalant, 1977). Many of the other signs displayed by withdrawing humans are also seen in other animal species. In one study, animals were trained in a drug discrimination paradigm to bar press on a specific lever when anxiogenic drugs were administered. When tested subsequently during ethanol withdrawal, the animals were shown to bar press on the anxiogenic drug lever (Lal et al., 1988). The elevated-plus-maze model of anxiety has also been tested during ethanol withdrawal. Withdrawing rats spend less time on the open arms of the maze, indicating an anxious response to withdrawal. Furthermore, anxiolytic drugs reverse this effect (Lal et al., 1991). Emmett-Oglesby et al. (1990) have reviewed rodent models of ethanol withdrawal-induced anxiety as an approach toward modeling psychological aspects of dependence. This chapter is devoted primarily to a review of the signs of physical dependence in genetic animal models.

1. Convulsions and Seizure Susceptibility

Because of their commonality across species, their severity, and their quantifiable nature, convulsion severity and seizure susceptibility measures have been prominent in ethanol withdrawal research. McQuarrie and Fingl (1958) measured pentylenetetrazol (PTZ)-induced seizure thresholds in mice following acute (4 g/kg) or chronic (6 g/kg/day for 14 days) ethanol injections. They found in either regimen that ethanol raised seizure thresholds (i.e., protected against seizures) shortly after dosage, but after elimination of ethanol, seizure thresholds were below those of the control group. This indicated that ethanol withdrawal was accompanied by a hyperexcitable state.

Goldstein and Pal (1971) introduced the handling-induced convulsion (HIC) as a method of quantifying withdrawal in mice after chronic administration of ethanol in a vapor inhalation chamber. The HIC was assessed by lifting the mouse by the tail and observing convulsive signs. A score was then assigned, based on the convulsion observed. If no convulsion occurred, the mouse was gently spun by the tail 180 degrees and observed and scored. The HIC was quantified on a scale ranging from 0 (no convulsion) to 4 (tonic-clonic convulsion induced without a spin). Goldstein (1972a,b, 1973a,b, 1974) subsequently demonstrated that the severity of withdrawal HICs was dose-dependent, heritable, and could be modified by a wide variety of drugs.

2. Other Withdrawal Signs

Other signs have been employed to index withdrawal in rodents, including increases in startle response (Pohorecky and Roberts, 1991), stereotypic wall climbing (Becker et al., 1987), audiogenic seizure susceptibility (e.g., Frye et al., 1983), hypothermia (e.g., Hutchins et al., 1981), reductions in open-field activity (e.g., Maier and Pohorecky, 1989), hole-in-wall activity, and vertical screen crossings (Hutchins et al., 1981). In rats undergoing withdrawal, general hyperactivity, "wet dog" shakes, tremor, and bizarre behavior that included uncontrolled running episodes and aggressiveness, as well as sound-induced and spontaneous convulsions, have been observed (Majchrowicz, 1975).

3. Electrophysiological Plasticity

Electroencephalographic (EEG) abnormalities were noted during ethanol withdrawal in the first controlled study using human subjects (Isbell et al., 1955). Results of several studies support the general hypothesis that ethanol withdrawal is accompanied by a hyperexcited neural state. Walker and Zornetzer (1974) examined EEG records while noting behavioral signs of withdrawal in C57BL/6J mice following chronic dietary ethanol. Significantly, observation of EEG epileptiform activity in dorsolateral thalamus, hippocampus, and dorsal cortex preceded observations of spontaneous tonic-clonic convulsions. EEG activity has also been investigated during ethanol withdrawal-potentiated audiogenic seizures. In contrast to the findings of Walker and Zornetzer (1974), an absence of abnormal cortical EEG activity and spontaneous convulsions during ethanol withdrawal in C57BL/6Bg mice was observed, although lesser ethanol consumption may be responsible (Maxson and Sze, 1976). Furthermore, cortical EEG activity was not epileptiform during audiogenic seizures. The authors concluded that audiogenic seizure responses, including those during ethanol withdrawal, are subcortical in origin (Maxson and Sze, 1976). Subsequent studies investigating the brain regions involved in seizure discharges and correlates of ethanol withdrawal convulsions in rats have hypothesized that an interaction of several subcortical loci, including the red nucleus and substantia nigra, with neostriatum, thalamus, limbic structures, and cerebellum may be necessary for the expression of behavioral convulsions (Hunter and Walker, 1978; Hunter et al., 1978).

Alterations in stimulus-evoked brain electrical potentials have also been demonstrated in several species following alcohol administration and withdrawal. The effects may be stimulus-dependent (cf. visual- vs. auditory "odd-ball"-evoked potentials in rats [Begleiter and Porjesz, 1979; Ehlers and Chaplin, 1991]). However, measured voltages of visual-evoked potentials were increased during ethanol withdrawal in alcoholic patients, monkeys, and rats, indicating reliability of the finding across species (Begleiter and Porjesz, 1979).

One hypothesis regarding the incidence of withdrawal seizures (measured by EEGs) and convulsions (directly observable) is that they occur as a result of kindling due to repeated intoxication-withdrawal episodes (Ballenger and Post, 1978). Kindling refers to the phenomenon wherein repeated administration of doses of drugs or amperages of electric shock, which initially are subconvulsant, eventually develop the ability to elicit a convulsion. Studies comparing previous detoxification histories in withdrawing alcoholics typically find that withdrawal seizure incidence correlates highly with the number of previous detoxifications (Ballenger and Post, 1978; Brown et al., 1988; Lechtenberg and Worner, 1991). Thus, the withdrawal episode can putatively act as a subthreshold excitatory stimulus that potentiates subsequent withdrawal responses. Many studies in rodents have examined the severity of withdrawal due to repeated intoxication-withdrawal cycles (e.g., Pohorecky and Roberts, 1991; McCown and Breese, 1993; Ulrichsen et al., 1992). Spontaneous convulsions and the abnormal EEGs preceding them were significantly more prominent after two episodes of withdrawal in mice (Walker and Zornetzer, 1974). Furthermore, the handling-induced convulsion has been shown to be enhanced for a longer time in mice treated with several episodes of vapor inhalation separated by periods of withdrawal than in mice treated for the same amount of time in the vapor chamber, but without intermittent withdrawal episodes (Goldstein, 1972a; Becker and Hale, 1993).

4. Neurochemical Plasticity

It is well-known that ethanol has effects on many neurotransmitter receptors, ion channels, and neuroactive peptides, and second-messenger systems, rather than interacting with a single specific receptor. Reviews of these effects are presented in several other chapters of this book. Neurochemical systems thought to be involved in ethanol withdrawal per se are discussed briefly here, and some specific findings in pharmacogenetic models are discussed later in the context of other findings using those models. In terms of the withdrawal-excitation model presented in Figure 16.1, chronic ethanol exposure induces the suppression of excitatory systems and enhancement of inhibitory systems; a reversal of these effects during withdrawal putatively underlies the observed electrophysiological changes and withdrawal seizures.

Two calcium ion channels have been clearly implicated in withdrawal severity. Activation of these channels acts to increase calcium ion flux into the neural cell, enabling neural firing, as well as phosphorylation events (see Chapters 3 and 7). Thus they mediate an excitatory function. The number of voltage-dependent dihydropyridine-sensitive (L-type) calcium channels is increased by chronic ethanol treatment (see Chapter 3). Modulators of this channel (e.g., agonist, BAY K 8644; antagonist, nitrendipine) have been found to affect ethanol withdrawal excitability *in vivo* and *in vitro* using a variety of paradigms (for review, see Little, 1991). Recently, it was shown that the upregulation of the dihydropyridine-sensitive calcium channel by chronic ethanol treatment requires the phosphorylation activity of protein kinase C isozymes (Roivainen et al., 1994).

The N-methyl-D-aspartate (NMDA) receptor complex associated with a voltage- and ligand-gated calcium ionophore is also increased in number after chronic ethanol treatment (see Chapters 3 and 5; also Hoffman and Tabakoff, 1994). The time course of upregulation in NMDA ligand-binding sites is paralleled by the time course of ethanol withdrawal convulsions (Hoffman and Tabakoff, 1994), and antagonists of the NMDA-gated calcium channel (e.g., dizocilpine) have repeatedly been shown to attenuate ethanol withdrawal (Hoffman et al., 1990; Buck and Harris, 1991; Tabakoff and Hoffman, 1993). As with the dihydropyridine-sensitive calcium channel, a role for protein-kinase C activity in ethanol modulation of calcium flux at the NMDA channel has recently been suggested. Either ethanol or activation of protein-kinase C reduces NMDA-stimulated calcium flux; furthermore, protein-kinase C inhibitors block the reduction of NMDA function by ethanol (Snell et al., 1994; Hoffman and Tabakoff, 1994).

An inhibitory role in neural function is played by the γ-aminobutyric acid (GABA)/benzodiazepine receptor/chloride ionophore complex, also called the $GABA_A$ receptor complex. Chloride flux into the neural cell decreases its ability to fire. This ion flux is influenced by a large variety of agonists (e.g., muscimol) and co-agonists (e.g., benzodiazepines, barbiturates, and neuroactive steroids), and is attenuated by a similarly large number of competitive (e.g., bicuculline) and noncompetitive antagonists (e.g., picrotoxin), as well as benzodiazepine inverse agonists (e.g., Ro15–4513 and β-carbolines). Antagonism of chloride flux is convulsant at moderate doses. Chronic ethanol treatment has sometimes, but not always, been shown to decrease the number of $GABA_A$ agonist and benzodiazepine binding sites and increase benzodiazepine inverse agonist binding sites (see Chapter 4; also Buck and Harris, 1991). Despite the inconsistency in binding site density findings, many studies have documented modulatory effects of ethanol on chloride flux (see reviews by Buck and Harris, 1991; Yu and Ho, 1990). Ethanol's modulatory role appears to occur via interactions at several of the binding sites on the complex. Certainly, ethanol withdrawal severity can be modulated by agonists and co-agonists (decreased severity), as well as antagonists and inverse agonists (increased severity) (e.g., Crabbe, 1992; Crabbe et al., 1993; Finn et al., 1995).

II. GENETIC STUDIES OF HUMAN WITHDRAWAL

Human genetic research on alcoholism has focused extensively on the possible existence of subtypes of alcoholism, on other issues related to genetic heterogeneity and the characteristics allowing subclassification (e.g., Devor and Cloninger, 1989), on the search for genetic markers (e.g., Thomasson et al., 1993), and on studies of twin concordance rates and incidence of alcoholism in children adopted away from alcoholics (e.g., Schuckit et al., 1985). Schuckit (1994) has recently reported that individuals with family histories of alcoholism have lesser sensitivity to some acute effects of ethanol and that such lowered sensitivity is a predictor of later propensity to develop alcoholism. Thus, it should be possible to link genetic differences in human susceptibility to alcoholism with particular responses. To our knowledge there are only two studies that have directly investigated genetic differences in ethanol withdrawal severity in humans. Neither study investigated severe withdrawal signs (e.g., convulsions or delirium tremens). This is to be expected because severe withdrawal is pharmacologically treated due to ethical considerations. Newlin and Pretorius (1990) first reported that adult male sons of alcoholic fathers gave self-reports of greater hangover symptom frequency than sons of nonalcoholic fathers. No mention of the alcoholic status of the mother was made. Hangover symptoms (including headache, nausea, vomiting, passing out, and blackouts) were rated on a 6-point scale indicating the percentage of times they occurred after drinking. Drinking quantity-frequency appeared not to differ between the two groups. In the second study, adult male subjects with a family history positive (FHP: alcoholic father and sometimes other relatives) or a family history negative (FHN: no first- or second-degree relatives alcoholic) for alcoholism were administered 1 g/kg ethanol or a placebo orally and monitored hourly between 3 and 8 hours and at 18 hours after ingestion (McCaul et al., 1991). Measures included physiological, subjective, psychomotor, and withdrawal indices. Compared to their FHN counterparts, FHP subjects had more severe total withdrawal scores beginning at hour three. Scores gradually declined to FHN levels by hour eight. In particular, sweatiness, shakiness, and tachycardia were reported to be elevated in FHP subjects. There was also a trend for greater craving. Furthermore, significant withdrawal after ethanol was seen only in the FHP subjects.

III. GENETIC STUDIES OF ANIMAL WITHDRAWAL

Many advances in ethanol withdrawal research have been made using genetic animal models, especially rats and mice (for reviews, see Crabbe and Li, 1995; Phillips and Crabbe, 1991). Advantages of rodent models include experimental control of matings (i.e., genotypes) and rapid generation cycle (from conception to adulthood is about 75 days for rats and mice). Also, a great deal of homology among rodent and human genomes exists, making rodents ideal subjects for identifying locations, and ultimately functions, of specific genes relevant to drug withdrawal responses.

A. COMPARISON OF INBRED STRAINS FOR WITHDRAWAL DIFFERENCES

Many inbred strains of mice exist. Several were originally developed by mouse fanciers for various traits, usually coat color (Morse, 1978). Because inbreds are developed by mating close relatives over 20 generations, members of any particular inbred strain are genetically identical to one another with the exception of sex-specific loci (Crabbe, 1989; McClearn, 1991). Therefore, differences in responses within an inbred strain must be due to environmental influences (including environment-genotype interactions). Furthermore, differences among several inbred strains can be attributed to genetic factors, given equivalent testing procedures. Goldstein and Kakihana (1974) first demonstrated that three inbred mouse strains — BALB/cJ, C57BL/6J, and DBA/2J — differed in ethanol withdrawal severity. Since then, it has become well-established that the various DBA/2 substrains of mice display severe ethanol withdrawal in comparison to almost any other inbred strain (Griffiths and Littleton, 1977; Grieve et al., 1979; Crabbe et al., 1983b; Roberts et al., 1992; Metten and Crabbe, 1994). The use of a new technique based on C57BL/6J and DBA/2J strain differences has recently led to exciting developments in the field of gene mapping (discussed in the last section).

The utility of inbred strains and methodological considerations in pharmacogenetic research have been discussed in detail elsewhere (Belknap, 1980; Deitrich and Spuhler, 1984; Crabbe et al., 1990b; McClearn, 1991). One important consideration is that the use of a relatively large panel of inbred strains (≥12 strains) is advisable when one is attempting to establish the range of genetic influences on a trait. This is because each strain represents a single genotype (i.e., the genetic sample size in the experiment equals the number of strains being tested). We caution that studies demonstrating a phenomenon in a single inbred strain, while adding valuable information on that strain to the inbred strain data base being compiled by several laboratories, are greatly restricting the generalizability of their findings. Because such studies are, by definition, not genetic studies, it is preferable to study heterogeneous animals in this situation (McClearn, 1991). Likewise, it is of questionable value to attempt to demonstrate genetic correlations in two or three inbred strains. The genes in inbred strains are fixed without respect to a particular phenotype; therefore, any pair of strains is likely to have similarities and differences on several traits that are genetically unrelated. The multiple-strain approach has been employed to investigate the influences of ethanol on several traits, including open-field activity, hypothermia, rotarod ataxia, loss of righting reflex (Crabbe, 1983; Crabbe et al., 1994b), ethanol preference drinking (Belknap et al., 1993a), and acute (Metten and Crabbe, 1994) and chronic (Crabbe et al., 1983b) ethanol withdrawal severity. Furthermore, we are compiling parallel data sets on several other drugs, including diazepam (Gallaher et al., submitted; Metten and Crabbe, 1994), pentobarbital (Crabbe et al., manuscript in preparation; Metten and Crabbe, 1994), and morphine (Belknap et al., 1995).

Another advantage to testing inbred strain panels is that it allows identification of genetically correlated responses. The fixed genetic identity of an inbred strain means that results of studies utilizing inbred strains can be compared across laboratories and time. We recently found that the magnitudes of withdrawal HIC severity following acute ethanol and pentobarbital are positively genetically correlated, whereas withdrawal from either drug was not significantly correlated with basal (predrug) HIC severity (Metten and Crabbe, 1994). Furthermore, withdrawal severity was not generally correlated with blood ethanol or brain pentobarbital drug levels, indicating that pharmacokinetic factors such as drug absorption or blood-brain barrier permeability are unlikely to explain the genetically correlated withdrawal severities to any significant degree. Chronic ethanol withdrawal severity has been shown to be weakly genetically correlated with blood ethanol concentrations and negatively correlated with ethanol hypothermia tolerance (Crabbe et al., 1983b); both acute and chronic withdrawal were genetically uncorrelated with ethanol-induced locomotor ataxia, decrease in open-field activity, loss of balance, and loss of righting reflex (correlation of data from Crabbe et al., 1983b; Crabbe et al., 1994b; Metten and Crabbe, 1994).

B. LINES SELECTED FOR WITHDRAWAL DIFFERENCES

Selective breeding for drug abuse-related traits has been a productive endeavor (Crabbe and Li, 1995). Selectively bred lines are developed by testing animals from a heterogeneous stock on the trait of interest and then mating together extreme-scoring animals. Usually, bidirectionally selected lines (high- and low-response lines) plus control (nonselected) lines are established. Divergence of the high and low lines on the selection trait in opposite directions from the control line is conclusive evidence that the trait is genetically influenced. Establishment of independently selected replicate lines assists the researcher with the interpretation of results suggesting correlated responses to selection. Similar line differences on a nonselected (correlated) trait in both replicates of the selected lines is strong evidence of pleiotropic influences of the genes fixed by selection. These issues have been discussed in detail elsewhere (Crabbe et al., 1990b).

Kosobud and Crabbe (1995) have most recently reviewed the ethanol withdrawal-related selectively bred lines (see also Phillips and Crabbe, 1991; Crabbe and Phillips, 1993). Genetic involvement in ethanol withdrawal HIC severity was initially demonstrated in a three-generation study that successfully bred lines of mice selected from the Swiss-Webster outbred stock (Goldstein, 1973a). Subsequent development of lines differing in ethanol withdrawal severity has been carried out by three laboratories; two of the four projects have been terminated (Phillips and Crabbe, 1991). All four projects used mice; to our knowledge, no selective breeding for withdrawal severity differences has been performed in rats. The four sets of withdrawal severity lines were each derived from the HS/Ibg foundation stock, and three were bred according to a within-family breeding scheme (Falconer, 1989). A brief historical overview of each project will be followed by new studies since the previous review (Kosobud and Crabbe, 1995). Presentation of preliminary results of a fifth selective breeding project (derived from a different foundation stock), the High Alcohol Withdrawal (HAW) and Low Alcohol Withdrawal (LAW) lines, will be made in the Section III.D.

1. Withdrawal Seizure Prone and Withdrawal Seizure Resistant Lines

The first and best characterized lines were bred following chronic ethanol treatment via vapor inhalation using a procedure slightly modified from the 3-day method of Goldstein and Pal (1971). Withdrawal Seizure Prone (WSP) and Withdrawal Seizure Resistant (WSR) mice were selectively bred for severe and mild HICs, respectively, following withdrawal from ethanol (Crabbe et al., 1983a,b, 1985). Two genetically independent replicate sets of lines were created, along with a replicate set of control lines (Withdrawal Seizure Control [WSC]) which are maintained without selection. Thus, there are four selected lines: WSP1, WSP2, WSR1, and WSR2; and two nonselected lines, WSC1 and WSC2. Mice from both replicates of the WSP line displayed 10-fold greater withdrawal severity than WSR mice by selection generation 11 (S_{11}), whereas WSC scores were intermediate (Crabbe et al., 1985). Selection pressure was applied for 26 generations; relaxation of the selection produced no drift in withdrawal scores for either line (i.e., WSP and WSR scores remained different to the same degree [Phillips et al., 1989a; Crabbe and Phillips, 1993]). This implies that the genes mediating the respective withdrawal intensities were fixed in the homozygous state by selection. The lines have been maintained under relaxed selection for 32 subsequent generations (i.e., S_{26} G_{58}) at the present writing. Sublines of the WSP and WSR lines are presently being inbred. The IWSP2 (Inbred WSP2), IWSR1, and IWSR2 strains have reached at least the 27th generation, and thus, can be considered to be inbred strains (Falconer, 1989). Inbreeding of the IWSP1 strain is also proceeding, but after approximately 10 generations of inbreeding, a reproduction bottleneck has twice led to loss of the IWSP1 strain. The current IWSP1 strain is in the seventh generation of inbreeding.

a. Characterization of the Selection Response

WSP and WSR mice were tested for selection by injection with loading doses of ethanol (1.5 g/kg) and 1 m*M* pyrazole, an alcohol dehydrogenase inhibitor, and by placement for 72 hours in an inhalation chamber containing ethanol vapor (about 5 to 12 mg ethanol per liter air [Crabbe et al., 1985]). Upon removal from the chamber, a blood sample (20 microliters) was taken from the tip of each mouse's tail, and withdrawal HICs were scored hourly for 15 hours, and at 24 and 25 hours. The selection index was the area under the withdrawal curve (Crabbe et al., 1985). About 26% of the genetic variability displayed in the selection trait has been shown to be heritable (Crabbe et al., 1985). This is a substantial amount for such a complex character. During early generations of selection, no differences among the lines were seen in blood ethanol concentration at removal from the chamber, or in ethanol metabolism following either an acute injection (3 g/kg, i.p.) in naive mice, or chronic vapor inhalation (Kosobud and Crabbe, 1986). Later in selection, WSP mice accumulated about 20% more ethanol during vapor inhalation than did WSR mice, a factor controllable by exposing them to different ethanol loading doses, pyrazole doses, and/or vapor concentrations (Terdal and Crabbe, 1994). Despite these pharmacokinetic differences that have arisen, the magnitude of withdrawal severity differences between the selected lines remains very large, even when blood ethanol concentrations are matched, indicating that the differences in blood ethanol concentration are not important in determining the differential susceptibility to withdrawal. This finding is supported by the studies reviewed earlier in inbred strains whose blood ethanol concentrations were, at best, weakly genetically correlated with withdrawal severity (Crabbe et al., 1983b; Metten and Crabbe, 1994).

b. Tests of the Lines for Correlated Responses to Selection

In any selective breeding project, identification of traits that are genetically correlated with the selection response can be a powerful tool in indicating potential mechanisms. If the genes fixed by selection influence other traits (pleiotropy), then the lines should display divergent responses to those traits, but not to genetically unrelated traits.

TABLE 16.1

Tests for Correlated Responses to Selection in WSP and WSR Mice	Sensitivity	Reference
Other Withdrawal Indices		
Correlated Responses		
↑Tremor	P>R	Kosobud and Crabbe, 1986
↑Backward Walking	R>P	Kosobud and Crabbe, 1986
Uncorrelated Responses		
↑Straub Tail, ↓Hole-in-Wall, ↓Vertical Screen Activity ↓Open-Field Activity	–	Kosobud and Crabbe, 1986, Crabbe et al., 1994b
Other Measures of Behavioral Sensitivity in Naive Mice		
Correlated Responses		
Ethanol Free-Choice Drinking	R>P	Kosobud et al., 1988
Uncorrelated Responses		
Ethanol-induced Hypothermia, Hypothermic Tolerance, Loss of Righting Reflex, Locomotor Activity	–	Crabbe and Kosobud, 1986, Crabbe et al., 1986, 1988
Withdrawal from Other Drugs (Increased HIC)		
Correlated Responses		
Chronic Phenobarbital, Diazepam	P>R	Belknap et al., 1988, 1989
Acute Diazepam, Pentobarbital, t-Butanol, Acetaldehyde	P>R	Crabbe et al., 1991a
Inhalation (1 h) of Nitrous Oxide Gas	P>R	Belknap et al., 1987
Morphine	R>P	Belknap, unpublished
Sensitivity to Effects of Pro- and Anti-Convulsant Treatments in Naive Mice		
Correlated Responses		
Pro-Convulsant Treatments:		
ED_{50}s for Tonic Hindlimb Extensor Seizures: NMDA	R>P	Kosobud and Crabbe, 1993
Latencies to clonus: Picrotoxin, CHEB, 4-Aminopyridine	P>R	Crabbe and Kosobud, 1990
Enhancement of HIC: Nicotine, 3-MPA, Strychnine, Bicuculline, TBPS, Kainic Acid, BAY K 8644, Ro15–4513, Pentylenetetrazol, Picrotoxin, NMDA	P>R	Crabbe et al., 1991b, 1993,[a] Feller et al., 1988
Audiogenic Seizure Sensitivity	P>R	Feller et al., 1994
Anti-convulsant Treatments:		
Drug vs. ECS: Phenytoin, Valproic acid, Diazepam, Barbital, Phenobarbital, Pentobarbital, C1–C5 straight-chain alcohols, Methyprylon, Ethchlorvynol, Dizocilpine	R>P	Crabbe et al., 1986, 1994c McSwigan et al., 1984
Ethanol vs. Convulsant: Strychnine, Flurothyl	R>P	McSwigan et al., 1984
Uncorrelated Responses		
ECS-CA_{50}[b] for Tonic Hindlimb Extensor Seizures	–	McSwigan et al., 1984, Crabbe et al., 1994c
ED_{50}[b]s for Tonic Hindlimb Extensor Seizures: Flurothyl, Strychnine, Bicuculline, Picrotoxin, Pentylenetetrazol	–	McSwigan et al., 1984
Latencies to clonus: Kainic Acid, Bicuculline, Pentylenetetrazol, TBPS, DMCM	–	Crabbe and Kosobud, 1990, Kosobud et al., 1992
Rate of Pentylenetetrazol kindling of HIC	–	Crabbe et al., 1990a
Ethanol vs. Convulsant: Pentylenetetrazol, Picrotoxin, Bicuculline	–	McSwigan et al., 1984
Sensitivity to Pro- and Anti-Convulsant Treatments During Withdrawal from Acute Ethanol		
Correlated Responses		
Pro-Convulsant Treatments:		
Corticosterone	P>R	Roberts et al., 1991, 1994[a]
NMDA,[a] Kainic Acid[a]	P>R[a]	Crabbe et al., 1993
Anticonvulsant Treatments:		
Aminoglutethimide[a]	P>R[a]	Roberts et al., 1991,
Dizocilpine[a]	P>R[a]	Crabbe et al., 1993,
$3\alpha,5\alpha$-P[a]	P>R[a]	Finn et al., 1995

TABLE 16.1 (continued)

Tests for Correlated Responses to Selection in WSP and WSR Mice	Sensitivity	Reference
Diazepam,[a] Abecarnil[a]	P>R[a]	Crabbe, 1992,
Nitrous Oxide[a]	P>R[a]	Belknap et al., 1987
Other Neurochemical Characters		
Correlated Responses		
Enhancement of nitrendipine binding after chronic ethanol	P>R	Brennan et al., 1990
Benzodiazepine inverse agonist modulation of Cl^- flux after ethanol	P>R	Buck et al., 1991b
Benzodiazepine agonist modulation of Cl^- flux after ethanol	R>P	Buck et al., 1991b
$GABA_A$ $\alpha1$ subunit whole brain mRNA down-regulation after ethanol	P>R	Buck et al., 1991a
$GABA_A$ $\alpha6$ subunit whole brain mRNA down-regulation after ethanol	R>P	Buck et al., 1991a
$GABA_A$ $\alpha3$ and $\alpha6$ subunit whole brain mRNA levels in naive mice	R>P	Buck et al., 1991a
$GABA_A$ $\alpha1$ and $\alpha6$ subunit cerebellar mRNA levels in naive mice	R>P	Keir and Morrow, in press
Hippocampal mossy fiber zinc content in naive mice	R>P	Feller et al., 1990
[c] Number of hippocampal NMDA binding sites (naive or after ethanol)	P>R	Valverius et al., 1990
Uncorrelated Responses		
TBPS, Flunitrazepam binding and affinity (naive)	–	Feller et al., 1988
Membrane fluidity (naive or after ethanol)	–	Harris et al., 1984
$GABA_A$ $\alpha1$ and $\beta2$ subunit cerebral cortex mRNA levels in naive mice	–	Keir and Morrow, in press
[c] Number of hippocampal NMDA binding sites (naive or after ethanol)	–	Carter et al., submitted

[a] Data collected in WSP mice only, because significant withdrawal is not obtained in WSR mice.
[b] CA_{50}: Convulsant amperage at which 50% of animals convulse; ED_{50}: Effective dose at which 50% of animals convulse.
[c] Note the appearance of this measure in both Correlated and Uncorrelated Responses.

The WSP and WSR lines have been tested for many traits besides chronic ethanol withdrawal (Table 16.1). These include:

1. ethanol withdrawal severity measures other than HIC
2. other behavioral measures of ethanol effects
3. withdrawal from other drugs
4. sensitivity to pro- and anti-convulsant treatments in the naive and/or ethanol-withdrawn states
5. other tests, including neurochemical differences

WSP mice have more severe tremors following chronic ethanol (Kosobud and Crabbe, 1986), as well as greater withdrawal severity than WSR mice following treatment with a variety of central nervous system depressants (see Table 16.1 [Belknap et al., 1987, 1988, 1989; Crabbe et al., 1991a]). In contrast, WSR mice voluntarily consume more ethanol and have more intense morphine withdrawal (Kosobud et al., 1988; Belknap et al., unpublished data). No differences between the lines in chronic ethanol withdrawal-induced Straub tail or reductions in hole-in-wall behaviors, vertical screen crossings, or open-field activity are seen, demonstrating a lack of pleiotropic effects on these indices of withdrawal (Kosobud and Crabbe, 1986). Furthermore, ethanol-induced hypothermia, hypothermic tolerance, loss of righting reflex, and locomotor activity have been shown to be similar in naive animals from the lines (Crabbe and Kosobud, 1986; Crabbe et al., 1986, 1988). These data provide independent, converging evidence with that derived from inbred strains (see above) about the relationships among these characters and imply that the genes controlling ethanol withdrawal are largely independent of those mediating other ethanol effects.

Much effort has been made to identify potential neurochemical differences in the WSP and WSR lines by attempting to find differential effects of pro- and anti-convulsant drugs with known neurochemical actions. Table 16.1 also summarizes the general findings of these studies. One interesting recent finding is that WSP mice are more sensitive to manipulations of the NMDA receptor complex during ethanol withdrawal than while naive (Crabbe et al., 1993). The converse appears to be the case for the GABAergic competitive antagonist, PTZ (Crabbe et al., 1993). In apparent agreement with these results, naive WSR mice are more sensitive to the convulsant effects of intravenous NMDA and the anticonvulsant effects of dizocilpine against electroconvulsive shock (ECS) than WSP mice (Kosobud and Crabbe, 1993; Crabbe et al., 1994c). Chronic ethanol treatment has been shown to upregulate dizocilpine receptors (Grant et al., 1990; Gulya et al., 1991; Valverius et al., 1990), but the original finding that naive and withdrawing WSP mice have more dizocilpine binding sites than WSR mice do (Valverius et al., 1990) has not been replicated (Carter et al., 1995). Further studies will need to clarify this issue; however, the convulsant sensitivity results would predict that, if anything, WSR mice may have more binding sites than WSP mice.

Another recent group of studies has shown that WSP mice are sensitive to steroid hormone manipulations of ethanol withdrawal. Roberts et al. (1991) showed that corticosterone enhanced acute ethanol withdrawal in WSP mice, while having no effect on WSR mice. The corticosterone synthesis inhibitor, aminoglutethimide, attenuated acute ethanol withdrawal (Roberts et al., 1991). Subsequent studies showed that corticosterone preferentially enhanced acute ethanol, pentobarbital, and precipitated diazepam withdrawal in WSP mice, while having no affect on HICs in naive mice (Roberts et al., 1994). In related studies, the neuroactive steroid, 3α-hydroxy-5α-pregnan-20-one (3α,5α-P) also dose-dependently attenuated ethanol withdrawal in WSP mice treated for 24 hours in the vapor inhalation chamber, whereas only high doses affected air-treated controls (Finn et al., 1995). This endogenous compound acts at the GABA/benzodiazepine receptor-chloride ionophore complex to potentiate GABA agonist-stimulated chloride conductance (Lambert et al., 1987; Belelli et al., 1990; Paul and Purdy, 1992). Other agents that manipulate chloride conductance at this complex have repeatedly been shown to modulate ethanol withdrawal (Goldstein, 1972b; Crabbe, 1992; Crabbe et al., 1993). Furthermore, benzodiazepine agonist and inverse agonist modulation of chloride flux was shown to be differentially effective in WSP and WSR mice only after chronic ethanol treatment (Buck et al., 1991b). Finally, genetic differences in alcohol withdrawal may be related to differential contributions of the different $GABA_A$ receptor subunits. $GABA_A$ receptor α1, α3, and α6 subunits are differentially regulated in naive WSP and WSR mice. Messenger RNA levels of α1 and α6 subunits in cerebellum were higher in WSR than WSP mice (Keir and Morrow, 1994), consistent with the earlier finding that α3 and α6 subunit whole-brain mRNA levels were higher in naive WSR than WSP mice (Buck et al., 1991a). Chronic ethanol treatment has been shown to decrease $GABA_A$ α1 subunit whole-brain mRNA levels in WSP, but not WSR mice; the converse was found for ethanol treatment effects on whole-brain α6 subunit mRNA levels (Buck et al., 1991a). Taken together, these studies clearly imply that receptors on the $GABA_A$ complex may play a role in withdrawal from central nervous system depressant drugs. Future studies will doubtless pursue these very interesting findings.

2. Severe Ethanol Withdrawal and Mild Ethanol Withdrawal Lines and High Addictability and Low Addictability Lines

The Severe Ethanol Withdrawal (SEW)/Mild Ethanol Withdrawal (MEW) (McClearn et al., 1982; Allen et al., 1983; Wilson et al., 1984) and High Addictability (HA)/Low

Addictability (LA) (Berta and Wilson, 1989, 1990, 1992) selective breeding projects have several features in common. As mentioned previously, both began with HS/Ibg foundation stock, and the breeding of replicate lines and controls proceeded using within-family selection. Unlike the methods in WSP/WSR selection, ethanol was administered via a liquid diet for 9 days to SEW and MEW mice and 11 days to HA and LA mice. Ethanol concentrations in the diets were increased from 10–35% of ethanol-derived calories for the SEW/MEW lines, and from 10–30% for the HA/LA lines. Body weight and diet consumption were recorded for each animal. Blood ethanol concentrations were not monitored routinely during selection. The selection indices for the two selection projects differed substantially.

The SEW/MEW selection index was a composite of seven different measures, including HIC, reduction in body temperature, reduced vertical screen activity, reduced rearings and crossings and spontaneous seizure incidence during hole-in-wall testing, and total ethanol consumption. Six to seven hours after removal of the ethanol diet on the 10th day, mice of both replicates, lines, and controls were serially tested on these measures, and the results were subjected to principal component analysis to derive a single index describing the first factor (McClearn et al., 1982). Divergence of the SEW from the MEW lines on this index of withdrawal was about 1.2 standard deviations by the 10th selected generation (Wilson et al., 1984). Naive animals from each line were examined in generation nine for correlated responses to selection. Loss of righting reflex duration was measured after an acute dose of ethanol (4.2 g/kg). During this test, body temperatures and blood ethanol concentrations were also measured. Only blood ethanol concentrations at hour one following injection was significantly different in both replicates of the lines: MEW > SEW. A number of difficulties, including escalated death rates in succeeding generations (reaching ~25%), resulted in the termination of the project. The death rates may have risen due to increases in ethanol consumption (by 35–50%, regardless of line) as selection progressed. The authors presented the intriguing hypothesis that natural selection in individuals for ability to tolerate more ethanol to enable the consumption of adequate non-ethanol calories may have interfered with the authors' artificial selection for differences in withdrawal severity (Wilson et al., 1984).

The HA/LA selection project attempted to correct some of the difficulties seen in the SEW/MEW project. The most significant difference in selection procedures was that of the criterion for selection, which was reduced to the single measure, HIC severity (Berta and Wilson, 1992). Using a 4-point scale, HICs were assessed at 0, 2, 4, 6, and 8 hours following removal of the ethanol diet on day 12, and the selection index was the sum of these scores. The scores of the initial breeders in all the lines were about 5.75. Over the succeeding seven generations, the scores of the LA lines decreased to about 2.4, giving a cumulated selection response of about 3.35. The scores of the HA lines increased by only about one unit overall, despite the fact that there was more scalar room for a selection response in the upward direction (scores could range from 0 to 20). In contrast to the increase in consumption seen in the SEW/MEW lines, a decrease in ethanol consumption in all lines during the sixth and seventh generations was reported in the HA/LA lines (Berta and Wilson, 1992). Furthermore, the initially greater consumption by the HA lines compared to the LA lines disappeared in generation seven. Blood ethanol concentrations at withdrawal were reportedly not different between the lines in generation seven, and the authors concluded that the differences in seizure scores were not attributable to blood ethanol concentration. However, it remains possible that consumption differences on earlier days, and thus different levels of exposure to ethanol between the lines, resulted in the differential seizure severity. Examination of ethanol preference and loss of righting reflex in naive mice of generation three did not detect correlated responses to selection. This project has also been terminated.

In two different experiments, the SEW/MEW and HA/LA lines were each tested with the WSP/WSR lines in the chronic ethanol vapor inhalation paradigm (Kosobud and Crabbe, 1995). Withdrawal HICs were significantly different (in the expected direction) in both pairs of WSP/WSR lines, but only one replicate of the SEW/MEW lines differed. A gender difference was found in the HA/LA test. Female LA1 mice had significantly lower, and male LA1 mice had significantly higher withdrawal HIC area scores than did the other lines (LA2, HA1, and HA2). Given the similarity in selection indices between the WSP/WSR and HA/LA selections, the lack of concordance in the results was initially surprising, but may be explained by the potential concomitant selection in the HA/LA lines for pharmacokinetic differences. Finally, the magnitudes of difference between high scoring lines and low scoring lines of each selection were greater for the WSP/WSR selection than for either of the other two (Kosobud and Crabbe, 1995). The line differences in withdrawal severity were not explained by differences in blood ethanol concentrations at withdrawal among mice from the three projects (SEW/MEW about three times higher than WSP/WSR, and WSR slightly lower than HA/LA/WSP [Kosobud and Crabbe, 1995]).

3. High Withdrawal and Low Withdrawal Lines

Replication of the WSP/WSR selection project via mass selection (Falconer, 1989), using the same foundation stock, and nearly the same ethanol administration methods and selection index, is underway in the laboratory of Dr. V. Gene Erwin (personal communication). These High Withdrawal (HW) and Low Withdrawal (LW) lines are currently in the second selected generation and show clear divergence of withdrawal severity. Testing of these lines for their divergence in traits suggested to be correlated with withdrawal severity by results from the WSP/WSR lines will be important for confirming the WSP/WSR differences. They will provide a powerful test of genetic correlation because such correlates can most clearly be ascertained during early generations of selective breeding.

C. OTHER GENETIC EVIDENCE

The inbred strain and selected line data reviewed in the previous sections suggest that there is little or no genetic relationship between ethanol withdrawal severity and ethanol sensitivity measured as loss of righting reflex, open-field activity, hypothermia, or ethanol preference. Another test of these hypotheses would be to test lines selectively bred for each of these traits for ethanol withdrawal severity. Some of these tests have been made. Long Sleep (LS) and Short Sleep (SS) mice, selectively bred for the long and short durations, respectively, of the loss of righting reflex after an acute 4.2 g/kg dose of ethanol (McClearn and Kakihana, 1981), were tested for ethanol withdrawal severity (Goldstein and Kakihana, 1975). LS mice showed significantly lower withdrawal HIC scores than did SS mice following 72 hours of chronic ethanol vapor inhalation. Thus, a significant negative genetic correlation between the two characters was posited (Goldstein and Kakihana, 1975). The mice were matched for blood ethanol concentrations to avoid potential ethanol exposure differences between the lines. One possible explanation might be that LS mice simply have less excitable central nervous systems than SS mice do. This possibility was addressed by testing for pentylenetetrazol-induced convulsion sensitivity, which revealed no difference between the lines. Examination of differential sensitivity to other convulsant drugs (bicuculline, picrotoxin, 3-carbomethoxy-β-carboline, caffeine, flurothyl, and strychnine) generally suggested that LS mice might be more resistant to excitation than are SS mice, but demonstrated no consistent neurochemical pattern of differences in the lines (McIntyre and Alpern, 1989; Phillips et al., 1989b).

These results appear to conflict with those in inbred strains and the lines selectively bred for withdrawal. However, closer examination of more technical genetic parameters in this experiment may help illuminate the nature of the relationship between the traits. First, the LS and SS lines were selected without replicate lines and control lines (McClearn and Kakihana, 1981). Therefore, any genetic correlation study using them is testing a two-data-point correlation. Second, the lines were tested for withdrawal late in selection when cumulative inbreeding rates are high and the fortuitous fixation of genes unrelated to the selection trait (i.e., loss of righting reflex duration) was a significant risk. In fact, Goldstein and Kakihana (1975) tested heterogeneous mice of the parental stock for the LS/SS selection (analogous to testing control lines) and found no phenotypic correlation between loss of righting reflex duration and withdrawal severity. They therefore concluded that the association seen in the selected lines was fortuitous; thus, these data do not support a relationship between the characters (Goldstein and Kakihana, 1975).

Mouse lines selectively bred for ethanol-stimulated (FAST) and -depressed (SLOW) locomotor activity in an open-field apparatus have also been tested for ethanol withdrawal (Crabbe et al., 1988). Briefly, FAST1 mice had significantly milder withdrawal HIC scores than did their SLOW1 counterparts, but the FAST2 and SLOW2 lines did not differ. Since the WSP and WSR lines did not differ in response to ethanol in the open-field activity test (see previous discussion), this suggests that there may be only a weak negative genetic correlation between the locomotor stimulant or depressant effects of ethanol and withdrawal severity. Inbred strain correlations between activity and acute or chronic ethanol withdrawal have both been demonstrated to be nonsignificant (correlation of data from Crabbe et al., 1983b, 1994b; Metten and Crabbe, 1994).

Similar lines and replicates, selectively bred for a large (COLD) and small (HOT) hypothermic response to an acute dose of ethanol, were tested for ethanol withdrawal after vapor inhalation treatment (Crabbe et al., 1988). Like the FAST/SLOW results, HOT and COLD mice of one replicate differed in response to chronic ethanol withdrawal (HOT1 > COLD1), but no line difference was seen in the other replicate. Similar sensitivity to acute ethanol hypothermia and hypothermic tolerance was seen in WSP and WSR mice (Crabbe and Kosobud, 1986; Crabbe et al., 1988). Neither acute nor chronic ethanol withdrawal severity was genetically correlated with hypothermia in inbred strains; however, tolerance to ethanol hypothermia was found to be negatively correlated with chronic ethanol withdrawal (analysis of Crabbe et al., 1983b, 1994b; Metten and Crabbe, 1994). Thus, the combined results suggest that there may be, at best, a weak genetic relationship between these characters when animals are chronically administered ethanol.

Finally, Alcohol-Preferring (P) and -Nonpreferring (NP) rats, selectively bred for their respective preference in a two-bottle free-choice paradigm, have been examined for ethanol withdrawal severity (Waller et al., 1982). After preference testing, the animals were food-restricted to 80% of their free-feeding weight and then given free access to a 10% ethanol solution and water for 8 weeks. Ethanol consumption was slightly less in the NP rats (~12 g/kg/day) than in P rats (~14 g/kg/day). Withdrawal signs monitored included Straub tail, hyperreactivity, wet-dog shakes, and audiogenic convulsion sensitivity. Rats of both lines displayed withdrawal signs, but no difference was seen between the lines in the maximum stage of withdrawal attained, despite the trend for less consumption in NP rats (Waller et al., 1982). These data suggest that there is no genetic correlation between preference and withdrawal severity. However, no studies have forcibly produced physical dependence in P and NP rats and then compared severity of withdrawal. Our results in inbred mouse strains suggest a negative correlation between chronic ethanol withdrawal severity and consumption of unsweetened 10% ethanol in a free-choice paradigm (correlation of data from Crabbe et al., 1983b; Belknap et al., 1993a). Furthermore, a negative relationship was also suggested by preference testing in WSP and WSR mice (Kosobud et al., 1988). Differences in method (forced vs. self-administration), species, and/or the withdrawal indices may be responsible for the disparity in results between the rat and mouse studies. Further studies will be necessary to clarify this relationship.

D. MAPPING PROTECTIVE AND RISK GENES FOR WITHDRAWAL

1. First Stage: Quantitative Trait Loci Analysis

A recent method, quantitative trait locus (QTL) gene mapping (Lander and Botstein, 1989), has been exploited in several laboratories to identify the chromosomal location of genes (QTL) that confer risk or protection to drug-related behaviors. Although a number of strategies have been developed, in our studies, we first analyze the responses of a relatively large number of recombinant inbred (RI) strains. By comparing the pattern of strain sensitivities to a data base that shows the genotype of each strain at nearly 800 genetic marker loci whose chromosomal locations are known, we can identify regions of the chromosome that appear to contain QTL influencing the trait of interest (Plomin et al., 1991). Use of this method has recently been reviewed (Crabbe and Belknap, 1992; Plomin and McClearn, 1993; Tanksley, 1993).

We employed a two-stage procedure to identify and map several QTL affecting acute alcohol withdrawal severity (Belknap et al., 1993b). In the first stage, a panel of RI strains inbred from F_2 offspring of an F_1 intercross of C57BL/6J and DBA/2J inbred progenitor strains (the BXD RI strain panel [Bailey, 1981; Taylor, 1978]) was tested for ethanol withdrawal. Withdrawal severity was indexed by handling-induced convulsion scores after injection of 4 g/kg ethanol (see Crabbe et al., in press). Strain mean ethanol withdrawal severity (the quantitative trait) was then correlated with the allelic status for each marker gene in the data base.

We identified five groups of markers on four chromosomes that were significantly associated with ethanol withdrawal severity ($P < 0.01$). These were the *Pmv-7/D2Mit9* and the *D2Mit17* regions of chromosome 2, the *Cph-1* region of chromosome 8, the *Tmpt* region of chromosome 13, and the *Prm-1* region of chromosome 16 (Crabbe et al., in press). Chromosomal regions near these markers are likely to contain a QTL that affects withdrawal severity, but some of the associations may represent false-positive (chance) associations (Belknap, 1992; see also other articles in the same issue of *Behavior Genetics* for further discussion). Therefore, it was necessary to verify these provisional QTL with further analyses.

2. Second Stage: Verification in Segregating F_2 Populations

Two QTL were selected for immediate further study. The first of these provisional QTL appeared to account for as much as 48% of the genetic variability in acute ethanol withdrawal severity. The markers most highly associated were *D2Mit9* and *Scn2a*, which mapped to a position 36–37 centiMorgans (cM) from the centromere of chromosome 2 (Silver et al., 1993). If this represented a true association, then individual mice with the DBA/2J allele at markers in this region should have higher withdrawal severity than mice with the C57BL/6J allele. To verify this, we bred more than 150 B6D2 F_2 mice (from F_1 crosses of C57BL/6J [B6] and DBA/2J [D2] mice) and tested them for ethanol withdrawal severity. All mice were then genotyped for six nearby markers, using PCR to determine whether they had the C57 or the DBA allele at each marker. Withdrawal was significantly associated with allelic status at the marker *D2Mit9* (Crabbe et al., in press). F_2 mice that were B6B6 homozygotes at *D2Mit9* had withdrawal that was significantly less pronounced and of shorter duration than that in F_2 mice that were D2D2 homozygotes (Crabbe et al., in press).

This result suggests that a gene conferring risk of DBA-like severe withdrawal resides in this region. Two plausible candidate genes are mapped near *D2Mit9:* (1) *Gad-1*, which codes for glutamic acid decarboxylase, a catalyzing synthesis of GABA, and (2) a cluster of genes coding for α-subunits of brain voltage-dependent sodium channels *(Scn1a, Scn2a,* and *Scn3a).* These genes seem likely candidates for enhancing withdrawal convulsion severity. This region of mouse chromosome 2 is homologous with human chromosome 2q24–q37. If a gene can be located that promotes withdrawal risk in mice, potential extrapolation of such a mouse gene to a human map site is possible due to the high degree of homology between mouse and human chromosome maps (Copeland et al., 1993). This is a very attractive feature of QTL mapping in mice.

A second chromosomal region somewhat more distally located on chromosome 2 was also associated with withdrawal severity, but in this case the association was negative. That is, B × D RI strains bearing DBA/2J alleles at markers *D2Mit17, Ebp4.2*, and *D2Mc1* had *less* severe acute ethanol withdrawal than mice possessing C57BL/6J alleles at these loci. We are presently attempting to verify this association in F_2 mice. If successful, verification would suggest that a locus-conferring protection against withdrawal is located in the region between 68–83 cM distal on mouse chromosome 2.

3. Future Directions

There are several obvious and interesting extensions of the QTL mapping approach. In a preliminary analysis, we compared the chromosomal distribution of QTL for many different responses to alcohol. The location of QTL underlying these traits, and several other drug response traits, were combined into a composite map. Since many potential candidate genes are being mapped in mice (e.g., neurotransmitter receptors and transporters, synthetic and metabolic enzymes, ion channels, etc.), this composite map also incorporated locations of some candidate genes. By analyzing the "genetic landscape" for drug-response genes, we found that there appear to be several regions of the genome where presumed QTL affecting multiple responses are located (Crabbe et al., 1994a). For example, many responses to multiple drugs suggested the presence of a QTL on mouse chromosome 9 in a region near the gene encoding the dopamine D2 receptor gene *(Drd2).* Although this composite map offers several interesting possibilities for future research, it represents results from Stage 1 analyses of BXD RI strain data only, so further verification of these associations still remains to be attempted.

Because individual mice can be genotyped for markers flanking a QTL of interest from a DNA sample derived from a tail snip, it is possible to breed mice selectively for the genotype of interest (Plomin and McClearn, 1993). Such a genotypic selection, which achieves fixation in a single generation, could be used to generate individuals possessing several risk-promoting or several protective QTL so that they could be studied for their drug responses. We are currently selectively breeding by genotype for both of the chromosome 2 QTL markers *(D2Mit9* and *D2Mit17).* Four lines of mice are being developed from F_2 mice of the two parental strains according to genotype at the two loci: D2D2/B6B6, B6B6/D2D2, B6B6/B6B6, and D2D2/D2D2. Mice of the first two lines (that is, recombinants) should display very severe and very mild acute withdrawal, respectively, whereas mice of the latter two lines should have scores intermediate to the extreme-scoring lines.

Genotypically selected mice could serve as the starting point for another purpose, moving small regions of DNA into one of the progenitor strain genomes through repetitive backcrossing. For example, such a "congenic" strain would have the C57BL/6J genotype in more than 99% of its genome, but would have introduced only the small region of chromosome 2 containing the DBA/2J withdrawal-promoting QTL described above (i.e., *D2Mit9*). These congenics would be excellent source material for molecular biological studies that could positionally clone the actual risk-promoting gene. Alternatively, a technique such as genetically directed representational difference analysis could be employed using the congenic versus the standard inbred strain to develop the map locations of a large number of previously unmapped polymorphisms in the small region of interest (Lisitsyn et al., 1994).

Finally, we are also selectively breeding High Alcohol Withdrawal (HAW) and Low Alcohol Withdrawal (LAW) lines from B6D2 F_2 mice for severe and mild ethanol withdrawal HICs, respectively, after 4 g/kg ethanol. Preliminary results after one generation of selection indicate that significant divergence in withdrawal severity has been achieved (HAW mice score approximately two units higher than do LAW mice), and estimated trait heritabilities were about 0.19 (Belknap et al., unpublished data). Genotype information in the lines is being collected for the relevant QTL markers to compare with withdrawal severity analogous to the verification step in gene mapping. If allele frequencies at the QTL markers change in parallel with the development of phenotypic differences in withdrawal, it will provide further evidence in support of the importance of QTL in those regions.

IV. CONCLUSIONS AND FUTURE DIRECTIONS

Since the first demonstration of genetic mediation of ethanol withdrawal convulsions in mice (Goldstein, 1973a), remarkable convergence has been found in data from different genetic animal models, including selectively bred lines and standard and recombinant inbred strains. These data clearly show that a substantial proportion of the variance in ethanol withdrawal severity is genetically mediated, and that withdrawal severity is largely genetically independent of other effects of ethanol.

Full advantage of the diversity of techniques for studying ethanol withdrawal is just beginning to be taken with the integration of behavioral and molecular pharmacogenetic research. We and others are using several approaches to identify genes involved in ethanol withdrawal severity in mice.

First, selective breeding for withdrawal severity (WSP and WSR lines) has clearly demonstrated genetic differences in a quantifiable measure of ethanol withdrawal (the HIC) and provided researchers with a valuable model for hypothesis testing. Table 16.1 summarized the results of many studies characterizing these lines.

Second, many of the significant findings in the WSP/WSR selected line studies have been replicated in panels of standard inbred strains. The advantages of inbred strain panels over studies with only a few strains include increased power for detection of correlated responses. More data in inbred strain panels are required, including genotyping of PCR-based markers, before the full potential of this technique can be reached.

Third, a two-stage gene mapping approach in two populations of mice derived from two extreme-scoring inbred strains, DBA/2J and C57BL/6J, has been undertaken. BXD RI strains and B6D2 F_2 mice were used to localize genes conferring ethanol withdrawal risk or protection, thus enabling selective breeding from B6D2 F_2 mice for particular genotypes identified by QTL analysis as important in withdrawal as well as for phenotypic differences in withdrawal severity.

Finally, the development of congenic strains and the use of other candidate-gene methods, such as production of "knockdown" or "knockout" transgenic animals, should eventually make possible identification of the function of genes involved in ethanol withdrawal.

REFERENCES

Allen, D.L., Petersen, D.R., Wilson, J.R., McClearn, G.E., and Nishimoto, T.K. (1983) Selective breeding for a multivariate index of ethanol dependence in mice: results from the first five generations. *Alcoholism: Clin. Exp. Res.*, **7**(4), 443–447.

Bailey, D.W. (1981) Recombinant inbred strains and bilineal congenic strains. In *The Mouse in Biomedical Research* (Eds: H.L. Foster, J.D. Small, and J.G. Fox), pp 223–239, Academic Press, New York.

Ballenger, J.C. and Post, R.M. (1978) Kindling as a model for alcohol withdrawal syndromes. *Br. J. Psychiatry*, **133**, 1–14.

Becker, H.C. and Hale, R.L. (1993) Repeated episodes of ethanol withdrawal potentiate the severity of subsequent withdrawal seizures: an animal model of alcohol withdrawal "kindling". *Alcoholism: Clin. Exp. Res.*, **17**(1), 94–98.

Becker, H.C., Anton, R.F., and Randall, C.L. (1987) Stereotypic wall climbing in mice during ethanol withdrawal: a new measure of physical dependence. *Alcohol*, **4**(6), 443–447.

Begleiter, H. and Porjesz, B. (1979) Persistence of a "subacute withdrawal syndrome" following chronic ethanol intake. *Drug Alcohol Dependence*, **4**, 353–357.

Belelli, D., Lan, N.C., and Gee, K.W. (1990) Anticonvulsant steroids and the GABA/benzodiazepine receptor-chloride ionophore complex. *Neurosci. Biobehav. Rev.*, **14**, 315–322.

Belknap, J.K. (1980) Genetic factors in the effects of alcohol: Neurosensitivity, functional tolerance and physical dependence. In *Alcohol Tolerance and Dependence* (Eds: H. Rigter and J.C. Crabbe, Jr.), 93–121, Elsevier/North-Holland Biomedical Press, Amsterdam.

Belknap, J.K. (1992) Empirical estimates of Bonferroni corrections for use in chromosome mapping studies with BXD recombinant inbred strains *Behav. Genet.*, **22**, 677–684.

Belknap, J.K., Crabbe, J.C., and Laursen, S.E. (1989) Ethanol and diazepam withdrawal convulsions are extensively codetermined in WSP and WSR mice. *Life Sci.*, **44**, 2075–2080.

Belknap, J.K., Crabbe, J.C., and Young, E.R. (1993a) Voluntary consumption of ethanol in 15 inbred mouse strains. *Psychopharmacology*, **112**, 503–510.

Belknap, J.K., Danielson, P.W., Lame, M., and Crabbe, J.C. (1988) Ethanol and barbiturate withdrawal convulsions are extensively codetermined in mice. *Alcohol*, **5**, 167–171.

Belknap, J.K., Laursen, S.E., and Crabbe, J.C. (1987) Ethanol and nitrous oxide produce withdrawal-induced convulsions by similar mechanisms in mice. *Life Sci.,* **41,** 2033–2040.

Belknap, J.K., Metten, P., Helms, M.L., O'Toole, L.A., Angeli-Gade, S., Crabbe, J.C., and Phillips, T.J. (1993b) Quantitative trait loci (QTL) applications to substances of abuse: physical dependence studies with nitrous oxide and ethanol in BXD mice. *Behav. Genet.,* **23**(2), 213–222.

Belknap, J.K., Riggan, J., Cross, S., Young, E.R., Gallaher, E., and Crabbe, J.C. (1995) Genetic determinants of morphine activity and thermal responses in inbred mice. *Psychopharmacology*, (accepted pending modification).

Berta, J. and Wilson, J.R. (1989) Selection in mice for alcohol withdrawal seizures. *Behav. Genet.,* **19**(6), 745.

Berta, J. and Wilson, J.R. (1990) Correlated responses to selection in mice for alcohol withdrawal seizures. *Behav. Genet.,* **20**(6), 703.

Berta, J. and Wilson, J.R. (1992) Seven generations of genetic selection for ethanol dependence in mice. *Behav. Genet.,* **22**(3), 345–359.

Brennan, C.H., Crabbe, J., and Littleton, J.M. (1990) Genetic regulation of dihydropyridine-sensitive calcium channels in brain may determine susceptibility to physical dependence on alcohol. *Neuropharmacology*, **29**(5), 429–432.

Brown, M.E., Anton, R.F., Malcolm, R., and Ballenger, J.C. (1988) Alcohol detoxification and withdrawal seizures: clinical support for a kindling hypothesis. *Biological Psychiatry*, **23,** 507–514.

Buck, K.J. and Harris, R.A. (1991) Neuroadaptive responses to chronic ethanol. *Alcoholism: Clin. Exp. Res.,* **15**(3), 460–470.

Buck, K.J., Hahner, L., Sikela, J., and Harris, R.A. (1991a) Chronic ethanol treatment alters brain levels of γ-aminobutyric acid$_A$ receptor subunit mRNAs: relationship to genetic differences in ethanol withdrawal seizure severity. *J. Neurochem.,* **57,** 1452–1455.

Buck, K.J., McQuilkin, S.J., and Harris, R.A. (1991b) Modulation of γ-aminobutyric acid$_A$ receptor-operated chloride channels by benzodiazepine inverse agonists is related to genetic differences in ethanol withdrawal seizure severity. *J. Neurochem.,* **57,** 2100–2105.

Cappell, H. and LeBlanc, A.E. (1979) Tolerance to, and physical dependence on, ethanol: why do we study them? *Drug Alcohol Dependence,* **4,** 15–31.

Carter, L.A., Belknap, J.K., Crabbe, J.C., and Janowsky, A. (1995) Allosteric regulation of the N-Methyl-D-Aspartate receptor-linked ion channel complex and effects of ethanol in ethanol Withdrawal Seizure-Prone and -Resistant mice. *J. Neurochem.,* 64(1), 213–219.

Cicero, T.H. (1980) Alcohol self-administration, tolerance and withdrawal in humans and animals: theoretical and methodological issues. In *Alcohol Tolerance and Dependence* (Eds: H. Rigter and J.C. Crabbe, Jr.), 1–51. Elsevier/North-Holland Biomedical Press, Amsterdam.

Copeland, N.G., Jenkins, N.A., Gilbert, D.J., Eppig, J.T., Maltais, L.J., Miller, J.C., Deitrich, W.F., Weaver, A., Lincoln, S.E., Steen, R.G., Stein, L.D., Nadeau, J.H., and Lander, E.S. (1993) A genetic linkage map of the mouse: current applications and future prospects. *Science*, **262,** 57–66.

Crabbe, J.C. (1983) Sensitivity to ethanol in inbred mice: genotypic correlations among several behavioral responses. *Behav. Neurosci.,* **97**(2), 280–289.

Crabbe, J.C. (1989) Genetic animal models in the study of alcoholism. *Alcoholism: Clin. Exp. Res.,* **13**(1), 120–127.

Crabbe, J.C. (1992) Antagonism of ethanol withdrawal convulsions in Withdrawal Seizure Prone mice by diazepam and abecarnil. *Eur. J. Pharmacol.,* **221,** 85–90.

Crabbe, J. and Li, T.-K. (1995) Genetic strategies in preclinical substance abuse research. In *Psychopharmacology: The Fourth Generation of Progress* (Eds: F.E. Bloom and D.J. Kupfer), pp 799–811, Raven Press, Ltd., New York.

Crabbe, J.C. and Belknap, J.K. (1992) Genetic approaches to drug dependence. *Trends Pharmacol. Sci.,* **13,** 212–219.

Crabbe, J.C. and Kosobud, A. (1986) Sensitivity and tolerance to ethanol in mice bred to be genetically prone or resistant to ethanol withdrawal seizures. *J. Pharmacol. Exp. Therap.,* **239**(2), 327–333.

Crabbe, J.C. and Kosobud, A. (1990) Alcohol withdrawal seizures: Genetic animal models. In *Alcohol and Seizures* (Eds: R.J. Porter, R.H. Mattson, J.A. Cramer, and I. Diamond), 126–139. F. A. Davis, Philadelphia.

Crabbe, J.C. and Phillips, T.J. (1993) Selective breeding for alcohol withdrawal severity. *Behav. Genet.,* **23**(2), 171–177.

Crabbe, J.C., Belknap, J.K., and Buck, K.J. (1994a) Genetic animal models of alcoholism and drug abuse. *Science*, **264,** 1715–1723.

Crabbe, J.C., Buck, K.J., Metten, P., and Belknap, J.K. Strategies for identifying genes underlying drug abuse susceptibility. In *Molecular Approaches to Drug Abuse Research, Volume III* (Ed: T.N.H. Lee), NIDA Research Monograph, No. XXX, (in press). USDHHS, Rockville, MD.

Crabbe, J.C., Gallaher, E.J., Phillips, T.J., and Belknap, J.K. (1994b) Genetic determinants of sensitivity to ethanol in inbred mice. *Behav. Neurosci.,* **108**(1), 186–195.

Crabbe, J.C., Kosobud, A., Feller, D.J., and Phillips, T.J. (1988) Use of selectively bred mouse lines to study genetically correlated traits related to alcohol. In *Biomedical and Social Aspects of Alcohol and Alcoholism* (Eds: K. Kuriyama, A. Takada, and H. Ishii), pp 427–430, Elsevier/North-Holland Biomedical Press, Amsterdam.

Crabbe, J.C., Kosobud, A., and Young, E.R. (1983a) Genetic selection for ethanol withdrawal severity: differences in replicate mouse lines. *Life Sci.,* **33,** 955–962.

Crabbe, J.C., Kosobud, A., Young, E.R., Tam, B.R., and McSwigan, J.D. (1985) Bidirectional selection for susceptibility to ethanol withdrawal seizures in *Mus musculus*. *Behav. Genet.,* **15,** 521–536.

Crabbe, J.C., Merrill, C., and Belknap, J.K. (1991a) Acute dependence on depressant drugs is determined by common genes in mice. *J. Pharmacol. Exp. Therap.,* **257**(2), 663–667.

Crabbe, J.C., Merrill, C.D., and Belknap, J.K. (1991b) Effects of convulsants on handling-induced convulsions in mice selected for ethanol withdrawal severity. *Brain Res.,* **550,** 1–6.

Crabbe, J.C., Merrill, C.M., and Belknap, J.K. (1993) Effect of acute alcohol withdrawal on sensitivity to pro- and anticonvulsant treatments in WSP mice. *Alcoholism: Clin. Exp. Res.,* **17**(6), 1233–1239.

Crabbe, J.C., Merrill, C.M., Kim, D., and Belknap, J.K. (1990a) Alcohol dependence and withdrawal: a genetic animal model. *Ann. Med.,* **22,** 259–263.

Crabbe, J.C., Phillips, T.J., Kosobud, A., and Belknap, J.K. (1990b) Estimation of genetic correlation: interpretation of experiments using selectively bred and inbred animals. *Alcoholism: Clin. Exp. Res.,* **14**(2), 141–151.

Crabbe, J.C., Young, E.R., and Dorow, J. (1994c) Effects of dizocilpine in Withdrawal Seizure-Prone (WSP) and Withdrawal Seizure-Resistant (WSR) mice. *Pharmacol. Biochem. Behav.,* **47**(3), 443–450.

Crabbe, J.C., Young, E.R., and Kosobud, A. (1983b) Genetic correlations with ethanol withdrawal severity. *Pharmacol. Biochem. Behav.,* **18**(Suppl. 1), 541–547.

Crabbe, J.C., Young, E.R., Tam, B., Kosobud, A., Belknap, J.K., and Laursen, S.E. (1986) Genetic differences in anticonvulsant sensitivity in mouse lines selectively bred for ethanol withdrawal severity. *J. Pharmacol. Exp. Therap.,* **239**(1), 154–159.

Deitrich, R.A. and Spuhler, K. (1984) Genetics of alcoholism and alcohol actions. In *Research Advances in Alcohol and Drug Problems, Vol. 8* (Eds: R.G. Smart, H.D. Cappell, F.B. Glazer, Y. Israel, H. Kalant, R. Popham, W. Schmidt, and E.M. Sellers), 47–98. Plenum Press, New York.

Devor, E.J. and Cloninger, C.R. (1989) Genetics of Alcoholism. *Ann. Rev. Genet.,* **23,** 19–36.

Ehlers, C.L. and Chaplin, R.I. (1991) EEG and ERP response to chronic ethanol exposure in rats. *Psychopharmacology,* **104,** 67–74.

Ellis, F.W. and Pick, J.R. (1970) Experimentally induced ethanol dependence in rhesus monkeys. *J. Pharmacol. Exp. Therap.,* **175**(1), 88–93.

Emmett-Oglesby, M.W., Mathis, D.A., Moon, R.T.Y., and Lal, H. (1990) Animal models of drug withdrawal symptoms. *Psychopharmacology,* **101,** 292–309.

Essig, C.F. and Lam, R.L. (1968) Convulsions and hallucinatory behavior following alcohol withdrawal in the dog. *Arch. Neurol.,* **18,** 626–632.

Falconer, D.S. (1989) *Introduction to Quantitative Genetics* (Third Edition) Longman Scientific and Technical, New York.

Feller, D.J., Bassir, J.M., Crabbe, J.C., and LeFevre, C.A. (1994) Audiogenic seizure susceptibility in WSP and WSR mice. *Epilepsia,* 35(4), 861–867.

Feller, D.J., Harris, R.A., and Crabbe, J.C. (1988) Differences in GABA activity between ethanol withdrawal seizure prone and resistant mice. *Eur. J. Pharmacol.,* **157,** 147–154.

Feller, D.J., Tso-Olivas, D.Y., and Savage, D.D. (1990) Hippocampal mossy fiber zinc deficit in mice genetically selected for ethanol withdrawal seizure susceptibility. *Brain Res.,* **545,** 73–79.

Finn, D.A., Roberts, A.J., and Crabbe, J.C. (1995) Neuroactive steroid sensitivity in Withdrawal Seizure Prone and Resistant mice. *Alcoholism: Clin. Exp. Res.,* 19, 410–415.

Freund, G. (1969) Alcohol withdrawal syndrome in mice. *Archives of Neurology,* **21**(September), 315–320.

Friedman, H.J. (1980) Assessment of physical dependence on and withdrawal from ethanol in animals. In *Alcohol Tolerance and Dependence* (Eds: H. Rigter and J.C. Crabbe, Jr.), 93–121. Elsevier/North-Holland Biomedical Press, Amsterdam.

Frye, G.D., McCown, T.J., and Breese, G.R. (1983) Characterization of susceptibility to audiogenic seizures in ethanol-dependent rats after microinjection of γ-aminobutyric acid (GABA) agonists into the inferior colliculus, substantia nigra or medial septum. *J. Pharmacol. Exp. Therap.,* **227,** 663–670.

Gallaher, E.J., Belknap, J.K., Jones, G., Cross, S.J., and Crabbe, J.C. Genetic determinants of sensitivity to diazepam in inbred mice. *Psychopharmacology,* (accepted pending revision).

Goldstein, D.B. (1972a) Relationship of alcohol dose to intensity of withdrawal signs in mice. *J. Pharmacol. Exp. Therap.,* **180,** 203–215.

Goldstein, D.B. (1972b) An animal model for testing effects of drugs on alcohol withdrawal reactions. *J. Pharmacol. Exp. Therap.,* **183,** 14–22.

Goldstein, D.B. (1973a) Inherited differences in intensity of alcohol withdrawal reactions in mice. *Nature,* **245,** 154–156.

Goldstein, D.B. (1973b) Alcohol withdrawal reactions in mice: effects of drugs that modify neurotransmission. *J. Pharmacol. Exp. Therap.,* **186**(1), 1–9.

Goldstein, D.B. (1974) Rates of onset and decay of alcohol physical dependence in mice. *J. Pharmacol. Exp. Therap.,* **190**(2), 377–383.

Goldstein, D.B. and Kakihana, R. (1974) Alcohol withdrawal reactions and reserpine effects in inbred strains of mice. *Life Sci.,* **15,** 415–425.

Goldstein, D.B. and Kakihana, R. (1975) Alcohol withdrawal reactions in mouse strains selectively bred for long or short sleep times. *Life Sci.,* **17,** 981–986.

Goldstein, D.B. and Pal, N. (1971) Alcohol dependence produced in mice by inhalation of ethanol: grading the withdrawal reaction. *Science*, **172,** 288–290.

Grant, K.A., Valverius, P., Hudspith, M., and Tabakoff, B. (1990) Ethanol withdrawal seizures and the NMDA receptor complex. *Eur. J. Pharmacol.,* **176,** 289–296.

Grieve, S.J., Griffiths, P.J., and Littleton, J.M. (1979) Genetic influences on the rate of development of ethanol tolerance and the ethanol physical withdrawal syndrome in mice. *Drug Alcohol Dependence,* **4,** 77–86.

Griffiths, P.J. and Littleton, J.M. (1977) Concentrations of free amino acids in brains of mice of different strains during the physical syndrome of withdrawal from ethanol. *Br. J. Exp. Pathol.,* **58,** 391–399.

Guerrero-Figueroa, R., Rye, M.M., Gallant, D.M., and Bishop, M.P. (1970) Electrographic and behavioral effects of diazepam during alcohol withdrawal state in cats. *Neuropharmacology*, **9,** 143–150.

Gulya, K., Grant, K.A., Valverius, P., Hoffman, P.L., and Tabakoff, B. (1991) Brain regional specificity and time-course of changes in the NMDA receptor-ionophore complex during ethanol withdrawal. *Brain Res.,* **547,** 129–134.

Harris, R.A., Crabbe, J.C., and McSwigan, J.D. (1984) Relationship of membrane physical properties to alcohol dependence in mice selected for genetic differences in alcohol withdrawal. *Life Sci.,* **35,** 2601–2608.

Hoffman, P.L. and Tabakoff, B. (1994) The role of the NMDA receptor in ethanol withdrawal. In *Toward a Molecular Basis of Alcohol Use and Abuse* (Eds: B. Jansson, H. Jörnvall, U. Rydberg, L. Terenius, and B.L. Vallee), 61–70. Birkhäuser Verlag Basel, Switzerland.

Hoffman, P.L., Rabe, C.S., Grant, K.A., Valverius, P., Hudspith, M., and Tabakoff, B. (1990) Ethanol and the NMDA receptor. *Alcohol,* **7,** 229–231.

Hunter, B.E. and Walker, D.W. (1978) Ethanol dependence in the rat: role of extrapyramidal motor systems in the withdrawal reaction. *Exp. Neurol.,* **62,** 374–393.

Hunter, B.E., Boast, C.A., Walker, D.W., Zornetzer, W.F., and Riley, J.N. (1978) Ethanol dependence in the rat: role of non-specific and limbic regions in the withdrawal reaction. *Electroencephalogr. Clin. Neurophysiol.,* **45,** 483–495.

Hutchins, J.B., Allen, D.L., Cole-Harding, L.S., and Wilson, J.R. (1981) Behavioral and physiological measures for studying ethanol dependence in mice. *Pharmacol. Biochem. Behav.,* **15,** 55–59.

Isbell, H., Fraser, H.F., Wikler, A., Belleville, M.A., and Eisenman, A.J. (1955) An experimental study of the etiology of "rum fits" and delirium tremens. *Q. J. Stud. Alcohol.,* **16,** 1–33.

Jaffe, J.H. (1985) Drug addiction and drug abuse. In *Goodman and Gilman's The Pharmacological Basis of Therapeutics* (Seventh Edition) (Eds: A.G. Gilman, L.S. Goodman, T.W. Rall, and F. Murad), 532–581. Macmillan Publishing Company, New York.

Jaffe, J.H. and Ciraulo, D.A. (1985) Drugs used in the treatment of alcoholism. In *The Diagnosis and Treatment of Alcoholism* (Eds: J.H. Mendelson and N.K. Mello), 355–389. McGraw-Hill Book Company, New York.

Kalant, H. (1977) Alcohol withdrawal syndromes in the human: Comparison with animal models. In *Alcohol Intoxication and Withdrawal-IIIb* (Ed: M. Gross), 57–64. Plenum Press, New York.

Kalant, H., LeBlanc, A.E., and Gibbins, R.J. (1971) Tolerance to, and dependence on, some non-opiate psychotropic drugs. *Pharmacol. Rev.,* 23(3), 135–191.

Keir, W.J. and Morrow, A.L. (1994) Differential expression of GABAA receptor subunit mRNAs in ethanol-naive Withdrawal Seizure Resistant (WSR) vs. Withdrawal Seizure Prone (WSP) mouse brain. *Mol. Brain Res.,* 25(3–4), 200–208.

Kosobud, A. and Crabbe, J.C. (1986) Ethanol withdrawal in mice bred to be genetically prone or resistant to ethanol withdrawal seizures. *J. Pharmacol. Exp. Therap.,* **238**(1), 170–177.

Kosobud, A.E. and Crabbe, J.C. (1993) Sensitivity to N-methyl-D-aspartic acid-induced convulsions is genetically associated with resistance to ethanol withdrawal seizures. *Brain Res.,* **610,** 176–179.

Kosobud, A.E. and Crabbe, J.C. (1995) Genetic influences on the development of alcohol physical dependence and withdrawal. In *The Genetics of Alcoholism* (Eds: H. Begleiter and B. Kissin), 221–256, Oxford University Press, Oxford, United Kingdom (in press).

Kosobud, A., Bodor, A.S., and Crabbe, J.C. (1988) Voluntary consumption of ethanol in WSP, WSC and WSR selectively bred mouse lines. *Pharmacol. Biochem. Behav.,* **29,** 601–607.

Kosobud, A.E., Cross, S.J., and Crabbe, J.C. (1992) Neural sensitivity to pentylenetetrazol convulsions in inbred and selectively bred mice. *Brain Res.,* **592,** 122–128.

Lal, H., Harris, C.M., Benjamin, D., Springfield, A.C., Bhadra, S., and Emmett-Oglesby, M.W. (1988) Characterization of a pentylenetetrazol-like interoceptive stimulus produced by ethanol withdrawal. *J. Pharmacol. Exp. Therap.,* **247,** 508–518.

Lal, H., Prather, P.L., and Rezazadeh, S.M. (1991) Anxiogenic behavior in rats during acute and protracted ethanol withdrawal: reversal by buspirone. *Alcohol,* **8,** 467–471.

Lambert, J.J., Peters, J.A., and Cottrell, G.A. (1987) Actions of synthetic and endogenous steroids on the GABAA receptor. *Trends Pharmacol. Sci.,* **8,** 224–227.

Lander, E.S. and Botstein, D. (1989) Mapping Mendelian factors underlying quantitative traits using RFLP linkage maps. *Genetics*, **121,** 185–199.

Lechtenberg, R. and Worner, T.M. (1991) Relative kindling effect of detoxification and non-detoxification admissions in alcoholics. *Alcohol Alcoholism*, **26**(2), 221–225.

Lisitsyn, N.A., Segre, J.A., Kusumi, K., Lisitsyn, N.M., Nadeau, J.H., Frankel, W.N., Wigler, M.H., and Lander, E.S. (1994) Direct isolation of polymorphic markers linked to a trait by genetically directed representational difference analysis. *Nat. Genet.,* **6,** 57–63.

Litten, R.Z. and Allen, J.P. (1991) Pharmacotherapies for alcoholism: promising agents and clinical issues. *Alcoholism: Clin. Exp. Res.,* **15**(4), 620–633.

Little, H.J. (1991) The role of neuronal calcium channels in dependence on ethanol and other sedatives/hypnotics. *Pharmacol. Ther.,* **50,** 347–365.

Maier, D.M. and Pohorecky, L.A. (1989) The effect of repeated withdrawal episodes on subsequent withdrawal severity in ethanol-treated rats. *Drug Alcohol Dependence,* **23,** 103–110.

Majchrowicz, E. (1975) Induction of physical dependence upon ethanol and the associated behavioral changes in rats. *Psychopharmacologia*, **43,** 245–254.

Maxson, S.C. and Sze, P.Y. (1976) Electroencephalographic correlates of audiogenic seizures during ethanol withdrawal in mice. *Psychopharmacology*, **47,** 17–20.

McCaul, M.E., Turkkan, J.S., Svikis, D.S., and Bigelow, G.E. (1991) Alcohol and secobarbital effects as a function of familial alcoholism: extended intoxication and increased withdrawal effects. *Alcoholism: Clin. Exp. Res.,* **15**(1), 94–101.

McClearn, G.E. (1991) The tools of pharmacogenetics. In *The Genetic Basis of Alcohol and Drug Actions* (Eds: J.C. Crabbe and R.A. Harris), 1–23. Plenum Press, New York.

McClearn, G.E. and Kakihana, R. (1981) Selective breeding for ethanol sensitivity: Short-Sleep and Long-Sleep mice. In *Development of Animal Models as Pharmacogenetic Tools,* (Eds: G.E. McClearn, R.A. Deitrich, and V.G. Erwin), 147–159. USDHHS-NIAAA Research Monograph No. 6, Washington.

McClearn, G.E., Wilson, J.R., Petersen, D.R., and Allen, D.L. (1982) Selective breeding in mice for severity of the ethanol withdrawal syndrome. *Substance Alcohol Actions/Misuse,* **3,** 135–143.

McCown, T.J. and Breese, G.R. (1993) A potential contribution to ethanol withdrawal kindling: reduced GABA function in the inferior collicular cortex. *Alcoholism: Clin. Exp. Res.,* **17**(6), 1290–1294.

McIntyre, T.D. and Alpern, H.P. (1989) Patterns of convulsive susceptibility in the long-sleep and short-sleep selected mouse lines. *Brain Res. Bull.,* **22**(5), 859–865.

McQuarrie, D.G. and Fingl, E. (1958) Effects of single doses and chronic administration of ethanol on experimental seizures in mice. *J. Pharmacol. Exp. Therap.,* **124,** 264–271.

McSwigan, J.D., Crabbe, J.C., and Young, E.R. (1984) Specific ethanol withdrawal seizures in genetically selected mice. *Life Sci.,* **35,** 2119–2126.

Metten, P. and Crabbe, J.C. (1994) Common genetic determinants of severity of acute withdrawal from ethanol, pentobarbital, and diazepam in inbred mice. *Behav. Pharmacol.,* **5**(4), 533–547.

Morse, H.C., III (1978) Introduction. In *Origins of Inbred Mice* (Ed: H.C. Morse, III), 3–21. Academic Press, New York.

Newlin, D.B. and Pretorius, M.B. (1990) Sons of alcoholics report greater hangover symptoms than sons of nonalcoholics: a pilot study. *Alcoholism: Clin. Exp. Res.,* **14**(5), 713–716.

Paul, S.M. and Purdy, R.H. (1992) Neuroactive steroids. *FASEB J.,* **6,** 2311–2322.

Phillips, T.J. and Crabbe, J.C., Jr. (1991) Behavioral studies of genetic differences in alcohol action. In *The Genetic Basis of Alcohol and Drug Actions* (Eds: J.C. Crabbe and R.A. Harris), 25–104. Plenum Press, New York.

Phillips, T.J., Feller, D.J., and Crabbe, J.C. (1989a) Selected mouse lines, alcohol and behavior. *Experientia*, **45,** 805–827.

Phillips, T.J., Kim, D., and Dudek, B.C. (1989b) Convulsant properties of GABA antagonists and anticonvulsant properties of ethanol in selectively bred Long- and Short-Sleep mice. *Psychopharmacology*, **98,** 544–548.

Pieper, W.A., Skeen, M.J., McClure, H.M., and Bourne, P.G. (1972) The chimpanzee as an animal model for investigating alcoholism. *Science*, **176,** 71–73.

Plomin, R. and McClearn, G.E. (1993) Quantitative trait loci (QTL) analysis and alcohol-related behaviors. *Behav. Genet.,* **23,** 197–212.

Plomin, R., McClearn, G.E., Gora-Maslak, G., and Neiderhiser, J. (1991) Use of recombinant inbred strains to detect quantitative trait loci associated with behavior. *Behav. Genet.,* **21,** 99–116.

Pohorecky, L.A. and Roberts, P. (1991) Development of tolerance to and physical dependence on ethanol: daily versus repeated cycles treatment with ethanol. *Alcoholism: Clin. Exp. Res.,* **15**(5), 824–833.

Roberts, A.J., Chu, H.-P., Crabbe, J.C., and Keith, L.D. (1991) Differential modulation by the stress axis of ethanol withdrawal seizure expression in WSP and WSR mice. *Alcoholism: Clin. Exp. Res.,* **15**(3), 412–417.

Roberts, A.J., Crabbe, J.C., and Keith, L.D. (1992) Genetic differences in hypothalamic-pituitary-adrenal axis responsiveness to acute ethanol and acute ethanol withdrawal. *Brain Res.,* **579,** 296–302.

Roberts, A.J., Crabbe, J.C., and Keith, L.D. (1994) Corticosterone increases severity of acute withdrawal from ethanol, pentobarbital, and diazepam in mice. *Psychopharmacology*, **115,** 278–284.

Roivainen, R., Hundle, B., and Messing, R.O. (1994) Protein kinase C and adaptation to ethanol. In *Toward a Molecular Basis of Alcohol Use and Abuse* (Eds: B. Jansson, H. Jörnvall, U. Rydberg, L. Terenius, and B.L. Vallee), 29–38. Birkhäuser Verlag Basel, Switzerland.

Schuckit, M.A. (1994) Low level of response to alcohol as a predictor of future alcoholism. *Am. J. Psychiatry,* **151,** 184–189.

Schuckit, M.A., Li, T.K., Cloninger, C.R., and Deitrich, R.A. (1985) University of California, Davis - Conference: Genetics of alcoholism. *Alcoholism: Clin. Exp. Res.,* **9**(6), 475–492.

Silver, L.M., Nadeau, J.H., and Goodfellow, P.N. (1993) Encyclopedia of the mouse genome III. *Mammalian Genome,* **4,** S1–S283.

Snell, L.D., Tabakoff, B., and Hoffman, P.L. (1994) Involvement of protein kinase C in ethanol-induced inhibition of NMDA receptor function in cerebellar granule cells. *Alcoholism: Clin. Exp. Res.,* **18**(1), 81–85.

Swift, R.M. (1994) Alcoholism and substance abuse. In *Clinical Psychiatry for Medical Students,* Second Edition (Ed: A. Stoudemire), 306–337. J. B. Lippincott Company, Philadelphia.

Tabakoff, B. and Hoffman, P.L. (1993) Ethanol, sedative hypnotics, and glutamate receptor function in brain and cultured cells. *Behav. Genet.,* **23**(2) 231–236.

Tanksley, S.D. (1993) Mapping polygenes. *Ann. Rev. Genet.,* **27,** 205–233.

Taylor, B.A. (1978) Recombinant inbred strains: Use in gene mapping. In *Origins of Inbred Mice* (Ed: H.C. Morse, III), pp 423–438, Academic Press, New York.

Terdal, E.S. and Crabbe, J.C. (1994) Indexing withdrawal in mice: matching genotypes for exposure in studies using ethanol vapor inhalation. *Alcoholism: Clin. Exp. Res.,* **18**(3), 542–547.

Thomasson, H.R., Crabb, D.W., Edenberg, H.J., and Li, T.-K. (1993) Alcohol and aldehyde dehydrogenase polymorphisms and alcoholism. *Behav. Genet.,* **23**(2), 131–136.

Ulrichsen, J., Clemmesen, L., and Hemmingsen, R. (1992) Convulsive behaviour during alcohol dependence: discrimination between the role of intoxication and withdrawal. *Psychopharmacology,* **107,** 97–102.

Valverius, P., Crabbe, J.C., Hoffman, P.L., and Tabakoff, B. (1990) NMDA receptors in mice bred to be prone or resistant to ethanol withdrawal seizures. *Eur. J. Pharmacol.,* **184,** 185–189.

Victor, M. and Adams, R.D. (1953) The effect of alcohol on the nervous system. *Assoc. Res. Nerv. Ment. Dis.,* **32,** 526–573.

Walker, D.W. and Zornetzer, S.F. (1974) Alcohol withdrawal in mice: electroencephalographic and behavioral correlates. *Electroencephalogr. Clin. Neurophysiol.,* **36,** 233–243.

Waller, M.B., McBride, W.J., Lumeng, L., and Li, T.-K. (1982) Induction of dependence on ethanol by free-choice drinking in alcohol-preferring rats. *Pharmacol. Biochem. Behav.,* **16,** 501–507.

Wilson, J.R., Erwin, V.G., DeFries, J.C., Petersen, D.R., and Cole-Harding, S. (1984) Ethanol dependence in mice: direct and correlated responses to ten generations of selective breeding. *Behav. Genet.,* **14**(3), 235–256.

Yu, S. and Ho, I.K. (1990) Effects of acute barbiturate administration, tolerance and dependence on brain GABA system: comparison to alcohol and benzodiazepines. *Alcohol,* **7,** 261–272.

Chapter 17

ETHANOL'S ACUTE EFFECTS ON THERMOREGULATION: MECHANISMS AND CONSEQUENCES

R.L. Alkana, D.L. Davies, and A.D. Lê

TABLE OF CONTENTS

0-8493-8389-7/95/$0.00+$.50

I. INTRODUCTION

This chapter describes the effects of acute ethanol on thermoregulation. It focuses on the mechanisms of ethanol-induced changes in thermoregulation and the consequences of these changes on ethanol sensitivity. The primary emphasis is on studies in laboratory animals. The chapter also presents variables that may influence the extent and measurement of ethanol-induced changes in body temperature. The effects of chronic ethanol on thermoregulation, particularly the development of acute and rapid tolerance to ethanol's hypothermic effects, are covered in Chapter 15. Similarly, the effects of ethanol exposure during the fetal and early postnatal period on development and adult thermoregulation and ethanol sensitivity are presented in Chapter 22. A number of excellent reviews of ethanol and thermoregulation are available (Freund, 1979; Lomax and Schönbaum, 1979, 1991; Crawshaw et al., 1992; Kalant and Lê, 1984, 1991; Pohorecky and Brick, 1988). Therefore, the present chapter concentrates on newer material and is not meant to be inclusive.

II. EFFECTS OF ETHANOL ON THERMOREGULATION

Ethanol alters thermoregulation in humans and laboratory animals. The effects on laboratory animals are more pronounced than on humans due to differences in size and thermoregulatory capacity.

A. HUMANS

1. Forensic

Forensic studies have suggested a link between ethanol consumption and an impairment of thermoregulation. High ethanol concentrations have been shown during post autopsy investigations of fatalities to be present in a majority of subjects due to hypothermia or accidental hypothermic death (Hirvonen, 1976; Albiin and Eriksson, 1984; Kortelainen, 1987). In 51 cases of accidental hypothermic death, high blood ethanol concentrations (with a mean average of 160 mg/dl) were observed in 33 cases (Albiin and Eriksson, 1984). A link between hyperthermic death and ethanol consumption has also been implicated. Brown (1981) reported that 12 of 30 residential hot tub deaths in the United States were related to ethanol consumption. Similarly, "sauna deaths" in Finland are often connected with ethanol consumption (Hirvonen and Huttunen, 1982; Kortelainen, 1987). In 26 cases of hyperthermic death, ethanol was present in most of the cases with an average blood alcohol concentration (BAC) of 140 mg/dl (Kortelainen, 1987). These studies suggest some linkage between high ethanol consumption and hypothermic or hyperthermic death. However, they do not permit any conclusion about a role of ethanol in the production of hypothermia or hyperthermia or any causal relationship between the effects of ethanol on thermoregulation and related deaths.

2. Experimental

The initial observations that ethanol ingestion can lower body temperature in humans at normal room temperature was documented as early as 1852 (Lichtenfels and Frohlich, 1852). Since then, a number of experimental studies have been conducted to investigate the effects of ethanol on thermoregulation in humans at various ambient temperatures (T_as).

At normal room T_a, the majority of studies have reported that consumption of ethanol in doses ranging from 0.3–1 g/kg causes a modest decrease in oral or rectal temperatures (Lichtenfels and Frohlich, 1852; Mullin et al., 1933; Risbo et al., 1981; Fellows et al., 1984). An absence of changes (Andersen et al., 1963; Martin and Cooper, 1964) or an increase (Marbach and Schwertz, 1964), however, has also been reported following consumption of similar doses of ethanol. Exposure to a cold T_a (–2°C) or to cold water (7.5–13°C) following ingestion of small doses of ethanol (0.3–0.5 g/kg) has also been shown to have no effect (Andersen et al., 1963; Risbo et al., 1981; Keatinge and Evans, 1960; Martin et al., 1977; Fox et al., 1979) or to cause only a small decrease in rectal temperature (Gupta, 1960; Graham and Baulk, 1980). However, when additional metabolic load, such as by exercising in the cold, is imposed on thermoregulatory mechanisms, ethanol has been shown to produce a small but consistent reduction in rectal or esophageal temperature (Haight and Keatinge, 1973; Fox et al., 1979; Graham, 1981a,b; 1983). No significant changes in body temperature were observed in subjects receiving ethanol (0.25–0.7 g/kg) and exposed to heat stress either by immersion in warm water of 30–40°C (Martin and Cooper, 1978; Allison and Reger, 1992) or warm air of 30°C (Fellows et al., 1984).

The results of these various studies showed that consumption of ethanol, in doses ranging from 0.25–1.2 g/kg, did not produce consistent effects on body temperature in human subjects at normal and low T_a. A modest but consistent reduction in temperature, however, was observed when the effects of ethanol were assessed in subjects exposed to strenuous exercise at a low T_a. The failure to observe consistent or significant changes in body temperature in human subjects might be due to the doses of ethanol used. The doses of ethanol used in these studies were quite moderate, and it is unlikely that blood ethanol concentration would ever exceed 100 mg/dl in the majority of these studies following oral administration. This is quite different from those forensic studies in which postmortem average blood ethanol concentrations were 160 (Albiin and Eriksson, 1984) or 140–150 mg/dl in hypo- and hyperthermic death. With a lower rate of absorption from oral administration, it is also likely that acute tolerance would develop quite rapidly and could mask the effect of ethanol on body temperature.

Studies with experimental animals have shown that the rate of absorption plays a critical role in determining the manifestation and/or degree of hypothermia (Gilliam, 1989). As discussed later, relatively high ethanol doses are required to modify temperature in experimental animals, even though they have a higher rate of heat loss and a larger surface:mass ratio than humans.

B. LABORATORY ANIMALS

Ethanol produces a consistent hypothermic effect at normal and low T_as in various species such as mice, rats, guinea pigs, rabbits, cats, dogs, and monkeys (see reviews by Kalant and Lê 1984, 1991 for complete citations). Mice and rats have been the most common subjects used for studying the effects of ethanol on thermoregulation.

At normal T_a (20–24°C), ethanol has been shown to produce a dose-dependent reduction in body temperature in mice (Freund, 1973; Crabbe et al., 1979) and rats (Lomax et al., 1980; Lê et al., 1981). Administration of ethanol by i.p. (Crabbe et al., 1979; Malcolm and Alkana, 1981; Lê et al., 1981), oral (Myers, 1981), or i.c.v. (Ritzmann and Tabakoff, 1976; Chai et al., 1971) routes have been shown to produce hypothermia in experimental rodents. The reduction in rectal temperature induced by ethanol is dose dependent. Ethanol given i.p. in doses ranging from 2–4 g/kg has been shown to reduce body temperature by 1.25–2.75°C in mice and rats (Crabbe et al., 1979; Lê et al., 1981). The maximum reduction in rectal temperature usually occurs at about 30–60 min after such administration (Crabbe et al., 1979; Malcolm and Alkana, 1981; Lê et al., 1981).

There is little information concerning the threshold or minimum dose of ethanol required to produce hypothermia. Most studies generally employed the hypothermic effect of ethanol as a dependent variable to evaluate ethanol effects and, therefore, commonly employed high doses of ethanol to produce a substantial degree of hypothermia. Lomax et al. (1980) examined the effects of various doses of ethanol ranging from 0.25–3.2 g/kg on temperature in the rat. They found that the minimum effective dose of ethanol required to lower rectal temperature was 0.5 g/kg. It should be mentioned that the drop in rectal temperature observed in this study (0.5°C) is comparable to that produced by doses of 1.0–1.2 g/kg of ethanol (i.p.) observed in other studies (Linakis and Cunningham, 1979; Mansfield and Cunningham, 1980). It is possible that a reduction in temperature such as that observed in the Lomax et al. study might be related to a lower T_a (18°C).

The effects of ethanol on body temperature have been shown to be dependent on the T_a. A low T_a enhances the hypothermic effect of ethanol, whereas a high T_a reduces or antagonizes such effects (Freund, 1973; Malcolm and Alkana, 1981; Myers, 1981; Lomax and Lee, 1982; Lê et al., 1986; Bejanian et al., 1987; Hirvonen and Huttunen, 1977; Huttunen et al., 1980). Lomax and Lee (1982) demonstrated that the drop in temperature induced by a dose of 1 g/kg in the rat is a logarithmic function of the T_a over the range of 0–18°C. Such doses of ethanol produce a fall in rectal temperature of 1.1°C and 0.6°C at T_as of 0°C and 18°C, respectively (Lomax et al., 1981). Such influences of T_a on ethanol-induced changes in body temperature have also been observed with higher doses of ethanol. A dose of 3.6 g/kg produces a 3°C drop in rectal temperature at a T_a of 24°C, but causes a reduction of 11°C at a T_a of 12°C (Malcolm and Alkana, 1981). Whether a low T_a can modify the threshold dose of ethanol required to produce hypothermia is not known. As mentioned above, Lomax et al. reported that an ethanol dose of 0.5 g/kg (i.p.) is the minimum effective dose to produce a significant drop in temperature at a T_a of 18°C. However, in their subsequent study (Lomax and Lee, 1982), similar doses of ethanol produced no significant change in temperature at a T_a of 4°C.

At a T_a higher than normal room temperature, the hypothermic effects induced by various doses of ethanol have been shown to be antagonized in mice (Ritzmann and Tabakoff, 1976; Malcolm and Alkana, 1981; Gordon and Stead, 1986; Bejanian et al., 1987) and rats (Myers, 1981; Pohorecky and Rizek, 1981; Gordon and Mohler, 1990). For example, Bejanian et al. (1987) showed that both brain and rectal temperatures of mice treated with various doses of ethanol were reduced and antagonized, respectively, at a T_a of 32°C and 35°C. At a T_a of 37–42°C, ethanol produced hyperthermia. Myers (1981) reported an increase in rectal temperature of 1°C in rats following oral administration of 4 g/kg of ethanol at T_a of 42°C. Similarly, an increase of 1–2°C in mice has been shown following the i.p. administration of 3.6 g/kg of ethanol in mice exposed to a T_a of 37°C (Malcolm and Alkana, 1981).

The available data, however, suggest that the hyperthermic effect of ethanol occurs only after administration of high doses of ethanol and at a T_a of 37°C or higher. In rats gavaged with 0, 2, 4, 6, and 8 g/kg of ethanol and exposed to a T_a of 37°C, hyperthermia was observed only in rats treated with doses of 4 g/kg or higher (Gordon and Mohler, 1990). However, there was no dose-related hyperthermic effects at these high doses of ethanol since the degree of hyperthermia was also essentially the same among those treated with 4, 6, or 8 g/kg, suggesting that the hypothermia may reflect impairment of behaviors by high-ethanol doses used for heat dissipation (see Section III.B. below). In another study with mice, doses of 0.3 and 1 g/kg given i.p. had no apparent effect on the heating rate of body temperature in mice placed in a T_a of 55°C (Gordon and Stead, 1988). Thus, it appears that, independent of the T_a employed, the dose of ethanol required to produce hypothermia is lower than those required to produce hyperthermia.

C. VARIABLES

The quantitative and qualitative effects of ethanol on thermoregulation and body temperature can be influenced by a number of factors. These include ethanol dose, ambient temperature (T_a), measurement techniques, timing of temperature measurement, stress and related factors, circadian rhythms, age, and genetic background. Failure to adequately control these variables can complicate the interpretation of individual studies or comparisons across studies that investigate ethanol's effects on thermoregulation or that use ethanol's effects on thermoregulation as a dependent variable for measuring ethanol sensitivity.

1. Dose and Ambient Temperature

The dose of ethanol and the T_a to which animals are exposed during intoxication can have marked effects on the manner in which ethanol alters thermoregulation as well as on the magnitude and direction of the thermic response to ethanol. Therefore, these factors represent important variables in studies investigating the effects of ethanol on thermoregulation. The roles of ethanol dose and T_a were discussed in detail in other sections of this chapter in the context of presenting the effects of ethanol on thermoregulation. Briefly, ethanol causes dose-dependent (Freund, 1973; Crabbe et al., 1979; Lomax et al., 1980; Lê et al., 1981; Myers, 1981; Finn et al., 1994; Lomax and Bajorek, 1980), rate-of-absorption-dependent (Gilliam, 1989), and concentration-dependent (Linakis and Cunningham, 1979) changes in thermoregulation that appear to result from both a lowering of the thermoregulatory set-point (behavioral thermoregulation) and an impairment of thermoregulatory ability (Kalant and Lê, 1991; Crawshaw et al., 1992). Impaired thermoregulation makes intoxicated animals behave somewhat like a poikilotherm, because their body temperatures can be manipulated by the T_a during intoxication (Freund, 1973; Myers, 1981; Malcolm and Alkana, 1981; Tabakoff et al., 1975). Collectively, the dose-response and T_a studies indicate that higher doses of ethanol are necessary to produce hyperthermia than are required to produce hypothermia. Although less well studied, available evidence suggests that ethanol-induced impairment of thermoregulation and change in set-point may also each have different ethanol concentration thresholds, dose-response curves, and time courses (O'Connor et al., 1989b). Therefore, when designing and comparing studies of ethanol's effects on thermoregulation, the individual and interactive effects of dose and T_a must be carefully considered.

2. Measurement Technique

Studies investigating the effects of ethanol on body temperature have used a variety of techniques to estimate core and brain temperatures during intoxication. These span from rectal probing (Lomax, 1966), skin temperature measurements with thermisters (Lomax et al., 1981), and infrared radiation detection (Brick and Pohorecky, 1983a) to surgically implanted miniature radiotelemetric probes (Cunningham and Peris, 1983; Gallaher and Egner, 1985; Bejanian et al., 1991b). However, the majority of studies of ethanol-induced changes in body temperature use rectal or colonic core temperature as the dependent measure.

Several technical variables can influence body temperature measured by rectal probes. Lomax (1966) demonstrated that the depth of probe insertion is important. He showed that rectal temperatures can provide reliable information if the probe is properly placed, but that measurements can be highly variable and artificially low if the probe is not inserted sufficiently deep. This work indicates that the minimum insertion depth for accurate and consistent body temperature data is 6 cm for rats, cats, and rabbits and 3 cm for guinea pigs. In mice, insertion depths greater than 1.9 cm provide consistent core temperatures (Malcolm and Alkana, 1981; Alkana et al., 1985a; Bejanian et al., 1991b; Melchior and Allen, 1993).

The duration of probe insertion can also influence body temperature. In experiments that involve a single acute body temperature measurement, the duration is controlled by the time constant of the probe (i.e., the time it takes the thermister to reach equilibrium) and can vary from 5 (Alkana et al., 1985a) to 40 s (Freund, 1973; York and Chan, 1993) with typically used temperature telemetry systems. Comparisons across studies in C57 mice exposed to typical laboratory temperatures suggest that i.p. injection of 3.6 g/kg ethanol induced 2–3°C greater hypothermia when measurements required 30 s (Malcolm and Alkana, 1981) compared to 5 s (Alkana et al., 1985a). As discussed in Section 3 later, the greater hypothermia accompanying the longer probe duration in these acute studies likely reflects the effects of restraint. Similarly, restraint associated with the use of rectal probes for continuous measurement of body temperature during time course studies can affect body temperature (Poole and Stephenson, 1977) and may influence the magnitude of ethanol's effect. Nevertheless, comparisons between similarly treated saline- and ethanol-injected mice and rats (Malcolm and Alkana, 1981; Myers, 1981) indicate that the qualitative effects of ethanol and the interaction with T_a in intoxicated mice cannot be explained by restraint in continuously probed animals (Malcolm and Alkana, 1981).

Overall, the evidence indicates that rectal temperature can provide useful information regarding the effects of ethanol on body temperature. However, differences in technique and equipment can alter the magnitude of ethanol's effect and complicate comparisons within and between studies.

Much of the technical variability associated with the rectal probe technique can be avoided by using miniature radiotelemetric surgical implants (Cunningham and Peris, 1983; Gallaher and Egner, 1985; Melchior and Allen, 1993). These systems transmit a pulse-modulated signal at a rate that is proportional to the local temperature. Radiotelemetry provides an accurate automated method of continuously assessing baseline temperature and ethanol's effects on body temperature in single or multiple test sessions without the need for restraint or for disturbing the animals. However, this technique requires surgery under anesthesia, a recovery period, and more extensive equipment than that needed for assessing temperature via rectal probes.

3. Brain versus Core Temperature

The brain in homeotherms is protected against heat stress by independent cooling mechanisms in nonintoxicated animals (Baker, 1982). This is achieved through a heat exchange process at the base of the brain between the arterial blood and the cooler venous blood coming from the upper respiratory mucosa (Baker, 1982). These thermoregulatory control systems presumably allow brain tissue, which is especially vulnerable to damage by heat (Burger and Fuhrman, 1964; Bowler and Tirri, 1974), to be kept within an optimal temperature range in the face of large body heat loads without expending energy and water to cool the bulk of the body (Bamford and Eccles, 1983). Further, several investigators report regional variation in brain temperature (Donhoffer et al., 1959; Lomax et al., 1964), time lags between rectal and brain temperatures under conditions of change (Nadel, 1977), and consistent differences between rectal and brain temperatures in animals experiencing heat stress (Taylor and Lyman, 1972; Baker, 1979). Collectively, these findings demonstrate that brain temperature can be controlled independently from deep core temperature and suggest that rectal and other core temperature measurements might give a false impression of brain temperature.

A few studies have investigated the relationship between brain and core temperatures in saline- and ethanol-injected animals exposed to different T_as after injection (Bejanian et al., 1987, 1991b; Finn et al., 1991a). The findings indicate that brain (third ventricle) and rectal temperature are highly correlated in saline- and ethanol-injected (3.2–4.0 g/kg) C57 mice exposed to ambients from 15–34°C for 30–120 min after injections (Bejanian et al., 1987). Brain temperatures were consistently lower than rectal temperatures in saline animals, regardless of the T_a. Ambient temperature challenge did not significantly affect brain or rectal temperature in saline controls. Ethanol significantly reduced brain and rectal temperature in mice exposed to 15°C and 22°C, but not in mice exposed to 34°C. Further, although correlated, brain and rectal temperatures did not parallel each other in temperature-challenged animals. Brain temperatures were lower than rectal temperatures in intoxicated mice exposed to 15°C and 22°C, but not in intoxicated mice exposed to 34°C.

Subsequent studies in C57 mice using radiotelemetric needle brain probes implanted in the lateral ventricle confirmed the differential between brain and rectal temperatures in controls and the high correlations between brain and rectal temperatures (Bejanian et al., 1991b). However, in contrast to the previous study (Bejanian et al., 1987), this work found that ethanol (3.6 g/kg) eliminated the temperature differential between brain and rectal temperatures and suggested that ethanol impairs the thermoregulatory system that maintains the differential between brain and core temperatures. The conflicting findings between these studies could result from technical differences. Additional studies, using the radiotelemetric needle brain probe technique in LS and SS mice, indicated that ethanol does not completely eliminate the differential between brain and rectal temperatures during intoxication in LS mice and that there are genotypic differences between mice in their abilities to differentially control brain and rectal temperature in the presence and absence of ethanol (Finn et al., 1991a).

Collectively, the consistent correlation between brain and rectal temperatures in control and ethanol mice exposed to different ambients indicates that rectal temperature can be used to estimate the qualitative and quantitative effects of ethanol on brain temperature. On the other hand, rectal temperature does not appear to provide an accurate estimate of brain temperature per se.

4. Time Post-Ethanol

The qualitative and quantitative effects of ethanol on core temperature and selected temperature — used as a measure of thermoregulatory set-point — depends on the timing of the measurement with respect to ethanol or saline administration (Lomax et al., 1980, 1981; O'Connor et al., 1989b). The time course and shape of the temperature response curve is also influenced by ethanol dose and T_a (Freund, 1973; Myers, 1981; Malcolm and Alkana, 1983, 1981, 1982; Finn et al., 1994; Maickel and Nash, 1985; O'Connor et al., 1989b). These responses appear to be relatively consistent across species (Lomax and Bajorek, 1980) and sexes (Alkana et al., 1985a), but can show quantitative differences between genotypes (Crabbe, 1983; Crabbe et al., 1989; Finn et al., 1990). Following i.p. injection of subhypnotic or hypnotic ethanol doses, the maximum hypothermia measured by rectal probes typically occurs around 30–60 min post injection and can last for more than 2 h (Freund, 1973; Lomax et al., 1981; Malcolm and Alkana, 1981). Higher, potentially lethal, i.p. ethanol doses generally caused peak hypothermia between 2 and 8 h after administration (Malcolm and Alkana, 1983; Finn et al., 1989). When ethanol was administered to rats via intragastric intubation, peak hypothermia occured between 2 and 4 h after intubation and lasted longer than 5 h (Myers, 1981).

Studies that employ peritoneally implanted radiotelemetry probes give a somewhat different picture. For example, O'Connor et al. (1989b) found that an i.p. injection of 2.25 g/kg ethanol caused a small decline in the core temperature of mice and in the temperature selected by the mice in a thermal gradient. The maximum decline in core and selected temperatures occurred approximately 20 min after injection and returned to near baseline levels by 40 min. The effect was much more pronounced after 2.6 g/kg ethanol, which produced a maximum decline in core temperature at about 25 min after injection, whereas selected temperature declined to its peak at 10 min postinjection. Both temperatures returned to near normal levels by 50 min postinjeciton. This radiotelemetric study indicates that the time course for ethanol-induced impairment of thermoregulation may differ from the time course for ethanol-induced reduction in set-point. Other work suggests that a rebound hyperthermia occurs, beginning at about the time of complete ethanol elimination, and that this hyperthermia persists for several days (Gallaher and Egner, 1987).

Taken together, standard and radiotelemetric studies indicate that both rectal probe and telemetry techniques agree with respect to the time of peak ethanol-induced hypothermia and that the use of rectal probes might artifically increase the magnitude and prolong the duration of ethanol's hypothermic effect.

5. Stress

Ethanol and stress interact in a complex manner. As discussed by Peris and Cunningham (1986), ethanol and stress can be mutually antagonistic in some situations in that each can counteract some effects of the other (Pohorecky, 1981, 1985; Brick and Pohorecky, 1983b; Wallgren and Tirri, 1963; Leikola, 1962). Ethanol alone can act as a stressor, whereas in some situations the combined effects of ethanol and stress can be additive rather than antagonistic (Peris and Cunningham, 1986; Pohorecky, 1981; Leikola, 1962). This complex interaction holds for the effects of stress on ethanol-induced hypothermia.

Several stress-related factors, including restraint, handling, rectal probing, and housing conditions, can influence body temperature and ethanol's effects on body temperature. Early work with laboratory thermometers inserted rectally found that restraining rats (by restrictive caging) impaired their thermoregulatory ability (Bartlett et al., 1954). Impairment was noted by increased susceptibility to hypothermia during cold T_a exposure (2°C) and to hyperthermia during warm T_a exposure (40–59°C) (Frankel, 1959). In a classical study, Poole and Stephenson

(1977) used probes implanted in the chests of rats to show that insertion of a rectal probe for 1 min increased core temperature for about 70 min after the probe was removed and that the elevation remained if the rectal probe was left *in situ.* Based on these findings, the authors concluded that, although rectal probes provide a good index of core temperature, they are only suitable for a single determination of body temperature when the subsequent elevation is unimportant. Peris and Cunningham (1986) extended this work using miniature radiotelemetric probes inserted surgically in the peritoneal cavity of rats. They found that a mild stressor (handling) caused hyperthermia. More important, they surprisingly showed that, despite its hyperthermic effect alone, handling before injection significantly increased the hypothermic effect of 2.0 g/kg ethanol. Further, in contrast to the nonadditive effects of handling on ethanol hypothermia, handling caused an additive increase in ethanol-induced tachycardia. Additional work indicated that the differences between handled and unhandled animals in thermic response to ethanol could not be explained by changes in ethanol's pharmacokinetics (Peris and Cunningham, 1986) or by differences in heat conservation secondary to periodically disturbing the animals (Cunningham and Bischof, 1987). Additional testing revealed that not all stressors that induce hyperthermia by themselves interact with ethanol in the same manner. This work found that intermittent exposure to a bright flashing light enhanced ethanol-induced hypothermia in rats, but that electric shock did not (Cunningham and Bischof, 1987).

The handling involved in rectal temperature measurement also causes hyperthermia in mice (Cabanac and Briese, 1991). Moreover, Melchior and Allen (1993), using surgically implanted miniature radiotelemetric peritoneal probes, demonstrated that rectal probing exacerbated ethanol-induced hypothermia in a dose-dependent manner on both the falling and rising phases of ethanol-induced changes in body temperature. The interaction was small at doses of 1 and 2 g/kg ethanol and strong after 3.5 g/kg. They also showed that the effect of probing on body temperature can be observed when the mice are under the hypnotic influence of ethanol, and that the regain of the righting reflex itself is followed by a fall in body temperature, regardless of whether the animals were or were not disturbed for temperature measurements. Using these findings, the authors suggest that the enhanced hypothermia seen in mice following the manipulations used to assess the thermal and hypnotic effects of ethanol is due to a facilitation of heat loss secondary to disruption of the local thermal environment and to a disruption of heat conservation rather than to an interaction with stress. Interestingly, treatment with salicylates attenuated the hyperthermic effect of emotionally induced hyperthermia in rats (Kluger et al., 1987; Singer et al., 1986), but not in mice, suggesting that rats and mice differ in the mechanisms underlying stress-induced hyperthermia (Melchior and Allen, 1993).

To further complicate the picture, changing the housing environment also appears to act as a stressor that can influence sensitivity to ethanol-induced hypothermia. York and Regan (1982) using standard telemetric rectal probes, found that moving a group of naive rats from their usual positions on a cage rack to the laboratory bench for 60 min caused hyperthermia (~0.85°C) and attenuated the hypothermia induced by 2.0 g/kg ethanol compared to controls. The controls were moved to the benchtop for baseline temperature measurement and injection and then were returned to the cage rack. A 1.0 g/kg dose of ethanol blocked the hyperthermic effect of changing the environment, but did not cause hypothermia in either environment. The effect of moving the rats was more pronounced in a group of chronically handled animals (i.e., animals that previously had been handled and injected with 1.0–3.0 g/kg ethanol twice weekly for eight weeks). In other studies, individual housing for 21–22 days, beginning at age 45 days, significantly reduced the hypothermic effect of 2.8 g/kg ethanol in LS mice and of 5.0 g/kg ethanol in SS mice 30 and 60 min after i.p. injection compared to respective mice housed five per cage (Jones et al., 1990). The effects in LS mice could not be attributed to changes in ethanol pharmacokinetics, but the role of ethanol pharmacokinetics was less clear in SS mice.

Together, these findings demonstrate the subtle and complicated interactions between stress and the assessment of ethanol's effects on body temperature and underline the importance of careful controls and proper selection of measurement techniques in conducting such studies and in comparing results across studies.

6. Circadian Rhythms

Circadian rhythms impact both baseline core temperature and ethanol's effects on core temperature. Under normal conditions, where the rhythms are well established, the body temperature cycle is syncronous with the daily sleep-activity cycle (Heller et al., 1983; Aschoff, 1970). That is, the maximum body temperature occurs during the active portion of the day, and the minium body temperature occurs during the inactive or sleep portion. Further, although there is a strong correlation between activity level and body temperature, the body temperature for any given activity level is higher during the active than during the inactive portion of the day, suggesting that the influence of activity metabolism is superimposed on a basic daily cycle of body temperature. Therefore, the link between activity and body temperature cannot be explained alone by the metabolic consequences of activity or rest. Similarly, diurnal declines in temperature cannot be solely attributed to sleep.

Circadian rhythms affect sensitivity to a number of drugs including ethanol. In general, rodents are more sensitive to ethanol's toxic, behavioral, and thermoregulatory effects during the dark phase of the 12 h:12 h light-dark cycle when the normal activity of the animal is the highest (Haus and Halberg, 1959; Lagerspetz, 1972; Deimling and Schnell, 1980; Brick et al., 1982; Pohorecky and Brick, 1988). In the most comprehensive study available, Deimling and Schnel (1980) measured rectal temperatures at 4-h intervals using standard telemetry probes inserted 3 cm into the rectums of male Swiss-Webster mice exposed to light from 800–1800 h in a room maintained at 22–24°C. They found that the normal rectal temperature varied throughout the day with a peak around 38.2°C at 2400 h (and possibly at 800 h) and a nadir of about 36.9°C at 1200–1600 h. They measured the effects of ethanol on body temperature using two criteria: (1) the lowest rectal temperature attained within 180 min of ethanol administration and (2) the area enclosed by the time-temperature response curves. The results indicated that ethanol-induced hypothermia, assessed by both criteria, varied in a circadian manner after ethanol doses of 4 and 5 g/kg i.p., with the greatest effect being observed at 400–800 h (body temperatures of about 32°C and 27°C, respective to dose) and the least effect at 1600–2000 h (body temperatures of about 33.5°C and 31.5°C, respective to dose). Ethanol-induced hypothermia did not vary with the time of day after 2 g/kg when measured as the lowest rectal temperature attained, but showed rhythmicity similar to the higher doses when the area under the effect-duration curve was used. Careful pharmacokinetic assessments indicated that these diurnal variations in ethanol's effects on body temperature did not result from changes in the *in vivo* rate of disappearance of ethanol from the blood or from the rate of ethanol metabolism by hepatic ethanol dehydrogenase measured *in vitro*. These factors did not change significantly during the 24-h period. Interestingly, the authors point out that previous work had found circadian changes in ethanol disposition in rats and humans, leading them to suggest that mice may differ from rats and humans with respect to daily variations in ethanol disposition. Additional studies reported in the same paper found diurnal variations in sensitivity to ethanol-induced spontaneous locomotor stimulation and depression, and a loss of the righting reflex and lethality that followed the classic pattern of highest sensitivity during the dark phase and lowest sensitivity during the light phase. Together, these findings strongly indicated that diurnal variations, including changes in hypothermic sensitivity, reflected temporal changes in the sensitivity of the central nervous system to ethanol.

Subsequent work found a similar diurnal pattern in baseline temperatures, but in contrast to Deimling and Schnell's findings, suggested that mice, housed at 21°C in the 12:12 light-dark cycle (800 on) and injected i.p. with 3 g/kg ethanol, were more sensitive to ethanol hypothermia during the light (maximum hypothermia about –2.4°C at 1000 h and –2.2°C at 1400 h) than during the dark cycle (maximum hypothermia about –1.8°C at 600 h) (Walker et al., 1982). However, recalculation to express Walker et al.'s data as the lowest rectal temperature during each period reveals a diurnal pattern very similar to that reported by Deimling and Schnell, with the time of maximium ethanol-induced hypothermia (lowest body temperatures) near the transition from dark to light (600–1400 h) and the minimal hypothermia (highest body temperatures) at the transition from light to dark (1400–600 h). Therefore, the discrepancy appears to reflect differences between studies in the operational definition of hypothermia, that is, the lowest rectal temperature (Deimling and Schnell, 1980) compared to the change in temperature from the baseline (Walker et al., 1982). More recent work suggests that circadian rhythms can also affect the development of tolerance to ethanol hypothermia (expressed as a change in temperature from the baseline), with tolerance developing when injections occur in the light, but not in the dark phase (Williams et al., 1993). Baseline data were not furnished in this study. Therefore, it would be interesting to see if the conclusions regarding the relationship between tolerance and diurnal rhythm hold when hypothermia is expressed as the lowest body temperature.

In conclusion, the available studies suggest that sensitivity to ethanol hypothermia is affected by diurnal rhythms, but that the pattern does not fit the pattern seen with most other ethanol-induced behaviors, which are more sensitive to ethanol during the dark cycle. With hypothermia, the maximal and minimal sensitivities to ethanol appear to revolve around the transitions from light to dark and dark to light.

7. Age

Age can have a major impact on sensitivity to ethanol including ethanol's thermoregulatory effects, with ethanol sensitivity generally increasing with age (Chesler et al., 1942; Wood et al., 1982; Wood and Ambrecht, 1982b; Hollstedt and Rydberg, 1985; Lable and Rydberg, 1982; Hunt et al., 1991). These age-related differences in sensitivity to ethanol's thermic effects extend from the time of birth through adulthood.

In rats, the thermoregulatory system matures rapidly during the first three weeks of postnatal development, with thermostability and independence from T_a occurring at about 21 days of age (Conklin and Hegggeness, 1971; Spiers and Adair, 1986). Early studies found that injection with 4 g/kg ethanol did not significantly affect colonic temperature in 6- to 16-day-old rats tested at an ambient temperture of 24°C (Kruckeburg et al., 1984). However, the lack of ethanol effect likely reflected the hypothermic state of the neonates at this T_a. Subsequent work, which tested at a thermoneutral T_a (defined as an ambient at which temperature regulation is achieved without regulatory changes in metabolic heat production or evaporative heat loss), found that the homeothermic ability of immature rats was affected by ethanol, with the magnitude of the effect being directly related to age (Spiers and Fusco, 1991). These studies, using copper-constantan thermocouples to measure colonic and skin temperatures at tail, abdominal, and back sites without a handling of the animals, showed that 4 g/kg ethanol i.p. did not alter colonic temperature in 2- to 3-day-old rats tested at a thermoneutral ambient of 36°C, whereas, the same dose decreased colonic temperature by 0.3°C and 0.6°C in 8- to 9- and 14- to 15-day-old animals tested at thermoneutral T_as of 35°C and 33°C, respectively. Further, injection of 2.0 g/kg ethanol did not significantly affect colonic temperature in 8- to 9-day-old rats, but produced a 0.6°C decrease in 14- to 15-day-old animals. Ethanol also depressed metabolic rates in an age-dependent manner, with 2- to 3-day-old rats showing a smaller magnitude and shorter duration response than older animals. These age-dependent differences in neonatal responsiveness to ethanol could not be explained by differences in ethanol pharmacokinetics because the older animals had greater ethanol metabolizing capacity than the younger animals. The authors (Spiers and Fusco, 1991) suggested that the differences in sensitivity to ethanol could reflect the rapid development of the CNS during this period (Dobbing and Sands, 1979) as well as the important relationship between CNS maturation and ethanol sensitivity (Lable and Rydberg, 1982; Kiianmaa and Sinclair, 1985).

Sensitivity to ethanol-induced impairment of thermoregulation also increases with age in adult rats. Although early work suggested that these age-related differences in thermic sensitivity to ethanol might be secondary to differences in ethanol pharmacokinetics or body composition (Abel and York, 1979; York, 1982), other work documented that increasing sensitivity of target tissues played a role in these age-related differences, particularly when animals were exposed to cold stress (York, 1983; Ott et al., 1985). In recent studies using standard telemetric rectal probes, York and Chan (1993) investigated sensitivity to ethanol hypothermia in female rats from ages 4–25 months. They found that the 4-month-old rats had significantly lower baseline temperatures than 13-month-old and 25-month-old animals and consequently used the change in temperature (predrug minus postdrug) to measure ethanol sensitivity. This approach revealed that the youngest rats had significantly less hypothermia from ethanol than the middle or older aged groups. The response of the middle aged group was more similar to that of old rats than to that of young rats. Interestingly, the response to ethanol was not different among the age groups if the absolute values of rectal temperature produced by ethanol were used as a measure. To eliminate baseline differences, and to further assess sensitivity to ethanol hypothermia, the authors matched baseline temperatures across age groups and measured sensitivity to hypothermia by measuring the blood ethanol concentration when a target level of hypothermia was reached (36°C). They also adjusted the dose of ethanol to compensate for pharmacokinetic differences between groups, with young and middle aged groups receiving 3.5 g/kg and the old aged group 3.0 g/kg. In partial agreement with the initial study, the blood ethanol concentration at the designated target level of hypothermia was higher in young vs. middle- and old-aged groups.

The interaction between aging and sensitivity to ethanol's thermic effects in adult mice appears to be more complicated. As with rats, sensitivity to ethanol, measured, for example, by a loss of the righting reflex, increased with age in animals from 2–24 months old (Abel, 1978; Ritzmann and Springer, 1980). However, Wood and Ambrecht (1982a) found that sensitivity to the hypothermia induced by 3 g/kg injected i.p., measured rectally as a change in temperature from the baseline 30 and 120 min after injection in mice exposed to a T_a of 25°C, was greater in young than in older animals (8 > 18 > 28 months). The maximum difference between groups was about 1.5°C. On the other hand, the duration of the hypothermic response was shorter in the young vs. older animals, suggesting greater sensitivity in old vs. young using this parameter. Baseline rectal temperatures did not differ significantly between the age groups. In contrast to their hypothermic responses, but consistent with previous work (Abel, 1978; Ritzmann and Springer, 1980), the older animals were more sensitive to ethanol-induced loss of the righting reflex than were the younger animals. The authors (Wood and Ambrecht, 1982a) suggest that differences between mice and rats in the effects of age on sensitivity to ethanol hypothermia between studies could reflect, in part, the relatively warm T_a to which the mice were exposed (25°C for mice compared to 17–19°C in the rats) (Abel and York, 1979).

Studies in nonhuman primates (squirrel monkeys) found that sensitivity to ethanol-induced hypothermia increased with age in animals from those less than 4 years old to those more than 9 years old (Murphy and Lipton, 1983). Ethanol (0.5–2.0 g/kg) given orally caused a dose-dependent hypothermia, which occurred in thermoneutral (25.5°C) and cool (18.5°C) environments and was augmented by exposure to the cooler T_a. The older primates not only developed greater decreases in core temperature (up to 3.3°C colder) than did younger animals, but the hypothermic effect of each ethanol dose was also more prolonged in the older animals. The authors conclude that the increased sensitivity to the temperature-altering effects of ethanol and the greater duration of the hypothermic effect of this drug in aged homeotherms suggest that elderly patients admitted to hospitals with any indication of recent ethanol ingestion should be carefully evaluated for the possibility of accidental hypothermia.

In summary, sensitivity to ethanol hypothermia increases with age. During neonatal development, the increased sensitivity to ethanol-induced hypothermia appears to reflect maturation of the CNS. In adult animals, ethanol sensitivity also increases with age, but the interaction is more complicated and the underlying mechanisms are not well understood. Nonetheless, aging appears to increase the risk from ethanol-related lethal hypothermia in humans.

8. Genetic Background

Genetic background influences baseline rectal temperature (Crabbe et al., 1994) and sensitivity to ethanol's thermoregulatory and other effects (Phillips and Crabbe, 1991; Crabbe, 1989). The evidence for a genetic influence of these factors comes from studies of inbred strains and from the successful development of selectively bred lines of animals that differ in sensitivity to ethanol's thermic effects (Phillips and Crabbe, 1991).

As discussed by McClearn (1991), the primary value of inbred strains is genetic uniformity. Each animal within a strain is essentially identical genetically to the others, with the exception of male-female differences in loci on the sex chromosomes. Therefore, if two inbred strains are reared and assessed under standard conditions, phenotypic differences between the strains can be interpreted as direct evidence of a genetic influence on the phenotype.

Previous work clearly demonstrated differences between inbred strains in sensitivity to ethanol hypothermia and established a genetic influence over sensitivity to ethanol-induced thermoregulatory changes (Crabbe, 1983). Crabbe et al. (1994) recently extended this investigation by studying the genetic determinants of sensitivity to ethanol-induced hypothermia and other behaviors (open field activity, rotarod ataxia, and anesthesia) in 15 inbred strains of mice. In these studies, mean baseline rectal temperatures ranged from 38.5°C in 129/J mice to 36.9°C in BALB/J, AKR/J, and C57BR/J mice. Due to these baseline differences, the researchers used change in temperature from the baseline to assess thermic sensitivity and response to ethanol. The results indicated that strains differed in sensitivity to 2 and 3 g/kg ethanol, but not after saline or 1 g/kg ethanol. Injection of 3 g/kg ethanol caused the maximal response to ethanol in all strains. The magnitude of the hypothermic response to ethanol varied by more than 4°C across strains — from a maximum –4.4°C in CBA/J mice to a minimum of –0.16°C in DBA/2J mice. Further, the extrapolated blood ethanol concentration required to reduce body temperature by 2°C ranged from 1.27 mg/ml of blood in the most sensitive strain (SWR/J) to 3.24 mg/ml in the least sensitive (BALB/J). This and other data in this study convincingly argued that the strain differences in sensitivity were mediated by pharmacodynamic rather than by pharmacokinetic factors. In addition, baseline rectal temperature was genetically correlated with sensitivity to ethanol-induced hypothermia, but in contrast to initial studies (Crabbe, 1983), baseline temperature was not related to ethanol-induced ataxia or loss of the righting reflex. Similarly, studies using selectively bred animals to investigate possible genetic correlations between sensitivity to hypothermia and other traits have not found a consistent relationship (Crabbe et al., 1990; Howerton et al., 1983). The considerable genetic independence in ethanol sensitivity between behaviors suggests that ethanol sensitivity for these behaviors is influenced by discrete populations of genes and emphasizes the potential importance of genetic correlations between behavioral sensitivities to ethanol when they do occur (Crabbe et al., 1994).

Selective breeding for a specific trait offers another method for (1) demonstrating genetic influence over a trait, (2) developing useful tools for identifying genetically correlated responses, and (3) investigating the biochemical and genetic bases underlying these traits. In contrast to inbreeding, which is nondirectional, selective breeding is directed toward a specific objective. As discussed by McClearn (1991), selective breeding can be used to manipulate genotypes toward specific preplanned objectives. The process begins with a genetically heterogeneous population having a normally distributed phenotype of the trait to be selected. The parents for the next generation are selected from the upper and lower tails of the phenotype distribution in the foundation population. The process is repeated over several generations. If the heritability of the trait is zero, the distribution of the trait is entirely due to environmental circumstances, that is, there are no genetic influences to be transmitted to the offspring. Therefore, selective breeding should not influence the population distribution. However, if heritability is present, then selective breeding will transmit to offspring more alleles contributing the selected expression (high or low). In this case, the offspring will have increased representation of the selected genes, and their phenotypic mean will be displaced from that of their parent generation. The achievement of a selection response is definitive proof of a significant genetic component to the variance for the phenotype in the foundation population.

To investigate the heritability of ethanol's hypothermic response, Crabbe and colleagues initiated a replicate selection (i.e., two independent selections) for sensitivity (COLD) and resistance (HOT) to ethanol-induced (3 g/kg, i.p.) hypothermia, measured rectally (Crabbe et al., 1987). They initiated the selection using genetically heterogeneous breeding pairs of HS/Ibg mice originally derived from crossing eight inbred strains. After five generations of selection, they demonstrated a difference in maximal hypothermic response to 3 g/kg i.p. ethanol between COLD and HOT lines of 1.6°C in the first replicate and 1.2°C in the second replicate. There were no significant differences between the lines in ethanol metabolism, indicating that the selection altered neurosensitivity to ethanol. The estimated heritability (h^2 = proportion of total response variance due to additive genetic sources) for the two lines combined was 0.17 for the two replicates, which indicated that 17% of the variance in ethanol-induced hypothermia in mice comes from additive genetic origin. Continued selection has driven the lines farther apart. In selected generation 14, the COLD and HOT lines differed by about 4°C in the selected hypothermic response, and heritability estimates suggested that approximately 20% of the variance in ethanol-induced hypothermic responses in mice is of additive genetic origin (Phillips et al., 1990). Interestingly, the response to selection has been unidirectional. That is, the COLD lines have gradually become more sensitive to ethanol-induced hypothermia than have the nonselected control lines or the HOT lines, whereas the HOT lines have not developed significant resistance to ethanol-hypothermia (maximum hypothermia of 5.6°C and 6.0°C in the COLD lines versus 2.2°C and 1.2°C in the HOT lines) (Phillips and Crabbe, 1991). The 4°C divergence in magnitude of hypothermic response between the selected lines matches closely the difference in sensitivity between inbred lines (Crabbe et al., 1994) and underscores the important role that genetic background can play as a variable in the study of ethanol's effects on hypothermia or in the use of ethanol's effects on hypothermia as a measure of ethanol sensitivity.

The COLD and HOT lines differed in sensitivity to the hypothermic effects of other drugs and to other behavioral effects of ethanol including the development of tolerance, loss of the righting reflex, locomotor activation, place and taste conditioning responses, preference for drinking ethanol-containing solutions, withdrawal intensity, and lethality. These potential correlated responses to selection for sensitivity and resistance to ethanol hypothermia, and the importance of these traits and selectively bred animals as tools for identifying underlying neurochemical and genetic mechanisms, are reviewed by Phillips and Crabbe (1991). It is beyond the scope of the present chapter to discuss these responses. However, recent investigation of ethanol's effects on thermoregulation in these lines provides insight into the central systems affected by the selection process. In this work, O'Connor et al. (1993) studied the effect of ethanol on regulated temperature in the COLD and HOT lines using miniature radiotelemetric peritoneal implants in mice trained in a temperature gradient (8–40°C). They found that ethanol caused a regulated decline in body temperature, as evidenced by a selection of low temperatures in the gradient at the same time internal temperatures were falling. Further, COLD mice were more sensitive than HOT mice to the regulated decline, indicating that the selection has differentially altered the CNS regulator of body temperature in these mice. These results were found in both replicates, suggesting that the differences represent pleiotropic effects of genes determining acute sensitivity to ethanol hypothermia. These findings led the authors to conclude that a major aspect of the way in which the COLD and HOT mice differ in body temperature response to ethanol involves a selection-induced difference in the sensitivity to ethanol of the CNS regulator of body temperature. From another perspective, this study shows a genetic influence over the thermoregulatory set-point and, thus, demonstrates the potential role that genetic background can play in studies investigating ethanol's effects on the thermoregulatory set-point.

In conclusion, studies using inbred and selectively bred animals clearly show that genetic background can influence sensitivity to ethanol's thermoregulatory effects, including the magnitude of body temperature change and alterations in the thermoregulatory set-point. The use of pharmacogenetic tools in research on ethanol's thermoregulatory effects and consequences are discussed in each relevant section of this chapter.

III. MECHANISMS AND SITES OF ETHANOL'S EFFECTS ON THERMOREGULATION

The thermoregulatory system is rather complex and involves various processes such as sensing, integrating, and responding to thermal challenges. Homeotherms achieve the constancy of their temperature by maintaining an equilibrium between heat production and heat loss. In response to changes in the temperature of the blood perfusing the brain and to information brought by thermosensory pathways from the skin or other organs, certain neurons in the hypothalamus that determine the set-point change their firing rate and bring about an appropriate compensatory response. The sensitivity of these hypothalamic neurons is controlled or can be modified by various neurotransmitters or hormones, ions, etc. (Boulant et al., 1989; Crawshaw et al., 1992).

In addition to physiologic responses, organisms can make behavioral responses that either modify the environment, or they can move to a more favorable one. In fact, the studies of ethanol effects on behavioral thermoregulation have provided much information concerning the possible mechanisms underlying the effects of ethanol on thermoregulation. A number of studies mentioned above have shown that the temperature effects of ethanol are quite dependent on T_a. The same dose of ethanol can produce various degrees of hypothermia or hyperthermia depending on T_a (For example, see Myers, 1981; Malcolm and Alkana, 1981). These studies have suggested that ethanol might abolish or broaden the thermoregulatory set-point and produce a poikilothermic-like response. Studies concerning the effects of ethanol on behavioral thermoregulation, however, have indicated that ethanol might affect temperature by lowering the set-point. In this section, evidence for the lowering of thermoregulatory set-point by ethanol is discussed.

A. EFFECTS OF ETHANOL ON THERMOREGULATORY SET-POINT

If ethanol acts by lowering the thermoregulatory set-point, one would expect that the organism would prefer a relatively cool T_a while undergoing a decrease in body temperature. A reduction in heat production and an increase in heat loss to the environment by stimulating vasodilation might accompany such behavioral responses.

Evidence for the possibility that ethanol might lower the set-point was first provided by Lomax et al. (1981). A dose of 1.5 g/kg i.p. ethanol was found to reduce the latency of escape from radiant heat in rats, even though the animals experienced hypothermia (Lomax et al., 1981).

A very thorough and direct examination of the effects of ethanol on thermoregulatory set-point was conducted by Gordon and co-workers (Gordon and Stead, 1986; Gordon and Mohler, 1990; Gordon et al., 1988a), and by O'Connor and co-workers (O'Connor et al., 1988, 1989, 1993). In a T_a gradient ranging from 18–36°C, mice or rats treated with i.p. or oral administration of 3 g/kg of ethanol, respectively, (Gordon and Stead, 1986; Gordon et al.,

1988a) voluntarily selected a lower T_a than saline controls did. Similarly, O'Connor et al. (1989) showed dose-related effects of ethanol on core and selected T_a. Mice receiving an i.p. injection of 2.25 g/kg had a lower core temperature and tended to select a lower T_a than did those receiving a saline injection. Mice injected with 2.6 g/kg ethanol had a greater fall in rectal temperature and selected a significantly lower T_a than did those receiving a saline injection or 2.25 g/kg of ethanol. Studies with selectively bred COLD and HOT mice also showed similar patterns of effect on selected T_a after ethanol administration (O'Connor et al., 1993). The COLD line of mice displayed hypothermia, but selected a lower T_a after receiving an i.p. administration of 2.65 g/kg of ethanol. It is of interest to note that the HOT line of mice, which are resistant to the hypothermic effect of ethanol, did not show any reduction in core temperature after receiving a similar dose of ethanol. Nevertheless, these mice selected a lower T_a after ethanol injection compared to the preinjection baseline. The reduction in selected T_a induced by ethanol was also observed in goldfish (O'Connor et al., 1988), which are ectothermic and do not have autonomic effectors (such as sweating, vasodilation, etc.) to regulate temperature. The temperature selected by goldfish exposed to a 1–1.5% ethanol solution was about 2°C lower than those exposed to saline (O'Connor et al., 1988). These studies thus pointed out that, despite experiencing the hypothermic effect of ethanol, animals were consistently selecting a lower T_a than were controls.

The selection of a lower T_a is a strong indication that ethanol does lower the thermoregulatory set-point. The lowering of set-point is consistent with the observations that ethanol delays the onset of shivering and reduces its duration during cold exposure. In fact, it might explain the paradoxical undressing observed in ethanol-related hypothermic deaths reported by Albiin and Eriksson (1984).

B. DOES ETHANOL ABOLISH OR LOWER THE THERMOREGULATORY SET-POINT?

Results from behavioral thermoregulation studies suggested that ethanol might lower the thermoregulatory set-point. On the other hand, the observed effects of T_a on ethanol-induced changes in body temperature indicated that ethanol might broaden or abolish the set-point and produce poikilothermia. Studies that support a lowering of thermoregulatory set-point by ethanol have utilized low to moderate doses ranging from 1.5–3 g/kg, whereas those that demonstrated a poikilothermic effect employed doses ranging from 3.6–8 g/kg of ethanol. It is possible that the nature of ethanol's effects on thermoregulatory set-point is dependent on the doses employed, with small to moderate doses lowering, and high doses of ethanol impairing, the set-point. It should be pointed out that the effect of ethanol on behavioral thermoregulation can be studied only at low to moderate doses, because higher doses of ethanol produce narcosis and incapacitate the animals. It is therefore difficult, if not impossible, to determine whether high doses of ethanol would lower the thermoregulatory set-point using behavioral thermoregulatory techniques. The absence of such data, however, does not negate the possibility that higher doses might also produce a lowering of the thermoregulatory set-point. It has been suggested that high doses of ethanol act as an anesthetic and impair various physiological functions including the processes involved in thermoregulation (Myers, 1981). However, one can argue that an impairment of autonomic effectors and behavioral thermoregulation might possibly contribute to the hyperthermic effect observed at a high T_a, even though the thermoregulatory set-point is lower.

At a T_a above the thermoneutral zone, metabolic rate increases as a direct consequence of the effect of rising tissue temperatures on metabolic processes as well as the use of increased ventilation to dissipate heat (Gordon et al., 1988b). A dose of 3 g/kg ethanol has been shown to reduce metabolic rate (MR) and evaporative water loss (EWL) at a T_a of 20°C, but not at 30°C or 35°C in mice (Gordon and Stead, 1986). Similarly, ethanol doses ranging from 4–6 g/kg have been shown to suppress EWL without altering the MR in rats at a high T_a. These findings indicate that high doses of ethanol impair heat dissipating mechanisms. The suppression of EWL induced by high doses of ethanol is predictable, because grooming of saliva onto the fur at a T_a of 37°C, which is a mechanism to accelerate evaporative heat loss, would be impaired by these doses of ethanol (Gordon and Mohler, 1990).

Whether high doses of ethanol lower or impair thermoregulatory set-point is not understood. It is, however, clear that high doses of ethanol impair both autonomic and thermoregulatory mechanisms. How such impairment might contribute to the observed "poikilothermia" remains to be examined.

C. ANATOMICAL SITES OF ACTION

It is well documented that neurons in the preoptic and anterior hypothalamic nuclei (POAH) play a major role in integrating the various thermal inputs and initiating the appropriate response. The thermoregulatory roles of POAH have been shown to be subserved by different types of neurons. Electrophysiologic studies have indicated that about 30% of the neurons are warm sensitive, and 10% are cold sensitive. These neurons show a rise in the firing rate in response to local warming or cooling, respectively (Boulant, 1981; Boulant et al., 1989). Although the response to thermal stimulation by POAH neurons has been studied extensively, the nature of the interaction among these neurons and the pharmacological characterization of these neurons are still unclear.

Little work has been conducted to examine the direct effects of ethanol on the activity of these thermosensitive neurons. An increase in skin temperature followed by a drop in rectal temperature has been demonstrated following perfusion of 2.75% or 5.5% ethanol into the POAH by push-pull cannulae over a period of 5 min (Huttunen and Myers, 1985). The same workers (Huttunen et al., 1988) also showed that the temperature response to the infusion of ethanol into the PAOH displays a very high anatomical specificity. Depending on the sites of perfusion, hypothermia or hyperthermia have been shown to occur following such infusion of ethanol. In fact, the neurons or nerve terminals in the POAH that mediate a hyperthermic response to ethanol infusion are separated from those underlying hypothermic responses by a few hundred microns (Huttunen and Myers, 1985). Recent work by Ticho et al. (1992) however, failed to demonstrate any significant changes in brain temperature following microinjection of similar concentrations of ethanol into the preoptic area. The failure to observe any significant change in temperature in the Ticho et al. study might be related to relative concentrations of ethanol achieved as well as to the site of infusion. Although the concentrations of ethanol per se were essentially the same, ethanol was infused at a rate of 1 μl/min over a period of 30 s, whereas in the studies by Huttunen and Myers (1985) and Huttunen et al. (1988), ethanol was perfused at a rate of 20 μl/min over a period of 5 min. In addition, the observed effects of ethanol on temperature in the studies by Huttunen and Myers (1985) were localized to a very circumscribed region of the preoptic area. The effects of direct administration of ethanol into the nucleus preopticus periventricualaris on behavioral thermoregulation has also been examined in goldfish. Microinjection of very low doses of ethanol (0.0475 ng) in this area led the fish to select a water temperature about 9°C lower than that selected by control fish (Crawshaw et al., 1989).

These studies together demonstrated that ethanol can produce an alteration in body temperature, at least in part, through its action on the neurons located in the POAH area. How ethanol exerts its action in this area to modify temperature is not clear. Changes in serotonin (5-HT) and norepinephrine (NE) released in this area have been implicated as possible mechanisms for such action of ethanol. These studies are discussed in the following section.

D. NEUROCHEMICAL CAUSES

1. Earlier Studies

The involvement of various neurochemical systems in the hypothermic effect of ethanol has been reviewed quite extensively by Kalant and Lê (1984, 1991). Generally, the evidence for the involvement of neurochemical systems in the effects of ethanol on body temperature have been derived from a number of experimental approaches. These studies:

- examined the effects of ethanol on the activity of neurochemical systems that have been shown to play a role in thermoregulation. For example, the effects of ethanol on brain levels, synthesis, or release of various neurotransmitters such as dopamine (DA), norepinephrine (NE), and serotonin (5-HT) have been examined extensively.
- determined how modification of the activity of these neurochemical systems would modify the effects of ethanol on temperature. These studies tested how specific lesioning of neurotransmitter systems or administration of receptor agonists or antagonists modifies the hypothermic effect of ethanol.
- determined how chronic ethanol treatment would modify the effects of neurotransmitter agonists or antagonists on body temperature. For example, chronic ethanol treatment has been shown to reduce the hypothermic effects of DA (Hoffman and Tabakoff, 1987) or NE agonists (Mullin and Ferko, 1981).

Based on the results from such experimental studies, a number of neuronal systems have been implicated in playing a role in the thermoregulatory effects of ethanol. These include acetylcholine, dopamine, GABA, histamine, norepinephrine, serotonin, various neuropeptides (e.g., β-endorphin, thyrotropin-releasing hormone, and vasopressin), prostaglandins, and calcium ions (see Kalant and Lê 1984, 1991 and Pohorecky and Brick 1988 for reviews). Evidence for the involvement of these neurochemical systems in such action of ethanol, however, has been rather conflicting. Depending on the species and experimental approaches employed, negative and positive data have been reported for most neurochemical systems. For example, depletion of 5-HT by the administration of p-chlorophenylalanine (Frankel et al., 1978) or p-chloroamphetamine (Pohorecky et al., 1976) has been shown to decrease or augment the hypothermic effect of ethanol, respectively, in rats. Moreover, destruction of brain 5-HT by the administration of 5,7-dihydroxytryptamine has been shown to have no effect on ethanol-induced hypothermia in both mice (Melchior and Tabakoff, 1981) and rats (Lê et al., 1980).

In the following section, we examine some of the recent work in this area as well as some of the work that was not covered in the earlier reviews (Kalant and Lê, 1984, 1991; Pohorecky and Brick, 1988). Readers are referred to these reviews for detailed information on the earlier work.

2. Recent Studies

The approaches used over the last several years to study the involvement of various neurochemical systems in the effects of ethanol on thermoregulation have been different from those employed earlier. Generally, they can be classified into four different areas:

1. examining the effects of direct administration of ethanol into the POAH on release of neurotransmitters
2. examining the differences in the binding or responses to different neurotransmitter agonists and antagonists in animals that have differences in sensitivity to the hypothermic effects of ethanol
3. examining the involvement of neurotransmitter receptor types or subtypes in ethanol mediated hypothermia
4. examining the involvement of other neurochemical systems

These studies have provided some important new information concerning the role of NE, 5-HT, and neuropeptides (opioids) in mediating and modulating the effects of ethanol on thermoregulation. These are described next.

a. NE and 5-HT

Probably the strongest evidence for the involvement of NE and 5-HT in the hypothermic effect of ethanol are derived from the studies by Huttennen and Myers (1985) and Huttunen et al. (1988). These workers demonstrated that the i.p. injection of ethanol causes an elevated efflux of [^{3}H] NE from POAH in a pattern parallel to that of the lower colonic temperature (Huttunen and Myers, 1985). Moreover, the direct infusion of ethanol within a very circumscribed area of the POAH caused an elevation in skin temperature and a hypothermia of short latency. The decrease in temperature was associated with a delay in the efflux of NE. Similar types of experimental studies have also shown an association between 5-HT release from the POAH and ethanol induced changes in body temperature (Huttunen et al., 1988). The effects on 5-HT release induced by ethanol, however, is quite dependent on the direction of ethanol effects on body temperature. In areas of the POAH where ethanol infusion caused a decrease in temperature, an inhibition of 5-HT efflux was observed, whereas in areas where ethanol caused an increase in temperature, an enhanced release of 5-HT was observed (Huttunen et al., 1988). Since a presynaptic release of 5-HT has been suggested to activate the triggering of pathway subserving heat loss (Myers, 1980), the results from these studies indicated that ethanol-hypothermia might be mediated by respective inhibition and stimulation of 5-HT and NE release from the POAH area.

Additional evidence for the involvement of brain 5-HT in ethanol-mediated hypothermia has been provided from studies with mice selectively bred for differences in the hypnotic or hypothermic effect of ethanol. Ethanol produces a greater inhibition of tryptophan hydroxylase (TpH) in the hypothalamic area and a greater degree of hypothermia in LS than in SS mice (French and Weiner, 1991). Pentobarbital, however, produces equivalent effects in LS and SS mice on TpH activity in the hypothalamus and on body temperature (French and Weiner, 1991). For this reason, an inhibition in 5-HT activity in the hypothalamus might play a role in ethanol-induced hypothermia. Similarly, COLD mice, which exhibit an increased hypothermic response to ethanol, are also more sensitive to the hypothermic effect of 5-HT than are HOT mice (O'Connor et al., 1990).

Although much knowledge has been gained over the last several years concerning the existence of various receptor types and subtypes of 5-HT receptors, little work has been done to examine the involvement of specific neurotransmitter receptor types or subtypes in ethanol-mediated hypothermia.

b. Neuropeptides

Orts et al. (1991) reported that pretreatment with naloxone in doses of 10, but not 1, mg/kg can partially reverse the hypothermic effect of ethanol. Since a very high dose of naloxone was required to reverse the effect of ethanol, it is questionable whether such a reversal is mediated by the action of naloxone on the opioid systems. However, it is possible that the requirement for higher doses of naloxone to antagonize ethanol-hypothermia might be related to its low affinity to kappa opioid receptors. A number of studies have suggested that kappa opioid receptors might play an important role in the hypothermic effect of ethanol. Pretreatment with kappa opioid receptor agonist (Pohorecky et al., 1989) enhances, whereas kappa opioid receptor antagonist (Pillai and Ross, 1986; Pohorecky et al., 1989) antagonizes ethanol-induced hypothermia. In addition, COLD mice were more sensitive to the hypothermic effect of kappa opioid receptor agonist than were HOT mice (Feller and Crabbe, 1991). Besides kappa opioid receptors, delta opioid receptors have also been implicated in ethanol-mediated hypothermia. Widdowson (1987) reported that intracranial injection of delta opioid receptor antagonists can significantly reduce the hypothermic effect of ethanol in rats. These studies thus indicate an involvement of dynorphin, enkephalins, and endorphin in ethanol-induced hypothermia.

Besides opioid peptides, the role of neurotensin and thyrotropin releasing hormone (TRH) in ethanol-mediated hypothermia has also been investigated. Intracranial administration of neurotensin or TRH into the nucleus accumbens or caudate nucleus has been shown to potentiate the hypothermic effect of ethanol in the rats (Widdowson, 1987). Similarly, doses of ethanol (1 g/kg) and neurotensin (0.005 μg) that have no effect on hypothermia when administered separately, produced a pronounced hypothermia when given together (Erwin and Su, 1989). In addition, differences in brain neurotensin receptor density have been reported in LS and SS mice which differ in their responses to the hypnotic and hypothermic effects of ethanol (Erwin and Jones, 1993). However, it is likely that neurotensin might alter the hypothermic effect of ethanol indirectly by enhancing the effects of ethanol, because neurotensin has been shown to potentiate the sedative effects of ethanol (Widdowson, 1987).

c. Others

Over the last few years, much attention has been paid to the involvement of NMDA receptor antagonists as well as some neurosteroids in ethanol's acute effects and ethanol tolerance. Although a number of studies have shown that MK-801, the NMDA receptor antagonist, can block tolerance to the hypothermic and hypnotic effects of ethanol (Khanna et al., 1991, 1992; Szabó et al., 1994), little is known about the acute interaction between MK-801 and ethanol-induced hypothermia.

How GABA receptors might be involved in mediating the hypothermic effect of ethanol is not well understood. Ethanol-hypothermia has been reported to be unaffected by pretreatment with benzodiazepine antagonist (Ro15–1788), inverse agonist (Ro15–4513), and the GABA antagonist, picrotoxin (Hoffman et al., 1987; Martz et al., 1983; Syapin et al., 1990, 1987; Wood et al., 1988). On the other hand, administration of Ro15–1788 at the nadir of the fall of body temperature induced by ethanol has been shown to potentiate ethanol hypothermia (Paez and Myers, 1990). In contrast, pretreatment with the $GABA_B$ receptor antagonist, phaclofen, significantly decreased the hypothermic effects of ethanol (Allan and Harris, 1989). In addition, benzodiazepine agonists such as diazepam and chlordiazepoxide reduce body temperature (Paez and Myers, 1990; Chan et al., 1985), and cross-tolerance between chlordiazepoxide and ethanol with respect to their hypothermic effects has also been demonstrated (Chan et al., 1985). Some neurostreroids such as dihydroepiandrosterone (DHEA) or pregnanolone (Melchior and Ritzmann, 1992; Melchior and Allen, 1992), which have been suggested to be active at the $GABA_A$ receptor complex, have been found to potentiate the hypothermic, as well as the hypnotic, effects of ethanol. The involvement of GABA receptors and the link between neurosteroids and $GABA_A$ receptors in mediating ethanol-induced hypothermia requires further investigation.

E. PHYSICO-CHEMICAL CAUSES

The effects of ethanol on biological membranes are similar to the effects of increasing the temperature of the membrane in that both ethanol and temperature cause disordering of membrane lipids (Bejanian et al., 1991a; Chin and Goldstein, 1981; Harris and Schroeder, 1981). Details regarding the membrane action of ethanol are presented in Chapter 2. The similarities between the effects of temperature and ethanol on brain cell membranes suggest that homeostatic mechanisms, perhaps in the POAH, may react to ethanol-induced membrane perturbations as if they were an increase in brain temperature and, in response, may set in motion adaptive mechanisms, including a reduction in set-point, to reduce brain temperature (Sinensky, 1974; Cossins et al., 1977; Bejanian et al., 1991; Kalant and Lê, 1991a; Crawshaw et al., 1992). These theories, dubbed homeoviscous adaptation by Sinesky (1974), have been discussed in detail by Kalant and Lê (1991) and Crawshaw et al. (1992). As discussed next in Section IV, the hypothermic effect of ethanol does produce a protective effect vs. ethanol at the behavioral (e.g., Malcolm and Alkana, 1981, 1983; Randall et al., 1988) and cellular levels (e.g., Bejanian et al., 1991a). Further support for the theory comes from studies of ethanol tolerance which found that chronic exposure to ethanol causes changes in membrane composition and other factors that render the membranes less susceptible to the fluidizing effects of ethanol (see reviews by Goldstein et al., 1980; Littleton et al., 1980a, 1980b; Kalant and Woo, 1981; Kalant and Lê, 1991). The theory is also supported by demonstrations that aliphatic ethanols enhanced the sensitivity of mammalian cells in tissue culture to thermal cytotoxicity with acute exposure, but increased resistance to thermal toxicity with repeated exposure (Li and Hahn, 1978).

IV. CONSEQUENCES

Ethanol-induced changes in body temperature represent an important variable in ethanol research. Body temperature changes can influence physiological and neurochemical function. Therefore, it is important to distinguish between the direct effects of ethanol on these systems in contrast to changes secondary to hypothermia. This topic has received considerable attention (Altland et al., 1970; Freund, 1973, 1979; Pohorecky and Rizek, 1981; Kalant and Lê, 1991; Rangaraj and Kalant, 1982; Tabakoff et al., 1975; Oliveira de Souza and Masur, 1984). In addition, ethanol-induced changes in body temperature can markedly affect ethanol sensitivity. The impact of ethanol-induced changes in body temperature on ethanol sensitivity is an often overlooked variable in ethanol research and represents the prime topic of this section.

A. PHARMACOKINETICS

Body temperature can influence the absorption, distribution, and elimination of drugs by altering their physical chemical properties and by changing physiological parameters such as gastric emptying time and rate of metabolism (Ballard, 1974). Consequently, it is not surprising to find that body temperature changes during intoxication can affect ethanol pharmacokinetics. Most early studies found that hypothermia reduced the rate of ethanol elimination (Koren et al., 1989; Dybing, 1945; Krarup and Larsen, 1972; Larsen, 1971; Munoz, 1937) or altered its distribution (MacGregor et al., 1965) in several species. However, other studies failed to find a difference in the rate of ethanol elimination between normothermic and hypothermic animals (Ferko and Bobyock, 1978; MacGregor et al., 1965).

Recently, more comprehensive studies have shown a consistent inverse relationship between body temperature and the rate of ethanol elimination. Romm and Collins (1987) found that increasing the dose of ethanol (1–8 g/kg, i.p.) given to LS, SS, and C57BL/6 mice increased the degree of hypothermia (body temperature range approximately 38–30°C) and decreased the rate of ethanol elimination in all three genotypes by 50–60% (LS and SS from 1.7–0.7 mg/dl/min, C57s from 1.4–0.6 mg/dl/min). The correlation coefficients between body temperature and ethanol elimination rates were 0.74, 0.82, and 0.75, respectively, for each genotype. The authors suggest that these temperature-related changes in ethanol elimination rates may help to explain controversies concerning the pharmokinetic models that best describe ethanol elimination. Interestingly, the SS mice eliminated ethanol more quickly than did the LS mice at most ethanol doses, but if elimination rates at equal body temperatures were measured, the two mouse lines did not differ in elimination rates.

Bejanian et al. (1990) studied the relationship between body temperature and ethanol pharmacokinetics by using a T_a challenge to manipulate body temperature following fixed ethanol doses. They found that exposing C57BL mice to T_as from 34°C to 4°C immediately after i.p. injection of 3.6 g/kg resulted in mean body temperatures from 38.2°C to 26°C during intoxication and decreased the rate of ethanol elimination by 50% (from 1.5– 0.7 mg/dl/min). The goodness of fit (r^2) for the linear model generated suggested that body temperature could account for 77% of the variability in ethanol elimination rates. These findings closely paralleled those of Romm and Collins (1987) and clearly demonstrated that their results were due to the increasing degree of hypothermia rather than the increased dose of ethanol. Bejanian et al. (1990) also found that delaying the temperature challenge until 30 min after ethanol administration did not alter the results, suggesting that changes in ethanol absorption did not mediate temperature-induced changes in the ethanol elimination rate. Further, they found that a T_a challenge caused a small, statistically significant change in body temperature in mice injected with 2.0 g/kg ethanol (body temperatures from 38°C to 36.1°C), but did not significantly affect the rate of ethanol elimination. These latter results are consistent with previous work in rats using subhypnotic ethanol doses where T_a had no effect on the ethanol elimination rate (Ferko and Bobyock, 1978) and with the results of Romm and Collins (1987) who found that the greater the decrease in body temperature during intoxication, the greater the decrease in the ethanol elimination rate. Collectively, these results suggest that some of the previous inconsistencies in the literature regarding the effects of temperature on ethanol elimination rates reflect differences in the extent of body temperature change between studies.

The mechanism by which body temperature change, during intoxication, alters ethanol elimination is not known. As discussed by Romm and Collins (1987), a decrease in body temperature should result in a number of physiogical changes that would decrease the ethanol elimination rate. These include

1. a decrease in the rate of oxidation of NADH, the purported rate-limiting step in ethanol oxidation (Meijer et al., 1975; Thurman et al., 1975; Videla and Israel, 1970)
2. a reduction in blood flow to the liver (Krarup and Larsen, 1972; Larsen, 1971)
3. possible changes in temperature-sensitive enzymes such as ethanol dehydrogenase or the P_{450}-linked MEOS

In conclusion, ethanol-induced hypothermia can cause marked changes in the rate of ethanol elimination. The magnitude of the change in elimination is proportional to the magnitude of the hypothermia. Consequently, studies of ethanol elimination rate should consider body temperature as an important variable in attempts to compare ethanol elimination rates across genotypes or in experiments investigating such things as metabolic tolerance and drug interactions.

B. BEHAVIORAL AND LETHAL EFFECTS

Ethanol-induced changes in body temperature can influence the sensitivity of an organism to the behavioral and lethal effects of ethanol. Perhaps the first evidence of this interaction was presented by Meyer (1899, 1901), who showed that ethanol narcosis in frogs was greater at 30°C than at 3°C. Since then, a number of laboratories have shown that offsetting ethanol hypothermia can strongly influence sensitivity to ethanol's acute and chronic effects. Overall, studies in mice and rats indicate that increasing body temperature between 32°C and 38°C increases ethanol sensitivity, measured by loss of the righting reflex (Alkana et al., 1985a; Finn et al., 1986, 1990, 1994; Malcolm and Alkana, 1981), impairment of gross motor activity and swim performance (Pohorecky and Rizek, 1981), conditioned taste aversion — see later section (Cunningham et al., 1992), and lethality (Dinh and Gailis, 1979; Grieve and Littleton, 1979; Malcolm and Alkana, 1983; Finn et al., 1989, 1991b; Gailis and Tourigny, 1984). This direct relationship between body temperature during intoxication and ethanol sensitivity also has been reported for fish (Ingram et al., 1982; Olofsson and Lindahl, 1977). On the other hand, as discussed later, there can be genetically determined differences in the interaction between temperature and ethanol sensitivity.

The magnitude of interaction between body temperature during intoxication and ethanol sensitivity can be large. Dose and temperature response studies found a significant negative correlation between ethanol sensitivity (measured by loss of righting reflex duration and blood and brain ethanol concentrations at the return of the righting reflex) and body temperature during intoxication in male and female C57 mice exposed to T_as from 13°C to 36°C during intoxication (Alkana et al., 1985a). Loss of righting reflex duration increased by more than 300% and blood ethanol concentrations at the return of the righting reflex decreased by more than 75% as rectal temperature was manipulated from 32°C to 41°C in these animals. Linear regression analysis indicated that rectal temperature at return of the righting reflex accounted for up to 71% of the variability in ethanol sensitivity in these mice. Similar relationships were found when the change in body temperature from the baseline was substituted for absolute rectal temperature.

The impact of offsetting ethanol hypothermia during intoxication following lethal ethanol doses can be even more profound (Malcolm and Alkana, 1983). This work showed that offsetting ethanol-induced hypothermia reduced the 24-h LD_{50} from 8.0 g/kg to 5.3 g/kg in C57 mice. Similarly, exposing C57, LS, SS, and 129/J mice to 34°C after injection of potentially lethal ethanol doses offset ethanol induced hypothermia and significantly increased the percentage of mortality in all four mouse genotypes (Finn et al., 1989). Exposure to 34°C increased mortality at 24 h after injection from 15% to 95% in SS mice, from 37.5% to 100% in 129/J mice, and from 50% to 100% in LS and C57 mice. Blood ethanol data indicated that the results could not be explained by temperature-related changes in ethanol elimination.

The mechanisms by which body temperature affects ethanol sensitivity are uncertain. The blood and brain ethanol concentration data indicate that the effects of temperature on the behavioral and toxic responses to ethanol are mediated via changing brain sensitivity to ethanol per se, and not by temperature-induced changes in ethanol pharmacokinetics (Bejanian et al., 1990; Alkana et al., 1985a; Malcolm and Alkana, 1981; Finn et al., 1994). The effects of temperature on ethanol sensitivity have been linked directly to brain temperature during intoxication (Bejanian et al., 1991b; Finn et al., 1991a). *In vitro* studies have shown that temperature manipulations change the extent of membrane perturbations (Chin and Goldstein, 1981; Harris and Schroeder, 1981; Bejanian et al., 1991a) and neurochemical effects (Hoffman et al., 1984; Rangaraj and Kalant, 1982) induced by ethanol. Therefore, temperature effects on brain sensitivity to ethanol may reflect direct biophysical interactions at the site or sites of ethanol action and/or indirect interactions at points downstream.

Taken together, this evidence indicates that the dependence of ethanol sensitivity on temperature is a general phenomenon extending across genetic backgrounds and measures of intoxication. Under most circumstances, the hypothermia induced by ethanol is protective in that it reduces the animal's sensitivity to ethanol's behavioral and lethal effects. Experimentally, these findings underline the importance of considering body temperature during intoxication as an important variable in ethanol studies. Clinically, these results suggest that exposure to warm environments can increase the toxic effects of ethanol and, at least in rodents, can transform a nonlethal ethanol dose to a lethal dose. Further, the results also suggest that holding body temperature constant at a subnormal level may represent a simple, noninvasive means of enhancing existing supportive measures and further reducing mortality from ethanol overdose.

C. TERATOGENIC EFFECTS

Ethanol is teratogenic and reportedly causes changes in the development of thermoregulatory systems and in the thermic responses to ethanol (e.g., Taylor et al., 1981; Zimmerberg et al., 1993). Details regarding this topic are presented in Chapter 22 of this volume. The effects of ethanol-induced thermoregulatory changes on fetal development has received little attention.

As discussed above, hypothermia per se has been shown to underlie some of the biochemical effects of ethanol. In addition, hypothermia itself can be teratogenic (Edwards and Wanner, 1977). Further, Henderson and colleagues (1980) found that attenuating ethanol-induced hypothermia in pregnant rats reversed an ethanol-induced decrease in fetal protein synthesis. Taken together, these findings suggest a direct causative link between ethanol-induced hypothermia and ethanol's teratogenic effects.

Randall and associates (1988) tested maternal hypothermia to see if it represents an underlying mechanism of the teratogenic effects of ethanol in mice. They found that the oral administration of 5.8 g/kg ethanol to pregnant mice on day 10 of gestation caused the expected decrease in birth weight and increase in limb and kidney malformations in the offspring when the pregnant animals were exposed to a T_a of 22°C after intubation. Surprisingly, offsetting ethanol hypothermia by exposing the pregnant animals to a T_a of 32°C for 6 h after ethanol intubation exacerbated the teratogenic effects of ethanol. Exposure to 32°C did not significantly affect fetal development in sucrose controls. These singular findings in mice must be evaluated cautiously. However, they suggest that maternal hypothermia may not represent an etiologic factor in animal models of fetal alcohol syndrome. From a clinical perspective, these findings suggest that exposure to warm environments during pregnancy, in addition to the potential for direct teratogenic effects from hyperthermia (Edwards, 1986), may increase the teratogenic effects of ethanol.

D. GENOTYPIC DIFFERENCES

Three lines of evidence discussed in previous sections of this chapter converge to suggest that body temperature during intoxication could play an important role in mediating genetically determined differences in sensitivity to ethanol. First, acute administration of ethanol impairs thermoregulation and induces a dose-dependent degree of hypothermia in laboratory animals exposed to normal T_as during intoxication (Freund, 1973; Ritzmann and Tabakoff, 1976; Lomax et al., 1980; Myers, 1981; Kalant and Lê, 1991). Second, the extent of the hypothermic response varies between species and between genotypes within species (Crabbe, 1983; Crabbe et al., 1994; Phillips and Crabbe, 1991; Tabakoff et al., 1980). Third, body temperature during intoxication can markedly affect sensitivity to the behavioral and lethal effects of ethanol in a variety of species (Dinh and Gailis, 1979; Pohorecky and Rizek, 1981; Malcolm and Alkana, 1981, 1983; Alkana et al., 1983, 1985a, 1985b, 1988; Wenger and Alkana, 1984; Ingram et al., 1982; Finn et al., 1986, 1994; Olofsson and Lindahl, 1977). Therefore, body temperature during intoxication is genetically determined and may influence sensitivity to other effects of ethanol.

Alkana and colleagues (1988) investigated the importance of body temperature during intoxication in mediating differences between five inbred strains of mice (C57BL/6J, BALB/cJ, DBA/2J, A/HeJ, and 129/J) in their acute sensitivity to the hypnotic effects of ethanol. They found that exposing the mice to a T_a of 34°C following ethanol injection markedly reduced strain-related differences in body temperature during intoxication and eliminated the statistically significant differences between strains in their ethanol concentrations at the return of the righting reflex that were evident in mice tested at 22°C. These results confirmed that body temperature can play a significant role in mediating differences in ethanol sensitivity between inbred strains.

Subsequent work extended the investigation to mice selectively bred for sensitivity or resistance to the ethanol-induced loss of the righting reflex (LS and SS, respectively), and hypothermia (COLD and HOT, respectively) (Finn et al., 1990). As with inbred strains, offsetting ethanol hypothermia markedly reduced differences in ethanol sensitivity between genotypes that were evident in mice tested at 22°C. Of particular interest was the absence of statistically significant differences in ethanol sensitivity between the 34°C-exposed LS, C57, and 129 mice. This pattern also extended to the COLD and HOT mice where the differences between COLD and HOT mice in brain ethanol concentrations at the return of the righting reflex in mice exposed to 22°C were eliminated in mice exposed to 34°C during intoxication. On the other hand, ethanol sensitivity in the SS mice exposed to 34°C remained markedly lower than in the other mice tested. In addition, the findings clearly demonstrated qualitative differences in the effects of temperature on ethanol sensitivity within genotypes. Overall comparison within genotypes indicated that C57, A/He, SS, HOT, and COLD mice became more sensitive to ethanol whereas 129/J and LS mice become less sensitive to ethanol, when hypothermia was offset. These qualitative differences in the effects of temperature on ethanol sensitivity within genotypes were not evident when lethal ethanol doses were administered. In this situation, offsetting hypothermia increased sensitivity to lethal doses of ethanol in all genotypes (Finn et al., 1989).

These results clearly demonstrate the important role body temperature during intoxication can play in mediating genetically determined differences in ethanol sensitivity. The pattern of the results indicates that there are qualitative and quantitative differences between genotypes in their responses to body temperature manipulation, making it difficult to anticipate the contribution of temperature to genotypic differences in ethanol sensitivity without testing. Nevertheless, *in vitro* experiments in LS and SS and other mouse genotypes leave little doubt that genetically determined differences in brain sensitivity to ethanol exist at the cellular level and that these differences are not dependent on body temperature (Sorensen et al., 1980; Palmer et al., 1982; Spuhler et al., 1982; Johnson et al., 1985). Therefore, collectively, the available evidence suggests that net sensitivity to ethanol's behavioral effects in an animal probably reflects genetically determined differences in absolute sensitivity to ethanol at the cellular level coupled with genetically determined differences in body temperature response during intoxication.

E. ETHANOL INTAKE

The wide range of factors that can influence ethanol intake are discussed in detail in Chapter 13. The present section concentrates on the relationship between body temperature, ethanol intake, and related behaviors.

Cunningham and colleagues (Cunningham et al., 1988, 1992; Cunningham and Niehus, 1989) have investigated the link between ethanol's hypothermic effect and ethanol intake. Specifically, they have tested the hypothesis that hypothermia produced by ethanol ingestion is either aversive or is highly correlated with the aversive effects of ethanol that ordinarily inhibit consumption of ethanol (Cunningham et al., 1992). The initiation of this work was based, in part, on early studies which found that exposing rats to warm environments (Eriksson, 1969; Myers, 1962) or microwave irradiation (Lai et al., 1984) increased ethanol preference and intake and on their own preliminary work in which rats appeared to regulate ethanol intake in order to prevent or minimize acute ethanol hypothermia (Cunningham and Bischoff, 1986). They found that exposure to a T_a of 32°C significantly decreased the magnitude of ethanol-induced taste aversion in rats (Cunningham et al., 1988). Subsequent work demonstrated that rats would learn to drink more ethanol and prefer an ethanol-paired flavor when ethanol's effects were experienced in a warm (32°C) environment that eliminated ethanol's hypothermic effect, than when the effects of ethanol were experienced at normal room temperature (21°C) (Cunningham and Niehus, 1989). In support of this conclusion, Marfaing-Jallat and Le Magnen (1985) found a highly significant negative correlation between initial sensitivity to ethanol hypothermia and the level of ethanol dependence produced by chronic ethanol treatment and manifested by a voluntary high ethanol intake. Further, Hunt et al. (1991) found that hypothermia was necessary for conditioning taste aversion in preweanling rats. More recently, Cunningham et al. (1992) conducted a comprehensive series of experiments which demonstrated that:

1. Exposure to a low T_a (5°C) facilitated, whereas exposure to a high T_a (32°C) retarded, the acquisition of an ethanol-induced taste aversion.
2. Ethanol-induced taste aversion was directly related to ethanol dose.
3. The strength of ethanol-induced conditioned taste aversion was negatively correlated with core body temperature after ethanol injection.
4. Alterations of the T_a alone did not cause conditioned taste aversion in the absence of drug indicating that altering the T_a does not have direct aversive or rewarding effects.
5. Alterations in the T_a did not appear to influence conditioned taste aversion by changing ethanol pharmacokinetics.

In addition, Cunningham and Niehus (1993) used a place preference paradigm to demonstrate that the T_a's ability to influence ethanol-induced aversive effects is not restricted to tasks involving ingestive behavior. Collectively, these findings consistently support the hypothesis that ethanol hypothermia is aversive and reduces ethanol consumption.

Cunningham et al. (1991) also used genotype, in place of T_a modification, to manipulate body temperature following ethanol. These studies investigated the hedonic effects of ethanol (measured by place conditioning, taste conditioning, and ethanol drinking) in replicate lines of mice selectively bred for sensitivity (COLD) or resistance (HOT) to the hypothermic effects of ethanol (Crabbe et al., 1987). In contrast to the hypothesis that ethanol hypothermia is aversive, HOT mice showed greater conditioned aversion for ethanol-paired flavor cues and drank less ethanol at high concentrations than did COLD mice. These findings are opposite to those in rats discussed earlier, in whom the strength of conditioned taste aversion was directly related to the magnitude of ethanol-induced hypothermia. Subsequent studies suggest that the differences between mice and rats may be, at least in part, attributed to species differences in initial sensitivity to ethanol's hedonic effects, measured by the place-preference paradigm (Cunningham et al., 1993). Cunningham and colleagues (1991) suggest that the differences may also reflect one or both of two alternate explanations:

1. The effects obtained in the rat studies may not be due to differences in ethanol hypothermia, but to effects of elevated ambient temperature on nonthermal responses to ethanol such as ataxia.
2. Taste conditioning in mice may have been affected by an unknown correlated response to selection that had a greater impact on behavior than did the difference in body temperature. For example, the apparent discrepancy between the rat and mouse studies may be resolved by focusing on differences in the glycemic response to ethanol (Cunningham et al., 1991). This suggestion was based on Resinger and Cunningham's (1991) recent report that COLD mice show a greater hyperglycemic response to ethanol than do HOT mice and on previous work showing that elevated T_a enhances the hyperglycemic response to ethanol in rats (Oliveira de Souza and Masur, 1982).

A third, related explanation would be that selection for resistance to ethanol-induced hypothermia in the HOT mice may have affected genes that sensitized these mice to the aversive effects of hypothermia. Consequently, the HOT mice would develop a larger aversive response to ethanol-induced hypothermia and would exhibit less ethanol intake at high concentrations, which produce hypothermia, than COLD mice, even though the HOT mice experience less ethanol-induced hypothermia than COLD mice do. On the other hand, the higher level of ethanol-induced place preference in the HOT versus COLD mice could result from an enhancement of the CNS effect of ethanol by the warmer body temperature in the HOT mice (Finn et al., 1990). That is, since mice exhibit a species specific place preference to ethanol (Cunningham et al., 1993), this behavioral effect of ethanol would be enhanced at the warmer body temperature. The same argument could be made in reverse for the relative responses of the COLD mice.

Studies in selectively bred animals provide insight into the relationship between genes influencing ethanol hypothermia and ethanol intake. As presented in the previous paragraph, selection for sensitivity or resistance to ethanol-induced hypothermia in mice also caused differences between the lines in sensitivity to ethanol's hedonic effects (Cunningham et al., 1991). These results provide strong evidence for a genetic correlation between ethanol's thermal effect and its effect on both place and taste conditioning in mice since the line differences occurred in both replicates. Evidence for a genetic correlation between ethanol's thermal effect and ethanol drinking, however, was weaker since it occurred in only one replicate. Stewart et al. (1992) found that selection for oral ethanol preference (P) or nonpreference (NP) in rats also appears to cause line differences in sensitivity to ethanol's thermic effects. This study showed that P rats had a greater hypothermic response to ethanol than NP rats did on test day 1 of ethanol exposure. However, the P rats developed tolerance to ethanol hypothermia beginning on day 2, whereas, the NP rats became more sensitive to the hypothermic effect of ethanol with repeated injections. In agreement with the hypothesis proposed by Cunningham and colleagues, Stewart et al. (1992) suggest that the sensitization to ethanol hypothermia in the NP rats may serve to limit ethanol intake, and that the more rapid tolerance to ethanol hypothermia in the P line contributes to this line's high ethanol intake. In agreement with this work, Lê and Kiianma (1988) found tolerance in only the alcohol-preferring AA, but not in the alcohol-nonpreferring ANA rat lines. Interestingly, rats selectively bred to be taste aversion prone (TAP) or taste aversion resistant (TAR) to cyclophsophamide injection were prone and resistant to the taste aversive effects of ethanol, but did not exhibit differences in sensitivity to ethanol-induced hypothermia or loss of the righting reflex (Elkins et al., 1992). These latter findings suggest that selective breeding for sensitivity or resistance to taste aversion does not affect the genes controlling ethanol sensitivity, including sensitivity to hypothermia.

Rezvani et al. (1992) investigated the effect of MDMA (ecstasy), which stimulates serotonin release), on ethanol preference and body temperature in two alcohol-preferring rats, the selectively bred P line and the Fawn-Hooded strain. MDMA reduced ethanol consumption in both rat genotypes causing hyperthermia in Fawn-Hooded rats and hypothermia in P rats. The opposing effects of MDMA on body temperature in the two genotypes suggest that the MDMA-induced reduction in ethanol consumption did not result from changes in body temperature. However, the P strain, which had a hypothermic response to MDMA, showed a much greater MDMA-induced reduction in ethanol consumption than did the Fawn-Hooded rats. Consequently, the hypothermic response in P rats may have played a role in reducing ethanol consumption. Therefore, it may be premature to conclude that body temperature change does not contribute to MDMA's negative effects on ethanol consumption.

Some studies have not found a relationship between ethanol hypothermia and ethanol intake. For example, Khanna and co-workers (1990) investigated ethanol consumption and sensitivity to ethanol, including its hypothermic effect, in four outbred rat strains: Fischer 344, Long-Evans, Sprague-Dawley, and Wistar. They found signficant differences in ethanol consumption between the strains with Long-Evans consuming the highest and Fischer 344 the lowest amount of ethanol, but did not find a significant correlation between consumption and ethanol hypothermia. However, the lack of correlation likely reflects the small difference between strains in the size of their hypothermic responses to ethanol (<1.0°C). Similarly, Bisaga and Kostowski (1993) did not find a relationship between ethanol preference and sensitivity to ethanol-induced hypothermia or other measures of ethanol sensitivity in 52 Wistar rats. However, the absence of a correlation probably reflects the very small variability in the hypothermic response to ethanol (ΔT 1 ± 0.1°C).

The mechanisms by which ethanol-induced hypothermia produces its aversive effects and reduces ethanol intake are unknown. As discussed by Cunningham et al. (1992), the ability of ethanol to produce taste aversion despite the complete blockade of hypothermia by exposure to warm T_as indicates that the thermal response to ethanol is not entirely responsible for its ability to condition taste aversion. On the other hand, the ability of a warm T_a to attenuate ethanol's aversive effects cannot be attributed to temperature-mediated alterations in ethanol pharmacokinetics or to mechanisms involving rewarding or aversive effects of altered thermal environments. Further, it is unlikely that hypothermia alone is the source of aversion because previous work found that whole-body cooling to colonic temperatures as low as 21°C failed to cause conditioned taste aversion in rats (Ionescu and Buresova, 1977). Possible mechanisms for the relationship between ethanol's aversive effects and temperature proposed by Cunningham and colleagues include temperature-related changes in neural sensitivity to ethanol, rate of tolerance development, and magnitude of ethanol-induced hyperglycemia (Cunningham et al., 1992; Risinger and Cunningham, 1991). These authors also point out the difficulties in reconciling the ability of warm environments to offset ethanol hypothermia and conditioned taste aversion with the behavioral thermoregulation studies which suggest that ethanol produces a decrease in the body temperature set-point of mice and rats (Crawshaw et al., 1992). The latter studies show that ethanol-treated animals prefer a cooler location in a temperature gradient than do saline controls. In contrast, the taste aversion results suggest that ethanol-injected animals should prefer a warm environment because it reduces the aversiveness of hypothermia. Interestingly, conditioned taste aversion and set-point studies are conducted under very different experimental conditions that impose different types of environmental constraints on the animals. Therefore, it would be interesting to conduct ethanol-induced taste aversion studies in a thermal gradient that would allow an animal to adjust its body temperature, and at least theoretically, minimize the aversiveness of ethanol's thermic effects.

In conclusion, the available evidence supports the hypothesis that ethanol hypothermia is aversive and that ethanol self-administration is influenced by variables which change ethanol's aversive motivational effects by altering the thermal response to ethanol. There is also evidence supporting genetic correlations between ethanol's thermal effect and its effects on place conditioning, taste conditioning, and ethanol drinking. The mechanisms by which ethanol-induced hypothermia produces its aversive effects and reduces ethanol intake remain unknown.

F. *IN VITRO* SENSITIVITY

Temperature can influence membrane biophysical parameters, enzymatic reactions, and physiologic function. Consequently, researchers are careful to control temperature in *in vitro* studies of ethanol. Temperature control usually refers to adjusting temperature to give optimal responses in the particular system, or it means adjusting temperature to be the same in nonexperimental (baseline) and experimental groups. However, a growing body of evidence indicates that test temperature itself can play a critical role in determining *in vitro* sensitivity to ethanol, and that testing at inappropriate temperatures may lead to erroneous conclusions regarding ethanol sensitivity (Hoffman et al., 1984; Rangaraj and Kalant, 1982; Chin and Goldstein, 1981; Harris and Schroeder, 1981; Bejanian et al., 1991a; Lovinger, 1993; Musleh et al., 1994; McQuilkin and Harris, 1990; Towell and Erwin, 1982; Eskuri and Pozos, 1987). The following paragraphs present selected examples of temperature's effects on *in vitro* sensitivity to ethanol measured by biophysical, receptor, and electrophysiologic techniques.

Several studies have shown that the test temperature markedly alters the magnitude of and sensitivity to ethanol's effects in *in vitro* membrane preparations. Early studies by Chin and Goldstein (1981) and Harris and Schroeder (1981) using high ethanol concentrations (>300 m*M*), showed that the magnitude of ethanol's response in brain synaptosomal membranes measured, respectively, by electron paramagnetic resonance and fluorescence polarization of DPH, increased as test temperatures were increased from 10°C to 48°C. More recently, Bejanian et al. (1991a) extended this work across a broad range of ethanol doses. They found that increasing the test temperature from 25°C to 37°C increased baseline fluidity and the magnitude of the relative response to ethanol-induced fluidization of mouse brain synaptic plasma membranes. More important, increasing the test temperature from 25°C to 37°C decreased the minimum concentration of ethanol required to cause a significant increase in membrane fluidity from 170.7 m*M* to 85.0 m*M*. Therefore, altering the test temperature changed the minimum effective concentration of ethanol from supralethal to sublethal in this preparation.

The sensitivity of the NMDA system to ethanol also appears to be temperature sensitive. Early *in vitro* studies demonstrated that changes in temperature of 5°C or 10°C can significantly alter the intrinsic and synaptic physiology of the hippocampal slice preparations used in electrophysiologic studies of ethanol's effects on NMDA function (Thompson et al., 1985) and on the response of the closely related quisqualate/AMPA subclass of glutamate receptor (Larson and Lynch, 1991). Lovinger (1993) measured the inhibition of NMDA receptor-mediated excitatory postsynaptic potentials (EPSPs) by 50 and 100 m*M* ethanol. He found that rat neostriatal and hippocampal preparations were much more sensitive to ethanol when tested at 35°C than when tested at room temperature (22°C). Musleh et al. (1994, 1995) investigated the effects of ethanol on two types of NMDA receptor-mediated responses in three different mouse genotypes (C57, LS, and SS). They found that ethanol produced a decrease in the amplitude of NMDA receptor-mediated potentials from the CA1 region of hippocampal slice preparations. The inhibitory effect of ethanol was more pronounced at 30°C than at 35°C only in C57 mice. Similarly, ethanol produced a decrease in glutamate-stimulated ^{3}H-MK-801 binding in membrane fractions with greater inhibition occurring at 30°C than at 30°C in all genotypes. The effect of ethanol was concentration dependent, and the sensitivity of the genotypes at 35°C was LS>SS>C57, whereas it was SS>LS>C57 at 30°C. At this time, it is not clear whether the differences between the results of Lovinger and Musleh et al. reflect differences in the temperature range studied, species differences, or other factors. In any case, these studies illustrate the important role temperature can play as a variable in the system.

McQuilkin and Harris (1990) investigated the effects of temperature and other factors on the sensitivity of GABA-activated chloride ion channels to ethanol in brain cerebrocortical membrane vesicle (microsac) preparations from LS and SS mice. They found that temperature affected both GABA-activated and direct effects of ethanol on the system. Specifically, they reported that ethanol-potentiation of GABA-activated (5 μM) $^{36}Cl^-$ uptake did not occur with an assay temperature of 0°C, was small at 22°C, but was reliably seen with an assay temperature of 34°C. Interestingly, assay temperature did not markedly alter the effects of GABA or (in those cases where they were tested) the effects of pentobarbital or flunitrazepam on $^{36}Cl^-$ uptake. In addition, ethanol did not alter $^{36}Cl^-$ uptake in the absence of exogenous GABA when the assay temperature was 34°C. However, at 0°C ethanol increased $^{36}Cl^-$ uptake in the absence of GABA. The effect of ethanol at 0°C appeared to be mediated through a $GABA_A$ channel since it was antagonized by picrotoxin. Preliminary studies suggest that temperature also affects the sensitivity of this system to neuroactive steroids (Finn, D.A., Davies, D.L., and Alkana, R.L., in preparation). Here, the potentiation of GABA-activated $^{36}Cl^-$ by 3α,5αP in microsacs from LS and SS mice decreased as the assay temperature was increased from 30°C to 38°C. This effect of temperature occurred in the presence and absence of 50–100 mM ethanol.

In conclusion, the studies presented above exemplify how failure to test at temperatures that match the physiologically relevant situation may lead to erroneous conclusions regarding the sensitivity of the given system to ethanol.

V. SUMMARY AND CONCLUSIONS

Ethanol affects thermoregulation in humans and laboratory animals. The net effect of ethanol intoxication on body temperature depends on the dose and the T_a to which the subject is exposed. In humans, ethanol does not cause major changes in body temperature when moderate doses are consumed during exposure to normal T_as. However, the intake of high ethanol doses in combination with exposure to high or low T_as appears to increase the risk of death from hyperthermia or hypothermia, respectively, particularly in the elderly. Ethanol produces a consistent, dose-dependent hypothermic effect in a range of laboratory animals when they are exposed to normal (20–24°C) or low T_as during intoxication. Exposure to warmer T_as during intoxication can offset the hypothermic effect of ethanol and can produce hyperthermia. The minimum dose of ethanol required to cause ethanol's thermic effects is not clear, but it appears that, independent of the T_a, the dose of ethanol required to produce hypothermia is lower than the dose required to produce hyperthermia.

In addition to the T_a and the ethanol dose, several variables have been shown to affect ethanol's thermic effects. These include measurement technique (the method of temperature assessment, the depth and duration of rectal probe placement, and the timing of temperature assessment with respect to the administration of ethanol), stress, circadian rhythms, age, and genetic background. Failure to control these variables can complicate the interpretation of individual studies or comparisons across studies that investigate ethanol's effects on thermoregulation or that use ethanol's effects on thermoregulation as a dependent variable for measuring ethanol sensitivity.

The mechanisms by which ethanol affects thermoregulation are not fully understood. Studies showing that the same dose of ethanol can produce hypothermia or hyperthermia, depending on the T_a, suggest that ethanol abolishes or broadens the thermoregulatory set-point, thus producing a poikilothermic-like state. However, in contrast to predictions based purely on a thermoregulatory impairment model, hypothermic ethanol-intoxicated animals select a cooler T_a than that selected by nonintoxicated, normothermic controls. The latter point provides convincing evidence that ethanol lowers the thermoregulatory set-point. It is possible that the nature of ethanol's effects on thermoregulatory set-point might be dependent on the doses employed, with small or moderate doses lowering and higher doses impairing the set-point. It is noteworthy that the high doses of ethanol required to produce hyperthermia also impairs heat dissipating capacity.

The anatomical and neurochemical systems underlying ethanol's effects on thermoregulation have received considerable experimental attention. It appears that ethanol affects body temperature regulation, at least in part, by acting on neurons located in the POAH area. Changes in 5-HT and NE neurotransmitter systems have been implicated in mediating the effects of ethanol in the POAH. In addition to NE and 5-HT, several other neurotransmitter systems have been implicated in mediating or modulating ethanol's thermic effects. These include acetylcholine, dopamine, GABA, histamine, neuropeptides (β-endorphin, TRH, and vasopressin), prostaglandins, and calcium ions. Recent studies have not provided definitive answers, but have further implicated the 5-HT system and suggest possible roles for dynorphin, enkephalins, endorphin, neurotensin, TRH, and neuroactive steroids in mediating ethanol's thermic effects. Little is known about the role the NMDA system plays in acute ethanol hypothermia, despite considerable evidence that the NMDA receptor antagonist, MK-801, can block tolerance to ethanol hypothermia. The failure of benzodiazepine antagonists and picrotoxin to alter ethanol hypothermia suggests that the GABA system does not play a critical role in mediating this effect of ethanol. However, newer work with neuroactive steroids and $GABA_B$ antagonists suggests that the involvement of GABA receptors and the link between neurosteroids and $GABA_A$ receptors in mediating ethanol-induced hypothermia requires further investigation.

It is also possible that biophysical changes underlie ethanol's effects on body temperature. Both ethanol and increases in temperature cause disordering of brain membrane lipids. This common effect of temperature and ethanol has led to suggestions that homeostatic mechanisms may react to ethanol-induced membrane perturbations as if they were an increase in brain temperature. Consequently, the reduction in set-point and other thermic effects of ethanol may reflect, in part, adaptive responses to the physico-chemical interaction between ethanol and brain cell membranes.

The body temperature change induced by ethanol has important pharmacokinetic, pharmcodynamic, and behavioral consequences. Ethanol-induced hypothermia can markedly slow the rate of ethanol elimination and, thus, can prolong ethanol's behavioral effects. More important, body temperature can affect brain sensitivity to ethanol. Under most circumstances, ethanol-induced hypothermia is protective in that it reduces the animal's sensitivity to ethanol as measured by motor performance, loss of righting reflex, and lethality. Further, limited information suggests that offsetting ethanol hypothermia may exacerbate the teratogenic effects of ethanol.

Considerable evidence indicates that body temperature during intoxication can play an important role in mediating genetically determined differences in sensitivity to other effects of ethanol. The pattern of results indicates that there are qualitative and quantitative differences between genotypes in the interaction between ethanol's effects on body temperature and ethanol sensitivity, making it difficult to anticipate the contribution of temperature to ethanol sensitivity without testing. Taken together, this work suggests that net sensitivity to ethanol's behavioral effects in an animal probably reflects a combination of genetically determined differences in absolute sensitivity to ethanol at the cellular level coupled with genetically determined differences in body temperature response during intoxication.

There also appears to be a link between ethanol's hypothermic effects and the willingness of animals to voluntarily drink ethanol. In general, findings support the hypothesis that the hypothermia produced by ethanol ingestion is aversive and that ethanol self-administration is influenced by variables that change ethanol's aversive motivational effects by altering the thermal response to ethanol. Further, there appears to be a genetic correlation between ethanol's effects on thermoregulation and its effects on place conditioning, taste conditioning, and ethanol intake.

Temperature plays a critical role in determining *in vitro* sensitivity to ethanol. Temperature can influence both the magnitude of ethanol's effects and the threshold sensitivity to ethanol. The findings presented exemplify how failure to test at temperatures that match the physiologically relevant situation may lead to erroneous conclusions regarding the sensitivity of a given *in vitro* model system to ethanol.

Overall, the experimental findings discussed in this chapter underline the importance of considering temperature as a critical variable in both *in vivo* and *in vitro* studies of ethanol's effects.

REFERENCES

Abel, E.L. (1978) Effects of ethanol and pentobarbital on mice of different ages. *Physiol. Psychol.,* 6:366–368.

Abel, E.L. and York, J.L. (1979) Age-related differences in the response to ethanol in the rat. *Physiol. Psychol.,* 7:391–395.

Albiin, N. and Eriksson, A. (1984) Fatal accidental hypothermia and alcohol. *Alcohol Alcoholism,* 19:13–22.

Alkana, R.L., Finn, D.A., and Malcolm, R.D. (1983) The importance of experience in the development of tolerance to ethanol hypothermia. *Life Sci.,* 32:2685–2692.

Alkana, R.L., Boone, D.C., and Finn, D.A. (1985a) Temperature dependence of ethanol depression: linear models in male and female mice. *Pharmacol. Biochem. Behav.,* 23:309–316.

Alkana, R.L., Finn, D.A., Galleisky, G.G., Bejanian, M., Boone, D.C., Jones, B.L., and Syapin, P.J. (1985b) Temperature modulates ethanol sensitivity in mice: generality across strain and sex. *Alcohol,* 2:281–285.

Alkana, R.L., Finn, D.A., Bejanian, M., and Crabbe, J.C. (1988) Genetically determined differences in ethanol sensitivity influenced by body temperature during intoxication. *Life Sci.,* 43:1973–1982.

Allan, A.M. and Harris, R.A. (1989) A new alcohol antagonist: phaclofen. *Life Sci.,* 45:1771–1779.

Allison, T.G. and Reger, W.E. (1992) Thermoregulatory, cardiovascular, and physiological response to alcohol in man. *J. Appl. Physiol.,* 72:2099–2107.

Altland, P.D., Highman, B., Parker, M.G., and Dieter, M.P. (1970) Serum enzyme, corticosterone and tissue changes in rats following a single oral dose of ethanol. *Q. J. Stud. Alcohol,* 31:281–287.

Andersen, K.L., Helstrom, B., and Lorentzen, F.V. (1963) Combined effect of cold and alcohol on heat balance in man. *J. Appl. Physiol.,* 18:975–982.

Aschoff, J. (1970) Circadian rhythm of activity and body temperature. In *Physiological and Behavioral Temperature Regulation.* J.D. Hardy, A.P. Gagge, and J.A.J. Stolwijk, Eds., Thomas, Springfield, IL, pp. 905–919.

Baker, M.A. (1979) A brain-cooling system in mammals. *Sci. Am.,* 240:130–139.

Baker, M.A. (1982) Brain cooling in endotherms in heat and exercise. *Ann. Rev. Physiol.,* 44:85–96.

Ballard, B.E. (1974) Pharmacokinetics and temperature. *J. Pharm. Sci.,* 63:1345–1358.

Bamford, O.S. and Eccles, R. (1983) The role of sympathetic efferent activity in the regulation of brain temperature. *Pflugers Arch.,* 396:138–143.

Bartlett, R.G., Jr., Bohr, B.C., Helmendach, R.H., Foster, G.L., and Miller, M.A. (1954) Evidence of an emotional factor in hypothermia produced by restraint. Am. *J. Physiol.,* 179:343–346.

Bejanian, M., Finn, D.A., Syapin, P.J., and Alkana, R.L. (1987) Rectal and brain temperatures in ethanol intoxicated mice. *Psychopharmacology,* 92:301–307.

Bejanian, M., Finn, D.A., Syapin, P.J., and Alkana, R.L. (1990) Body temperature and ethanol pharmacokinetics in temperature-challenged mice. *Alcohol,* 7:331–337.

Benjanian, M., Alkana, R.L., Von Hungen, K., Baxter, C., and Syapin, P.J. (1991a) Temperature alters ethanol-induced fluidization of C57 mouse brain membranes. *Alcohol,* 8:117–121.

Bejanian, M., Jones, B.L., Syapin, P.J., Finn, D.A., and Alkana, R.L. (1991b) Brain temperature and ethanol sensitivity in C57 mice: a radiotelemetric study. *Pharmacol. Biochem. Behav.,* 39:457–463.

Bisaga, A. and Kostowski, W. (1993) Individual behavioral differences and ethanol consumption in Wistar rats. *Physiol. Behav.,* 54:1125–1131.

Boulant, J.A. (1981) Hypothalamic mechanism in thermoregulation. *Fed. Proc.,* 40:2843–2850.

Boulant, J.A., Curras, M.C., and Dean, J.B. (1989) Neurophysiological aspects of thermoregulation. In *Advances of Comparative and Enviromental Physiology.* L.C.H. Wang, Ed., Springer-Verlag, Berlin, pp. 117–134.

Bowler, K. and Tirri, R. (1974) The temperature characteristics of synaptic membrane ATPases from immature and adult rat brain. *J. Neurochem.,* 23:611–613.

Brick, J., Pohorecky, L.A., Faulkner, W., and Adams, M.N. (1982) Circadian variations in behavioral and biological sensitivity to ethanol. *Alcoholism: Clin. Exp. Res.,* 8:204–211.

Brick, J. and Pohorecky, L.A. (1983a) Effect of ethanol on surface body temperature as measured by infrared radiation detection. *Psychopharmacology,* 81:244–246.

Brick, J. and Pohorecky, L.A. (1983b) The neuroendocrine response to stress and the effect of ethanol. In *Stress and Alcohol Use.* L.A. Pohorecky and J. Brick, Eds., Elsevier, Amsterdam, pp. 389–402.

Brown, V. (1981) Spa associated hazards. An update and summary. US Consumer Product and Safety Commission, Washington D.C.

Burger, F.J. and Fuhrman, F.A. (1964) Evidence of injury by heat in mammalian tisssues. *Am. J. Physiol.,* 44:1057–1061.

Cabanac, A. and Briese, E. (1991) Handling elevates the colonic temperature of mice. *Physiol. Behav.,* 51:95–98.

Chai, C.Y., Chen, H.I., and Yin, T.H. (1971) Central cites of action and effects of acetylstrophanthidine on body temperature in monkeys. *Expl. Neurol.,* 33:618–628.

Chan, A.W.K., Schanley, D.L., Aleo, M.D., and Leong, F.W. (1985) Cross-tolerance between ethanol and chlordiazepoxide. *Alcohol,* 2:209–213.

Chesler, A., LaBelle, G.G., and Himwich, H.E. (1942) The relative effects of toxic doses of alcohol on fetal, newborn and adults rats. *J. Stud. Alcohol,* 3:1–4.

Chin, J.H. and Goldstein, D.B. (1981) Membrane-disordering action of ethanol. Variation with membrane cholesterol content and depth of the spin label probe. *Mol. Pharmacol.,* 19:425–431.

Conklin, P. and Hegggeness, F.W. (1971) Maturation of temperature homeostasis in the rat. *Am. J. Physiol.,* 220:333–336.

Cossins, A.R., Friedlander, M.J., and Prosser, C.L. (1977) Correlations between behavioral temperature adaptations of goldfish and the viscosity and fatty acid composition of their synaptic membranes. *J. Comp. Physiol.,* 120:109

Crabbe, J.C. (1983) Sensitivity to ethanol in inbred mice: genotypic correlations among several behavioral responses. *Behav. Neurosci.,* 97:280–289.

Crabbe, J.C., Kosobud, A., Tam, B.R., Young, E.R., and Deutsch, C.M. (1987) Genetic selection of mouse lines sensitive (COLD) and resistant (HOT) to acute ethanol hypothermia. *Alcohol Drug Res.,* 7:163–174.

Crabbe, J.C. (1989) Genetic aminal models in the study of alcoholism. *Alcoholism: Clin. Exp. Res.,* 13:120–127.

Crabbe, J.C., Feller, D.J., and Dorow, J.S. (1989) Sensitivity and tolerance to ethanol-induced hypothermia in genetically selected mice. *J. Pharmacol. Exp. Ther.,* 249(2):456–461.

Crabbe, J.C., Phillips, T.J., Kosobud, A., and Belknap, J.K. (1990) Estimation of genetic correlation: interpretation of experiments using selectively bred and inbred animals. *Alcoholism: Clin. Exp. Res.,* 14:141–151.

Crabbe, J.C., Gallaher, E.S., Phillips, T.J., and Belknap, J.K. (1994) Genetic determinants of sensitivity to ethanol in inbred mice. *Behav. Neurosci.,* 108:186–195.

Crabbe, J.D., Rigter, H., Uijlen, J., and Strijbos, C. (1979) Rapid developement of tolerance to the hypothermic effect of ethanol in mice. *J. Pharmacol. Exp. Ther.,* 208:128–133.

Crawshaw, L.I., O'Connor, C.S., and Wollmuth, L.P. (1992) Ethanol and the neurobiology of temperature regulation. In *Alcohol and Neurobiology: Brain Development and Hormone Regulation.* R.R. Watson, Ed., Boca Raton, FL, CRC Press, pp. 341–360.

Crawshaw, L.I., Wollmuth, L.P., and O'Connor, C.S. (1989) Intracranial ethanol and ambient anoxia elicit selection of cooler water by goldfish. *Am. J. Physiol.,* 256: R133–R137.

Cunningham, C.L., Hawks, D.M., and Niehus, D.R. (1988) Role of hypothermia in ethanol-induced conditioned taste aversion. *Psychopharmacology,* 95:318–322.

Cunningham, C.L., Hallett, C.L., Niehus, D.R., Hunter, J.S., Nouth, L., and Risinger, F.O. (1991) Assessment of ethanol's hedonic effects in mice selectively bred for sensitivity to ethanol-induced hypothermia. *Psychopharmacology,* 105:84–92.

Cunningham, C.L., Niehus, J.S., and Bachtold, J.F. (1992) Ambient temperature effects on taste aversion conditioned by ethanol: contribution of ethanol-induced hypothermia. *Alcoholism: Clin. Exp. Res.,* 16:1117–1124.

Cunningham, C.L., Niehus, J.S., and Noble, D. (1993) Species difference in sensitivity to ethanol's hedonic effects. *Alcohol,* 10:97–102.

Cunningham, C.L. and Bischof, L.L. (1987) Stress and ethanol-induced hypothermia. *Physiol. Behav.,* 40:377–382.

Cunningham, C.L. and Bischoff, L.L. (1986) Failure to induce tolerance to ethanol's hypothermic effect by oral self-administration in rats. *Alcoholism: Clin. Exp. Res.,* 10:108.

Cunningham, C.L. and Niehus, D.R. (1989) Effect of ingestion-contingent hypothermia on ethanol self-administration. *Alcohol,* 6:377–380.

Cunningham, C.L. and Niehus, J.S. (1993) Drug-induced hypothermia and conditioned place aversion. *Behav. Neurosci.,* 107:468–479.

Cunningham, C.L. and Peris, J. (1983) A microcomputer system for temperature biotelemetry. *Behav. Res. Methods Instrum.,* 15:598–603.

Deimling, M.J. and Schnell, R.C. (1980) Circadian rhythms in the biological response and disposition of ethanol in the mouse. *J. Pharmacol. Exp. Ther.,* 213:1–8.

Dinh, T.K.H. and Gailis, L. (1979) Effect of body temperature on acute ethanol toxicity. *Life Sci.,* 25:547–552.

Dobbing, J. and Sands, J. (1979) Comparative aspects of the brain growth spurt. *Early Human Dev.,* 3:70–83.

Donhoffer, S., Szegvari, G.Y., Jarai, I., and Farkas, M. (1959) Thermoregulatory heat production in the brain. *Nature,* 184:993–994.

Dybing, F. (1945) The blood alcohol curve in hypothermia. *Acta Pharmacol.,* 1:77–81.

Edwards, M.J. (1986) Hyperthermia as a teratogen: a review of experimental studies and their clinical significance. *Teratogen Carcinogen Mutagen,* 6:563–582.

Edwards, M.J. and Wanner, R.A. (1977) Extremes of temperature. In *Handbook of Teratology*. Anonymous. Plenum Press, New York, Wilson, J.G. and Fraser, F.C. (Eds.) Vol. 1, pp. 421–444.

Elkins, R.L., Walters, P.A., and Orr, T.E. (1992) Continued development and unconditioned stimulus characterization of selectively bred lines of taste aversion prone and resistant rats. *Alcoholism: Clin. Exp. Res.,* 16:928–934.

Eriksson, K. (1969) Factors affecting voluntary alcohol consumption in the albino rat. *Ann. Zool. Fennici.,* 6:227–265.

Erwin, V.G. and Jones, B.C. (1993) Genetic correlations among ethanol-related behaviors and neurotensin receptors in long sleep (LS) × short sleep (SS) recombinant inbred strains of mice. *Behav. Genet.,* 23:191–196.

Erwin, V.G. and Su, N.C. (1989) Neurotensin and ethanol interactions on hypothermia and locomotor activity in LS and SS mice. *Alcoholism: Clin. Exp. Res.,* 13:91–94.

Eskuri, S.A. and Pozos, R.S. (1987) The effect of ethanol and temperature on calcium dependent sensory neuron action potentials. *Alcohol Drug Res.,* 7:153–162.

Feller, D.J. and Crabbe, J.C. (1991) Effect of neurotransmitter-selective drugs in mice selected for differential sensitivity to the hypothermic actions of ethanol. *J. Pharmacol. Exp. Ther.,* 256:954–958.

Fellows, I.W., MacDonald, I.A., and Bennet, T. (1984) The influence of enviromental temperature upon thermoregulatory responses to ethanol in man. *Clin. Sci.,* 66:733–739.

Ferko, A.P. and Bobyock, E. (1978) Physical dependence on ethanol. Rate of ethanol clearance from the blood and effect of ethanol on body temperature in rats. *Toxicol. Appl. Pharmacol.,* 46:235–248.

Finn, D.A., Boone, D.C., and Alkana, R.L. (1986) Temperature dependence of ethanol depression in rats. *Psychopharmacology,* 90:185–189.

Finn, D.A., Bejanian, M., Jones, B.L., Syapin, P.J., and Alkana, R.L. (1989) Temperature affects ethanol lethality in C57BL/6, 129, LS and SS mice. *Pharmacol. Biochem. Behav.,* 34:375–380.

Finn, D.A., Bejanian, M., Jones, B.L., McGivern, R.F., Syapin, P.J., Crabbe, J.C., and Alkana, R.L. (1990) Body temperature differentially affects ethanol sensitivity in both inbred strains and selected lines of mice. *J. Pharmacol. Exp. Ther.,* 253:1229–1235.

Finn, D.A., Bejanian, M., Jones, B.L., Babbini, M., Syapin, P.J., and Alkana, R.L. (1991a) The relationship between brain temperature during intoxication and ethanol sensitivity in LS and SS mice. *Alcoholism: Clin. Exp. Res.,* 15:717–724.

Finn, D.A., Syapin, P.J., Bejanian, M., Jones, B.L., and Alkana, R.L. (1991b) Body temperature influences ethanol and ethanol/pentobarbital lethality in mice. *Alcohol,* 8:39–41.

Finn, D.A., Syapin, P.J., Bejanian, M., Jones, B.L., and Alkana, R.L. (1994) Temperature dependence of ethanol depression in mice: dose response. *Alcoholism: Clin. Exp. Res.,* 18(2):382–386.

Fox, G.R., Hayward, J.S., and Hobson, G.N. (1979) Effect of alcohol on thermal balance of man in cold water. *Can. J. Physiol. Pharmacol.,* 57:860–865.

Frankel, D., Khanna, J.M., Kalant, H., and LeBlanc, A.E. (1978) Effect of p-chlorophenylalanine on the acquisition of tolerance to the hypothermic effects of ethanol. *Psychopharmacology,* 57:239–242.

Frankel, H.M. (1959) Effects of restraint on rats exposed to high temperature. *J. Appl. Physiol.,* 14:997–999.

French, T.A. and Weiner, N. (1991) Serotoninergic involvement in ethanol-induced alterations of thermoregulation in long-sleep and short-sleep mice. *J. Pharmacol. Exp. Ther.,* 259:833–840.

Freund, G. (1973) Hypothermia after acute ethanol and benzyl alcohol administration. *Life Sci.,* 13:345–349.

Freund, G. (1979) Ethanol-induced changes in body temperature and their neurochemical consequences. In *Biochemistry and Pharmacology of Ethanol.* E. Majchrowicz and E.P. Noble, Eds., Plenum Press, New York, pp. 439–452.

Gailis, L. and Tourigny, A. (1984) Chlorpromazine and dithioerythritol protection against acute ethanol toxicity. *Alcoholism: Clin. Exp. Res.,* 8:808–813.

Gallaher, E.J. and Egner, D.A. (1985) Ethanol induced temperature disturbances measured by automated radiotelemetry. *Proc. West. Pharmacol. Soc.,* 28:221–223.

Gallaher, E.J. and Egner, D.A. (1987) Rebound hyperthermia follows ethanol-induced hypothermia in rats. *Psychopharmacology,* 91:34–39.

Gilliam, D.M. (1989) Alcohol absorption rate affects hypothermic response in mice: evidence for rapid tolerance. *Alcohol,* 6(5):357–362.

Goldstein, D.B., Chin, J.H., McComb, J.M., and Parsons, L.M. (1980) Chronic effects of alcohols on mouse biomembranes. *Adv. Exp. Med. Biol.,* 126:1–5.

Gordon, C.J., Fogelson, L., Mohler, F., Stead, A.G., and Rezvani, A.H. (1988a) Behavioral thermoregulation in the rat following the oral administration of ethanol. *Alcohol Alcoholism,* 23:383–390.

Gordon, C.J., Mohler, F.S., Watkinson, W.P., and Rezvani, A.H. (1988b) Temperature regulation in laboratory mammals following acute toxic insult. *Toxicol.,* 53:161–178.

Gordon, C.J. and Mohler, F.S. (1990) Thermoregulation at a high ambient temperature following the oral administration of ethanol in the rat. *Alcohol,* 7:551–555.

Gordon, C.J. and Stead, A.G. (1986) Effect of alcohol on behavioral and autonomic thermoregulation in mice. *Alcohol,* 3:339–343.

Gordon, C.J. and Stead, A.G. (1988) Effect of ethyl alcohol on thermoregulation in mice following the induction of hypothermia or hyperthermia. *Pharmacol. Biochem. Behav.,* 29(4):693–698.

Graham, T.E. (1981a) Alcohol ingestion and man's ability to adapt to exercise in cold enviroment. *Can. J. Appl. Sport Sci.,* 6:27–31.

Graham, T.E. (1981b) Thermal and glycemic responses during mild exercise +5 to –15°C environments following alcohol ingestion. *Aviat. Space Environ. Med.,* 52:517–522.

Graham, T.E. (1983) Alcohol ingestion and sex differences on the thermal responses to mild exercise in a cold enviroment. *Human Biol.,* 55:463–476.

Graham, T.E. and Baulk, K. (1980) Effect of alcohol ingestion on man's thermoregulatory responses during cold water immersion. *Aviat. Space Environ. Med.,* 51:155–159.

Grieve, S.J. and Littleton, J.M. (1979) Ambient temperatures and the development of functional tolerance to ethanol by mice. *J. Pharm. Pharmacol.,* 31:707–708.

Gupta, K.K. (1960) Effect of alcohol in cold climate. *J. Indian Med. Assoc.,* 35:211–212.

Haight, J.S.J. and Keatinge, W.R. (1973) Failure of thermoregulation in the cold during hypoglycemia induced by exercise and ethanol. *J. Physiol.,* 229:87–97.

Harris, R.A. and Schroeder, F. (1981) Ethanol and the physical properties of brain membranes: fluorescence studies. *Mol. Pharmacol.,* 20:128–137.

Haus, E. and Halberg, F. (1959) 24-hour rhythm in susceptibility of C mice to a toxic dose of ethanol. *J. Appl. Physiol.,* 14:878–880.

Heller, H.C., Graf, R., and Rautenberg, W. (1983) Circadian and arousal state influences on thermoregulation in the pigeon. *Am. J. Physiol.,* 245:R321–R328.

Henderson, G.I., Hoympa, A.M., Rothschild, M.A., and Schenker, S. (1980) Effect of ethanol and ethanol-induced hypothermia on protein synthesis in pregnant and fetal rats. *Alcoholism: Clin. Exp. Res.,* 4:165–177.

Hirvonen, J. (1976) Necropsy findings in fatal hypothermia cases. *Forens. Sci.,* 8:155–164.

Hirvonen, J. and Huttunen, P. (1977) The effect of ethanol on the ability of guinea pigs to withstand severe cold exposure. In *Drugs, Biogenic Amines and Body Temperature,* Third symposium on the Pharmacology of Thermoregulation, Banff, Alberta. K.E. Cooper, P. Lomax, and E. Schonbaum, Eds., Karger, Basel, Switzerland, pp. 230–232.

Hirvonen, J. and Huttunen, P. (1982) Increased urinary catecholamines in hypothermia deaths. *J. Forens. Sci.,* 27:264–271.

Hoffman, P.L., Chung, C.T., and Tabakoff, B.T. (1984) Effects of ethanol, temperature and endogenous regulatory factors on the characteristics of striatal opiate receptors. *J. Neurochem.,* 43:1003–1010.

Hoffman, P.L., Tabakoff, B., Szabó, G., Suzdak, P.D., and Paul, S.M. (1987) Effect of an imidazobenzodiazepine, Ro15–4513, on the incoordination and hypothermia produced by ethanol and pentobarbital. *Life Sci.,* 41:611–619.

Hoffman, P.L. and Tabakoff, B. (1987) Alterations in dopamine receptor sensitivity by chronic ethanol treatment. *Nature,* 268:551–553.

Hollstedt, C. and Rydberg, U.E. (1985) Postnatal effects of alcohol on the developing rat. In *Alcohol and The Developing Brain.* U. Rydberg, C. Alling, J. Engel, B. Pernow, L.A. Pellborn, and S. Rossner, Eds., Raben Press, New York, pp. 69–84.

Howerton, T.C., O'Connor, M.F., and Collins, A.C. (1983) Differential effects of long-chain alcohols in long- and short-sleep mice. *Psychopharmacology,* 79:313–317.

Hunt, P.S., Spear, L.P., and Spear, N.E. (1991) An ontogenic comparison of ethanol-mediated taste aversion learning and ethanol-induced hypothermia in preweanling rats. *Behav. Neurosci.,* 105: 971–983.

Huttunen, P., Penttinen, J., and Hirvonen, J. (1980) The effect of ethanol and cold-adaptation on the survival of guinea pigs in severe cold. *Z. Rechtmed.,* 85:289–294.

Huttunen, P., Lapinlampi, T., and Myers, R.D. (1988) Temperature-related release of serotonin from unrestrained rat's preoptic area perfused with ethanol. *Alcohol,* 5:189–193.

Huttunen, P. and Myers, R.D. (1985) Release of norepinephrine from the rat's hypothalamus perfused with alcohol in relation to body temperature. *Alcohol,* 2:683–691.

Ingram, L.O., Carey, V.C., and Dombek, K.M. (1982) On the relationship between alcohol narcosis and membrane fluidity. *Subst. Alcohol Actions Misuse,* 2:213–224.

Ionescu, E. and Buresova, O. (1977) Effects of hypothermia on the acquisition of conditioned taste aversion in rats. *J. Comp. Physiol. Psychol.,* 91:1297–1307.

Johnson, S.W., Hoffer, B.J., Baker, R., and Friedman, R. (1985) Correlation of purkinje neuron depression and hypnotic effects of ethanol in inbred strains of rat. *Alcoholism: Clin. Exp. Res.,* 9:56–58.

Jones, B.C., Connell, J.M., and Erwin, V.G. (1990) Isolate housing alters ethanol sensitivity in long-sleep and short-sleep mice. *Pharmacol. Biochem. Behav.,* 35:469–472.

Kalant, H. and Lê, A.D. (1984) Effects of ethanol on thermoregulation. *Pharmacol. Ther.,* 23:313–364.

Kalant, H. and Lê, A.D. (1991) Effects of ethanol on thermoregulation. In *Thermoregulation: Pathology, Pharmacology and Therapy.* E. Schönbaum and P. Lomax, Eds., Pergamon Press, New York, pp. 561–617.

Kalant, H. and Woo, N. (1981) Electrophysiological effects of ethanol on the nervous system. *Pharm. Ther.,* 14:431–457.

Keatinge, W.R. and Evans, M. (1960) Effect of food, alcohol and hyoscine on body-temperature and reflex responses of men immersed in cold water. *Lancet,* 2:176–178.

Khanna, J.M., Kalant, H., Shah, G., and Sharma, H. (1990) Comparison of sensitivity and alcohol consumption in four outbred strains of rats. *Alcohol,* 7(5):429–434.

Khanna, J.M., Wu, P.H., Weiner, J., and Kalant, H. (1991) NMDA antagonist inhibits rapid tolerance to ethanol. *Brain Res. Bull.,* 26:643–645.

Khanna, J.M., Kalant, H., Shah, G., and Chau, A. (1992) Effect of (+)MK-801 and ketamine on rapid tolerance to ethanol. *Brain Res. Bull.,* 28:311–314.

Kiianmaa, K. and Sinclair, J.D. (1985) Physiology of the young rat brain and alcohol. In *Alcohol and The Developing Brain.* U. Rydberg, C. Alling, J. Engel, B. Pernow, L.A. Pellborn, and S. Rossner, Eds., Raben Press, New York, pp. 11–18.

Kluger, M.J., O'Reilly, B., Shope, T.R., and Vander, A.J. (1987) Further evidence that stress hypertermia is a fever. *Physiol. Behav.,* 39:763–766.

Koren, G., Barker, C., Bohn, D., Kent, G., McGuigan, M., and Biggar, D. (1989) Effect of hypothermia on the pharmacokinetics of ethanol in piglets. *Ann. Emerg. Med.,* 18(2):118–121.

Kortelainen, M.L. (1987) Drugs and alcohol in hypothermia and hyperthermia related deaths: a retrospective study. *J. Forensic Sci.,* 32(6):1704–1712.

Krarup, N. and Larsen, J.A. (1972) The effect of slight hypothermia on liver function as measured by the elimination rate of ethanol, the hepatic uptake and excretion of indocyanine green and bile formation. *Acta Physiol. Scand.,* 84:396–407.

Kruckeburg, T.W., Gaetano, P.K., Burns, E.M., Stibler, H., Cerven, E., and Borg, S. (1984) Ethanol in preweanling rats with dams: body temperature unaffected. *Neurobehav. Toxicol. Teratol.,* 6:307–312.

Lable, R. and Rydberg, U. (1982) Effects of ethanol on locomotor activity in rats of different ages. *Acta Pharmacol. Toxicol.,* 50:246–250.

Lagerspetz, K.Y.H. (1972) Diurnal variation in the effects of alcohol and in the brain 5-hydroxytryptamine metabolism in mice. *Acta Pharmacol. Toxicol.,* 31:509–520.

Lai, H., Horita, A., Chou, C.K., and Guy, A.W. (1984) Ethanol-induced hypothermia and ethanol consumption in the rat are affected by low-level microwave irradiation. *Bioelectromagnetics,* 5:213–220.

Larsen, J.A. (1971) The effect of cooling on liver function in cats. *Acta Physiol. Scand.,* 81:197–207.

Larson, J. and Lynch, G. (1991) A test of the spine resistence hypothesis for LTP expression. *Brain Res.,* 538:347–350.

Leikola, A. (1962) Influence of stress on alcohol intoxication in rats. Q. *J. Stud. Alcohol,* 23:369–379.

Lê, A.D., Khanna, J.M., Kalant, H., and LeBlanc, A.E. (1980) Effect of 5,7-dihydroxytryptamine on the development of tolerance to ethanol. *Psychopharmacology,* 67:143–146.

Lê, A.D., Khanna, J.M., Kalant, H., and LeBlanc, A.E. (1981) The effect of lesions in the dorsal, median and magnus raphe nuclei on the development of tolerance to alcohol. *J. Pharmacol. Exp. Ther.,* 218:525–529.

Lê, A.D., Kalant, H., and Khanna, J.M. (1986) Influence of ambient temperature on the development and maintenance of tolerance to ethanol-induced hypothermia. *Pharmacol. Biochem. Behav.,* 25:667–672.

Lê, A.D. and Kiianma, K. (1988) Characteristics of ethanol tolerance in alcohol drinking (AA) and alcohol avoiding (ANA) rats. *Psychopharmacology,* 94:479–483.

Li, G.C. and Hahn, G.M. (1978) Ethanol-induced tolerance to heat and adriamycin. *Nature,* 274:699–701.

Lichtenfels, R. and Fröhlich, R. (1852) Bebachtugen uber die Gestze des Granges der Pulsfrequenz und Korperworme in den normalen Zustanden sowie unter dem Einflusse bestimmter Ursachen. *Denschr. Akad. Wiss. Wiem.,* 3:113–154.

Linakis, J.G. and Cunningham, C.L. (1979) Effects of concentration of ethanol injected intraperitoneally on taste aversion, body temperature, and activity. *Psychopharmacology,* 64:61–65.

Littleton, J.M., Grieve, S.J., Griffiths, P.J., and John, G.R. (1980a) Ethanol-induced alteration in membrane phospholipid composition: possible relationship to development of tolerance to ethanol. *Adv. Exp. Med. Biol.,* 126:7–19.

Littleton, J.M., John, G.R., Jones, P.A., and Grieve, S.J. (1980b) The rapid onset of functional tolerance to ethanol — role of different neurotransmitters and synaptosomal membrane lipids. *Acta Psychiatr. Scand. Suppl.,* 286:137–151.

Lomax, P., Malveaux, E., and Smith, R.E. (1964) Brain temperatures in the rat during exposure to low environmental temperatures. *Am. J. Physiol.,* 207:736–739.

Lomax, P. (1966) Measurement of core temperature in the rat. *Nature,* 210:854–855.

Lomax, P., Bajorek, J.G., Chesarek, W.A., and Chaffee, R.R.J. (1980) Ethanol induced hypothermia in the rat. *Pharmacology,* 21:288–294.

Lomax, P., Bajorek, J.G., Bajorek, T.A., and Chaffee, R.R.J. (1981) Thermoregulatory mechanisms and ethanol hypothermia. *Eur. J. Pharmacol.,* 71:483–487.

Lomax, P. and Bajorek, J.G. (1980) Comparative thermoregulatory effects of ethanol in rats, mice and gerbils. *Proc. West. Pharmacol. Soc.,* 23:219–223.

Lomax, P. and Lee, R.J. (1982) Cold acclimation and resistance to ethanol-induced hypothermia. *Eur. J. Pharmacol.,* 84:87–91.

Lomax, P. and Schönbaum, E. (1979) Body temperature: Regulation, drug effects and therapeutic implications. In *Biochemistry and Pharmacology of Ethanol.* E. Majchrowicz and E.P. Noble, Eds., Dekker, New York, pp. 439–452.

Lomax, P. and Schönbaum, E. (1991) *Thermoregulation: Pathology, Pharmacology and Therapy.* Pergamon Press, New York.

Lovinger, D.M. (1993) Temperature dependence of selective ethanol inhibition of NMDA receptor-mediated synaptic transmission. *Alcoholism: Clin. Exp. Res.,* 17:472(Abstract).

MacGregor, D.C., Schonbaum, E., and Bigelow, W.G. (1965) Effects of hypothermia on disappearance of ethanol from arterial blood. *Am. J. Physiol.,* 208:1016–1020.

Maickel, R.P. and Nash, J.F. (1985) Differing effects of short-chain alcohols on body temperature and coordinated muscular activity in mice. *Neuropharmacology,* 24:83–89.

Malcolm, R.D. and Alkana, R.L. (1981) Temperature dependence of ethanol depression in mice. *J. Pharmacol. Exp. Ther.,* 217:770–775.

Malcolm, R.D. and Alkana, R.L. (1982) Hyperbaric ethanol antagonism: role of temperature, blood and brain ethanol concentrations. *Pharmacol. Biochem. Behav.,* 16:341–346.

Malcolm, R.D. and Alkana, R.L. (1983) Temperature dependence of ethanol lethality in mice. *J. Pharm. Pharmacol.,* 35:306–311.

Mansfield, J.G. and Cunningham, C.L. (1980) Conditioning and extinction of tolerance to the hypothermic effect of ethanol in rats. *J. Comp. Physiol. Psychol.,* 94:962–969.

Marbach, G. and Schwertz, M.T. (1964) Effects physiologiques de l'alcool et de la cafeine au cours du sommeil chez l'homme. *Arch. Sci. Physiol.,* 18:163–210.

Marfaing-Jallat, T. and Le Magnen, J. (1985) Relationship between initial sensitivity to ethanol and the high alcohol intake in dependent rats. *Pharmacol. Biochem. Behav.,* 22:19–23.

Martin, S., Diegold, R.J., and Cooper, K.E. (1977) Alcohol, respiration, skin and body temperature during cold water immersion. *J. Appl. Physiol.* (Respir. Environ. Exercise Physiol.), 43:211–215.

Martin, S. and Cooper, K.E. (1964) Alcohol and respiratory and body temperature changes during tepid water immersion. *J. Appl. Physiol.,* 44:683–689.

Martz, A., Deitrich, R.A., and Harris, R.A. (1983) Behavioral evidence for the involvement of γ-aminobutyric acid in the actions of ethanol. *Eur. J. Pharmacol.,* 89:53–62.

McClearn, G.E. (1991) The tools of pharmacogenetics. In *The Genetic Basis of Alcohol and Drug Actions.* J.C. Crabbe and R.A. Harris, Eds., Plenum Press, New York, pp. 1–23.

McQuilkin, S.J. and Harris, R.A. (1990) Factors affecting actions of ethanol on GABA-activated chloride channels. *Life Sci.,* 46:527–541.

Meijer, A.J., Van Woerkom, G.M., Williamson, J.R., and Tager, J.M. (1975) Rate-limiting factors in oxidation of ethanol by isolated rat liver cells. *Biochem. J.,* 150:205–209.

Melchior, C.L. and Allen, P.M. (1992) Interaction of pregnanolone and prenenolone sulfate with ethanol and pentobarbital. *Pharmacol. Biochem. Behav.,* 42:605–611.

Melchior, C.L. and Allen, P.M. (1993) Temperature in mice after ethanol: effect of probing and regain of righting reflex. *Alcohol,* 10:17–20.

Melchior, C.L. and Ritzmann, R.F. (1992) Dehydroepiandrosterone enhances the hypnotic and hypothermic effects of ethanol and pentobarbital. *Pharmacol. Biochem. Behav.,* 43:223–227.

Melchior, C.L. and Tabakoff, B. (1981) Modification of environmentally cued tolerance to ethanol in mice. *J. Pharmacol. Exp. Ther.,* 219:175–180.

Meyer, H.H. (1899) Zur theorie der alkoholnarkose. *Arch. Exp. Path. Pharmakol.,* 42:109–118.

Meyer, H.H. (1901) Zur theorie der alkoholnarkose. Der einfluss wechselnder temperatur auf wirkingsstarke und theilungskoefficient der narcotica. *Arch. Exp. Path. Pharmakol.,* 46:338–346.

Mullin, F.J., Kleitman, N., and Cooperman, N.R. (1933) Studies on the physiology of sleep. X. The effects of alcohol and caffein on motility and body temperature during sleep. *Am. J. Physiol.,* 106:478–487.

Mullin, M.J. and Ferko, A.P. (1981) Ethanol and functional tolerance: interactions with pimozide and clonidine. *J. Pharmacol. Exp. Ther.,* 216:459–464.

Munoz, J.M. (1937) Influencia del alcohol sobre la resistencia a las temperaturas bajas o altas. *Rev. Soc. Argentina Biol.,* 13:244–247.

Murphy, M.T. and Lipton, J.M. (1983) Effects of alcohol on thermoregulation in aged monkeys. *Exp. Geron.,* 18:19–27.

Musleh, W., Alvarez, S., Baudry, M., and Alkana, R.L. (1994) Effects of ethanol and temperature on NMDA receptor-mediated responses in different mouse genotypes. *Alcoholism: Clin. Exp. Res.,* 18:444(Abstract).

Musleh, W., Alvarez, S., Baudry, M., and Alkana, R.L. (1995) Effects of ethanol and temperature on NMDA receptor function in different mouse genotypes. (in preparation).

Myers, A.K. (1962) Alcohol choice in Wistar and G-4 rats as a function of environmental temperature and alcohol concentration. *J. Comp. Physiol. Psychol.,* 55:606–609.

Myers, R.D. (1980) Hypothalamic control of thermoregulation: Neurochemical mechanisms. In *Handbook of the Hypothalamus.* P. Morgane and J. Panksepp, Eds., Marcel Dekker, New York, pp. 83–210.

Myers, R.D. (1981) Alcohol's effect on body temperature: hypothemia, hyperthermia or poikilothermia? *Brain Res. Bull.,* 7:209–220.

Nadel, E.R. (1977) *Problems with Temperature Regulation during Excercise.* Academic Press, San Francisco.

O'Connor, C.S., Crawshaw, L.I., Bedichek, R.C., and Crabbe, J.C. (1988) The effect of ethanol on temperature selection in the goldfish, Carassius auratus. *Pharmacol. Biochem. Behav.,* 29:243–248.

O'Connor, C.S., Crawshaw, L.I., Kosobud, A., Bedichek, R.C., and Crabbe, J.C. (1989) The effect of ethanol on behavioral temperature regulation in mice. *Pharmacol. Biochem. Behav.,* 33:315–319.

O'Connor, C.S., Hayteas, D.L., Crawshaw, L.I., and Crabbe, J.C. (1990) Mice selected for a difference in hypothermic response to ethanol also show a difference in body temperature response to i.c.v. 5-HT. *Fed. Am. Soc. Exp. Biol.,* 4:A989.

O'Connor, C.S., Crawshaw, L.I., and Crabbe, J.C. (1993) Genetic selection alters thermoregulatory response to ethanol. *Pharmacol. Biochem. Behav.,* 44:501–508.

Oliveira de Souza, M.L. and Masur, J. (1982) Does hypothermia play a role in the glycemic alterations induced by ethanol? *Pharmacol. Biochem. Behav.,* 16:903–908.

Oliveira de Souza, M.L. and Masur, J. (1984) Ethanol induces hyper and hypoglycemia in both fasted and nonfasted rats dependent on the ambient temperature. *Pharmacol. Biochem. Behav.,* 20:649–652.

Olofsson, S. and Lindahl, P.E. (1977) Temperature dependence of the effects of ethanol, as a model substance, on cod. *Environ. Res.,* 14:22–29.

Orts, A., Alcaraz, C., Goldfrank, L., Turndorf, H., and Puig, M.M. (1991) Morphine-ethanol interaction on body temperature. *Gen. Pharmacol.,* 22:111–116.

Ott, J.F., Hunter, B., and Walker, D. (1985) The effect of age on ethanol metabolism and on the hypotthermic and hypnotic responses to ethanol in the Fischer 344 rat. *Alcoholism: Clin. Exp. Res.,* 9:59–65.

Paez, X. and Myers, R.D. (1990) Differential actions of RO 15–1788 and diazepam on poikilothermia, motor impairment and sleep produced by ethanol. *Pharmacol. Biochem. Behav.,* 36:915–922.

Palmer, M.R., Sorensen, S.M., Freedman, R., Olson, L., Hoffer, B., and Seiget, A. (1982) Differential ethanol sensitivity of intraocular cerebellar grafts in long sleep and short sleep mice. *J. Pharmacol. Exp. Ther.,* 222:480–487.

Peris, J. and Cunningham, C.L. (1986) Handling-induced enhancement of alcohol's acute physiological effects. *Life Sci.,* 38:273–279.

Phillips, T.J., Terdal, E.S., and Crabbe, J.C. (1990) Response to selection for sensitivity to ethanol hypothermia: genetic analyses. *Behav. Genet.,* 20(4):473–480.

Phillips, T.J. and Crabbe, J.C., Jr. (1991) Behavioral studies of genetic differences in alcohol action. In *The Genetic Basis of Alcohol and Drug Actions.* J.C. Crabbe and R.A. Harris, Eds., Plenum Press, New York, pp. 25–104.

Pillai, N.P. and Ross, D.H. (1986) Ethanol-induced hypothermia in rats: possible involvement of opiate kappa receptors. *Alcohol,* 3:249–253.

Pohorecky, L.A. (1981) Interaction of ethanol and stress: a review. *Neurosci. Biobehav. Rev.,* 5:209–229.

Pohorecky, L.A. (1985) Ethanol diuresis in rats: possible modifying factors. *J. Phar. Pharmac.,* 37:271–273.

Pohorecky, L.A., Patel, V.A., and Roberts, P. (1989) Effects of ethanol on plasma corticosterone and rectal temperature modification by U50488H and WIN 44441–3. *Life Sci.,* 44:1637–1643.

Pohorecky, L.A. and Brick, J. (1988) Pharmacology of ethanol. *Pharmacol. Ther.,* 36:335–427.

Pohorecky, L.A. and Rizek, A.E. (1981) Biochemical and behavioral effects of acute ethanol in rats at different enviromental temperatures. *Psychopharmacology,* 72:205–209.

Pohorecky, L.A., Brick, J., and Sun, J.Y. (1976) Serotonergic involvement in the effect of ethanol on body temperature in rats. *J. Phar. Pharmac.,* 28:157–159.

Poole, S. and Stephenson, J.D. (1977) Core temperature: some shortcomings of rectal temperature measurements. *Physiol. Behav.,* 18:203–205.

Randall, C.L., Anton, R.F., and Becker, H.C. (1988) Role of alcohol-induced hypothermia in mediating the teratogenic effects of alcohol in C57BL/6J mice. *Alcoholism: Clin. Exp. Res.,* 12:412–416.

Rangaraj, N. and Kalant, H. (1982) Effect of chronic ethanol treatment on temperature dependence and on norepinephrine sensitization of rat brain ($Na^+ + K^+$)-adenosine triphosphatase. *J. Pharmacol. Exp. Ther.,* 223:536–539.

Rezvani, A.H., Garges, P.L., Miller, D.B., and Gordon, C.J. (1992) Attenuation of alcohol consumption by MDMA (ecstasy) in two strains of alcohol-preferring rats. *Pharmacol. Biochem. Behav.,* 43:103–110.

Risbo, A., Hagelsten, J.O., and Jessen, K. (1981) Human body temperature and controlled exposure during moderate and severe experimental alcohol-intoxication. *Acta Anesthesiol. Scand.,* 25:215–218.

Risinger, F.O. and Cunningham, C.L. (1991) The relationship between ethanol-induced hyperglycemia and hypothermia: evidence of genetic correlation. *Alcoholism: Clin. Exp. Res.,* 15:730–733.

Ritzmann, R.F. and Springer, A. (1980) Age-differences in brain sensitivity and tolerance to ethanol in mice. *Ageing,* 36:15–17.

Ritzmann, R.F. and Tabakoff, B. (1976) Body temperature in mice: a quantitative measure of alcohol tolerance and physical dependence. *J. Pharmacol. Exp. Ther.,* 199:158–170.

Romm, E. and Collins, A.C. (1987) Body temperature influences on ethanol elimination rate. *Alcohol,* 4:189–198.

Sinensky, M. (1974) Homeoviscous adaptation — A homeostatic process that regulates the viscosity of membrane lipids in *Escherichia coli. Proc. Natl. Acad. Sci. U.S.A.,* 71:522–525.

Singer, R., Harker, C.T., Vander, A.J., and Kluger, M.J. (1986) Hyperthermia induced by open-field stress is blocked by salicylate. *Physiol. Behav.,* 36:1179–1182.

Sorensen, S., Palmer, M., Dunwiddie, T., and Hoffer, B. (1980) Electrophysiological correlates of ethanol-induced sedation in differentially sensitive lines of mice. *Science,* 210:1143–1145.

Spiers, D.E. and Adair, E.R. (1986) Ontogeny of homeothermy in the immature rat: metabolic and thermal responses. *J. Appl. Physiol.,* 60:1190–1197.

Spiers, D.E. and Fusco, L.E. (1991) Age-dependent differences in thermoregulatory response of the immature rat to ethanol. *Alcoholism: Clin. Exp. Res.,* 15:23–28.

Spuhler, K., Hoffer, B., Weiner, N., and Palmer, M. (1982) Evidence for genetic correlation of hypnotic effects and cerebellar purkinje neuron depression response to ethanol in mice. *Pharmacol. Biochem. Behav.,* 17:569–578.

Stewart, R.B., Kurtz, D.L., Zweifel, M., Li, T.K., and Froehlich, J.C. (1992) Differences in the hypothermic response to ethanol in rats selectively bred for oral preference and nonpreference. *Psychopharmacology,* 106(2):169–174.

Syapin, P.J., Gee, K.W., and Alkana, R.L. (1987) Ro15–4513 differentially affects ethanol-induced hypnosis and hypothermia. *Brain Res. Bull.,* 19:603–605.

Syapin, P.J., Jones, B.L., Kobayashi, L.S., Finn, D.A., and Alkana, R.L. (1990) Interactions between benzodiazepine antagonists, inverse agonists, and acute behavioral effects of ethanol in mice. *Brain Res. Bull.,* 24:705–709.

Szabó, G., Tabakoff, B., and Hoffman, P.L. (1994) The NMDA receptor antagonist dozocilpine differentially affects enviromental-dependent and environment-independent ethanol tolerance. *Psychopharmacology,* 113:511–517.

Tabakoff, B., Ritzmann, R.F., and Boggan, W.O. (1975) Inhibition of the transport of 5-hydroxyindoleacetic acid from brain by ethanol. *J. Neurochem.,* 24:1043–1051.

Tabakoff, B., Ritzmann, R.F., Raju, T.S., and Deitrich, R.A. (1980) Characterization of acute and chronic tolerance in mice selected for inherent differences in sensitivity to ethanol. *Alcoholism: Clin. Exp. Res.,* 4:70–73.

Taylor, A.N., Branch, B.J., Liu, S.H., Wiechmann, A.F., Kill, M., and Kokka, N. (1981) Fetal exposure to ethanol enhances pituitary-adrenal and temperature responses to ethanol in adult rats. *Alcoholism: Clin. Exp. Res.,* 5:237–246.

Taylor, C.R. and Lyman, C.P. (1972) Heat storage in running antelopes: independence of brain and body temperatures. *Am. J. Physiol.,* 222:114–117.

Thompson, S.M., Masukawa, L.M., and Prince, D.A. (1985) Temperature dependence of intrinsic membrane properties and synaptic potentials in hippocampal CA1 neurons *in vitro. J. Neurosci.,* 5:817–824.

Thurman, R.G., McKenna, W.R., Brentzel, H.J., and Hesse, S. (1975) Significant pathways of hepatic ethanol metabolism. *Fed. Proc.,* 34:2075–2081.

Ticho, S.R., Stojanovic, M., Lekovic, G., and Radulovacki, M. (1992) Effects of ethanol injection to the preoptic area on sleep and temperature in rats. *Alcohol,* 9:275–278.

Towell, J.F., III and Erwin, G. (1982) Effects of ethanol and temperature on glucose utilization in the *in vivo* and isolated perfussed mouse brain. *Alcoholism: Clin. Exp. Res.,* 6:110–116.

Videla, L. and Israel, Y. (1970) Factors that modify the metabolism of ethanol in rat liver and adaptive changes produced by its chronic administration. *Biochem. J.,* 118:275–281.

Walker, P.Y., Soliman, K.F.A., and Walker, C.A. (1982) Diurnal rhythm of ethanol hypothermic action in mice. *Res. Commun. Subst. Abuse,* 3:503–506.

Wallgren, H. and Tirri, R. (1963) Studies on the mechanism of stress-induced reduction of alcohol intoxication in rats. *Acta Pharmacol. Toxicol.,* 20:27–38.

Wenger, J.R. and Alkana, R.L. (1984) Temperature dependence of ethanol depression in C57BL/6 and BALB/c mice. *Alcohol,* 1:297–303.

Widdowson, P.S. (1987) The effect of neurotensin, TRH and the delta-opioid receptor antagonist ICI 174864 on alcohol-induced narcosis in rats. *Brain Res.,* 424:281–289.

Williams, R.L., Soliman, K.F., and Mizinga, K.M. (1993) Circadian variation in tolerance to the hypothermia action of CNS drugs. *Pharmacol. Biochem. Behav.,* 46:283–288.

Wood, A.L., Healey, P.A., Menendez, J.A., Verne, S.L., and Atrens, D.M. (1989) The intrinsic and interactive effects of Ro15–4513 and ethanol on locomotor activity, body temperature, and blood glucose concentration. *Life Sci.,* 45:1467–1473.

Wood, W., Armbrecht, H., and Wise, R. (1982) Ethanol intoxication and withdrawal among three age groups of C57BL/6NNIA mice. *Pharmacol. Biochem. Behav.,* 17:1037–1041.

Wood, W.G. and Ambrecht, H. (1982a) Age differences in ethanol-induced hypothermia and impairment in mice. *Neurobiol. Aging,* 3:243–246.

Wood, W.G. and Ambrecht, H. (1982b) Behavioral effects of ethanol in animals: age differences and age changes. *Alcoholism: Clin. Exp. Res.,* 6:3–12.

York, J.L. (1982) Body water content, ethanol pharmacokinetics and the responsiveness to ethanol in young and old rats. *Dev. Pharmacol. Ther.,* 4:106–116.

York, J.L. (1983) Increased responsiveness to ethanol with advancing age in rats. *Pharmacol. Biochem. Behav.,* 19:687–691.

York, J.L. and Chan, A.W. (1993) Age-related differences in sensitivity to alcohol in the rat. *Alcoholism: Clin. Exp. Res.,* 17:864–869.

York, J.L. and Regan, S.G. (1982) Conditioned and unconditioned influences on body temperature and ethanol hypothermia in laboratory rats. *Pharmacol. Biochem. Behav.,* 17:119–124.

Zimmerberg, B., Tomlinson, T.M., Glaser, J., and Beckstead, J.W. (1993) Effects of prenatal alcohol exposure on the developmental pattern of temperature preference in a thermocline. *Alcohol,* 10:403–408.

Chapter 18

FREE RADICAL MECHANISMS AND ETHANOL-INDUCED BRAIN INJURY

Roger Nordmann and Hélène Rouach

TABLE OF CONTENTS

I. INTRODUCTION

We shall recall that free radicals are chemical species possessing an unpaired electron. They are generally very reactive and characterized by a short lifetime. The most important free radical in aerobic cells are radical derivatives of oxygen. Dioxygen (O_2) leads by successive addition of one electron to the superoxide radical ($O_2^{\bullet-}$), hydrogen peroxide (H_2O_2), the hydroxyl radical ($^{\bullet}OH$), and water. Among these oxygen derivatives, the hydroxyl radical appears as an exceptionally unpleasant species. Due to its high prooxidant activity, once generated it reacts with almost every molecule found in its immediate surrounding and thus has potentially devastating effects. It is mainly generated from the superoxide radical and hydrogen peroxide by the Haber-Weiss reaction:

$$O_2^{\bullet-} + H_2O_2 \longrightarrow {}^{\bullet}OH + OH^- + O_2$$

which is catalyzed in biological systems by transition metals, especially iron. Only a small fraction of the iron present in the cells acts as a catalyst in this reaction. It is represented by

0-8493-8389-7/95/$0.00+$.50

low-molecular-weight chelatable iron derivatives (also referred to as "free" iron or decompartmentalized iron) in opposition to the bulk of cellular iron, which is safely sequestered in storage proteins, mainly ferritin (Aust et al., 1985).

In order to avoid the damaging effects of prooxidant radicals, cells have efficient defenses aimed to prevent the generation of these radicals or to intercept the radicals that are formed. These defenses, which can be enzymes or nonenzymes, are located in both the aqueous and membranous compartments of cells (Cheeseman and Slater, 1993; Niki, 1993; Sies, 1993a). Due to the presence of such antioxidants, the steady-state level of prooxidant radicals inside the various cell compartments is kept very low and tightly controlled.

However, a disturbance in the prooxidant/antioxidant systems in favor of the former is often found in pathology. Such an imbalance may be denoted as "oxidative stress" (Sies, 1985), that is likely implicted in the pathogenesis of various diseases (Halliwell and Gutteridge, 1989; McCord, 1993).

A major target of oxidative stress is represented by the polyunsaturated fatty acid side-chains of membranous phospholipids, which may be converted into lipid peroxides through excessive lipid peroxidation, a self-perpetuating chain reaction. In the presence of transition metal catalysis, the lipid peroxides are themselves broken down to yield lipid peroxyl and alkoxyl radicals, as well as aldehydes, comprising biologically active hydroxyalkenals such as 4-hydroxynonenal (Esterbauer et al., 1991). The mechanism, measurement, and pathological significance of lipid peroxidation have been recently reviewed by Halliwell and Chirico (1993).

Among the various tissues, the brain appears to be particularly sensitive to the damaging effects of an oxidative stress on membranous lipids (Cohen, 1985; Prilipko, 1992). The reasons for such a high vulnerability have been recently reviewed by Evans (1993). Among the factors quoted, the most important are the following:

- high rate of oxidative metabolic activity
- high concentration of readily oxidizable substrates, in particular membranous polyunsaturated fatty acids
- low level of some protective enzymes, namely catalase and glutathione peroxidase, enzymes that both contribute to the decomposition of peroxides (the former acting on hydrogen peroxide, the latter on both hydrogen peroxide and fatty acid hydroperoxides).

These various characteristics are likely responsible for the high proneness of the central nervous system to lipid peroxidation.

Besides unsaturated fatty acids, oxidative stress can damage many other biological molecules. Indeed, proteins and DNA are often more significant targets of injury than are lipids (Halliwell and Chirico, 1993). Free radicals can thus modify and fragment cellular proteins. They are also able to accelerate proteolysis inside and outside of cells (Wolff et al., 1986). Many proteins, especially some enzymes, are sensitive to oxygen-derived radicals (Davies, 1987). Free radical-induced oxidation of –SH groups in key enzymes may contribute to the perturbation of Ca^{2+} homeostasis (Boobis et al., 1989), which can mediate cell death by causing disruption of the cytoskeleton, DNA fragmentation, and extensive damage to other cell components (Orrenius et al., 1989; Orrenius, 1993).

Rises in intracellular free calcium are particularly damaging to neurons (Halliwell, 1992). Free radicals could also induce an enhanced release of excitatory aminoacids such as glutamate (Pellegrini-Giampietro et al., 1988). Furthermore, one should also consider the putative role of oxidative damage of key enzymes such as brain glutamine synthetase, which may be inactivated by oxygen radicals (Schor, 1988). Still other factors, such as an increase in "free" iron, could also be involved in free radical-mediated brain injury. Lipid peroxidation and probably also protein oxidation induce changes in membrane fluidity, receptor function, and ion permeability, which can eventually lead to neuronal cell death (Cini et al., 1994). The dopaminergic system seems particularly susceptible to oxidative damage (Zaleska et al., 1989). The degree of alcohol consumed may therefore constitute one of the significant variables involved in the decline of this system with ageing (Bondy, 1992).

It is therefore not surprising that free radical mechanisms have been implicated in the pathogenesis of a number of pathological conditions of the brain. These conditions include brain trauma and stroke (Braughler and Hall, 1989; see Chapter 21), as well as Parkinson's disease, Alzheimer's dementia, and lipofuscinosis (Evans, 1993).

As emphasized by Cheeseman and Slater (1993), it remains essential that the role of free radicals in the *causation* of disorders and their production as a *consequence* of disorders be clearly distinguished. To do this one should understand the time course of the free radical production and the injury. Preventing free radical formation should decrease both the detectable free radicals and the severity of the injury. These criteria have not yet been achieved for the various brain pathological conditions listed above. This achievement is especially difficult for the brain, due to the heterogeneity of the central nervous system and the limitation of its accessibility *in vivo*. Spin trapping techniques adapted to the detection of free radical generation *in vivo* (Knecht and Mason, 1993) appear promising in the future. However, free radical mechanisms are presently generally assessed by indirect methods such as measurement of the various end products of reactions with lipids, proteins, and DNA (Holley and Cheeseman, 1993).

Other studies have been undertaken to assess disturbances in the antioxidant defense inside various areas of the central nervous system in the presence of brain diseases. They include the study of enzymes participating in the antioxidant defense, especially superoxide dismutases involved in the generation of hydrogen peroxide from superoxide radicals, catalase and glutathione peroxidase involved in the destruction of hydrogen peroxide, and glutathione S-transferase that contributes to the removal of products resulting from lipid peroxidation.

The disturbances of the most important nonenzymatic antioxidants have also been extensively studied. α-tocopherol (vitamin E) is known to be the major chain-breaking lipid-soluble antioxidant in biological membranes (Packer and Kagan, 1993). This general statement holds true for neural membranes (LeBel et al., 1989; Kagan et al., 1992). The antioxidant property of α-tocopherol is linked to its ability to donate a hydrogen atom to a peroxy radical and thereby interfere with the chain-reaction of lipid peroxidation. This leaves behind an unpaired electron on α-tocopherol, but the resulting radical can either degrade harmlessly or be reduced back to α-tocopherol. Ascorbate and glutathione interact with α-tocopherol in the protection against oxidative damage by contributing to the regeneration of α-tocopherol itself from its free radical derivative (Reed, 1993). However, ascorbate often acts as a prooxidant at low concentrations and as an antioxidant at higher concentrations. Its concentration in the brain being the highest for any tissue next to the adrenal gland (Kovachich and Mishra, 1983), ascorbate appears to have a protective role against brain lipid peroxidation. Besides its intervention in the regeneration of α-tocopherol, glutathione may also directly scavenge free radicals or contribute to the antioxidant defense as substrate of glutathione peroxidase and reductase as well as of glutathione S-transferases.

Due to these general considerations, disturbances in the level of α-tocopherol, ascorbate, and/or glutathione in the brain have frequently been interpreted as indirect indices of the occurrence of oxidative stress. Although evidence of primary involvement of free radicals in the pathogenesis of many brain diseases is generally lacking, free radical scavengers and antioxidants appear promising in their prophylaxis and treatment (Halliwell, 1992; Packer, 1992).

Therefore, it appears of large interest to address the question whether free radical mechanisms participate to the well-known disturbances induced by ethanol administration in the central nervous system. This appears especially difficult since some brain areas are more affected by ethanol than others. Furthermore, the conditions of acute or chronic ethanol administration are not yet standardized, even during experimental studies. Most research reported is concerned with rodents. We shall review, successively, the disturbances induced by an acute ethanol load and by long-term ethanol administration.

II. ETHANOL-INDUCED OXIDATIVE STRESS IN THE BRAIN OF RODENTS FOLLOWING AN ACUTE ETHANOL LOAD

A. CELLULAR TARGETS

1. Lipid Peroxidation

Reviewing recently the numerous experimental data suggesting that free radical mechanisms are contributing to ethanol-induced liver injury (Nordmann et al., 1992), we recalled the pioneer role of Di Luzio's report (1963) showing that the acute ethanol-induced fatty liver can be prevented in rats by the administration of some antioxidants. Di Luzio and Hartman (1967) later reported an increase in lipid peroxides in rat liver homogenates after the *in vitro* addition of ethanol as well as after acute ethanol administration. However, no ethanol-induced increase in the peroxide level of brain homogenates was observed in the same conditions (Hartman et al., 1967). This negative finding seems to have discouraged investigators from studying the effects of alcohol intoxication on lipid peroxidation and other free radical-mediated processes in the brain for many years.

However, Seligman et al. (1977) observed that lipid peroxidation was enhanced in the spinal cord following minimal physical trauma in ethanol-pretreated cats, whereas the same trauma did not induce any change in lipid peroxidation in control cats not pretreated with ethanol. Our group reported later that acute ethanol without any associated trauma can induce an enhanced lipid peroxidation (2 and 4 h after the i.p. administration of 2.3 g ethanol/kg b. wt.) in rat cerebellum (Nordmann et al., 1985; Rouach et al., 1987a, 1987b). The cerebellum was selected in these studies because of its well-known sensitivity to ethanol-induced dysfunction. Furthermore, among all the brain regions, the cerebellum shows the lowest concentration of α-tocopherol (Meydani et al., 1986), the most important chain-breaking antioxidant.

Besides the cerebellum, lipid peroxidation may also be enhanced by an acute ethanol load in other parts of the central nervous system. Acute ethanol administration elicits indeed an increase in the lipid peroxide level in whole-rat-brain homogenates without the cerebellum (Uysal et al., 1989). This increase was apparent from 1 to 4 h after the administration of 5 g ethanol/kg b. wt., i.p. Increased levels of thiobarbituric acid-reactive substances likely representing products formed through lipid peroxidation have also been reported in the cerebellum, cerebral cortex, and brain stem following acute ethanol administration to rats (Nadiger et al., 1988).

2. Other Cellular Targets

A decrease in glutathione has been reported after an acute ethanol load in mouse brain (Guerri and Grisolia, 1980) as well as in rat cerebellum (Nordmann, 1987) and brain (Uysal, 1989; Bondy and Guo, 1994). The decrease of this substrate, which has a pivotal antioxidant role, can either result from oxidative stress or contribute to its establishment. However, it should be pointed out that the concentration of glutathione is high in the brain (about 1–2 m*M*) (Marcus et al., 1994; Montoliu et al., 1994) and that the magnitude of the decrease induced by acute ethanol does not exceed 25%. The contribution of this decrease to oxidative damage is therefore not yet ascertained. Further studies aimed to determine glutathione concentrations in the various subcellular compartments (especially the mitochondria) may reveal selective changes at sites predisposed to oxidative damage.

Glutamine synthetase, which is especially susceptible to oxidative degradation, was surprisingly slightly increased in cerebral cytosol 2 h after the administration of ethanol (4.5 g/kg b. wt., i.p.) (Bondy and Guo, 1994).

B. MECHANISMS INVOLVED

1. Disturbances in the Generation of Prooxidant Free Radical Species

Because one of the most important sources of superoxide in the cell is the mitochondria, Ribière et al. (1987) studied the influence of an acute ethanol load (2.3 g/kg b. wt., i.p., 2 h before sacrifice) on rat brain mitochondrial superoxide production. They reported that the $O_2\bar{\bullet}$ generation was not enhanced, although a marked inhibition of state 3 respiratory activities was noticeable. These results were in contrast with the increased $O_2\bar{\bullet}$ production observed in the liver mitochondria of the same animals.

Another possible source of free radicals is represented by cytochrome P 450 IIE1. Ingelman-Sundberg (1993) has recently reviewed the experimental data emphasizing the role of this ethanol-inducible enzyme in the increased generation of reactive oxygen species, as well as in the biosynthesis of the 1-hydroxyethyl radical derived from ethanol itself in liver microsomes. The same radical was identified by electron-spin-resonance spectroscopy in brain microsomes incubated in the presence of ethanol (Gonthier et al., 1991).

Cytochrome P 450 IIE1 was shown to be constitutively expressed in the rat brain (Tindberg et al., 1989; Hansson et al., 1990). Although the level of P 450 is only 1–3% of that in the liver, Warner and Gustafsson (1994) recently made the interesting observation that a single dose of ethanol (0.8 ml/kg) administered either intraperitoneally or orally induces P 450 IIE1 as well as other P 450 isoforms in the brain. It, therefore, appears likely that even a small acute ethanol load elicits an increased production of oxygen- and/or ethanol-derived radicals at the microsomal level in the rat brain.

Whereas xanthine oxidase activity is very low in rat brain (Nihei et al., 1989), aldehyde oxidase appears to be involved in the generation of reactive oxygen species in the brain (Bondy and Orozco, 1994). The Km of aldehyde oxidase for acetaldehyde (1 m*M*) is much lower than that of xanthine oxidase with this substrate (cf. Nordmann et al., 1992). Focal accumulation of acetaldehyde may result from the oxidation of ethanol through ADH, as suggested by Bühler et al. (1983) and by Kerr et al. (1989). It could also result from ethanol oxidation through cytochrome P 450, as already mentioned, or through catalase, as reported by Cohen et al. (1980), Aragon et al. (1991), and Gill et al. (1992).

The production of free radicals is likely favored by the increase in the low-molecular-weight chelatable (LMWC) iron derivatives that we observed in the rat cerebellum following acute ethanol administration (2.3 g/kg, i.p.) (Rouach et al., 1990). Such an administration elicited in the cerebellum an increased total nonheme iron content together with an increase in the percentage of total nonheme iron represented by LMWC-iron inside the cytosol. The increase in the total cerebellar nonheme iron content appears to be, at least partly, linked to the increase in cerebellar iron uptake resulting from acute ethanol administration (Rouach et al., 1994). The increase in the percentage of LMWC-iron in the cytosol may be mediated by an enhanced production of reducing equivalents, such as NADH or superoxide, favoring the liberation of LMWC-iron from ferritin. Kerr et al. (1989) have, therefore, suggested that regional increases in the concentration of NADH could account for brain cell injury. These last authors detected type 1 ADH in a limited number of neurons. At the cerebellum level, it was present only in Purkinje cell cytoplasm.

Whatever the mechanism(s) involved in the increase in the LMWC-iron content and their putative links to local ethanol metabolism, this increase may favor the biosynthesis of aggressive prooxidant radicals in the cerebellum after an acute ethanol load.

2. Disturbances in the Antioxidant Systems

Disturbances in the antioxidant systems that could contribute to acute ethanol-induced oxidative stress have been especially studied at the cerebellar level in our laboratory. We observed that an acute ethanol load elicits a decrease in the cerebellar concentrations of α-tocopherol and ascorbate (Rouach et al., 1987a). We have already recalled the major role of α-tocopherol as an antioxidant in the brain, a role that has been confirmed in studies on the effects of vitamin E deficiency on postischemic cerebral lipid peroxidation (Yoshida et al., 1985). We have also recalled that ascorbate appears to have a protective role against brain lipid peroxidation. Changes in glutathione concentration, which we have already mentioned, may also contribute to the defect in antioxidant defense.

Furthermore, we have reported an acute ethanol-induced decrease in the cerebellar concentration of selenium, zinc, and copper (Houzé et al., 1991). The reported decrease in selenium may contribute to oxidative stress by affecting cerebellar selenoproteins, such as glutathione peroxidase, involved in the antioxidant defense. The decrease in zinc and copper concentrations is also a likely contributing factor, since zinc itself has an antioxidant function, whereas copper is involved in the activity of cytosolic Cu,Zn-superoxide dismutase. The activity of the latter enzyme is decreased in the brain following acute ethanol administration (Lédig et al., 1981; Saffar et al., 1988).

C. PREVENTION TRIALS

Only a few of the numerous antioxidants that could be efficient to prevent oxidative stress apparently induced in the brain by acute ethanol administration have been tested.

It was reported that increased lipid peroxidation, apparent in the cerebellum, cerebral cortex, and brain stem following an acute ethanol administration to rats fed a normal laboratory diet, was prevented when the same ethanol load was administered to rats fed a vitamin E-supplemented diet resulting in a daily intake higher than the normal rat requirement (Nadiger et al., 1988).

Allopurinol administered prior to an acute ethanol load has been shown to provide protection against the enhanced cerebellar lipid peroxidation observed after such a load in nonallopurinol pretreated animals (Park et al., 1988). The pretreatment with allopurinol also prevented the ethanol-induced cerebellar changes in nonheme iron, zinc, copper, and selenium (Rouach et al., 1989; Houzé et al., 1991) as well as the ethanol-induced decrease in Cu,Zn-superoxide dismutase activity in the brain (Saffar et al., 1988). The preventative effect of allopurinol could be linked to its activity as a xanthine-oxidase inhibitor, hydroxyl-radical scavenger, electronic-transfer activator, or adenosine-release enhancer (Houzé et al., 1991). Studying the influence of allopurinol on lipid peroxidation in rat brain homogenates, Cini et al. (1994) did not observe any effect that could be distinct from xanthine-oxidase inhibition. This enzyme is present in the capillaries of the central nervous system. Furthermore, its level is increased in rat blood plasma after acute ethanol administration (Zima et al., 1993).

III. ETHANOL-INDUCED OXIDATIVE STRESS IN THE BRAIN OF RODENTS FOLLOWING CHRONIC ALCOHOL ADMINISTRATION

A. CELLULAR TARGETS

1. Lipid Peroxidation

The effects of chronic ethanol administration on brain lipid peroxides estimated as thiobarbituric acid reactive substances (TBARS) are controversial. Uysal et al. (1986) observed no change in rats given 20% (v/v) ethanol in the drinking fluid for 2 months, whereas Marcus et al. (1988) reported an increase in TBARS in cerebral cortex, cerebellum, and brain stem from rats fed ethanol for 30 days.

Surprisingly, Rouach et al. (1993) reported no increase in lipid peroxidation during incubation (without adding any inducer) of cerebellar homogenates from rats given a 10% (v/v) aqueous ethanol solution as their sole drinking fluid during 4 weeks. Montoliu et al. (1994) made the same observation in brain homogenates from rats fed a Lieber-De Carli diet containing 5% (wt/v) ethanol for 8 weeks. However, these last authors reported in the same animals an increase in the generation of TBARS in synaptosomal preparations incubated during 2 h. This last finding could be related to the presence in the homogenates of a yet unidentified protective agent eliminated during the preparation of the synaptosomal fraction.

2. Other Cellular Targets

The reported changes in the brain glutathione level following chronic ethanol administration are controversial. Uysal et al. (1989) didn't observe any change in that level in whole brain. Marcus et al. (1994) also reported unchanged glutathione levels in cerebral cortex, cerebellum, and brain stem. On the other hand, Montoliu et al. (1994) and Bondy and Guo (1994) observed a decreased brain glutathione level. More interestingly, Montoliu et al. (1994) reported a highly significant increase in the GSSG:GSH ratio in brain homogenates from chronic alcohol-fed rats and suggested that this change could be the consequence of increased generation of prooxidant free radicals, leading to oxidative stress in brain tissue.

It has been recently emphasized that some proteins are particularly vulnerable to oxidative damage (Carney et al., 1991). Brain glutamine synthetase appears highly sensitive to inactivation by oxygen radicals (Schoz, 1988). A significant decrease in the activity of this enzyme in the cerebellum has been reported by Rouach et al. (1993) in chronic ethanol-fed rats. A similar decrease is also present in whole brain (Bondy and Guo, 1994). Due to the pivotal role of glutamine synthetase in the brain (Oliver et al., 1990), the decrease in its activity after chronic ethanol administration may represent a critical factor in ethanol-induced neurotoxicity. It might indeed result in glutamate accumulation leading to increased calcium flux through receptor-gated and/or voltage-sensitive calcium channels. These changes in calcium flux may activate nitric oxide formation, contributing to free radical-induced neuronal toxicity, as discussed in the last part of the present chapter.

Another finding that strengthens the hypothesis that chronic ethanol feeding results in oxidative stress is represented by the acceleration of lipofuscin formation in the cerebellum, hippocampus, and prefrontal cortex after long-term alcohol treatment in rats (Tavares and Paula-Barbosa, 1983; Paula-Barbosa et al., 1991). Lipofuscin pigments are insoluble conglomerates of polymerized and oxidized fatty acids, mainly formed by oxygen-derived free radicals oxidizing the unsaturated fatty acids of the phospholipids of the internal cellular membranes. An increase in the lipofuscin content in nerve cells is observed during senescence. This increase, which represents one of the most consistent cytological changes during ageing, supports the role of free radical mechanisms in brain damage observed in older animals or humans (Nohl, 1993a), as well as following chronic alcohol intake. However, the acceleration by chronic alcohol consumption of age-related impairment of avoidance learning in mice does not appear to be related to an increased brain lipofuscin deposition (Freund, 1979).

B. MECHANISMS INVOLVED

1. Disturbances in the Generation of Prooxidant Free Radical Species

No evidence of disturbed free radical generation was found in brain extracts from rats fed a high-fat ethanol-containing diet for 2 weeks (Reinke et al., 1987). Similarly no change in the total production of reactive oxygen species was reported in a crude cerebellar P2 synaptosomal-mitochondrial fraction after chronic ethanol administration (Bondy and Pearson, 1993). However, a significant increase in the formation of oxygen reactive species was reported in isolated cerebral synaptosomes from chronic ethanol-fed rats (Montoliu et al., 1994). Interestingly, lipid peroxidation was enhanced in these same synaptosomal preparations.

Besides, a significant increase in superoxide production was apparent in brain submitochondrial particles isolated from rats having received an aqueous ethanol solution (10% v/v) as their sole drinking fluid for 4 weeks (Ribière et al., 1994). Such an increase was not apparent in the mitochondria isolated from the liver and heart of the same animals. The increase in brain mitochondrial superoxide generation following a moderate alcohol intoxication model suggests that these mitochondria represent an early target for free radical attack. This suggestion is in accordance with the finding that the brain is the first organ to exhibit an increase in mitochondrial superoxide production during ageing when compared to the liver and heart (Sawada and Carlson, 1987). Since cytochrome P 450 IIE1 is induced in the brain by chronic ethanol treatment (Anandatheerthavarada et al., 1993; Montoliu et al., 1994), an increased free radical generation may also occur at the microsomal level. Another disturbance that may contribute to an enhanced generation of aggressive prooxidant radicals is the increase in low-molecular-weight iron species that was apparent in the cerebellar cytosolic fractions after long-term ethanol administration (Nordmann et al., 1990b; Rouach et al., 1991).

2. Disturbances in the Antioxidant Systems

Various, but controversial, disturbances have been reported in the antioxidant systems of the central nervous system following chronic ethanol intake. When rats were given an ethanol solution as their sole drinking fluid, a progressive decrease in brain superoxide dismutase activity was apparent, reaching a plateau after 6 weeks of treatment, and returning to control values within 48 h after withdrawal (Lédig et al., 1981, 1988). On the contrary, Montoliu et al. (1994) recently reported a significant increase in that activity in the brain following chronic alcohol intake. Catalase activity was also increased in the brain of the same animals.

Changes in some essential antioxidant substrates have also been reported in the rat central nervous system after long-term ethanol administration. Glutathione has thus been reported in some reports to be lowered in the brain as already mentioned, whereas significant decreases in the level of α-tocopherol and of selenium were apparent at cerebellar level (Rouach et al., 1991). However, the reduction in the selenium level was not accompanied by changes in the activity of the cerebellar selenium-dependent glutathione peroxidase (Rouach et al., 1993). The activity of glutathione peroxidase in chronic ethanol-fed rats was reported to be unchanged in the brain by Montoliu et al. (1994), but found to be significantly decreased in the cerebral cortex, cerebellum, and brain stem by Marcus et al. (1994). A significant increase in total glutathione-S-transferase activity (using 1-chloro-2,4-dinitrobenzene as substrate) was apparent in the cerebellum of long-term ethanol-treated rats (Rouach et al., 1993). This increase may represent an adaptative response protecting against an ethanol-generated free radical challenge and contributing to the lack of increased lipid peroxidation observed in the cerebellum of these same animals. A similar increase in glutathione-S-transferase activity was reported in the cerebellum and the cerebral cortex by Marcus et al. (1994). However, no changes were apparent in rat brain in the experimental conditions used by Montoliu et al. (1994).

C. PREVENTION TRIALS

Lipofuscin accumulation in the cerebellar Purkinje and hippocampal CA3 pyramidal cells was markedly decreased in long-term alcohol-treated rats when piracetam was added to the alcohol drinking solution (Paula-Barbosa et al., 1991). This effect may be related, at least partly, to the antioxidant properties of piracetam. Supplementation of diets with vitamin E was found to decrease the brain lipofuscin content in control and long-term ethanol-treated mice, but failed to prevent the ethanol-induced learning deficit (Freund, 1979).

IV. HUMAN STUDIES

Only a few investigations have been undertaken to study disturbances in the brain antioxidant defense in alcoholics. Marklund et al. (1983) reported that the Cu,Zn-SOD activities were slightly lower and the Mn-SOD activities slightly higher in brain pieces from chronic alcoholics compared to those from controls. They concluded, however, that the slight differences found can hardly be assigned etiological importance in the degenerative process connected in brain with chronic alcoholism. Kasarskis et al. (1985) reported a decrease in the zinc and copper concentration in brains from alcoholics.

Contrasting with the rarity of reports directly concerned with the brain, many studies have shown that the antioxidant status is frequently disturbed in the blood plasma of chronic alcoholics. Most of the reported data concerning vitamin E show significant lower-than-normal mean circulating level of α-tocopherol in alcoholics, at least in those who suffer from severe liver alcoholic disease (Nordmann and Rouach, 1992). Blood selenium levels are also generally lowered in alcoholics (Dworkin et al., 1984). These disturbances in blood plasma antioxidants could reflect a decreased antioxidant defense in many tissues, including the brain.

V. CONCLUSION AND FUTURE PROSPECTS

We stated some years ago (Nordmann et al., 1990a) that further research is necessary to ascertain whether ethanol intoxication causes oxidative stress in the central nervous system. This statement is still valid. However, the experimental data presently reviewed favor the hypothesis that free radical-mediated disturbances are induced by acute or chronic ethanol administration, at least in rat cerebellum. The involvement of these disturbances in the neurotoxicity of ethanol is presently not clearly assessed. Nevertheless, it appears likely that they contribute to the onset and/or the spreading of ethanol-induced neuronal damage. The evidence recently reviewed by Pellmar (1993) suggests that *acute* neuronal exposure to either ethanol or free radical-generating systems generally suppresses action potentials. Furthermore, both ethanol and radicals have been shown to decrease calcium currents in some preparations. The comparison of free radical- and ethanol-induced actions on electrophysiologic properties of neurons reveals that many similarities and some significant differences are also present. In the conclusion of his review, Pellmar (1993) suggests, therefore, that ethanol-induced synaptic damage may be linked to free radical and additional nonradical mechanisms.

Following *long-term exposure to ethanol,* the experimental data reported also suggest the involvement of free radical-mediated neuronal disturbances. However, they do not show conclusively that these disturbances lead to cell damage result of it (Hunt, 1993). Nevertheless, it appears likely that some of the disturbances reviewed earlier may have an important role in the adverse effects of ethanol on the brain. The effects of desferrioxamine on the withdrawal score in mice previously submitted to chronic ethanol vapor inhalation (Abu-Murad and Nordmann, 1983) are suggestive of an involvement of iron-catalyzed free radical mechanisms in alcohol dependence. This score was significantly reduced by daily administration of the iron chelator, desferrioxamine, a powerful inhibitor of free radical biosynthesis and lipid peroxidation, during the ethanol inhalation period. On the contrary, the withdrawal score was increased in mice receiving an iron load prior to, as well as during, ethanol exposure.

It also appears likely that the disturbances induced by chronic ethanol administration on enzymes especially sensitive to an oxidative stress contribute to neuronal damage, and possibly neuronal death. This could be the case for glutamine synthetase (Rouach et al., 1993).

The intensive research recently devoted to the role of nitric oxide in the brain (Snyder, 1992; Bruhwyler et al., 1993; Nohl, 1993b) may also produce new insights concerning the role of free radical mechanisms in alcoholic brain injury. The hypothesis that nitric oxide and ethanol-induced brain damage may be connected was put forward by Lancaster (1992; 1993).

The reduction of neuronal density associated with chronic ethanol exposure is well documented. A recent review by Crews and Chandler (1993) suggests that the slow progressive neuronal degeneration may involve depolarization of neurons by glutamate followed by the increased cellular entry of calcium through receptor-gated ion channels [especially N-methyl-D-aspartate (NMDA) receptors] and voltage-sensitive calcium channels. This amino acid-mediated excitotoxicity is a pathological mechanism unique to the brain that may be involved in a number of neurodegenerative processes, and especially in alcohol-related brain damage. Hyperexcitability following chronic alcohol exposure appears thus to be linked to enhanced activation of glutamatergic synapses. Prolonged or repetitive bouts of enhanced excitatory transmission during withdrawal may destroy central neurons via "excitotoxic" mechanisms that might be enhanced by thiamine deficiency (Lovinger, 1993).

Nitric oxide (•NO), a short-lived free radical, may play an important role in excitotoxicity. The neuronal nitric oxide synthase that generates •NO is calcium/calmoduline dependent and activated by increased intracellular calcium. The biosynthesis of •NO in the brain appears, therefore, to be strongly linked to NMDA-receptor activation (cf. Crews and Chandler, 1993; Lancaster, 1993). As an example, •NO was shown to mediate glutamate neurotoxicity in primary cortical cultures (Dawson et al., 1991). Moreover, astroglia possess an inducible form of nitric oxide synthase that may have important implications in terms of the glial-neuronal response to NMDA receptor-mediated neurotoxicity (Chandler et al., 1992).

An excess of •NO generation may be cell damaging through its own free radical properties and/or through formation of additional free radicals. •NO is thus able to combine with $O_2\bar{\bullet}$ to form peroxynitrite anion ($ONOO^-$). Since neuronal cells are able to generate both •NO and $O_2\bar{\bullet}$, peroxynitrite may be generated in excess during alcohol exposure. Peroxynitrite is potentially cytotoxic since it is a potent oxidant that mediates oxidation of both nonprotein and protein sulfhydryls (Radi et al., 1991a). Furthermore, the conjugate acid of peroxynitrite, peroxynitrous acid (ONOOH), and/or its decomposition products, i.e., •OH and nitrogen dioxide (•NO_2), are able to initiate lipid peroxidation (Radi et al., 1991b).

All these properties of •NO, and particularly the observation that ethanol and •NO induce neuronal injury in association with an NMDA response (Lancaster, 1993), suggest that nitric oxide could be highly involved in ethanol-induced brain injury. Presently, this involvement remains speculative. However, Khanna et al. (1993) have recently reported that the acquisition of rapid tolerance to the motor-incoordinating effects of ethanol was impaired in rats administered the nitric oxide synthase inhibitor, L-nitroarginine. They further suggested that the role of •NO in ethanol tolerance may be similar to its role in memory and learning, involving facilitation of transmission in certain NMDA synapses. This report shows that •NO may be effectively involved in some of the ethanol effects on the brain.

The uncertainty about the precise involvement of •NO and other free radicals in alcoholic brain damage does not preclude the possibility that free radical scavengers and antioxidants may be able to prevent, at least partly, brain damage linked to alcohol consumption in humans (Nordmann, 1994). One could consider various pharmacological antioxidants able to reduce free radical production (e.g., iron chelators), trap free radicals themselves, interrupt the peroxidation process, or reinforce the natural antioxidant defense (Sies, 1993a; 1993b). Supplementation of the diet with natural antioxidants may also be used. Due to its prominent role as a major membranous antioxidant, vitamin E should be specially considered in prophylaxy and treatment of brain diseases (Packer, 1992). However supplementation with a single, natural antioxidant might not always be beneficial for the patients. As a matter of fact, the various cellular antioxidants act in coordination and often in synergy. Long-term epidemiological studies are therefore needed to address the question of whether supplementation with a single antioxidant or a "cocktail" of antioxidant mixtures may reduce the severity of alcohol-induced brain damage.

REFERENCES

Abu-Murad, C. and Nordmann, R. (1983) Effect of two iron-chelators, desferrioxamine and diethylenetriaminepentaacetic acid, on the development of tolerance to and physical dependence on ethanol in mice. *Drug Alcohol Dependence,* 12, 371–379.

Anandatheerthavarada, H. K., Shankar, S. K., Bhamre, S., Boyd, M. R., Song, B. J., and Ravindranath, V. (1993) Induction of brain cytochrome P-450IIE1 by chronic ethanol treatment. *Brain Res.,* 601, 279–285.

Aragon, C. M. G., Stotland, L. M., and Amit, Z. (1991) Studies on ethanol-brain catalase interaction: evidence for central ethanol oxidation. *Alcoholism: Clin. Exp. Res.,* 15, 165–169.

Aust, S. D., Morehouse, L. A., and Thomas, C. E. (1985) Role of metals in oxygen radical reactions. *Free Radic. Biol. Med.,* 1, 3–25.

Bondy, S. C. (1992) Ethanol toxicity and oxidative stress. *Toxicol. Lett.,* 63, 231–241.

Bondy, S. C. and Pearson, K. R. (1993) Ethanol-induced oxidative stress and nutritional status. *Alcoholism: Clin. Exp. Res.,* 17, 651–654.

Bondy, S. C. and Guo, S. X. (1994) Effect of ethanol treatment on indices of cumulative oxidative stress. *Eur. J. Pharmacol.,* 270, 349–355.

Bondy, S. C. and Orozco, J. (1994) Effects of ethanol treatment upon sources of reactive oxygen species in brain and liver. *Alcohol Alcoholism,* 29, 375–383.

Boobis, A. R., Fawthrop, D. J., and Davies, D. S. (1989) Mechanisms of cell death. *Trends Pharm. Sci.,* (TIPS), 10, 275–280.

Braughler, J. M. and Hall, E. D. (1989) Central nervous system trauma and stroke: biochemical considerations for oxygen radical formation and lipid peroxidation. *Free Radic. Biol. Med.,* 6, 289–301.

Bruhwyler, J., Chleide, E., Liégeois, J. F., and Carreer, F. (1993) Nitric oxide: a new messenger in the brain. *Neurosci. Biobehav. Rev.,* 17, 373–384.

Bühler, R., Pestalozzi, D., Hess, M., and Von Wartburg J. P. (1983) Immunohistochemical localization of alcohol dehydrogenase in human kidney, endocrine organs and brain. *Pharm. Biochem. Behav.,* 18, Suppl. 1, 55–59.

Carney, J. M., Starke-Reed, P. E., Oliver, C. N., Landum, R. W., Cheng, M. S., Wu, J. F., and Floyd, R A. (1991) Reversal of age-related charges in brain protein oxidation, decrease in enzyme activity, and loss, in temporal and spatial memory by chronic administration of the spin trapping compound N-tert-butyl-α-phenylnitrone. *Proc. Natl. Acad. Sci. U.S.A.,* 88, 3633–3636.

Chandler, L. J., Sumners, C., Crews, F. T., and Guzman, N. (1992) Induction of nitric oxide synthase in astroglial cultures. *Alcoholism: Clin. Exp. Res.,* 16, 365.

Cheeseman, K. H. and Slater, T.F. (1993) An introduction to free radical biochemistry. *Br. Med. Bull.,* 49, 481–493.

Cini, M., Fariello, R. G., Bianchetti, A., and Moretti, A. (1994) Studies on lipid peroxidation in the rat brain. *Neurochem. Res.,* 19, 283–288.

Cohen, G., Sinet, P. M., and Heikkila, R. (1980) Ethanol oxidation by rat brain in vivo. *Alcoholism: Clin. Exp. Res.,* 4, 366–370.

Cohen, G. (1985) Oxidative stress in the nervous system. In *Oxidative Stress,* Sies, H., Ed., Academic Press, London, 383–402.

Crews, F. and Chandler, L. J. (1993) Excitotoxicity and the neuropathology of ethanol. In *Alcohol-induced Brain Damage,* Hunt, W. A. and Nixon, S. J., Eds., NIAAA Monograph, No. 22, NIH Publication No. 93–3549. Rockville, MD, NIAAA, 355–371.

Davies, K. J. A. (1987) Protein damage and degradation by oxygen radicals. I. General aspects. *J. Biol. Chem.,* 262, 9895–9901.

Dawson, V. L., Dawson, T. M., London, E. D., Bredt, D. S., and Snyder, S. H. (1991) Nitric oxide mediates glutamate neurotoxicity in primary cortical cultures. *Proc. Natl. Acad. Sci. U.S.A.,* 88, 6368–6371.

Di Luzio, R. (1963) Prevention of the acute ethanol-induced fatty liver by antioxidants. *Physiologist,* 6, 169–173.

Di Luzio, N. R. and Hartman, A. D. (1967) Role of lipid peroxidation in the pathogenesis of ethanol-induced fatty liver. *Fed. Proc.,* 26, 1436–1442.

Dworkin, B. B., Rosenthal, W. S., Gordon, G. G., and Jankowski, R. H. (1984) Diminished blood selenium levels in alcoholics. *Alcoholism: Clin. Exp. Res.,* 8, 535–538.

Esterbauer, H., Schaur, R. J., and Zollner, H. (1991) Chemistry and biochemistry of 4-hydroxynonenal, malondialdehyde and related aldehydes. *Free Radic. Biol. Med.,* 11, 81–128.

Evans, P. H. (1993) Free radicals in brain metabolism and pathology. *Br. Med. Bull.,* 49, 577–587.

Freund, G. (1979) The effects of chronic alcohol and vitamin E consumption on aging pigments and learning performance in mice. *Life Sci.,* 24, 145–151.

Gill, K., Ménez, J. F., Lucas, D., and Deitrich, R. A. (1992) Enzymatic production of acetaldehyde from ethanol in rat brain tissue. *Alcoholism: Clin. Exp. Res.,* 16, 910–915.

Gonthier, B., Jeunet, A., and Barret, L. (1991) Electron spin resonance study of free radicals produced from ethanol and acetaldehyde after exposure to a Fenton system or to brain and liver microsomes. *Alcohol,* 8, 369–375.

Guerri, C. and Grisolia, S. (1980) Changes in glutathione in acute and chronic alcohol intoxication. *Pharmacol. Biochem. Behav.,* 13, Suppl. 1, 53–61.

Halliwell, B. and Gutteridge, J. M. C. (1989) *Free Radicals in Biology and Medicine,* Clarendon Press, Oxford, 543.

Halliwell, B. (1992) Reactive oxygen species and the central nervous system. In *Free Radicals in the Brain,* Packer, L., Prilipko, L., and Christen, Y., Eds., Springer-Verlag, Berlin, 21–40.

Halliwell, B. and Chirico, S. (1993) Lipid peroxidation: its mechanism, measurement, and significance. *Am. J. Clin. Nutr.,* 57 (suppl.), 715S-725S.

Hansson, T., Tindberg, N., Ingelman-Sundberg, M., and Köhler, C. (1990) Regional distribution of ethanol-inducible cytochrome P450 IIE1 in the rat central nervous system. *Neuroscience,* 34, 451–463.

Hartman, A. D., Comporti, M., and Di Luzio, N. R. (1967) Peroxidation of lipids in the pathogenesis of acute ethanol-induced liver injury. *Gastroenterology* 52, 316 (abstract).

Holley, A. E. and Cheeseman, K. H. (1993) Measuring free radical reactions *in vivo. Br. Med. Bull.,* 49, 494–505.

Houzé, P., Rouach, H., Gentil, M., Orfanelli, M. T., and Nordmann, R. (1991) Effect of allopurinol on the hepatic and cerebellar iron, selenium, zinc and copper status following acute ethanol administration to rats. *Free Radic. Res. Commun.,* 12–13, 663–668.

Hunt, W. A. (1993) Role of free radical reactions in ethanol-induced brain damage: an introduction. In *Alcohol-induced Brain Damage,* Hunt, W. A. and Nixon, S. J., Eds., NIAAA Monograph, No. 22, NIH Publication No. 93–3549. Rockville, MD, NIAAA, 327–338.

Ingelman-Sundberg, M. (1993) Ethanol-inducible cytochrome P450 2E1. Regulation, radical formation and toxicological importance. In *Free Radicals: From Basic Science to Medicine,* Poli, G., Albano, E., and Dianzani, M. U., Eds., Birkhäuser Verlag Basel, Switzerland, 287–301.

Kagan, V. E., Bakalova, R. A., Koynova, G. M., Tyurin, V. A., Serbinova, E. A., Petkov, V. V., Petkov, V. D., Staneva, D. S., and Packer, L. (1992) Antioxidant protection of the brain against oxidative stress. In *Free Radicals in the Brain,* Packer, L., Prilipko, L., and Christen, Y., Eds., Springer-Verlag, Berlin, 49–61.

Kasarskis, E. J., Manton, W. I., Devenport, L. D., Kirkpatrick, J. B., Howell, G. A., Klitenick, M. A., and Frederickson, C. J. (1985) Effects of ethanol ingestion on zinc content of human and rat central nervous systems. *Exp. Neurol.,* 90, 81–95.

Kerr, T., Maxwell, D. S., and Crabb, D. W. (1989) Immunocytochemistry of alcohol dehydrogenase in the rat central nervous system. *Alcoholism: Clin. Exp. Res.,* 13, 730–736.

Khanna, J. M., Morato, G. S., Shah, G., Chau, A., and Kalant, H. (1993) Inhibition of nitric oxide synthesis impairs rapid tolerance to ethanol. *Brain Res. Bull.,* 32, 43–47.

Knecht, K. T. and Mason, R. P. (1993) *In vivo* spin trapping of xenobiotic free radical metabolites. *Arch. Biochem. Biophys.,* 303, 185–194.

Kovachich, G. B. and Mishra, O. P. (1983) The effect of ascorbic acid on malonaldehyde formation, K^+, Na^+ and water content of brain slices. *Exp. Brain Res.,* 50, 62–68.

Lancaster, F. E. (1992) Alcohol, nitric oxide, and neurotoxicity: is there a connection? a review. *Alcoholism: Clin. Exp. Res.,* 16, 539–541.

Lancaster, F. E. (1993) Nitric oxide and ethanol-induced brain damage. A hypothesis. In *Alcohol-induced Brain Damage,* Hunt, W. A. and Nixon, S. J., Eds., NIAAA Monograph, No. 22, NIH Publication No. 93–3549. Rockville, MD, NIAAA, 373–386.

LeBel, C. P., Odunze, I. N., Adams, J. D., and Bondy, S. C. (1989) Perturbations in cerebral oxygen radical formation and membrane order following vitamin E deficiency. *Biochem. Biophys. Res. Commun.,* 163, 860–866.

Lédig, M., M'Paria, J-R., and Mandel, P. (1981) Superoxide dismutase activity in rat brain during acute and chronic alcohol intoxication. *Neurochem. Res.,* 6, 385–391.

Lédig, M., Kopp, P., and Mandel, P. (1988) Alcohol effect on superoxide dismutase activity in nervous tissue. In *Alcohol Toxicity and Free Radical Mechanisms. Advances in the Biosciences,* Nordmann, R., Ribière, C., and Rouach, H., Eds., Vol. 71, Pergamon Press, Oxford, 79–86.

Lovinger, D. M. (1993) Excitotoxicity and alcohol-related brain damage. *Alcoholism: Clin. Exp. Res.,* 17, 19–27.

Marcus, S. R., Chandrakala, M. V., and Nadiger, H. A. (1994) Effect of chronic ethanol administration on glutathione levels and its metabolising enzymes in rat brain. *Med. Sci. Res.,* 22, 731–732.

Marklund, S. L., Oreland, L., Perdahl, E., and Winblad, B. (1983) Superoxide dismutase activity in brains from chronic alcoholics. *Drug Alcohol Dependence,* 12, 209–215.

McCord, J. M. (1993) Human disease, free radicals, and the oxidant/antioxidant balance. *Clin. Biochem.,* 26, 351–357.

Meydani, M., Macauley, J. B., and Blumberg, J. B. (1986) Influence of dietary vitamin E, selenium and age on regional distribution of α-tocopherol in the rat brain. *Lipids,* 21, 786–791.

Montoliu, C., Vallés, S., Renau-Piqueras, J., and Guerri, C. (1994) Ethanol-induced oxygen radical formation and lipid peroxidation in rat brain: effect of chronic alcohol consumption. *J. Neurochem.,* 63, 1855–1862.

Nadiger, H. A., Marcus, S. R., and Chandrakala, M. V. (1988) Lipid peroxidation and ethanol toxicity in rat brain: effect of vitamin E deficiency and supplementation. *Med. Sci. Res.,* 16, 1273–1274.

Nihei, H., Kanemitsu, H., Tamura, A., Oka, H., and Sano, K. (1989) Cerebral uric acid, xanthine, and hypoxanthine after ischemia: the effect of allopurinol. *Neurosurgery,* 25, 613–617.

Niki, E. (1993) Antioxidant defenses in eukariotic cells: an overview. In *Free Radicals: From Basic Science to Medicine,* Poli, G., Albano, E., and Dianzani, M. U., Eds., Birkhäuser Verlag Basel, Switzerland, 365–373.

Nohl, H. (1993a) Involvement of free radicals in ageing: a consequence or cause of senescence. *Br. Med. Bull.,* 49, 653–667.

Nohl, H. (1993b) Nitric oxide and related radicals. In *Free Radicals: From Basic Science to Medicine,* Poli, G., Albano, E., and Dianzani, M. U., Eds., Birkhäuser Verlag Basel, Switzerland, 38–46.

Nordmann, R., Ribière, C., Rouach, H., Sinaceur, J., and Sabourault, D. (1985) Alcool et radicaux libres. *Bull. Acad. Natle. Méd.,* 169, 1201–1206.

Nordmann, R. (1987) Oxidative stress from alcohol in the brain. *Alcohol Alcoholism,* Suppl. 1, 75–82.

Nordmann, R., Ribière, C., and Rouach, H. (1990a) Ethanol-induced lipid peroxidation and oxidative stress in extrahepatic tissues. *Alcohol Alcoholism,* 25, 231–237.

Nordmann, R., Rouach, H., and Houzé, P. (1990b) Alcool, fer et stress oxydatif. *Bull. Acad. Natle. Méd.,* 174, 95–104.

Nordmann, R., Ribière, C., and Rouach, H. (1992) Implication of free radical mechanisms in ethanol-induced cellular injury. *Free Radic. Biol. Med.,* 12, 219–240.

Nordmann, R. and Rouach, H. (1992) Vitamin E disturbances during alcohol intoxication. In *Vitamin E in Health and Disease,* Packer, L. and Fuchs, J., Eds., Marcel Dekker, New York, 935–945.

Nordmann, R. (1994) Alcohol and antioxidant systems. *Alcohol Alcoholism,* 29, 513–522.

Oliver, C. N., Starke-Reed, P. E., Stadtman, E. R., Liu, G. J., Carney, J. M., and Floyd, R A. (1990) Oxidative damage to brain proteins, loss of glutamine synthetase activity, and production of free radicals during ischemia/reperfusion-induced injury to gerbil brain. *Proc. Natl. Acad. Sci. U.S.A.,* 87, 5144–5147.

Orrenius, S., McConkey, D. J., Bellomo, G., and Nicotera, P. (1989) Role of Ca^{2+} in toxic cell killing. *Trends Pharm. Sci. (TIPS),* 10, 281–285.

Orrenius, S. (1993) Mechanisms of oxidative cell damage. In *Free Radicals: From Basic Science to Medicine,* Poli, G., Albano, E., and Dianzani, M. U., Eds., Birkhäuser Verlag Basel, Switzerland, 47–64.

Packer, L. (1992) Free radical scavengers and antioxidants in prophylaxy and treatment of brain diseases. In *Free Radicals in the Brain,* Packer, L., Prilipko, L., and Christen, Y., Eds., Springer-Verlag, Berlin, 1–20.

Packer, L. and Kagan, V. E. (1993) Vitamin E: The antioxidant harvesting center of membranes and lipoproteins. In *Vitamin E in Health and Disease,* Packer, L. and Fuchs, J., Eds., Marcel Dekker, New York, 179–192.

Park, M. K., Rouach, H., Orfanelli, M. T., Janvier, B., and Nordmann, R. (1988) Influence of allopurinol and desferrioxamine on the ethanol-induced oxidative stress in rat liver and cerebellum. In *Alcohol Toxicity and Free Radical Mechanisms,* (Adv. Bio., Vol. 71), Nordmann, R., Ribière, C., and Rouach, H., Eds., Pergamon Press, Oxford, 135–140.

Paula-Barbosa, M. M., Brandao, F., Pinho, M.C., Andrade, J. P., Madeira, M. D., and Cadete-Leite, A. (1991) The effects of piracetam on lipofuscin of the rat cerebellar and hippocampal neurons after long-term alcohol treatment and withdrawal: a quantitative study. *Alcoholism: Clin. Exp. Res.,* 15, 834–838.

Pellegrini-Giampietro, D. E., Cherichi, G., Alesiani, M., Carla, V., and Moroni, F. (1988) Excitatory amino acid release from rat hippocampal slices as a consequence of free-radical formation. *J. Neurochem.,* 51, 1961–1963.

Pellmar, T. C. (1993) Do free radicals contribute to ethanol-induced synaptic damage? In *Alcohol-induced Brain Damage,* Hunt, W. A. and Nixon, S. J., Eds., NIAAA Monograph, No. 22, NIH Publication No. 93–3549. Rockville, MD, NIAAA, 339–354.

Prilipko, L. (1992) The possible role of lipid peroxidation in the pathophysiology of mental disorders. In *Free Radicals in the Brain,* Packer, L., Prilipko, L., and Christen, Y., Eds., Springer-Verlag, Berlin, 146–152.

Radi, R., Beckman, J. S., Bush, K. M., and Freeman, B. A. (1991a) Peroxynitrite oxidation of sulfhydryls. The cytotoxic potential of superoxide and nitric oxide. *J. Biol. Chem.,* 266, 4244–4250.

Radi, R., Beckman, J. S., Bush, K. M., and Freeman, B. A. (1991b) Peroxynitrite-induced membrane lipid peroxidation: the cytotoxic potential of superoxide and nitric oxide. *Arch. Biochem. Biophys.,* 288, 481–487.

Reed, D. J. (1993) Interaction of vitamin E, ascorbic acid, and glutathione in protection against oxidative damage. In *Vitamin E in Health and Disease,* Packer, L. and Fuchs, J., Eds., Marcel Dekker, New York, 269–281.

Reinke, L. A., Lai, E. K., DuBose, C. M., and McCay, P. B. (1987) Reactive free radical generation *in vivo* in heart and liver of ethanol-fed rats: correlation with radical formation *in vitro. Proc. Natl. Acad. Sci. U.S.A.,* 84, 9223–9227.

Ribière, C., Sabourault, D., Saffar, C., and Nordmann, R. (1987) Mitochondrial generation of superoxide free radicals during acute ethanol intoxication in the rat. *Alcohol Alcoholism,* Suppl. 1, 241–244.

Ribière, C., Hininger, I., Saffar-Boccara, C., Sabourault, D., and Nordmann, R. (1994) Mitochondrial respiratory activity and superoxide radical generation in the liver, brain and heart after chronic ethanol intake. *Biochem. Pharmacol.,* 47, 1827–1833.

Rouach, H., Ribière, C., Park, M. K., Saffar, C., and Nordmann, R. (1987a) Lipid peroxidation and brain mitochondrial damage induced by ethanol. *Bioelectrochem. Bioenerg.,* 18, 211–217.

Rouach, H., Park, M. K., Orfanelli, M. T., Janvier, B., and Nordmann, R. (1987b) Ethanol-induced oxidative stress in the rat cerebellum. *Alcohol Alcoholism,* Suppl. 1, 207–211.

Rouach, H., Houzé, P., Park, M. K., and Nordmann, R. (1989) Altérations du métabolisme cérébelleux du fer et leur prévention par l'allopurinol lors du stress oxydatif lié à l'administration aiguë d'éthanol chez le rat. *Compt. Rend. Soc. Biol.,* 183, 40–47.

Rouach, H., Houzé, P., Orfanelli, M. T., Gentil, M., Bourdon, R., and Nordmann, R. (1990) Effect of acute ethanol administration on the subcellular distribution of iron in rat liver and cerebellum. *Biochem. Pharmacol.,* 39, 1095–1100.

Rouach, H., Houzé, P., Orfanelli, M. T., Gentil, M., and Nordmann, R. (1991) Effects of chronic ethanol intake on some anti- and pro-oxidants in rat cerebellum. *Alcohol Alcoholism,* 26, 257 (abstract).

Rouach, H., Orfanelli, M. T., Gentil, M., and Nordmann, R. (1993) Cerebellar disturbances in relation to free radical mechanisms during chronic ethanol administration. *Alcohol Alcoholism,* 28, 222 (abstract).

Rouach, H., Houzé, P., Gentil, M., Orfanelli, M. T., and Nordmann, R. (1994) Effects of acute ethanol administration on the uptake of ^{59}Fe-labeled transferrin by rat liver and cerebellum. *Biochem. Pharmacol.,* 47, 1835–1841.

Saffar, C., Ribière, C., Sabourault, D., and Nordmann, R. (1988) Prevention by allopurinol of the ethanol-induced disturbances in brain mitochondrial electron transport chain and superoxide dismutase activity. In *Alcohol Toxicity and Free Radical Mechanisms,* (Adv. Bio., Vol. 71), Nordmann, R., Ribière, C., and Rouach, H., Eds., Pergamon Press, Oxford, 147–151.

Sawada, M. and Carlson, J. C. (1987) Changes in superoxide radical and lipid peroxide formation in the brain, heart and liver during the lifetime of the rat. *Mech. Ageing Dev.,* 41, 125–137.

Schor, N. F. (1988) Inactivation of mammalian brain glutamine synthetase by oxygen radicals. *Brain Res.,* 456, 17–21.

Seligman, M. L., Flamm, E. S., Goldstein, B. D., Poser, R. G., Demopoulos, H. B., and Ransohoff, J. (1977) Spectrofluorescent detection of malonaldehyde as a measure of lipid free radical damage in response to ethanol potentiation of spinal cord trauma. *Lipids,* 12, 945–950.

Schor, N. F. (1988) Inactivation of mammalian brain glutamine synthetase by oxygen radicals. *Brain Res.,* 456, 17–21.

Sies, H. (1985) Oxidative stress: introductory remarks. In *Oxidative Stress,* Sies, H., Ed., Academic Press, London, 1–8.

Sies, H. (1993a) Strategies of antioxidant defense. *Eur. J. Biochem.,* 215, 213–219.

Sies, H. (1993b) Medical applications of antioxidants: an update of current problems. In *Free Radicals: From Basic Science to Medicine,* Poli, G., Albano, E., and Dianzani, M. U., Eds., Birkhäuser Verlag Basel, Switzerland, 419–423.

Snyder, S. H. (1992) Nitric oxide: first in a new class of neurotransmitters? *Science,* 257, 494–496.

Tavares, M. A. and Paula-Barbosa, M. M. (1983) Lipofuscin granules in Purkinje cells after long term alcohol consumption in rats. *Alcoholism: Clin. Exp. Res.,* 7, 302–306.

Tindberg, N., Hansson, T., Köhler, C., and Ingelman-Sundberg, M. (1989) Ethanol-inducible cytochrome P-450 in the rat central nervous system. *Alcohol Alcoholism,* 24, 389 (abstract).

Uysal, M., Keyer-Uysal, M., Kocak-Toker, N., and Aykac, G. (1986) The effect of chronic ethanol ingestion on brain lipid peroxide and glutathione levels in rats. *Drug Alcohol Dependence,* 18, 73–75.

Uysal, M., Kutalp, G., Ozdemirler, G., and Aykac, G. (1989) Ethanol-induced changes in lipid peroxidation and glutathione content in rat brain. *Drug Alcohol Dependence,* 23, 227–230.

Warner, M. and Gustafsson, J. A. (1994) Effect of ethanol on cytochrome P450 in the rat brain. *Proc. Natl. Acad. Sci. U.S.A.,* 91, 1019–1023.

Wolff, S. P., Garner, A., and Dean, R. T. (1986) Free radicals, lipids and protein degradation. *Trends Biochem. Sci. (TIBS),* 11, 27–31.

Yoshida, S., Busto, R., Watson, B. D., Santiso, M., and Ginsberg, M. D. (1985) Postischemic cerebral lipid peroxidation *in vitro*: modification by dietary vitamin E. *J. Neurochem.,* 44, 1593–1601.

Zaleska, M. M., Nagy, K., and Floyd, R. A. (1989) Iron-induced lipid peroxidation and inhibition of dopamine synthesis in striatum synaptosomes. *Neurochem. Res.,* 14, 597–605.

Zima, T., Novak, L., and Stipek, S. (1993) Plasma xanthine oxidase level and alcohol administration. *Alcohol Alcoholism,* 28, 693–694.

Chapter 19

TOXIC EFFECTS OF ETHANOL ON THE FETAL BRAIN

Wei-Jung A. Chen, Susan E. Maier, and James R. West

TABLE OF CONTENTS

I. INTRODUCTION

The term "fetal alcohol syndrome" (FAS) was first introduced about 20 years ago by Jones and colleagues (1973) to describe a constellation of physical anomalies, physiological dysfunctions, and psychological aberrations observed among children born to mothers who abused ethanol during pregnancy. The minimal diagnostic criteria for FAS are (1) pre- and postnatal somatic growth deficiency in height, weight, and head circumference; (2) central nervous system (CNS) dysfunctions in intellectual and motor performance; and (3) distinctive facial characteristics taking the form of midfacial hypoplasia. Two related terms, "fetal alcohol effect" (FAE) (Clarren and Smith, 1978) and "alcohol-related birth defect" (ARBD) (Sokol and Clarren, 1989) have been introduced to refer to characteristics in those born with only some of the features of FAS, but whose mothers had a history of ethanol abuse during pregnancy.

In this chapter, we begin by considering several risk factors that can exacerbate the danger to the fetus from ethanol exposure during gestation. Then we summarize key reports from the literature encompassing clinical and experimental research on FAS. Finally, we consider the

0-8493-8389-7/95/$0.00+$.50

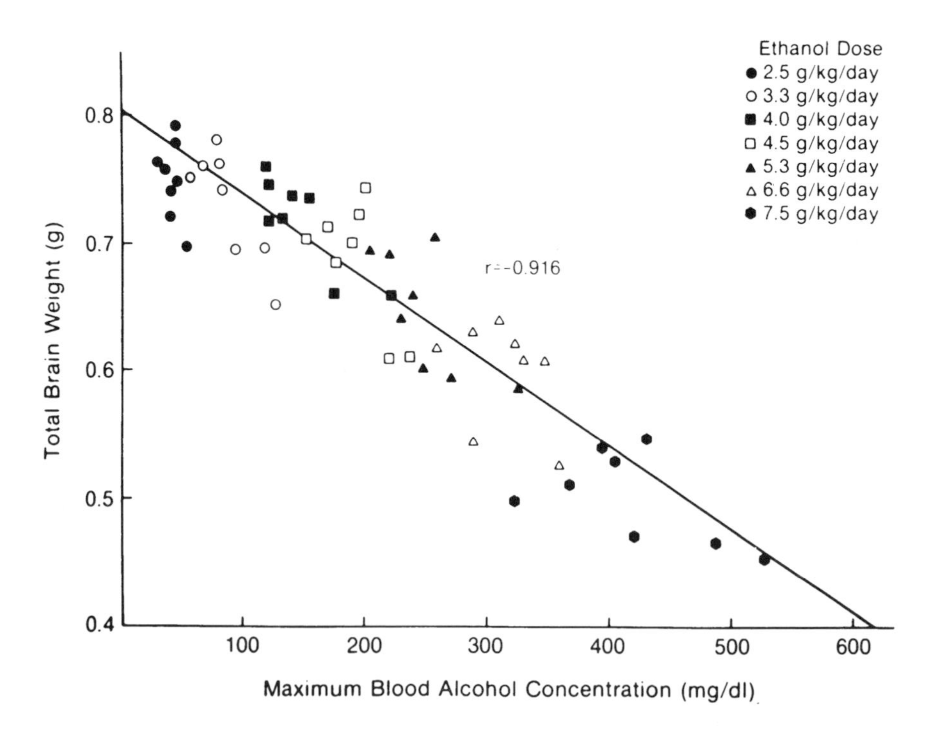

FIGURE 19.1. Brain growth restriction, as indicated by lower total brain weight, plotted as a function of the peak blood alcohol concentration obtained on postnatal day 6. Briefly, artificially reared rat pups were given their assigned dose of ethanol in 4 of 12 daily feedings (administered via an implanted gastric cannulae) from postnatal days 4 through 9. Brain weights were obtained following perfusion with 4% paraformaldehyde on postnatal day 10. The regression equation describing this function is the following: Brain weight (g) = $[-6.6 \times 10^{-4}$ (g/mg/dl)] [BAC(mg/dl)] + 0.803. Correlation coefficient (r) = –.916. (From Bonthius and West, *Teratol.*, **37,** 223–231, 1988, with permission.)

toxic effects of ethanol on the developing brain and discuss several putative mechanisms associated with ethanol-induced fetal brain damage.

II. RISK FACTORS

There is considerable diversity in outcome among children born to mothers who abuse ethanol during pregnancy. Some women may report drinking the same amount of ethanol during gestation, yet their babies are not affected equally by the *in utero* ethanol exposure. This suggests that there are risk factors that may influence the consequences of fetal ethanol exposure.

A. BLOOD ALCOHOL CONCENTRATION

Studies of the relationship between blood alcohol concentration (BAC) and ethanol-induced brain malformations have demonstrated convincingly that the peak BAC is a reliable predictor of risk and severity of damage, including microencephaly (Bonthius and West, 1988), neuronal loss (West et al., 1990), and neurobehavioral impairments (Goodlett et al., 1987, 1991b). Figure 19.1 demonstrates that higher BACs produce greater brain growth restrictions (a parameter of microencephaly) during the brain growth spurt. From these data, it is reasonable to speculate that any manipulation that increases peak BAC may increase the risk and severity of fetal brain damage. These manipulations include the concurrent use of over-the-counter (Bonthius and West, 1989) and prescription drugs (Caballeria et al., 1989a), coadministration of pharmacological agents that alter the metabolism of ethanol (Chen et al.,

1995; Ukita et al., 1993), disease states in various organs (Caballeria et al., 1989b; Lieber, 1993), ethanol concentration (Maier et al., in press; Roine et al., 1991), and drinking pattern (Bonthius and West, 1990).

As an example, Bonthius and West (1990) have shown that rat pups exposed to a condensed pattern of ethanol during the brain growth spurt exhibited more severe cerebellar weight restrictions than did those administered the identical daily dose in feedings spread evenly throughout the day. The detrimental consequences of this condensed "binge-like" drinking pattern are not restricted to short-term deficits in cerebellar weights. Goodlett et al. (1987, 1994) demonstrated long-term neurobehavioral and neuroanatomical deficits from this binge-like drinking pattern. It is important to note that clinical studies show that many pregnant women who abuse ethanol follow a binge-like drinking pattern when they engage in ethanol consumption (Stephens, 1985). If the peak BAC is a reliable predictor of adverse fetal outcome following perinatal ethanol exposure, then factors that influence BAC must be considered carefully.

B. TEMPORAL VULNERABILITY

When interpreting the effects of ethanol exposure on the developing organism, it is important to consider the issue of temporal vulnerability (West, 1987). Different brain structures develop uniquely according their genetically programmed sequences at various developmental stages. The presence of a neuroteratogen during a critical (vulnerable) developmental stage will result in more detrimental consequences compared with exposure during a less sensitive stage. One focus of our laboratory has been to determine the periods of vulnerability of neuronal populations in the cerebellum to ethanol exposure. The results offer clues to the mechanisms that underlie fetal-ethanol-induced brain damage. For example, rat pups given ethanol during the third trimester equivalent (the period of most rapid brain growth) exhibited differential Purkinje cell loss depending on the timing of ethanol exposure, with the most dramatic reduction in neuronal numbers observed following exposure on postnatal days four to five (Goodlett et al., 1990a; Hamre and West, 1993). Furthermore, other studies suggest that Purkinje cells are much less vulnerable following *prenatal* ethanol exposure (during the period of neurogenesis) than they are following *postnatal* exposure (see Figure 19.2) (Marcussen et al., 1994), a period after the Purkinje cells have been formed and are starting to grow (Phillips and Cragg, 1982). According to these findings, it is clear that there is a temporal window of vulnerability for Purkinje cells in response to ethanol insults, which occurs postnatally. Specific windows of vulnerability are likely to exist for each neuronal population in the brain.

C. GENETIC FACTORS

In humans, genotype appears to contribute to the variation in progeny outcome after heavy maternal ethanol consumption during pregnancy, even when other influential factors are controlled (Christoffel and Salafsky, 1975). Focusing on the effects of ethanol exposure during the brain growth spurt among several inbred strains of rats, the Marshall (M520) and Maudsley Reactive (MR) strains characterize the two opposing ends of a continuum in terms of susceptibility to cerebellar weight restriction (Goodlett et al., 1990b) and Purkinje cell loss (Goodlett et al., in preparation; Maier et al., 1994). Figure 19.3 illustrates the differences in Purkinje cell survival in the MR and M520 strains. Interestingly, the M520 ethanol-treated group also showed a loss of Purkinje cells, yet the *magnitude* of the decrement was not nearly as substantial as in the MR strain. These differential responses to ethanol exposure are independent of strain differences in pharmacokinetics (Goodlett et al., 1991a) or BAC (Goodlett et al., 1989). Taken together with other data, this is a clear demonstration of genetic-associated differences in susceptibility to ethanol-induced brain injuries during development.

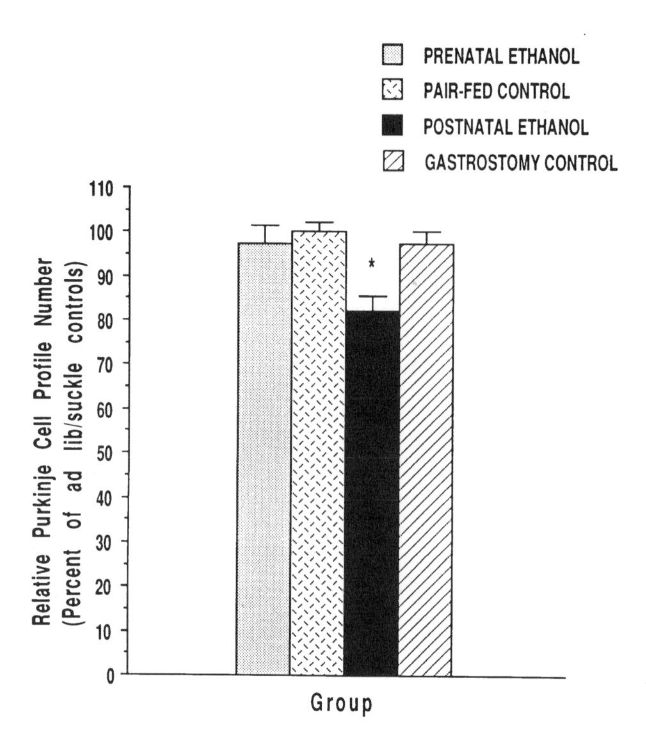

FIGURE 19.2. Comparison of Purkinje cell profile numbers from the single (2 μm) vermal sections of the cerebellum of prenatal and postnatal ethanol-treated rat pups groups expressed as a percent of the ad lib/suckle control group mean. Prenatally treated animals were offspring of dams given a daily oral dosing of 5 g/kg/day of 15% (w/v) ethanol solution (prenatal ethanol group) or a maltose dextrin solution (pair-fed control group) from gestation days 13 through 18. Postnatally treated animals were artificially reared from postnatal days 4 through 9 and received daily intragastric infusions of 2.5 g/kg/day of ethanol (postnatal ethanol group) or maltose dextrin solution (gastrostomy control group) via indwelling gastric cannulae. All pups were perfused and had their cerebella removed on postnatal day 10. Values presented are means (as percent) ± standard error of the mean. (*) significantly different compared to all other groups ($p < 0.05$, Scheffé test). (From Marcussen et al., *Alcohol,* **1,** 147–156, 1994, with permission.)

D. POLYDRUG ABUSE

The increasing incidence of polydrug use has raised concerns regarding the extent of damage due to the interactive effects of ethanol and other licit and illicit drugs on fetal brain development (Day and Richardson, 1994). Ethanol is often abused concomitantly with other drugs (Chasnoff et al., 1990; Kokotailo et al., 1992). There is an urgent need, therefore, to understand the interactive effects of ethanol with these drugs during pregnancy.

There is a remarkable decline in brain weights (microencephaly) in neonatal rats exposed to aspirin and ethanol during the brain growth spurt period (see Figure 19.4) (Bonthius and West, 1989). In contrast, ethanol-induced brain growth restrictions were not potentiated by concurrent exposure to cocaine during the brain growth spurt. However, the administration of ethanol and cocaine significantly increased mortality, suggesting an additive or a synergistic action with the combination of both drugs in this setting (Chen et al., 1994). In addition, results from other laboratories (e.g., Church et al., 1988; Coles et al., 1992; Hearn et al., 1991) support the contention that polydrug use during pregnancy increases the risk of damage from ethanol exposure to the fetus.

E. MATERNAL NUTRITIONAL STATUS

Perinatal malnutrition has been well recognized as an adverse factor in fetal brain development, regardless of whether it occurs prior to or during the brain growth spurt period (for an excellent recent review, see Morgane et al., 1993). Disturbance in nutritional status appears

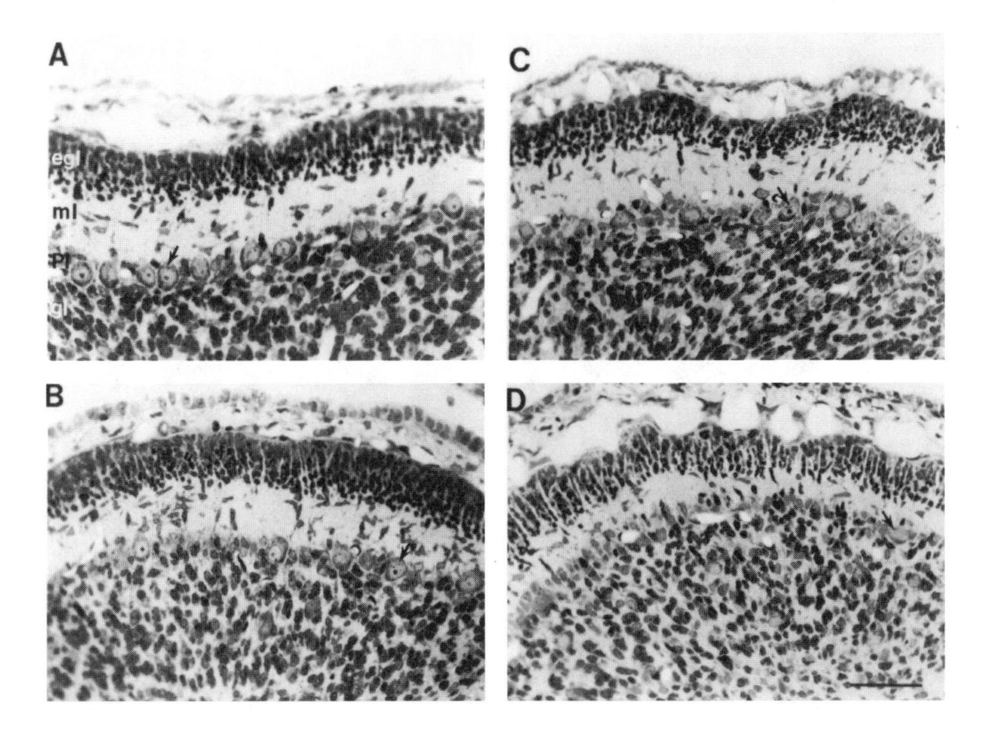

FIGURE 19.3. Purkinje cell profiles in 2 μm vermal sections near the primary fissure of the cerebellum in the M520 (A and B) and MR (C and D) strains of 10-day-old rat pups. All pups were artificially reared from postnatal days 5 through 9, were perfused on postnatal day 10, and had their cerebella removed and weighed. During artificial rearing, the ethanol treated groups of each strain received 6.6 g/kg/day of ethanol in 4 of 12 consecutive daily feedings, and the gastrostomized control animals received an isocaloric maltose dextrin solution substituted for ethanol in the same four feedings. Note the extensive vacuoles, previously occupied by Purkinje cells, within the Purkinje cell line from the MR ethanol group (D). The M520 ethanol group (B) also shows Purkinje cell loss, but to a lesser degree than in the MR ethanol group. The Purkinje cell line appears normal in the gastrostomized MR (C) and M520 (A) control cerebella. The arrow indicates a typical Purkinje cell. Abbreviation: egl = external granule layer, ml = molecular layer, Pl = Purkinje layer, gl = granule layer. Bar = 50 μm.

to interfere with chronologically programmed brain development. Chronic ethanol exposure results in a suppression of food intake and dysfunction in digestion, including interference in nutrient absorption and utilization (Thomson and Pratt, 1992), yet there is no systematic research or empirical evidence demonstrating how nutritional status or health condition during pregnancy interacts with ethanol to induce fetal brain damage.

III. CLINICAL STUDIES

Among the reported primary features of FAS, CNS dysfunction is considered the most serious. However, in the past, human research on FAS describing CNS anomalies has been limited to autopsy cases and less specific, noninvasive techniques such as electroencephalogram (EEG). Not until recently has the state-of-the-art magnetic resonance imaging (MRI) technique allowed any meaningful assessment of neuroanatomical anomalies in FAS cases.

A. NEUROANATOMICAL MODIFICATIONS

A common characteristic of CNS damage associated with FAS is microencephaly, indicating structural alterations within the brain in response to ethanol insults. Recent MRI studies examining structural defects on brain development as a result of prenatal ethanol exposure have identified reductions in the volumes of the corpus collosum, basal ganglia, cerebrum, and cerebellum (Mattson et al., 1992, 1994). Autopsy data on FAS brains documented various

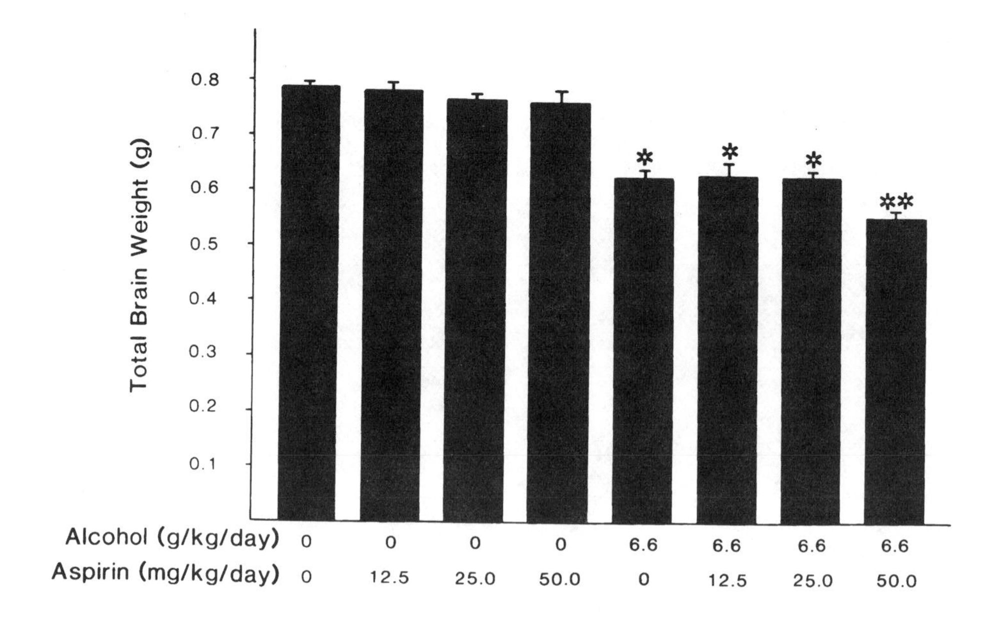

FIGURE 19.4. Average total brain weight of 10-day-old rat pups exposed to different daily doses of ethanol and aspirin. From postnatal days 4 through 9, artificially reared rat pups received 6.6 g/kg/day of ethanol (ethanol treated groups) or isocaloric maltose dextrin (gastrostomized control groups) in 4 of 12 daily intragastric infusions. Aspirin was administered through the diet in all 12 daily feedings. Pups were perfused on postnatal day 10 and had their brains removed and weighed. Data are presented as means ± standard error of the mean. (*) significantly different from all nonethanol treated groups ($p < 0.001$). (**) significantly different from all other groups ($p < 0.01$). (From Bonthius and West, *Neurotoxicol. Teratol.*, **11**, 135–143, 1989, with permission.)

degrees in the severity of malformations, such as agenesis of the corpus collosum, cerebellar vermis, and olfactory bulb (Clarren et al., 1978; Coulter et al., 1993; Majewski, 1981; Peiffer et al., 1979). From a microscopic viewpoint, brain damage from *in utero* ethanol exposure ranges from disorganized temporal and frontal cortical layers, to depletion of neurons in the dentate gyrus, a significant loss of Purkinje cells in the cerebellum, and other general anomalies in cellular structures, including dendritic spines (Coulter et al., 1993; Ferrer and Galofré, 1987; Majewski, 1981).

B. NEUROPSYCHOLOGICAL ASSESSMENTS

Neurobehavioral dysfunction, including cognitive impairments, appears to be one of the most recognizable characteristics among children, adolescents, and adults with FAS or FAE. IQ scores from affected individuals encompass the range from severe mental retardation to normal, regardless of the ages tested (Conry, 1990; Streissguth et al., 1978, 1991). Nevertheless, low average IQ remains a distinctive feature of FAS (Aronson et al., 1985; Robinson et al., 1987). In terms of other neuropsychological measures, individuals born with FAE or FAS exhibit sensory and motor deficiencies as well as impairments in learning and memory (Barr et al., 1990; Brown et al., 1991; Conry, 1990; Mattson et al., 1992; Streissguth et al., 1983, 1990).

The long-term consequences of ethanol exposure *in utero* has become one of the main issues in FAS research, since children previously diagnosed with FAS only recently have reached adolescence and young adulthood. Reports indicate that the distinctive facial features of FAS become less apparent during and beyond the periadolescent period. However, the somatic growth deficiency remains apparent (Streissguth et al., 1991). Recent papers from Streissguth's research group (1994a,b) reported a long-term deficiency in cognitive functions assessed by attention/memory and other academic performance tests that are appropriate for the corresponding developmental stage (age). It is quite clear that even though cognitive performance in some of the children with FAS or FAE may show some improvement with age, in general, the deficiencies are long-lasting.

C. LIMITATIONS

Although retrospective studies involving the teratogenetic effects of ethanol on the developing fetus have several advantages, there are many factors that remain unaccounted for in studies of FAS. As mentioned earlier, the peak BAC is a major determinant in assessing risk and severity of brain injury resulting from fetal ethanol exposure. Personal reports of drinking during pregnancy are notoriously unreliable. Moreover, as is often the case with ethanol abuse, women may be reluctant to report polydrug use, greatly complicating the interpretation of the effects of gestational ethanol exposure.

IV. EXPERIMENTAL STUDIES

Due to the inherent limitations of clinical FAS studies, animal research is providing valuable information toward understanding the mechanisms underlying the detrimental effects of ethanol on the developing fetus. Prior to extrapolating data from animal studies to humans, the equivalent periods of brain development must be considered. All mammals pass through the same stages of brain development. In humans, the brain growth spurt, which is the most dynamic period of CNS development, occurs *in utero* throughout the third trimester of gestation and continues into postnatal life, whereas in rats, for example, it occurs entirely postnatally (Dobbing and Sands, 1973, 1979). Therefore, comparisons of the timing of ethanol exposure relative to the stages of brain development in animal model systems need to be considered carefully in order to extrapolate accurately the results from animal studies to humans.

A. NEUROANATOMICAL MODIFICATIONS

Ethanol exposure during brain development produces neuroanatomical alterations ranging from gross changes (e.g., microencephaly and agenesis of the corpus callosum) to microscopic changes (e.g., neuronal depletion and subtle changes in cytoarchitecture). Numerous studies have demonstrated that ethanol exposure during the brain growth spurt period produces severe microencephaly and significant neuronal loss in brain regions such as the cerebellum (granule and Purkinje cells) (Phillips and Cragg, 1982; Hamre and West, 1993; Pierce et al., 1989, 1993), hippocampus (pyramidal cells in field CA1) (Bonthius and West, 1990, 1991; Pierce et al., 1989), and olfactory bulb (granule and mitral cells) (Bonthius et al., 1992). In addition to those involving the postnatal rat, many studies have employed prenatal model systems with similar findings, including depleted pyramidal cells in field CA1 of the hippocampus (Barnes and Walker, 1981), fewer axons in the pyramidal tract (Miller and Al-Rabiai, 1994), reduced Purkinje cells in the cerebellum (Bonthius et al., 1991), and a deficit in the number of retinal ganglion cells (Clarren et al., 1990). Recently, a modern three-dimensional stereological technique, which permits more accurate quantitative estimations of the total cell number, has been used to identify deficits in neuronal numbers in the olfactory bulb (Bonthius et al., 1992), cerebellum (Maier et al., 1994), and inferior olivary nucleus (Napper and West, submitted) resulting from ethanol exposure.

Aside from inducing cell death in existing developing neurons, there is convincing evidence that ethanol alters cell proliferation and induces errors in neuronal migration, the first two of four ontogenetic phases of neuronal development. Miller and Muller (1989) reported that ethanol exposure throughout the entire gestational period significantly reduced the number of small neurons in the principal sensory nucleus (PSN) of the trigeminal nerve and produced a delay in the neurogenesis of large PSN neurons. They suggested that the deficits in number and the delay in neurogenesis in PSN neurons could have resulted from a decrease in the number of developing neurons leaving the proliferative zone and a prolonged S-phase (synthesis of DNA) within a cell cycle. The cerebral cortex is another brain region where the effects of prenatal exposure to ethanol on cell proliferation and migration have been identified. Miller (1986, 1988, 1993) and Miller and Robertson (1993) have demonstrated that prenatal ethanol exposure alters the cell proliferative activity and migration by affecting the synthesis of DNA, RNA, and the maturity of supportive glial cells.

In addition to producing deficits in neuron numbers, perinatal ethanol exposure can adversely affect the development of neuronal circuitry. The hippocampal formation is an attractive model for studying effects of ethanol on neuronal circuitry because of the orderly development and lamination of its afferents. One important pathway, the hippocampal mossy fiber system, forms the major excitatory input from the dentate gyrus to the pyramidal cells in field CA3 of the hippocampus. Normally, mossy fibers traverse a precise course making multiple connections with pyramidal cell apical dendrites in the *stratum lucidum*. Various developmental manipulations, including increased levels of thyroxine (Lauder and Mugnaini, 1980) and hippocampal lesions (Laurberg and Zimmer, 1980), can alter the distribution of mossy fiber. Gestational exposure to ethanol also can produce an aberrant distribution of mossy fibers (West et al., 1981). Furthermore, ethanol exposure during the brain growth spurt resulted in a greater aberrant mossy fiber terminal field (West and Hamre, 1985), suggesting that mossy fibers are vulnerable over a wide period of development, but that they are most vulnerable during the brain growth spurt. More recently, Miller and Al-Rabiai (1994) found that prenatal exposure to ethanol caused a disruption in the structure of the pyramidal tract, but that the number of fibers actually increased with ethanol treatment.

Synaptic networks also appear to be affected by ethanol exposure during brain development. Mohamed et al. (1987) observed altered climbing fiber maturation with a significant increase in atypical climbing fiber synapses on Purkinje cell somas, rather than on dendrites, suggesting a delay in developmental sequence. Moreover, Tanaka et al. (1991) reported a marked decline in the number of synaptic junctions in CA3 of the hippocampus in response to prenatal ethanol exposure. With regard to dendritic structure, Miller et al. (1990) showed that adult rats exposed to ethanol prenatally exhibited long-term changes in the altered distribution of cortical pyramidal cells, and that these ectopic cells had more complex dendritic trees and a higher density of dendritic spines. In contrast, others have reported that perinatal ethanol exposure leads to a less extensive dendritic branching and fewer dendritic spines on pyramidal cells of the sensorimotor cortex (Hammer and Schiebel, 1981; Reyes et al., 1983) and hippocampal pyramidal cells (Smith and Davies, 1990) among younger animals. The discrepancy may be a function of variables such as regional differences or the ages of the animals when they were examined. Perinatal exposure to ethanol may lead to a decrease in spine regression during adulthood. In other words, the increase in dysmorphic and dysfunctional spines may contribute to the overall increase in spine density. Such alterations in the components of synaptic networks may interfere with functional intercellular communications, mediating some of the behavioral and cognitive deficits associated with FAS.

Nonneuronal elements such as glial cells also are affected by ethanol exposure during brain development. Ethanol exposure during the brain growth spurt period produced transient cortical astrogliosis (Goodlett et al., 1993), identified by the increased expression of glial fibrillary acidic protein (GFAP), a potential marker for neuronal injuries (O'Callaghan, 1993; Stone, 1993). The intensive immunoreactivity of GFAP, located mostly in cortical layer V and around the blood vessels, suggested ethanol-induced injuries to pyramidal cells or axon terminals within layer V and the developing blood-brain barrier, respectively. The increase in cortical GFAP immunoreactivity may be a function of the direct action of ethanol on the rate of transcription of GFAP mRNA (Fletcher and Shain, 1993). Furthermore, Miller and Robertson (1993) reported that prenatal exposure to ethanol accelerated the early postnatal transformation of radial glia to astrocytes. Since radial glia participate in guiding neuronal migration, the early transformation of radial glia into astrocytes may interfere with normal neuronal migration in the cortex, accounting for the observed alterations in cortical organization. The effects of ethanol on oligodendrocytes, another glial cell type, was reported by Phillips and Krueger (1990). They found that postnatal ethanol exposure in the rat markedly retards the maturation of developing oligodendrocytes, which may indirectly delay the myelination process in the CNS. Furthermore, Zoeller et al. (1994) reported that the mRNAs encoding specific myelin basic protein and myelin-associated glycoprotein are significantly reduced in the cerebellum following postnatal ethanol exposure. Together, these studies have demonstrated that ethanol exposure during development produces deficits in both neuronal and nonneuronal cell types in the developing brain.

B. NEUROCHEMICAL ALTERATIONS

Many neurobehavioral alterations or developmental anomalies associated with perinatal ethanol exposure may be associated with alterations in neurochemical status, such as the disruption in development of neurotransmitter, hormonal, and neurotrophic systems, whether or not there are concomitant structural changes in neuronal organization. Studies of neurotransmitters during brain development have demonstrated that serotonin serves as a trophic factor during neurogenesis (Lauder et al., 1983). Prenatal exposure to ethanol significantly decreased serotonin concentration and binding sites, and altered serotonergic projections to the terminals as well (Druse and Paul, 1989; Druse et al., 1991; Tajuddin and Druse, 1993). In addition, exposing cultured fetal astroglia to ethanol significantly reduced the number of serotonin uptake sites (Lokhorst and Druse, 1993).

Cholinergic systems are also affected by ethanol exposure. Chick embryos exposed to chronic ethanol showed a significant reduction in forebrain choline acetyltransferase (ChAT) activity (Swanson et al., 1994). However, Light and associates (1989a,b) exposed neonatal rat pups to ethanol during a portion of the brain growth spurt and found no significant change in ChAT activities in the cortex, hippocampus, or cerebellum on postnatal day 20. The reason for these differences is unclear, and comparisons between these two studies are complicated by different model systems (avian vs. mammalian), different brain regions assessed, different ethanol doses used, and temporal variations in the ethanol exposure. Nevertheless, data from our laboratory suggest a long-lasting increase in the number of muscarinic cholinergic receptors in the hippocampus of adult rats exposed to ethanol during the brain growth spurt period (Kelly et al., 1989). Studies involving other neurotransmitter systems (catecholaminergic, glutaminergic, etc.) also have demonstrated adverse effects from perinatal ethanol exposure. For example, Druse et al. (1990) reported decreased dopamine concentrations in the corpus striatum and the striatal D1 receptors following *in utero* ethanol exposure, and Savage et al. (1991) showed that prenatal ethanol exposure decreases hippocampal N-methyl-D-aspartate-sensitive glutamate receptor binding-site density. Taken together, the data suggest that perinatal ethanol treatment affects fundamental neurotransmitter balances pertaining to normal brain development.

Neurotrophic support is important for inducing neuronal differentiation and for promoting neuronal survival during brain development. Recent studies investigating the interactive actions between ethanol and neurotrophic factors indicate that ethanol inhibits neurotrophic activity in the rat hippocampus (Walker et al., 1990, 1992) and chick embryonic forebrain (Heaton et al., 1992). It is thought that ethanol reduces the levels of nerve growth factor (NGF) receptors on responsive cells, which subsequently inhibits the neurotrophic function of NGF. Furthermore, it has been reported that the supplementation of NGF or basic fibroblast growth factor (bFGF) prevents neuronal degeneration in cultured cerebellar granule cells (Pantazis et al., 1994) and embryonic chick dorsal root ganglion cells (Heaton et al., 1993). These results suggest that ethanol may change the concentrations of available neurotrophic factors in addition to its effects on reducing the number of neurotrophic factor receptors.

The adenylate cyclase/cyclic AMP second-messenger system appears to be a target of ethanol exposure during development. Pennington (1988) reported that chick embryos exposed to ethanol result in decreased responsiveness of brain adenylate cyclase. Moreover,

Pennington (1990) also found that cyclic AMP levels and the binding activity of cyclic AMP to the regulatory subunit of cyclic AMP-dependent protein kinase were both decreased following ethanol exposure during the embryonic period. Similarly, Beeker et al. (1988) reported that ethanol exposure significantly reduced the cyclic AMP-stimulated autophosporylation of cyclic AMP-dependent protein kinase in embryonic chick brain. It is known that maintaining a functional cyclic AMP-dependent protein kinase-induced phosphorylation is essential for many physiological responses, including neurotransmitter release (Nestler and Greengard, 1989). Fetal ethanol exposure may disrupt the activities of cyclic AMP and cyclic AMP-dependent protein kinase, and subsequently interfere with specific protein phosphorylation activities that are uniquely associated with CNS development (Pennington, 1992).

V. PUTATIVE MECHANISMS

A. ETHANOL VS. ACETALDEHYDE-MEDIATED TOXICITY

A key step in understanding the damage mechanism of any teratogen is determining the specific teratogenic agent or metabolite responsible for the damage. An unresolved debate concerns whether the proximate teratogen in ethanol-mediated fetal brain damage is ethanol, per se, or its primary metabolite, acetaldehyde.

1. Ethanol

A recent study from our laboratory suggests that ethanol is the key teratogen in mediating fetal ethanol-induced brain injury. The administration of an ADH inhibitor, 4-methylpyrazole, in combination with ethanol significantly augmented ethanol-induced brain growth restrictions compared with that of ethanol treatment alone (Chen et al., 1995). Like most ADH inhibitors, 4-methylpyrazole prevents the conversion of ethanol into acetaldehyde, which results in a higher peak of BAC and an increase in the duration at peak BAC level due to its immediate action in blocking ethanol metabolism. This recent finding supports the hypothesis that peak BAC is a better predictor of fetal brain damage than ethanol dose. Furthermore, Blakley and Scott (1984) reported that the cotreatment of ethanol and 4-methylpyrazole significantly increased embryolethality and teratogenicity compared with ethanol treatment alone. Taken together, these findings support the notion that ethanol is the key component in producing microencephaly, because the action of the ADH inhibitor ensures that only trace amounts of acetaldehyde are generated. Nevertheless, it should be noted that 4-methylpyrazole does not completely inhibit the formation of acetaldehyde, since 4-methylpyrazole blocks ADH-dependent, but only part of the ADH-independent metabolic pathways.

A possible role of ethanol in mediating fetal brain damage lies in its interaction with the synthesis of retinoic acid from retinol. Retinoic acid is an endogenous vitamin A metabolite that has been identified in modulating CNS development and limb morphogenesis. Ethanol is hypothesized to interfere with the synthesis of retinoic acid by directly disrupting the activity of a rate-limiting enzyme, retinol dehydrogenase (RDH), which is identical to class I ADH, an enzyme that converts ethanol to acetaldehyde (Shean and Duester, 1993). It appears that RDH (class I ADH) exhibits a higher affinity for ethanol than for retinol, which inhibits the synthesis of retinoic acid. This model can explain the actions of ADH inhibitors in magnifying ethanol-mediated fetal brain injuries. The presence of an ADH inhibitor would not only impede the metabolism of ethanol, but also prevent the production of retinoic acid, which would lead to more severe impairments of fetal CNS development.

Fetal ethanol-induced brain damage also has been proposed to be mediated through the activities of gangliosides. Gangliosides are enriched in brain tissues and are specific for maintaining the functional and structural integrity of the neuronal membrane. Recently, a "dehydration theory" associated with ganglioside activity was hypothesized to account for the intoxicating effects of ethanol (Klemm, 1990). According to Klemm, ethanol molecules displace water molecules that normally link the gangliosides to membrane proteins. This disrupts the stable hydrogen bonding status between gangliosides and membrane proteins, while concurrently freeing the oligosaccharide moieties of the gangliosides into their intramembrane microenvironments. Such modifications in the physical constraints of the neuronal membrane may subsequently lead to membrane fluidization and alterations in the functions of the adjacent membrane proteins or receptors, particularly those concerned with neurotrophic activities. Presently, it is unclear whether acetaldehyde molecules are capable of exerting actions similar to those of the ethanol molecules to replace water molecules and free the oligosaccharide moieties of the gangliosides.

Additional evidence for the role of ethanol as a primary teratogen in fetal ethanol-induced toxicity comes from an interesting study of wild-type fruit flies *(Drosophila melanogaster).* This species is especially useful for answering questions regarding ethanol/acetaldehyde toxicity due to a unique characteristic in its ethanol metabolic pathway. *D. melanogaster* is capable of displaying a dual function for ADH, in which ADH metabolizes ethanol to acetaldehyde and further converts acetaldehyde into acetate without an accumulation of acetaldehyde in the physiological system (Miller et al., 1992). Ranganathan et al. (1987) observed dose-dependent gross malformations among fruit flies exposed to ethanol. In summary, although the current literature does not provide an absolute link of the direct effects of ethanol in neuroteratogenicity, it corroborates the hypothesis that ethanol, per se, is the causative agent in fetal ethanol teratogenicity.

2. Acetaldehyde

Despite convincing evidence demonstrating that ethanol is the proximate candidate for producing ARBDs, assessing the role of acetaldehyde in mediating ethanol teratogenicity has attracted considerable attention. The direct administration of acetaldehyde appears to be a more accurate method for examining the possible involvement of acetaldehyde in fetal ethanol damage since it is easier to control the amount of acetaldehyde. However, the experimental procedures employed to manipulate acetaldehyde levels are problematic partly due to its low boiling point (bp 21°C), and because the accurate quantitative measurement of trace amounts of acetaldehyde formed *in vivo* systems is difficult (Eriksson and Fukunaga, 1993), particularly in developing organisms (Zorzano and Herrera, 1989).

Disulfiram (Antabuse®) is an aldehyde dehydrogenase (ALDH) inhibitor that prevents the oxidation of acetaldehyde to acetate while increasing circulating levels of acetaldehyde. Disulfiram has been the primary focus of research assessing the effects of acetaldehyde on ethanol-induced neurotoxicity, since it results in the accumulation of toxic acetaldehyde following ethanol use. There is little information available concerning the use of disulfiram treatment in pregnant women and its subsequent effects on developing fetuses. However, one report suggests that Antabuse® does exacerbate the detrimental effects of ethanol on the conceptus (Jones et al., 1991), suggesting an influential role of acetaldehyde in ARBDs.

Several studies have assessed the direct role of acetaldehyde as a mediator in FAS by the administration of acetaldehyde, rather than ethanol, to pregnant rats and mice (Ali and Persaud, 1988; O'Shea and Kaufman, 1979). These studies demonstrated various malformations among exposed fetuses and implicate acetaldehyde as responsible for FAS. Although acetaldehyde-induced teratogenicity has been documented, the effects of acetaldehyde on brain development requires more focused research to further compare the severity and pattern of brain injuries with the effects of ethanol exposure.

The role of acetaldehyde in producing the teratogenic effects of fetal ethanol exposure is substantiated by evidence that toxic free radicals are formed during the oxidation of acetaldehyde (Fridovich, 1989; Oei et al., 1982). During the oxidative process, the univalent and divalent reduction of oxygen occurs as well as a releasing of radical superoxide (O_2^-), which is a toxic oxygen metabolite. It is known that glutathione (GSH), a tripeptide consisting of glutamate, cysteine, and glycine, is an antioxidant capable of scavenging toxic free radicals.

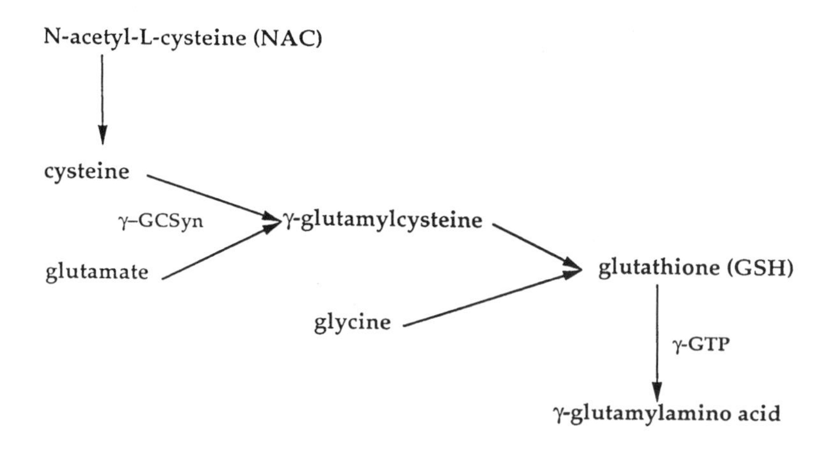

FIGURE 19.5. N-acetyl-L-cysteine (NAC) derives its protective effect on ethanol-induced fetal toxicity by increasing the levels of glutathione (GSH) in the system. Ethanol given during gestation decreases GSH levels and increases GSH turnover rates. By giving the precursor of GSH, cysteine, the levels of GSH are increased and the toxic effects of ethanol on the developing brain are averted.

Reports regarding acute and chronic effects of ethanol on GSH indicate a decrease in GSH levels or an increase in GSH turnover rates (Boyer and Peterson, 1990), which may in turn facilitate the toxicity of acetaldehyde. Recent studies by Reyes and associates (1989, 1993) have suggested that prenatal ethanol significantly decreases the levels of γ-glutamylcysteine synthetase (γ-GCSyn), an enzyme necessary for the synthesis of GSH, and increases the concentrations of γ-glutamyltranspeptidase (γ-GTP), an enzyme involving GSH metabolism (Reyes et al., 1989, 1993, 1994). In addition, the administration of N-acetyl-L-cysteine (NAC), a precursor of GSH, during pregnancy, significantly reduces ethanol-induced teratogenicity in terms of body and brain growth restrictions (see Figure 19.5) (Reyes et al., 1991).

Another interpretation of acetaldehyde's role in fetal ethanol-induced brain damage is derived from studies of prostaglandins. Prostaglandins are involved in many biological functions including cell differentiation in the CNS (Wolfe, 1989), and an imbalance of prostaglandin levels during pregnancy may lead to various abnormalities among fetuses (Horrobin, 1980). Evidence has shown that acute ethanol exposure during gestation in mice elevates prostaglandin levels, and fetal malformations were ameliorated by pretreatment with a prostaglandin synthetase inhibitor (Randall et al., 1991). On the other hand, chronic ethanol exposure induces a depletion of prostaglandins. Our laboratory has shown that the brain growth restrictions resulting from a decline in prostaglandin levels may be augmented by the administration of aspirin, a prostaglandin synthetase inhibitor (Bonthius and West, 1989); however, the prostaglandin levels were not measured.

Another piece of evidence implicating acetaldehyde toxicity in FAS comes from studies of ethanol-mediated vitamin B6 (pyridoxal-5′-phosphate, 5-PLP) deficiency, since vitamin B6 is the coenzyme involved in the biosynthesis of neurotransmitters and in myelin formation. The decrease in 5-PLP following ethanol exposure has proven to be due to the interference of acetaldehyde, rather than ethanol, with the metabolic pathway of vitamin B6, inhibiting pyridoxine kinase from producing 5-PLP and stimulating pyridoxal phosphate phosphatase, thus catalyzing 5-PLP destruction (Lumeng, 1978; Ryle and Thomson, 1984). These studies all suggest that acetaldehyde is a responsible agent in fetal ethanol damage.

In sum, it appears that both ethanol and acetaldehyde are capable of producing teratogenic actions through some of the same as well as some different mechanisms. With this caveat, it appears that minimizing the presence of both ethanol and acetaldehyde may be a solution to reducing the severity of ethanol-mediated brain damage.

B. INDIRECT VS. DIRECT EFFECTS OF ETHANOL ON BRAIN DAMAGE

Beyond the discussion of whether ethanol or acetaldehyde is the proximate teratogen responsible for ethanol-induced teratogenicity, another important issue related to the putative mechanisms of damage concerns whether the detrimental effects are a result of direct or indirect actions. The definitions of "direct" and "indirect" are somewhat confusing at times. For example, the effects of ethanol on membrane fluidization can be interpreted as either "direct," when ethanol participates in disordering membrane integrity when ethanol penetrates through the bilayer membrane, or "indirect," when ethanol interacts with ganglioside molecules. However, in this context and in the following section these two terms are defined in a more traditional manner. "Direct" refers to the primary effects on the fetus resulting from a direct action of ethanol or acetaldehyde via infusion through placental and blood brain barriers, and diffusion into all aqueous compartments of the fetuses. "Indirect" refers to the secondary effects on fetuses as a consequence of the primary actions of ethanol on pregnant mothers.

Ethanol-induced hypoxia has been hypothesized to be a major indirect cause of ethanol teratogenicity (Abel, 1984). Fetal hypoxia, resulting from an interruption in blood or blood oxygen supply during placental transport, induces a variety of dysmorphogenesis and alters many biochemical and physiological functions of cells. Altura et al. (1983) demonstrated that exposing human umbilical arteries and veins to ethanol *in vitro* results in vasospasms that could lead to a decrease in oxygenated blood supply to the fetus. However, fetal hypoxia-induced neuroteratogenicity has not been examined comprehensively. In contrast to the findings of McDonald and colleagues (1987) of a reduction in neonatal brain weights, our data (unpublished) did not support the neuroteratogenic role of hypoxia. However, the methodological procedures used in these two studies were quite different; the McDonald study ligated the right carotid in addition to the hypoxic treatment. Considering all of the existing facts, hypoxia may either enhance or act concurrently with other factors, rather than act alone in mediating brain damage. Given that the occurrence of fetal hypoxia is unavoidably confounded by other injurious factors, such as an interruption in nutrient transport across the placenta, the potential role of ethanol-induced hypoxia in FAS merits a further investigation.

In addition to hypoxia, malnutrition is considered a possible secondary cause of ethanol-induced fetal brain damage. Fetal ethanol malnutrition can originate from two sources: (1) poor maternal nutritional status from chronic ethanol abuse and (2) disruption in placental-fetal nutrient transport system. Amankwah and Kaufmann (1984) found ultrastructural malformations in placentas obtained from mothers who abused ethanol during gestation. Overwhelming evidence indicates that ethanol-induced deficits in the placental transport of various amino acids (e.g., leucine and valine), vitamins (e.g., folate, thiamin, and B6) and trace elements (e.g., selenium and copper) to the fetus has profoundly pernicious effects on fetal brain development (Thomson and Pratt, 1992).

Another potentially significant indirect action of ethanol implicated in FAS is zinc deficiency. Zinc deficiency during development is teratogenic, interfering with RNA, DNA, and protein synthesis, resulting in fetal growth retardation, brain dysmorphogenesis, and long-term cognitive dysfunction (Dreosti, 1986). Prenatal exposure to ethanol has been shown to depress the placental transport of zinc and the uptake of zinc by the fetus, resulting a fetal zinc deficiency. Of related interest, the inhibition of placental zinc transport was not reversible by exogenous zinc supplementation (Ghishan and Greene, 1983), suggesting some level of ethanol-induced damage in the ultrastructure of the placenta. Several experimental studies have further strengthened the hypothetical zinc model for FAS by demonstrating that ethanol interacts synergistically with zinc deficiency in producing teratogenicity (Dreosti, 1986, 1993; Keppen et al., 1985). This interactive effect between ethanol and zinc deficiency bridges the "direct (ethanol)" and "indirect (ethanol-induced zinc deficiency)" effects of ethanol in a manner similar to the idea posed earlier that ethanol-induced hypoxia may work in concert with the direct action of ethanol to produce neuroteratogenic effects.

VI. CONCLUSIONS

The consequences of ethanol exposure on the neuromorphological parameters of brain development range from gross (microencephaly) to microscopic (cytoarchitectural abnormalities), and this wide range of effects contributes in different ways to the severity of FAS. Although knowledge regarding the toxicity of ethanol on the fetal brain has increased significantly in recent years, it is presently unclear just how much ethanol is harmful to the developing fetus. This is a difficult question to answer, because there are many factors that interact with ethanol to produce subtle brain damage.

The evidence for ethanol neuroteratogenicity is convincing from both clinical and experimental perspectives (West, 1986). However, it is noteworthy that a majority of the research within each specific domain has been limited in focus. For example, the majority of neuroanatomical data comes from assessing the toxic effects of ethanol on hippocampal formation, the cortex, and the cerebellum. This limited focus is due largely to the relative simplicity of the cytoarchitectural organization of these regions. Although it is known that these three regions are vulnerable to ethanol insults, there is a need to focus on other brain regions. It is quite possible that other developing brain regions are even more vulnerable to ethanol insult. The same analogy can be applied to other perspectives as well. The discussion of potential mechanisms regarding the possible causative agents, ethanol or acetaldehyde, suggests that fetal ethanol-induced brain damage occurs through multiple mechanisms. It is premature to conclude that fetal alcohol damage is associated solely with either ethanol or acetaldehyde. At the present time, the existing evidence supports the idea that both indirect and direct actions of ethanol are involved in mediating FAS. What remains to be determined is how the indirect actions of ethanol may interact with its direct actions in modifying the risk and severity of brain damage.

ACKNOWLEDGMENTS

This work was supported by Grants AA05523 and AA07313 from the National Institute on Alcohol Abuse and Alcoholism. We thank Dr. Charles R. Goodlett for his generosity and assistance in providing and producing the photomicrographs.

REFERENCES

Abel, E.L. (1984). Mechanisms of action. *Fetal Alcohol Syndrome and Fetal Alcohol Effects*. New York, Plenum Press. pp. 207–212.

Ali, F. and Persaud, T.V.N. (1988). Mechanisms of fetal alcohol effects: role of acetaldehyde. *Exp. Pathol.*, **33:** 17–21.

Altura, B.M., Altura, B.T., Carella, A., Halevy, S., and Tejani, N. (1983). Alcohol produces spasms of human umbilical blood vessels: relationship to fetal alcohol syndrome (FAS). *Eur. J. Pharmacol.*, **86:** 311–312.

Amankwah, K.S. and Kaufmann, R.C. (1984). Ultrastructure of human placenta: effects of maternal drinking. *Gynecol. Obstet. Invest.*, **18:** 311–316.

Aronson, M., Kyllerman, M., Sabel, K.G., Sandin, B., and Olegard, R. (1985). Children of alcoholic mothers: outcome in relation to the social environment in which the children were brought up. *Rep. Depart. Appl. Psychol.*, **9:** 3–7.

Barnes, D.E. and Walker, D.W. (1981). Prenatal alcohol exposure permanently reduces the number of pyramidal neurons in the rat hippocampus. *Dev. Brain Res.*, **1:** 333–340.

Barr, H.M., Streissguth, A.P., Darby, B.L., and Sampson, P.D. (1990). Prenatal exposure to alcohol, caffeine, tobacco and aspirin: effects on fine and gross motor performance in 4-year-old children. *Dev. Psychol.,* **26:** 339–348.

Beeker, K., Deans, D., Elton, C., and Pennington, S.N. (1988). Ethanol-induced growth inhibition in embryonic chick brain is associated with changes in cyclic AMP-dependent protein kinase regulatory subunit. *Alcohol Alcohol.,* **23:** 477–482.

Blakley, P.M. and Scott, W.J. (1984). Determination of the proximate teratogen of the mouse fetal alcohol syndrome. 1. Teratogenicity of ethanol and acetaldehyde. *Toxicol. Appl. Pharmacol.,* **72:** 355–363.

Bonthius, D.J., Bonthius, N.E., Napper, R.M.A., Astley, S.J., Clarren, S.K., and West, J.R. (1991). Purkinje cell deficits in the pig-tailed macaque following gestational alcohol exposure. *Alcoholism: Clin. Exp. Res.,* **15:** 340.

Bonthius, D.J., Bonthius, N.E., Napper, R.M.A., and West, J.R. (1992). A stereological study of the effect of early postnatal alcohol exposure on the number of granule cells and mitral cells in the rat olfactory bulb. *J. Comp. Neurol.,* **324:** 557–566.

Bonthius, D.J. and West, J.R. (1988). Blood alcohol concentration and microencephaly: a dose-response study in the neonatal rat. *Teratology,* **37:** 223–231.

Bonthius, D.J. and West, J.R. (1989). Aspirin augments alcohol in restricting brain growth in the neonatal rat. *Neurotoxicol. Teratol.,* **11:** 135–143.

Bonthius, D.J. and West, J.R. (1990). Alcohol-induced neuronal loss in developing rats: increased brain damage with binge exposure. *Alcoholism: Clin. Exp. Res.,* **14:** 107–118.

Bonthius, D.J. and West, J.R. (1991). Permanent neuronal deficits in rats exposed to alcohol during the brain growth spurt. *Teratology,* **44:** 147–163.

Boyer, C.S. and Peterson, D.R. (1990). Potentiation of cocaine-mediated hepatotoxicity by acute and chronic ethanol. *Alcoholism: Clin. Exp. Res.,* **14:** 28–31.

Brown, R.T., Coles, C.D., Smith, I.E., Platzman, K.A., Silverstein, J., Erickson, S., and Falek, A. (1991). Effects of prenatal alcohol exposure at school age: II. Attention and behavior. *Neurotoxicol. Teratol.,* **13:** 369–376.

Caballeria, J., Baraona, E., Rodamilans, M., and Lieber, C.S. (1989a). Effects of cimetidine on gastric alcohol dehydrogenase activity and blood ethanol levels. *Gastroenterology,* **96:** 388–392.

Caballeria, J., Frezza, M., Hernández-Muñoz, DiPadova, C., Korsten, M.A., Baraona, E., and Lieber, C.S. (1989b). Gastric origin of the first-pass metabolism of ethanol in humans: Effect of gastrectomy. *Gastroenterology,* **97:** 1205–1209.

Chasnoff, I.J., Harvey, J., Landress, A.C.S.W., and Barrett, M.E. (1990). The prevalence of illicit-drug or alcohol use during pregnancy and discrepancies in mandatory reporting in Pinellas County, Florida. *N. Engl. J. Med.,* **322:** 1202–1206.

Chen, W.J., McAlhany, R.E., Jr., and West, J.R. (1995). 4-methylpyrazole, an alcohol dehydrogenase inhibitor, exacerbates alcohol-induced microencephaly during the brain growth spurt. *Alcohol,* **12:** 351–355.

Chen, W.J., Andersen, K.H., and West, J.R. (1994). Alcohol-induced brain growth restrictions (microencephaly) were not affected by concurrent exposure to cocaine during the brain growth spurt. *Teratology,* **50:** 250–255.

Christoffel, K.K. and Salafsky, I. (1975). Fetal alcohol syndrome in dizygotic twins. *J. Pediatr.,* **87:** 963–967.

Church, M.W., Dintcheff, B.A., and Gessner, R.K. (1988). The interactive effects of alcohol and cocaine on maternal and fetal toxicity in Long-Evans rat. *Neurotoxicol. Teratol.,* **10:** 355–361.

Clarren, S.K., Alvord, E.C., Jr., Sumi, S.M., Streissguth, A.P., and Smith, D.W. (1978). Brain malformations related to prenatal exposure to alcohol. *J. Pediatr.,* **92:** 64–67.

Clarren, S.K. and Smith, D.W. (1978). The fetal alcohol syndrome. *N. Engl. J. Med.,* **298:** 1063–1067.

Clarren, S.K., Astley, S.J., Bowden, D.M., Lai, H., Milam, A.H., Rudeen, P.K., and Shoemaker, W.J. (1990). Neuroanatomic and neurochemical abnormalities in non-human primate infants exposed to weekly doses of ethanol during gestation. *Alcoholism: Clin. Exp. Res.,* **14:** 674–683.

Coles, C.D., Platzman, K.A., Smith, I., James, M.E., and Falek, A. (1992). Effects of cocaine and alcohol use in pregnancy on neonatal growth and neurobehavioral status. *Neurotoxicol. Teratol.,* **14:** 23–33.

Conry, J. (1990). Neuropsychological deficits in fetal alcohol syndrome and fetal alcohol effects. *Alcoholism: Clin. Exp. Res.,* **14:** 650–655.

Coulter, C.L., Leech, R.W., Schaefer, G.B., Scheithauer, B.W., and Brumback, R.A. (1993). Midline cerebral dysgenesis, dysfunction of the hypothalamic-pituitary axis and fetal alcohol effects. *Arch. Neurol.,* **50:** 771–775.

Day, N.L. and Richardson, G.A. (1994). Comparative teratogenicity of alcohol and other drugs. *Alcohol Health Res. World,* 18, 42–48.

Dobbing, J. and Sands, J. (1979). Comparative aspects of the brain growth spurt. *Early Hum. Develop.,* **3:** 79–83.

Dobbing, J. and Sands, J. (1973). The quantitative growth and development of the human brain. *Arch. Dis. Child.,* **48:** 757–767.

Dreosti, I.E. (1986). Zinc-alcohol interactions in brain development. In West, J.R. (Ed.). *Alcohol and Brain Development,* Oxford University Press, New York. pp. 373–405.

Dreosti, I.E. (1993). Nutritional factors underlying the expression of the fetal alcohol syndrome. *Ann. N.Y. Acad. Sci.,* **678:** 193–204.

Druse, M.J., Kuo, A., and Tajuddin, N. (1991). Effects of *in utero* ethanol exposure on the developing serotonergic system. *Alcoholism: Clin. Exp. Res.,* **15:** 678–684.

Druse, M.J. and Paul, L.H. (1989). Effects of *in utero* ethanol exposure on serotonin uptake in cortical regions. *Alcoholism,* **5:** 455–459.

Druse, M.J., Tajuddin, N., Kuo, A.P., and Connerty, M. (1990). Effects of *in utero* ethanol exposure on the developing dopaminergic system in rats. *J. Neurosci. Res.,* **27:** 233–240.

Eriksson, C.J.P. and Fukunaga, T. (1993). Human blood acetaldehyde (update 1992). *Alcohol Alcohol.,* Suppl. 2, 9–25.

Ferrer, I. and Galofré, E. (1987). Dendritic spine anomalies in fetal alcohol syndrome. *Neuropediatrics,* **18:** 161–163.

Fletcher, T.L. and Shain, W. (1993). Ethanol-induced changes in astrocyte gene expression during rat central nervous system development. *Alcoholism: Clin. Exp. Res.,* **17:** 993–1001.

Fridovich, I. (1989). Oxygen radicals from acetaldehyde. *Free Rad. Biol. Med.,* **7:** 557–558.

Ghishan, F.K. and Greene, H.L. (1983). Fetal Alcohol Syndrome: failure of zinc supplementation to reverse the effect of ethanol on placental transport of zinc. *Pediatr. Res.,* **17:** 529–531.

Goodlett, C.R., Gilliam, D.M., Nichols, J.M., and West, J.R. (1989). Genetic influences on brain growth restriction induced by developmental exposure to alcohol. *Neurotoxicology,* **10:** 321–334.

Goodlett, C.R., Kelly, S.J., and West, J.R. (1987). Early postnatal alcohol exposure that produces high blood alcohol levels impairs development of spatial learning. *Psychobiology,* **15:** 64–74.

Goodlett, C.R., Leo, J.T., O'Callaghan, J.P., Mahoney, J.C., and West, J.R. (1993). Transient cortical astrogliosis induced by alcohol exposure during the neonatal brain growth spurt in rats. *Dev. Brain Res.,* **72:** 85–97.

Goodlett, C.R., Marcussen, B.L., and West, J.R. (1990a). A single day of alcohol exposure during the brain growth spurt induces brain weight restriction and cerebellar Purkinje cell loss. *Alcohol,* **7:** 107–114.

Goodlett, C.R., Nichols, J.M., and West, J.R. (1990b). Dose-dependent deficits in cerebellar growth induced by early postnatal alcohol exposure: differences among inbred strains of rats. *Soc. Neurosci. Abs.,* **16:** 33.

Goodlett, C.R., Nichols, J.M., and West, J.R. (1991a). Strain differences in susceptibility to alcohol-induced brain growth restriction are not accounted for by differences in alcohol elimination rates. *Alcoholism: Clin. Exp. Res.,* **15:** 343.

Goodlett, C.R. and Peterson, S.D. (1994). The role of duration and timing of binge-like neonatal alcohol exposure in determining the extent of alcohol-induced deficits in spatial navigation and motor learning. *Alcoholism: Clin. Exp. Res.,* **18:** 501.

Goodlett, C.R., Thomas, J.D., and West, J.R. (1991b). Long-term deficits in cerebellar growth and rotarod performance of rats following "binge-like" alcohol exposure during the neonatal brain growth spurt. *Neurotoxicol. Teratol.,* **13:** 69–74.

Hammer, R.P., Jr. and Scheibel, A.B. (1981). Morphologic evidence for a delay of neuronal maturation in fetal alcohol exposure. *Exp. Neurol.,* **74:** 587–596.

Hamre, K.M. and West, J.R. (1993). The effects of the timing of ethanol exposure during the brain growth spurt on the number of cerebellar Purkinje and granule cell nuclear profiles. *Alcoholism: Clin. Exp. Res.,* **17:** 610–622.

Hearn, W.L., Rose, S., Wagner, J., Ciarleglio, A., and Mash, D.C. (1991). Cocaethylene is more potent than cocaine in mediating lethality. *Pharmacol. Biochem. Behav.,* **39:** 531–533.

Heaton, M.B., Paiva, M., Swanson, D.J., and Walker, D.W. (1993). Modulation of ethanol neurotoxicity by nerve growth factor. *Brain Res.,* **620:** 78–85.

Heaton, M.B., Swanson, D.J., Paiva, M., and Walker, D.W. (1992). Ethanol exposure affects trophic factor activity and responsiveness in chick embryo. *Alcohol,* **9:** 161–166.

Horrobin, D.F. (1980). A biochemical basis for alcoholism and alcohol-induced damage including the fetal alcohol syndrome and cirrhosis: interference with essential fatty acid and prostaglandin metabolism. *Med. Hypotheses,* **6:** 929–942.

Jones, K.L., Chambers, C.C., and Johnson, K.A. (1991). The effect of disulfiram on the unborn baby. *Teratology,* **43:** 438.

Jones, K.L., Smith, D.W., Ulleland, C.N., and Streissguth, A.P. (1973). Pattern of malformation in offspring of chronic alcoholic mothers. *Lancet,* **1:** 1267–1269.

Kelly, S.J., Black, A.C., Jr., and West, J.R. (1989). Changes in the muscarinic cholinergic receptors in the hippocampus of rats exposed to ethyl alcohol during the brain growth spurt. *J. Pharm. Exp. Thera.,* **249:** 798–804.

Keppen, L.D., Pysher, T., and Rennert, O.M. (1985). Zinc deficiency acts as a co-teratogen with alcohol in fetal alcohol syndrome. *Pediatr. Res.,* **19:** 944–947.

Klemm, W.R. (1990). Dehydration: a new alcohol theory. *Alcohol,* **7:** 49–59.

Kokotailo, P.K., Adger, H., Jr., Duggan, A.K., Repke, J., and Joffe, A. (1992). Cigarette, alcohol and other drug use by school-age pregnant adolescents: Prevalence, detection and associated risk factors. *Pediatrics* **90:** 328–334.

Lauder, J.M. and Mugnaini, E. (1980). Infrapyramidal mossy fibers in the hippocampus of the hyperthyroid rat. *Dev. Neurosci.,* **3:** 248–265.

Lauder, J.M., Wallace, J.A., Wilkie, M.B., DiNome, A., and Krebs, H. (1983). Roles for serotonin in neurogenesis. *Monogr. Neural Sci.,* **9:** 3–10.

Laurberg, S. and Zimmer, J. (1980). Lesion-induced rerouting of hippocampal mossy fibers in developing but not in adult rats. *J. Comp. Neurol.,* **190:** 627–650.

Lieber, C.S. (1993). Biochemical factors in alcoholic liver disease. *Semi. Liver Dis.,* **13:** 136–153.

Light, K.E., Serbus, D.C., and Santiago, M. (1989a). Exposure of rats to ethanol from postnatal days 4 to 8: alterations of cholinergic neurochemistry is the cerebral cortex and corpus striatum at day 20. *Alcoholism: Clin. Exp. Res.,* **13:** 29–35.

Light, K.E., Serbus, D.C., and Santiago, M. (1989b). Exposure of rats to ethanol from postnatal days 4 to 8: alterations of cholinergic neurochemistry in the hippocampus and cerebellum at day 20. *Alcoholism: Clin. Exp. Res.,* **13:** 686–692.

Lokhorst, D.K. and Druse, M.J. (1993). Effects of ethanol on cultured fetal astrogliosis. *Alcoholism: Clin. Exp. Res.,* **17:** 810–815.

Lumeng, L. (1978). The role of acetaldehyde in mediating the deterious effect of alcohol on pyridoxal 5′-phosphate metabolism. *J. Clin. Invest.,* **62:** 286–293.

Maier, S.E., Mahoney, J.C., Goodlett, C.R., and West, J.R. (1994). Strain differences in susceptibility to alcohol-induced Purkinje cell loss demonstrated using unbiased stereological counting methods. *Alcoholism: Clin. Exp. Res.,* **18:** 437.

Maier, S.E., Strittmatter, M.A., Chen, W.-J.A., and West, J.R. (in press). Changes in blood alcohol levels as a function of alcohol concentration and repeated alcohol exposure in adults female rats: potential risk factors for alcohol-induced fetal brain injury. *Alcoholism: Clin. Exp. Res.,*

Majewski, F. (1981). Alcohol embryopathy: some facts and speculations about pathogenesis. *Neurobehav. Toxicol. Teratol.,* **3:** 129–144.

Marcussen, B.L., Goodlett, C.R., Mahoney, J.C., and West, J.R. (1994). Developing rat Purkinje cells are more vulnerable to alcohol-induced depletion during differentiation than during neurogenesis. *Alcohol,* **11:** 147–156.

Mattson, S.N., Riley, E.P., Jernigan, T.L., Ehlers, C.L., Delis, D.C., Jones, K.L., Stern, C., Johnson, K.A., Hesselink, J.R., and Bellugi, U. (1992). Fetal alcohol syndrome: a case report of neuropsychological, MRI and EEG assessment of two children. *Alcoholism: Clin. Exp. Res.,* **16:** 1001–1003.

Mattson, S.N., Riley, E.P., Jernigan, T.L., Garcia, A., Kaneko, W.M., Ehlers, C.L., and Jones, K.L. (1994). A decrease in the size of the basal ganglia following prenatal alcohol exposure: a preliminary report. *Neurotoxicol. Teratol.,* **16:** 283–289.

McDonald, J.W., Silverstein, F.S., and Johnston, M.V. (1987). MK-801 protects the neonatal brain from hypoxic-ischemic damage. *Eur. J. Pharmacol.,* **140:** 359–361.

Miller, M.W. (1986). Effects of alcohol on the generation and migration of cerebral cortical neurons. *Science,* **233:** 1308–1311.

Miller, M.W. (1988). Effect of prenatal exposure to ethanol on the development of cerebral cortex: I. neuronal generation. *Alcoholism: Clin. Exp. Res.,* **12:** 440–449.

Miller, M.W. (1993). Migration of cortical neurons is altered by gestational exposure to alcohol. *Alcoholism: Clin. Exp. Res.,* **17:** 304–314.

Miller, M.W. and Al-Rabiai, S. (1994). Effects of prenatal exposure to ethanol on the number of axons in the pyramidal tract of the rat. *Alcoholism: Clin. Exp. Res.,* **18:** 346–354.

Miller, M.W., Chiaia, N.L., and Rhoades, R.W. (1990). Intracellular recording and injection study of corticospinal neurons in the rat stomatosensory cortex: effect of prenatal exposure to ethanol. *J. Comp. Neurol.,* **297:** 91–105.

Miller, M.W. and Muller, S.J. (1989). Structure and histogenesis of the principal sensory nucleus of the trigeminal nerve: effects of prenatal exposure to ethanol. *J. Comp. Neur.,* **282:** 570–580.

Miller, M.W. and Robertson, S. (1993). Prenatal exposure to ethanol alters the postnatal development and transformation of radial glia to astrocytes in the cortex. *J. Comp. Neurol.,* **337:** 253–266.

Miller, R.R. Jr., Heinstra, P.W.H., and Geer, B.W. (1992). Metabolic flux and the role of lipids in alcohol tolerance in *Drosophila.* In Watson, R.R. and Watzl, B. (Eds.). *Nutrition and Alcohol.* CRC Press, Boca Raton, FL. pp. 205–242.

Mohamed, S.A., Nathaniel, E.J., Nathaniel, D.R., and Snell, L. (1987). Altered Purkinje cell maturation in rats exposed prenatally to ethanol. *Exp. Neurol.,* **97:** 35–52.

Morgane, P.J., Austin-LaFrance, R., Bronzino, J., Tonkiss, J., Díaz-Cintra, S., Cintra, L., Kemper, T., and Galler, J.R. (1993). Prenatal malnutrition and development of the brain. *Neurosci. Biobehav. Rev.,* **17:** 91–128.

Napper, R.M.A. and West, J.R. (submitted). Early postnatal alcohol exposure permanently reduces the number of neurons in the inferior olivary nucleus. *Alcoholism: Clin. Exp. Res.*

Nestler, E.J. and Greengard, P. (1989). Protein phosphorylation and the regulation of neuronal function. In Siegel, G.J., Agranoff, B.W., Albers, R.W., and Molinoff, P.B. (Eds.). *Basic Neurochemistry,* 4th Ed. Raven Press, New York. pp. 373–398.

Oei, H.H.H., Stroo, W.E., Burton, K.P., and Schaffer, S.W. (1982). A possible role of xanthine oxidase in producing oxidative stress in the heart of chronically alcohol treated rats. *Res. Commun. Chem. Path. Pharmacol.,* **38:** 453–461.

O'Callaghan, J.P. (1993). Quantitative features of reactive gliosis following toxicant-induced damage of the CNS. In Johannessen, J.N. (Ed.), *Markers of Neuronal Injury and Degeneration.* New York Academy of Science, pp. 195–210.

O'Shea, K.S. and Kaufman, M.H. (1979). The teratogenic effect of acetaldehyde: implications for the study of the fetal alcohol syndrome. *J. Anat.,* **128:** 65–76.

Pantazis, N.J., Luo, J., and West, J.R. (1994). Growth factor-mediated neuroprotection against alcohol-induced death of cerebellar granule cells. *Alcoholism: Clin. Exp. Res.,* **18:** 443.

Peiffer, J., Majewski, F., Fishbach, H., Bierich, J.R., and Volk, B. (1979). Alcohol embryo- and fetopathy. *J. Neurol. Sci.,* **41:** 125–137.

Pennington, S.N. (1988). Ethanol-induced growth inhibition: the role of cyclic AMP dependent protein kinase. *Alcoholism: Clin. Exp. Res.,* **12:** 125–130.

Pennington, S.N. (1990). Molecular changes associated with ethanol-induced growth suppression in chick embyro. *Alcoholism: Clin. Exp. Res.,* **14:** 832–837.

Pennington, S.N. (1992). Ethanol-induced teratology and second messenger signal transduction. In Miller, M.W. (Ed.). *Development of the Central Nervous System: Effects of Alcohol and Opiates.* Wiley-Liss, Inc., New York. pp. 189–207.

Phillips, D.E. and Krueger, S.K. (1990). Effects of postnatal ethanol exposure on glial cell development in rat optic nerve. *Exp. Neurol.,* **107:** 97–105.

Phillips, S.C. and Cragg, B.G. (1982). A change in susceptibility of rat cerebellar Purkinje cells to damage by alcohol during fetal, neonatal and adult life. *Neuropathol. Appl. Neurobiol.,* **8:** 441–454.

Pierce, D.R., Goodlett, C.R., and West, J.R. (1989). Differential neuronal loss following early postnatal alcohol exposure. *Teratology,* **40:** 113–126.

Pierce, D.R., Serbus, D.C., and Light, K.E. (1993). Intragastric intubation of alcohol during postnatal development of rats results in selective cell loss in the cerebellum. *Alcoholism: Clin. Exp. Res.,* **17:** 1275–1280.

Randall, C.L., Anton, R.F., Becker, H.C., Hale, R.L., and Ekblad, U. (1991). Aspirin dose-dependently reduces alcohol-induced birth defects and prostaglandin E levels in mice. *Teratology,* **44:** 521–529.

Ranganathan, S., Davis, D.G., and Hood, R.D. (1987). Developmental toxicity of alcohol in *Drosophila melanogaster. Teratology,* **36:** 45–49.

Reyes, E., Contreras, R., Montoya, R., and Ott, S. (1994). Effects of *in utero* administration of alcohol on enzymes of the g-glutamyl cycle. *Alcoholism: Clin. Exp. Res.,* **18:** 434.

Reyes, E., Lucero, M.M., Ott, S., and Coker, H. (1991). Effects of the *in utero* administration of N-Acetyl-L-Cysteine and alcohol on fetal development. *Alcoholism: Clin. Exp. Res.,* **15:** 343.

Reyes, E., Ott, S., and Robinson B. (1993). Effects of *in utero* administration of alcohol on glutathione levels in brain and liver. *Alcoholism: Clin. Exp. Res.,* **17:** 877–881.

Reyes, E., Rivera, J.M., Saland, L.C., and Murray, H.M. (1983). Effects of maternal administration of alcohol on fetal brain development. *Neurobehav. Toxicol. Teratol.,* **5:** 263–267.

Reyes, E., Wolfe, J., and Marquez, M. (1989). Effects of prenatal alcohol on g-glutamyl transpeptidase in various brain regions. *Phys. Behav.,* **46:** 49–53.

Robinson, G.C., Conry, J.L., and Conry, R.F. (1987). Clinical profile and prevalence of fetal alcohol syndrome in an isolated community in British Columbia. *Can. Med. Assoc. J.,* **137:** 203–207.

Roine, R.P., Gentry, R.T., Lim, R.T., Jr., Baraona, E., and Lieber, C.S. (1991). Effect of concentration of ingested ethanol on blood alcohol levels. *Alcoholism: Clin. Exp. Res.,* **15:** 734–738.

Ryle, P.R. and Thomson, A.D. (1984). Nutrition and vitamins in Alcoholism. In Rosalki, S.B. (Ed.). *Clinical Biochemistry of Alcoholism.* Churchill Livingstone, New York, pp. 188–224.

Savage, D.D., Montano, C.Y., Otero, M.A., and Paxton, L.L. (1991). Prenatal ethanol exposure decreases hippocampal NMDA-sensitive [^{3}H]glutamate binding site density in 45-day-old rats. *Alcohol,* **8:** 193–201.

Shean, M.L. and Duester, G. (1993). The role of alcohol dehydrogenase in retinoic acid homeostasis and fetal alcohol syndrome. *Alcohol Alcohol.,* Suppl 2, 51–56.

Smith, D.E. and Davies, D.L. (1990). Effect of perinatal administration of ethanol on the CA1 pyramidal cell of the hippocampus and Purkinje cell of the cerebellum: an ultrastructural survey. *J. Neurocytol.,* **19:** 708–717.

Sokol, R.J. and Clarren, S.K. (1989). Guidelines for use of terminology describing the impact of prenatal alcohol on the offspring. *Alcoholism: Clin. Exp. Res.,* **13:** 597–598.

Stephens, C.J. (1985). Alcohol consumption during pregnancy among southern city women. *Drug Alcohol Depend.,* **9:** 339–350.

Stone, R. (1993). New marker for nerve damage. *Science,* **259:** 1541.

Streissguth, A.P., Aase, J.M., Clarren, S.K., Randalls, S.P., LaDue, R.A., and Smith, D.W. (1991). Fetal Alcohol Syndrome in adolescents and adults. *JAMA,*. **265:** 1961–1967.

Streissguth, A.P., Barr, H.M., and Martin, D.C. (1983). Maternal alcohol and neonatal habituation assessed with the Brazelton Scale. *Child Develop.,* **54:** 1109–1118.

Streissguth, A.P., Barr, H.M., Olson, H.C., Sampson, P.D., Bookstein, F.L., and Burgess, D.M. (1994a). Drinking during pregnancy decreases word attack and arithmetic scores on standardized tests: adolescent data from a population-based prospective study. *Alcoholism: Clin. Exp. Res.,* **18:** 248–254.

Streissguth, A.P., Barr, H.M., and Sampson, P.D. (1990). Moderate prenatal alcohol exposure: effects on child IQ and learning problems at age $7^1/_2$ years. *Alcoholism: Clin. Exp. Res.,* **14:** 662–669.

Streissguth, A.P., Herman, C.S., and Smith, D.W. (1978). Intelligence, behavior and dysmorphogenesis in the fetal alcohol syndrome: a report on 20 patients. *J. Pediatr.,* **92:** 363–367.

Streissguth, A.P., Sampson, P.D., Olson, H.C., Bookstein, F.L., Barr, H.M., Scott, M., Feldman, J., and Mirsky, A.F. (1994b). Maternal drinking during pregnancy. Attention and short-term memory in 14-year-old offspring: a longitudinal prospective study. *Alcoholism: Clin. Exp. Res.,* **18:** 202–218.

Swanson, D.J., Daniels, H., Meyer, E.M., Walker, D.W., and Heaton, M.B. (1994). Chronic ethanol alters CNS cholinergic and cerebellar development in chick embryos. *Alcohol,* **11:** 187–194.

Tajuddin, N.F. and Druse, M.J. (1993). Treatment of pregnant alcohol-consuming rats with buspirone: effects on serotonin and 5-hydroxyindoleacetic acid content in offspring. *Alcoholism: Clin. Exp. Res.,* **17:** 110–114.

Tanaka, H., Nasu, F., and Inomata, K. (1991). Fetal alcohol effects: decreased synaptic formations in the field CA3 of fetal hippocampus. *Int. J. Devl. Neurosci.,* **9:** 509–517.

Thomson, A.D. and Pratt, O.E. (1992). Interaction of nutrients and alcohol: Absorption, transport, utilization and metabolism. In: Watson, R.R. and Watzl, B. (Eds.). *Nutrition and Alcohol.* CRC Press, Boca Raton, FL. pp. 75–99.

Ukita, K., Fukui, Y., and Shiota, K. (1993). Effects of prenatal alcohol exposure in mice: influence of an ADH inhibitor and a chronic inhalation study. *Reproduct. Toxicol.,* **7:** 273–281.

Walker, D.W., Heaton, M.B., Smothers, C.T., and Hunter, B.E. (1990). Chronic ethanol ingestion reduces the neurotrophic activity contained in rat hippocampus. *Alcoholism: Clin. Exp. Res.,* **14:** 350.

Walker, D.W., Lee, N., Heaton, M.B., King, M.A., and Hunter, B.E. (1992). Chronic ethanol consumption reduces the neurotrophic activity in rat hippocampus. *Neurosci. Lett.,* **147:** 77–80.

West, J.R. (1986). (Ed.). *Alcohol and Brain Development.* Oxford University Press, New York.

West, J.R. (1987). Fetal alcohol-induced brain damage and the problem of determining temporal vulnerability: a review. *Alcohol Drug Res.,* **7:** 423–441.

West, J.R., Goodlett, C.R., Bonthius, D.J., Hamre, K. M., and Marcussen, B.L. (1990). Cell population depletion associated with fetal alcohol brain damage: mechanisms of BAC-dependent cell loss. *Alcoholism: Clin. Exp. Res.,* **14:** 813–818.

West, J.R. and Hamre, K.M. (1985). Effects of alcohol exposure during different periods of development: changes in hippocampal mossy fibers. *Develop. Brain Res.,* **17:** 280–284.

West, J.R., Hodges, C.A., and Black, A.C., Jr. (1981). Prenatal exposure to alcohol alters the organization of hippocampal mossy fibers in rats. *Science,* **211:** 957–959.

Wolfe, L.S. (1989). Eicosanoids. In Siegel, G.J., Arganoff, B.W., Alberts, R. W., and Molinoff, P.B. (Eds.). *Basic Neurochemistry.* 4th ed. Raven Press, New York. pp. 399–414.

Zoeller, R.T., Butnariu, O.V., Fletcher, D.L., and Riley, E.P. (1994). Limited postnatal ethanol exposure permanently alters the expression of mRNA encoding myelin basic protein and myelin-associated glycoprotein in cerebellum. *Alcoholism: Clin. Exp. Res.,* **18:** 909–916.

Zorzano, A. and Herrera, E. (1989). Decreased *in vivo* rate of ethanol metabolism in the suckling rat. *Alcoholism: Clin. Exp. Res.,* **13:** 527–532.

Chapter 20

CENTRAL NERVOUS SYSTEM PATHOLOGY IN ALCOHOLISM

Michael E. Charness

TABLE OF CONTENTS

I. INTRODUCTION

Alcoholism is complicated by a variety of neurological disorders (Charness et al., 1989), most of which produce characteristic brain lesions (Charness, 1993; Victor et al., 1989). Malnutrition is common in alcoholics, who may derive the majority of their calories from nonnutritive ethanol (7 kcal/gm). Thiamine deficiency is particularly frequent and leads to Wernicke's encephalopathy. Thiamine deficiency may also contribute to alcoholic cerebellar degeneration and alcoholic peripheral neuropathy. Alcoholics are subject to repeated episodes of head trauma, which may cause contusions of the frontal and temporal poles as well as subdural, subarachnoid, and parenchymal hemorrhage. Alcoholic liver disease may cause hepatocerebral degeneration. Prenatal exposure to alcohol causes a variety of brain lesions associated with cognitive dysfunction. Finally, alcohol may directly damage the mature nervous system. In this chapter, I review the neuropathologic findings in alcoholism, emphasizing clinical, radiographic, and neuropathologic correlations in humans.

II. WERNICKE'S ENCEPHALOPATHY

Wernicke's encephalopathy is a common neurological disorder caused by thiamine deficiency, and alcoholics account for most cases in the Western world (Victor et al., 1989). Thiamine deficiency in alcoholics results from a combination of inadequate dietary intake, reduced gastrointestinal absorption, decreased hepatic storage, and impaired utilization (Thomson et al., 1983). Only a subset of thiamine-deficient alcoholics develop Wernicke's encephalopathy, perhaps because they have inherited (Blass and Gibson, 1977; Mukherjee

0-8493-8389-7/95/$0.00+$.50

et al., 1987) or acquired (Jeyasingham et al., 1987) abnormalities of the thiamine-dependent enzyme, transketolase, that reduce its affinity for thiamine. No mutations in the gene encoding transketolase were identified in Korsakoff patients with the low-affinity variant of this enzyme (McCool et al., 1993). It remains possible that some Korsakoff patients inherit an abnormality that affects the assembly of transketolase into a holoenzyme.

A. NEUROPATHOLOGY

The lesions of Wernicke's encephalopathy occur in a characteristic, symmetrical distribution in structures surrounding the third ventricle, Aqueduct of Sylvius, and the fourth ventricle (see Figure 20.1) (Victor et al., 1989). The mamillary bodies are involved in virtually all cases, and the dorsomedial thalamus, locus ceruleus, periaqueductal gray, ocular motor nuclei, and vestibular nuclei are commonly affected. Lesions occur less frequently in the colliculi, fornices, septal region, hippocampus, and cerebral cortex, which may show patchy, diffuse neuronal loss and astrocytic proliferation. In about half of the cases, midsagittal sections through the cerebellum reveal a selective loss of Purkinje cells at the tips of the folia of the anterior superior cerebellar vermis. These changes are identical to those found in alcoholic cerebellar degeneration, where they can occur in the absence of other Wernicke lesions.

Acute Wernicke lesions can be identified by endothelial prominence, microglial proliferation, and occasional petechial hemorrhages. In chronic cases, there is demyelination, gliosis, and loss of neuropil with a relative preservation of neurons. Neuronal loss is most prominent in the relatively unmyelinated medial thalamus (Torvik, 1985; Victor et al., 1989). Atrophy of the mamillary bodies is a highly specific finding in chronic Wernicke's encephalopathy and is present in up to 80% of the cases (Victor et al., 1989).

B. PATHOGENESIS

It is unclear how thiamine deficiency causes brain lesions. Thiamine is a cofactor for transketolase, alpha-ketoglutarate dehydrogenase, and pyruvate dehydrogenase, (Victor et al., 1989), and may also function in axonal conduction and synaptic transmission (Iwata, 1982). Decreased activity of the thiamine-dependent enzymes has been measured at autopsy in the cerebellar vermis of alcoholics with clinical and pathologic evidence of Wernicke's encephalopathy as compared with alcoholics without Wernicke's lesions (Butterworth et al., 1993). Because thiamine-dependent enzymes play an important role in cerebral energy utilization, it has been proposed that thiamine deficiency initiates neuronal injury by inhibiting metabolism in brain regions with high metabolic requirements and high thiamine turnover (Butterworth et al., 1993). Thiamine deficiency has been shown to cause a diffuse decrease in cerebral glucose utilization (Hakim and Pappius, 1983). Shortly before the development of structural brain lesions, vulnerable brain areas exhibit a burst of metabolic activity accompanied by a local production of lactate (Hakim, 1984; Hakim and Pappius, 1983).

Animal studies suggest that excitotoxicity may be part of the final pathway leading from thiamine deficiency to neuronal injury. Extracellular concentrations of glutamate increase up to 7-fold in the medial thalamus and 3-fold in the hippocampus following seizures in thiamine-deficient rats (Langlais and Zhang, 1993). This burst of glucose utilization could represent a shift from aerobic metabolism to rapid glycolysis due to reduced pyruvate dehydrogenase activity (Hakim and Pappius, 1983). MK-801, an NMDA receptor antagonist (Choi, 1988), reduces the neurologic signs and the severity and extent of lesions in experimental thiamine deficiency induced by the administration of the thiamine antagonist, pyrithiamine (Langlais and Mair, 1990). Wernicke lesions can also be prevented by dialysing away the glutamate that accumulates in selective brain regions following pyrithiamine-associated seizures (Langlais and Zhang, 1993). The observation that chronic ethanol treatment increases the density of NMDA receptors in specific brain regions (Grant et al., 1990; Gulya et al., 1991) suggests a possible mechanism by which chronic ethanol ingestion could potentiate the excitotoxicity of

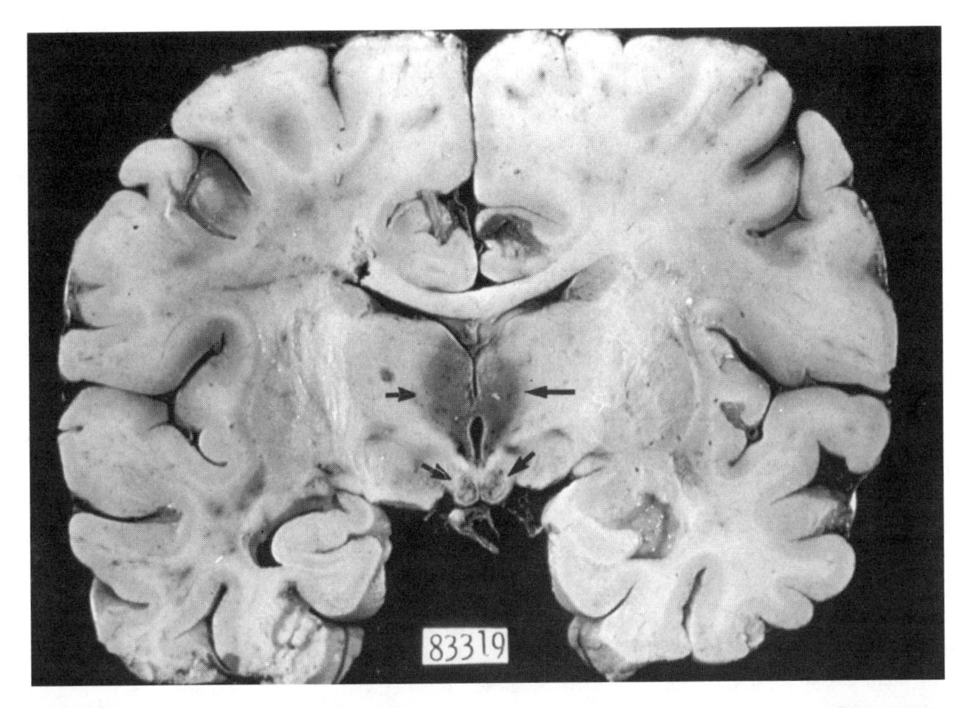

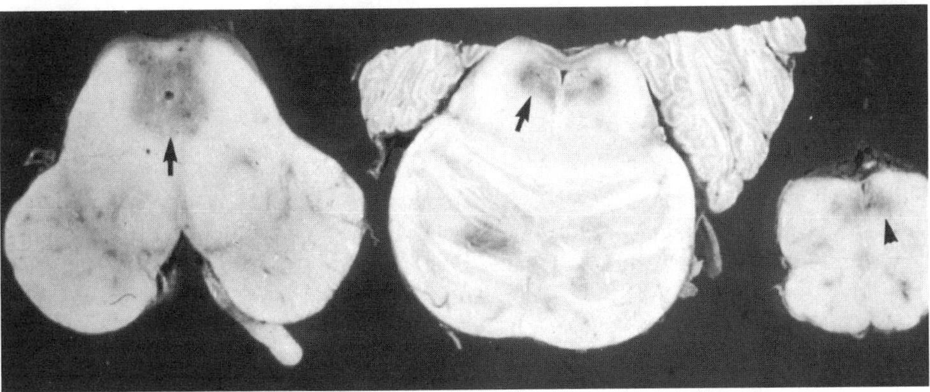

Figure 20.1. Wernicke's encephalopathy. Areas of discoloration (arrows) and petechial hemorrhage are seen in a symmetrical distribution in structures surrounding (A) the third ventricle, (B) the aqueduct and the fourth ventricle. (Courtesy of Dr. Richard L. Davis, Department of Pathology, University California San Francisco.)

glutamate in Wernicke's encephalopathy (Lovinger, 1993). It will be important to determine whether glutamate release in pyrithiamine-treated rats is mediated by thiamine deficiency or associated seizures, because thiamine deficiency alone does not generally induce seizures in humans.

C. CLINICAL PRESENTATION

The distinctive nature of the pathological findings in Wernicke's encephalopathy permits subclinical cases to be diagnosed postmortem. Autopsy studies have consistently revealed a higher incidence of Wernicke lesions in the general population (0.8–2.8%) than is predicted by clinical studies (0.04–0.13%) (Harper, 1983; Victor et al., 1989). The observation that acute Wernicke's encephalopathy was correctly diagnosed before death in only 1 of 22 patients (Torvik et al., 1982) suggests that the classic clinical triad of encephalopathy, opthalmoplegia, and ataxia is either surprisingly rare (Harper, 1983; Reuler et al., 1985;

Torvik et al., 1982) or is not properly recognized by clinicians (Harper et al., 1986; Victor et al., 1989).

In clinical studies, only one-third of patients with acute Wernicke's encephalopathy present with the classic clinical triad (Victor et al., 1989). A majority of patients are profoundly disoriented, indifferent, and inattentive, and some exhibit an agitated delirium related to ethanol withdrawal (delirium tremens). Although fewer than 5% of patients present with a depressed level of consciousness, the course in untreated patients may progress through stupor and coma to death (Victor et al., 1989; Wallis et al., 1978). Ocular motor abnormalities including nystagmus, lateral rectus palsy, and conjugate gaze palsies occur in 96%, reflecting lesions of the oculomotor, abducens, parabducens, and vestibular nuclei. Gait ataxia occurs in 87% and is likely due to a combination of polyneuropathy, noted in 82% of cases, cerebellar involvement, and vestibular paresis (Ghez, 1969; Victor et al., 1989). In keeping with the restriction of cerebellar pathology to the anterior and superior vermis, ataxia of the arms and dysarthria (scanning speech) are each observed in less than 20% of cases (Victor et al., 1989).

Autopsy-based series also suggest a very high incidence (82%) of mental status abnormalities, but much lower incidences of ataxia (23%), ocular motor abnormalities (29%), and polyneuropathy (11%) (Harper et al., 1986). The classic triad of clinical findings was identified retrospectively in only 10% of autopsy cases, and 19% showed none of the classic elements. Stupor or (coma), hypotension, and hypothermia were predominant findings in unsuspected cases (Harper, 1983; Lindboe and Loberg, 1989; Torvik et al., 1982). The discrepancy between prospective clinical and retrospective, autopsy-based descriptions of Wernicke's encephalopathy is likely due to the exclusion of atypical presentations in clinical series and the underestimation in autopsy series of classic signs that were not properly elicited, recognized, or recorded.

D. NEUROIMAGING

Neuroimaging studies may help bridge the gap between the clinical and neuropathological studies of Wernicke's encephalopathy. Abnormalities on computerized axial tomography (CT) scans or magnetic resonance imaging (MRI) have now been reported in small numbers of patients with acute Wernicke's encephalopathy (Warach and Charness, 1994). CT scanning may show symmetric, low-density abnormalities in the diencephalon and periventricular regions (Hofmann et al., 1988; McDowell and LeBlanc, 1984; Mensing et al., 1984) that are enhanced after the injection of contrast (Kitaguchi et al., 1987). Gross hemorrhages are uncommon in acute Wernicke's encephalopathy, but have been detected by CT (Roche et al., 1988). Symmetric areas of increased or decreased signal in the diencephalon, midbrain, and periventricular regions are relatively uncommon in other disorders, and when present in alcoholics, should strongly suggest the diagnosis of acute Wernicke's encephalopathy. However, the paucity of reports of such findings would suggest that CT is an insensitive method for their detection. Hence, a normal CT scan does not exclude the diagnosis of acute Wernicke's encephalopathy.

In general, MRI appears to be more sensitive than CT scans in detecting acute diencephalic and periventricular lesions (Gallucci et al., 1990; Ostertun et al., 1990). Patients with acute Wernicke's encephalopathy have exhibited areas of increased T2 signal surrounding the aqueduct and third ventricle and within the medial thalamus and mamillary bodies (see Figure 20.2) consistent with the localization of the pathologic lesions (Gallucci et al., 1990; Yokote et al., 1991). Corresponding areas of decreased T1 signal have also been detected (Gallucci et al., 1990). The alterations in T1 and T2 signals have been shown to resolve over months, leaving an enlarged aqueduct and third ventricle (see Figure 20.2) (Gallucci et al., 1990). These MR abnormalities have been reported in small numbers of patients with acute Wernicke's encephalopathy. The sensitivity of MRI in detecting these lesions has not been determined; hence a normal MRI does not rule out acute Wernicke's encephalopathy. We did not observe any abnormalities in the brainstem, thalamus, or aqueduct on transaxial T2-weighted

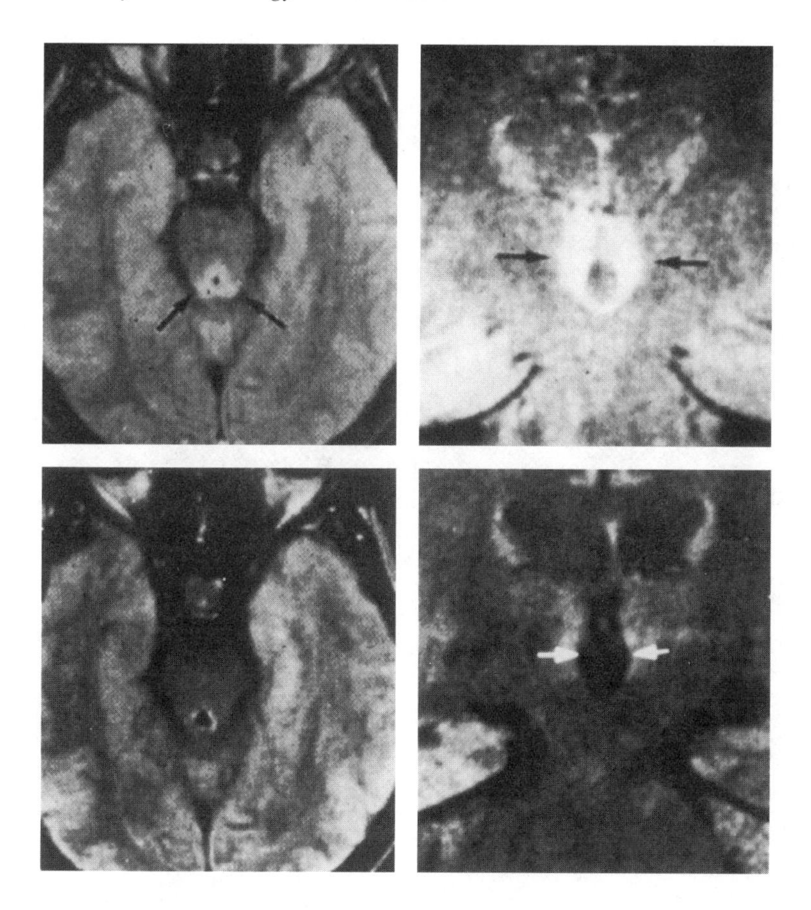

FIGURE 20.2. MRI in acute Wernicke's encephalopathy. Increased T2 signal is seen surrounding the aqueduct (arrows, top left) and third ventricle (arrows, top right). A later study shows resolution of these abnormalities and enlargement of the aqueduct (lower left) and third ventricle (white arrows, lower right). (From Gallucci et al., *AJR*, 155, 1309–1314, 1990. With permission.)

images of two patients presenting with the classical triad of Wernicke's encephalopathy (Charness and DeLaPaz, 1987).

Atrophy of the mamillary bodies can be identified by MRI in approximately 80% of alcoholics with a history of classic Wernicke's encephalopathy (Charness and DeLaPaz, 1987) (see Figure 20.3) and is not found in control subjects, Alzheimer's patients, or alcoholics without a history of Wernicke's encephalopathy (Charness and DeLaPaz, 1987, 1988). Alcoholics with the Wernicke-Korsakoff syndrome have small mamillary bodies and normal hippocampal volumes, as determined by MRI, whereas the converse is true in nonalcoholic amnestic patients (Squire et al., 1990). A decrease in mamillary body volume may be found months after acute Wernicke's encephalopathy in patients who do not develop Korsakoff's amnestic syndrome; in contrast, one of our patients with marked memory loss had normal mamillary body volume (Charness and DeLaPaz, 1987). The ability to detect specific Wernicke lesions by MRI should prove valuable in studying the cognitive disorders of alcoholics (Charness and DeLaPaz, 1987). The finding of small mamillary bodies in a mentally impaired patient will indicate that nutritional deficiency has likely played a role in the cognitive disorder. This technique may also prove useful in the diagnosis of atypical chronic cases of Wernicke's encephalopathy (Charness and DeLaPaz, 1988) and in elucidating, prospectively, the full clinical spectrum of the disorder.

MRI studies comparing amnestic alcoholics, nonamnestic alcoholics, and controls identified increases in CSF volume, ventricular enlargement, and reductions in cortical and subcortical

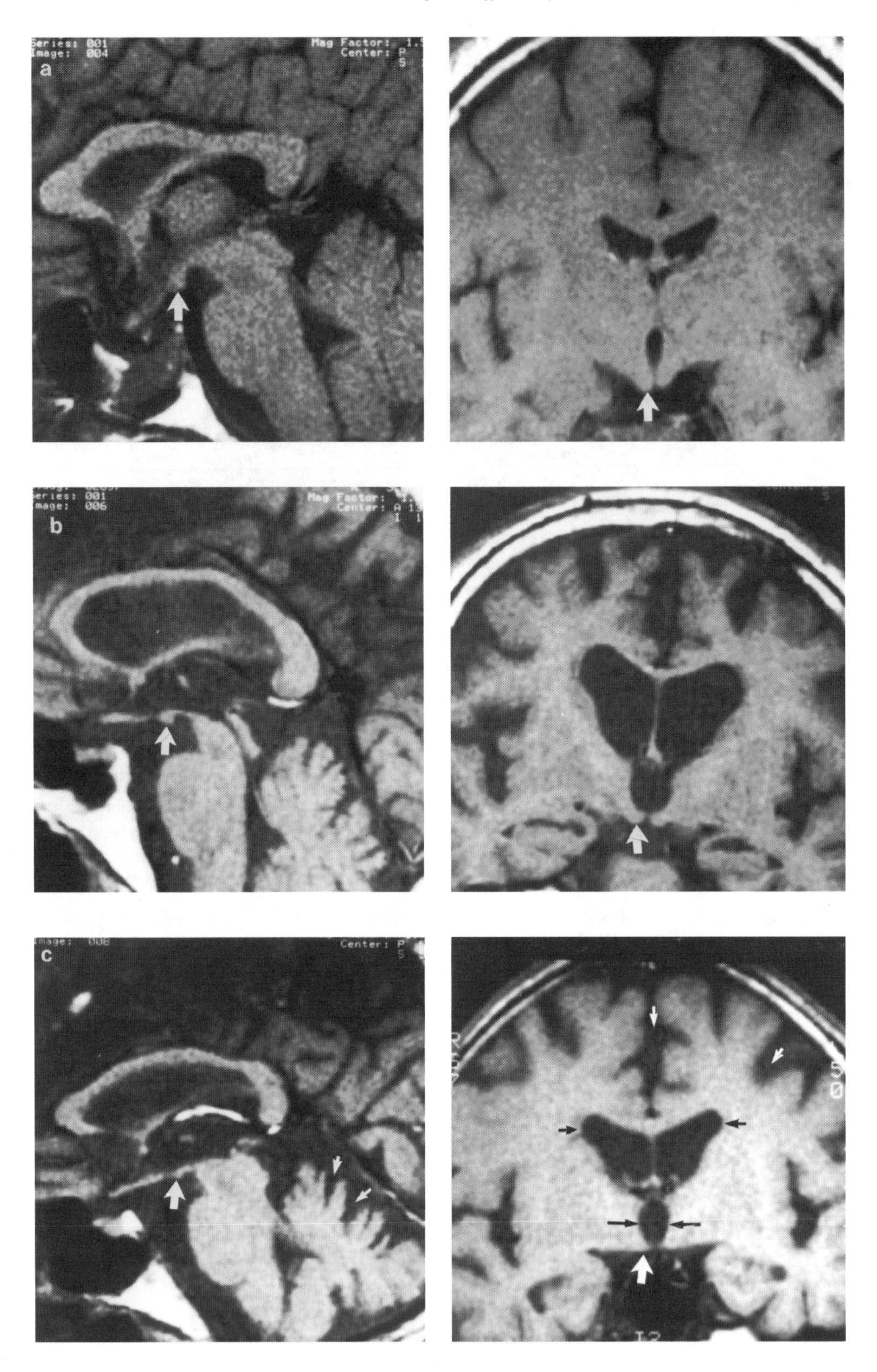
Series: 001
Image: 004
Mag Factor:
Center:
a
Series: 001
Image: 006
Mag Factor:
Center:
b
Center:
c

FIGURE 20.3

grey matter structures in both alcoholic groups (Jernigan et al., 1991a, 1991b). The amnestic alcoholics showed particularly large decreases of grey matter volume in the anterior diencephalon, mesial temporal, and orbitofrontal regions (Jernigan et al., 1991b). Neuropsychological measures in the nonamnesic alcoholics did not correlate with grey matter volume in a variety of brain regions.

E. RESPONSE TO TREATMENT

With prompt administration of thiamine, ocular signs improve within hours to days, and ataxia and confusion within days to weeks (Victor et al., 1989). This early response likely represents recovery of a biochemical rather than a structural lesion. A majority of patients are left with horizontal nystagmus, ataxia, and sometimes a disabling memory disorder known as the Korsakoff amnestic syndrome. These sequelae may result from the accumulation of lesions during repeated subclinical episodes of thiamine deficiency (Bowden, 1990; Harper, 1983; Lishman, 1981; Witt and Goldman-Rakic, 1983), rapid development of irreversible lesions during a single acute episode (Witt and Goldman-Rakic, 1983), or inadequate treatment of patients with low-affinity transketolase variants (Jeyasingham et al., 1987).

III. KORSAKOFF'S AMNESTIC SYNDROME

Approximately 80% of alcoholic patients recovering from classic Wernicke's encephalopathy exhibit the selective memory disturbance of Korsakoff's amnestic syndrome (Victor et al., 1989), which is characterized by marked deficits in anterograde and retrograde memory, apathy, an intact sensorium, and relative preservation of other intellectual abilities (Ekhardt and Martin, 1986; Victor et al., 1989). Korsakoff's amnestic syndrome may also appear without an antecedent episode of Wernicke's encephalopathy (Blansjaar and Van Dijk, 1992; Harper, 1983; Victor et al., 1989). Acute histopathologic lesions may be superimposed on chronic lesions, suggesting that subclinical episodes of Wernicke's encephalopathy may culminate in Korsakoff's amnestic syndrome (Bowden, 1990; Harper, 1979; Lishman, 1981). The memory disorder correlates best with the presence of histopathologic lesions (Victor et al., 1989) and areas of decreased density on CT scan (Shimamura et al., 1988) in the dorsomedial thalamus. However, Wernicke lesions have been confined to the mamillary bodies in one reported case of Korsakoff's amnestic syndrome (Pittella and de Castro, 1990).

Although Korsakoff's amnestic syndrome is most readily recognized as a relatively selective disorder of anterograde and retrograde memory, some alcoholics with Wernicke lesions exhibit a more global abnormality of higher cognitive function (Bowden, 1990). Torvik et al. (1982) identified Wernicke's lesions at autopsy in 20 alcoholics who had been evaluated by psychiatrists during the course of their illness. Fifteen of the 20 patients were felt to have a global dementia and only 5 had a circumscribed memory disorder; consequently, Korsakoff's amnestic syndrome was diagnosed in only 3 of 20 patients, whereas Alzheimer's disease was diagnosed in 8. Twelve of 15 patients diagnosed with dementia showed only the lesions of Wernicke's encephalopathy at autopsy, although subtle structural abnormalities were not sought (Torvik et al., 1982). The occurrence of global dementia in Wernicke's encephalopathy and the absence of non-Wernicke lesions in demented alcoholics has led some investigators

---

FIGURE 20.3. MRI of chronic Wernicke's encephalopathy. (A) normal control, (B) Alzheimer's disease, and (C) chronic Wernicke's encephalopathy. T1-weighted sagittal (left) and coronal (right) images in the plane of the mamillary bodies (arrows). The Wernicke patient shows atrophy of the mamillary bodies (arrows) and anterior superior cerebellar vermis (arrowheads) and enlargement of the third ventricle (black arrows), lateral ventricles, interhemispheric fissure (white arrowhead), and cerebral sulci (white arrowhead). The Alzheimer patient exhibits greater cerebral atrophy and ventricular enlargement than does the Wernicke patient, yet shows larger mamillary bodies. Images were acquired using TR 600, TE 25. (From Charness and DeLaPaz, *Ann. Neurol.*, 22, 595–600, 1987. With permission.)

to conclude that most cases of dementia in alcoholics are nutritional in origin (Lishman, 1986; Torvik et al., 1982; Victor et al., 1989). Whether alcohol neurotoxicity plays a role in these cases can only be determined by correlating computerized brain morphometry with premorbid neuropsychiatric examination.

One of 3 patients in Wernicke's original report was a nonalcoholic (Victor et al., 1989). Wernicke's encephalopathy has since been described in nonalcoholic patients with malnutrition due to hyperemesis, starvation, gastric plication, renal dialysis, malignancy, and AIDS (Acker et al., 1982; Bjorneboe et al., 1988; Davtyan and Vinters, 1987; Victor et al., 1989). Indeed, in one autopsy series, nonalcoholics accounted for 12 of 52 cases of Wernicke's encephalopathy (Lindboe and Loberg, 1989).

Korsakoff's amnestic syndrome is an infrequent sequela of Wernicke's encephalopathy in nonalcoholics (Freund, 1973). This observation has led to speculation that ethanol neurotoxicity is a contributing factor in the memory disorders of alcoholics (de Wardener and Lennox, 1947). Although it is possible that neurotoxic effects of ethanol worsen the cerebral disorder of thiamine deficiency, it is also clear that Korsakoff's amnestic syndrome can occur in the absence of ethanol ingestion. Accounts of thiamine deficiency in prisoners of war include descriptions of some individuals with enduring disorders of mental function following treatment with thiamine (Cruickshank, 1950; de Wardener and Lennox, 1947). More recent cases provide clearer descriptions of Korsakoff's amnestic syndrome following Wernicke's encephalopathy in nonalcoholics (Beatty et al., 1988; Becker et al., 1990; Engel et al., 1991; Parkin et al., 1991; Pittella and de Castro, 1990). Quantitative brain morphometry (see later) in these uncommon patients will indicate whether ventricular enlargement, shrinkage of cerebral white matter, selective neuronal loss, and simplification of dendritic arbors are specific lesions of ethanol neurotoxicity or previously unrecognized manifestations of thiamine deficiency.

Quantitative morphometric analysis has demonstrated a selective loss of serotonergic cells in the brainstem of alcoholics with Wernicke's encephalopathy as compared with nonalcoholic controls (Halliday et al., 1993). A similar degree of serotonergic cell loss was observed in Wernicke patients with and without Korsakoff's amnestic syndrome, suggesting that serotonergic cell loss does not account for memory loss in this disorder. Interestingly, the brain of a single alcoholic without Wernicke lesions also exhibited a striking loss of serotonergic neurons, raising the possibility that these lesions are related to alcohol neurotoxicity.

Involvement of the locus ceruleus in Wernicke's encephalopathy has led to the hypothesis that noradrenergic deficits underlie the memory disorder of Korsakoff's amnestic syndrome (McEntee and Mair, 1990). However, analysis of cerebrospinal fluid has not consistently revealed reductions in catecholamine metabolites in Korsakoff patients (Martin et al., 1984; McEntee and Mair, 1990). Using quantitative morphometry, Halliday and colleagues did not find loss of pigmented neurons in the locus coeruleus in three groups of alcoholics: those with Wernicke lesions and no memory loss, those with Wernicke lesions and Korsakoff's amnestic syndrome, and those without Wernicke lesions (Halliday et al., 1992). This finding reduces the likelihood that the memory loss of Korsakoff's amnestic syndrome is caused by a noradrenergic deficit.

IV. CEREBELLAR DEGENERATION

Alcoholic patients may be afflicted by a chronic cerebellar syndrome related to the degeneration of Purkinje cells in the cerebellar cortex (Victor et al., 1956). Midline cerebellar structures — especially the anterior and superior vermis — are predominantly affected (see Figure 20.4), a pattern that resembles the distribution of cerebellar pathology in Wernicke's encephalopathy (Victor et al., 1989). In humans, the loss of Purkinje cells is more prominent than the loss of granule cells or other cellular elements. The cause of alcoholic cerebellar degeneration is not known with certainty, but its similarity to the cerebellar lesion in Wernicke's encephalopathy suggests that thiamine deficiency is likely to be an important factor (Mancall

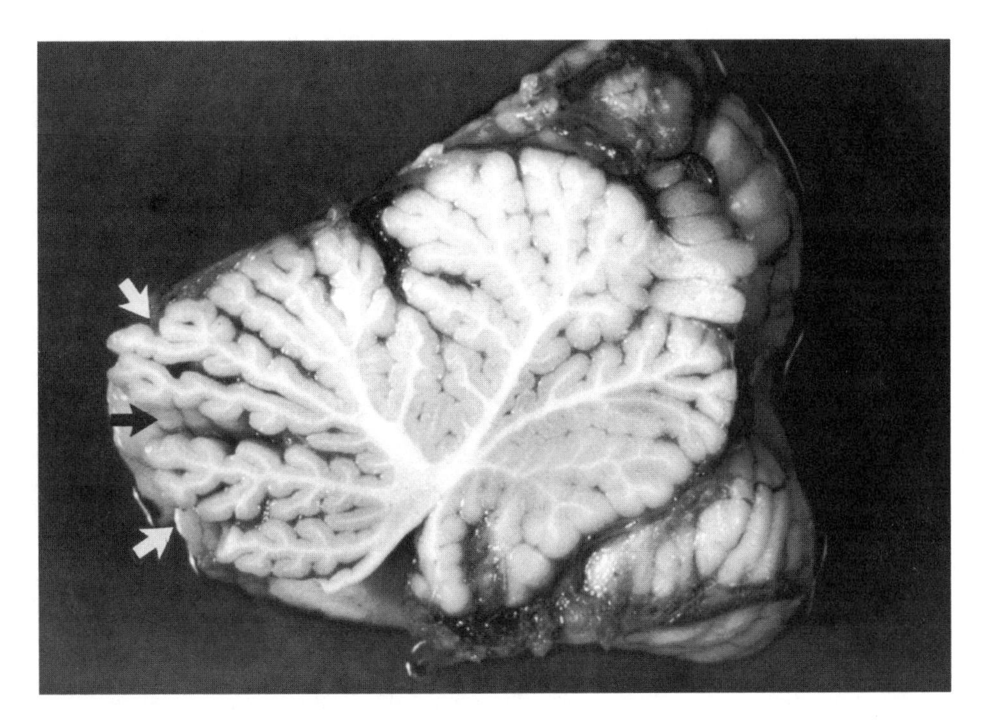

FIGURE 20.4. Alcoholic cerebellar degeneration. Sagittal section through the cerebellum discloses selective atrophy of the anterior superior vermis (arrows).

and McEntee, 1965; Victor et al., 1956). Evidence for a direct toxic effect of ethanol as the cause is poor: cerebellar ataxia in alcoholics does not correlate with daily, annual, or lifetime consumption of ethanol (Estrin, 1987), and animal models of cerebellar degeneration induced by ethanol in the absence of nutritional deficiency show a somewhat different pattern of cerebellar pathology, in which the granule cells and molecular layer interneurons are more vulnerable than are Purkinje cells (Tavares et al., 1987).

Alcoholic cerebellar degeneration typically occurs only after 10 or more years of excessive ethanol use. It is usually a gradual, progressive disorder that develops over weeks to months, but may also evolve over years or commence abruptly (Victor et al., 1956). Mild and apparently stable cases may become suddenly worse. As in Wernicke's encephalopathy, ataxia affects the gait most severely. Limb ataxia and dysarthria occur more often than in Wernicke's encephalopathy, whereas nystagmus is rare (Victor et al., 1956).

The diagnosis of alcoholic cerebellar ataxia is based on the clinical history and neurologic examination. MRI or CT scans may show cerebellar cortical atrophy (see Figure 20.3), but one-half of alcoholic patients with this finding are not ataxic on examination (Hillbom et al., 1986). Whether these represent subclinical cases in which symptoms will develop subsequently is unclear.

V. HEPATOCEREBRAL DEGENERATION

Hepatic encephalopathy develops in many alcoholics with liver disease, and is characterized by altered sensorium, frontal release signs, asterixis, hyperreflexia, extensor plantar responses, and occasional seizures. Whereas some patients progress from stupor to coma and then death, others recover and suffer recurrent episodes. Brains of patients with hepatic encephalopathy show enlargement and proliferation of protoplasmic astrocytes in the basal ganglia, thalamus, red nucleus, pons, and cerebellum, in the absence of neuronal loss or other glial changes (Victor et al., 1965).

Occasional patients do not recover fully after an episode of hepatic encephalopathy, but rather, go on to develop a progressive syndrome of tremor, choreoathetosis, dysarthria, gait ataxia, and dementia. Hepatocerebral degeneration may progress in step-wise fashion with incomplete recovery after each episode of hepatic encephalopathy, or proceed slowly and inexorably, without a discrete episode of encephalopathy. Brain examination reveals astrocytic proliferation (as described), laminar necrosis in the cortex, patchy loss of neurons throughout the cortex, basal ganglia, and cerebellum, and cavitation of the cortico-subcortical junction and superior pole of the putamen (Victor et al., 1965, 1989).

VI. MARCHIAFAVA-BIGNAMI SYNDROME

Marchiafava-Bignami syndrome is a rare disorder of demyelination or necrosis of the corpus callosum and adjacent subcortical white matter, which occurs predominantly in malnourished alcoholics (Brion, 1976). In some cases there are associated lesions of Wernicke's encephalopathy or selective neuronal loss and gliosis in the third cortical layer. A few cases have been described in nonalcoholics (Kosaka et al., 1984; Leong, 1979), demonstrating that ethanol alone is not responsible for the lesion. The course may be acute, subacute, or chronic, and is marked by dementia, spasticity, dysarthria, and inability to walk. Patients may lapse into coma and die, survive for many years in a demented condition, or occasionally recover (Delangre et al., 1986).

Marchiafava-Bignami disease was formerly diagnosed only at autopsy, but lesions can now be imaged using CT or MRI (Kawamura et al., 1985). CT may demonstrate hypodense areas in portions of the corpus callosum. MRI typically shows cystic areas of decreased T1 signal or areas of increased T2 signal (see Figure 20.5) within the same regions (Kawamura et al., 1985). The abnormalities may be restricted to one region of the corpus callosum or may be present diffusely throughout the corpus callosum. The CT and MRI findings may persist after the resolution of clinical signs (Delangre et al., 1986).

VII. CENTRAL PONTINE MYELINOLYSIS

Central pontine myelinolysis is a disorder of cerebral white matter that usually affects alcoholics, but also occurs in nonalcoholics with liver disease, including Wilson's disease, malnutrition, anorexia, burns, cancer, Addison's disease, and severe electrolyte disorders, such as thiazide-induced hyponatremia (Adams et al., 1959; Victor et al., 1989). Central pontine myelinolysis is frequently associated with a rapid correction of hyponatremia; however, the majority of cases occur in alcoholics, suggesting that alcoholism may contribute to the genesis of central pontine myelinolysis in as yet undefined ways.

The most common macroscopic lesion is a triangular region of pallor in the base of the pons (see Figure 20.6). Approximately 10% of human cases also have symmetric extrapontine lesions, most frequently in the striatum, thalamus, cerebellum, and cerebral white matter (Wright et al., 1979). Microscopic examination reveals demyelinated axons with preserved cell bodies except in the center of lesions, which may show cavitation. Myelinolytic lesions can be induced experimentally by rapid correction of chronic hyponatremia (Illowsky and Laureno, 1987; Norenberg and Papendick, 1984). In rats, the lesions are primarily extrapontine in location (Kleinschmidt-DeMasters and Norenberg, 1981), whereas in dogs, a mixture of pontine and extrapontine lesions are observed (Laureno, 1983). The marked species differences in the distribution of lesions in central pontine myelinolysis illustrates one of the difficulties in generalizing pathologic findings from experimental animals to humans.

Symptoms and signs of central pontine myelinolysis may be absent (Adams et al., 1959) or obscured by associated conditions, such as ethanol withdrawal, Wernicke's encephalopathy, or hepatic encephalopathy. Treatment of these disorders may lead to an initial improvement

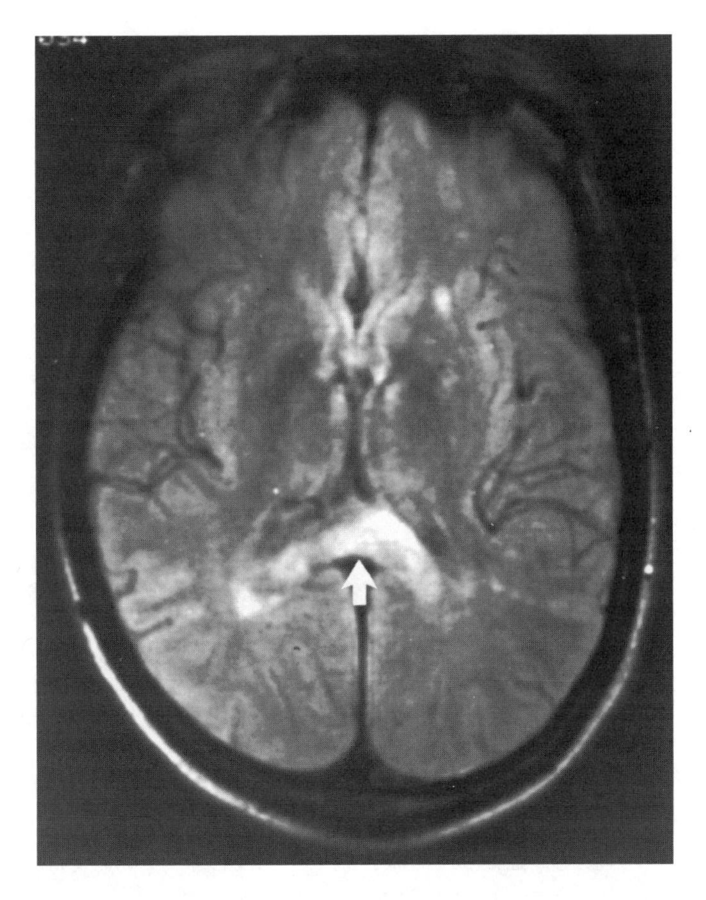

FIGURE 20.5. MRI of Marchiafava-Bignami Syndrome. The splenium of the corpus callosum shows an area of increased T2 signal (arrow). (TR 2000, TE 40). (From Charness, M.E., *Alcoholism: Clin. Exp. Res.*, 17, 2–11, 1993. With permission.)

in mental status, followed within days by confusion, lethargy, and coma due to central pontine myelinolysis. Involvement of the corticospinal tracts causes paraparesis or quadriparesis, and demyelination of the corticobulbar tracts leads to dysarthria, dysphagia, and inability to protrude the tongue. The tendon reflexes may be increased, decreased, or normal, and Babinski signs may be present. Disorders of conjugate eye movement occur occasionally and may reflect extension of the lesion in the pons or associated Wernicke lesions. Disproportionate involvement of motor function may produce the "locked-in" syndrome, with only limited ability to move the limbs or face despite a normal level of consciousness (Messert et al., 1979).

The lesions of central pontine myelinolysis can be visualized using CT scanning or MRI. MRI is more sensitive than CT in imaging the pontine lesions; however, even MRI may be unremarkable early in the course of central pontine myelinolysis (Miller et al., 1988). The most common MRI finding is an area of decreased T1 signal (see Figure 20.7A) or increased T2 signal (see Figure 20.7B) within the basis pontis (Miller et al., 1988; Ostertun et al., 1990). Symmetrical extrapontine abnormalities have also been detected by CT and MRI in patients suspected of having central pontine myelinolysis (Dickoff et al., 1988; Miller et al., 1988). Other disorders, such as multiple sclerosis, multiinfarct dementia, and encephalitis, may produce areas of increased T2 signal in the pons that resemble those of central pontine myelinolysis, but also can cause significant periventricular abnormalities and a distinctive clinical picture (Miller et al., 1988). Serial CT or MRI studies indicate that the radiographic lesions of central pontine myelinolysis may resolve in parallel with patient recovery (Miller et al., 1988); hence, the absence of lesions on MRI does not exclude a past episode of central

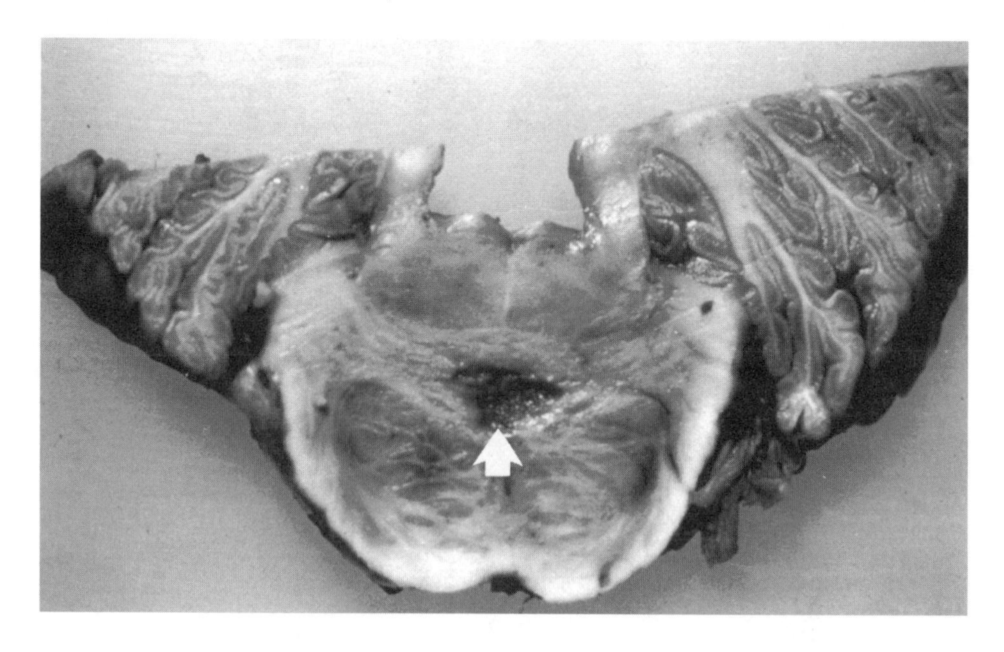

FIGURE 20.6. Central pontine myelinolysis. Arrow indicates a large area of demyelination and necrosis in the midpons. (Courtesy of Dr. Richard L. Davis, Department of Pathology, University California San Francisco.)

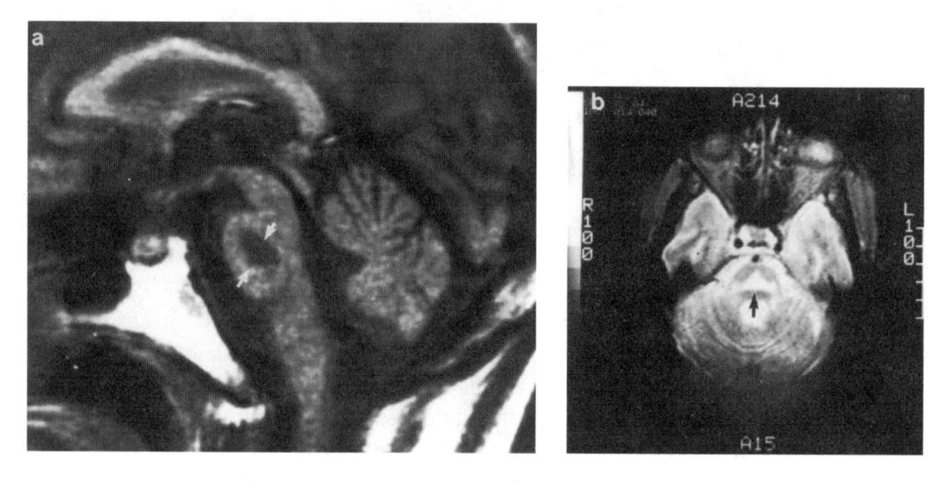

FIGURE 20.7. MRI of central pontine myelinolysis in an alcoholic man. (A) T1-weighted sagital image (TR 600, TE 25) reveals a sharply delineated zone of decreased signal in the midpons. (B) An axial image through the same region shows a triangular area of increased T2 signal (TR 2000, TE 70). (From Charness, M.E., *Alcoholism: Clin. Exp. Res.*, 17, 2–11, 1993. With permission.)

pontine myelinolysis. Because small lesions of central pontine myelinolysis may be asymptomatic, typical MRI abnormalities can be an incidental finding in seriously ill patients.

VIII. ETHANOL NEUROTOXICITY IN THE DEVELOPING NERVOUS SYSTEM

Brain lesions in alcoholics may in some instances antedate birth. Many alcoholics have been exposed to high concentrations of alcohol during critical stages of brain development, leading in some cases to the fetal alcohol syndrome (FAS). This disorder is characterized by pre- and postnatal growth retardation, microcephaly, neurologic abnormalities, facial

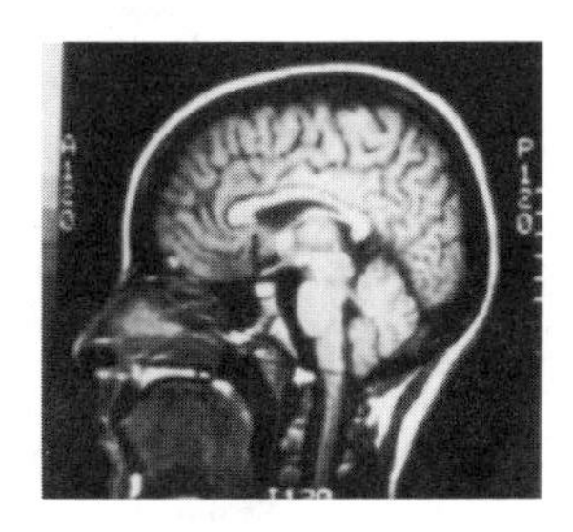
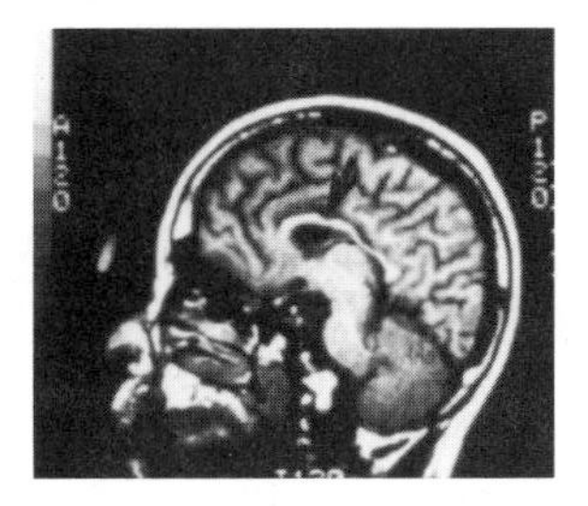
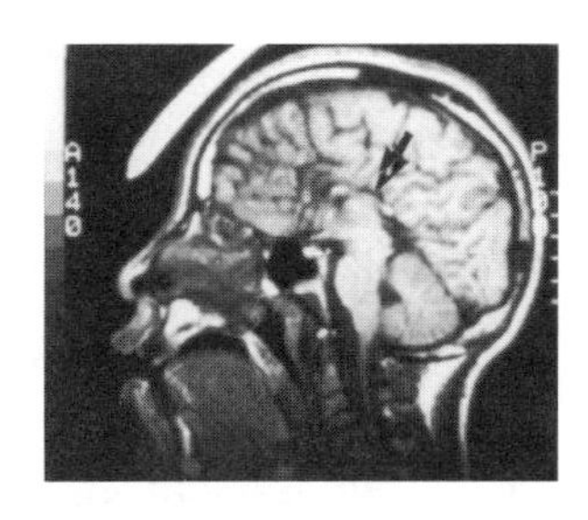

FIGURE 20.8. MRI showing agenesis of the corpus callosum after gestational exposure to alcohol. (A) Normal 13-year-old female, (B) 13-year-old male with fetal alcohol syndrome and focal thinning (arrow) of the corpus callosum, (C) 14-year-old male with fetal alcohol syndrome and agenesis of the corpus callosum. (Courtesy of Dr. Edward P. Riley, San Diego State University.)

dysmorphology, and other congenital anomalies (Rosett and Weiner, 1984; Streissguth et al., 1980). The full syndrome occurs in approximately 0.33 cases per 1000 of live births (Abel and Sokol, 1991). More often, heavy intrauterine exposure to ethanol is associated with a constellation of less severe "fetal alcohol effects," including mental retardation, intrauterine growth retardation, and minor, ungrouped congenital anomalies involving the cutaneous, genitourinary, musculoskeletal, and cardiac systems (Rosett and Weiner, 1984). Some dysmorphic features appear to recede with postnatal development (Spohr and Steinhausen, 1987), however, learning disabilities and hyperactivity persist (Streissguth et al., 1984, 1986, 1989).

Neuropathologic examination in FAS reveals microcephaly, cerebellar dysplasia, hydrocephalus, agenesis of the corpus callosum, and neuronal-glial heterotopias (Clarren et al., 1978; Jones and Smith, 1973; Siebert et al., 1991) — lesions consistent with the decreased proliferation and disordered migration of neurons that have been demonstrated by *in vivo* [^{3}H]thymidine labeling studies in rats (Miller, 1986). In animal studies, gestational exposure to ethanol also causes microcephaly (Diaz and Samson, 1980) and disrupts the cytoarchitecture of cerebral cortex, hippocampus, cerebellum, and other subcortical structures (Kotkoskie and Norton, 1988; Miller, 1986, 1987, 1992; Pentney and Miller, 1992; West et al., 1981; West and Hodges, 1983). The pathological basis for mental retardation and learning disabilities may reside in more subtle structural abnormalities, such as the simplification of hippocampal dendrites seen in animals exposed to ethanol in utero (Davies and Smith, 1981). Prenatal ethanol exposure alters synaptogenesis and dendritic structure (reviewed in Pentney and Miller, 1992) and reduces the density of parallel fibers in the molecular layer of the cerebellum (Smith and Davies, 1990).

MRI in a small number of children with FAS has identified agenesis or hypoplasia of the corpus callosum (see Figure 20.8) (Mattson et al., 1994) and selective reductions in the volume of the cerebrum, cerebellum, and basal ganglia (Gabrielli et al., 1990; Mattson et al., 1992). Interestingly, agenesis of the corpus callosum and hydrocephalus are also features of X-linked hydrocephalus, a genetic disorder associated with mutations in the gene for the neural cell adhesion molecule, L1 (Rosenthal et al., 1992). The recent finding that alcohol inhibits cell-cell adhesion mediated by L1 and N-CAM (Charness et al., 1994) raises the possibility that the common developmental abnormalities of X-linked hydrocephalus and FAS are caused by a disruption of cell-cell interactions.

IX. ETHANOL NEUROTOXICITY IN THE MATURE NERVOUS SYSTEM

Neuroimaging studies, neuropathological observations, and experiments in animals and cultured cells provide evidence that ethanol is neurotoxic. CT and MRI show enlargement of the cerebral ventricles and sulci in a majority of alcoholics. However, when corrected for the effects of aging, the radiographic indices do not correlate consistently with either the duration of drinking or the severity of cognitive impairment (Victor et al., 1989). The ventricles and sulci become significantly smaller within about one month of abstinence, (Carlen et al., 1978, 1984; Schroth et al., 1988; Zipursky et al., 1989), whereas brain water, estimated by MRI (Schroth et al., 1988) or chemical analysis (Harper et al., 1988) does not change consistently. According to these findings, it has been hypothesized that changes in brain parenchyma, but not brain water, may account for the reversible radiographic and cognitive abnormalities of alcoholics (Carlen et al., 1984; Schroth et al., 1988). In support of this hypothesis are reports that chronic ingestion of ethanol by well-nourished animals can reversibly reduce the complexity of dendritic arborization (McMullen et al., 1984) and alter the density of dendritic spines (King et al., 1988) in the hippocampus.

Pathologic studies of the brains of alcoholics have provided mixed evidence for ethanol-induced cerebral atrophy (Victor et al., 1989). Brain weight in alcoholics is reduced only slightly as compared with that in nonalcoholics (Harper et al., 1985, 1988; Harper and Kril, 1985; Torvik et al., 1982), and some workers find no differences at all (de la Monte, 1988; Victor et al., 1989). Brain volume, estimated by the volume of the pericerebral space — the CSF-filled region between the brain and skull — is reduced in alcoholics compared to that in controls, but this indirect measure of cerebral atrophy is most abnormal in alcoholics with liver disease or Wernicke's encephalopathy (Harper and Kril, 1985). Quantitative morphometry suggests that alcoholics, including those with liver disease and Wernicke's encephalopathy, lose a disproportionate amount of subcortical white matter as compared with cortical gray matter (Harper et al., 1985). This loss of cerebral white matter is also apparent when brains of nondemented alcoholics with liver disease are compared with those from patients with nonalcoholic liver disease; hence, liver disease cannot be the sole cause of this selective loss of brain tissue. The loss of cerebral white matter is evident across a wide range of ages, is not accentuated in the frontal lobes (Harper et al., 1985), and is of sufficient magnitude (6–17%) to account for the associated ventricular enlargement (de la Monte, 1988).

The identification of selectively affected neurotransmitter pathways could have important implications for the management of dementia in alcoholics. Cholinergic neurons in the nucleus basalis of the basal forebrain, which innervate much of the cerebral cortex and are preferentially depleted in dementia due to Alzheimer's disease, have also been reported to be lost in three patients with Korsakoff's syndrome (Arendt et al., 1983). Levels of the acetylcholine-synthesizing enzyme, choline acetyltransferase (CAT), which are reduced in the cerebral cortex in Alzheimer's disease, are also depleted in the brains of alcoholics, as reported in some (Antuono et al., 1980; Nordberg et al., 1980), but not all (Smith et al., 1988) studies. Whether putative cholinergic deficits in alcoholics are a consequence of thiamine deficiency or direct ethanol neurotoxicity is unclear.

Long-term administration of ethanol to well-nourished rats causes memory deficits, reductions in CAT levels and choline uptake, and a slight (17%) loss of neurons in the nucleus basalis (Arendt et al., 1988a, 1988b). Transplantation of cholinergic neurons into the hippocampus and neocortex corrects both the cholinergic deficits and memory abnormalities, suggesting that, at least in rats, ethanol can directly damage cholinergic projection neurons (Arendt et al., 1988; Arendt et al., 1983).

The recent use of computerized morphometry has revealed alterations in neuronal size, number, architecture, and synaptic complexity in alcoholics. Neuronal density in the superior frontal cortex was reduced by 22% in alcoholics compared to nonalcoholic controls (Harper et al., 1987). Neuronal loss was accompanied by selective glial proliferation in the superior frontal cortex and was more pronounced in a subgroup of alcoholics with cirrhosis or Wernicke's encephalopathy. By contrast, neuronal counts in motor, temporal, or cingulate cortex did not differ between the two groups (Harper et al., 1987; Kril and Harper, 1989). A decrease in neuronal area was also observed in the superior frontal, cingulate, and motor cortices (Harper et al., 1987; Kril and Harper, 1989). The complexity of basal dendritic arborization of layer III pyramidal cells in both superior frontal and motor cortices was significantly reduced in a group of 15 alcoholics compared to controls (Harper and Corbett, 1990). Similarly, a significant reduction in dendritic arborization of Purkinje cells in the anterior superior vermis was observed in four alcoholics, three of whom also had lesions of Wernicke's encephalopathy or pellagra (Ferrer et al., 1984). Finally, a group of five alcoholics without lesions of nutritional deficiency showed a decrease in the density of synaptic spines in layer V cortical pyramidal cells when compared to control patients (Ferrer et al., 1986).

These data demonstrate that there is selective neuronal loss, dendritic simplification, and reduction of synaptic complexity in different brain regions of alcoholics; however, it remains uncertain how these cellular lesions relate to the selective loss of white matter described earlier. The etiology of these cellular lesions is also unclear, because patients often had coincident lesions of nutritional deficiency or cirrhosis. In fact, the cellular lesions were often most severe in the alcoholics with liver disease or Wernicke lesions. There are three possible explanations for these findings:

1. Liver disease and Wernicke lesions are markers of greater ethanol intake and more severe ethanol neurotoxicity.
2. Liver disease and thiamine deficiency potentiate the neurotoxic actions of ethanol.
3. Liver disease and thiamine deficiency cause these lesions.

The fact that the neurological syndromes of liver disease and thiamine deficiency can develop slowly suggests that lesions in these disorders may also arise insidiously. Indeed, it is conceivable that some of the subtle, nonspecific lesions identified in alcoholic brains are early manifestations of malnutrition and liver disease that precede more recognizable, specific lesions of Wernicke's disease and hepatocerebral degeneration. This hypothesis can be tested by using computerized morphometry to seek similar, subtle lesions in nonalcoholic patients with Wernicke's disease and hepatocerebral degeneration. To date, all of the Wernicke patients studied by computerized morphometry have been alcoholics, some of whom have also had liver disease. In the meantime, evidence that ethanol is neurotoxic must come from animal studies where nutrition can be better controlled. Here, the data suggest that ethanol is neurotoxic in various murine species, in some instances producing lesions similar to those observed in humans. Thus, with respect to hippocampal neurons, chronic ethanol treatment reduces (Walker et al., 1980) or does not change (McMullen et al., 1984) cell number, reversibly decreases (McMullen et al., 1984) or increases (Durand et al., 1989) dendritic complexity, and reversibly decreases or increases the density of synaptic spines (Durand et al., 1989), depending on the cell type.

REFERENCES

Abel, E.L. and Sokol, R.J. (1991) A revised conservative estimate of the incidence of FAS and its economic impact. *Alcoholism: Clin. Exp. Res.,* 15: 514–524.

Acker, W., Aps, E.J., Majumdar, S.K., and Shaw, G.K. (1982) The relationship between brain and liver damage in chronic alcoholic patients. *J. Neurol. Neurosurg. Psychiat.,* 45: 984–987.

Adams, R.D., Victor, M., and Mancall, E.L. (1959) Central pontine myelinolysis. A hitherto undescribed disease occurring in alcoholic and malnourished patients. *Arch. Neurol.,* 81: 154–172.

Antuono, P., Sorbi, S., Bracco, L., Fusco, T., and Amaducci, L. (1980). A discrete sampling technique in senile dementia of the Alzheimer type and alcoholic dementia: study of the cholinergic system. *Aging of the Brain and Dementia.* New York, Raven Press.

Arendt, T., Allen, Y., Sinden, J., Schugens, M.M., Marchbanks, R.M., Lantos, P.L., and Gray, J.A. (1988a) Cholinergic-rich brain transplants reverse alcohol-induced memory deficits. *Nature,* 332: 448–450.

Arendt, T., Henning, D., Gray, J.A., and Marchbanks, R. (1988b) Loss of neurons in the rat basal forebrain cholinergic projection system after prolonged intake of ethanol. *Brain Res. Bull.,* 21: 563–569.

Arendt, T., Bigl, V., Arendt, A., and Tennstedt, A. (1983) Loss of neurons in the nucleus basalis of Meynert in Alzheimer's disease, paralysis agitans and Korsakoff's disease. *Acta Neuropathol.* (Berlin), 61: 101–108.

Beatty, W.W., Bailly, R.C., and Fisher, L. (1988) Korsakoff-like amnesic syndrome in a patient with anorexia and vomiting. *Int. J. Clin. Neuropsychol.,* 11: 55–65.

Becker, J.T., Furman, J.M., Panisset, M., and Smith, C. (1990) Characteristics of the memory loss of a patient with Wernicke-Korsakoff's syndrome without alcoholism. *Neuropsychologia,* 28: 171–179.

Bjorneboe, G.-E.A., Johnsen, J., Bjorneboe, A., Marklund, S.F., Skylv, N., Hoiseth, A., Bache-Wiig, J.-E., Morland, J., and Drevon, C.A. (1988) Some aspects of antioxidant status in blood from alcoholics. *Alcoholism: Clin. Exper. Res.,* 12: 806–810.

Blansjaar, B.A. and Van Dijk, J.G. (1992) Korsakoff minus Wernicke syndrome [see comments]. *Alcohol Alcoholism,* 27: 435–437.

Blass, J.P. and Gibson, G.E. (1977) Abnormality of a thiamine-requiring enzyme in patients with Wernicke-Korsakoff syndrome. *N. Engl. J. Med.,* 297: 1367–1370.

Bowden, S.C. (1990) Separating cognitive impairment in neurologically asymptomatic alcoholism from Wernicke-Korsakoff syndrome: is the neuropsychological distinction justified? *Psychol. Bull.,* 107: 355–366.

Brion, S. (1976) Marchiafava-Bignami syndrome. *Metabolic and Deficiency Diseases of the Nervous System, Part 2.* Amsterdam, North-Holland Publishing Company.

Butterworth, R.F., Kril, J.J., and Harper, C.G. (1993) Thiamine-dependent enzyme changes in the brains of alcoholics: relationship to the Wernicke-Korsakoff syndrome. *Alcoholism,* 17: 1084–1088.

Carlen, P.L., Wilkinson, D.A., Wortzman, G., and Holgate, R. (1984) Partially reversible cerebral atrophy and fuctional improvement in recently abstinent alcoholics. *Can. J. Neurol. Sci.,* 11: 441–446.

Carlen, P.L., Wortzman, G., Holgate, R.C., Wilkinson, D.A., and Rankin, J.G. (1978) Reversible cerebral atrophy in recently abstinent chronic alcoholics measured by computed tomography scans. *Science,* 200: 1076–1078.

Charness, M.E. (1993) Brain lesions in alcoholics. *Alcoholism: Clin. Exp. Res.,* 17: 2–11.

Charness, M.E. and DeLaPaz, R.L. (1987) Mamillary body atrophy in Wernicke's encephalopathy: antemortem identification using magnetic resonance imaging. *Ann. Neurol.,* 22: 595–600.

Charness, M.E. and DeLaPaz, R.L. (1988) Periodic alternating nystagmus in an alcoholic with small mamillary bodies. *Neurology,* 38 (suppl): 421.

Charness, M.E., Safran, R.M., and Perides, G. (1994) Ethanol inhibits neural cell-cell adhesion. *J. Biol. Chem.,* 269: 9304–9309.

Charness, M.E., Simon, R.P., and Greenberg, D.A. (1989) Ethanol and the nervous system. *N. Engl. J. Med.,* 321: 442–454.

Choi, D.W. (1988) Glutamate neurotoxicity and diseases of the nervous system. *Neuron,* 1: 623–634.

Clarren, S.K., Alvord, E.J., Sumi, S.M., Streissguth, A.P., and Smith, D.W. (1978) Brain malformations related to prenatal exposure to ethanol. *J. Pediatr.,* 92: 64–67.

Cruickshank, E.K. (1950) Wernicke's encephalopathy. *Q. J. Med.,* 19: 327–338.

Davies, D.L. and Smith, D.E. (1981) A Golgi study of mouse hippocampal CA1 pyramidal neurons following perinatal ethanol exposure. *Neurosci. Lett.,* 26: 49–54.

Davtyan, D.G. and Vinters, H.V. (1987) Wernicke's encephalopathy in AIDS patient treated with zidovudine. *Lancet,* 1: 919–920.

de la Monte, S.M. (1988) Disproportionate atrophy of cerebral white matter in chronic alcoholics. *Arch. Neurol.,* 45: 990–992.

de Wardener, H.E. and Lennox, B. (1947) Cerebral beriberi (Wernicke's encephalopathy). *Lancet,* 1: 11–17.

Delangre, T., Hannequin, D., Clavier, E., Denis, P., Mihout, B., and Samson, M. (1986) Maladie de Marchiafava-Bignami d'evolution favorable. *Rev. Neurol.* (Paris), 142: 933–936.

Diaz, J. and Samson, H.H. (1980) Impaired brain growth in neonatal rats exposed to ethanol. *Science,* 208: 751–753.

Dickoff, D.J., Raps, M., and Yahr, M.D. (1988) Striatal syndrome following hyponatremia and its rapid correction. *Arch. Neurol.,* 45: 112–114.

Durand, D., Saint Cyr, J.A., Gurevich, N., and Carlen, P.L. (1989) Ethanol-induced dendritic alterations in hippocampal granule cells. *Brain Res.,* 477: 373–377.

Ekhardt, M.J. and Martin, P.R. (1986) Clinical assessment of cognition in alcoholism. *Alcoholism: Clin. Exp. Res.,* 10: 123–127.

Engel, P.A., Grunnet, M., and Jacobs, B. (1991) Wernicke-Korsakoff syndrome complicating T-cell lymphoma: unusual or unrecognized? *South. Med. J.,* 84: 253–256.

Estrin, W.J. (1987) Alcoholic cerebellar degeneration is not a dose-dependent phenomenon. *Alcoholism: Clin. Exp. Res.,* 11: 372–375.

Ferrer, I., Fabregues, I., Pineda, M., Gracia, I., and Ribalta, T. (1984) A Golgi study of cerebellar atrophy in human chronic alcoholism. *Neuropathol. Appl. Neurobiol.,* 10: 245–253.

Ferrer, I., Fabregues, I., Rairiz, J., and Galofre, E. (1986) Decreased numbers of dendritic spines on cortical pyramidal neurons in human chronic alcoholism. *Neurosci. Lett.,* 69: 115–119.

Freund, G. (1973) Chronic central nervous system toxicity of alcohol. *Ann. Rev. Pharmacol.,* 13: 217–227.

Gabrielli, O., Salvolini, U., Coppa, G.V., Catassi, C., Rossi, R., Manca, A., Lanza, R., and Giorgi, P.L. (1990) Magnetic resonance imaging in the malformative syndromes with mental retardation. *Pediatr. Radiol.,* 21: 16–19.

Gallucci, M., Bozzao, A., Splendiani, A., Masciocchi, C., and Passariello, R. (1990) Wernicke encephalopathy: MR findings in five patients. *AJR,* 155: 1309–1314.

Ghez, C. (1969) Vestibular paresis: a clinical feature of Wernicke's disease. *J. Neurol. Neurosurg. Psychiat.,* 32: 134–139.

Grant, K.A., Valverius, P., Hudspith, M., and Tabakoff, B. (1990) Ethanol withdrawal seizures and the NMDA receptor complex. *Eur. J. Pharmacol.,* 176: 289–296.

Gulya, K., Grant, K.A., Valverius, P., Hoffman, P.L., and Tabakoff, B. (1991) Brain regional specificity and time-course of changes in the NMDA receptor-ionophore complex during ethanol withdrawal. *Brain Res.,* 547: 129–134.

Hakim, A.M. (1984) The induction and reversibility of cerebral acidosis in thiamine deficiency. *Ann. Neurol.,* 16: 673–679.

Hakim, A.M. and Pappius, H.M. (1983) Sequence of metabolic, clinical, and histological events in experimental thiamine deficiency. *Ann. Neurol.,* 13: 365–375.

Halliday, G., Ellis, J., Heard, R., Caine, D., and Harper, C. (1993) Brainstem serotonergic neurons in chronic alcoholics with and without the memory impairment of Korsakoff's psychosis. *J. Neuropath. Exper. Neurol.,* 52: 567–579.

Halliday, G., Ellis, J., and Harper, C. (1992) The locus coerulus and memory: a study of chronic alcoholics with and without the memory impairment of Korsakoff's psychosis. *Brain Res.,* (1–2): 33–37.

Harper, C. and Corbett, D. (1990) Changes in the basal dendrites of cortical pyramidal cells from alcoholic patients — a quantitative Golgi study. *J. Neurol. Neurosurg. Psychiat.,* 53: 856–861.

Harper, C., Kril, J.J., and Daly, J.M. (1988) Brain shrinkage in alcoholics is not caused by changes in hydration: a pathological study. *J. Neurol. Neurosurg. Psychiat.,* 51: 124–127.

Harper, C.G. (1979) Wernicke's encephalopathy: a more common disease than realised. *J. Neurol. Neurosurg. Psychiat.,* 42: 226–231.

Harper, C.G. (1983) The incidence of Wernicke's encephalopathy in Australia — a neuropathological study of 131 cases. *J. Neurol. Neurosurg. Psychiat.,* 46: 593–598.

Harper, C.G., Giles, M., and Finlay-Jones, R. (1986) Clinical signs in the Wernicke-Korsakoff complex: a retrospective analysis of 131 cases diagnosed at necropsy. *J. Neurol. Neurosurg. Psychiat.,* 49: 341–345.

Harper, C.G., Kril, J., and Daly, J. (1987) Are we drinking our neurons away. *Br. Med. J.,* 294: 534–536.

Harper, C.G. and Kril, J.J. (1985) Brain atrophy in chronic alcoholic patients: a quantitative pathological study. *J. Neurol. Neurosurg. Psychiat.,* 48: 211–217.

Harper, C.G., Kril, J.J., and Daly, J. (1988) Does a "moderate" alcohol intake damage the brain? *J. Neurol. Neurosurg. Psychiat.,* 51: 909–913.

Harper, C.G., Kril, J.J., and Holloway, R.L. (1985) Brain shrinkage in chronic alcoholics: a pathological study. *Br. Med. J.,* 290: 501–504.

Hillbom, M., Muuronen, A., Holm, L., and Hindmarsh, T. (1986) The clinical versus radiological diagnosis of alcoholic cerebellar degeneration. *J. Neurol. Sci.,* 73: 45–53.

Hofmann, E., Friedburg, H., Rasenack, J., Ott, D., and Wimmer, B. (1988) Wernicke's encephalopathy on CT and MR. [German]. *Rofo: Fortschritte Auf Dem Gebiete Der Rontgenstrahlen Und Der Nuklearmedizin,* 148: 97–98.

Illowsky, B.P. and Laureno, R. (1987) Encephalopathy and myelinolysis after rapid correction of hyponatremia. *Brain,* 110: 855–867.

Iwata, H. (1982) Possible role of thiamine in the nervous system. *Trends Pharmacol. Sci.,* 3: 171–173.

Jernigan, T.L., Butters, N., DiTraglia, G., Schafer, K., Smith, T., Irwin, M., Grant, I., Schuckit, M., and Cermak, L.S. (1991a) Reduced cerebral grey matter observed in alcoholics using magnetic resonance imaging. *Alcoholism: Clin. Exp. Res.,* 15: 418–427.

Jernigan, T.L., Schafer, K., Butters, N., and Cermak, L.S. (1991b) Magnetic resonance imaging of alcoholic Korsakoff patients. *Neuropsychopharmacology,* 4: 175–186.

Jeyasingham, M.D., Pratt, O.E., Burns, A., Shaw, G.K., Thomson, A.D., and Marsh, A. (1987) The activation of red blood cell transketolase in groups of patients especially at risk from thiamin deficiency. *Psychol. Med.,* 17: 311–318.

Jones, K.L. and Smith, D.W. (1973) Recognition of the fetal alcohol syndrome in early infancy. *Lancet,* 2: 999–1001.

Kawamura, M., Shiota, J., Yagishita, T., and Hirayama, K. (1985) Marchiafava-Bignami disease: computed tomographic scan and magnetic resonance imaging. *Ann. Neurol.,* 18: 103–104.

King, M.A., Hunter, B.E., and Walker, D.W. (1988) Alterations and recovery of dendritic spine density in rat hippocampus following long-term ethanol ingestion. *Brain Res.,* 459: 381–385.

Kitaguchi, T., Kobayashi, T., Tobimatsu, S., Goto, I., and Kuroiwa, Y. (1987) Computed tomography and magnetic resonance imaging in a young patient with Wernicke's encephalopathy. *J. Neurol.,* 234: 449–450.

Kleinschmidt-DeMasters, B.K., and Norenberg, M.D. (1981) Rapid correction of hyponatremia causes demyelination: relation to central pontine myelinolysis. *Science,* 211: 1068–1070.

Kosaka, K., Aoki, M., Kawasaki, N., Adachi, Y., Konuma, I., and Iizuka, R. (1984) A nonalcoholic Japanese patient with Wernicke's encephalopathy and Marchiafava-Bignami disease. *Clin. Neuropathol.,* 3: 231–236.

Kotkoskie, L.A. and Norton, S. (1988) Prenatal brain malformations following acute ethanol exposure in the rat. *Alcoholism: Clin. Exp. Res.,* 12: 831–836.

Kril, J.J. and Harper, C.G. (1989) Neuronal counts from four cortical regions of alcoholic brains. *Acta Neuropathol.,* 79: 200–204.

Langlais, P.J. and Mair, R.G. (1990) Protective effects of the glutamate antagonist MK-801 on pyrithiamine-induced lesions and amino acid changes in rat brain. *J. Neurosci.,* 10: 1664–1674.

Langlais, P.J. and Zhang, S.X. (1993) Extracellular glutamate is increased in thalamus during thiamine deficiency-induced lesions and is blocked by MK-801. *J. Neurochem.,* 61: 2175–2182.

Laureno, R. (1983) Central pontine myelinolysis following rapid correction of hyponatremia. *Ann. Neurol.,* 13: 232–242.

Leong, A.S.Y. (1979) Marchiafava-Bignami disease in a non-alcoholic Indian male. *Pathology,* 11: 241–249.

Lindboe, C.F. and Loberg, E.M. (1989) Wernicke's encephalopathy in non-alcoholics: an autopsy study. *J. Neurol. Sci.,* 90: 125–129.

Lishman, W.A. (1981) Cerebral disorder in alcoholism. Syndromes of impairment. *Brain,* 104: 1–20.

Lishman, W.A. (1986) Alcoholic dementia: a hypothesis. *Lancet,* 1: 1184–1186.

Lovinger, D.M. (1993) Excitotoxicity and alcohol-related brain damage. *Alcoholism: Clin. Exp. Res.,* 17: 19–27.

Mancall, E.L. and McEntee, W.J. (1965) Alterations of the cerebellar cortex in nutritional encephalopathy. *Neurology,* 15: 303–313.

Martin, P.R., Weingartner, H., Gordon, E.K., Burns, R.S., Linnoila, M., Kopin, I., and Ebert, M.H. (1984) Central nervous system catecholamine metabolism in Korsakoff's psychosis. *Ann. Neurol.,* 15: 184–187.

Mattson, S.N., Jernigan, T.L., and Riley, E.P. (1994) MRI and prenatal alcohol exposure. *Alcohol Health Res. World,* 18: 49–52.

Mattson, S.N., Riley, E.P., Jernigan, T.L., and Jones, E.K. (1992) Magnetic resonance imaging of children with fetal alcohol syndrome. *Alcoholism: Clin. Exp. Res.,* 16: 369.

McCool, B.A., Plonk, S.G., Martin, P.R., and Singleton, C.K. (1993) Cloning of human transketolase cDNAs and comparison of the nucleotide sequence of the coding region in Wernicke-Korsakoff and non-Wernicke-Korsakoff individuals. *J. Biol. Chem.,* 268: 1397–1404.

McDowell, J.R. and LeBlanc, H.J. (1984) Computed tomographic findings in Wernicke-Korsakoff syndrome. *Arch. Neurol.,* 41: 453–454.

McEntee, W.J. and Mair, R.G. (1990) The Korsakoff syndrome: a neurochemical perspective. *Trends Neurosci.,* 13: 340–344.

McMullen, P.A., Saint-Cyr, J.A., and Carlen, P.L. (1984) Morphological alterations in rat CA1 hippocampal pyramidal cell dendrites resulting from chronic ethanol consumption and withdrawal. *J. Comp. Neurol.,* 225: 111–118.

Mensing, J.W.A., Hoogland, P.H., and Sloof, J.L. (1984) Computed tomography in the diagnosis of Wernicke's encephalopathy: a radiological-neuropathological correlation. *Ann. Neurol.,* 16: 363–365.

Messert, B., Orrison, W.W., Hawkins, M.J., and Quaglieri, C.E. (1979) Central pontine myelinolysis. Considerations on etiology, diagnosis, and treatment. *Neurology,* 29: 147–160.

Miller, G.M., Baker, H.L., Okazaki, H., and Whisnant, J.P. (1988) Central pontine myelinolysis and its imitators: MR findings. *Radiology,* 168: 795–802.

Miller, M.W. (1986) Effects of alcohol on the generation and migration of cerebral cortical neurons. *Science,* 233: 1308–1311.

Miller, M.W. (1987) Effect of prenatal exposure to alcohol on the distribution and time of origin of corticospinal neurons in the rat. *J. Comp. Neurol.,* 257: 372–382.

Miller, M.W. (1992). Effect of ethanol on cell proliferation and neuronal migration. *Development of the Central Nervous System: Effects of Ethanol and Opiates.* New York, Wiley-Liss, Inc.

Mukherjee, A.B., Svoronos, S., Ghazanfari, A., Martin, P.R., Fisher, A., Roecklein, B., Rodbard, D., Staton, R., Behar, D., Berg, C.J., and Manjunath, R. (1987) Transketolase abnormality in cultured fibroblasts from familial chronic alcoholic men and their offspring. *J. Clin. Invest.,* 79: 1039–1043.

Nordberg, A., Adolfsson, R., Aquilonius, S.M., Marklund, S., Oreland, L., and Wimblad, B. (1980). Brain enzymes and Ach. receptors in senile dementia of Alzheimer type and chronic alcohol abuse. *Aging of the Brain and Dementia.* New York, Raven Press.

Norenberg, M.D. and Papendick, R.E. (1984) Chronicity of hyponatremia as a factor in experimental myelinolysis. *Ann. Neurol.,* 15: 544–547.

Ostertun, B., Dewes, W., Hanisch, E., and Harder, T. (1990) [MR tomography and computer tomography of alcohol-induced brain tissue changes]. *Rofo: Fortschritte Auf Dem Gebiete Der Rontgenstrahlen Und Der Nuklearmedizin,* 152: 87–90.

Parkin, A.J., Blunden, J., Rees, J.E., and Hunkin, N.M. (1991) Wernicke-Korsakoff syndrome of nonalcoholic origin. *Brain Cogn.,* 15: 69–82.

Pentney, R.J. and Miller, M.W. (1992) The effects of ethanol on neuronal morphogenesis. *Development of the Central Nervous System: Effects of Ethanol and Opiates.* New York, Wiley-Liss, Inc.

Pittella, J.E. and de Castro, L.P. (1990) Wernicke's encephalopathy manifested as Korsakoff's syndrome in a patient with promyelocytic leukemia. *South. Med. J.,* 83: 570–573.

Reuler, J.B., Girard, D.E., and Cooney, T.G. (1985) Wernicke's encephalopathy. *N. Engl. J. Med.,* 312: 1035–1039.

Roche, S.W., Lane, R.J., and Wade, J.P. (1988) Thalamic hemorrhages in Wernicke-Korsakoff syndrome demonstrated by computed tomography [letter]. *Ann. Neurol.,* 23: 312.

Rosenthal, A., Jouet, M., and Kenwrick, S. (1992) Aberrant splicing of neural cell adhesion molecule L1 mRNA in a family with X-linked hydrocephalus. *Nature Genet.,* 2: 107–112.

Rosett, H.L. and Weiner, L. (1984) *Alcohol and the Fetus.* New York, Oxford University Press.

Schroth, G., Naegele, T., Klose, U., Mann, K., and Petersen, D. (1988) Reversible brain shrinkage in abstinent alcoholics, measured by MRI. *Neuroradiology,* 30: 385–389.

Shimamura, A.P., Jernigan, T.L., and Squire, L.R. (1988) Korsakoff's syndrome: radiological (CT) findings and neuropsychological correlates. *J. Neurosci.,* 8: 4400–4410.

Siebert, JR,. Astley, S.J., and Clarren, S.K. (1991) Holoprosencephaly in a fetal macaque (Macaca nemestrina) following weekly exposure to ethanol. *Teratology,* 44: 29–36.

Smith, C.J., Perry, E.K., Perry, R.H., Candy, J.M., Johnson, M., Bonham, J.R., Dick, D.J., Fairbairn, A., Blessed, G., and Birdsall, N. (1988) Muscarinic cholinergic receptor subtypes in hippocampus in human cognitive disorders. *J. Neurochem.,* 50: 847–856.

Smith, D.E. and Davies, D.L. (1990) Effect of perinatal administration of ethanol on the C1 pyramidal cell of the hippocampus and the Purkinje cells of the cerebellum: an ultrastructural survey. *J. Neurocytol.,* 19: 708–717.

Spohr, H.L. and Steinhausen, H.C. (1987) Follow-up studies of children with fetal alcohol syndrome. *Neuropediatrics,* 18: 13–17.

Squire, L.R., Amaral, D.G., and Press, G.A. (1990) Magnetic resonance imaging of the hippocampal formation and mammillary nuclei distinguish medial temporal lobe and diencephalic amnesia. *J. Neurosci.,* 10: 3106–3117.

Streissguth, A.P., Barr, H.M., Sampson, P.D., Darby, B.L., and Martin, D.C. (1989) IQ at age 4 in relation to maternal alcohol use and smoking during pregnancy. *Dev. Psychol.,* 25: 3–11.

Streissguth, A.P., Barr, H.M., Sampson, P.D., Parrish-Johnson, J.C., Kirschner, G.L., and Martin, D.C. (1986) Attention, distraction and reaction time at age 7 years and prenatal alcohol exposure. *Neurobehav. Toxicol. Teratol.,* 8: 717–725.

Streissguth, A.P., Landesman-Dwyer, S., Martin, J.C., and Smith, D.W. (1980) Teratogenic effects of alcohol in humans and laboratory animals. *Science,* 209: 353–361.

Streissguth, A.P., Martin, D.C., Barr, H.M., MacGregor, B., Kirchner, G.L., and Darby, B.L. (1984) Intrauterine alcohol and nicotine exposure: attention and reaction time in 4-year-old children. *Dev. Psychol.,* 20: 533–541.

Tavares, M.A., Paula-Barbosa, M.M., and Cadete-Leite, A. (1987) Chronic alcohol consumption reduces the cortical layer volumes and the number of neurons of the rat cerebellar cortex. *Alcoholism: Clin. Exp. Res.,* 11: 315–319.

Thomson, A.D., Ryle, P.R., and Shaw, G.K. (1983) Ethanol, thiamine, and brain damage. *Alcohol Alcoholism,* 18: 27–43.

Torvik, A. (1985) Two types of brain lesions in Wernicke's encephalopathy. *Neuropath. Appl. Neurobiol.,* 11: 179–190.

Torvik, A., Lindboe, C.F., and Rodge, S. (1982) Brain lesions in alcoholics. A neuropathological study with clinical correlation. *J. Neurol. Sci.,* 56: 233–248.

Victor, M., Adams, R.A., and Collins, G.H. (1989) *The Wernicke-Korsakoff Syndrome and Related Disorders Due to Alcoholism and Malnutrition.* Philadephia, F.A. Davis.

Victor, M., Adams, R.D., and Cole, M. (1965) The acquired (non-Wilsonian) type of hepatocerebral degeneration. *Medicine,* 44: 345–395.

Victor, M., Adams, R.D., and Mancall, E.L. (1956) A restricted form of cerebellar cortical degeneration occurring in alcoholic patients. *Arch. Neurol.,* 76: 579–688.

Walker, D.W., Barnes, D.E., Zornetzer, S.F., Hunter, B.E., and Kubanis, P. (1980) Neuronal loss in hippocampus induced by prolonged ethanol consumption in rats. *Science,* 209: 711–713.

Wallis, W.E., Willoughby, E., and Baker, P. (1978) Coma in the Wernicke-Korsakoff syndrome. *Lancet,* 2: 400–401.

Warach, S.J. and Charness, M.E. (1994). Imaging the brain lesions of alcoholics. *Neuroimaging: A Companion to Adams and Victor's Principles of Neurology.* New York, McGraw-Hill.

West, J.R., Hodges, C.A., and Black, A.J. (1981) Prenatal exposure to ethanol alters the organization of hippocampal mossy fibers in rats. *Science,* 211: 957–959.

West, J.R. and Hodges, S.C. (1983) Permanent hippocampal mossy fiber hyperdevelopment following prenatal ethanol exposure. *Neurobehav. Toxicol. Teratol.,* 5: 139–150.

Witt, E.D. and Goldman-Rakic, P.S. (1983) Intermittent thiamine deficiency in the rhesus monkey. I. Progression of neurological signs and neuroanatomical lesions. *Ann. Neurol.,* 13: 376–395.

Wright, D.G., Laureno, R., and Victor, M. (1979) Pontine and extrapontine myelinolysis. *Brain,* 102: 361–385.

Yokote, K., Miyagi, K., Kuzuhara, S., Yamanouchi, H., and Yamada, H. (1991) Wernicke encephalopathy: follow-up study by CT and MR. *J. Comput. Assist. Tomogr.,* 15: 835–858.

Zipursky, R.B., Lim, K.O., and Pfefferbaum, A. (1989) MRI study of brain changes with short-term abstinence from alcohol. *Alcoholism: Clin. Exp. Res.,* 13: 664–666.

Chapter 21

NEUROLOGICAL DAMAGE — STROKE

Matti E. Hillbom

TABLE OF CONTENTS

I. INTRODUCTION

Stroke and alcoholism are major causes of illness and death in Western industrialized countries. Stroke is a very costly disease and the leading cause of neurological disability. It is also a very complex disease having multiple subtypes with different causes and risk factors. Alcohol and stroke became a focus of epidemiological research about 15 years ago, although several papers had previously pointed out a possible relationship. Experimental studies have revealed both beneficial and untoward actions of alcohol with respect to the development of stroke, and the relationship between alcohol and stroke is still intriguing (Ashley, 1982; Altura & Altura, 1984; Wolf, 1986; Hillbom, 1987; Gorelick, 1987, 1989).

The two main mechanisms of stroke are infarction and hemorrhage. Cerebral infarction (CI) is the most common subtype of stroke. A clinical stroke results from cerebral ischemia of insufficient duration, long enough to cause infarction, that is, tissue destruction. Shorter periods of ischemia cause transient ischemic attacks (TIAs). TIAs often precede infarction, particularly when the underlying disease is advanced atherosclerosis of the extracranial arteries. Although atherosclerosis is the major cause of ischemic cerebral infarction among elderly people, it is only one of the many causes of ischemic stroke.

Hemorrhagic stroke is also a multifactorial disease. Subarachnoid hemorrhage (SAH) is mainly caused by a rupture of a saccular arterial aneurysm, but it can also be caused by trauma and a rupture of an arteriovenous malformation. Occasionally, no underlying vascular pathology

0-8493-8389-7/95/$0.00+$.50

is found. If bleeding occurs within cerebral tissue, an intracerebral hemorrhage (ICH) develops. The occurrence of ICH strongly increases with age, as does that of brain infarction. Also ICH can be either traumatic or spontaneous. The most powerful risk factor for ICH is chronic hypertension (high blood pressure). Among elderly people, a degenerative vessel disease, amyloid angiopathy, is often the causative disease underlying spontaneous ICH.

Among the numerous risk factors for stroke, alcohol is a debated one. Hypertension is a key risk factor for both ICH and CI, and habitual heavy drinking of alcohol is clearly linked with elevated blood pressure (Keil et al., 1991). Although long-term heavy alcohol use increases the risk for all types of stroke according to most epidemiological surveys, there is evidence to suggest that light-to-moderate alcohol use might even protect against CI caused by atherosclerotic disease (Friedman & Klatsky, 1993). The following review highlights the most recent observations and suggests new avenues for future research.

II. EPIDEMIOLOGIC STUDIES

Alcohol-related diseases are diseases in which alcohol use is the major risk factor. Liver cirrhosis and Wernicke's encephalopathy are good examples of such diseases. Stroke is not a typical alcohol-related disease because alcohol use has not been shown to be a major risk factor for stroke. However, alcohol has recently been claimed to be an independent risk factor for several subtypes of stroke, although a causal relationship remains to be proved. Heavy drinking of alcohol has long been known to carry an increased risk for death due to hemorrhagic stroke (Klatsky et al., 1992).

A. INTERNATIONAL COMPARISONS

There is no evidence to suggest that the occurrence of stroke among different populations is directly proportional to the per capita alcohol consumption. For example, stroke is not more common among French than Finnish people, although alcohol consumption is. In fact, the incidence of stroke is very high in Finland compared to several other European countries where alcohol consumption has long been higher. Many similar examples can be found outside of Europe. The drinking habits, however, are also different in France and Finland. Heavy binge drinking is common in Finland, but regular daily drinking is common in several other European countries, for example.

As far as I know, controlled comparisons between heavy drinking populations and populations avoiding alcohol drinking have not been carried out with respect to the incidence and prevalence of stroke. Black Americans, for example, have a higher mortality from stroke than white Americans (Gillum, 1988), but the difference has neither been attributed to alcohol consumption nor other life-style factors. The report of Gillum shows that the excess mortality of black Americans is very conspicuous in people of working age, but nonexistent in elderly people with low risk-factor profiles (Manton et al., 1987). This could suggest that life-style factors, such as physical inactivity, heavy alcohol drinking, and cigarette smoking may play a role. Interestingly, papers cited by Gillum indicate that such risk factors as hypertension, diabetes, and smoking, which are often associated with heavy alcohol drinking, are more prevalent in blacks than in whites. However, the incidence of stroke has been found to be higher in blacks after adjustment for age, hypertension, and diabetes, and differences in compliance or medical care related to hypertension and diabetes were not found to explain the excess stroke risk among blacks (Kittner et al., 1990).

Mortality from stroke has decreased remarkably in many countries during recent decades, but this has not been shown to be influenced by any change in alcohol consumption. More powerful factors, such as improvements in the treatment of both hypertension and acute stroke, have been recognized. Comparisons between various populations are hampered by the great variability in the etiology and risk factors of stroke in different countries and also by the different facilities for diagnosis and treatment.

B. DRINKING HABITS

Accurate identification of stroke subtypes and reliable measurement of alcohol consumption have been and still are the main problems in the research aimed at assessing the relationship between alcohol use and stroke. The reader will find valuable discussions of several methodologic problems in the excellent review of previous (1966–1989) epidemiological reports by Camargo (1989). In his paper, the need for further studies addressing the role of different drinking habits was strongly emphasized.

The simple use of the categories, drinkers and nondrinkers, to represent the drinking pattern of the study population is a poor design for epidemiological studies, although it has been used widely. Continuous alcohol intake produces effects different from those of occasional drinking. Adaptive changes will take place and lead to a situation where even high doses of alcohol cause only minor effects. On the other hand, an abrupt cessation of chronic heavy drinking precipitates a withdrawal state that may create conditions provoking stroke. Since tolerance to ethanol develops within a couple of weeks, it has been suggested that periodic heavy drinking might be the most dangerous pattern of drinking, particularly if one considers that heavy alcohol drinking and its withdrawal may be able to trigger stroke. Periodic heavy drinking is not an equally common drinking habit in all countries and among all populations.

To prove causality, a positive dose-response association between alcohol consumption and stroke occurrence has been sought. However, such an association disregards the fact that a similar dose of alcohol may not cause the same effect in all individuals. Habitual daily drinkers with good tolerance do not get the same degree of intoxication as occasional moderate drinkers from a bottle of wine. Particularly, if we consider that alcohol intoxication could trigger a stroke via several different mechanisms, and take into account the tolerance that develops to many physiological effects of alcohol, it could well be difficult to find a linear dose-response relationship between the risk of stroke and alcohol intake.

An ideal epidemiologic design to investigate the problem is the one that has been applied to investigations of whether physical exertion triggers acute myocardial infarction (Curfman, 1993). This is the case-crossover method (Maclure, 1991), which, when applied to alcohol research, means that the patient's consumption during a few hours or a day before the stroke onset will be compared with the consumption of normal control subjects at the same time of the day on the same day of the week. No such studies with sufficiently large populations have been published yet, although the method has been applied in a few studies reviewed in this chapter that used hospital-based controls.

As a matter of fact, it is reasonable to ask whether alcohol consumption is beneficial — or hazardous — for your stroke risk. The majority of epidemiological evidence published thus far suggests that drinkers as a group have an equal risk for total stroke compared with nondrinkers (Camargo, 1989), but these studies have rarely used any other parameters of drinking habits except the classification of drinkers vs. nondrinkers. Although heavy drinking seems to carry a slightly greater risk than light and moderate drinking, the latter seem to carry a slightly lower risk than nondrinking (Goldberg et al., 1994). Many open issues are still to be answered before we are prepared to suggest safe doses and patterns of alcohol drinking for the public.

C. STROKE SUBTYPES

1. Hemorrhagic Stroke

Numerous reports have suggested that alcohol abuse increases the risk for intracranial hemorrhage. The Honolulu Heart Study demonstrated an increased risk even after hypertension had been controlled (Kagan et al., 1980; Donahue et al., 1986). Accordingly, alcohol-induced hypertension cannot be the only factor that increases the risk. The risk for hemorrhagic stroke among drinkers compared to nondrinkers was significant, irrespective of whether the subjects reported a consumption level of less than one drink per day or more, but the role of binge drinking was not investigated. The risk for SAH appeared more significant than that for ICH. The authors also reported that a reduction of alcohol intake resulted in a decrease of the risk, suggesting reversibility, a finding that was simultaneously suggested by another report (Kono et al., 1986). Later on, Klatsky et al. (1989) found in a large cohort study of white and black Americans that the alcohol-induced risk for ICH was greater than that for SAH and confirmed the role of habitual heavy drinking as a risk factor for hemorrhagic stroke, although no causal relationship was demonstrated.

Cohort studies usually include too few subjects with end-points to demonstrate the role of alcohol consumption. A good example is a study from Sweden where alcohol consumption was defined as abuse, carrying legal consequences (Harmsen et al., 1990). That study included too few cases of both SAH and ICH to give any reasonable conclusion of the role of alcohol as a risk factor. In addition, the use of registered alcoholics as selected representatives of heavy drinkers excluded from the study all the other heavy drinkers.

2. Intracerebral Hemorrhage

ICH is divided into spontaneous and traumatic types. Although traumatic intracranial hemorrhage has frequently been shown to associate with alcohol drinking, in this review I concentrate on spontaneous ICH, which accounts for about 10–15% of all cases of spontaneous stroke. Reports dealing with spontaneous ICH usually exclude cases of brain tumor, aneurysm, and vascular malformation as well as cases of trauma. The known risk factors for spontaneous ICH include arterial hypertension, anticoagulant treatment, hepatic disease, alcoholism, and various associated laboratory parameters.

The role of hypertension has been established convincingly (Caplan, 1992), but alcoholism is frequently mentioned as another main risk factor. Hepatic disease seems to be associated with alcoholism (Niizuma et al., 1988). Both severe hepatic disease and alcoholism are known to elevate blood pressure, yet there are data to suggest that alcoholism, independently of hypertension, increases the risk for spontaneous ICH, but the mediating mechanism is unknown (Calandre et al., 1986).

Among young and middle-aged people, lobar rather than basal ganglia and thalamic hematomas were found to be associated with alcoholism, but whether minor undetected head traumas contributed to this finding remains unclear, since cases presenting with both alcohol withdrawal and seizures were included in the study (Monforte et al., 1990). However, more than one-third of the putaminal hematomas in men were found in alcoholics in another study (Schutz et al., 1990), although such a location is generally typical of elderly people with chronic arterial hypertension rather than alcoholics. In the study by Calandre et al. (1986) no clear association between alcohol and location of hematoma was found. Taken together, the previously mentioned observations suggest that habitual heavy alcohol intake may increase the risk of ICH both because of and independently of vascular degeneration caused by hypertension. Even the outcome of spontaneous ICH is aggravated by chronic alcoholism, according to studies excluding subjects who die before admission to a hospital (Schutz et al., 1990). These new reports are in line with the earlier observations and confirm that habitual heavy drinking of alcohol is a risk factor for spontaneous ICH, but they do not discuss the role of binge drinking.

In a case-control study, Juvela et al. (1995) were able to demonstrate that very recent heavy alcohol intake (i.e., intake within the 24 hours preceding the onset of illness) is a significant independent risk factor for spontaneous ICH after adjustment for age, hypertension, body mass index, smoking status, and alcohol consumption during the week before the stroke (see Table 21.1). Alcohol consumption within one week (after an exclusion of the most recent 24-hour consumption) before the stroke onset also increased the risk of hemorrhage in men, but more previous heavy alcohol consumption did not increase the risk. These observations suggest that recent moderate and heavy drinking, including binge drinking, may provoke ICH, whereas recent light drinking may associate with a decreased risk. They support the earlier

TABLE 21.1
The risk of intracerebral hemorrhage (ICH), subarachnoid hemorrhage (SAH), and cerebral infarction (CI) in men and women by recent alcohol consumption after adjustment for age, hypertension, and smoking status.

Recent alcohol intake in grams per preceding 24 hours	ICH	SAH	CI
Men			
0	1.0	1.0	1.0
1–40	0.2[b]	0.3[a]	0.9
41–120	5.1[c]	2.5[a]	6.0[b]
>120	11.1[c]	4.5[b]	7.1[a]
Women			
0	1.0	1.0	1.0
1–40	0.7	0.4[a]	1.4
>40	5.3[b]	6.4[c]	7.8[a]

[a] $p < 0.05$
[b] $p < 0.01$
[c] $p < 0.001$

Note: The figures represent odds ratios as estimates of relative risks after adjustment of possible confounding variables (other risk factors) calculated by multiple logistic regression separately in men and women with ICH and SAH (16–60 years) and CI (16–40 years).

For details see: Juvela et al. 1995 (ICH), Juvela et al. 1993 (SAH), and Hillbom et al. 1995 (CI).

observations that alcohol consumption has a relatively short-term association with the risk for hemorrhagic stroke, and that the risk is reversible (Kono et al., 1986).

Several issues remain to be answered. Among them is the question of whether alcohol withdrawal seizures and delirium carry an increased risk for spontaneous ICH as does intoxication. This could be investigated by using animal models, since rats easily get intracerebral hemorrhage during severe tonic-clonic seizures. The hitherto reported studies have not addressed the question of whether the regular use of alcohol is less detrimental than long-term periodic heavy drinking. Although alcoholism impairs the outcome for hospitalized patients, we lack data to show whether acute mortality (before hospital admission) is influenced by alcohol drinking. The effects of age, race, type of ingested beverage, and the mixed use of drugs and alcohol also require further exploration.

In conclusion, the most recent clinical investigations using head CT scans to verify the type of acute stroke indicate a relationship between the heavy drinking of alcohol and spontaneous ICH. The risk for ICH is increased by recent drinking, including occasional drinking, but not by previous problem drinking, suggesting that alcohol per se is at least partly responsible for it. Alcohol-induced hypertension and hepatic disease are additional factors that contribute to the risk. The mechanisms by which alcohol per se may promote the occurrence of cerebral hemorrhage are largely unknown, but are discussed later in this chapter.

3. Subarachnoid Hemorrhage

Spontaneous nontraumatic SAH results in most (80%) cases from a rupture of a saccular arterial aneurysm. Minor entities include those in which the bleeding results from a ruptured arterio-venous malformation (5%) and those that demonstrate no vessel abnormality at autopsy or by angiography (15%). Ageing, female gender, cigarette smoking, hypertension, and the use of oral contraceptives have been suggested to be risk factors for SAH (Longstreth et al., 1985). Interestingly, SAH has recently also been reported to be a complication of recreational drug use (Levine et al., 1991). Aneurysmal SAH is a very serious disease carrying a 40–50% acute mortality. It constitutes about 5–10% of all spontaneous strokes.

Most previous studies have investigated hemorrhagic stroke by combining SAH with ICH, although these two conditions have different pathophysiologies and risk factors. Whereas hypertension is the dominant risk factor for ICH, cigarette smoking is the dominant one for SAH (Bell & Symon, 1979; Longstreth et al., 1992; Juvela et al., 1993). Two recent case-control studies have confirmed the role of alcohol as an independent risk factor for SAH and have emphasized the role of binge drinking.

Longstreth et al. (1992) defined a binge to consist of more than 5 drinks (or units) of alcohol per day and demonstrated that if a subject had at least one such binge within a week, the relative risk for SAH was 4.3 (95% confidence interval 1.5–12.3) compared to those who were nondrinkers and those who had 1 to 11 binges within a year. The risk was not significantly increased until the number of binges amounted to more than 4 per month (Longstreth et al., 1992). The control subjects were drawn from the general population. This might have resulted in an underrepresentation of heavy drinkers and smokers among the controls.

The other study used hospital-based control subjects matched by age, sex, and acuteness of the disease onset, as well as the day of onset of the stroke (i.e., was it a holiday or a working day), so that the confounding factors influencing alcohol consumption could be eliminated (Juvela et al., 1993). After adjustment for age, hypertension, and smoking status, the results (see Table 21.1) were strikingly similar to those of Longstreth et al. (1992). Previous problem drinking did not increase the risk, and the bleedings occurred most frequently during the hangover period. Assuming that recent heavy drinking and current smoking are etiological risk factors, the authors calculated that approximately 12 and 48% of the hemorrhages were attributable to recent drinking and current smoking, respectively.

The issues remaining to be answered include whether periodic heavy drinking with repeated intoxications and withdrawals is a more deleterious drinking pattern than habitual heavy drinking, whether light and moderate habitual drinking influences the risk at all, whether the subjects are mainly stricken during intoxication or withdrawal, and whether there are other activities that associated with drinking precipitate SAH (i.e., trauma).

It is well-known that traumatic SAH with brain contusion is strongly associated with alcohol abuse. The rare occasion of traumatic SAH without brain contusion due to a rupture of a vertebral artery against the bony processes in the neck is almost exclusively seen in alcoholics (Harland et al., 1983; Simonsen, 1984). Intoxicated subjects are very prone not to control their neck movements when falling, fighting, etc. Although mortality from this condition approaches 100%, its rarity makes it less important as a public health problem. The condition in which the vertebral arteries do not rupture completely, but are damaged by intimal tears, is much more common. This condition, which is called vertebral arterial dissection, causes brain ischemia and will be discussed together with the risk of ischemic stroke caused by alcohol.

In summary, both traumatic and spontaneous occurrences of SAH associate significantly with recent heavy, and probably even moderate, drinking of alcohol. Binge drinking or drinking for intoxication (more than 5 units per day) seems to be a precipitating factor, provided that the drinking episodes amount to one per week. Drinking itself or other associated activities may trigger SAH. By contrast, former problem drinking has not been found to increase the risk, suggesting a reversibility of the effect after prolonged abstinence. The finding that even a moderate intake of alcohol might provoke SAH (Stampfer et al., 1988) remains an open issue, since drinking habits may vary over time, and cohort studies investigating the usual level of alcohol intake cannot control for risky behavior very close to the onset of the disease. It is reasonable to think that this finding is due to undetected binge drinking. There is no conclusive evidence yet to indicate that habitual light drinking influences the risk for SAH.

4. Ischemic Stroke

ICH and SAH are relatively simple entities etiologically compared to ischemic stroke, which has very many known etiological causes and precipitating factors making it very difficult in this complex disease group to define the role of alcohol drinking. Since the invention of new brain imaging techniques, it has become a simple task to confirm the presence or absence of local ischemic lesions in the brain and to separate them from other cerebral lesions. Therefore, the most recent reports discussing the risk factors for stroke are much more reliable in excluding hemorrhage, for example, than many of the earlier ones. Yet the effort to reach an accurate etiological diagnosis of ischemic stroke continues to be a very difficult task. Strokes caused by emboli of cardiogenic origin, extracranial large-vessel disease, and intracranial small-vessel disease, including lacunar infarction, cervicocerebral arterial dissection, arteritis, vasospasm, paradoxal embolism, infectious, hematologic and metabolic disorders, etc., require very great efforts by the clinician to clarify the true etiology. Consequently, the real underlying cause remains unrevealed in 15 to 25% of cases. In 4 to 66% of young stroke patients in reported series, no cause was found (van den Berg & Limburg, 1993). The reader should keep this in mind when considering the role of alcohol in ischemic stroke.

Ischemic strokes account for about 80% of all spontaneous strokes. Some previous reports cited in other reviews (Gorelick, 1987, 1989; Camargo, 1989) suggest that excessive alcohol consumption increases the risk of ischemic stroke, whereas light and moderate consumption may have a protective effect similar to that frequently proposed for coronary artery disease. Analyses of case histories indicate that alcohol may precipitate ischemic stroke via different mechanisms (Hillbom & Kaste, 1990). Most of the papers cited below discuss CI verified by imaging techniques, but some include even TIAs.

Cohort studies usually do not differentiate between subgroups of ischemic stroke and cannot address recent drinking variables. A female nurse study showed that average alcohol intakes of 5–14 g and >14 g of ethanol per day carried relative risks of 0.3 (95% confidence intervals 0.1–0.7) and 0.5 (0.2–1.1), respectively, for ischemic stroke (Stampfer et al., 1988). Another study categorizing participants into those who were daily drinkers and those who were not, and excluding verified SAHs from the stroke category, revealed a relative risk of 0.7 caused by alcohol consumption in women, but this was not significant (Lindenström et al., 1993). In another study (Klatsky et al., 1989), in which even the type of beverage, but not eating habits, was taken into account, participants were classified into abstainers, former drinkers, and current drinkers of alcohol with varying daily intakes, and alcohol use was shown to be associated with lower hospitalization rates for occlusive cerebrovascular disease. An inverse relationship was present in both sexes, whites and blacks, and for extracranial and intracerebral occlusive lesions. The end-point admission of the subjects in this study was not always as first-ever stroke victims because many of them were hospitalized for diagnostic purposes and probably for surgical treatment of an established extracranial atherosclerotic disease. The results suggest, however, that there is an inverse relationship between habitual daily alcohol consumption and the risk for atherosclerotic cerebrovascular disease. The risk for former drinkers was at the same level as that for lifelong abstainers, which is higher than that for current drinkers. This study did not address the question of whether binge drinking is able to trigger acute ischemic episodes among subjects with different background diseases predisposing to ischemic stroke. Finally, a recent study from Finland demonstrates, that compared with nondrinkers, light-to-moderate consumption of alcohol is negatively associated with the risk of CI in men if the pattern of drinking is regular, but not if it is irregular (Palomäki & Kaste, 1993a). The beneficial effect of regular drinking was greatest when the reported average weekly alcohol intake ranged between 50 and 150 grams. In the same study, however, heavy drinking was found to be an independent risk factor for CI. According to this study, alcohol consumption appears to exhibit a nonlinear relationship to ischemic stroke in young and middle-aged men.

5. Cerebral Infarction

In the studies by Gorelick et al. (1987, 1989) young adults and cases of nonatherosclerotic ischemic stroke were excluded. Although the mean weekly alcohol consumption was greater among the stroke index cases than the controls, and a dose-response relationship between alcohol intake and CI was observed, matched multiple logistic regression analysis showed that weekly alcohol consumption was not an independent risk factor for ischemic stroke when hypertension and smoking were taken into account (Gorelick, 1989). Acute alcohol ingestion was not found to be an independent risk factor for CI in middle-aged and elderly patients, who frequently have advanced carotid and cerebral arterial atherosclerosis (Gorelick, 1987). The finding suggests that binge drinking does not trigger ischemic episodes in subjects with advanced atherosclerotic lesions, but it needs to be confirmed in a larger population, preferably using the case-crossover method (Maclure, 1991). We have also been unable to prove that binge drinking significantly precipitates CI in elderly subjects (Hillbom & Kaste, 1990).

The methods of the previous case-control studies have been criticized in some studies, which did not find heavy drinking to be a risk factor for CI (Henrich & Horwitz, 1989; Ben-Shlomo et al., 1992; Shinton et al., 1993). One of these reports ends in a recommendation that future studies must attempt to overcome the various biases associated with control selection and employ a sample size large enough to detect even a modestly increased risk. The study behind this report had insufficient power to significantly detect a modestly increased risk associated with alcohol consumption and suffered from an underrepresentation of young male subjects with CI (Ben-Shlomo et al., 1992). The authors also stated that the "true risk" may have been underestimated by the use of hospital-based control groups and overestimated by the use of community-based controls. Another work demonstrated that proper hospital-based controls should not include patients admitted into the hospital for elective surgical treatments because such patients may reduce their usual alcohol intake shortly preceding the admission, which certainly is true (Henrich & Horwitz, 1989). The third study used community-based controls, but still did not find recent alcohol intake to increase the risk for CI (Shinton et al., 1993). In this study, the prevalence of heavy drinkers among the controls (12%) suggests that the controls represented well the general population. However, the stroke subjects were rather old (median age 66 years) and few in number (81 patients with CI).

Several investigations have suggested that habitual light-to-moderate daily alcohol consumption may protect against ischemic cerebrovascular disease (Bogousslavsky et al., 1990; Palomäki & Kaste, 1993a), strokes other than primary SAH (Rodgers et al., 1993), and all types of stroke (Jamrozik et al., 1994). The observations are in line with the evidence for the inverse association between alcohol consumption and coronary heart disease attributed to alcohol-induced increases in HDL cholesterols (Gaziano et al., 1993). Among elderly people, atherosclerosis and coronary heart disease are the main underlying conditions that predispose to ischemic stroke. Cervical arterial stenosis and recent myocardial infarction are well-known risk factors for stroke because they may release emboli into the cerebral circulation. A significant part (36–45%) of the elderly subjects with ischemic stroke have a carotid artery disease (Zhu & Norris, 1990), which is detected in clinical investigations of stroke patients more often than intracerebral arterial or aortic atherosclerosis. Age, male gender, hypertension, cigarette smoking, diabetes, hyperlipidemias, and homocystinuria are known to be risk factors for cervicocerebral atherosclerosis (Palomäki et al., 1993b; Fryer et al., 1993). Smoking seems to be the most powerful predictor of severe extracranial carotid artery atherosclerosis (Whisnant et al., 1990).

In the study by Bogousslavsky et al. (1990), regular light-to-moderate alcohol intake showed an inverse linear relationship to the degree of internal carotid artery stenosis in subjects >50 years of age with first-ever CI, that is, the tighter the stenosis, the less was the usual weekly alcohol intake. In this study, the weekly alcohol intake of the subjects varied between 0 and 250 g of ethanol. No conclusions could be drawn concerning heavier drinkers because there were too few of them. Some medico-legal autopsy series have suggested that coronary and aortic atherosclerosis is not less advanced in chronic alcoholics than in the general population (Viel et al., 1966; Rissanen et al., 1974), but the confounding effects of smoking and periodic drinking were not taken into account. However, many autopsy studies have demonstrated a remarkable absence of atherosclerosis in the arteries of regular heavy drinkers who have liver cirrhosis (Hall et al., 1953; Creed et al., 1955; Hirst et al., 1965).

Interestingly, in a British study, lifelong abstention from alcohol was found to be associated with an increased risk of stroke, presumably ischemic stroke, but brain imaging techniques were not used routinely to verify the type of stroke (Rodgers et al., 1993). However, current male heavy drinkers had an increased risk of stroke, whereas those who had formerly been regular heavy drinkers, but who were currently nondrinkers, did not have an increased risk. If moderate alcohol intake does not protect against stroke, an alternative explanation for the finding will be that lifelong abstainers are at increased risk because of other reasons, such as diet, personality, etc. In this context, it should be mentioned that alcohol consumption has been suggested to be especially protective against ischemic heart disease in a group of subjects who are genetically prone to the disease (Hein et al., 1993). However, this finding needs to be confirmed for possible selection bias.

Finally, consumption of 1 to 20 g of ethanol per day during the preceding week of stroke onset was associated with a significant reduction in the risk of all strokes, all ischemic strokes, and primary intracerebral hemorrhage in an Australian study where even dietary factors were taken into account (Jamrozik et al., 1994). Results suggesting that the lower the alcohol intake, the better the protective effect against stroke are hard to understand on the basis of the biological actions of alcohol. Such observations rather suggest the presence of some bias due to a healthy cohort effect, which means an unintended selection of subjects with very good health in the category of those who use very small amounts of alcohol. Interestingly, in our case-control studies we repeatedly found a lower relative risk for all types of stroke among those who had recently consumed very small amounts of alcohol compared to those who were nondrinkers or were drinking moderate and heavy doses (Juvela et al., 1993, 1995; Hillbom et al., 1995). We believe that such a finding is due to selection bias and does not represent a true protective action of ethanol or some other component of the beverages. Other investigators have found an association between nondrinking and a history of overweight and inactivity and suggest that this could explain the apparent protective effect of lighter alcohol consumption on the risks of both stroke and coronary heart disease (Shinton et al., 1993).

The role of recent drinking as a triggering factor for CI has not been established in case-control studies of elderly people, who mainly have an atherosclerotic predisposition to stroke (Gorelick et al., 1987; Hillbom & Kaste, 1990; Shinton et al., 1993). One reason for the failure not mentioned in the reports, which cover a rather small number of subjects, might be that occasional heavy drinking is relatively uncommon among elderly people, who usually avoid drinking for intoxication. The effect will not be evident in a study, unless it includes enough subjects who have been drinking for intoxication just prior to the onset of the illness. For the same reason, it has been difficult to prove that women have an increased risk. Assuming that other factors predisposing to stroke in addition to severe intoxication are needed, one would expect that the susceptible subjects would already be stricken at an early age. In addition, since the drinking habits of young adults are more exaggerated, such an effect will be more easily detected by studying young adults.

In fact, some observations indicate that very recent drinking, particularly the drinking of large amounts of alcohol, may trigger the onset of brain infarction in young adults (Hillbom et al., 1995). By comparing the recent drinking of alcohol by 75 consecutive first-ever CI subjects aged 16 to 40 years to that of 133 carefully matched hospital-based control subjects, we found alcohol intake exceeding 40 g of ethanol within the 24 hours to precede the onset of stroke significantly more often among both men and women than it preceded the onset of other acute illnesses, believed to be neither positively nor negatively correlated with alcohol consumption (see Table 21.1). Alcohol intoxication did not seem to precipitate migrainous infarction, but in some cases it was associated with the onset of trauma leading to cervicocerebral arterial dissection.

The largest published case-control study of ischemic stroke in young adults included 135 subjects with TIA and 173 subjects with CI aged 15–44 years as well as 308 hospital-based and 308 community-based controls (Marini et al., 1993). Patients were excluded from the study if they had or once had chronic alcoholism, acute psychosis, cancer, pregnancy, conventional embolic heart diseases, or an ischemic event that had occurred during a migrainous attack, cardiac or carotid surgery, general anesthesia, or carotid angiography after delivery or after head trauma. Accordingly, some obvious causes of ischemic stroke were excluded from the series along with chronic alcoholics. On the basis of their alcohol consumption, the subjects were categorized into two groups: (1) those who had consumed at least 100 g of ethanol daily over the past two months and (2) those who had consumed less or were nonconsumers. About 13% of the subjects belonged to the heavy drinking group, whereas 32% reported no consumption at all. In univariate analyses, habitual heavy alcohol intake (>100 g daily) increased the risk for cerebral ischemic events (CI and TIA together) significantly when the heavy drinkers were compared to both hospital and population controls, the odds ratios being 4.4 (95% confidence intervals 2.0–9.4) and 6.4 (2.6–15.7), respectively. Even moderate alcohol consumption tended to increase the risk slightly, though insignificantly. The significance of the findings remained when the data were processed by multivariate statistics, suggesting that habitual heavy drinking of alcohol may operate through a mechanism that is independent of smoking, hypertension, blood lipids, and all the other parameters used in the model. In a separate report, the authors stated that acute alcohol intoxication occurred as a precipitating factor in three male patients, one of whom had left atrial enlargement and the other two atherothrombotic disease (Carolei et al., 1993).

In conclusion, the recent literature emphasizes that moderate and light regular drinking of alcohol may protect against ischemic stroke. If we are going to explain this by some known metabolic effect of alcohol itself, we should keep in mind that the greater the dose of alcohol, the stronger its effect. Light regular drinking of alcohol exerts effects that are hardly long-lasting and marked enough on HDL cholesterols and related parameters to provide true protection. Such effects usually show a linear dose-response relationship to alcohol consumption (Flegal & Cauley, 1985). However, if the other effects of heavy drinking precipitate stroke, the protecting effect of heavy drinking via HDL cholesterol may be cancelled out. In fact, heavy drinking seems itself to be a risk factor for ischemic stroke. Some new studies using laboratory markers of alcohol drinking to confirm self-reported alcohol use support the former observations of alcohol being a risk factor (Gill et al., 1986, 1988, 1991). Former heavy drinkers do not run an increased risk for ischemic stroke, whereas current heavy drinkers do. It remains to be shown whether recent heavy drinking could precipitate stroke, even among infrequent drinkers, and which underlying conditions, if any, predispose to alcohol-provoked stroke.

III. EXPERIMENTAL STUDIES, AUTOPSY STUDIES, AND CASE REPORTS

Just as cohort studies cannot demonstrate the role of drinking habits, which vary over time, so the more sensitive case-control studies are unable to detect causal relationships. It might well be that the associations found in case-control studies are due to chance and do not represent real effects of alcohol. A risk factor shown to be independent of the other risk factors does not guarantee causality. Therefore, studies of another type are needed to support and confirm what has been found by epidemiologic investigations.

Surprisingly few experimental and autopsy studies address the problem of alcohol intake in relation to stroke. The roles of intoxication, alcohol withdrawal, binge drinking, and regular drinking could be easily tested by animal models. However, stroke in experimental animals does not always occur in similar conditions as in humans. One study explored the effect of alcohol withdrawal on the extent of ischemic stroke in a gerbil model. The study showed that experimentally induced brain infarctions were larger in volume and led more often to a fatal outcome if the onset occurred while the animal was in florid alcohol withdrawal state

(Mandybur & Mendenhall, 1983). This is in line with clinical observations suggesting that higher levels of alcohol consumption are associated with increased stroke mortality (Semenciew et al., 1988), although we do not know whether the increased mortality of heavy drinkers is due to hemorrhagic or to ischemic stroke (Klatsky et al., 1992). The model should be further applied to test whether the outcome will be better if the animals are habituated to small daily doses of alcohol or if they are kept intoxicated for some time during and after the onset of stroke. The actions of alcohol have also been tested in an animal model of SAH in order to prove that alcohol may ameliorate the vasospastic reaction frequently following the bleeding (Barry & Scott, 1979).

One autopsy study particularly designed to prove an association between stroke and alcoholism has been reported. The major observation of the study suggests that heavy drinkers are likely to be stricken by ischemic stroke at an earlier age than other drinkers and nondrinkers (Walbran et al., 1981). This finding supports the concept that heavy drinking is an important risk factor for ischemic stroke in young adults. Subjects who engage in heavy drinking as young adults may damage their vascular systems by hemodynamic stress and the other effects of alcohol caused by repeated severe intoxications. Those who are susceptible to be stricken will have a stroke at an early age. Another finding of the autopsy study was that heavy drinkers more often have multiple infarctions. Whether the infarctions were silent or not could not be demonstrated, but the observation warrants further study to discover whether heavy drinkers are more prone to get repeated infarctions than are other drinkers and nondrinkers. As a matter of fact, one recent investigation has already produced evidence to suggest that alcohol abuse may be a determinant of ischemic stroke recurrence (Sacco et al., 1994).

Mechanisms that mediate the onset of stroke during alcoholic intoxication or shortly thereafter are difficult to demonstrate without thorough analyses of individual cases of alcohol-related strokes. Some published case reports describe subjects stricken during severe intoxication or shortly thereafter and emphasize that despite thorough etiological investigations, no other precipitating factor apart from the excessive consumption of alcohol was found. Wilkins and Kendall (1985) reported on two young men who, after drinking heavily during the previous night, woke up with ischemic stroke the next morning, and despite a thorough cardiac examination and angiography of the cervicocerebral arteries, showed no evident cause of the stroke. Likewise, Weisberg (1988) found no autopsy evidence of aneurysm, vascular malformation, neoplasm, or amyloid angiopathy in half of his alcoholic subjects who had succumbed to spontaneous intracerebral hemorrhage. In the absence of a mediating mechanism, however, it is hard to believe that alcohol was responsible for the occurrence of these strokes.

On the other hand, several other case reports describe how some known causes of stroke have been associated with the onset of stroke during heavy drinking. Alcoholism is a known cause of dilative cardiomyopathy (Ahmed et al., 1980), and such cardiomyopathies are well-known potential sources of cardiogenic embolism. Cases of cerebral embolism due to alcoholic cardiomyopathy have been described (Gonzalez et al., 1988). Alcoholic cardiomyopathy may well be one significant factor that increases the risk of ischemic stroke in heavy drinkers, since it is certainly a disease that precipitates stroke. However, it is not likely to be responsible for the association of binge drinking and stroke among very young adults.

Other cardiac sources of embolism have also been related to alcoholic intoxication. A 37-year-old man sustained two ischemic strokes during acute alcoholic intoxication, and the autopsy indicated no cause for the episodes. A careful inspection of the case history, however, revealed that the man had suffered from an alcohol-induced atrial flutter (Gras et al., 1992). Accordingly, paroxysmal cardiac arrhythmias can be added to the list of alcohol effects that are able to precipitate ischemic stroke. We know of a 32-year-old man who had a congenital cardiac conduction block and suffered two ischemic strokes five years apart, both of which occurred when the subject was binge drinking. We are also aware of a heavy drinker who got paradoxical embolism from deep vein thrombosis via an atrial septal defect while he was severely intoxicated (Hillbom & Kaste, 1990). Finally, we can add the history of a heavily drinking man, who was treated in our hospital because of an acute myocardial infarction and did not receive anticoagulants. The day when the man was permitted to go home, exactly one month after the onset of the illness, he purchased a bottle of spirits and drank it the same night. The next morning he was stricken by an ischemic stroke due to cardiac embolism propagated either by a rapid heart rate or an unnoticed cardiac arrhythmia provoked by the alcohol and the recent myocardial infarction together.

Rebound thrombocytosis, which frequently accompanies withdrawal from prolonged heavy drinking of alcohol, was speculated by Neiman (1988) to have provoked a TIA in a 48-year-old male alcoholic. Although cerebral infarctions are not frequently encountered in detoxification units, about one-fourth of the clients do develop rebound thrombocytosis after they stop drinking. Surprisingly few alcoholics with rebound thrombocytosis at the onset of an ischemic stroke are seen in neurological units. This might mean either that rebound thrombocytosis produced by alcohol withdrawal is not a potential factor to provoke ischemic stroke or that alcoholics have a mechanism that compensates for the effect of thrombocythemia.

Trauma is certainly a mechanism linking alcoholic intoxication to stroke. Traumatic SAH almost always occurs in subjects who are heavily drunk. Traumatic ICH is also a well-known alcohol-related disease, which frequently complicates the falls of intoxicated heavy drinkers. Cervicocerebral arterial dissection can be either spontaneous or traumatic, leading to cerebral ischemia via emboli. Although a common cause of CI in the young (Hilton-Jones & Warlow, 1985), it easily remains undetected, particularly when located in the vertebral arteries, which are seldom investigated by angiography. The role of alcohol drinking in carotid and vertebral dissection is unclear and warrants further exploration.

A case history presented a hypertensive 41-year-old man with carotid dissection who did not have a trauma, but who had a definite history of alcohol abuse (Hess et al., 1990). On the other hand, we have seen several cases of traumatic cervical arterial dissections resulting from violence due to alcoholic intoxication (Hillbom & Kaste, 1990). Sometimes the dissection does not lead directly to CI, which occurs a few days later when emboli from the lesion are released into circulation. An example of such a case was a 36-year-old man who, while intoxicated, wrestled with another intoxicated man. Finally, when he lost the game, the other man compressed him around the neck too forcefully. A week later, after having felt neck pain for several days, he developed a brain infarction due to emboli propagated from a traumatic carotid arterial dissection. Accordingly, it is wise to inquire about trauma and violence of all kinds as well as alcohol drinking in the cases of young subjects with first-ever cervical dissections to determine whether alcoholic intoxication is a significant risk factor for this type of stroke, but we feel that it probably is not. However, cases of traumatic dissection occurring because of violence prompted by alcoholic intoxication are not infrequent. It has also been suggested that extracranial vessel compression due to unusual posturing during alcoholic stupor should be considered (Prendes, 1979), and we have seen one such case.

IV. ACTIONS OF ALCOHOL AND MECHANISMS OF STROKE

1. Primary Intracerebral Hemorrhage

Drinking may promote ICH and SAH partly via the same mechanisms. The traumatic varieties are certainly precipitated by the association between accidents and alcoholic intoxication (see Table 21.2). Another common link may be ethanol-induced hypertension. Alcohol consumption elevates blood pressure in a dose-dependent way and, after the cessation of drinking, blood pressure usually normalizes (Puddey et al., 1985, 1986). The effect of alcohol on blood pressure is independent of sodium, potassium, body mass index, and smoking; seems to be more prominent among periodic than regular heavy drinkers (Marmot et al., 1994); and may be prevented by magnesium (Hsieh, 1992). Heavy drinkers have higher blood pressure

TABLE 21.2
Factors Which Possibly Link Alcohol Consumption to Occurrence of Stroke

Hemorrhagic Stroke	
Predisposing factor	Enhanced formation of hypertensive arterial lesions
Precipitating factors	Head trauma
	Acute increase in the systolic blood pressure
	Acute increase in the cerebral blood flow
	Cerebral arterial vasoconstriction and vasodilation
Contributing factors	Liver disease
	Impairment of the primary hemostasis
	Deficiency of the clotting factors
	Increase of the fibrinolytic activity
Ischemic Stroke	
Predisposing factors	Ehanced formation of hypertensive arterial lesions
	Alcoholic cardiomyopathy
Precipitating factors	Neck trauma
	Cardiac arrhythmias
	Alcohol-induced postprandial metabolic effects
Contributing factors	The sleep apnea syndrome
	Hyperhomocystinemia
	Rebound thrombocytosis
	Decrease of the fibrinolytic activity
	Activation of the clotting cascade

regardless of whether they have consumed alcohol over the previous 24 hours or not, which implies a sustained effect of alcohol on blood pressure (Marmot et al., 1994). We do not know the precise mechanism of alcohol-induced hypertension, but it may involve a direct effect of alcohol on the vascular smooth muscle, possibly mediated by calcium (Hsieh et al., 1992).

Heavy drinking can be assumed to promote hypertension-induced pathological lesions in the small cerebral arteries, such as lipohyalinosis, microaneurysms, and fibrinoid necrosis. The untoward effect may be eliminated if prolonged abstention ensues, provided that the subject does not have hypertension for reasons other than alcohol consumption. This may contribute to the observation that former heavy drinking does not carry an increased risk for hemorrhagic stroke.

Interestingly, intoxicating doses of alcohol have been reported to induce contractions of the cerebral arteries. These occur independent of endothelial cells, require calcium, and are potentiated by magnesium deficiency (Zhang et al., 1993). Alcohol consumption may deplete intracellular free magnesium in the cerebral vascular smooth muscle cells (Altura et al., 1993). If such contractions were produced by alcohol in the human brain, they could explain the precipitation of both ICH and SAH during acute intoxication. In addition to cerebral ischemia, a severe vasospasm could cause an arterial rupture or, after the vasoconstriction subsides, an acute increase in blood flow could rupture an artery. Interestingly, cocaine may also precipitate vasoconstriction, and it is known to precipitate both hemorrhagic and ischemic strokes (Levine et al., 1991; Sloan et al., 1991). Contractions of cerebral arteries can even be caused by endothelins, a family of three peptides with extremely potent and characteristically sustained vasoconstrictor and vasopressor actions (Greenberg et al., 1992), but there are no data available to indicate alcohol-induced activation of endothelin release in cerebral arteries.

Observations of regional cerebral food flow measurements have shown that alcohol has influence (Schwartz et al., 1993). A transient reduction in cerebral blood flow has been described during alcohol withdrawal (Berglund & Risberg, 1981), and moderate doses of alcohol seem to provoke cerebral vasodilation (Mathew & Wilson, 1986). The autoregulation of cerebral blood flow is usually relatively resistant to external agents, but may be more vulnerable in subjects having chronic hypertension. Accordingly, hypertensives could be more prone to strokes related to alcohol consumption, but this has not yet been proved. By contrast, it has been shown that regular alcohol intake causes a prolonged lifetime in spontaneously hypertensive rats (Schlicht et al., 1992).

Effects of alcohol on fibrinolysis, blood clotting, and primary hemostasis may contribute to the risk of bleeding and thrombosis. It is a common clinical experience that disturbances in hemostasis, blood clotting, or fibrinolysis must be grave before a major bleeding, such as a hemorrhagic stroke, can result from them (Olson, 1993). There are no data available to indicate that the hemorrhagic stroke of heavy drinkers is associated with impaired hemostasis, deficient blood clotting, or enhanced fibrinolytic activity.

Heavy drinking has been reported to be associated with reduced (Lee et al., 1980), normal (Wallerstedt et al., 1977), or elevated (Meade et al., 1979) fibrinolytic activity. In heavy drinkers, fibrinolytic activity may vary according to whether it is measured during drinking or afterwards. In alcoholic liver disease, fibrinolysis may be enhanced (Cowan, 1980).

Recent studies indicate that there is a strong correlation of fibrinolytic variables with alcohol intake. Heavy drinking seems to have a deleterious (Iso et al., 1993), but regular moderate alcohol consumption a beneficial effect (Ridker et al., 1994). Acute drinking of alcohol produces a biphasic effect. During the period that alcohol is present in the body, fibrinolytic activity is significantly impaired (Hillbom et al., 1983; Hendriks et al., 1994), but shortly after the elimination of alcohol, a rebound effect, that is, increased activity, is apparent (Hendriks et al., 1994). *In vitro,* alcohol has been shown to augment plasminogen activator secretion by endothelial cells (Lang, 1983), a finding that contrasts with the decreased fibrinolytic activity found *in vivo.*

Coagulation factors are usually normal in alcoholics without significant liver disease (Wallerstedt et al., 1977). In the presence of liver cirrhosis and hepatitis, however, they may be deficient (Ragni et al., 1982). Acute alcohol intake transiently activates the coagulation cascade (Hillbom et al., 1983).

Occasionally, platelet count is severely decreased and platelet function is impaired in chronic alcoholics (Ballard, 1989). Prolonged heavy drinking of alcohol has been found to decrease platelet aggregability and to prolong bleeding time (Mikhailidis et al., 1986). The prolongation of bleeding time seems to be associated with alcoholic liver disease (Hillbom & Neiman, 1988), but it has also been considered to result from the degree of recent alcohol consumption (Mikhailidis et al., 1986). These phenomena may contribute to the bleeding diathesis associated with alcoholism. Two extensive epidemiological studies indicated that platelet aggregability is decreased by regular alcohol drinking, but the amount of alcohol and the duration of drinking needed for the effect were not given (Meade et al., 1985; Renaud et al., 1992). Reports of acute alcohol intake on platelet function suggest either no significant effect or impairment (Mikhailidis et al., 1987, 1990; Rubin & Rand, 1994).

The alcohol-induced increased risk for spontaneous ICH may include predisposing, precipitating, and contributing factors. A predisposing condition may include damage to the penetrating small arteries of the brain. Prolonged hypertension caused by heavy drinking may produce such damage. A greater risk of former heavy drinkers might indicate that alcohol consumption predisposes to ICH via this mechanism. Accordingly, the risk for spontaneous ICH would be expected to increase along with the intensity and duration of alcohol-induced hypertension. Epidemiological studies have not yet specifically addressed this question, and we lack the final answer to it, but some data suggest that former heavy drinkers do not have a greater risk than light drinkers and nondrinkers (Juvela et al., 1995).

Acute increases in blood pressure (Potter et al., 1986), particularly systolic blood pressure, and cerebral blood flow (Mathew & Wilson, 1986) have been described to follow acute moderate and heavy drinking of alcohol. Such phenomena may act to trigger bleeding from degenerated arteries. The most recent clinical data indicate that acute heavy drinking triggers spontaneous ICH (Juvela et al., 1995).

Finally, disturbances in primary hemostasis, blood coagulation, and fibrinolysis, whether associated with liver disease or not, are common conditions among heavy drinkers. Their contributory role is clear if the subject already has degenerated small arteries in the brain and continues heavy drinking of alcohol. In concert with the predisposing and triggering factors, these disturbances certainly influence both the outcome and the frequency of ICH, but further studies of their significance are still needed.

2. Aneurysmal Subarachnoid Hemorrhage

While hypertension and alcoholism seem to be important risk factors for ICH, cigarette smoking and binge drinking have been emphasized in connection with aneurysmal SAH. It has been speculated that the formation of saccular aneurysms in cerebral arteries results from high blood flow and velocity, which implicates an increased shear stress against the vessel walls, particularly at the bifurcations of the large cerebral arteries (Stehbens, 1989). Aneurysmal SAH includes a distinct familial pattern, and hereditary factors contribute to the development of aneurysms. The frequent formation of aneurysms in the cerebral arteries may be due to their lack of external elastic lamina. Other structural characteristics deviant from the other arteries have also been described (Östergaard, 1989).

The incidence of saccular aneurysms increases steadily with age, although age itself is not a risk factor for the rupture of an aneurysm (Wiebers et al., 1987). Women are more susceptible to aneurysm formation than men, but aneurysm rupture seems to occur more frequently in men during the working age, probably reflecting some life-style factors of men (see Figure 21.1). Chronic hypertension may promote aneurysm formation and act as a predisposing factor, but an occasional increase in systolic blood pressure up to very high levels may be the most potent triggering factor of aneurysmal rupture (Torner et al., 1981). Such an increase occurs occasionally during a Valsalva maneuver, for example. With this in mind, it is not surprising that bleeding can occur during sexual intercourse and defecation (Matsuda et al., 1993). Hypertension is frequently associated with heavy drinking, particularly periodic heavy drinking (Marmot et al., 1994), which certainly causes enhanced hemodynamic stress on the vessel walls of the cerebral arteries. Sudden increases of systolic blood pressure may occur during intoxication. The following hangover can thus trigger an aneurysmal rupture. It has not yet been demonstrated by clinical studies that alcohol-induced contractions of the cerebral arteries (Zhang et al., 1993) are responsible for the rupture of an aneurysm.

The role of smoking is interesting because although it may enhance hemodynamic stress by causing vasospasm (Caralis et al., 1992), it may also contribute to the formation of saccular aneurysms via mechanisms that are different from those of alcohol drinking. This could happen via a release of proteolytic enzymes from lung tissue into the systemic circulation, as has been speculated by Fogelholm and Murros (1987). This, however, remains to be proved, and the mechanisms that might relate smoking to SAH are not yet clear (Higa & Davanipour, 1991).

Apart from hypertension, the other contributing factors to aneurysm formation include inherited diseases such as certain connective tissue disorders (Stehbens, 1989; Östergaard, 1989). Theoretically, impaired hemostasis and enhanced fibrinolytic activity might worsen the outcome of aneurysmal bleeding, which not infrequently (20%) first occurs as a minor leak or warning bleed (Versari et al., 1993). In fact, the clinical evidence available indicates that heavy drinkers do have a poorer outcome after SAH than do nondrinkers, but this is mainly due to the more severe complications, such as cerebral vasospasm and rebleeding (Juvela, 1992). There is no evidence yet to prove that early mortality, that is, dying before admission into hospital, is higher in alcoholics and heavy drinkers than in infrequent drinkers.

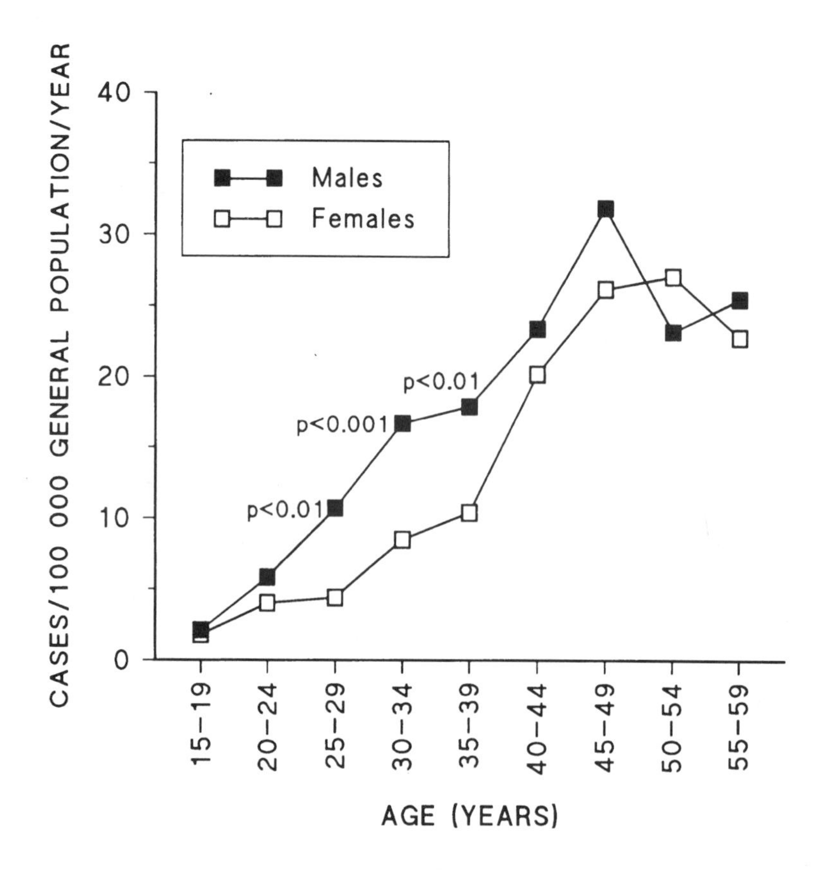

FIGURE 21.1 Incidence of nonhypertensive aneurysmal subarachnoid hemorrhage among working-age people in Finland during 1978–1983 according to official statistics based on discharge information.

3. Infarctions of Atherothrombotic Origin

As already mentioned, there are abundant data to suggest that moderate alcohol consumption protects against ischemic stroke of atherothrombotic origin. In particular, regular daily moderate drinking has often been found to associate with a decreased risk for cervical arterial atherosclerosis, which is a common cause of ischemic stroke among elderly people. However, this does not exclude the possibility that heavy drinking may promote atherogenesis and that acute alcohol might precipitate stroke in subjects with advanced atherosclerosis.

Atherosclerosis, which is a well-known predisposing condition for TIA and CI, commonly occurs at the bifurcations of carotid arteries, but may also occur in other arteries, including the aortic arch and the intracranial arteries. Recent observations suggest that stenoses and thrombi attached on the ulcerative atherosclerotic plaques can undergo episodes of formation and dissolution (Svindland & Torvik, 1988). Symptoms of cerebral ischemia arise when the oxygen demand of brain tissue is compromised by the occlusion of a stenotic artery or when emboli from mural thrombi are dispatched into the distal circulation. If the stenosis is tight and includes ulcerative plaques, the likelihood that it will cause a stroke is great. Alcohol ingestion could act on this disease process in several ways.

First, atherogenesis seems to be prevented by regular daily moderate alcohol drinking. This has frequently been ascribed to alcohol-induced elevated high density lipoprotein cholesterol (Langer et al., 1992; Gaziano et al., 1993; Srivastava et al., 1994). Indeed, even experimental findings indicate that moderate regular doses of alcohol (12% of total calories from ethanol) produce an antiatherogenic lipoprotein profile (increased HDL, normal LDL levels, and lecithin-cholesterol acyltransferase activity), whereas higher doses (24–36% of calories) produce both antiatherogenic (increased HDL) and atherogenic (increased LDL) responses

(Hojnacki et al., 1992). Alcoholics usually have high levels of plasma cholesterol, LDL cholesterol, and HDL cholesterol while drinking, whereas their HDL cholesterol often rapidly decreases afterwards, and the lipid and lipoprotein profiles change to become more atherogenic (Clemens et al., 1986; Hurt et al., 1986). These observations are interesting since light drinkers have a lower atherosclerosis risk than either abstainers or heavy drinkers (Kiechl et al., 1994), and since the inverse association of HDL cholesterol and coronary mortality is less marked at higher levels of alcohol intake (Paunio et al., 1995).

There is no evidence to suggest that periodic heavy drinking protects against the development of atherosclerosis. In addition, since heavy drinkers are frequently heavy smokers (Kaprio et al., 1982), it is not surprising that the possible beneficial effect of alcohol is not detectable in heavy drinkers (Baldwa, 1983). These data do not, however, disprove the assumption that moderate regular alcohol consumption may protect against atherosclerosis by inhibiting atheromatous plaque formation.

Alcohol can also be assumed to promote atherogenesis via its effects on blood pressure and plasma homocysteine concentration. Alcoholics do have elevated plasma homocysteine during their drinking bouts (Hultberg et al., 1993), and hyperhomocysteinemia is a known risk factor for carotid arterial atherosclerosis (Fryer et al., 1993). Alcohol may also have different effects depending on whether it is ingested alone or together with a meal. If taken together with a meal, alcohol has been shown to increase postprandial lipemia and atherogenic lipoprotein particles (Superko, 1992), to increase platelet aggregation, and to decrease fibrinolytic activity (Veenstra et al., 1990).

Alcohol and thrombogenesis have also been of great interest. Platelets play the key role in primary hemostasis. They adhere to dysfunctional endothelial cells and their underlying tissues. Such adherence takes place in severely injured vessel walls exposed by a rupture of atherosclerotic plaques (Buchanan, 1988). Alcohol might enhance or alleviate this process and thereby contribute to stroke occurrence, but the issue is still unclear, and further studies are needed.

Two excellent reviews are available concerning the effects of alcohol on platelet function (Mikhailidis et al., 1990; Rubin and Rand, 1994). A majority of the data published previously suggests that alcohol inhibits platelet aggregability *in vitro* (Mikhailidis et al., 1987; Rand et al., 1988). Increased platelet aggregability after ingestion of intoxicating doses of alcohol has also been observed (Hillbom et al., 1985a), but the effect might have been artifactual and due to hemolysis (Abi-Younes et al., 1991) or an excessively alkaline pH of the platelet-rich plasma. It is unlikely that alcohol activates platelets by interfering with the binding of the platelet-activating factor to its platelet receptor, since alcohol has been found to inhibit the process of platelet activation by the platelet-activating factor (Baker et al., 1989).

Chronic short-term administration of a moderate amount of ethanol has been shown to inhibit some platelet functions in cholesterol-fed rabbits (Latta et al., 1994a, 1994b), suggesting mechanisms for ethanol-induced alleviation of thrombogenesis *in vivo.* Acute ethanol has also been demonstrated to inhibit experimentally induced thrombosis in rabbits (Rand et al., 1990, 1991). Although these data need not necessarily reflect phenomena that take place in a living human organism, they support the hypothesis that alcohol could protect against thrombogenesis-mediated cerebrovascular disease.

In humans, platelet hypoaggregability *in vitro* is seen during prolonged alcohol drinking (Mikhailidis et al., 1986), whereas the cessation of drinking is usually followed by platelet hyperaggregability (Hutton et al., 1981). The appearance of sticky platelets is a transient phenomenon and seems to be associated with a shortening of the bleeding time (Hillbom et al., 1985b), alterations of plasma lipoprotein concentrations (Desai et al., 1986; Beitz et al., 1986), and possibly a resolution of alcoholic fatty liver (Hillbom & Neiman, 1988). It is unclear whether the sticky platelet phenomenon observed *in vitro* represents an artifact as suggested by some later observations (Neiman et al., 1989), or whether it occurs *in vivo* and is significant for thrombogenesis. It is also difficult to say whether the effects of alcohol on platelet function measured *in vitro* and *ex vivo* are important with regard to the occurrence of stroke.

While being activated, platelets release thromboxane, a potent vasoconstrictor and platelet-aggregating agent. On the other hand, vascular endothelial cells release prostacyclin, a potent vasodilator and antiaggregating agent, which neutralizes the action of thromboxane. Prostacyclin synthesis by endothelial cells may be decreased in patients with advanced atherosclerosis (Kyrle et al., 1989). Because there is a local decrement in prostacyclin production at the site of an atherosclerotic plaque (Rush et al., 1988), the potential of alcohol to enhance the actions of prostacyclin (Jakubowski et al., 1988) may be decreased, and the net result caused by an increase in the velocity of blood flow during alcoholic intoxication will hence be an enhancement of shear stress-induced platelet activation and thrombus formation. Since platelet activation caused by shear stress against a rough vessel wall may include mechanisms that differ from those previously studied, further investigations of platelet function in intoxicated human volunteers are needed to prove or disprove the hypothesis.

The effects of alcohol on hemostasis and platelet function can also be studied indirectly by measuring the skin bleeding time and the plasma levels as well as the urinary excretion of substances released during platelet aggregation. Prolonged bleeding time is often encountered in alcoholics during a drinking bout (Mikhailidis et al., 1986; Hillbom & Neiman, 1988). After alcohol withdrawal, the skin bleeding time is normalized or may even become transiently shorter than in healthy controls (Hillbom et al., 1985b). The effects of acute alcohol ingestion have been controversial. Alcohol has repeatedly been shown to potentiate the aspirin-induced prolongation of bleeding time, which suggests that alcohol may ameliorate rather than enhance thrombogenesis and is consistent with the alcohol potentiation of the inhibitory effect of prostacyclin on platelet aggregation (Jakubowski et al., 1988). The shortening of the skin bleeding time, which was observed in healthy nonalcoholic volunteers after the ingestion of a single large dose of alcohol (Hillbom et al., 1983), may have resulted from a change in vascular tone instead of enhanced platelet aggregability, since the volunteers were supine and sleepy when the shortening of bleeding time occurred. The skin bleeding time is influenced by central vasomotor control mechanisms that also have been shown to be modulated by alcohol ingestion (Malpas et al., 1990).

Urinary excretion of prostanoid metabolites has shown that withdrawal from prolonged heavy drinking is followed by a decrease in the systemic production of prostacyclin and by unchanged thromboxane production, a shift in balance that is favorable for thrombus formation (Neiman et al., 1987, 1994; Försterman & Feuerstein, 1987). Acute ingestion of alcohol in amounts sufficient to cause a significant increase in the velocity of circulation might enhance the urinary excretion of both prostacyclin and thromboxane. Whether there is a shift in balance after acute alcohol intake remains to be proved.

Finally, it is also relevant to consider the effect of alcohol on systemic circulation in general. Very small amounts of alcohol do not increase the heart rate significantly, but amounts that produce moderate to severe intoxication elevate the heart rate markedly (Stott et al., 1987). Ingestion of large amounts of alcohol is followed by tachycardia, which prevails until most of the alcohol has been eliminated. In addition to causing hemodynamic stress, tachycardia helps to dispatch already existing thrombi from diseased vessel walls and the heart into circulation. This action of alcohol may be significant enough to trigger ischemic stroke both among subjects having advanced atherosclerosis and those with cardiac disease.

4. Infarctions of Cardioembolic Origin

The lack of validated clinical criteria for the diagnosis of cardiogenic embolism makes it difficult to discriminate between ischemic stroke cases of cardiac origin and those with other causes. However, the recent advances in diagnostic techniques, including detection of intracardiac sources of cerebral emboli by transesophageal echocardiography, have made the task somewhat easier. On the other hand, clinical reports have not specifically addressed the role of alcohol consumption in ischemic stroke of cardiac origin. The few available reports indicate that heavy drinking might be associated with cerebral infarctions caused by cardiac emboli (Gonzalez et al., 1988; Hillbom & Kaste, 1990; Gras, 1992).

The most convincing link is that between alcoholic cardiomyopathy and ischemic stroke. Dilated cardiomyopathies have long been known to predispose to intracardiac thrombus formation. Alcoholic cardiomyopathy is a type of dilated cardiomyopathy usually diagnosed by the presence of heavy drinking and an ameloriation of the condition after prolonged abstention (Teragaki et al., 1993). Propagation of thrombi is certainly enhanced by cardiac arrhythmias, and alcohol, even in modest doses, has the potential to produce arrhythmias in patients with a history of chronic alcohol consumption and heart disease (Greenspon & Schaal, 1983). Alcoholic cardiomyopathy presents a broad spectrum of clinical conditions from an asymptomatic preclinical state with impaired early filling of the left ventricle due to delayed relaxation (Kupari et al., 1980) to severe cardiac insufficiency. It is unclear at what stage of the disease the risk for thrombus formation emerges, but it may already be present in less advanced cardiomyopathies.

Cardiac arrhythmias, particularly atrial fibrillation, are well-known factors that predispose to cerebral emboli of cardiac origin. Arrhythmias frequently occur in association with heart disease, particularly recent myocardial infarction, but also in the absence of overt cardiac disease. It is of interest that binge drinking can trigger episodes of atrial fibrillation in apparently healthy people who are moderate or even infrequent drinkers (Thornton, 1984) and that subjects with recurrent atrial fibrillation are frequently recent moderate-to-heavy drinkers or problem drinkers (Koskinen et al., 1990).

In considering alcohol and the risk for cerebral emboli of cardiac origin, of special interest are the sleep apnea syndromes, which are characterised by repeated long apneic episodes, arterial hypoxia, cardiac dysfunction, and arrhythmias (Miller, 1982; Motta & Guilleminault, 1985; Palomäki, 1991). Some observations suggest that alcohol ingestion may aggravate sleep apneas in subjects with such syndromes (Block et al., 1986; Mitler et al., 1988; Vitiello, 1990; Dawson, 1993). If this is true, it is reasonable to suppose that intracardiac thrombi may be formed during the arrhythmias provoked by sleep apneas and, when the person wakes up in the morning, dispatch into circulation.

Finally, prolonged heavy drinking of alcohol frequently followed by the syndrome of acute alcohol withdrawal may create conditions that facilitate the entrance of paradoxical emboli into the cerebral circulation. A paradoxical embolus most frequently results from deep venous thrombosis of the leg veins. As a matter of fact, some clinicians consider it necessary for the diagnosis that a deep venous thrombosis is simultaneously verified. Emboli of venous origin reach arterial circulation via a patent foramen ovale with a frequency of about 20% in the normal population. Blood will usually not shunt from the right to the left side of the heart across very small defects, but such shunting may occasionally occur if the subject performs a Valsalva maneuver, as when vomiting, which frequently occurs with a troublesome hangover, for example. Certain inherited conditions, systemic diseases, bed rest, surgical operations, alcoholism (Nylander et al., 1977), and decreased fibrinolytic activity (Isacson & Nilsson, 1972) are known to be risk factors for the formation of deep venous thromboses, and acute alcohol intake powerfully decreases fibrinolytic activity, although this may be followed by a rebound activation (Hendriks et al., 1994).

5. Other Subtypes of Cerebral Ischemia

The other subtypes of ischemic stroke include those due to vasospasm, hypotension, blood cell abnormalities, etc. Compared to the previously mentioned varieties, these are relatively rare. Among young adults, however, migrainous strokes may be more frequent (Carolei et al., 1993), and it has been speculated that vasospasm of the cerebral arteries may play a role in their pathogenesis (Rothrock et al., 1988). The pathogenesis of migrainous attacks has remained unclear, and, as far as I know, no reports attributing the onset of migrainous strokes to acute alcohol ingestion have been published.

Cerebral ischemia due to systemic hypotension may be due to poor cardiac performance in subjects with cardiac arrhythmias and insufficiency. Ingestion of even moderate doses of alcohol may impair cardiac performance (Kelbaek et al., 1988), but large doses of alcohol may be more dangerous. Alcohol-induced hypotension has been observed in severely intoxicated rats (Brackett et al., 1994). If it also occurs in humans, it might exacerbate even focal cerebral ischemia in subjects who have poor focal circulation of cerebral tissue due to local vascular lesions or abnormalities and/or poor cardiac performance.

Thrombocytosis, leukocytosis, and polycythemias are known to be associated with ischemic stroke. An increased platelet count has been reported to provoke TIA and brain infarction, presumably as a result of an obstruction of the cerebral microcirculation (Preston et al., 1979). There is no convincing evidence yet to indicate that rebound thrombocytosis after the cessation of prolonged heavy drinking really provides a significant risk for ischemic stroke among alcoholics, although a case report suggests so (Neiman, 1988). Alcohol consumption by subjects with other blood abnormalities and dyscrasias has not been reported to precipitate ischemic stroke.

V. CONCLUSIONS

Clinical epidemiological data on the association between alcohol consumption and stroke have substantially increased during the recent years. New observations suggest that particularly recent heavy drinking of alcohol carries an increased risk for stroke. The risk for intracranial hemorrhage seems to be greater than that for ischemic infarction.

The status of alcohol as a risk factor for ischemic stroke is a complex one. Moderate regular consumption seems to be associated with a decreased risk, whereas recent ingestion of large doses of alcohol associates with an increased risk. In spontaneous infarctions, alcohol may act via a diversity of mechanisms, including alcohol-induced changes in blood pressure, heart rate, vascular tone, hemostasis, blood clotting, fibrinolytic activity, and metabolic effects. Prolonged drinking of alcohol elevates blood pressure in a dose-dependent manner, which can result in a degeneration of the vessel walls. Therefore, alcohol can even be thought to predispose for cerebrovascular disease in the same way as other factors causing chronic hypertension. Acute heavy drinking may be particularly dangerous for those who already have some predisposing disease.

Alcoholic intoxication promotes the occurrence of traumas of all kinds, which explains the frequency of traumatic ICH and SAH in intoxicated subjects. Traumatic dissection of the neck arteries and the consequent CI have been insufficiently studied to prove a causal relationship. Acute alcoholism may produce cerebral vasospasm via actions that are independent of cerebral arterial endothelial cells and prostanoids, but human studies are needed to confirm this. Severe constriction and the resulting vasodilation of the cerebral arteries, high systolic blood pressure, or both may trigger aneurysmal rupture and produce SAH, which may also contribute to the onset of spontaneous ICH.

REFERENCES

Abi-Younes, S.A., Ayers, M.L., and Myers, A.K. Mechanism of ethanol-induced aggregation in whole blood. *Thromb. Res.*, 1991;63:481–489.

Ahmed, S.S., Levinson, G.E., Fiore, J.J., and Regan, T.J. Spectrum of heart muscle abnormalities related to alcoholism. *Clin. Cardiol.*, 1980;3:335–341.

Altura, B.M. and Altura, B.T. Alcohol, the cerebral circulation and strokes. *Alcohol*, 1984;1:325–331.

Altura, B.M., Zhang, A., Gheng, T.P.-O., and Altura, B.T. Ethanol promotes rapid depletion of intracellular free Mg in cerebrovascular smooth muscle cells: possible relation to alcohol-induced behavioral and stroke-like effects. *Alcohol,* 1993;10:563–566.

Ashley, M.J. Alcohol consumption, ischemic heart disease and cerebrovascular disease. An epidemiologic perspective. *J. Stud. Alcohol,* 1982; 43:869–887.

Baker, R.C., Fish, H.T., and Fitzpatrick, F.A. Effects of ethanol on human platelets stimulated with platelet-activating factor, a biologically active ether phospholipid. *Alcoholism,* 1989;13:824–828.

Baldwa, V.S., Gupta, M.C., Maheshwari, V.D., and Bhansali, A. Effect of prolonged smoking and alcohol on lipid profile, separately and in combination. *JAPI,* 1983;31:573–575.

Ballard, H.S. Hematological complications of alcoholism. *Alcoholism,* 1989;13:706–720.

Barry, K.J. and Scott, R.M. Effect of intravenous ethanol on cerebral vasospasm produced by subarachnoid blood. *Stroke,* 1979;10:535–537.

Beitz, J., Block, H.-U., Beitz, A., Muller, G., Winkler, L., Dargel, R., and Mest, H.-J. Endogenous lipoproteins modify the thromboxane formation capacity of platelets. *Atherosclerosis,* 1986;60:95–99.

Bell, B.A. and Symon, L. Smoking and subarachnoid hemorrhage. *Br. Med. J.,* 1979;1:577–578.

Ben-Shlomo, Y., Markowe, H., Shipley, M., and Marmot, M.G. Stroke risk from alcohol consumption using different control groups. *Stroke,* 1992;23:1093–1098.

van den Berg, J.S.P. and Limburg, M. Ischemic stroke in the young: Influence of diagnostic criteria. *Cerebrovasc. Dis.,* 1993;3:227–230.

Berglund, M. and Risberg, J. Regional cerebral blood flow during alcohol withdrawal. *Arch. Gen. Psychiatry,* 1981;38:351–355.

Block, A.J., Hellard, D.W., and Slayton, P.C. Effect of alcohol ingestion on breathing and oxygenation during sleep. *Am. J. Med.,* 1986;80:595–600.

Bogousslavsky, J., Van Melle, G., Despland, P.A., and Regli, F. Alcohol consumption and carotid atherosclerosis in the Lausanne stroke registry. *Stroke,* 1990;21:715–720.

Brackett, D.J., Gauvin, D.V., Lerner, M.R., Holloway, F.A., and Wilson, M.F. Dose- and time-dependent cardiovascular responses induced by ethanol. *J. Pharmacol. Exp. Ther.,* 1994;268:78–84.

Buchanan, M.R., Mechanisms of pathogenesis of arterial thrombosis: potential sites of inhibition by therapeutic compounds. *Semin. Thromb. Hemostasis,* 1988;14:33–40.

Calandre, L., Arnal, C., Fernandez Ortega, J., Bermejo, F., Felgeroso, B., del Ser, T., and Vallejo, A. Risk factors for spontaneous cerebral hematomas. Case-control study. *Stroke,* 1986;17:1126–1128.

Camargo, C.A. Moderate alcohol consumption and stroke. The epidemiologic evidence. *Stroke,* 1989;20:1611–1626.

Caplan, L.R. Intracerebral hemorrhage. *Lancet,* 1992;339:656–658.

Caralis, D.G., Deligonui, U., Kern, M.J., and Cohen, J.D. Smoking is a risk factor for coronary spasm in young women. *Circulation,* 1992;85:905–909.

Carolei, A., Marini, C., Ferranti, E., Frontoni, M., Prencipe, M., Fieschi, C., and the National Research Council Study Group. A prospective study of cerebral ischemia in the young. Analysis of pathogenic determinants. *Stroke,* 1993;24:362–367.

Clemens, M.R., Schied, H.W., and Waller, H.D. Serum lipids of alcoholics before and after withdrawal. Importance for coronary risk. *J. Clin. Chem. Clin. Biochem.,* 1986;24:369–374.

Cowan, D.H. Effects of alcoholism on hemostasis. *Semin. Hematol.,* 1980;17:137–147.

Creed, D.L., Baird, W.F., and Fisher, E.R. The severity of aortic atherosclerosis in certain diseases. A necropsy study. *Amer. J. Med. Sci.,* 1955;230:385–391.

Curfman, G.D. Is exercise beneficial — or hazardous — to your heart? *N. Engl. J. Med.,* 1993;329:1730–1731.

Dawson, A., Lehr, P., Bigby, B.G., and Mitler, M.M. Effect of bedtime ethanol on total inspiratory resistance and respiratory drive in normal nonsnoring men. *Alcoholism,* 1993;17:256–262.

Desai, K., Owen, J.S., Wilson, D.T., and Hutton, R.A. Platelet aggregation and plasma lipoproteins in alcoholics during alcohol withdrawal. *Thromb. Haemost.,* 1986;55:173–177.

Donahue, R.P., Abbott, R.D., Reed, D.M., and Yano, K. Alcohol and hemorrhagic stroke: the Honolulu Heart Study. *JAMA,* 1986;255:2311–2314.

Flegal, K.M. and Cauley, J.A. Alcohol consumption and cardiovascular risk factors. *Rec. Dev. Alcohol,* 1985;3:165–180.

Fogelholm, R. and Murros, K. Cigarette smoking and subarachnoid haemorrhage: a population-based case-control study. *J. Neur. Neurosurg. Psychiatry,* 1987;50:78–80.

Friedman, G.D. and Klatsky, A.L. Is alcohol good for your health? *N. Engl. J. Med.,* 1993;329:1882–1883.

Fryer, R.H., Wilson, B.D., Gubler, D.B., Fitzgerald, L.A., and Rodgers, G.M. Homocysteine, a risk factor for premature vascular disease and thrombosis, induces tissue factor activity in endothelial cells. *Arterioscl. Thromb.,* 1993;13:1327–1333.

Försterman, U. and Feuerstein, T.J. Decreased systemic formation of prostaglandin E and prostacyclin, and unchanged thromboxane formation, in alcoholics during withdrawal as estimated from metabolites in urine. *Clin. Sci.,* 1987;73:277–283.

Gaziano, J.M., Buring, J.E., Breslow, J.L., Goldhaber, S.Z., Rosner, B., VanDenburgh, M., Willett, W., and Hennekens, C.H. Moderate alcohol intake, increased levels of high-density lipoprotein and its subfractions, and decreased risk of myocardial infarction. *N. Engl. J. Med.,* 1993;329:1829–1834.

Gill, J.S., Zezulka, A.V., Shipley, M.J., Gill, S.K., and Beevers, D.G. Stroke and alcohol consumption. *N. Engl. J. Med.,* 1986;315:1041–1046.

Gill, J.S., Shipley, M.J., Hornby, R.H., Gill, S.K., and Beevers, D.G. A community case-control study of alcohol consumption in stroke. *Int. J. Epidemiol.,* 1988;17:542–547.

Gill, J.S., Shipley, M.J., Tsementzis, S.A., Hornby, R.S., Gill, S.K., Hitchcock, E.R., and Beevers, D.G. Alcohol consumption, a risk factor for hemorrhagic and non-hemorrhagic stroke. *Am. J. Med.,* 1991;90:489–497.

Gillum, R.F. Stroke in blacks, *Stroke,* 1988;19:1–9.

Goldberg, R.J., Burchfiel, C.M., Reed, D.M., Wergowske, G., and Chiu, D. A prospective study of the health effects of alcohol consumption in middle-aged and elderly men. The Honolulu Heart Program. *Circulation,* 1994;89:651–659.

Gonzalez, M.R.M., Donderis, A.C., Sanhauja, J.J., and Rieger, J.S. Cardiomyopathy, alcoholism and cerebral embolism. *Rev. Clin. Esp.,* 1988;182:79–80.

Gorelick, P.B. Alcohol and stroke. *Stroke,* 1987;18:268–271.

Gorelick, P.B., Rodin, M.B., Langenberg, P., Hier, D.B., Costigan, J., Gomez, I., and Spontak, S. Is acute alcohol ingestion a risk factor for ischemic stroke? Results of a controlled study in middle-aged and elderly stroke patients at three urban medical centers. *Stroke,* 1987;18:359–364.

Gorelick, P.B. The status of alcohol as a risk factor for stroke. *Stroke,* 1989;20:1607–1610.

Gorelick, P.B., Rodin, M.B., Langenberg, B., Hier, D.B., and Costigan, J. Weekly alcohol consumption, cigarette smoking, and the risk of ischemic stroke: results of a case-control study at three urban medical centers in Chicago, Illinois. *Neurology,* 1989;39:339–343.

Gras, P., Karim, A.A., Grosmaire, N., Borsotti, J.P., Giroud, M., Blettery, B., and Dumas, R. Multiple brain embolism in acute alcohol intoxication. A clinico-pathological case. *Rev. Neurol.,* 1992;148:215–217.

Greenberg, D.A., Chan, J., and Sampson, H.A. Endothelins and the nervous system. *Neurology,* 1992;42:25–31.

Greenspon, A.J. and Schaal, S.F. The "holiday heart": electrophysiologic studies of alcohol effects in alcoholics. *Ann. Int. Med.,* 1983;98:135–139.

Hall, E.M., Olsen, A.Y., and Davis, F.E. Portal cirrhosis. *Am. J. Pathol.,* 1953;29:993–1027.

Harland, W.A., Pitts, J.F., and Watson, A.A. Subarachnoid haemorrhage due to upper cervical trauma. *J. Clin. Pathol.,* 1983;36:1335–1341.

Harmsen, P., Rosengren, A., Tsipogianni, A., and Wilhelmsen, L. Risk factors for stroke in middle-aged men in Göteborg, Sweden. *Stroke,* 1990;21:223–229.

Hein, H.O., Sörensen, H., Suadicani, P., and Gyntelberg, F. Alcohol consumption, Lewis phenotypes, and risk of ischaemic heart disease. *Lancet,* 1993;341:392–396.

Hendriks, H.F.J., Veenstra, J., Velthuis-te Wierik, E.J.M., Schaafsma, G., and Kluft, C. Effect of moderate dose of alcohol with evening meal on fibrinolytic factors. *Br. Med. J.,* 1994;308:1003–1006.

Henrich, J.B. and Horwitz, R.I. Evidence against the association between alcohol use and ischemic stroke risk. *Arch. Intern. Med.,* 1989;149:1413–1416.

Hess, D.C., Sethi, K.D., and Nichols, F.T. Carotid dissection: a new false localizing sign. *J. Neurol. Neurosurg. Psychiatry,* 1990;53:804–805.

Higa, M. and Davanipour, Z. Smoking and stroke. *Neuroepidemiology,* 1991;10:211–222.

Hillbom, M., Kaste, M., and Rasi, V. Can ethanol intoxication affect hemocoagulation to increase the risk of brain infection in young adults? *Neurology,* 1983;33:381–384.

Hillbom, M., Kangasaho, M., Kaste, M., Numminen, H., and Vapaatalo, H. Acute alcohol ingestion increases platelet reactivity: is there a relationship to stroke? *Stroke,* 1985a;16:19–23.

Hillbom, M., Muuronen, A., Löwbeer, C., Änggård, E., Beving, H., and Kangasaho, M. Platelet thromboxane formation and bleeding time is influenced by ethanol withdrawal but not by cigarette smoking. *Thromb. Haemost.,* 1985b;53:419–422.

Hillbom, M.E. What supports the role of alcohol as a risk factor for stroke? *Acta Med. Scand.,* 1987;Suppl. 717:93–106.

Hillbom, M. and Neiman, J. Platelet thromboxane formation capacity after ethanol withdrawal in chronic alcoholics. *Haemostasis,* 1988;18:170–178.

Hillbom, M. and Kaste, M. Alcohol abuse and brain infarction. *Ann. Med.,* 1990;22:347–352.

Hillbom, M., Haapaniemi, H., Juvela, S., Palomäki, H., Numminen, H., and Kaste, M. Recent alcohol consumption, cigarette smoking and cerebral infarction in young adults. *Stroke,* 1995;26:40–45.

Hilton-Jones, D. and Warlow, C.P. Non-penetrating arterial trauma and cerebral infarction in the young. *Lancet,* 1985;1:1435–1448.

Hirst, A.E., Hadley, G.G., and Gore, I. The effect of chronic alcoholism and cirrhosis of the liver on atherosclerosis. *Am. J. Med. Sci.,* 1965;249:143–149.

Hojnacki, J.L., Cluette-Brown, J.E., Dechenes, R.N., Mulligan, J.J., Osmolski, T.V., Rencricca, N.J., and Barboriak, J.J. Alcohol produces dose-dependent antiatherogenic and atherogenic plasma lipoprotein responses. *Proc. Soc. Exp. Biol. Med.,* 1992;200:67–77.

Hsieh, S.-T., Sano, H., Saito, K., Kubota, Y., and Yokoyama, M. Magnesium supplementation prevents the development of alcohol-induced hypertension. *Hypertension,* 1992;19:175–182.

Hultberg, B., Berglund, M., Andersson, A., and Frank, A. Elevated plasma homocysteine in alcoholics. *Alcoholism,* 1993;17:687–689.

Hurt, R.D., Briones, E.R., Offord, K.P., Patton, J.G., Mao, S.J.T., Morse, R.M., and Kottke, B.A. Plasma lipids and apolipoprotein A-I and A-II levels in alcoholic patients. *Am. J. Clin. Nutr.,* 1986;43:521–529.

Hutton, R.A., Fink, R., Wilson, D.T., and Marjot, T.H. Platelet hyperaggregability during alcohol withdrawal. *Clin. Lab. Haematol.,* 1981;3:223–229.

Isacson, S. and Nilsson, I.M. Defective fibrinolysis in blood and vein walls in recurrent "idiopathic" venous thrombosis. *Acta Chir. Scand.,* 1972;138:313–319.

Iso, H., Folsom, A.R., Koike, K.A., Sato, S., Wu, K.K., Shimamoto, T., Iida, M., and Komachi, Y. Antigens of tissue plasminogen activator and plasminogen activator inhibitor 1: correlates in nonsmoking Japanese and Caucasion men and women. *Thromb. Hemostas.,* 1993;70:475–480.

Jakubowski, J.A., Vaillancourt, R., and Deykin, D. Interaction of ethanol, prostacyclin, and aspirin in determining human platelet reactivity *in vitro. Arteriosclerosis,* 1988;8:436–441.

Jamrozik, K., Broadhurst, R.J., Anderson, C.S., and Stewart-Wynne, E.G. The role of lifestyle factors in the etiology of stroke. A population-based case-control study in Perth, Western Australia. *Stroke,* 1994;25:51–59.

Juvela, S. Alcohol consumption as a risk factor for poor outcome after aneurysmal subarachnoid hemorrhage. *Br. Med. J.,* 1992;304:1663–1667.

Juvela, S., Hillbom, M., Numminen, H., and Koskinen, P. Cigarette smoking and alcohol consumption as risk factors for aneurysmal subarachnoid hemorrhage. *Stroke,* 1993;24:639–646.

Juvela, S., Hillbom, M., and Palomäki, H. Risk factors for spontaneous intracerebral hemorrhage. *Stroke,* 1995; 26.

Kagan, A., Popper, J.S., and Rhoads, G.G. Factors related to stroke incidence in Hawaii Japanese men: The Honolulu Heart Study. *Stroke,* 1980;11:14–21.

Kaprio, J., Hammar, N., Koskenvuo, M., Floderus-Myrhed, B., Langinvainio, H., and Sarna, S. Cigarette smoking and alcohol use in Finland and Sweden: a cross-national twin study. *Int. J. Epidemiol.,* 1982;11:378–386.

Keil, U., Chambless, L., Filipiak, B., Härtel, U. Alcohol and blood pressure and its interaction with smoking and other behavioural variables: results from the MONICA Augsburg Survey 1984–1985. *J. Hypertens.,* 1991;9:491–498.

Kelbaek, H., Heslet, L., Skagen, K., Christensen, N.J., Godtfredsen, J., and Munck, O. Hemodynamic effects of alcohol at rest and during upright exercise in coronary artery disease. *Am. J. Cardiol.,* 1988;61:61–64.

Kiechl, S., Willeitt, J., Egger, G., Oberhollenzer, M., and Aichner, F. Alcohol consumption and carotid atherosclerosis: evidence of dose-dependent atherogenic and antiatherogenic effects. Results from the Bruneck Study. *Stroke,* 1994;25:1593–1598.

Kittner, S.J., White, L.R., Losonczy, K.G., Wolf, P.A., and Hebel, J.R. Black-white differences in stroke incidence in a national sample. The contribution of hypertension and diabetes mellitus. *JAMA,* 1990;264:1267–1270.

Klatsky, A.I., Armstrong, M.A., and Friedman, G.D. Alcohol use and subsequent cerebrovascular disease hospitalizations. *Stroke,* 1989;20:741–746.

Klatsky, A.L., Armstrong, M.A., and Friedman, G.D. Alcohol and mortality. *Ann. Int. Med.,* 1992;117:646–654.

Kono, S., Ikeda, M., Tokudome, S., Nishizumi, M., and Kuratsune, M. Alcohol and mortality: a cohort study of male Japanese physicians. *Int. J. Epidemiol.,* 1986;15:527–532.

Koskinen, P., Kupari, M., and Leinonen, H. Role of alcohol in recurrences of atrial fibrillation in persons <65 years of age. *Am. J. Cardiol.,* 1990;66:954–958.

Kupari, M., Koskinen, P., Suokas, A., and Ventilä, M. Left ventricular filling impairment in asymptomatic chronic alcoholics. *Am. J. Cardiol.,* 1990;66:1473–1477.

Kyrle, P.A., Minar, E., Brenner, B., Eichler, H.G., Heistinger, M., Marosi, L., and Lechner, K. Thromboxane A_2 and prostacyclin generation in the microvasculature of patients with atherosclerosis. Effect of low dose aspirin. *Thromb. Haemost.,* 1989;61:374–377.

Lang, W.E. Ethyl alcohol enhances plasminogen activator secretion by endothelial cells. *JAMA,* 1983;250:772–776.

Langer, R.D., Criqui, M.H., and Reed, D.M. Lipoproteins and blood pressure on biological pathways for effect of moderate alcohol consumption on coronary heart disease. *Circulation,* 1992;85:910–915.

Latta, E.K., Packham, M.A., DaCosta, S.M., and Rand, M.L. Effects of chronic administration of ethanol on platelets from rabbits with diet-induced hypercholesterolemia. Unchanged characteristics and responses to ADP but reduction of enhanced thrombin-induced, TXB_2-independent platelet response. *Arterioscler. Thromb.,* 1994a;14:1372–1377.

Latta, E.K., Packham, M.A., Gross, P.L., and Rand, M.L. Enhanced collagen-induced responses of platelets from rabbitrs with diet-induced hypercholesterolemia are due to increased sensitivity to TXA_2. Response inhibition by chronic ethanol administration in hypercholesterolemia is due to reduced TXA_2 formation. *Arterioscler. Thromb.,* 1994b;14:1379–1385.

Lee, K., Nielsen, J.D., Zeeberg, I., and Gormsen, J. Platelet aggregation and fibrinolytic activity in young alcoholics. *Acta Neurol. Scand.,* 1980;62:287–292.

Levine, S.R., Brust, J.C.M., Futrell, N., Brass, L.M., Blake, D., Fayad, P., Schultz, L.R., Millikan, C.H., Ho, K.-L., and Welch, K.M.A. A comparative study of the cerebrovascular complications of cocaine: alkaloidal versus hydrochloride — a review. *Neurology,* 1991;41:1173–1177.

Lindenström, E., Boysen, G., and Nyboe, J. Lifestyle factors and risk of cerebrovascular disease in women. The Copenhagen City Heart Study. *Stroke,* 1993;24:1468–1472.

Longstreth, W.T., Koepsell, T.D., Yerby, M.S., and van Belle, G. Risk factors for subarachnoid hemorrhage. *Stroke,* 1985;16:377–385.

Longstreth, W.T., Jr., Nelson, L.M., Koepsell, T.D., and van Belle, G. Cigarette smoking, alcohol use, and subarachnoid hemorrhage. *Stroke,* 1992;23:1242–1249.

Maclure, M. The case-crossover design: a method for studying transient effects on the risk of actue events. *Am. J. Epidemiol.,* 1991;133:144–153.

Malpas, S.C., Robinson, B.J., and Maling, T.J.B. Mechanism of ethanol-induced vasodilation. *J. Appl. Physiol.,* 1990;68:731–734.

Mandybur, T.I. and Mendenhall, C.L. The effects of chronic alcoholism on development of ischemic cerebral infarcts following unilateral carotid artery ligation in gerbils. *Alcoholism,* 1983;7:357–361.

Manton, K.G., Patrick, C.H., and Johnson, K.W. Health differentials between blacks and whites: recent trends in mortality and morbidity. *Milbank Q,* 1987;65 (suppl 1):129–199.

Marini, C., Carolei, A., Roberts, R.S., Prencipe, M., Gandolfo, C., Inzitari, D., Landi, G., De Zanche, L., Scoditti, U., Fieschi, C., and the National Research Council Study Group. Focal cerebral ischemia in young adults: a collaborative case-control study. *Neuroepidemiology,* 1993;12:70–81.

Marmot, M.G., Elliott, P., Shipley, M.J., Dyer, A.R., Ueshima, H., Beevers, D.G., Stamler, R., Kesteloot, H., Rose, G., and Stamler, J. Alcohol and blood pressure: the INTERSALT study. *Br. Med. J.,* 1994;308:1263–1267.

Mathew, R.J. and Wilson, W.H. Regional cerebral blood flow changes associated with ethanol intoxication. *Stroke,* 1986;17:1156–1159.

Matsuda, M., Ohashi, M., Shiino, A., Matsumura, K., and Handa, J. Circumstances precipitating aneurysmal subarachnoid hemorrhage. *Cerebrovasc. Dis.,* 1993;3:285–288.

Meade, T.W., Chakrabarti, R., Haines, A.P., North, W.R.S., and Stirling, Y. Characteristics affecting fibrinolytic activity and plasma fibrinogen concentrations. *Br. Med. J.,* 1979;1:153–156.

Meade, T.W., Vickers, M.V., Thompson, S.G., Stirling, Y., Haines, A.P., and Miller, G.J. Epidemiological characteristics of platelet aggregability. *Br. Med. J.,* 1985;290:428–432.

Mikhailidis, D.P., Jenkins, W.J., Barradas, M.A., Jeremy, J.Y., and Dandona, P. Platelet function defects in chronic alcoholism. *Br. Med. J.,* 1986;293:715–718.

Mikhailidis, D.P., Barradas, M.A., Epemolu, O., and Dandona, P. Ethanol ingestion inhibits human whole blood platelet impendance aggregation. *Am. J. Clin. Pathol.,* 1987;88:342–345.

Mikhailidis, D.P., Barradas, M.A., and Jeremy, J.Y. The effect of ethanol on platelet function and vascular prostanoids. *Alcohol,* 1990;7:171–180.

Miller, W.P. Cardiac arrhythmias and conduction distrubances in the sleep apnea syndrome. *Am. J. Med.,* 1982;73:317–321.

Mitler, M.M., Dawson, A., Henriksen, S.J., Sobers, M., and Bloom, F.E. Bedtime ethanol increases resistance of upper airways and produces sleep apneas in asymptomatic snorers. *Alcoholism,* 1988;12:801–805.

Monforte, R., Estruch, R., Graus, F., Nicolas, J.M., and Urbano-Marquez, A. High ethanol consumption as risk factor for intracerebral hemorrhage in young and middle-aged people. *Stroke,* 1990;21:1529–1532.

Motta, J. and Guilleminault, C. Cardiac dysfunction during sleep. *Ann. Clin. Res.,* 1985;17:190–198.

Neiman, J., Hillbom, M., Benthin, G., and Ânggård, E.E. Urinary excretion of 2,3-dinor-6-keto prostaglandin F_1a and platelet thromboxane formation during ethanol withdrawal in alcoholics. *J. Clin. Pathol.,* 1987;40:512–515.

Neiman, J. Association of transient ischemic attack in alcohol withdrawal with changes in haemostasis. *Br. J. Addict.,* 1988;83:1457–1459.

Neiman, J., Rand, M.L., Jakowec, D.M., and Packham, M.A. Platelet responses to platelet activating factor are inhibited in alcoholics undergoing alcohol withdrawal. *Thromb. Res.,* 1989;56:399–405.

Neiman, J., Nowak, J., Benthin, G., Numminen, H., and Hillbom, M. Increased urinary excretion of a major thromboxane metabolite in early alcohol withdrawal. *Clin. Physiol.,* 1994;14:405–409.

Niizuma, H., Suzuki, J., Yonemitsu, T., and Otsuki, T. Spontaneous intracerebral hemorrhage and liver dysfunction. *Stroke,* 1988;19:852–856.

Nylander, G., Olivecrona, H., and Hedner, U. Earlier and concurrent morbidity of patients with acute lower leg thrombosis. *Acta Chir. Scand.,* 1977;143:425–429.

Olson, J.D. Mechanisms of hemostasis. Effect on intracerebral hemorrhage. *Stroke,* 1993;24(suppl 1):1-109–1-114.

Östergaard, J.R. Risk factors in intracranial saccular aneurysms. Aspects on the formation and rupture of aneurysms, and development of cerebral vasospasm. *Acta Neurol. Scand.,* 1989;80:81–98.

Palomäki, H. Snoring and the risk of ischemic brain infarction. *Stroke,* 1991;22:1021–1025.

Palomäki, H. and Kaste, M. Regular light-to-moderate intake of alcohol and the risk of ischemic stroke. Is there a beneficial effect? *Stroke,* 1993a;24:1828–1832.

Palomäki, H., Kaste, M., Raininko, R., Salonen, O., Juvela, S., and Sarna, S. Risk factors for cervical atherosclerosis in patients with transient ischemic attack or minor ischemic stroke. *Stroke,* 1993b;24:970–975.

Paunio, M., Heinonen, O.P., Virtamo, J., Klag, M.J., Manninen, V., Albanes, D., and Comstock, G.W. HDL cholesterol and mortality in Finnish men with special reference to alcohol intake. *Circulation,* 1994;90:2909–2918.

Potter, J.F., Macdonald, I.A., and Beevers, D.G. Alcohol raises blood pressures in hypertensive patients. *J. Hypertens.*, 1986;4:435–441.

Prendes, J.L. Cerebral ifnarction and alcohol. *Lancet,* 1979;1:219.

Preston, F.E., Martin, J.F., Stewart, R.M., and Davies-Jones, G.A.B. Thrombocytosis, circulating platelet aggregates, and neurological dysfunction. *Br. Med. J.,* 1979;2:1561–1563.

Puddey, I.B., Beilin, L.J., Vandongen, R., Rouse, I.L., and Rogers, P. Evidence for a direct effect of consumption in normotensive men: a randomized controlled trial. *Hypertension,* 1985;7:707–713.

Puddey, I.B., Beilin, L.J., and Vandongen, R. Effect of regular alcohol use on blood pressure control in treated hypertensive subjects: a controlled study. *Clin. Exp. Pharmacol. Physiol.,* 1986;13:315–318.

Ragni, M.V., Lewis, J.H., Spero, J.A., and Hasiba, U. Bleeding and coagulation abnormalities in alcoholic cirrhotic liver disease. *Alcoholism,* 1982;6:267–274.

Rand, M.L., Packham, M.A., Kinlough-Rathbone, R.L., and Mustard, J.F. Effects of ethanol on pathways of platelet aggregation *in vitro. Thromb. Haemost.,* 1988;59:383–387.

Rand, M.L., Groves, H.M., Packham, M.A., Mustard, J.F., and Kinlough-Rathbone, R.L. Acute administration of ethanol to rabbits inhibits thrombus formation induced by indwelling aortic catherters. *Lab. Invest.,* 1990;63:742–745.

Rand, M.L., Groves, H.M., Packham, M.A., Mustard, J.F., and Kinlough-Rathbone, R.L. Inhibition by ethanol of thrombus formation induced by indwelling aortic catheters in rabbits. *Thromb. Haemost.,* 1991;65:454–455.

Renaud, S.C., Beswick, A.D., Fehily, A.M., Sharp, D.S., and Elwood, P.C. Alcohol and platelet aggregation: the Caerphilly prospective heart disease study. *Am. J. Clin. Nutr.,* 1992;55:1012–1017.

Ridker, P.M., Vaughan, D.E., Stampfer, M.J., Glynn, R.J., and Hennekens, C.H. Association of moderate alcohol consumption and plasma concentration of endogenous tissue-type plasminogen activator. *JAMA,* 1994;272:929–933.

Rissanen, V. Coronary and aortic atherosclerosis of chronic alcoholics. *Z. Rechtsmed.,* 1974;75:183–189.

Rodgers, H., Aitken, P.D., French, J.M., Curless, R.H., Bates, D., and James O.F.W. Alcohol and stroke. A case-control study of drinking habits past and present. *Stroke,* 1993;24:1473–1477.

Rothrock, J.F., Walicke, P., Swenson, M.R., Lyden, P.D., and Logan, W.R. Migrainous stroke. *Arch. Neurol.,* 1988;45:63–67.

Rubin, R. and Rand, M.L. Alcohol and platelet function. *Alcoholism,* 1994;18:105–110.

Rush, D.S., Kerstein, M.D., Bellan, J.A., Knoop, S.M., Mayeux, P.R., Hyman, A.L., Kadowitz, P.J., and McNamara, D.B. Prostacyclin, thromboxane A_2 and prostaglandin E_2 formation in atherosclerotic human carotid artery. *Arteriosclerosis,* 1988;8:73–78.

Sacco, R.L., Shi, T., Zamanillo, M.C., and Kargman, D.E. Predictors of mortality and recurrence after hosptilalized cerebral infarction in an urban community: the Northern Manhattan Stroke Study. *Neurology,* 1994;44:626–634.

Schlicht, I., Falk, S., Krössen, A., and Mohnhaupt, A. Lifetime prolongation in voluntary alcohol-consuming rats (SHR) treated with clofibrate. *Alcohol,* 1992;9:139–148.

Schutz, H., Bödeker, R.-H., Damian, M., Krack, P., and Dorndorf, W. Age-related spontaneous intracerebral hematoma in a German community. *Stroke,* 1990;21:1412–1418.

Schwartz, J.S., Speed, N.M., Gross, M.D., Lucey, M.R., Bazakis, A.M., Harihara, M., and Beresford, T.P. Acute effects of alcohol administration on regional cerebral blood flow: the role of acetate. *Alcoholism,* 1993;17:1119–1123.

Semenciew, R.M., Morrison, M.I., Mao, Y., Johansen, H., Davies, J.W., and Wigle, D.T. Major risk factors for cardiovascular disease mortality in adults: results from the Nutrition Canada Survey Study. *Int. J. Epidemiol.,* 1988;17:317–324.

Shinton, R., Sagar, G., and Beevers, G. The relation of alcohol consumption to cardiovascular risk factors and stroke. The west Birmingham stroke project. *J. Neurol. Neurosurg. Psychiatry.,* 1993;56:458–462.

Simonsen, J. Fatal subarachnoid hemorrhages in relation to minor injuries in Denmark from 1967 to 1981. *Forensic Sci. Int.,* 1984;24:57–63.

Sloan, M., Kittner, S.J., Rigamonti, D., and Price, T.R. Occurrence of stroke associated with use/abuse of drugs. *Neurology,* 1991;41:1358–1364.

Srivastava, L.M., Vasisht, S., Agarwal, D.P., and Goedde, H.W. Relation between alcohol intake, lipoproteins and coronary heart disease: the interest continues. *Alcohol Alcoholism,* 1994;29:11–24.

Stampfer, M.J., Colditz, G.A., Willett, W.C., Speizer, F.E., and Hennekens, C.H. A prospective study of moderate alcohol consumption and the risk of coronary disease and stroke in women. *N. Engl. J. Med.,* 1988;319:267–273.

Stehbens, W.E. Etiology of intracranial berry aneurysms. *J. Neurosurg.,* 1989;70:823–831.

Stott, D.J., Ball, S.G., Inglis, G.C., Davies, D.L., Fraser, R., Murray, G.D., and McInnes, G.T. Effects of a single moderate dose of alcohol on blood pressure, heart rate and associated metabolic and endocrine changes. *Clin. Sci.,* 1987;73:411–416.

Superko, H.R. Effects of acute and chronic alcohol consumption on postprandial lipemia in healthy normotriglyceridemic men. *Am. J. Cardiol.,* 1992;69:701–704.

Svindland, A. and Torvik, A. Atherosclerotic carotid disease in asymptomatic individuals. A histological study of 53 cases. *Acta Neurol. Scand.,* 1988;78:506–517.

Teragaki, M., Takeuchi, K., and Takeda, T. Clinical and histologic features of alcohol drinkers with congestive heart failure. *Am. Heart J.,* 1993;125:808–817.

Thornton, J.R. Atrial fibrillation in healthy nonalcoholic people after an alcoholic binge. *Lancet,* 1984;2:1013–1014.

Torner, J.C., Kassels, N.F., Wallace, R.B., and Adams, H.P. Preoperative prognostic factors for rebleeding and survival of aneurysm patients receiving antifibrinolytic therapy: report on the operative aneurysm study. *Neurosurgery,* 1981;9:506–513.

Veenstra, J., Kluft, C., Ockhuizen, T., Pol, H., Wedel, M., and Schaafsma, G. Effects of moderate alcohol consumption on platelet function, tissue-type plasminogen activator and plasminogen activator inhibitor. *Thromb. Haemost.,* 1990;63:345–348.

Versari, P. Bassi, P., Limoni, P., D'Aliberti, G., Loiero, M., Levati, A., and Mangoni, A. Unrecognized warning leak in ruptured intracranial aneurysm. *Cerebrovasc. Dis.,* 1993;3:289–294.

Viel, B., Donoso, S., Salcedo, D., Rojas, P., Varela, A., and Alessandri, R. Alcoholism and socioeconomic status, hepatic damage, and arterioslcerosis. Study of 777 autopsied men in Santiago, Chile. *Arch. Intern. Med.,* 1966;117:84–91.

Vitiello, M.V., Prinz, P.N., Personius, J.P., Vitaliano, P.P., Nuccio, M.A., and Koerker, R. Relationship of alcohol abuse history to nighttime hypoxemia in abstaining chronic alcoholic men. *J. Stud. Alcohol,* 1990;51:29–33.

Walbran, B.B., Nelson, J.S., and Taylor, J.R. Association of cerebral infarction and chronic alcoholism: an autopsy study. *Alcoholism,* 1981;5:531–535.

Wallerstedt, S., Cederblad, G., Korsan-Bengtsen, K., and Olsson, R. Coagulation factors and other plasma proteins during abstinence after heavy alcohol consumption in chronic alcoholics. *Scand. J. Gastroenterol.,* 1977;12:649–656.

Weisberg, L.A. Alcoholic intracerebral hemorrhage. *Stroke,* 1988;19:1565–1569.

Whisnant, J.P., Homer, D., Ingall, T.J., Baker, H.L., Jr., O'Fallon, W.M., and Wiebers, D.O. Duration of cigarette smoking is the strongest predictor of severe extracranial carotid artery atherosclerosis. *Stroke,* 1990;21:707–714.

Wiebers, D.O., Whisnant, J.P., Sundt, T.M., and O'Fallon, M. The significance of unruptured intracranial saccular aneurysms. *J. Neurosurg.,* 1987;66:23–29.

Wilkins, M.R. and Kendall, M.J. Stroke affecting young men after alcoholic binges. *Br. Med. J.,* 1985;291:1342.

Wolf, P.A. Cigarettes, alcohol, and stroke. *N. Engl. J. Med.,* 1986;315:1087–1089.

Zhang, A., Altura, B.T., and Altura, B.M. Ethanol-induced contraction of cerebral arteries in diverse mammals and its mechanism of action. *Eur. J. Pharmacol.,* 1993;248:229–236.

Zhu, C.Z. and Norris, J.W. Role of carotid stenosis in ischemic stroke. *Stroke,* 1990;21:1131–1134.

Chapter 22

EFFECTS OF ETHANOL ON THE CENTRAL NERVOUS SYSTEM: FETAL DAMAGE – NEUROBEHAVIORAL EFFECTS

Howard C. Becker

TABLE OF CONTENTS

I. INTRODUCTION

Just over two decades ago, in 1973, a description of a common pattern of birth defects observed in children born to alcoholic mothers was published (Jones and Smith, 1973; Jones et al., 1973). The distinct cluster of symptoms was labeled Fetal Alcohol Syndrome (FAS). Since these initial reports, a great deal of clinical and basic research has ensued, firmly establishing ethanol as a teratogen (i.e., an agent that produces defects in offspring *in utero*). As a result of this research, it soon became apparent that the deleterious effects of *in utero* ethanol exposure exist on a continuum, ranging from gross morphological defects to more subtle cognitive-behavioral dysfunctions. The latter effects have established ethanol as a behavioral teratogen. Although the behavioral teratogenic actions of ethanol may be more subtle in nature (from a diagnostic perspective), this is not to suggest that these effects are any less clinically significant. In fact, neurobehavioral deficits may very well represent the most pernicious consequences of *in utero* exposure to ethanol in that many of these effects appear to persist beyond infancy and, unfortunately, are the most difficult aspects of the syndrome to effectively treat. This point is further underscored by the fact that prenatal ethanol exposure is one of the leading known causes of mental retardation in the Western world (Abel and Sokol, 1986).

The minimal criteria needed for a diagnosis of FAS include: (1) prenatal and postnatal growth retardation, (2) a characteristic constellation of craniofacial anomalies, and (3) central nervous system (CNS) dysfunction (Clarren and Smith, 1978; Rosett, 1980). The terms "alcohol related birth defects" (ARBD) or "Fetal Alcohol Effects" (FAE) are appropriate when only some of the criteria for FAS are met (Clarren and Smith, 1978; Sokol and Clarren, 1989). Thus, the behavioral teratogenic properties of ethanol represent a hallmark feature of both full (FAS) and partial (FAE) expression of the syndrome.

Estimates of the world-wide incidence of FAS have been reported in the range of 0.33 to 1.9 cases per 1000 live births (Abel and Sokol, 1987, 1991). The variance in estimated incidence rates is most likely due to a number of factors including methodological differences in epidemiological surveillance, underdiagnosis of FAS, underreporting of FAS diagnoses, and inclusion-exclusion of some high-risk groups (Cordero et al., 1994; Little et al., 1990). In addition, the incidence rates are much greater when one considers only the population of heavy drinking women. Moreover, given the less restrictive diagnostic criteria, the incidence of FAE has been estimated to be at least three to four times greater than that for FAS (Abel, 1984).

Both FAS and FAE represent major public health problems, with treatment costs for FAS escalating in recent years in the United States (Abel and Sokol, 1987, 1991; Sokol and Abel, 1988). Given the enormous medical, social, and economic burden FAS/FAE costs society, it is not surprising that a great deal of research has focused on characterizing the adverse effects of *in utero* ethanol exposure. Much of this research effort has been directed toward addressing key issues that are of interest to scientists and the lay public alike. These issues include the identification of a threshold dose; critical periods of greater vulnerability; the role of other factors such as smoking, undernutrition, and other drug use; the permanence of the adverse consequences; and whether ethanol-induced fetal damage can be prevented or reversed by therapeutic intervention. The immense complexity of these issues cannot be overstated, and for a variety of reasons, including the need for greater experimental control as well as ethical considerations, it is not surprising that clinicians have turned to basic research to address many of these questions. This chapter reviews recent findings on the behavioral teratogenic effects of ethanol, as well as possible neural mechanisms that might underlie the observed behavioral abnormalities. Although this report focuses on basic research with animals, parallel findings in the clinical literature are raised throughout the discussion.

II. ANIMAL MODELS AND METHODOLOGICAL ISSUES

The development of animal models has provided an invaluable research tool for characterizing and advancing the study of ethanol's adverse effects on the developing fetus. Employing several species (including subhuman primates) as well as a variety of methodologies and experimental approaches, animal research has played a major role in advancing our knowledge of the many immediate and long-term deleterious consequences that follow prenatal ethanol exposure. Most importantly, the use of animal models allows for the control of a variety of potentially confounding factors that are commonly associated with chronic alcohol use in humans. Some of these confounding variables include malnutrition, poor rearing (environmental) conditions, disease, smoking, and other drug use.

To address these potential problems, a variety of techniques and procedures have been developed to allow for analysis of effects that may be attributed to prenatal ethanol exposure, per se. For example, by using pair-feeding techniques and adjusting for the caloric value of ethanol, problems associated with undernutrition can be addressed to a great extent. Through the use of fostering manipulations, the role of pre- and postnatal influences on pregnancy outcome following maternal ethanol consumption can be dissociated. In addition, rigorous control can be exerted over such variables as dose and pattern of ethanol consumption, as well as duration and timing of exposure. The potential interactive effects of other drug use can also be examined in a controlled fashion. Further, the relative role of maternal variables and genetic factors governing susceptibility to prenatal ethanol effects can be best realized in research employing animal models (e.g., Abel and Dintcheff, 1984, 1985; Gilliam and Irtenkauf, 1990; Goodlett et al., 1989; Vorhees, 1989).

In addition, as observed clinically, the deleterious effects of *in utero* ethanol exposure in animal models have been shown to exist on a continuum. In fact, there is a remarkable similarity and correspondence of findings in animal models with that observed in the clinical setting (Driscoll et al., 1990). This applies to both physical anomalies and behavioral teratogenic effects. For example, mouse models have been shown to be sensitive to the full spectrum of prenatal ethanol effects ranging from dysmorphology to cognitive-behavioral deficits in the absence of physical abnormalities (Becker, 1992; Becker and Randall, 1989; Becker et al., 1989). Moreover, most of the deleterious effects of prenatal ethanol exposure observed in animals have been identified following maternal ethanol exposure (blood ethanol levels) that approximates typical levels of exposure reported in pregnant alcohol-consuming women (Driscoll et al., 1990).

In some cases, findings generated from animal laboratory studies have helped guide clinicians in identifying defects previously not documented in children of alcohol-dependent mothers. For instance, after first observing auditory defects in laboratory rodents prenatally exposed to ethanol, hearing impairment was searched for and found in children with FAS (Church, 1987; Church and Gerkin, 1988). A similar sequence of events applies to the discovery of renal defects in animals (Randall et al., 1977) and then in FAS children (Debeukelaer et al., 1977).

Finally, and perhaps of greatest significance, basic research with animals has allowed for the investigation of mechanisms underlying the teratogenic actions of ethanol. Inherent in studies focused on the behavioral teratogenic properties of ethanol is a presumed underlying perturbation of brain structure and function. The complexity of the brain, as well as the dynamic nature of its development, has presented some unique challenges to basic researchers. For most organ systems, the period of greatest vulnerability to the action of a teratogen corresponds to the organ's major period of organogenesis. This typically coincides with the first trimester of pregnancy. However, in comparison to other organs, the brain is unique in that it is one of the first to begin to develop and the last to be completed (West, 1987). Thus, the brain appears to be sensitive to the teratogenic actions of ethanol throughout its development, which encompasses all three trimesters of fetal development (West and Goodlett, 1990). In addition, even though all mammals pass through the same stages of brain development, the timing of those stages relative to birth varies considerably (Dobbing and Sands, 1979; West, 1987). For example, the brain is less mature at birth in rats than it is in humans. Thus, with regard to brain development, the full gestation period in the rat is equivalent to the first two trimesters in humans, and the first 10 days of postnatal life in the rat may be considered the third trimester equivalent.

The brain is particularly sensitive to the adverse effects of ethanol during the period of time when it is undergoing rapid growth, referred to as the brain growth spurt. Since this corresponds to the third trimester equivalent, an "artificial-rearing" procedure was developed to administer ethanol to rat neonates (Diaz and Samson, 1980; West et al., 1984). This model involves implanting an intragastric feeding tube in the rat pup so that it may be raised apart from its mother by having a milk formula containing ethanol infused directly into the neonate's stomach. This method allows for external control over the dose of ethanol administered. Such neonatal ethanol exposure (during the brain growth spurt) has been shown to result in a variety of structural and functional (behavioral) abnormalities (e.g., Barron and Riley, 1990; Bonthius and West, 1988; Goodlett et al., 1987, 1990; Kelly et al., 1988, 1990; Meyer et al., 1990a, 1990b; Pierce and West, 1986; West, Goodlett, Bonthius et al., 1990). Thus, both prenatal and early postnatal ethanol exposure have been employed in animal models to demonstrate the behavioral teratogenic effects of ethanol, as well as to probe for mechanisms underlying the observed abnormalities. Some of the more recent research findings generated from animal studies on the behavioral teratogenic effects of ethanol are outlined in the following sections.

III. SENSORIMOTOR EFFECTS

A variety of sensorimotor deficits have been identified in children with FAS/FAE and then examined in greater detail in animal studies. These include visual, auditory, vestibular, and motor coordination problems.

A. VISUAL SYSTEM

Visual system anomalies are very common in FAS/FAE. Aside from the facial and morphologic malformations that comprise some of the craniofacial features used for FAS diagnosis (microphthalmia, ptosis, and short palpebral fissures), disorders of ocular muscle coordination and defects of several intraocular structures have been reported as well. Strabismus or esotropia (crossed eyes), optic nerve hypoplasia (a reduction in the number of optic nerve axons), and abnormal vasculature in the retina have been most frequently observed (Stromland, 1987). All of these defects contribute to compromised visual acuity, typically myopia (nearsightedness). A similar pattern of ocular facial features has been obtained in animal models (Sulik and Johnston, 1983), and recently the pathogenesis of these ophthalmologic abnormalities have been examined. Embryonic ethanol exposure has been found to increase neuronal cell death in the retina as well as alter the cell cycle (Kennedy and Elliot, 1986). In another study, the rate of myelination of optic nerve axons was shown to be retarded in rats whose mothers consumed ethanol during pregnancy (Ashwell and Zhang, 1994; Samorajski et al., 1986). Similar results have been observed following early postnatal ethanol exposure (Phillips, 1989; Phillips and Krueger, 1990). Taken together, these results suggest that an increase in retinal cell loss along with impaired cell replication due to prolongation of the cell cycle and hypomyelination of optic nerve axons may contribute to the pathogenesis of ocular abnormalities seen in FAS/FAE.

B. AUDITORY SYSTEM

The incidence of hearing impairment in FAS/FAE children has been reported be quite high (Church, 1987). It also appears that prenatal ethanol exposure can have a negative impact at various levels of the auditory system. Evidence of sensorineural hearing loss in rats prenatally exposed to ethanol has been obtained from studies demonstrating abnormal electrical activity in the brainstem following the presentation of high-frequency sounds, and disruptions in auditory processing at the level of the cortex have been detected, as measured by cortical auditory-evoked potentials (Berman et al., 1992; Church, 1987; Church and Holloway, 1984).

These results have important clinical implications because good hearing is essential for normal speech and language development in children.

C. VESTIBULAR AND MOTOR SYSTEMS

A variety of vestibular and motor coordination problems have been associated with FAS/FAE. For example, Marcus (1987) reported on five FAS children who showed signs of axial ataxia (particularly problems with balance and gait) and kinetic tremor. Abnormalities in motor development and performance have been noted during infancy and well beyond birth in children who were exposed to "moderate" amounts of alcohol during gestation (Conry, 1990; Streissguth et al., 1980, 1990). Deficits in both gross and fine motor function were observed in 4-year-old children of social drinking mothers (Barr et al., 1990).

Animal models have also revealed developmental abnormalities in vestibular and motor function as a result of ethanol exposure. Rats prenatally exposed to ethanol have shown delayed development of motor reflexes (Lee et al., 1980), deficits in performance of tasks that require motor coordination (Abel and Dintcheff, 1978), and alterations in walking pattern or gait (Hannigan and Riley, 1988). Similar findings were reported following neonatal ethanol exposure. That is, these animals showed shorter stride lengths and more open step angles (Meyer et al., 1990a), as well as inferior performance on tasks requiring balance and motor coordination (Goodlett et al., 1991; Meyer et al., 1990b). All of these effects resemble the motor dysfunctions that occur following the disruption of cerebellar maturation, and as described later, there is evidence to suggest that ethanol-induced cerebellar damage may underlie, at least in part, many of the observed motor problems in FAS/FAE offspring.

IV. NEONATAL AND REGULATORY BEHAVIORAL EFFECTS

A. FETAL MOVEMENT

Although the deleterious effects of prenatal ethanol exposure have been well characterized in offspring after birth, very little is known about how fetuses respond to ethanol in the womb. However, a procedure that allows direct observation of rat fetuses *in utero* has been employed to study the effects of ethanol administered to the mothers (Smotherman et al., 1986). Following ethanol administration, a cesarean section was performed, and the mother's uterus was externalized and allowed to float in a warm bath so that fetal activity could be observed. Maternal ethanol administration produced a 51% reduction in overall fetal activity. More recently, spontaneous and stress-induced (cold saline applied on the skin) forelimb movements were examined in near-term fetal sheep (Leader et al., 1990). Maternal intravenous ethanol infusion that resulted in fetal sheep attaining blood ethanol levels of about 85 mg/dl decreased spontaneous forelimb movements. In contrast, stress-induced increases in forelimb movements were exaggerated in the ethanol-exposed fetal sheep. In addition, treatment of pregnant ewes with ethanol resulted in a suppression of breathing movements and brain activity in the near-term fetal sheep (Smith, Brien, Carmichael et al., 1989; Smith, Brien, Homan et al., 1989). These results provide the most direct evidence that maternal ingestion of ethanol influences fetal behavior *in utero*. They also are consistent with a study indicating altered human fetal "breathing movements" monitored indirectly in mothers who had just consumed alcohol (McLeod et al., 1983).

Taken together, these results related to fetal activity have important implications because reduced fetal movement *in utero* has been shown to have profound effects on morphological development in animals and humans. In addition, direct observation of fetal behavior in rats revealed that the adaptive response patterns to brief, experimentally induced hypoxia were diminished in intensity and duration among fetuses exposed to ethanol *in utero* (Smotherman and Robinson, 1987). Thus, altered fetal activity as a result of *in utero* ethanol exposure may be a contributing factor in the etiology of FAS/FAE.

B. FEEDING BEHAVIOR

Human newborns prenatally exposed to ethanol have been characterized as poor feeders, exhibiting a weak sucking response and irregular sucking patterns early in life. These infants are typically described as easily distracted and fatigued when suckling (Martin et al., 1979; Van Dyke et al., 1982). Similarly, feeding abnormalities have been found in rodents prenatally exposed to ethanol. In comparison to controls, rat pups exposed to ethanol *in utero* took longer to attach to the nipple of a test dam, spent less time suckling, and displayed weaker suckling pressure with an altered pattern of suckling (Chen et al., 1982; Rockwood and Riley, 1986; Subramanian, 1992). It was suggested that the deficits in nipple attachment behavior may be due to an impaired ability to detect olfactory attachment cues, difficulty in coordinating motor behaviors and orienting responses compatible with locating a nipple, or a dampened arousal level to significant stimuli related to nipple attachment (Rockwood and Riley, 1990). Similar deficits in suckling behavior have been noted in rat neonates exposed to ethanol during the early postnatal period (Barron et al., 1991). In addition, it recently has been hypothesized that growth retardation in rodents prenatally exposed to ethanol may be related to disturbances in feeding behavior that result from ethanol-induced alterations in developing monoamine systems such as dopamine (Middaugh and Boggan, 1991). These systems are clearly involved in food regulation (Stricker and Verbalis, 1990), and have been shown to be altered in animals prenatally exposed to ethanol (Druse et al., 1990). Taken together, these human and animal data suggest that inferior feeding behavior results from *in utero* ethanol exposure and, in turn, may contribute importantly to the postnatal growth retardation commonly seen in FAS and FAE.

C. REGULATORY BEHAVIOR

Human neonates born to mothers who consumed ethanol during pregnancy have been described as displaying poor state regulation. Sleep state disturbances, tremulousness, and jitteriness have been reported in infants exposed prenatally to ethanol (Coles et al., 1984; Landesman-Dwyer et al., 1978; Rosett et al., 1979; Scher et al., 1988; Streissguth, 1986). They also were observed to exhibit poor habituation, as measured on the Brazelton Neonatal Behavior Scale, and frequent state changes at the low end of the arousal continuum, primarily alterations between awake and drowsy (Streissguth et al., 1983).

Assessment of state regulation has been studied in animal models as well. For example, prenatal ethanol exposure has been shown to cause a delay in the development of thermoregulation, and the difference relative to controls is not simply a function of their smaller body size (Zimmerberg, Ballard, and Riley, 1987). Similar findings were obtained in mice (Gilliam and Kotch, 1992). In addition, ethanol-exposed rat pups moved farther along a thermal gradient to maintain body temperature equivalent to that of controls, suggesting that ethanol-exposed pups compensated behaviorally for the greater heat loss they experienced (Zimmerberg, Beckstead, and Riley, 1987). These data indicate that prenatal ethanol exposure impairs the ability of rat neonates to defend their body temperature, and hence, may underlie the altered thermal response to drug challenges observed in these animals later in adulthood (e.g., Abel, Bush et al., 1981; Nelson, Lewis et al., 1986; Taylor et al., 1987). In addition, altered thermoregulatory capacity may have an impact on drug responsiveness since this variable has been shown to influence sensitivity to a number of behavioral and physiological actions of ethanol and related drugs (Finn et al., 1994).

V. EFFECTS ON ACTIVITY AND EXPLORATORY BEHAVIOR

One of the most common behavioral effects observed following prenatal exposure to ethanol is an alteration in spontaneous locomotor activity. Increased open-field activity in ethanol-exposed offspring relative to controls has been demonstrated in rats (Abel, 1982; Bond and DiGiusto, 1976, 1977a; Branchey and Friedhoff, 1976; Caul et al., 1979; Fernandez, Caul, Haelein et al., 1983; Fernandez, Caul, Osborne et al., 1983; Osborne et al., 1980; Shaywitz et al., 1976) and mice (Becker and Randall, 1989; Randall et al., 1986). Similar findings have been reported following neonatal ethanol exposure in rats as well (Grant et al., 1983; Kelly, Hulsether et al., 1987; Kelly, Pierce et al., 1987; Riley et al., 1993). In many cases, the hyperactivity was found to be transient, with the effect waning as the animal matured (Bond and DiGiusto, 1977a; Shaywitz et al., 1976). However, other studies have shown this effect to persist, as evidenced by hyperactivity in adult offspring (Caul et al., 1979; Kelly, Pierce et al., 1987; Osborne et al., 1980; Randall et al., 1986). There are some reports that have not demonstrated an effect of prenatal ethanol exposure on spontaneous locomotor activity (Abel, 1979b; Abel and York, 1979; Abel, Dintcheff et al., 1981; Gallo and Weinberg, 1982; Shaywitz et al., 1979). Inconsistent results also have been obtained from studies assessing activity in running wheels (Abel, Dintcheff et al., 1981; Martin et al., 1978). Further, in one mouse strain, a significant reduction in open-field activity has been noted in prenatal ethanol-exposed offspring (Becker et al., 1993). A number of factors may account for the discrepancy in results, including species, strain, ethanol dosage, and timing of exposure, as well as various testing parameters (Meyer and Riley, 1986a).

A measure somewhat related to open-field activity involves the natural tendency for rodents to engage in exploratory behavior. Using experimental procedures that allow for exploratory behavior to be assessed independent of ambulatory tendencies, prenatal ethanol exposure was found to increase nose-poking and head-dipping behavior in rat offspring (Abel and Dintcheff, 1986b; Plonsky and Riley, 1983; Riley, Shapiro et al., 1979). Taken together, alterations in activity and exploratory behavior reported in animal models appears to be congruent with hyperactivity commonly noted in FAS/FAE children and adults (Streissguth et al., 1991).

VI. EFFECTS ON LEARNING AND MEMORY

Compromised mental abilities represents perhaps the most devastating consequence of prenatal ethanol exposure. Mental retardation is a common affliction in FAS/FAE patients. Clinical studies have revealed a number of cognitive disabilities that result from ethanol exposure during fetal development, including psychomotor dysfunction, attentional problems, and learning deficiencies. As is the case for many of the physical and physiological abnormalities associated with *in utero* ethanol exposure, animal models have been extensively used to study the cognitive deficits that result from prenatal ethanol exposure. This work has involved the use of a variety of testing situations, including both appetitively and aversively motivated tasks.

A. ATTENTION

The ability to attend to salient stimuli in the environment is a crucial component of the learning process. A number of clinical studies have reported attentional deficits in FAS/FAE children (Nanson and Hiscock, 1990; Streissguth et al., 1984, 1986), although this finding is not universal and in some cases has been found to depend on the postnatal (rearing) conditions of the patients (Boyd et al., 1991; Brown et al., 1991; Greene et al., 1991). In one study, attentional deficits were observed in rat offspring prenatally exposed to ethanol (Strupp et al., 1989). However, in another more recent study with rats, prenatal ethanol exposure did not alter attention, as measured by the cardiac orienting response (bradycardia) to a novel, innocuous, olfactory stimulus in preweanlings (Hayne et al., 1992). The orienting response is thought to index central processes involved in attention. Thus, at present there is only a limited amount of animal research addressing the issue of attentional deficits.

B. DISCRIMINATION AND REVERSAL LEARNING

A number of studies have demonstrated that prenatal ethanol exposure results in performance deficits in tasks involving discrimination learning. The deficits are particularly apparent under reversal learning conditions in which a previously rewarded response is subsequently no longer appropriate (not rewarded). For example, prenatal ethanol-exposed rats were found to exhibit impaired escape latencies (Shaywitz et al., 1976, 1979) and spatial discrimination (Lee et al., 1980; Lochry and Riley, 1980) in aversively-motivated T-maze tasks. Deficits were also noted in reversal learning when the arm of the T-maze that was previously associated with electroshock was switched and movement to that arm subsequently became the appropriate response (Riley, Lochry, Shapiro et al., 1979; Lochry and Riley, 1980). In a similar task involving brightness discrimination, these deficits were not observed in prenatal ethanol-exposed rats (Abel, 1978). Deficits in reversal learning in a water T-maze have been reported in mice exposed prenatally to ethanol (Wainwright et al., 1990). In addition, others have reported deficits in appetitively-motivated discrimination learning tasks (Anandam and Stern, 1980; Lee et al., 1980).

C. AVOIDANCE LEARNING

In passive avoidance tests, the subject is required to withhold responding in order to avoid receiving a noxious stimulus (typically electroshock). Relative to control offspring, deficits in passive avoidance learning have been reported in rats (Abel, 1982; Driscoll et al., 1982; Gallo and Weinberg, 1982; Lochry and Riley, 1980; Riley, Lochry and Shapiro, 1979) and mice (Becker and Randall, 1989; Becker et al., 1989) prenatally exposed to ethanol. In one study, gestational exposure to ethanol during the second trimester equivalent rather than the first trimester equivalent was found to be critical in producing impaired passive avoidance performance in rats (Driscoll et al., 1982). Deficits in passive avoidance learning also have been reported in rats exposed to ethanol during the early postnatal period, which corresponds to the third trimester equivalent (Barron and Riley, 1990).

Active avoidance tasks require the subject to make a response in order to avoid receiving shock. In one-way avoidance paradigms, the animal learns to move to another compartment of the test chamber to avoid electroshock. In two-way shuttle avoidance tasks, shock is delivered to each side of the test apparatus on an alternating basis. Hence, following the presentation of a warning cue (light or tone), the subject must learn to return to an area of the test chamber in which it was previously shocked. The effects of prenatal ethanol exposure on one-way avoidance learning have been inconsistent (Abel, 1978; Harris and Case, 1979; Shaywitz et al., 1979), whereas deficits in shuttle avoidance learning have been more consistently reported in rats (Abel, 1979a; Bond, 1981; Bond and DiGiusto, 1977b, 1978) and mice (Becker et al., 1989; Randall et al., 1986). In contrast to these reported deficits in shuttle avoidance performance, prenatal ethanol-exposed rat offspring have been shown to exhibit enhanced performance in a Y-maze discrimination avoidance task (Caul et al., 1979; Osborne et al., 1980), as well as in an unsignaled (Sidman) avoidance task (Riley et al., 1982). Thus, depending on the demands of a particular test situation, abnormal behavior in prenatal ethanol-exposed animals has been shown to result in altered performance in avoidance learning paradigms.

D. CLASSICAL CONDITIONING

Most of the studies examining the influence of prenatal ethanol exposure on classical conditioning have employed the conditioned taste aversion paradigm. In this situation, animals learn an association between the orosensory qualities of a novel fluid or food and subsequent illness. The strength of the association is assessed by measuring the extent to which subjects refrain from consuming the substance that was previously paired with illness. Prenatal ethanol exposure has been shown to impair such conditioned taste aversions in rats

(Riley, Lochry and Shapiro, 1979; Riley et al., 1984). Moreover, the effect was dose-dependent in that an inverse relationship was demonstrated between the level of exposure to ethanol *in utero* and the degree of consummatory avoidance behavior (learned aversion). In another study, prenatal ethanol exposure was found to influence performance in a conditioned lick suppression task (Caul et al., 1983). In this paradigm, a tone is first paired with the occurrence of foot shock. As a result of this classical conditioning association, licking behavior is suppressed when the tone is subsequently presented to thirsty subjects along with water. In comparison to control offspring, prenatal ethanol-exposed rats exhibited a more enduring lick suppression response over days (greater resistance to extinction). This pattern of results is similar to that observed in operant conditioning paradigms, as well.

E. OPERANT CONDITIONING

Prenatal ethanol exposure has been shown to produce impaired performance in a number of operant conditioning situations. For example, in comparison to controls, a greater number of trials were required for ethanol-exposed offspring to learn the response requirements in a fixed-ratio (FR) schedule of reinforcement (Riley et al., 1980). Further, once the response was learned, prenatal ethanol-exposed offspring exhibited greater resistance to extinction. Similarly, retarded acquisition of FR responding has been reported in mice, with the effect being greatest when the demands of the schedule were high (e.g., FR100) (Gentry and Middaugh, 1988; Middaugh and Ayers, 1988). Moreover, the deficit in FR responding was demonstrated in mice when *in utero* ethanol exposure was limited to only the second trimester equivalent (Middaugh and Gentry, 1992). Deficits in operant behavior also have been demonstrated when the response requirements for reinforcement are low. Accordingly, prenatal ethanol exposure has been shown to produce impaired performance in diffential reinforcement of low rates of responding (DRL) and differential reinforcement of response omission (DRO) schedules of reinforcement (Driscoll et al., 1980; Gentry and Middaugh, 1988). In both situations, reinforcement was contingent upon the animal refraining from responding until a cue was presented that signaled the availability of a reward. The inferior performance in operant behavior displayed by prenatal ethanol-exposed offspring has been attributed to reduced efficacy of reward (Gentry and Middaugh, 1988; Middaugh and Gentry, 1992) as well as impaired response inhibition (Riley, Barron, and Hannigan, 1986; Riley, Lochry, and Shapiro, 1979).

F. SPATIAL LEARNING AND DISCRETE-TRIALS INSTRUMENTAL LEARNING

A number of studies have demonstrated that prenatal ethanol exposure results in impaired performance in tasks that require animals to utilize spatial cues to navigate to a particular place or goal. For example, in the Morris water maze procedure, prenatal ethanol-exposed rats exhibited delays in finding a hidden platform (Blanchard, Riley et al., 1987; Blanchard et al., 1990; Gianoulakis, 1990). Similar effects were observed in rats exposed to ethanol as neonates (during the third trimester equivalent) (Goodlett et al., 1987; Kelly et al., 1988). Prenatal ethanol exposure also has been shown to produce impaired radial arm maze performance (Reyes et al., 1989), as well as deficits in operant tasks involving spatial learning (Zimmerberg et al., 1991). One study using mice failed to find an effect of prenatal ethanol exposure on Morris water maze performance (Wainwright et al., 1993).

A recent study examined the influence of prenatal ethanol exposure on serial-pattern learning (La Fiette et al., 1994). This type of task is thought to involve the cognitive processing of information such that "rules" are generated on the basis of patterns and relationships among serially presented stimuli or events. In this study the number of food pellets was varied in the goalbox of a straight alley. As expected, the latency to reach the goalbox increased as the number of anticipated pellets decreased. However, prenatal ethanol-exposed rats exhibited less change in performance as the number of food pellets in the goalbox changed. This deficit in serial-pattern performance is similar to that observed in children with FAS (La Fiette et al., 1994).

Another study examined the influence of prenatal ethanol exposure on the response to an abrupt, unexpected reduction in reward (Becker et al., 1988). In an appetitively-motivated task (consummatory negative contrast), animals typically consume substantially less of a small (sucrose) reward if they had previous experience with a larger reward. Although not statistically significant, prenatal ethanol-exposed rats exhibited less change in consummatory behavior following reward reduction (Becker et al., 1988). Thus, behavior following complete reward reduction (extinction) appears to be more sensitive to prenatal ethanol exposure than procedures involving partial reward reduction.

G. PERSEVERATIVE TENDENCIES AND RESPONSE DISINHIBITION

Many of the behavioral abnormalities noted in animals prenatally and neonatally exposed to ethanol have been characterized as being perseverative in nature, as well as reflective of a general deficit in response inhibition (Meyer and Riley, 1986a; Riley, Barron, and Hannigan, 1986). Perseverative tendencies in prenatal ethanol-exposed offspring have been revealed in tasks involving spontaneous alternations of responses (Abel, 1982; Riley, Lochry, Shapiro et al., 1979) and operant situations in which alternating responses between two levers was a requirement for reinforcement (Zimmerberg et al., 1989). Increased resistance to extinction displayed by prenatal ethanol-exposed offspring in classical conditioning (Caul et al., 1983) and operant conditioning (Riley et al., 1980) situations may be also taken as evidence for perseverative tendencies in these animals.

Riley and his colleagues have advanced the notion that many of the behavioral alterations exhibited in animals prenatally exposed to ethanol may be accounted for by a general deficit in response inhibition (Meyer and Riley, 1986a; Riley, Barron, and Hannigan, 1986). Poorer performance on tasks, such as passive avoidance and conditioned taste aversion, that involve withholding, or inhibiting, a response most clearly support this hypothesis. In addition, increased spontaneous activity and exploratory behavior, deficits in reversal learning and spontaneous alternation, and increased resistance to extinction all lend support to the response-inhibition-deficit hypothesis. Furthermore, lack of response inhibition may contribute to enhanced performance in tasks that require the subject to make an active response. Of course, other cognitive disabilities such as impaired discrimination learning cannot be easily explained by a general deficit in inhibitory responding.

Nevertheless, behavioral abnormalities in prenatal ethanol-exposed animals characterized as perseverative and disinhibitory may reflect many of the behavioral problems observed in FAS/FAE children and adults. These include impulsivity, restlessness, hyperactivity, attentional deficits and distractibility, and poor adaptive or coping responses to relatively normal daily challenges (Streissguth et al., 1991). Further, as discussed later, many of these behavioral effects in prenatal ethanol-exposed animals that reflect a response inhibition deficit are similar to those observed following hippocampal damage. This, along with the recognized role of the hippocampus in spatial learning and memory function, have served to focus attention on the relationship between hippocampal damage and cognitive-behavioral dysfunction in offspring prenatally exposed to ethanol (Riley, Barron, and Hannigan, 1986).

VII. PERMANENCE OF BEHAVIORAL EFFECTS

One of the many advantages of using laboratory animals in prenatal alcohol research is that the long-term effects of such treatment can be studied in a relatively shorter period of time than would be possible in humans. As previously described, prenatal ethanol exposure has been shown to have adverse effects on a variety of neonatal behaviors that are vital for normal growth and development in animals and humans. These include impaired feeding and thermoregulation, motor dysfunctions, and learning disabilities (Barron et al., 1987; Coles et al., 1985; Martin et al., 1977; 1979; Riley et al., 1984; Smith et al., 1986; Streissguth et al., 1983; Zimmerberg, 1989). In fact, prenatal ethanol exposure has been shown to produce deficits in associative learning in rats as early as 3 days of age (Barron, Gagnon et al., 1988). Many of the behavioral teratogenic effects of ethanol have been reported to be transient. That is, the effects appear to diminish with age, suggesting that prenatal ethanol exposure produces only a developmental delay in CNS maturation (Abel, 1982). Alternatively, or in addition to a simple delay in development, some of this "normalization" that comes with age may be due to the development of independent compensatory strategies designed to cope with and overcome the disabilities. Recent data showing the reemergence of some of these deficits in older animals suggests that the compensatory systems break down with age and/or the recruitment of such compensatory mechanisms compromises the ability to perform other more complex functions later in life (Riley, 1990). For example, under particular testing conditions, behavioral abnormalities such as hyperactivity, tendency toward perseveration, and impaired performance in learning and memory tasks have been observed in adult animals that were prenatally exposed to ethanol (Abel, 1979a; Abel and Dintcheff, 1986a; Becker et al., 1989; Gianoulakis, 1990; Middaugh and Ayers, 1988; Plonsky and Riley, 1983; Randall et al., 1986; Reyes et al., 1989; Zimmerberg et al., 1989, 1991).

In addition, behavioral deficits not evident under normal unchallenged conditions may become "unmasked" when the complexity of the testing situation increases or when testing is conducted under stress and drug challenge conditions (Riley, 1990). For instance, adult rats prenatally exposed to ethanol evidenced deficits in performance in an operant behavioral situation that required response inhibition and visual discrimination, but did not differ from control offspring on a component of the operant task that required relatively simple nondiscrimination responding (Vigliecca et al., 1989). Likewise, learning to alternate responding between two levers in order to receive a reward was impaired in ethanol-exposed rats only when visual cues were removed so that the subjects had to rely solely on spatial cues (Zimmerberg et al., 1989). Finally, behavioral abnormalities have been noted in prenatal ethanol-exposed animals when the adult offspring are tested under a stressful or drug-challenged state (Becker et al., 1993; Hannigan et al., 1987; Means et al., 1984; Nelson, Lewis et al., 1986).

Thus, animal research has revealed that many of the behavioral dysfunctions observed in young offspring are permanent, persisting into adulthood. That some of these effects have been noted in rats more than a year old (Abel and Dintcheff, 1986a, 1986b) further attests to the permanence of these functional deficits. These findings also complement those of clinical studies that have revealed persistent attentional, cognitive, and neurobehavioral deficits in children of alcohol-abusing mothers (Aronson et al., 1985; Spohr and Steinhausen, 1987; Streissguth et al., 1984, 1986). Most recently, the long-term manifestations of heavy alcohol consumption during pregnancy have been documented in adolescent and adult FAS/FAE cases (Streissguth et al., 1991). Taken together, both basic and clinical studies have clearly established the potential for prenatal ethanol exposure to produce long-lasting cognitive-behavioral disabilities. Studies aimed at distinguishing between those deleterious effects of *in utero* ethanol exposure that wane with maturity from those that are enduring, as well as identifying the conditions under which such deficits may be "unveiled" or expressed are of particular clinical relevance because the information gained from such studies would undoubtedly be important for targeting treatment and therapy for the affected children.

VIII. EFFECTS ON SEXUALLY DIMORPHIC BEHAVIORS

Sexual differentiation of the rat brain is both organizationally and functionally dependent upon the hormonal milieu of the fetus during early development. This critical period in the rat corresponds to the last week of gestation through the first week of postnatal life. Testosterone biosynthesis and secretion by the fetal and neonatal testes is the primary hormonal influence acting during this period to promote masculinization of the brain. Sexually dimorphic patterns of behavior represent some of the functional manifestations of these early hormonal forces. A number of studies have demonstrated that prenatal ethanol exposure alters these behaviors. For example, adult male rats prenatally exposed to ethanol have been shown to exhibit demasculinized sexual behavior (Barron, Treman et al., 1988; Hard et al., 1984; Parker et al., 1984; Rudeen et al., 1986) and a feminized pattern of maze learning (McGivern et al., 1984), saccharin preference (McGivern et al., 1984, 1987), and rough and tumble play (Meyer and Riley, 1986b). Adult prenatal ethanol-exposed male C3H/He mice exhibited a demasculinized pattern of scent marking behavior (Hale et al., 1992). Female offspring prenatally exposed to ethanol have displayed a masculinization of saccharin consumption, maze learning performance, and play behavior, as well as a marked reduction in maternal behavior (Barron and Riley, 1985; McGivern et al., 1984; Meyer and Riley, 1986b). Additionally, sex differences in sensitivity to prenatal ethanol effects on a variety of behavioral, pharmacological, immune, and stress responses have been noted as well (Weinberg, 1988; Weinberg and Jerrells, 1991). Some studies have not found an effect of prenatal ethanol exposure on sexually dimorphic behaviors (Abel and Dintcheff, 1986c; Becker et al., 1985; Blanchard and Hannigan, 1994). Nevertheless, as discussed later, there is some evidence to suggest that these functional alterations may result from ethanol-induced perturbation of the developing hypothalamic-pituitary-gonadal axis.

IX. EFFECTS ON STRESS RESPONSIVENESS

Following exposure to a stressful situation, a number of physiological and behavioral events are set in motion to ready the body for its response. One of the physiological components of the stress response is the stimulation of brain structures (hypothalamus and pituitary gland) that, in turn, activate the adrenal cortex to release corticosteroid hormones. Alcohol consumption during pregnancy has been shown to have profound effects on both maternal and fetal hypothalamic-pituitary-adrenocortical (HPA) axes (Weinberg et al., 1986). Indeed, several studies have demonstrated that adult rats prenatally exposed to ethanol exhibit a heightened stress response (elevated plasma corticosteroid levels) in comparison to control offspring. This augmented stress response was observed after exposure to a variety of stressors such as noise, shaking, electric footshock, physical restraint, and placement in a novel environment (Nelson, Taylor et al., 1986; Taylor, Branch, and Kokka, 1981; Taylor, Branch, Liu et al., 1982; Weinberg, 1988). Similar results were obtained following challenges with drugs such as ethanol and morphine (Nelson, Lewis et al., 1986; Taylor, Branch, and Kokka, 1981; Taylor, Branch, Liu et al., 1981). Moreover, recovery from the stress response was found to be retarded in offspring prenatally exposed to ethanol. These animals also were shown to be less sensitive than control offspring to manipulations that typically dampen the stress response, suggesting that prenatal ethanol exposure results in a deficit in HPA response inhibition or recovery from stress (Weinberg, 1988). Furthermore, prenatal ethanol exposure has been shown to result in a suppressed β-endorphin response to stress in neonatal rats, but a potentiated stress response at later stages of development (Angelogiani and Gianoulakis, 1989).

In addition to the enhanced hormonal (corticosteroid) and opioid (β-endorphin) response to stress, adult rats prenatally exposed to ethanol also demonstrate an augmented behavioral response to stress (Hannigan et al., 1987; Nelson et al., 1985). For example, after exposure to intermittent footshock, ethanol-exposed offspring displayed greater analgesia (reduced pain sensitivity) than did controls (Nelson et al., 1985). This type of stress-induced analgesia can be blocked by administering a narcotic antagonist (naloxone), indicating that the analgesia is mediated by an endogenous opioid mechanism. In contrast, when an equivalent amount of footshock is presented continuously rather than intermittently, the resultant analgesia cannot be blocked by opiate antagonists, and is hence, thought to be mediated by nonopioid mechanisms. The nonopioid form of stress-induced analgesia (that which follows continuous shock treatment) is not altered by prenatal ethanol exposure (Nelson et al., 1985). Whether this distinction is related to alterations in brain opioid systems (β-endorphins?) as a function of *in utero* ethanol exposure remains to be determined.

The mechanism whereby ethanol exposure during fetal life heightens the behavioral and HPA response to certain stressors later in adulthood is unclear at present. As discussed later, studies have shown that prenatal ethanol exposure alters the HPA axis at several levels. Regardless of the mechanism(s), however, it is clear that fetal exposure to ethanol has long-term effects, altering both physiological and behavioral responsiveness to stress in adulthood. These perturbations in stress responsiveness in offspring prenatally exposed to ethanol may be of great significance because the ability to respond to and cope with a constantly changing and challenging (stressful) environment is crucial for survival and the maintenance of normal life in animals and humans. Furthermore, alterations in the HPA system may represent a contributing etiologic factor in the behavioral teratogenic actions of ethanol, and might impact on the integrity of immune responsiveness, which has been shown to be compromised in offspring prenatally exposed to ethanol (Gottesfeld and Abel, 1991; Lee and Rivier, 1993).

X. EFFECTS ON ETHANOL PREFERENCE, SENSITIVITY, AND TOLERANCE

A number of studies have examined whether *in utero* exposure to ethanol influences subsequent responsiveness to the drug later in life. Given the prevalence of "social" drinking during pregnancy, this variable clearly has the potential of contributing to the myriad of factors that influence the propensity to use and/or abuse ethanol. Unfortunately, results from these studies have been generally inconsistent. For example, some studies have shown prenatal ethanol exposure to increase ethanol intake in preference tests (Bond and DiGiusto, 1976; Phillips and Stainbrook, 1976), whereas others have not confirmed this finding (Abel and York, 1979). In some cases the increase in ethanol preference was only transient (Randall et al., 1983), whereas others observed the effect in adult offspring (Bond and DiGiusto, 1976). Whereas the above studies involved chronic administration of ethanol during the gestation period, Molina et al. (1985; 1987) demonstrated that even an acute exposure to ethanol early in gestation (day 8) can influence (increase) later postnatal ethanol intake.

More recently, studies have demonstrated that the rat in both fetal and infancy stages of development is capable of attending to and processing ethanol-related orosensory cues. That is, prenatal and early postnatal nonassociative and associative experience with these cues subsequently influenced responsiveness to the drug. For example, direct administration of ethanol into amniotic fluid just prior to delivery (gestation day 21) resulted in an enhanced neonatal bradycardiac response to the odor of ethanol in comparison to control rat neonates (Chotro and Molina, 1992). This treatment also resulted in an increase in ethanol intake in preweanling rat offspring (Chotro and Molina, 1990). Further, when the administration of ethanol into amniotic fluid was paired with an aversive stimulus (electroshock), the neonatal response to the odor of ethanol was greatly inhibited. These results suggest that learning relative to ethanol cues occurs *in utero* (Chotro and Molina, 1992). Early postnatal experience with ethanol cues also has been shown to influence later ethanol intake (Hunt et al., 1993; Molina et al., 1986). However, inconsistent results have been obtained with regard to the influence of postweanling ethanol experience on later ethanol selection (Ho et al., 1989; Tolliver and Samson, 1991). Whether exposure to ethanol during the perinatal period has a lasting influence on ethanol preference-intake remains unclear.

Inconsistent results have been reported regarding other ethanol responses, as well. For example, relative to control offspring, rats prenatally exposed to ethanol have been shown to be more (Reyes et al., 1993; Taylor, Branch, and Kokka, 1981; Taylor, Branch, Liu et al., 1981) or less (Abel, Bush et al., 1981; Anandam et al., 1980; Molina et al., 1987) sensitive to the hypothermic actions of ethanol. Studies with mice have found no effect of prenatal ethanol exposure on this measure (Gilliam et al., 1990). Likewise, prenatal ethanol exposure has been reported to either increase (Reyes et al., 1993), decrease (Fulginiti et al., 1989), or not influence (Abel, 1979b; Gilliam et al., 1990; Harris and Case, 1979; Perez et al., 1983; Randall et al., 1983) sensitivity to the hypnotic effects of ethanol, as measured by "sleeptime" duration. Offspring exposed prenatally to ethanol also have been shown to exhibit greater sensitivity to the locomotor stimulant properties of ethanol than do control offspring (Becker et al., 1993; Rockman et al., 1989). This effect in mice was found to vary as a function of age, gender, and dose of ethanol (Becker et al., 1993). A similar pattern of results was obtained in a study measuring low-dose ethanol-induced dopamine release in the nucleus accumbens in rat offspring (Blanchard et al., 1993). In contrast, reduced sensitivity to the motor incoordinating actions of ethanol (Reyes et al., 1993) and the drug's disruptive effects on operant responding (Middaugh and Ayers, 1988) have been noted in offspring exposed prenatally to ethanol. Explanation for these discrepant findings has yet to be specifically identified, although differences in species and strain, method of ethanol administration to the pregnant dams, dose of the ethanol challenge, and gender and age of the tested offspring are all viable possibilities.

In a drug discrimination paradigm, Middaugh and Ayers (1988) reported that mice exposed prenatally to ethanol exhibited reduced sensitivity to the discriminative stimulus properties of ethanol. In drug discrimination studies, interoceptive cues of a drug are associated with a response-contingent reward on one lever whereas the subjective effects of a placebo are associated with a response on an alternative lever. Following such discrimination training, prenatal ethanol exposure was found to impair generalization to an ethanol cue. That is, ethanol-exposed mice were less able to discriminate lower doses of ethanol from a placebo than were control offspring (Middaugh and Ayers, 1988).

Although a number of studies have addressed the potential influence of perinatal ethanol exposure on later acute sensitivity to the drug, very little attention has focused on chronic ethanol effects. In one study, prenatal ethanol exposure was shown to retard the development of tolerance to the motor incoordinating effects of ethanol in adult mouse offspring (Becker and Hale, 1991). Interestingly, development of ethanol dependence was not influenced by prenatal ethanol exposure (Becker, unpublished data).

XI. EFFECTS ON RESPONSIVENESS TO OTHER PHARMACOLOGIC CHALLENGES

Psychopharmacologic assessment of ethanol-exposed offspring has proven to be a valuable tool in studies on the behavioral teratogenic properties of ethanol. As indicated by Hannigan and Blanchard (1988), this experimental approach has provided researchers with a unique opportunity to examine the functional integrity of various brain neurotransmitter systems. This work also plays an important role in assessing the efficacy of potential pharmacologic treatment strategies designed to mitigate behavioral-cognitive impairments that result from *in utero* ethanol exposure.

A variety of behavioral-cognitive testing situations have been used to show that prenatal ethanol exposure alters sensitivity to numerous drug challenges targeted at various brain neurochemical systems. For example, sensitivity to morphine-induced hypothermia and analgesia was found to be enhanced in offspring exposed prenatally to ethanol (Nelson, Lewis et al., 1986; Nelson, Taylor et al., 1986). In another study, adult rats prenatally exposed to ethanol exhibited reduced sensitivity to the hypothermic effects of pentobarbital and diazepam, but did not differ from control offspring in response to the thermic actions of morphine, chlorpromazine, and amphetamine (Abel, Bush et al., 1981).

In an attempt to reveal disturbances in cholinergic systems, sensitivity to cholinergic agonists (physostigmine) and antagonists (scopolamine) have been examined in a number of behavioral situations. Physostigmine was reported to attenuate hyperactivity in young ethanol-exposed offspring (Riley, Barron, Driscoll et al., 1986), but in another study the drug exacerbated this response (Bond, 1988). The muscarinic cholinergic antagonist, scopolamine, decreased overactivity in ethanol-exposed rat offspring (Bond, 1986). In contrast, prenatal ethanol exposure did not influence sensitivity of adult offspring to physostigmine effects on shuttle avoidance performance (Blanchard and Riley, 1988) or scopolamine effects on performance in Morris maze and shuttle avoidance tasks (Hannigan et al., 1993; Rockwood and Riley, 1985). These studies suggest that perturbation of cholinergic systems may be only transient in offspring exposed prenatally to ethanol.

A number of studies have examined the effects of *in utero* exposure to ethanol on later sensitivity to catecholaminergic agents. Enhanced sensitivity to the locomotor stimulant properties of dopamine agonists such as apomorphine (Hannigan et al., 1990), methylphenidate (Means et al., 1984; Ulug and Riley, 1983), and amphetamine (Blanchard, Hannigan et al., 1987; Hannigan and Pilati, 1991) have been demonstrated in rats exposed prenatally to ethanol. Reduced sensitivity to the cataleptic effects of the D1 dopamine receptor antagonist, SCH 23390, was noted as well in rats exposed prenatally to ethanol (Hannigan, 1990). These effects have been primarily observed in male rat offspring. In adult male mouse offspring, prenatal ethanol exposure has been shown to dampen sensitivity to apomorphine, particularly its ability to attenuate ethanol-stimulated activity (Becker et al., 1994). On the other hand, adult male mice exposed prenatally to ethanol exhibited enhanced sensitivity to the ability of the catecholamine synthesis blocker, alpha-methyl-p-tyrosine, to attenuate the stimulant properties of ethanol (Becker et al., 1993). In general, these studies suggest that prenatal ethanol exposure alters nigrostriatal and mesotelencephalic dopamine systems in rat and mouse offspring.

Reduced sensitivity to the locomotor suppressant effects of the alpha-2 adrenoceptor agonist, clonidine, given alone as well as in combination with a stimulant dose of ethanol has been observed in adult mice exposed prenatally to ethanol (Weathersby et al., 1994). A study with rats found no alteration in sensitivity to the locomotor and thermal effects of higher doses of clonidine (Hannigan et al., 1988), but altered hormonal responses to clonidine challenge has been reported in rats exposed prenatally to ethanol (McGivern et al., 1986). Taken together, these psychopharmacologic studies have indicated that prenatal ethanol exposure produces functional disturbances in a number of neurotransmitter systems. Further support of these findings has been obtained through more direct examination of neurochemical abnormalities that follow prenatal ethanol exposure.

XII. POSSIBLE MECHANISMS UNDERLYING BEHAVIORAL ABNORMALITIES

As noted earlier, one of the major advantages of using animals in fetal alcohol research is that it allows for the identification of CNS defects that may underlie observed cognitive-behavioral deficits. The following is an overview of the diverse adverse effects ethanol has been shown to exert on the developing brain, with particular emphasis on the possible relationship of such effects to observed behavioral abnormalities.

A. NEUROANATOMICAL EFFECTS

Adverse effects of ethanol on brain growth and development have been documented in a number of regions spanning all levels of the nervous system. These include development of the neural tube (Ross and Persaud, 1989), peripheral nerves (Baruah and Kinder, 1989), cranial nerves (Miller and Muller, 1989), structural components of the extrapyramidal motor system (Zajac et al., 1989), the hypothalamus (Fresnillo et al., 1989), the hippocampus (Barnes and Walker, 1981; West and Pierce, 1986), the cerebellum (Pierce et al., 1989, 1993; West, Goodlett, Bonthius et al., 1990), and the cerebral cortex (Kotkoskie and Norton, 1990; Miller, 1993; Miller and Nowakowski, 1991; Miller and Potempa, 1990). In addition, glial cells (which play an important role in CNS development and function) also have been shown to be sensitive targets to ethanol-induced damage when exposure occurs during gestation and/or postnatally (Fletcher and Shain, 1993; Goodlett et al., 1993; Lokhorst and Druse, 1993a; Phillips and Krueger, 1992; Shetty and Phillips, 1992). These neuroanatomical anomalies have been observed in rats and mice following chronic ethanol exposure, as well as following acute exposure during early gestation (Kotch and Sulik, 1992; Sulik et al., 1984), late gestation (Kotkoskie and Norton, 1988, 1990), and the early postnatal period (Goodlett et al., 1990). Furthermore, it has become apparent that different regions of the developing brain may have different thresholds and perhaps different critical periods of susceptibility to ethanol injury. Thus, in considering the issues of threshold and critical periods in relation to ethanol-induced brain and behavior abnormalities, it appears that the teratogenic outcome depends not only on the dose, pattern, and timing of the ethanol exposure, but on the brain region as well (West and Goodlett, 1990).

In more recent years, attention has focused on examining whether ethanol-induced brain defects are, indeed, related to observed behavioral dysfunctions (Riley and Barron, 1989; West, Goodlett, Bonthius et al., 1990). For example, Riley and his colleagues (Riley, 1990; Riley and Barron, 1989; Riley, Barron, and Hannigan, 1986) have suggested that many of the behavioral abnormalities that result from ethanol exposure during development bear remarkable similarity to effects observed following hippocampal damage. Indeed, response-inhibition deficits along with impaired learning and memory performance and spatial navigational dysfunction have all been demonstrated in animals exhibiting ethanol-induced hippocampal damage (Abel et al., 1984; West, Goodlett, Bonthius et al., 1990; Wigal and Amsel, 1990). In some cases, the magnitude of the behavioral deficit was correlated with the degree of hippocampal perturbation.

Similarly, motor deficits have been linked to structural defects in the cerebellum. In fact, many of the balance and motor dysfunctions observed in children (Barr et al., 1990; Conry, 1990; Streissguth et al., 1990) and animals (Goodlett et al., 1991; Meyer et al., 1990b) prenatally exposed to ethanol resemble those that occur following disruption of cerebellar maturation, and the developing cerebellum has been shown to be particularly vulnerable to ethanol-induced insult (Clarren, 1986; Phillips, 1986; West, Goodlett, Bonthius et al., 1990). Further, adult animals exposed to ethanol during the neonatal brain growth spurt that exhibited deficits in performance on several tasks requiring balance and motor coordination, also evidenced retarded cerebellar growth and dysmorphology (Goodlett et al., 1991; Meyer et al., 1990b). Alterations in brain structures that are components of the extrapyramidal motor system may also contribute to motor deficits (Mattson et al., 1992; Zajac et al., 1989).

Another intensely studied brain region that has been shown to be sensitive to prenatal ethanol exposure is the cerebral cortex. Miller and his colleagues have demonstrated that the generation, proliferation, and migration of cerebral cortical neurons are altered following gestational ethanol exposure (Miller, 1986, 1987, 1988, 1989, 1993; Miller and Potempa, 1990). Consequently, a reduction in the number of cortical neurons as well as an abnormal pattern of distribution and organization of these neurons has been observed in several subregions of the cortex (Miller, 1986, 1988; Miller and Potempa, 1990). It has been suggested that these anomalies may be, at least in part, related to an alteration in the number of cells cycling and the cell cycle length of proliferating cells (Miller and Nowakowski, 1991), as well as delays in the migration of cortical neurons (Miller, 1993). Abnormalities in cortical neurons that project to the spinal cord have been noted as well (Miller, 1987; Miller and Al-Rabiai, 1994; Miller et al., 1990). Thus, whereas an ethanol exposure during the early postnatal brain growth spurt has been shown to result in neuronal cell death among relatively mature neuronal populations in certain brain regions such as prenatally derived hippocampal pyramidal neurons (West and Pierce, 1986) and cerebellar Purkinje cells (Pierce et al., 1989; West, Goodlett, Bonthius et al., 1990), a prenatal exposure to ethanol exerts deleterious effects on cortical neurons during their proliferative stage of development. Further, it is certainly conceivable that aberrant cortical development may underlie impaired sensory and motor functions, as well as disabilities involving higher cognitive processing noted in FAS/FAE clinical populations.

B. NEUROCHEMICAL EFFECTS

In addition to morphologic anomalies, ethanol exposure during CNS development has been shown to produce alterations in brain microcircuitry, including neurochemical aberrations, as well as disturbances in the synaptic integrity of neuronal connections. Whereas many earlier studies examined neurotransmitter levels in whole-brain tissue, more recent work has focused on steady-state levels as well as the utilization and receptor function of transmitter systems in more specific brain regions (Druse, 1986). In addition, effects of prenatal ethanol exposure on neurotransmitter receptor-effector mechanisms have been recently reported (Balduini and Costa, 1989, 1990; Queen et al., 1993; Zimmerman and Collins, 1989). Most of this neurochemistry work has focused on the monoamines (dopamine, norepinephrine, and serotonin), although effects on other systems such as acetylcholine (Brodie and Vernadakis, 1990, 1991; Kelly et al., 1989; Light et al., 1989a, 1989b; Schambra et al., 1990; Serbus et al., 1986; Wigal et al., 1990) and amino acid neurotransmitters including gamma-aminobutyric acid (GABA) (Zhulin and Zabludovskii, 1989; Zimmerman et al., 1990) and glutamate (Farr et al., 1988; Martin et al., 1992; Morrisett et al., 1989; Noble and Richie, 1989; Savage et al., 1991) have been reported.

Prenatal exposure to ethanol has been shown to markedly affect the development of the dopamine system. A greater than 40% reduction in striatal dopamine (DA) was evident in ethanol-exposed rats at 19 days of age, and a 25% reduction in DA uptake sites was observed at 35 days of age. Further, these animals showed a 20 to 25% reduction in the number of D1 receptor sites at 19 and 37 days of age, whereas D2 receptor sites were not significantly influenced by *in utero* ethanol exposure (Druse et al., 1990). A 30% reduction in DA was observed in the cortex as early as five days of age. This, along with a deficiency in cortical D1 receptors, indicates that prenatal ethanol exposure alters the development of the dopamine system in the frontal cortex as well as in the striatum (Druse et al., 1990). Given the role of dopamine in a variety of neurobehavioral events, such as regulatory and maintenance functions, motor activity, and arousal, these effects on the developing dopamine system may be related to feeding, motor, and attentional deficits observed in children born to mothers who consumed alcohol during pregnancy. In addition, such perturbations of the DA system may underlie altered responsiveness to drugs acting on the dopamine system (Becker et al., 1993; Becker et al., 1994; Blanchard et al., 1987; Hannigan, 1990; Hannigan et al., 1990) as well as suppressed lesion-induced plasticity of dopaminergic neurons (Gottesfeld et al., 1989) in ethanol-exposed offspring.

There is some evidence for ethanol-induced alteration of central and peripheral noradrenergic systems. Deficiencies in norepinephrine levels have been noted in the hypothalamus, septal area, and cortex (Cooper and Rudeen, 1988; Detering et al., 1980, 1981). These changes may underlie altered sensitivity to noradrenergic drugs (Weathersby et al., 1994). In addition, prenatal ethanol exposure has been shown to influence the sympathetic nervous system regulation of immune function. That is, rats prenatally exposed to ethanol exhibited enhanced norepinephrine turnover, lower steady-state norepinephrine levels, and a reduced number of beta-adrenoceptor binding sites in lymphoid tissues (Gottesfeld, Christie et al., 1990; Gottesfeld, Morgan et al., 1990). There is also evidence that impaired thermoregulation in rats prenatally exposed to ethanol may be related to an alteration in the noradrenergic innervation of brown adipose tissue (Zimmerberg et al., 1993).

Druse and her colleagues have demonstrated large (50%) reductions in serotonin (5HT) and its major metabolite (5HIAA) in the cortex of rats exposed prenatally to ethanol in comparison to control offspring (Rathbun and Druse, 1985). Decreased levels of 5HT were also detected in the cerebellum and brainstem, but not in other brain regions, including the hippocampus, hypothalamus, and striatum. Further, the deficiency of 5HT and 5HIAA in the brainstem and cortex was found as early as gestation days 15 and 19, respectively (Druse et al., 1991). In addition, prenatal ethanol exposure was shown to reduce the number of cortical 5HT uptake sites (15 to 30% decrease) and $5HT_1$ receptor sites (25% decrease), but not alter the number of $5HT_2$ receptor sites (Druse and Paul, 1989; Tajuddin and Druse, 1989a, 1989b). The reduction in $5HT_1$ receptors was found to be due to a decrease in cortical $5HT_{1A}$ receptor subtype binding sites, but not in $5HT_{1B}$ (Druse et al., 1991). Interestingly, maternal treatment during pregnancy with the $5HT_{1A}$ agonist, buspirone, was found to reverse the ethanol-induced deficiency of 5HT and 5HIAA in certain brain regions (Tajuddin and Druse, 1993).

In contrast to the findings of these *in vivo* ethanol exposure studies, 5HT content and uptake in cultured fetal serotonergic neurons was not found to be influenced by four days of *in vitro* ethanol exposure (Lokhorst and Druse, 1993b). The reason for a difference in results obtained from *in vivo* and *in vitro* exposure studies is not clear at present. In any event, the reported effects of *in vivo* ethanol exposure on brain monoamine systems are particularly significant because dopamine and serotonin not only are implicated in mediating a variety of neurobehavioral functions, but they are also thought to play an important role in neuronal maturation and differentiation during embryonic development (Lankford et al., 1988; Lauder et al., 1983).

Neurochemical and electrophysiological analysis of the hippocampus also has revealed significant alterations following prenatal exposure to ethanol. For example, 45-day-old rats that were exposed to relatively low levels of ethanol *in utero* (maternal blood ethanol levels were in the range of 30–40 mg/dl) evidenced a 20–36% reduction in mossy fiber zinc concentrations, a trace element thought to be involved in hippocampal synaptic neurotransmission (Savage et al., 1989). In addition, a marked reduction in the number of NMDA-sensitive [^{3}H]glutamate binding sites was observed in the dorsal hippocampal formation (Farr et al., 1988; Savage et al., 1991). The metabotropic glutamate receptor subtype also has been shown to be influenced by prenatal ethanol treatment, but this effect may require a higher level of ethanol exposure (Noble and Richie, 1989; Queen et al., 1993). In contrast, prenatal ethanol exposure did not alter [^{3}H]AMPA and [^{3}H]vinylidene kainate binding in the same hippocampal regions where NMDA binding was affected (Farr et al., 1989; Martin et al., 1992). Further, the littermates of animals that displayed a decreased number of NMDA binding sites also exhibited reduced sensitivity to NMDA and a decreased capacity to generate long-term potentiation, as measured electrophysiologically in hippocampal slices (Morrisett et al., 1989; Swartzwelder et al., 1988). Long-term potentiation is thought to represent the major biological substrate underlying learning and memory consolidation. In another recent electrophysiological study, abnormal synaptic potentials were recorded from the CA1 area of hippocampal slices derived from adult prenatal ethanol-exposed rats (Tan et al., 1990). These results indicate that prenatal ethanol exposure can result in long-term subtle abnormalities in the function of the hippocampus, a brain region important in the process of learning and memory consolidation.

Finally, there is evidence to suggest that the above mentioned neuroanatomical and neurochemical aberrations may result in, or be related to, an alteration in brain activity (Miller and Dow-Edwards, 1988; Vingan et al., 1986). These studies employed an autoradiographic 2–[^{14}C] deoxyglucose technique to examine the rate of glucose use in several brain regions of adult prenatal ethanol-exposed and control offspring. Prenatal ethanol exposure was found to significantly decrease glucose utilization in sensory and motor cortex regions as well as in numerous limbic system structures including the hippocampus (Miller and Dow-Edwards, 1988; Vingan et al., 1986).

C. NEUROENDOCRINE EFFECTS

Prenatal ethanol exposure disturbs the normal functioning of several neuroendocrine systems, including prolactin secretion, thyroid function, and gonadal and adrenocortical hormonal control. The clinical impact of these changes may include feeding deficits; poor state and metabolic regulation; altered sexual maturation; and a host of sensory, motor, and cognitive impairments. For example, maternal ethanol consumption both prior to and following birth has been shown to produce alterations in the structure and function of mammary gland tissue (Steven et al., 1989; Vilaro et al., 1989), as well as inhibited suckling-induced prolactin release and milk yield (Subramanian and Abel, 1988; Subramanian et al., 1990). Although the mechanism is unknown, it has been suggested that ethanol may disrupt the transmission of neural impulses from the nipples to the CNS and/or directly influence central control of prolactin release via action at the hypothalamus or pituitary gland (Subramanian et al., 1990). This latter possibility is intriguing because dopamine is known to control prolactin production in the anterior pituitary gland, and this neurotransmitter is sensitive to the adverse effects of ethanol during development (Druse et al., 1990). Moreover, aside from the possible dangers of ethanol delivered to babies through breast milk (Little et al., 1989), these data suggest that ethanol consumption may hinder the ability of a mother to adequately nourish her child after delivery.

Many physical, sensorimotor, and cognitive-behavioral abnormalities associated with prenatal ethanol exposure are similar to those produced by perinatal hypothyroidism. Fetal exposure to ethanol has been shown to result in abnormal thyroid function, including a 10–15% reduction in serum thyroxine (T4) concentrations (Hannigan and Bellisario, 1990; Lee and Wakabayashi, 1986; Portoles et al., 1988). Ethanol may act directly on the fetus where it has been shown to have profound effects on the thyroid axis (Yamamoto et al., 1989), or indirectly via the mother since T4 is known to cross the placenta (Vulsma et al., 1989). Since thyroid hormones are critical trophic factors for normal somatic and neural maturation, an alteration in the hypothalamic-pituitary-thyroid axis may result in impaired trophic regulation of growth, differentiation, and general metabolic activity. Interestingly, a recent study indicated the early thyroid hormone treatment was effective in reversing some of the developmental delays observed in rats exposed prenatally to ethanol (Gottesfeld and Silverman, 1990).

Prenatal ethanol exposure has been shown to influence the hypothalamic-pituitary-gonadal (HPG) axis at all levels in males and females. For example, in female rats prenatally exposed to ethanol, hypothalamic luteinizing hormone-releasing hormone (LHRH) and plasma luteinizing hormone (LH) were found to be significantly reduced in comparison to those of controls (Morris et al., 1989). Pituitary tissue from female rats exposed prenatally to ethanol was found to exhibit a blunted response to LHRH and estradiol-17β, as measured by the reduced expression of LHβ-mRNA (Creighton-Taylor and Rudeen, 1991a). In addition, ethanol has been shown to decrease the *in vitro* LH-induced production of estrogens and progesterone in cultured human ovarian granulosa cells (Saxema et al., 1990). Taken together, these results suggest that the adverse effects of ethanol on hypothalamic, pituitary, and ovarian hormonal activity may underlie some of the reproductive abnormalities observed in alcoholic women. Indeed, animal research has demonstrated delayed sexual maturation, reproduction dysfunction, and altered sexual behavior in female offspring prenatally exposed to ethanol (Becker and Randall, 1987; Boggan et al., 1979; Creighton-Taylor and Rudeen, 1991b; Hard et al., 1984; Parker et al., 1984).

Likewise, the HPG axis in males is influenced at all levels by fetal ethanol exposure. As previously indicated, the development of sexual dimorphic neuroanatomical and behavioral patterns is dependent on the androgenic milieu of the fetus during the critical period of brain sexual differentiation. Leydig cells in the interstitial compartment of the fetal rat testes are responsible for the production of testosterone commencing around the sixteenth day of gestation. Prenatal ethanol exposure has been shown to suppress testicular steroidogenesis (Kelce et al., 1989; Kelce et al., 1990a), as well as to blunt the normal testosterone surge that occurs on gestation days 18 and 19 (McGivern, Raum et al., 1988) and immediately following birth (McGivern et al., 1993). In addition, fetal testes from rats prenatally exposed to ethanol were relatively insensitive to *in vitro* LH-induced stimulation of testosterone secretion as compared to controls (McGivern, Raum et al., 1988). However, this effect was not replicated in a subsequent study (McGivern et al., 1993). Further, Rudeen and his co-workers (Kelce et al., 1989; Kelce et al., 1990a) found that the inhibition of testicular steroidogenesis was due to a specific reduction (36% decrease) in the activity of the catalytic enzyme 17α-hydroxylase (i.e., the activity of other testicular steroidogenic microsomal enzymes was not similarly influenced). Interestingly, fetal Leydig cells are apparently more sensitive to this ethanol effect than are the structurally and functionally distinct adult-type Leydig cells, since inhibition of testicular steroidogenic enzyme activity was not evident in rats on postnatal days 20, 40, or 60 (Kelce et al., 1989; Kelce et al., 1990a).

In addition to the direct effects of ethanol on testicular steroidogenesis, another mechanism by which fetal ethanol exposure might alter sexual differentiation of the CNS is through the altered metabolism of androgens within the brain itself. Local formation of testosterone metabolites within the hypothalamic preoptic area is essential for normal sexual differentiation of the male central nervous system. Both androgens and estrogens are involved in brain sexual differentiation through the conversion of circulating testosterone to more potent androgens (5α-dihydroxytestosterone) via the 5α-reductase pathway and to estrogens (17β-estradiol) via the aromatase pathway. Studies on whether prenatal ethanol exposure influences hypothalamic activity of 5α-reductase and/or aromatase have yielded conflicting results (Kelce et al., 1990b; McGivern, Roselli et al., 1988). Nevertheless, other studies have shown that prenatal ethanol exposure results in decreased testicular weight, decreased plasma testosterone levels, and decreased responsiveness to steroid feedback regulation in the hypothalamus of adult male rats (Jungkuntz-Burgett et al., 1990; Parker et al., 1984; Rudeen, 1986; Rudeen et al., 1986; Udani et al., 1985).

Of particular significance is the finding that these hormonal perturbations produced by perinatal ethanol exposure may underlie the altered gender differences in brain structure and function (behavior) commonly observed in ethanol-exposed offspring. For example, in addition to alterations in sexually dimorphic patterns of behavior mentioned earlier, male offspring exposed prenatally to ethanol have been shown to exhibit a demasculinization of the sexual dimorphic nucleus of the hypothalamic preoptic area (Barron, Treman et al., 1988; Rudeen, 1986; Rudeen et al., 1986) and the corpus callosum (Zimmerberg and Mikus, 1990), as well as brain asymmetries or cerebral laterality (Zimmerberg and Reuter, 1989). Thus, the ethanol-induced suppression of perinatal testicular steroidogenesis and decreased circulating androgen levels may have permanent consequences because *in utero* ethanol exposure occurred during the critical period for brain sexual differentiation. Alterations in the sexual dimorphic nature of brain structure and behavioral patterns following prenatal ethanol exposure support this contention.

Finally, both maternal and fetal hypothalamic-pituitary-adrenocortical (HPA) axes have been shown to be influenced by ethanol (Weinberg et al., 1986). For example, elevated plasma corticosteroid levels have been noted in pregnant females (Weinberg and Bezio, 1987). Likewise, plasma, brain, and adrenal corticosteroid levels were found to be elevated in rodent neonates exposed prenatally to ethanol (Kakihana et al., 1980; Taylor, Branch, Cooley-Matthews et al., 1982; Taylor et al., 1986; Weinberg, 1989). These neonates were also found to have higher hypothalamic corticotropin-releasing factor (CRF) content and higher pituitary adrenocorticotropin hormone (ACTH) content (Redei et al., 1989). Further, during the first week of life, ethanol-exposed pups exhibited a blunted response to several stressors (Taylor et al., 1986; Weinberg, 1989). Although this reduced responsiveness is a transient phenomenon (the stress response normalized by the second week of life), the perturbation in basal and stress-related HPA activity early in life may have long-term consequences on the manner in which these animals respond to stress in adulthood. Indeed, as noted earlier, prenatal ethanol exposure has been shown to result in a heightened and prolonged HPA response to certain stressors in adult offspring (Nelson, Taylor et al., 1986; Taylor, Branch, and Kokka, 1981; Taylor, Branch, Liu et al., 1982; Weinberg, 1988).

Fetal ethanol exposure could perturb the HPA axis at several levels, including the hypothalamus (where CRF is produced), pituitary corticotrophs (which upon CRF stimulation, secrete ACTH), or the adrenal cortex (which following ACTH stimulation, produces and releases corticosteroids into circulation). Some data have suggested that the effect of prenatal ethanol exposure may be centrally mediated because adrenocortical sensitivity to ACTH is not altered in ethanol-exposed offspring (Taylor, Branch, Liu et al., 1982). On the other hand, elevated plasma ACTH levels were noted concomitant with higher plasma corticosteroid levels following footshock stress in rats prenatally exposed to ethanol (Nelson, Taylor et al., 1986).

Another possible explanation for the HPA hyperresponsiveness in ethanol-exposed offspring is that *in utero* ethanol exposure may produce deficits in feedback control of the HPA system. Support for this hypothesis is provided by data indicating a faster rebound from dexamethasone suppression in ethanol-exposed rats compared to that in control offspring (Weinberg et al., 1986). A more recent study addressed the possibility that alterations in brain receptors for glucocorticoids may underlie the proposed deficit in feedback regulation of HPA activity (Weinberg and Petersen, 1991). The hippocampus contains the highest concentration of glucocorticoid receptors in the brain and is the principal brain target site for adrenocortical steroid hormones (McEwen et al., 1986). Two major receptor types have been identified. The type I receptor is concentrated in the hippocampus and appears to be involved in corticosteroid modulation of hippocampus-associated brain functions, whereas the type II receptor (which exists in higher concentrations and is more widely distributed) is thought to be involved in mediating feedback actions of corticosteroids on stress-activated brain mechanisms (DeKloet and Reul, 1987). Prenatal ethanol exposure was found to have no effect on maximal binding or binding affinity of either type I or type II glucocorticoid receptors in the hippocampus under nonstressed (basal) conditions (Weinberg and Petersen, 1991). Whether prenatal treatment differences in these receptors will emerge following stress remains to be determined. Of course, it is possible that even under nonstressed conditions, prenatal ethanol exposure may alter glucocorticoid receptors in brain regions other than the hippocampus (Weinberg and Petersen, 1991). In any case, it is clear that prenatal ethanol exposure has both immediate and long-term effects on the HPA axis, which in turn, may be manifested as altered physiological and behavioral responsiveness to stress.

XIII. SUMMARY

Ethanol is widely recognized as a teratogen, exerting deleterious effects on many major organ systems in the developing fetus (USDHHS, 1993). Adverse effects of ethanol on the developing CNS may represent some of the most devastating consequences of excessive ethanol use during pregnancy. Both clinical and animal research has demonstrated that the teratogenic properties of ethanol exist on a continuum, with effects ranging from gross physical malformations and severe mental retardation to more subtle cognitive-behavioral dysfunctions. These latter effects have been shown to occur in the absence of overt physical anomalies, and hence, define ethanol as a classic behavioral teratogen.

Animal research has played a major role in advancing our knowledge of the myriad immediate and long-lasting adverse consequences that follow prenatal ethanol exposure, including the behavioral teratogenic effects of the drug. Use of animal models has allowed researchers to exert rigorous control over such variables as dose, pattern, timing, and duration of ethanol exposure, as well as a number of potentially confounding factors commonly associated with chronic alcohol use, such as malnutrition, poor rearing (environmental) conditions, disease, smoking, and other drug use. Importantly, the plethora of adverse consequences following prenatal ethanol exposure documented in animals bears remarkable similarity to the neurobehavioral abnormalities and mental disabilities observed in humans afflicted with FAS/FAE.

More specifically, animal studies have demonstrated impaired sensory function in several modalities (e.g., vision, audition), a number of motor and balance abnormalities, and a host of learning and memory deficits. These cognitive-behavioral deficiencies have been demonstrated using a wide variety of methodologies and test situations, including acute and chronic models of gestational ethanol exposure, as well as neonatal exposure models. Some of the behavioral teratogenic effects of ethanol appear to be transient, with the deficit gradually diminishing as the animal matures. This has led to the suggestion that at least some of these effects may be reflective of a developmental delay (Abel, 1982). However, a number of cognitive-behavioral abnormalities have been shown to persist into adulthood. Further, many of the behavioral abnormalities observed in animals (and humans) prenatally and neonatally exposed to ethanol have been hypothesized to result from a general deficit in response inhibition (Riley, Barron, and Hannigan, 1986).

In addition, and perhaps of most significance, the use of animal models has allowed for a more mechanistic analysis of ethanol's behavioral teratogenic properties. For example, neuroanatomical, neurochemical, and neuroendocrine substrates of observed behavioral dysfunction can be elucidated and studied more directly in animals, whereas such studies in humans are obviously limited for a variety of reasons including ethical considerations. Indeed, with the teratogenic properties of ethanol firmly established, more recent experimental attention has been focused on underlying mechanisms. For example, neuroanatomical studies have demonstrated that prenatal-neonatal ethanol exposure alters the cytoarchitectural structure, organization, and metabolic activity of numerous brain regions. Electrophysiological and neurochemical investigations have revealed disturbances in the integrity of synaptic neurotransmission, as well as perturbation of several neurotransmitter systems. Neuroendocrine studies have demonstrated alterations in many hormonal systems, including dysregulation of nervous system control of endocrine function. These effects have been primarily identified in animals that exhibit no external physical abnormalities, and thus, lend support to the notion that such alterations in CNS function may represent the neurobiological substrates for the behavioral teratogenic actions of ethanol. Continued refinement of animal model systems along with the use of more sophisticated molecular biological techniques and experimental approaches hold promise for the further advancement of the field. In addition, the recruitment of other more sophisticated neurobiological techniques in clinical investigations, such as nuclear magnetic resonance (NMR) brain imaging procedures, as well as the application of more specific neuropsychological test batteries to clinical FAS/FAE populations, will provide further insight into the relationship between structural and functional deficits that result from prenatal ethanol exposure (West, Goodlett, and Brandt, 1990). These research efforts ultimately may contribute to the development of therapeutic interventions or prevention strategies.

ACKNOWLEDGMENTS

This work was partially supported by grants from the National Institute on Alcohol Abuse and Alcoholism (AA07791) and the Medical Research Division of the Department of Veterans Affairs.

REFERENCES

Abel, E.L. Effects of ethanol on pregnant rats and their offspring. *Psychopharmacology*, 57:5–11, 1978.

Abel, E.L. Prenatal effects of alcohol on adult learning in rats. *Pharmacol. Biochem. Behav.*, 10:239–243, 1979a.

Abel, E.L. Prenatal effects of alcohol on open-field behavior, step-down latencies and "sleep time." *Behav. Neural Biol.*, 25:406–410, 1979b.

Abel, E.L. *In utero* alcohol exposure and developmental delay of response inhibition. *Alcoholism: Clin. Exp. Res.,* 6:369–376, 1982.

Abel, E.L. *Fetal Alcohol Syndrome and Fetal Alcohol Effects.* New York: Plenum Press, 1984.

Abel, E.L., Bush, R., and Dintcheff, B.A. Exposure of rats to alcohol *in utero* alters drug sensitivity in adulthood. *Science,* 212:1531–1533, 1981.

Abel, E.L. and Dintcheff, B.A. Effects of prenatal alcohol exposure on growth and development in rats. *J. Pharmacol. Exp. Therap.,* 207:916–921, 1978.

Abel, E.L. and Dintcheff, B.A. Factors affecting the outcome of maternal alcohol exposure: I. Parity. *Neurobehav. Toxicol. Teratol.,* 6:373–377, 1984.

Abel, E.L. and Dintcheff, B.A. Factors affecting the outcome of maternal alcohol exposure: II. Maternal age. *Neurobehav. Toxicol. Teratol.,* 7:263–266, 1985.

Abel, E.L. and Dintcheff, B.A. Effects of prenatal alcohol exposure on behavior of aged rats. *Drug Alcohol Dependence,* 16:321–330, 1986a.

Abel, E.L. and Dintcheff, B.A. Effects of prenatal alcohol exposure on nose poking in year-old rats. *Alcohol,* 3:210–214, 1986b.

Abel, E.L. and Dintcheff, B.A. Saccharin preference in animals prenatally exposed to alcohol: no evidence of altered sexual dimorphism. *Neurobehav. Toxicol. Teratol.,* 8:521–523, 1986c.

Abel, E.L., Dintcheff, B.A., and Bush, R. Effects of beer, wine, whiskey, and ethanol on pregnant rats and their offspring. *Teratology,* 23:217–222, 1981.

Abel, E.L., Jacobson, S., and Sherwin, B.T. *In utero* alcohol exposure produced functional and structural damage. *Neurobehav. Toxicol. Teratol.,* 5:363–366, 1984.

Abel, E.L. and Sokol, R.J. Fetal alcohol syndrome is now leading cause of mental retardation. *Lancet,* 2:1222, 1986.

Abel, E.L. and Sokol, R.J. Incidence of fetal alcohol syndrome and economic impact of FAS-related anomalies. *Drug Alcohol Dependence,* 19:51–70, 1987.

Abel, E.L. and Sokol, R.J. A revised conservative estimate of the incidence of FAS and its economic impact. *Alcoholism: Clin. Exp. Res.,* 15:514–524, 1991.

Abel, E.L. and York, J.L. Absence of effect of prenatal ethanol on adult emotionality and ethanol in rats. *J. Stud. Alcohol,* 40:547–553, 1979.

Anandam, N. and Stern, J.M. Alcohol *in utero*: effects on preweanling appetitive learning. *Neurobehav. Toxicol.,* 2:199–205, 1980.

Anandam, N., Strait, T., and Stern, J.M. *In utero* ethanol retards early discrimination learning and decreases adult responsiveness to ethanol. *Teratology,* 21:25–26A, 1980.

Angelogianni, P. and Gianoulakis, C. Prenatal exposure to ethanol alters the ontogeny of the β-endorphin response to stress. *Alcoholism: Clin. Exp. Res.,* 13:564–571, 1989.

Aronson, M., Kyllerman, M., Sabel, K.G., Sandin, B., and Olegard, R. Children of alcoholic mothers: developmental, perceptual, and behavioural characteristics as compared to matched controls. *Acta Paediatrica Scandinavica,* 74:27–35, 1985.

Ashwell, K.W.S. and Zhang, L.-L. Optic nerve hypoplasia in an acute exposure model of the fetal alcohol syndrome. *Neurotoxicol. Teratol.,* 16:161–167, 1994.

Balduini, W. and Costa, L.G. Effects of ethanol on muscarinic receptor-stimulated phosphoinositide metabolism during brain development. *J. Pharmacol. Exp. Therap.,* 250:541–547, 1989.

Balduini, W. and Costa, L.G. Developmental neurotoxicity of ethanol: *in vitro* inhibition of muscarinic receptor-stimulated phosphoinositide metabolism in brain from neonatal but not adult rats. *Brain Res.,* 512:248–252, 1990.

Barnes, D.E. and Walker, D.W. Prenatal ethanol exposure permanently reduces the number of pyramidal neurons in rat hippocampus. *Dev. Brain Res.,* 1:333–340, 1981.

Barr, H.M., Streissguth, A.P., Darby, B.L., and Sampson, P.D. Prenatal exposure to alcohol, caffeine, tobacco, and aspirin: effects on fine and gross motor performance in 4-year-old children. *Dev. Psychol.,* 26:339–348, 1990.

Barron, S., Gagnon, W.A., Mattson, S.N., Kotch, L.E., Meyer, L.S., and Riley, E.P. The effects of prenatal alcohol exposure on odor associative learning in rats. *Neurotoxicol. Teratol.,* 10:333–339, 1988.

Barron, S., Kelly, S.J., and Riley, E.P. Neonatal alcohol exposure alters suckling behavior in neonatal rat pups. *Pharmacol. Biochem. Behav.,* 39:423–427, 1991.

Barron, S. and Riley, E.P. Pup-induced maternal behavior in adults and juvenile rats exposed to alcohol prenatally. *Alcoholism: Clin. Exp. Res.,* 9:360–365, 1985.

Barron, S. and Riley, E.P. Passive avoidance performance following neonatal alcohol exposure. *Neurotoxicol. Teratol.,* 12:135–138, 1990.

Barron, S., Treman, S.B., and Riley, E.P. Effects of prenatal alcohol exposure on the sexually dimorphic nucleus of the preoptic area of the hypothalamus in male and female rats. *Alcoholism: Clin. Exp. Res.,* 12:59–64, 1988.

Barron, S., Zimmerberg, B., Rockwood, G.A., and Riley, E.P. Prenatal alcohol exposure: recent work on behavioral dysfunctions in pre-weaning rats. *Adv. Alcohol Substance Abuse,* 6:105–118, 1987.

Baruah, J.K. and Kinder, D. Pathological changes in peripheral nerves in experimental fetal alcohol syndrome. *Alcoholism: Clin. Exp. Res.,* 13:547–548, 1989.

Becker, H.C. The effects of alcohol on fetal and postnatal development in mice. *Special Report to the National Institute on Alcohol Abuse and Alcoholism*, 1992.

Becker, H.C. and Hale, R.L. Effect of prenatal ethanol exposure on ethanol tolerance in adult offspring is dependent on blood level after chronic ethanol exposure. *Soc. Neurosci. Abstr.*, 17:1504, 1991.

Becker, H.C., Hale, R.L., Boggan, W.O., and Randall, C.L. Effects of prenatal ethanol exposure on later sensitivity to the low-dose stimulant actions of ethanol in mouse offspring: possible role of catecholamines. *Alcoholism: Clin. Exp. Res.*, 17:1325–1336, 1993.

Becker, H.C. and Randall, C.L. Two generations of maternal alcohol consumption in mice: effect on pregnancy outcome. *Alcoholism: Clin. Exp. Res.*, 11:240–242, 1987.

Becker, H.C. and Randall, C.L. Effects of prenatal ethanol exposure in C57BL mice on locomotor activity and passive avoidance behavior. *Psychopharmacology*, 97:40–44, 1989.

Becker, H.C., Randall, C.L., and Anton, R.F. Prenatal alcohol exposure does not influence sexual dimorphism of saccharin preference in C57 mice. *Soc. Neurosci. Abstr.*, 11:293, 1985.

Becker, H.C., Randall, C.L., and Middaugh, L.D. Behavioral teratogenic effects of ethanol in mice. *Ann. N.Y. Acad. Sci.*, 562:340–341, 1989.

Becker, H.C., Randall, C.L., and Riley, E.P. Effect of prenatal ethanol exposure on response to abrupt reward reduction. *Neurotoxicol. Teratol.*, 10:121–125, 1988.

Becker, H.C., Weathersby, R.T., and Hale, R.L. Prenatal ethanol exposure alters sensitivity to the effects of apomorphine given alone and in combination with ethanol on locomotor activity in adult male mouse offspring. *Neurotoxicol. Teratol.*, 17:57–64, 1994.

Berman, R.F., Beare, D.J., Church, M.W., and Abel, E.L. Audiogenic seizure susceptibility and auditory brainstem responses in rats prenatally exposed to alcohol. *Alcoholism: Clin. Exp. Res.*, 16:490–498, 1992.

Blanchard, B.A. and Hannigan, J.H. Prenatal ethanol exposure: Effects on androgen and nonandrogen dependent behaviors and on gonadal development in male rats. *Neurotoxicol. Teratol.*, 16:31–39, 1994.

Blanchard, B.A., Hannigan, J.H., and Riley, E.P. Amphetamine-induced activity after fetal alcohol exposure and undernutrition in rats. *Neurotoxicol. Teratol.*, 9:113–119, 1987.

Blanchard, B.A., Pilati, M.L., and Hannigan, J.H. The role of stress and age in spatial navigation deficits following prenatal exposure to ethanol. *Psychobiology*, 18:48–54, 1990.

Blanchard, B.A. and Riley, E.P. Effects of physostigmine on shuttle avoidance in rats exposed prenatally to ethanol. *Alcohol*, 5:27–31, 1988.

Blanchard, B.A., Riley, E.P., and Hannigan, J.H. Deficits on a spatial navigation task following prenatal exposure to ethanol. *Neurotoxicol. Teratol.*, 9:253–258, 1987.

Blanchard, B.A., Steindorf, S., Wang, S., Le Fevre, R., Mankes, R.F., and Glick, S.D. Prenatal ethanol exposure alters ethanol-induced dopamine release in nucleus accumbens and striatum in male and female rats. *Alcoholism: Clin. Exp. Res.*, 17:974–981, 1993.

Boggan, W.O., Randall, C.L., and Dodds, H.M. Delayed sexual maturation in female C57BL/6J mice prenatally exposed to alcohol. *Res. Commun. Chem. Pathol. Pharmacol.*, 23:117–125, 1979.

Bond, N.W. Effects of prenatal alcohol exposure on avoidance conditioning in high- and low-avoidance rat strains. *Psychopharmacology*, 74:177–181, 1981.

Bond, N.W. Prenatal alcohol exposure and offspring hyperactivity: effects of scopolamine and methylscopolamine. *Neurobehav. Toxicol. Teratol.*, 8:287–292, 1986.

Bond, N.W. Prenatal alcohol exposure and offspring hyperactivity: effects of physostigmine and neostigmine. *Neurotoxicol. Teratol.*, 10:59–63, 1988.

Bond, N.W. and DiGiusto, E.L. Effects of prenatal alcohol consumption on open-field behaviour and alcohol preference in rats. *Psychopharmacology*, 46:163–168, 1976.

Bond, N.W. and DiGiusto, E.L. Prenatal alcohol consumption and open-field behaviour in rats: effects of age at time of testing. *Psychopharmacology*, 52:311–312, 1977a.

Bond, N.W. and DiGiusto, E.L. Effects of prenatal alcohol consumption on shock avoidance learning in rats. *Psychological Rep.*, 41:1269–1270, 1977b.

Bond, N.W. and DiGiusto, E.L. Avoidance conditioning and Hebb-Williams maze performance in rats treated prenatally with alcohol. *Psychopharmacology*, 58:69–71, 1978.

Bonthius, D.J. and West, J.R. Blood alcohol concentration and microencephaly: a dose-response study in the neonatal rat. *Teratology*, 37:223–231, 1988.

Boyd, T.A., Ernhart, C.B., Greene, T.H., Sokol, R.J., and Martier, S. Prenatal alcohol exposure and sustained attention in the preschool years. *Neurotoxicol. Teratol.*, 13:49–55, 1991.

Branchey, L. and Friedhoff, A.J. Biochemical and behavioral changes in rats exposed to ethanol *in utero*. *Ann. N.Y. Acad. Sci.*, 273:328–330, 1976.

Brodie, C. and Vernadakis, A. Critical periods to ethanol exposure during early neuroembryogenesis in the chick embryo: cholinergic neurons. *Dev. Brain Res.*, 56:223–228, 1990.

Brodie, C. and Vernadakis, A. Muscle-derived factors reverse the cholinotoxic effects of ethanol during early neuroembryogenesis in the chick embryo. *Dev. Brain Res.*, 61:183–188, 1991.

Brown, R.T., Coles, C.D., Smith, I.E., Platzman, K., Silverstein, J., Erickson S., and Falek A. Effects of prenatal alcohol exposure at school age. II. Attention and behavior. *Neurotoxicol. Teratol.,* 13:369–376, 1991.

Caul, W.F., Fernandez, K., and Michaelis, R.C. Effects of prenatal ethanol exposure on heart rate, activity, and response suppression. *Neurobehav. Toxicol. Teratol.,* 5:461–464, 1983.

Caul, W.F., Osborne, G.L., Fernandez, K., and Henderson, G.I. Open-field and avoidance performance of rats as a function of prenatal ethanol treatment. *Addictive Behav.,* 4:311–322, 1979.

Chen, J.S., Driscoll, C.D., and Riley, E.P. The ontogeny of suckling behavior in rats prenatally exposed to alcohol. *Teratology,* 26:145–153, 1982.

Chotro, M.G. and Molina, J.C. Acute ethanol contamination of the amniotic fluid during gestational day 21: postnatal changes in alcohol responsiveness in rats. *Dev. Psychobiol.,* 23:535–547, 1990.

Chotro, M.G. and Molina, J.C. Bradycardiac responses elicited by alcohol odor in rat neonates: influence of *in utero* experience with ethanol. *Psychopharmacology,* 106:491–496, 1992.

Church, M.W. Chronic *in utero* alcohol exposure affects auditory function in rats and humans. *Alcohol,* 4:231–239, 1987.

Church, M.W. and Gerkin, K.P. Hearing disorders in children with Fetal Alcohol Syndrome: findings from case reports. *Pediatrics,* 82:147–154, 1988.

Church, M.W. and Holloway, J.A. Effects of prenatal ethanol exposure on the postnatal development of the brainstem auditory evoked potential in the rat. *Alcoholism: Clin. Exp. Res.,* 8:258–265, 1984.

Clarren, S.K. Neuropathology in fetal alcohol syndrome. In West, J.R. (Ed.) *Alcohol and Brain Development.* New York: Oxford University Press, 1986, 158–166.

Clarren, S.K. and Smith, D.W. The fetal alcohol syndrome: a review of the world literature. *N. Engl. J. Med.,* 298:1063–1067, 1978.

Coles, C.D., Smith, I.E., Fernhoff, P.M., and Falek, A. Neonatal ethanol withdrawal: characteristics in clinically normal, nondysmorphic neonates. *J. Pediatr.,* 105:445–451, 1984.

Coles, C.D., Smith, I.E., Fernhoff, P.M., and Falek, A. Neonatal neurobehavioral characteristics as correlates of maternal alcohol use during gestation. *Alcoholism: Clin. Exp. Res.,* 9:454–460, 1985.

Conry, J. Neuropsychological deficits in Fetal Alcohol Syndrome and Fetal Alcohol Effects. *Alcoholism: Clin. Exp. Res.,* 14:650–655, 1990.

Cooper, J.D. and Rudeen, P.K. Alterations in regional catecholamine content and turnover in the male rat brain in response to *in utero* ethanol exposure. *Alcoholism: Clin. Exp. Res.,* 12:282–285, 1988.

Cordero, J.F., Floyd, R.L., Martin, M.L., Davis, M., and Hymbaugh, K. Tracking the prevalence of FAS. *Alcohol Health Res. World.,* 18:82–85, 1994.

Creighton-Taylor, J.A. and Rudeen, P.K. Fetal alcohol exposure and effects of LHRH and PMA on LHβ-mRNA expression in the female rat. *Alcoholism: Clin. Exp. Res.,* 15:1031–1035, 1991a.

Creighton-Taylor, J.A. and Rudeen, P.K. Prenatal ethanol exposure and opiatergic influence on puberty in the female rat. *Alcohol,* 8:187–191, 1991b.

Debeukelaer, M.M., Randall, C.L., and Stroud, D.R. Renal anomalies in the fetal alcohol syndrome. *J. Pediatr.,* 91:759–760, 1977.

Detering, N., Collins, R.M., Hawkins, R.L., Ozand, P.T., and Karahasan, A.M. The effects of ethanol on developing catecholamine neurons. *Adv. Exp. Med. Biol.,* 132:721–727, 1980.

Detering, N., Collins, R.M., Hawkins, R.L., Ozand, P.T., and Karahasan, A.M. Comparative effects of ethanol and malnutrition on the development of catecholamine neurons: a long lasting effect in the hypothalamus. *J. Neurochem.,* 36:2094–2096, 1981.

DeKloet, E.R. and Reul, J.M.H.M. Feedback action and tonic influence of corticosteroids on brain function: a concept arising from the heterogeneity of brain receptor systems. *Psychoneuroendocrinology* 12:83–105, 1987.

Diaz, J. and Samson, H.H. Impaired brain growth in neonatal rat pups exposed to ethanol. *Science,* 208:751–753, 1980.

Dobbing, J. and Sands, J. Comparative aspects of the brain growth spurt. *Early Human Dev.,* 3:79–83, 1979.

Driscoll, C.D., Chen, J.S., and Riley, E.P. Operant DRL performance in rats following prenatal alcohol exposure. *Neurobehav. Toxicol.,* 2:207–211, 1980.

Driscoll, C.D., Chen, J.S., and Riley, E.P. Passive avoidance performance in rats prenatally exposed to alcohol during various periods of gestation. *Neurobehav. Toxicol. Teratol.,* 4:99–103, 1982.

Driscoll, C.D., Streissguth, A.P., and Riley, E.P. Prenatal alcohol exposure: comparability of effects in humans and animal models. *Neurotoxicol. Teratol.,* 12:231–237, 1990.

Druse, M.J. Effects of perinatal alcohol exposure on neurotransmitters, membranes, and proteins. In West, J.R. (Ed.) *Alcohol and Brain Development.* New York: Oxford University Press, 1986, pp. 343–372.

Druse, M.J., Ku, A., and Tajuddin, N. Effects of *in utero* ethanol exposure on the developing serotonergic system. *Alcoholism: Clin. Exp. Res.,* 15:678–684, 1991.

Druse, M.J. and Paul, L.H. Effects of *in utero* ethanol exposure on serotonin uptake in cortical regions. *Alcohol,* 5:455–459, 1989.

Druse, M.J., Tajuddin, N., Kuo, A., and Connerty, M. Effects of *in utero* ethanol exposure on the developing dopaminergic system in rats. *J. Neurosci. Res.,* 27:233–240, 1990.

Farr, K.L., Montano, C.Y., Paxton, L.L., and Savage, D.D. Prenatal ethanol exposure decreases hippocampal ^{3}H-glutamate binding in 45-day-old rats. *Alcohol,* 5:125–133, 1988.

Farr, K.L., Montano, C.Y., Paxton, L.L., and Savage, D.D. Prenatal ethanol exposure decreases hippocampal ^{3}H-vinylidene kainic acid binding in 45-day-old rats. *Neurotoxicol. Teratol.,* 10:563–568, 1989.

Fernandez, K., Caul, W.F., Haelein, M., and Vorhees, C.V. Effects of prenatal alcohol on homing behavior, maternal responding and open-field activity in rats. *Neurobehav. Toxicol. Teratol.,* 5:351–356, 1983.

Fernandez, K., Caul, W.F., Osborne, G.L., and Henderson, G.I. Effects of chronic alcohol exposure on offspring activity in rats. *Neurobehav. Toxicol. Teratol.,* 5:135–137, 1983.

Finn, D.A., Syapin, P.J., Bejanian, M., Jones, B.L., and Alkana, R.L. Temperature dependence of ethanol depression in mice: dose response. *Alcoholism: Clin. Exp. Res.,* 18:382–386, 1994.

Fletcher, T.L. and Shain, W. Ethanol-induced changes in astrocyte gene expression during rat central nervous system development. *Alcoholism: Clin. Exp. Res.,* 17:993–1001, 1993.

Fresnillo, M., Villa-Elizaga, I., Olazabal, A., Frizell, E., de-Ochoa, J.A.F., Ballesteros, A., and Sierrasesumaga, L. Alterations in the nuclear volume of neurones from the hypothalamic magnocellular nuclei of fetuses born to rats subject to chronic alcoholism. *Revista Espanola de Fisiologia,* 45 (Suppl):43–48, 1989.

Fulginiti, S., Artinian, J., Cabrera, R., and Contreras, P. Response to an ethanol challenge dose on sleep time and blood alcohol level in Wistar rats prenatally exposed to ethanol during gestational day 8. *Alcohol,* 6:253–256, 1989.

Gallo, P.V. and Weinberg, J. Neuromotor development and response inhibition following prenatal ethanol exposure. *Neurobehav. Toxicol. Teratol.,* 4:505–513, 1982.

Gentry, G.D. and Middaugh, L.D. Prenatal ethanol weakens the efficacy of reinforcers for adult mice. *Teratology,* 37:135–144, 1988.

Gianoulakis, C. Rats exposed prenatally to alcohol exhibit impairment in spatial navigation test. *Behavioral Brain Res.,* 36:217–228, 1990.

Gilliam, D.M., Dudek, B.C., and Riley, E.P. Responses top ethanol challenge in long- and short-sleep mice prenatally exposed to alcohol. *Alcohol,* 7:1–5, 1990.

Gilliam, D.M. and Irtenkauf, K.T. Maternal genetic effects on ethanol teratogenesis and dominance of relative embryonic resistance to malformations. *Alcoholism: Clin. Exp. Res.,* 14:539–545, 1990.

Gilliam, D.M. and Kotch, L.E. Developmental thermoregulatory deficits in prenatal ethanol exposed long- and short-sleep mice. *Dev. Psychobiol.,* 25:365–373, 1992.

Goodlett, C.R., Gilliam, D.M., Nichols, J.M., and West, J.R. Genetic influences on brain growth restriction induced by developmental exposure to alcohol. *Neurotoxicology,* 10:321–334, 1989.

Goodlett, C.R., Kelly, S.J., and West, J.R. Early postnatal alcohol exposure that produces high blood alcohol levels impairs development of spatial navigation learning. *Psychobiology,* 15:64–74, 1987.

Goodlett, C.R., Leo, J.T., O'Callaghan, J.P., Mahoney, J.C., and West, J.R. Transient cortical astrogliosis induced by alcohol exposure during the neonatal brain growth spurt in rats. *Dev. Brain Res.,* 72:85–97, 1993.

Goodlett, C.R., Marcussen, B.L., and West, J.R. A single day of alcohol exposure during the brain growth spurt induces brain weight restriction and cerebellar purkinje cell loss. *Alcohol,* 7:107–114, 1990.

Goodlett, C.R., Thomas, J.D., and West, J.R. Long-term deficits in cerebellar growth and rotarod performance of rats following "binge-like" alcohol exposure during the neonatal brain growth spurt. *Neurotoxicol. Teratol.,* 13:69–74, 1991.

Gottesfeld, Z. and Abel, E.L. Maternal and paternal alcohol use: effects of the immune system of the offspring. *Life Sci.,* 48:1–8, 1991.

Gottesfeld, Z., Christie, R., Felten, D.L., and LeGrue, S.J. Prenatal ethanol exposure alters immune capacity and noradrenergic synaptic transmission in lymphoid organs of the adult mouse. *Neuroscience,* 35:185–194, 1990.

Gottesfeld, Z., Garcia, C.J., Lingham, R.B., and Chronister, R.B. Prenatal ethanol exposure impairs lesion-induced plasticity in a dopaminergic synapse after maturity. *Neuroscience,* 29:715–723, 1989.

Gottesfeld, Z., Morgan, B., and Perez-Polo, J.R. Prenatal alcohol exposure alters the development of sympathetic synaptic components and of nerve growth factor receptor expression selectivity in lymphoid organs. *J. Neurosci. Res.,* 26:308–316, 1990.

Gottesfeld, Z. and Silverman, P.B. Developmental delays associated with prenatal alcohol exposure are reversed by thyroid hormone treatment. *Neurosci. Lett.,* 109:42–47, 1990.

Grant, K.A., Choi, E.Y., and Samson, H.H. Neonatal ethanol exposure: effects on adult behavior and brain growth patterns. *Pharmacol. Biochem. Behav.,* 18(Suppl 1): 331–336, 1983.

Greene, T., Ernhart, C.B., Sokol, R.J., Martier, S., Marler, M.R., Boyd, T.A., and Age, J. Prenatal alcohol exposure and preschool physical growth: A longitudinal analysis. *Alcoholism: Clin. Exp. Res.,* 15:905–913, 1991.

Hale, R.L., Randall, C.L., Becker, H.C., and Middaugh, L.D. The effect of prenatal ethanol exposure on scentmarking in the C57BL/6J and C3H/He mouse strains. *Alcohol,* 9:287–292, 1992.

Hannigan, J.H. The ontogeny of SCH 23390-induced catalepsy in male and female rats exposed to ethanol in utero. *Alcohol,* 7:11–16, 1990.

Hannigan, J.H. and Bellisario, R.L. Lower serum thyroxine levels in rats following prenatal exposure to ethanol. *Alcoholism: Clin. Exp. Res.,* 14:456–460, 1990.

Hannigan, J.H. and Blanchard, B.A. Commentary: psychopharmacological assessment in neurobehavioral teratology. *Neurotoxicol. Teratol.,* 10:143–145, 1988.

Hannigan, J.H., Blanchard, B.A., Horner, M.P., Riley, E.P., and Pilati, M.L. Apomorphine-induced motor behavior in rats exposed prenatally to alcohol. *Neurobehav. Toxicol. Teratol.,* 12:79–84, 1990.

Hannigan, J.H., Blanchard, B.A., and Riley, E.P. Altered grooming responses to stress in rats exposed prenatally to ethanol. *Behav. Neural Biol.,* 47:173–185, 1987.

Hannigan, J.H., Cortese, B.M., DiCerbo, J.A., and Radford, L.D. Scopolamine does not differentially affect Morris maze performance in adult rats exposed prenatally to alcohol. *Alcohol,* 10:529–535, 1993.

Hannigan, J.H., Fitzgerald, L.W., Blanchard, B.A., and Riley, E.P. Absence of differential motoric and thermic responses to clonidine in young rats exposed prenatally to alcohol. *Alcohol,* 5:431–436, 1988.

Hannigan, J.H. and Pilati, M.L. The effects of chronic postweaning amphetamine on rats exposed to alcohol *in utero:* weight gain and behavior. *Neurotoxicol. Teratol.,* 13:649–656, 1991.

Hannigan, J.H. and Riley, E.P. Prenatal ethanol alters gait in rats. *Alcohol,* 5:451–454, 1988.

Hard, E., Dahlgren, I., Engel, J., Larsson, K., Liljequist, S., Linde, A., and Musi, B. Development of sexual behavior in prenatally ethanol exposed rats. *Drug Alcohol Dependence,* 14:51–61, 1984.

Harris, R.A. and Case, J. Effects of maternal consumption of ethanol, barbital, or chlordiazepoxide on the behavior of the offspring. *Behav. Neural Biol.,* 26:234–247, 1979.

Hayne, H., Hess, M., and Campbell, B.A. The effect of prenatal alcohol exposure on attention in the rat. *Neurotoxicol. Teratol.,* 14:393–398, 1992.

Ho, A., Chin, A.J., and Dole, V.P. Early experience and the consumption of alcohol by adult C57BL/6J mice. *Alcohol,* 6:511–515, 1989.

Hunt, P.S., Kraebel, K.S., Rabine, H., Spear, L.P., and Spear, N.E. Enhanced ethanol intake in preweanling rats following exposure to ethanol in a nursing context. *Dev. Psychobiol.,* 26:133–153, 1993.

Jones, K.L. and Smith, D.W. Recognition of the fetal alcohol syndrome in early infancy. *Lancet,* 2:999–1001, 1973.

Jones, K.L., Smith, D.W., Ulleland, C.N., and Streissguth, A.P. Pattern of malformation in offspring of chronic alcoholic mothers. *Lancet,* 1:1267–1271, 1973.

Jungkuntz-Burgett, L., Paredez, S., and Rudeen, P.K. Reduced sensitivity of hypothalamic-preoptic area norepinephrine and dopamine to testosterone feedback in adult fetal alcohol-exposed male rats. *Alcohol,* 7:513–516, 1990.

Kakihana, R., Butte, J.C., and Moore, J.A. Endocrine effects of maternal alcoholization: plasma and brain testosterone, dihydrotestosterone, estradiol, and corticosterone. *Alcoholism: Clin. Exp. Res.,* 4:57–61, 1980.

Kelce, W.R., Ganjam, V.K., and Rudeen, P.K. Inhibition of testicular steroidogenesis in the neonatal rat following acute ethanol exposure. *Alcohol,* 7:75–80. 1990a.

Kelce, W.R., Ganjam, V.K., and Rudeen, P.K. Effects of fetal alcohol exposure on brain 5a-reductase/aromatase activity. *J. Steroid Biochem.,* 35:103–106, 1990b.

Kelce, W.R., Rudeen, P.K., and Ganjam, V.K. Prenatal ethanol exposure alters steroidogenic enzyme activity in newborn rat testes. *Alcoholism: Clin. Exp. Res.,* 13:617–621, 1989.

Kelly, S.J., Black, A.C., Jr., and West, J.R. Changes in the muscarinic cholinergic receptors in the hippocampus of rats exposed to ethyl alcohol during the brain growth spurt. *J. Pharmacol. Exp. Therap.,* 249:798–804, 1989.

Kelly, S.J., Goodlett, C.R., Hulsether, S.A., and West, J.R. Impaired spatial navigation in adult female but not adult male rats exposed to alcohol during the brain growth spurt. *Behav. Brain Res.,* 27:247–257, 1988.

Kelly, S.J., Hulsether, S.A., and West, J.R. Alterations in sensorimotor development: relationship to postnatal alcohol exposure. *Neurotoxicol. Teratol.,* 9:243–251, 1987.

Kelly, S.J., Mohonney, J.C., and West, J.R. Changes in brain microvasculature resulting from early postnatal alcohol exposure. *Alcohol,* 7:43–47, 1990.

Kelly, S.J., Pierce, D.R., and West, J.R. Microencephaly and hyperactivity in adult rats can be induced by neonatal exposure to high blood alcohol concentrations. *Exp. Neurol.,* 96:580–593, 1987.

Kennedy, L.A. and Elliot, M.J. Ocular changes in the mouse embryo following acute maternal ethanol intoxication. *Int. J. Dev. Neurosci.,* 4:311–317, 1986.

Kotch, L.E. and Sulik, K.K. Patterns of ethanol-induced cell death in the developing nervous system of mice: neural fold states through the time of anterior neural tube closure. *Int. J. Dev. Neurosci.,* 10:273–279, 1992.

Kotkoškie, L.A. and Norton, S. Prenatal brain malformations following acute ethanol exposure in the rat. *Alcoholism: Clin. Exp. Res.,* 12:831–836, 1988.

Kotkoskie, L.A. and Norton, S. Acute response of the fetal telencephalon to short-term maternal exposure to ethanol in the rat. *Acta Neuropathologica,* 79:513–519, 1990.

La Fiette, M.H., Carlos, R., and Riley, E.P. Effects of prenatal alcohol exposure on serial pattern performance in the rat. *Neurotoxicol. Teratol.,* 16:41–46, 1994.

Landesman-Dwyer, S., Keller, L.S., and Streissguth, A.P. Naturalistic observations of newborns: Effects of maternal alcohol intake. *Alcoholism: Clin. Exp. Res.,* 2:171–177, 1978.

Lankford, K.L., DeMello, F.G., and Klein, W.L. D_1-type dopamine receptors inhibit growth cone motility in cultured retina neurons: Evidence that neurotransmitters act as morphogenic growth regulators in the developing central nervous system. *Proc. Natl. Acad. Sci.,* 85:2839–2843, 1988.

Lauder, J.M., Wallace, J.A., Wilkie, M.B., Dinome, A., and Krebs, H. Role for serotonin in neurogenesis. *Monogr. Neural Sci.,* 9:3–10, 1983.

Leader, L.R., Smith, F.G., and Lumbers, E.R. Effect of ethanol on habituation and the cardiovascular response to stimulation in fetal sheep. *Eur. J. Obstet. Gynecol. Reproductive Biol.,* 36:87–95, 1990.

Lee, M.H., Haddad, R., and Rabe, A. Developmental impairment in the progeny of rats consuming ethanol during pregnancy. *Neurobehav. Toxicol.,* 2:189–198, 1980.

Lee, S. and Rivier, C. Prenatal alcohol exposure blunts interleukin-1 induced ACTH and beta-endorphin secretion by immature rats. *Alcoholism: Clin. Exp. Res.,* 17:940–945, 1993.

Lee, M. and Wakabayashi, K. Pituitary and thyroid hormones in pregnant alcohol-fed rats and their fetuses. *Alcoholism,* 10:428–431, 1986.

Light, K.E., Serbus, D.C., and Santiago, M. Exposure of rats to ethanol from postnatal days 4 to 8: alterations of cholinergic neurochemistry in the hippocampus and cerebellum at day 20. *Alcoholism: Clin. Exp. Res.,* 13:686–692, 1989a.

Light, K.E., Serbus, D.C., Santiago, M. Exposure of rats to ethanol from postnatal days 4 to 8: alterations of cholinergic neurochemistry in the cerebral cortex and corpus striatum at day 20. *Alcoholism: Clin. Exp. Res.,* 13:29–35, 1989b.

Little, R.E., Anderson, K.W., Ervin, C.H., Worthington-Roberts, B., and Clarren, S.K. Maternal alcohol use during breast-feeding and infant mental and motor development at one year. *N. Engl. J. Med.,* 321:425–430, 1989.

Little, B.B., Snell, L.M., Rosenfeld, C.R., Gilstrap, L.C., III, and Gant, N.F. Failure to recognize fetal alcohol syndrome in newborn infants. *Am. J. Dis. Children,* 144:1142–1146, 1990.

Lochry, E.A. and Riley, E.P. Retention of passive avoidance and T-maze escape in rats exposed to alcohol prenatally. *Neurobehav. Toxicol.,* 2:107–115, 1980.

Lokhorst, D.K. and Druse, M.J. Effects of ethanol on cultured astroglia. *Alcoholism: Clin. Exp. Res.,* 17:810–815, 1993a.

Lokhorst, D.K. and Druse, M.J. Effects of ethanol on cultured fetal serotonergic neurons. *Alcoholism: Clin. Exp. Res.,* 17:86–93, 1993b.

Marcus, J.C. Neurological findings in the fetal alcohol syndrome. *Neuropediatrics,* 18:158–160, 1987.

Martin, D.C., Martin, J.C., Lund, C.A., and Streissguth, A.P. Maternal alcohol ingestion and cigarette smoking and their effects on newborn conditioning. *Alcoholism,* 1:243–247, 1977.

Martin, J.C., Martin, D.C., Sigman, G., and Radow, B. Maternal ethanol consumption and hyperactivity in cross-fostered offspring. *Physiological Psychol.,* 6:362–365, 1978.

Martin, D.C., Martin, J.C., Streissguth, A.P., and Lund, C.A. Sucking frequency and amplitude in newborns as a function of maternal drinking and smoking. In Galanter, M. (Ed.) *Currents in Alcoholism.* Vol. V. New York: Grune and Stratton, 1979, pp. 359–366.

Martin, D., Savage, D.D., and Swartzwelder, H.S. Effects of prenatal ethanol exposure on hippocampal ionotropic-quisqualate and kainate receptors. *Alcoholism: Clin. Exp. Res.,* 16:816–821, 1992.

Mattson, S.N., Riley, E.P., Jernigan, T.L., Ehlers, C.L., Delis, D.C., Jones, K.L., Stern, C., Johnson, K.A., Hesselink, J.R., and Bellugi, U. Fetal alcohol syndrome: a case report of neuropsychological, MRI and EEG assessment of two children. *Alcoholism: Clin. Exp. Res.,* 116:1001–1003, 1992.

McEwen, B.S., DeKloet, E.R., and Rostene, W. Adrenal steroid receptors and actions in the nervous system. *Physiological Rev.,* 66:1121–1188, 1986.

McGivern, R.F., Clancy, A.N., Hill, M.A., and Noble, E.P. Prenatal alcohol exposure alters adult expression of sexually dimorphic behavior in the rat. *Science,* 224:896–898, 1984.

McGivern, R.F., Handa, R.J., and Redei, E. Decreased postnatal testosterone surge in male rats exposed to ethanol during the last week of gestation. *Alcoholism: Clin. Exp. Res.,* 17:1215–1222, 1993.

McGivern, R.F., Holcomb, C., and Poland, R.E. Effects of prenatal testosterone propionate treatment on saccharin preference of adult rats exposed to ethanol *in utero. Physiol. Behav.,* 39:241–246, 1987.

McGivern, R.F., Poland, R.E., Noble, E.P., and Lane, L.A. Influence of prenatal ethanol exposure on hormonal responses to clonidine and naloxone in prepubescent male and female rats. *Psychoneuroendocrinology,* 11:105–110, 1986.

McGivern, R.F., Raum, W.J., Salido, E., and Redei, E. Lack of prenatal testosterone surge in fetal rats exposed to alcohol: Alterations in testicular morphology and physiology. *Alcoholism: Clin. Exp. Res.,* 12:243–247, 1988.

McGivern, R.F., Roselli, C.E., and Handa, R.J. Perinatal aromatase activity in male and female rats: Effect of prenatal alcohol exposure. *Alcoholism: Clin. Exp. Res.,* 12:769–772, 1988.

McLeod, W.J., Brien, C., Loomis, L., Carmichael, L., Probert, C., and Patrick, J. Effect of maternal ethanol ingestion on fetal breathing movements, gross body movements, and heart rate at 37 to 40 weeks' gestational age. *Am. J. Obstet. Gynecol.,* 145:251–257, 1983.

Means, L.W., Medlin, C.W., Hughes, V.D., and Gray, S.L. Hyperresponsiveness to methylphenidate in rats following prenatal ethanol exposure. *Neurobehav. Toxicol. Teratol.,* 6:187–192, 1984.

Meyer, L.S., Kotch, L.E., and Riley, E.P. Alterations in gait following ethanol exposure during the brain growth spurt in rats. *Alcoholism: Clin. Exp. Res.,* 14:23–27, 1990a.

Meyer, L.S., Kotch, L.E., and Riley, E.P. Neonatal ethanol exposure: Functional alterations associated with cerebellar growth retardation. *Neurotoxicol. Teratol.,* 12:15–22, 1990b.

Meyer, L.S. and Riley, E.P. Behavioral teratology of alcohol. In Riley, E.P. and Vorhees, C.V. (Eds.) *Handbook of Behavioral Teratology,* New York: Plenum Press, 1986a, pp. 101–140.

Meyer, L.S. and Riley, E.P. Social play in juvenile rats prenatally exposed to alcohol. *Teratology,* 34:1–7, 1986b.

Middaugh, L.D. and Ayers, K.L. Effects of ethanol on mature offspring of mice given ethanol during pregnancy. *Alcoholism: Clin. Exp. Res.,* 12:388–393, 1988.

Middaugh, L.D. and Boggan, W.O. Postnatal growth deficits in prenatal ethanol exposed mice: Characteristics and critical periods. *Alcoholism: Clin. Exp. Res.,* 15:919–926, 1991.

Middaugh, L.D. and Gentry, G.D. Prenatal ethanol effects on reward efficacy for adult mice are gestation stage specific. *Neurotoxicol. Teratol.,* 14:365–370, 1992.

Miller, M.W. Effects of alcohol on the generation and migration of cerebral cortical neurons. *Science,* 233:1308–1311, 1986.

Miller, M.W. Effect of prenatal exposure to alcohol on the distribution and time of origin of corticospinal neurons in the rat. *J. Comp. Neurol.,* 257:372–382, 1987.

Miller, M.W. Effect of prenatal exposure to ethanol on the development of cerebral cortex: I. Neuronal generation. *Alcoholism: Clin. Exp. Res.,* 12:440–449, 1988.

Miller, M.W. Effects of prenatal exposure to ethanol on neocortical development: II. Cell proliferation in the ventricular and subventricular zones of the rat. *J. Comp. Neurol.,* 287:326–338, 1989.

Miller, M.W. Migration of cortical neurons is altered by gestational exposure to ethanol. *Alcoholism: Clin. Exp. Res.,* 17:304–314, 1993.

Miller, M.W. and Al-Rabiai, S. Effects of prenatal exposure to ethanol on the number of axons in the pyramidal tract of the rat. *Alcoholism: Clin. Exp. Res.,* 18:346–354, 1994.

Miller, M.W., Chiaia, N.L., and Rhoades, R.W. Intracellular recording and injection study of corticospinal neurons in the rats somatosensory cortex: effect of prenatal exposure to ethanol. *J. Comp. Neurol.,* 297:91–105, 1990.

Miller, M.W. and Dow-Edwards, D.L. Structural and metabolic alterations in rat cerebral cortex induced by prenatal exposure to ethanol. *Brain Res.,* 474:316–326, 1988.

Miller, M.W. and Muller, S.J. Structure and histogenesis of the principal sensory nucleus of the trigeminal nerve: effects of prenatal exposure to ethanol. *J. Comp. Neurol.,* 282:570–580, 1989.

Miller, M.E. and Nowakowski, R.S. Effect of prenatal exposure to ethanol on the cell cycle kinetics and growth fraction in the proliferative zones of fetal rat cerebral cortex. *Alcoholism: Clin. Exp. Res.,* 15:229–232, 1991.

Miller, M.E. and Potempa, G. Numbers of neurons and glia in mature rat somatosensory cortex: Effects of prenatal exposure to ethanol. *J. Comp. Neurol.,* 293:92–102, 1990.

Molina, J.C., Hoffmann, H., Spear, L.P., and Spear, N.E. Sensorimotor maturation and alcohol responsiveness in rats prenatally exposed to alcohol during gestational day 8. *Neurobehav. Toxicol. Teratol.,* 9:121–128, 1987.

Molina, J.C., Moyano, H.F., Spear, L.P., and Spear, N.E. Acute alcohol exposure during gestational day 8 in the rat: effect upon physical and behavioral parameters. *Alcohol,* 1:459–464, 1985.

Molina, J.C., Serwatka, J., and Spear, N.E. Alcohol drinking patterns of young adult rats as a function of infantile aversive experiences with alcohol odor. *Behav. Neural Biol.,* 46:257–271, 1986.

Morris, D.L., Harms, P.G., Petersen, H.D., and McArthur, N.H. LHRH and LH in peripubertal female rats following prenatal and/or postnatal ethanol exposure. *Life Sci.,* 44:1165–1171, 1989.

Morrisett, R.A., Martin, D., Wilson, W.A., Savage, D.D., and Swartzwelder, S.H. Prenatal exposure to ethanol decreases the sensitivity of the adult rat hippocampus to N-Methyl-D-Aspartate. *Alcohol,* 6:415–420, 1989.

Nanson, J.L. and Hiscock, M. Attention deficits in children exposed to alcohol prenatally. *Alcoholism: Clin. Exp. Res.,* 14:656–661, 1990.

Nelson, L.R., Lewis, J.W., Kokka, N., Branch, B.A., and Taylor, A.N. Prenatal exposure to ethanol potentiates morphine-induced hypothermia in adult rats. *Neurobehav. Toxicol. Teratol.,* 8:469–474, 1986.

Nelson, L.R., Taylor, A.N., Lewis, J.W., Branch, B.J., Liebeskind, J.C. Opioid but not nonopioid stress-induced analgesia is enhanced following prenatal exposure to ethanol. *Psychopharmacology,* 85:92–96, 1985.

Nelson, L.R., Taylor, A.N., Lewis, J.W., Poland, R.E., Redei, E., and Branch, B.J. Pituitary-adrenal responses to morphine and footshock stress are enhanced following prenatal alcohol exposure. *Alcoholism: Clin. Exp. Res.,* 10:397–402, 1986.

Noble, E.P. and Richie, T. Prenatal ethanol exposure reduces the effects of excitatory amino acids in the rat hippocampus. *Life Sci.,* 45:803–810, 1989.

Osborne, G.L., Caul, W.F., and Fernandez, K. Behavioral effects of prenatal ethanol exposure and differential early experience in rats. *Pharmacol. Biochem. Behav.,* 12:393–401, 1980.

Parker, S., Udani, M., Gavaler, J.S., and Van Thiel, D.H. Adverse effects of ethanol upon the adult sexual behavior of male rats exposed *in utero*. *Neurobehav. Toxicol. Teratol.,* 6:289–293, 1984.

Perez, V.J., Gonzalez, G.E., and Smith, C.J. Exposure to ethanol during pregnancy in mice — potential importance of dose for the development of tolerance in offspring. *Physiol. Behav.,* 30:485–488, 1983.

Phillips, S.C. Alcohol and histology of the developing cerebellum. In West, J.R. (Ed.) *Alcohol and Brain Development.* New York: Oxford University Press, 1986, pp. 204–224.

Phillips, D.E. Effects of limited postnatal ethanol exposure on the development of myelin and nerve fibers in rat optic nerve. *Exp. Neurol.,* 103:90–100, 1989.

Phillips, D.E. and Krueger, S.K. Effects of postnatal ethanol exposure on glial cell development in rat optic nerve. *Exp. Neurobiol.,* 107:97–105, 1990.

Phillips, D.E. and Krueger, S.K. Effects of combined pre- and postnatal ethanol exposure (three trimester equivalency) on glial cell development in rat optic nerve. *Int. J. Dev. Neurosci.,* 10:197–206, 1992.

Phillips, D.S. and Stainbrook, G.L. Effects of early alcohol exposure upon adult learning ability and taste preferences. *Physiological Psychol.,* 4:473–475, 1976.

Pierce, D.R., Goodlett, C.R., and West, J.R. Differential neuronal loss following early postnatal alcohol exposure. *Teratology,* 40:113–126, 1989.

Pierce, D.R., Serbus, D.C., and Light, K.E. Intragastric intubation of alcohol during postnatal development of rats results in selective cell loss in the cerebellum. *Alcoholism: Clin. Exp. Res.,* 17:1275–1280, 1993.

Pierce, D.R. and West, J.R. Alcohol-induced microencephaly during the third trimester equivalent: Relationship to dose and blood alcohol concentration. *Alcohol,* 3:185–191, 1986.

Plonsky, M. and Riley, E.P. Head-dipping behaviors in rats exposed to alcohol prenatally as a function of age at testing. *Neurobehav. Toxicol. Teratol.,* 5:309–314, 1983.

Portoles, M., Sanchis, R., and Guerri, C. Thyroid hormone levels in rats exposed to alcohol during development. *Horm. Metab. Res.,* 20:267–270, 1988.

Queen, S.A., Sanchez, C.F., Lopez, S.R., Paxton, L.L., and Savage, D.D. Dose- and age-dependent effects of prenatal ethanol exposure on hippocampal metabotropic-glutamate receptor-stimulated phosphoinositide hydrolysis. *Alcoholism: Clin. Exp. Res.,* 17:887–893, 1993.

Randall, C.L., Becker, H.C., and Middaugh, L.D. Effect of prenatal ethanol exposure on activity and shuttle avoidance behavior in adult C57 mice. *Alcohol Drug Res.,* 6:351–360, 1986.

Randall, C.L., Hughes, S.S., Williams, C.K., and Anton, R.F. Effect of prenatal alcohol exposure on consumption of alcohol and alcohol-induced sleep-time in mice. *Pharmacol. Biochem. Behav.,* 18:325–329, 1983.

Randall, C.L., Taylor, W.J., and Walker, D.W. Ethanol-induced malformations in mice. *Alcoholism: Clin. Exp. Res.,* 1:219–223, 1977.

Rathbun, W.E. and Druse, M.J. Dopamine, serotonin and acid metabolites in brain regions from the developing offspring of ethanol-treated rats. *J. Neurochem.,* 44:57–62, 1985.

Redei, E., Clark, W.R., and McGivern, R.F. Alcohol exposure in utero results in diminished T-cell function and alterations in brain corticotropin-releasing factor and ACTH content. *Alcoholism: Clin. Exp. Res.,* 13:439–443, 1989.

Reyes, E., Duran, E., and Switzer, S.H. Effects of in utero administration of alcohol on alcohol sensitivity in adult rats. *Pharmacol. Biochem. Behav.,* 44:307–312, 1993.

Reyes, E., Wolfe, J., and Savage, D.D. Effects of prenatal alcohol exposure on radial arm maze performance in adult rats. *Physiol. Behav.,* 46:45–48, 1989.

Riley, E.P. The long term behavioral effects of prenatal alcohol exposure in rats. *Alcoholism: Clin. Exp. Res.,* 14:670–673, 1990.

Riley, E.P. and Barron, S. The behavioral and neuroanatomical effects of prenatal alcohol exposure in animals. *Ann. N.Y. Acad. Sci.,* 562:173–177, 1989.

Riley, E.P., Barron, S., Driscoll, C.D., and Chen, J.S. Taste aversion learning in preweanling rats exposed to alcohol prenatally. *Teratology,* 29:325–331, 1984.

Riley, E.P., Barron, S., Driscoll, C.D., and Hamlin, R.T. The effects of physostigmine on open-field behavior in rats exposed to alcohol prenatally. *Alcoholism: Clin. Exp. Res.,* 10:50–53, 1986.

Riley, E.P., Barron, S., and Hannigan, J.H. Response inhibition deficits following prenatal alcohol exposure: A comparison to the effects of hippocampal lesions in rats. In West, J.R. (Ed.) *Alcohol and Brain Development.* New York: Oxford University Press, 1986, pp. 71–102.

Riley, E.P., Barron, S., Melcer, T., and Gonzalez, D. Alterations in activity following alcohol administration during the third trimester equivalent in P and NP rats. *Alcoholism: Clin. Exp. Res.,* 17:1240–1246, 1993.

Riley, E.P., Lochry, E.A., and Shapiro, N.R. Lack of response inhibition in rats prenatally exposed to alcohol. *Psychopharmacology,* 62:47–52, 1979.

Riley, E.P., Lochry, E.A., Shapiro, N.R., and Baldwin, J. Response perseveration in rats exposed to alcohol prenatally. *Pharmacol. Biochem. Behav.,* 11:513–519, 1979.

Riley, E.P., Plonsky, M.Z., and Rosellini, R.A. Acquisition of an unsignalled avoidance task in rats exposed to alcohol prenatally. *Neurobehav. Toxicol. Teratol.,* 4:525–530, 1982.

Riley, E.P., Shapiro, N.R., and Lochry, E.A. Nose-poking and head-dipping behaviors in rats prenatally exposed to alcohol. *Pharmacol. Biochem. Behav.,* 11:513–519, 1979.

Riley, E.P., Shapiro, N.R., Lochry, E.A., and Broida, J. Fixed-ratio performance and subsequent extinction in rats prenatally exposed to alcohol. *Physiological Psychol.,* 8:47–50, 1980.

Rockman, G.E., Markert, L., and Delrizzo, M. Effects of prenatal ethanol exposure on ethanol-induced locomotor activity in rats. *Alcohol,* 6:353–356, 1989.

Rockwood, G.A. and Riley, E.P. Effects of scopolamine on spontaneous alternation and shuttle avoidance in rats exposed to alcohol *in utero. Alcohol,* 2:575–579, 1985.

Rockwood, G.A. and Riley, E.P. Suckling deficits in rat pups exposed to alcohol *in utero. Teratology,* 33:145–151, 1986.

Rockwood, G.A. and Riley, E.P. Nipple attachment behavior in rat pups exposed to alcohol *in utero*. *Neurotoxicol. Teratol.,* 12:383–389, 1990.

Rosett, H.L. A clinical perspective of the fetal alcohol syndrome. *Alcoholism: Clin. Exp. Res.,* 4:119–122, 1980.

Rosett, H.L., Snyder, R., Sander, L.N., Lee, A., Cook, P., Weiner, L., and Gould, J. Effects of maternal drinking on neonatal state regulation. *Dev. Med. Child Neurol.,* 21:464–473, 1979.

Ross, C.P. and Persaud, T.V.N. Neural tube defects in early rat embryos following maternal treatment with ethanol and caffeine. *Anatomischer Anzeiger,* 169:247–252, 1989.

Rudeen, P.K. Reduction of the volume of the sexually dimorphic nucleus of the preoptic area by in utero ethanol exposure in male rats. *Neurosci. Lett.,* 72:363–368, 1986.

Rudeen, P.K., Kappel, A., and Lear, K. Postnatal or in utero ethanol exposure reduction of the volume of the sexually dimorphic nucleus of the preoptic area in male rats. *Drug Alcohol Dependence,* 18:247–252, 1986.

Samorajski, T., Lancaster, F., and Wiggins, R.C. Fetal ethanol exposure: a morphometric analysis of myelination in the optic nerve. *Int. J. Dev. Neurosci.,* 4:369–374, 1986.

Savage, D.D., Montano, C.Y., Otero, M.A., and Paxton, L.L. Prenatal ethanol exposure decreases hippocampal NMDA-sensitive 3H-glutamate binding site density in 45-day-old rats. *Alcohol,* 8:193–201, 1991.

Savage, D.D., Montano, C.Y., Paxton, L.L., and Kasarskis, E.J. Prenatal ethanol exposure decreases hippocampal mossy fiber zinc in 45-day-old rats. *Alcoholism: Clin. Exp. Res.,* 13:588–593, 1989.

Saxema, S., Meehan, D., Coney, P., and Wimalasena, J. Ethanol has direct inhibitory effects on steroidogenesis in human granulosa cells: Specific inhibition of LH action. *Alcoholism: Clin. Exp. Res.,* 14:522–527, 1990.

Schambra, U.B., Lauder, J.M., Petrusz, P., and Sulik, K.K. Development of neurotransmitter systems in the mouse embryo following acute ethanol exposure. *Int. J. Dev. Neurosci.,* 8:507–522, 1990.

Scher, M.S., Richardson, G.A., Coble, P.A., Day, N.L., and Stoffer, D.S. The effects of prenatal alcohol and marijuana exposure: disturbances in neonatal sleep cycling and arousal. *Ped. Res.,* 24:101–105, 1988.

Serbus, D.C., Stull, R.E., and Light, K.E. Neonatal ethanol exposure to rat pups: Resultant alterations of cortical muscarinic and cerebellar H_1-histaminergic receptor binding dynamics. *Neurotoxicology,* 7:257–278, 1986.

Shaywitz, B.A., Griffieth, G.G., and Warshaw, J.B. Hyperactivity and cognitive deficits in developing rat pups born to alcoholic mothers: an experimental model of the expanded fetal alcohol syndrome (EFAS). *Neurobehav. Toxicol.,* 1:113–122, 1979.

Shaywitz, B.A., Klopper, J.H., and Gordon, J.W. A syndrome resembling minimal brain dysfunction (MBD) in rat pups born to alcoholic mothers. *Ped. Res.,* 10:451, 1976.

Shetty, A.K. and Phillips, D.E. Effects of prenatal ethanol exposure on the development of Bermann glia and astrocytes in the rat cerebellum: an immunohistochemical study. *J. Comp. Neurol.,* 321:19–32, 1992.

Smith, G.N., Brien, J.F., Carmichael, L., Homan, J., Clarke, D.W., and Patrick, J. Development of tolerance to ethanol-induced suppression of breathing movements and brain activity in the near-term fetal sheep during short-term maternal administration of ethanol. *J. Dev. Physiol.,* 11:189–197, 1989.

Smith, G.M., Brien, J.F., Homan, J., Carmichael, L., and Patrick, J. Indomethacin antagonizes the ethanol-induced suppression of breathing activity but not the suppression of brain activity in the near-term fetal sheep. *J. Dev. Physiol.,* 12:69–75, 1989.

Smith, I.E., Coles, C.D., Lancaste, J.S., Fernhoff, P.M., and Falek, A. The effect of volume and duration of prenatal ethanol exposure on neonatal physical and behavioral development. *Neurobehav. Toxicol. Teratol.,* 8:375–381, 1986.

Smotherman, W.P. and Robinson, S.R. Stereotypic behavioral response of rat fetuses to acute hypoxia is altered by maternal alcohol consumption. *Am. J. Obstet. Gynecol.,* 157:982–986, 1987.

Smotherman, W.P., Woodruff, K.S., Robinson, S.R., Del Real, C., Barron, S., and Riley, E.P. Spontaneous fetal behavior after maternal exposure to ethanol. *Pharmacol. Biochem. Behav.,* 24:165–170, 1986.

Sokol, R.J. and Abel, E.L. Alcohol-related birth defects: Outlining current research opportunities. *Neurotoxicol. Teratol.,* 10:183–186, 1988.

Sokol, R.J. and Clarren, S.K. Guidelines for use of terminology describing the impact of prenatal alcohol on the offspring. *Alcoholism: Clin. Exp. Res.,* 13:597–598, 1989.

Spohr, H.L. and Steinhausen, H.C. Follow-up studies of children with fetal alcohol syndrome. *Neuropediatrics,* 18:13–17, 1987.

Steven, W.M., Bulloch, B., and Seelig, L.L., Jr. A morphometric study of the effects of ethanol consumption on lactating mammary glands of rats. *Alcoholism: Clin. Exp. Res.,* 13:209–212, 1989.

Streissguth, A.P. The behavioral teratology of alcohol: Performance, behavioral, and intellectual deficits in prenatally exposed children. In West, J.R. (Ed.) *Alcohol and Brain Development.* New York: Oxford University Press, 1986, pp. 3–44.

Streissguth, A.P., Aase, J.M., Clarren, S.K., Randels, S.P., LaDue, R.A., and Smith, D.F. Fetal Alcohol Syndrome in adolescents and adults. *J. Am. Med. Assoc.,* 265:1961–1967, 1991.

Streissguth, A.P., Barr, H.M., Martin, D.C., and Herman, C. Effects of maternal alcohol, nicotine, and caffeine use during pregnancy on infant mental and motor development at eight months. *Alcoholism: Clin. Exp. Res.,* 4:152–164, 1980.

Streissguth, A.P., Barr, H.M., and Sampson, P.D. Moderate prenatal alcohol exposure: effects on child IQ and learning problems at age $7^1/_2$ years. *Alcoholism: Clin. Exp. Res.,* 14:662–669, 1990.

Streissguth, A.P., Barr, H.M., Sampson, P.D., Parrish-Johnson, J.C., Kirchner, G.L., and Martin, D.C. Attention, distraction and reaction time at age 7 years and prenatal alcohol exposure. *Neurobehav. Toxicol. Teratol.,* 8:717–725, 1986.

Streissguth, A.P., Martin, D.C., and Barr, H.M. Maternal alcohol use and neonatal habituation assessed with the Brazelton Scale. *Child Dev.,* 54:1109–1118, 1983.

Streissguth, A.P., Martin, D.C., Barr, H.M., Sandman, B.M., Kirchner, G.L., and Darby, B.L. Intrauterine alcohol and nicotine exposure: Attention and reaction time in four-year-old children. *Dev. Psychol.,* 20:533–541, 1984.

Stricker, E.M. and Verbalis, J.G. Control of appetite and satiety: Insights from biologic and behavioral studies. *Nutr. Rev.,* 48:49–56, 1990.

Stromland, K. Ocular involvement in the fetal alcohol syndrome. *Surv. Opthalmol.,* 31:277–284, 1987.

Strupp, B.J., Korahais, J., Levitsky, D.A., and Ginsberg, S. Attentional impairment in rats exposed to alcohol prenatally: lack of hypothesized masking by food deprivation. *Ann. N.Y. Acad. Sci.,* 562:380–382, 1989.

Subramanian, M.G. Lactation and prolactin release in foster dams suckling prenatally ethanol exposed pups. *Alcoholism: Clin. Exp. Res.,* 16:891–894, 1992.

Subramanian, M.G. and Abel, E.L. Alcohol inhibits suckling-induced prolactin release and milk yield. *Alcohol,* 5:95–96, 1988.

Subramanian, M.G., Chen, X.G., and Bergeski, B.A. Pattern and duration of the inhibitory effect of alcohol administered acutely on suckling-induced prolactin in lactating rats. *Alcoholism: Clin. Exp. Res.,* 14:771–775, 1990.

Sulik, K.K. and Johnston, M.C. Sequence of developmental alterations following acute ethanol exposure in mice: craniofacial features of the fetal alcohol syndrome. *Am. J. Anat.,* 166:257–269, 1983.

Sulik, K.K., Lauder, J.M., and Dehart, D.B. Brain malformations in prenatal mice following acute maternal ethanol administration. *Int. J. Dev. Neurosci.,* 2:203–214, 1984.

Swartzwelder, H.S., Farr, K.L., Wilson, W.A., and Savage, D.D. Prenatal exposure to ethanol decreases hippocampal plasticity in the adult rat. *Alcohol,* 5:121–124, 1988.

Tajuddin, N. and Druse, M.J. Chronic maternal ethanol consumption results in decreased serotonergic 5-HT_1 sites in cerebral cortical regions from offspring. *Alcohol,* 5:465–470, 1989a.

Tajuddin, N. and Druse, M.J. Effects of in utero ethanol exposure on cortical 5-HT_2 binding sites. *Alcohol,* 5:461–464, 1989b.

Tajuddin, N. and Druse, M.J. Treatment of pregnant alcohol-consuming rats with buspirone: effects on serotonin and 5-hydroxyindoleacetic acid content in offspring. *Alcoholism: Clin. Exp. Res.,* 17:110–114, 1993.

Tan, S.E., Berman, R.F., Abel, E.L., and Zajac, C.S. Prenatal alcohol exposure alters hippocampal slice electrophysiology. *Alcohol,* 7:507–511, 1990.

Taylor, A.N., Branch, B.J., Cooley-Matthews, B., and Poland, R.E. Effects of maternal ethanol consumption on basal and rhythmic pituitary-adrenal function in neonatal offspring. *Psychoneuroendocrinology,* 7:49–58, 1982.

Taylor, A.N., Branch, B.J., and Kokka, N. Neuroendocrine effects of fetal alcohol exposure. *Prog. Biochem. Pharmacol.,* 18:99–100, 1981.

Taylor, A.N., Branch, B.J., Liu, S., and Kokka, N. Long-term effects of fetal ethanol exposure on pituitary-adrenal responses to stress. *Pharmacol. Biochem. Behav.,* 16:585–589, 1982.

Taylor, A.N., Branch, B.J., Liu, S., Weichmann, A.F., Hill, M.A., and Kokka, N. Fetal exposure to ethanol enhances pituitary-adrenal and temperature responses to ethanol in adult rats. *Alcoholism: Clin. Exp. Res.,* 5:237–246, 1981.

Taylor, A.N., Branch, B.J., Nelson, L.R., Lane, L.A., and Poland, R.E. Prenatal ethanol and ontogeny of pituitary-adrenal responses to ethanol and morphine. *Alcohol,* 3:255–259, 1986.

Taylor, A.N., Branch, B.A., Randolph, D., Hill, M.A., and Kokka, N. Prenatal ethanol exposure affects temperature responses of adult rats to pentobarbital and diazepam alone and in combination with ethanol. *Alcoholism: Clin. Exp. Res.,* 9:355–359, 1987.

Tolliver, G.A. and Samson, H.H. The influence of early postweaning ethanol exposure on oral self-administration behavior in the rat. *Pharmacol. Biochem. Behav.,* 38:575–580, 1991.

Udani, M., Parker, S., Gavaler, J., and Van Thiel, D.H. Effects of *in utero* exposure to alcohol upon male rats. *Alcoholism: Clin. Exp. Res.,* 9:355–359, 1985.

Ulug, S. and Riley, E.P. The effect of methylphenidate on overactivity in rats prenatally exposed to alcohol. *Neurobehav. Toxicol. Teratol.,* 5:35–39, 1983.

U.S. Department of Health and Human Services. *Eighth Special Report to the U.S. Congress on Alcohol and Health.* DHSS Publication No. (ADM) 94–3699. U.S. Government Printing Office, Washington, D.C., 1993.

Van Dyke, D.C., MacKay, L., and Ziaylek, E.N. Management of severe feeding dysfunction in children with fetal alcohol syndrome. *Clin. Pediatr.,* 21:336–339, 1982.

Vigliecca, N.S., Fulginiti, S., and Minetti, S.A. Acute ethanol exposure during pregnancy in rats: Effects upon a multiple learning task. *Alcohol,* 6:363–368, 1989.

Vilaro, S., Vinas, O., and Remesar, X. Altered ultrastructure of lactating rat mammary epithelial cells induced by chronic ethanol ingestion. *Alcoholism: Clin. Exp. Res.,* 13:128–136, 1989.

Vingan, R.D., Dow-Edwards, D.L., and Riley, E.P. Cerebral metabolic alterations in rats following prenatal alcohol exposure: A deoxyglucose study. *Alcoholism: Clin. Exp. Res.,* 10:22–26, 1986.

Vorhees, C.V. Concepts in teratology and developmental toxicology derived from animal research. *Ann. N.Y. Acad. Sci.,* 562:31–41, 1989.

Vulsma, T., Gons, M.H., and Vijlder, J.J.M. Maternal-fetal transfer of thyroxine in congenital hypothyroidism due to a total organification defect or thyroid agenesis. *N. Engl. J. Med.,* 321:13–16, 1989.

Wainwright, P.E., Levesque, S., Krempulec, L., Bulman-Fleming, B., and McCutcheon, D. Effects of environmental enrichment on cortical depth and Morris-maze performance in B6D2F_2 mice exposed prenatally to ethanol. *Neurotoxicol. Teratol.,* 15:11–20, 1993.

Wainwright, P.E., Ward, R.P., and McCutcheon, D. Effects of prenatal ethanol and long-chain n-3 fatty acid supplementation on development in mice: body and brain growth, sensorimotor development, and water T-maze reversal learning. *Alcoholism: Clin. Exp. Res.,* 14:405–411, 1990.

Weathersby, R.T., Becker, H.C., and Hale, R.L. Reduced sensitivity to the effects of clonidine on ethanol-stimulated locomotor activity in adult mouse offspring prenatally exposed to ethanol. *Alcohol,* 11:517–522, 1994.

Weinberg, J. Hyperresponsiveness to stress: differential effects of prenatal ethanol on males and females. *Alcoholism: Clin. Exp. Res.,* 12:647–652, 1988.

Weinberg, J. Prenatal ethanol exposure alters adrenocortical development in offspring. *Alcoholism: Clin. Exp. Res.,* 13:73–83, 1989.

Weinberg, J. and Bezio, S. Alcohol-induced changes in pituitary-adrenal activity during pregnancy. *Alcoholism: Clin. Exp. Res.,* 11:274–280, 1987.

Weinberg, J. and Jerrells, T.R. Suppression of immune responsiveness: Sex differences in prenatal ethanol effects. *Alcoholism: Clin. Exp. Res.,* 15:525–531, 1991.

Weinberg, J., Nelson, L.R., and Taylor, A.N. Hormonal effects of fetal alcohol exposure. In West, J.R. (Ed.) *Alcohol and Brain Development.* New York: Oxford University Press, 1986, pp. 310–342.

Weinberg, J. and Petersen, T.D. Effects of prenatal ethanol exposure on glucocorticoid receptors in rat hippocampus. *Alcoholism: Clin. Exp. Res.,* 15:711–716, 1991.

West, J.R. Fetal alcohol-induced brain damage and the problem of determining temporal vulnerability: a review. *Alcohol Drug Res.,* 7:423–441, 1987.

West, J.R. and Goodlett, C.R. Teratogenic effects of alcohol on brain development. *Ann. Med.,* 22:319–325, 1990.

West, J.R., Goodlett, C.R., Bonthius, D.J., Hamre, K.M., and Marcussen, B.L. Cell population depletion associated with fetal alcohol brain damage: mechanisms of BAC-dependent cell loss. *Alcoholism: Clin. Exp. Res.,* 14:813–818, 1990.

West, J.R., Goodlett, C.R., and Brandt, J.P. New approaches to research on the long-term consequences of prenatal exposure to alcohol. *Alcoholism: Clin. Exp. Res.,* 14:684–689, 1990.

West, J.R., Hamre, K.M., and Pierce, D.R. Delay in brain development induced by alcohol in artificially reared rat pups. *Alcohol,* 1:213–222, 1984.

West, J.R. and Pierce, D.R. Perinatal alcohol exposure and neuronal damage. In West, J.R. (Ed.) *Alcohol and Brain Development.* New York: Oxford University Press, 1986, pp. 120–157.

Wigal, T. and Amsel, A. Behavioral and neuroanatomical effects of prenatal, postnatal, or combined exposure to ethanol in weanling rats. *Behav. Neurosci.,* 104:116–126, 1990.

Wigal, T., Amsel, A., and Wilcox, R.E. Fetal ethanol exposure diminishes hippocampal beta-adrenergic receptor density while sparing muscarinic receptors during development. *Dev. Brain Res.,* 55:161–169, 1990.

Yamamoto, M., Toguchi, M., Arisima, K., Eguchi, Y., Leichter, J., and Lee, M. Effect of maternal alcohol consumption on fetal thyroid in the rat. *Proc. Soc. Exp. Biol. Med.,* 191:382–386, 1989.

Zajac, C.S., Bunger, P.C., and Moore, J.C. Changes in red nucleus neuronal development following maternal alcohol exposure. *Teratology,* 40:567–570, 1989.

Zhulin, V.V. and Zabludovskii, A.L. Binding of ^{3}H-muscimol by neurocortical membranes of rats exposed prenatally to ethanol. *Bull. Exp. Biol. Med.,* 106:1456–1457, 1989.

Zimmerberg, B. Thermoregulatory deficits following prenatal alcohol exposure: Structural correlates. *Alcohol,* 6:389–393, 1989.

Zimmerberg, B., Ballard, G.A., and Riley, E.P. The development of thermoregulation after prenatal exposure to alcohol in rats. *Psychopharmacology,* 91:478–489, 1987.

Zimmerberg, B., Beckstead, J.W., and Riley, E.P. Prenatal alcohol exposure and thermotaxic behavior in neonatal rats. *Neurotoxicol. Teratol.,* 9:283–286, 1987.

Zimmerberg, B., Carson, E.A., Kaplan, L.J., Zuniga, J.A., and True, R.C. Role of noradrenergic innervation of brown adipose tissue in thermoregulatory deficits following prenatal alcohol exposure. *Alcoholism: Clin. Exp. Res.,* 17:418–422, 1993.

Zimmerberg, B., Mattson, S., and Riley, E.P. Impaired alternation test performance in adult rats following prenatal alcohol exposure. *Pharmacol. Biochem. Behav.,* 32:293–299, 1989.

Zimmerberg, B. and Mikus, L.A. Sex differences in corpus callosum: influence of prenatal alcohol exposure and maternal undernutrition. *Brain Res.,* 537:115–122, 1990.

Zimmerberg, B. and Reuter, J.M. Sexually dimorphic behavioral and brain asymmetries in neonatal rats: effects of prenatal alcohol exposure. *Dev. Brain Res.,* 46:281–290, 1989.

Zimmerberg, B., Sukel, H.L., and Stekler, J.D. Spatial learning of adult rats with fetal alcohol exposure: Deficits are sex-dependent. *Behav. Brain Res.,* 42:49–56, 1991.

Zimmerman, E.F. and Collins, M. Chloride transport in embryonic cells: effect of ethanol and GABA. *Teratology,* 40:593–601, 1989.

Zimmerman, E.F., Scott, W.J., Jr., and Collins, M.D. Ethanol-induced limb defects in mice: effect of strain and Ro15-4513. *Teratology,* 41:453–462, 1990.

INDEX

A

C

D

E

F

G

H

I

K

L

M

N

O

P

Q

R

S

T

U

V

W

X

X

Z